S0-AUV-614

Y. Chen

Power System Control and Stability

VOLUME I

Power System Control and Stability

P. M. ANDERSON
A. A. FOUAD

THE IOWA STATE UNIVERSITY PRESS, AMES, IOWA, U.S.A.

P. M. Anderson is Program Manager for Research in Power System Planning, Security, and Control with the Electric Power Research Institute, Palo Alto, California. Prior to 1975 he was Professor of Electrical Engineering at Iowa State University where he received the Ph.D. degree in 1961. His industrial experience includes six years with the Iowa Public Service Company and a university leave for research with the Pacific Gas and Electric Company, San Francisco. He is author of numerous journal articles and the book *Analysis of Faulted Power Systems.*

A. A. Fouad is Professor of Electrical Engineering at Iowa State University. He received the B.Sc. degree from Cairo University and the Ph.D. degree from Iowa State University in 1956 and has extensive overseas experience in Egypt, Brazil, and the Philippines. He is active in the Power Engineering Society of the Institute of Electrical and Electronics Engineers where he serves as a member of the Power System Engineering Committee, System Control Subcommittee, and the working Group on Dynamic System Performance.

© 1977 The Iowa State University Press
Ames, Iowa 50010. All rights reserved

Composed and printed by Science Press, Ephrata, Pa. 17522

First edition, 1977

Second printing, 1980
Third printing, 1982

Library of Congress Cataloging in Publication Data

Anderson, Paul M 1926–
Power system control and stability.

Includes bibliographical references.
1. Electric power systems. 2. System analysis.
I. Fouad, Abdel-Aziz A., joint author. II. Title.
TK1005.A7 621.3 76-26022
ISBN 0-8138-1245-3

To Our Families

Contents

Part II The Electromagnetic Torque

Chapter 4. The Synchronous Machine

Chapter 5. The Simulation of Synchronous Machines

Chapter 6. Linear Models of the Synchronous Machine

Chapter 7. Excitation Systems

Preface

This book was written for the practicing engineer or the advanced student who is interested in power system dynamic behavior. The work grew out of the need for a modern course on the subject at Iowa State University, which would take advantage of prior training in elementary power system analysis, electric machines, circuit theory, control systems, and computer applications. The intent is to examine the detailed mathematical models of system components and to analyze the system behavior using the necessary computational tools. Thus digital or analog computations are utilized where either method seems appropriate.

Fine books have been written on the subject of power system stability; for example, the excellent volumes by E. W. Kimbark and by S. B. Crary. We believe, however, that there is a need for a book that stresses detailed modeling for computer solution and takes into account the massive amount of new information that has appeared in the technical literature in the last twenty years.

This book, which is the first of a planned two-volume series, is organized in two parts. Part I is a review of the theory of power system stability in common use in the power industry in the 1950s. The electrical torque, including a detailed treatment of the excitation system, is covered in Part II. A chapter dealing with multimachine systems with loads represented by constant impedances concludes this volume. Further development of a more complete tutorial treatment of power system stability and control will be undertaken in the second volume, with special attention given to the production of the mechanical torque.

The material in the book is presented in sufficient detail to satisfy the needs of the advanced student and to be used in the classroom. It is also intended to be useful to the practicing engineer. Modern methods of mathematical formulation of system equations have been adopted whenever possible, for example, in the use of state-space equations, matrices, etc. A significant departure from previous work is the use of a modified version of Park's transformation. We feel that the advantages to be gained far outweigh the occasional inconvenience suffered by giving up something familiar.

Only two authors are named on the cover of this book, but they know fully the generous contribution of many others to this work. Our students, too many to be named here, are the principal collaborators in a sense, as they offered an endless flow of corrections and suggestions to the developing manuscript. Our colleagues, particularly David D. Robb, Kenneth C. Kruempel, and John R. Pavlat contributed many valuable suggestions and computer checks of problems and examples. Associates in industry, particularly Clifford C. Young, have been very helpful, not only in offering corrections, but in educating the authors to many of the finer points of power system analysis. We acknowledge their generous contributions with thanks.

Part I Introduction

mechanical torque is adjusted accordingly. This occurs under normal operating conditions and during disturbances.

To be stable under normal conditions, the torque speed characteristic of the turbine speed control system should have a "droop characteristic"; i.e., a drop in turbine speed should accompany an increase in load. Such a characteristic is shown in Figure 2.3(b). A typical "droop" or "speed regulation" characteristic is 5% in the United States (4% in Europe). This means that a load pickup from no load (power) to full load (power) would correspond to a speed drop of 5% if the speed load characteristic is assumed to be linear. The droop (regulation) equation is derived as follows: from Figure 2.3(b), $T_m = T_{m0} + T_{m\Delta}$, and $T_{m\Delta} = -\omega_\Delta / R$, where R is the regulation in rad/N·m·s. Thus

$$T_m = T_{m0} - (\omega - \omega_R)/R \quad \text{N·m} \tag{2.25}$$

Multiplying (2.25) by ω_R, we can write

$$P_m \cong T_m \omega_R = P_{m0} - (\omega_R / R)\omega_\Delta \tag{2.26}$$

Let P_{mu} = pu mechanical power on machine VA

$$P_{mu} \triangleq P_m / S_B = P_{m0}/S_B - (\omega_R / S_B R)\omega_\Delta$$

or

$$P_{mu} = P_{m0u} - \omega_R^2 \omega_{\Delta u} / S_B R \quad \text{pu} \tag{2.27}$$

Since $P_{m\Delta} = P_m - P_{m0}$,

$$P_{m\Delta u} = -\omega_R^2 \omega_{\Delta u} / S_B R = -\omega_{\Delta u} / R_u \quad \text{pu} \tag{2.28}$$

where the pu regulation R_u is derived from (2.28) or

$$R_u \triangleq S_B R / \omega_R^2 \quad \text{pu} \tag{2.29}$$

As previously mentioned, R_u is usually set at 0.05 in the United States.

We also note that the "effective" regulation in a power system could be appreciably different from the value 0.05 if some of the machines are not under active governor control. If $\sum S_B$ is the sum of the ratings of the machines under governor control, and $\sum S_{sB}$ is the sum of the ratings of all machines, then the effective pu regulation is given by

$$R_{u\text{eff}} = R_u (\textstyle\sum S_B / \sum S_{sB}) \tag{2.30}$$

Similarly, if a system base other than that of the machine is used in a stability study, the change in mechanical power in pu on the system base $P_{m\Delta su}$ is given by

$$P_{m\Delta su} = -(S_B \omega_{\Delta u} / S_{sB} R_u) \quad \text{pu} \tag{2.31}$$

A block diagram representing (2.28) and (2.31) is shown in Figure 2.4 where

$$K = S_B / S_{sB}$$

The droop characteristic shown in Figure 2.3(b) is obtained in the speed control system with the help of feedback. It will be shown in Volume 2 that without feedback the speed control mechanism is unstable. Finally we should point out that the steady-state regulation characteristic determines the ultimate contribution of each machine to a change in load in the power system and fixes the resulting system frequency error.

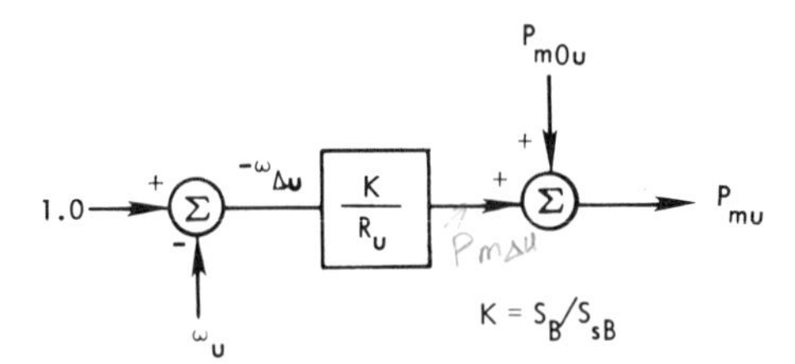

Fig. 2.4 Block diagram representation of the droop equation.

During transients the discrepancy between the mechanical and electrical torques for the various machines results in speed changes. The speed control mechanism for each machine under active governor control will attempt to adjust its output according to its regulation characteristic. Two points can be made here:

1. For a particular machine the regulation characteristic for a small (and sudden) change in speed may be considerably different in magnitude from its overall average regulation.
2. In attempting to adjust the mechanical torque to correspond to the speed change, time lags are introduced by the various delays in the feedback elements of the speed control system and in the steam paths; therefore, the dynamic response of the turbine could be appreciably different from that indicated by the steady-state regulation characteristic. This subject will be dealt with in greater detail in Volume 2.

2.4 Electrical Torque

In general, the electrical torque is produced by the interaction between the three stator circuits, the field circuit, and other circuits such as the damper windings. Since the three stator circuits are connected to the rest of the system, the terminal voltage is determined in part by the external network, the other machines, and the loads. The flux linking each circuit in the machine depends upon the exciter output voltage, the loading of the magnetic circuit (saturation), and the current in the different windings. Whether the machine is operating at synchronous speed or asynchronously affects all the above factors. Thus a comprehensive discussion of the electrical torque depends upon the synchronous machine representation. If all the circuits of the machine are taken into account, discussion of the electrical torque can become rather involved. Such a detailed discussion will be deferred to Chapter 4. For the present we simply note that the electrical torque depends upon the flux linking the stator windings and the currents in these windings. If the instantaneous values of these flux linkages and currents are known, the correct instantaneous value of the electrical torque may be determined. As the rotor moves, the flux linking each stator winding changes since the inductances between that winding and the rotor circuits are functions of the rotor position. These flux linkage relations are often simplified by using Park's transformation. A modified form of Park's transformation will be used here (see Chapter 4). Under this transformation both currents and flux linkages (and hence voltages) are transformed into two fictitious windings located on axes that are 90° apart and fixed with respect to the rotor. One axis coincides with the center of the magnetic poles of the rotor and is called the direct axis. The other axis lies along the magnetic neutral axis and is called the quadrature axis. Expressions for the electrical quantities such as power and torque are developed in terms of the direct and quadrature axis voltages (or flux linkages) and currents.

A simpler mathematical model, which may be used for stability studies, divides the electrical torque into two main components, the synchronous torque and a second component that includes all other electrical torques. We explore this concept briefly as an aid to understanding the generator behavior during transients.

2.4.1 Synchronous torque

The synchronous torque is the most important component of the electrical torque. It is produced by the interaction of the stator windings with the fundamental component of the air gap flux. It is dependent upon the machine terminal voltage, the rotor angle, the machine reactances, and the so-called quadrature axis EMF, which may be thought of as an effective rotor EMF that is dependent on the armature and rotor currents and is a function of the exciter response. Also, the network configuration affects the value of the terminal voltage.

2.4.2 Other electrical torques

During a transient, other extraneous electrical torques are developed in a synchronous machine. The most important component is associated with the damper windings. While these asynchronous torques are usually small in magnitude, their effect on stability may not be negligible. The most important effects are the following.

1. *Positive-sequence damping* results from the interaction between the positive-sequence air gap flux and the rotor windings, particularly the damper windings. In general, this effect is beneficial since it tends to reduce the magnitude of the machine oscillations, especially after the first swing. It is usually assumed to be proportional to the slip frequency, which is nearly the case for small slips.
2. *Negative-sequence braking* results from the interaction between the negative-sequence air gap flux during asymmetrical faults and the damper windings. Since the negative-sequence slip is $2 - s$, the torque is always retarding to the rotor. Its magnitude is significant only when the rotor damper winding resistance is high.
3. *The* dc *braking* is produced by the dc component of the armature current during faults, which induces currents in the rotor winding of fundamental frequency. Their interaction produces a torque that is always retarding to the rotor.

It should be emphasized that if the correct expression for the instantaneous electrical torque is used, all the above-mentioned components of the electrical torque will be included. In some studies approximate expressions for the torque are used, e.g., when considering quasi-steady-state conditions. Here we usually make an estimate of the components of the torque other than the synchronous torque.

2.5 Power-Angle Curve of a Synchronous Machine

Before we leave the subject of electrical torque (or power), we return momentarily to synchronous power to discuss a simplified but very useful expression for the relation between the power output of the machine and the angle of its rotor.

Consider two sources $\bar{V} = V\angle 0$ and $\bar{E} = E\angle\delta$ connected through a reactance x as shown in Figure 2.5(a).[3] Note that the source $\bar{V}$ is chosen as the reference. A current

3. A phasor is indicated with a bar above the symbol for the rms quantity. For example if I is the rms value of the current, $\bar{I}$ is the current phasor. By *definition* the phasor $\bar{I}$ is given by the transformation $\mathcal{P}$ where $\bar{I} \triangleq Ie^{j\theta} = I(\cos\theta + j\sin\theta) = \mathcal{P}[\sqrt{2}\, I\cos(\omega t + \theta)]$. A phasor is a complex number related to the corresponding time quantity $i(t)$ by $i(t) = \mathcal{R}e\,(\sqrt{2}\,\bar{I}e^{j\omega t}) = \sqrt{2}I\cos(\omega t + \theta) = \mathcal{P}^{-1}(Ie^{j\theta})$.

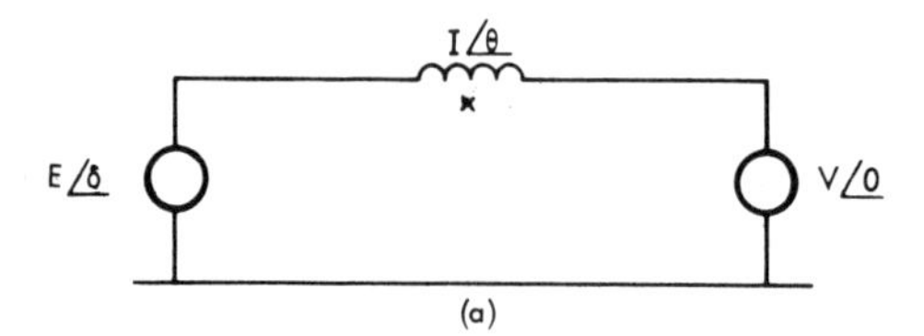

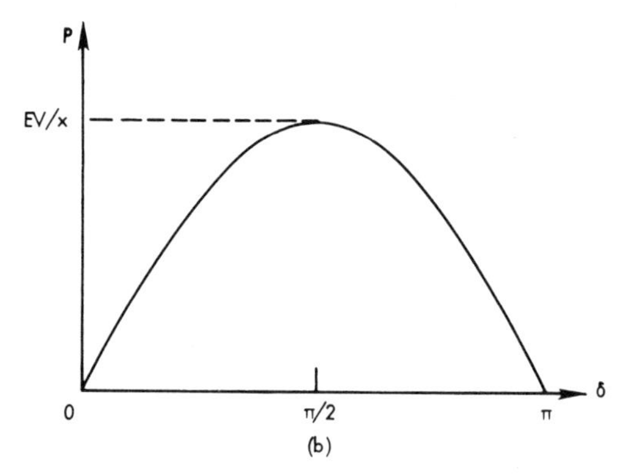

Fig. 2.5 A simple two-machine system: (a) schematic representation, (b) power-angle curve.

$\bar{I} = I\underline{/\theta}$ flows between the two sources. We can show that the power P is given by

$$P = (EV/x) \sin \delta \tag{2.32}$$

Since E, V, and x are constant, the relation between P and δ is a sine curve, as shown in Figure 2.5(b). We note that the same power is delivered by the source $\bar{E}$ and received by the source $\bar{V}$ since the network is purely reactive.

Consider a round rotor machine connected to an infinite bus. At steady state the machine can be represented approximately by the above circuit if V is the terminal voltage of the machine, which is the infinite bus voltage; x is the direct axis synchronous reactance; and E is the machine excitation voltage, which is the EMF along the quadrature axis. We say approximately because such factors as magnetic circuit saturation and the difference between direct and quadrature axis reluctances are overlooked in this simple representation. But (2.32) is essentially correct for a round rotor machine at steady state. Equation (2.32) indicates that if E, V, and x are constant, EV/x is a constant that we may designate as P_M to write $P = P_M \sin \delta$; and the power output of the machine is a function only of the angle δ associated with E. Note that E can be chosen to be any convenient EMF, not necessarily the excitation voltage; but then the appropriate x and δ must be defined accordingly.

2.5.1 Classical representation of a synchronous machine in stability studies

The EMF of the machine (i.e., the voltage corresponding to the current in the main field winding) can be considered as having two components: a component E' that corresponds to the flux linking the main field winding and a component that counteracts the armature reaction. The latter can change instantaneously because it corresponds to currents, but the former (which corresponds to flux linkage) cannot change instantly.

When a change in the network occurs suddenly, the flux linkage (and hence E') will not change, but currents will be produced in the armature; hence other currents will be induced in the various rotor circuits to keep this flux linkage constant. Both the armature and rotor currents will usually have ac and dc components as required to match the ampere-turns of various coupled coils. The flux will decay according to the effective time constant of the field circuit. At no load this time constant is on the order of several seconds, while under load it is reduced considerably but still on the order of one second or higher.

From the above we can see that for a period of less than a second the natural characteristic of the field winding of the synchronous machine tends to maintain constant flux linkage and hence constant E'. Exciters of the conventional type do not usually respond fast enough and their ceilings are not high enough to appreciably alter this picture. Furthermore, it has been observed that during a disturbance the combined effect of the armature reaction and the excitation system is to help maintain constant flux linkage for a period of a second or two. This period is often considered adequate for determining the stability of the machine. Thus in some stability studies the assumption is commonly made that the main field flux linkage of a machine is constant.

The main field-winding flux is almost the same as a fictitious flux that would create an EMF behind the machine direct axis transient reactance. The model used for the synchronous machine is shown in Figure 2.6, where x'_d is the direct axis transient reactance.

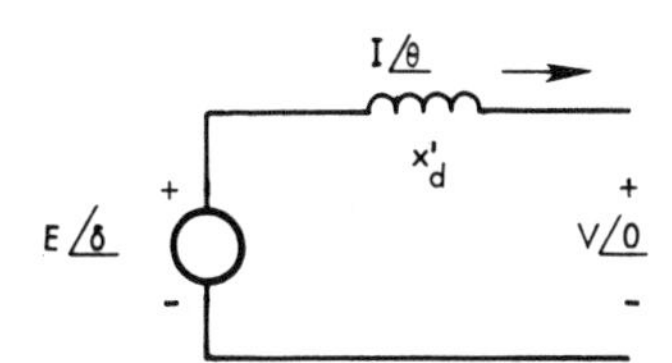

Fig. 2.6 Representation of a synchronous machine by a constant voltage behind transient reactance.

The constant voltage source $E\angle\delta$ is determined from the initial conditions, i.e., pretransient conditions. During the transient the magnitude E is held constant, while the angle δ is considered as the angle between the rotor position and the terminal voltage V.

Example 2.1

For the circuit of Figure 2.6 let $V = 1.0$ pu, $x'_d = 0.2$ pu, and the machine initially operating at $P = 0.8$ pu at 0.8 PF.

Solution

Using $\overline{V}$ as reference, $\overline{V} = 1.0\angle 0$

$$\overline{I}_0 = 1.0\angle -36.9^\circ = 0.8 - j0.6$$
$$\overline{E} = E\angle\delta = 1.0 + j0.2(0.8 - j0.6)$$
$$= 1.12 + j0.16 = 1.1314\angle 8.06^\circ$$

The magnitude of E is 1.1314. This will be held constant during the transient, although δ may vary. The initial value of δ, called δ_0, is 8.06°.

During the transient period, assuming that V is held constant, the machine power as a function of the angle δ is also given by a power-angle curve. Thus

$$P = (EV/x_d') \sin \delta = P_M \sin \delta \tag{2.33}$$

For the example given above $P_M = 1.1314/0.2 = 5.657$.

2.5.2 Synchronizing power coefficients

Consider a synchronous machine the terminal voltage of which is constant. This is the case when the machine is connected to a very large power system (infinite bus). Let us assume that the machine can be represented by a constant voltage magnitude behind a constant reactance, as shown in Figure 2.6. The power is given by (2.32). Let the initial power delivered by the machine be P_0, which corresponds to a rotor angle δ_0 (which is the same as the angle of the EMF E). Let us assume that δ changes from its initial value δ_0 by a small amount δ_Δ; i.e., $\delta = \delta_0 + \delta_\Delta$. From (2.32) P also changes to $P = P_0 + P_\Delta$. Then we may write

$$P_0 + P_\Delta = P_M \sin(\delta_0 + \delta_\Delta) = P_M(\sin \delta_0 \cos \delta_\Delta + \cos \delta_0 \sin \delta_\Delta) \tag{2.34}$$

If δ_Δ is small then, approximately, $\cos \delta_\Delta \cong 1$ and $\sin \delta_\Delta \cong \delta_\Delta$, or

$$P_0 + P_\Delta \cong P_M \sin \delta_0 + (P_M \cos \delta_0)\delta_\Delta$$

and since $P_0 = P_M \sin \delta_0$,

$$P_\Delta = (P_M \cos \delta_0)\delta_\Delta \tag{2.35}$$

The quantity in parentheses in (2.35) is defined to be the *synchronizing power coefficient* and is sometimes designated P_s. From (2.35) we also observe that

$$P_s \triangleq P_M \cos \delta_0 = \left.\frac{\partial P}{\partial \delta}\right]_{\delta = \delta_0} \tag{2.36}$$

Equation (2.35) is sometimes written in one of the forms

$$P_\Delta = P_s \delta_\Delta = \frac{\partial P}{\partial \delta} \delta_\Delta \tag{2.37}$$

(Compare this result with dP, the differential of P.)

In the above analysis the appropriate values of x and E should be used to obtain P_M. In dynamic studies x_d' and the voltage E' are used, while in steady-state stability analysis a saturated steady-state reactance x_d is used. If the control equipment of the machine is slow or inoperative, it is important that the machine be operating such that $0 \leq \delta \leq \pi/2$ for the operating point to be stable in the static or steady-state sense. This is the same as having a positive synchronizing power coefficient. This criterion was used in the past to indicate the so-called "steady-state stability limit."

2.6 Natural Frequencies of Oscillation of a Synchronous Machine

A synchronous machine, when perturbed, has several modes of oscillation with respect to the rest of the system. There are also cases where coherent groups of machines oscillate with respect to other coherent groups of machines. These oscillations cause fluctuations in bus voltages, system frequencies, and tie-line power flows. It is important that these oscillations should be small in magnitude and should be damped if the system is to be stable in the sense of the definition of stability given in Section 1.2.1.

In this section we will illu ronous
machine by the following exam

Example 2.2

A two-pole synchronous tage $\overline{V}$ through a reactance x as in Fig a small change in speed is given to the machine (the rotor is given a small twist); i.e., $\omega = \omega_0 + \epsilon u(t)$, where $u(t)$ is a unit step function. Let the resulting angle change be δ_Δ. Let the damping be negligible. Compute the change in angle as a function of time and determine its frequency of oscillation.

Solution

From (2.10) we write $M\ddot{\delta}/S_{B3} + P_e = P_m$. But we let $\delta = \delta_0 + \delta_\Delta$ such that $\ddot{\delta} = \ddot{\delta}_\Delta$ and $P_e = P_{e0} + P_{e\Delta}$; P_m is constant. Then $M\ddot{\delta}_\Delta/S_{B3} + P_{e\Delta} = P_m - P_{e0} = 0$ since $\ddot{\delta}_0 = 0$. From (2.37) for small δ_Δ we write $P_{e\Delta} = P_s\delta_\Delta$, where from (2.36) P_s is the synchronizing power coefficient. Then the swing equation may be written as

$$M\ddot{\delta}_\Delta/S_{B3} + P_s\delta_\Delta = 0$$

which has the solution of the form

$$\delta_\Delta(t) = \epsilon\sqrt{M/P_s}\,\sin\sqrt{P_sS_{B3}/M}\,t \quad \text{elect rad} \tag{2.38}$$

Equation (2.38) indicates that the angular frequency of oscillation of the synchronous machine with respect to the rest of the power system is given by $\sqrt{P_sS_{B3}/M}$. This frequency is usually referred to as the natural frequency of the synchronous machine.

It should be noted that P_s is a function of the operating point on the power-angle characteristic. Different machines, especially different machine types, have different inertia constants. Therefore, the different machines in a power system may have somewhat different natural frequencies.

We now estimate the order of magnitude of this frequency. From (2.6) and (2.16) we write $M/S_{B3} = 2H/\omega_m$ or $P_sS_{B3}/M = P_s\omega_m/2H$ where P_s is in pu, ω_m is in rad/s, and H is in s. Now P_s is the synchronizing power coefficient in pu (on a base of the machine three-phase rating). If the initial operating angle δ is small, P_s is approximately equal to the amplitude of the power-angle curve. We must also be careful with the units.

For example, a system having $P_s/S_{B3} = 2$ pu, $H = 8$,

$$\omega_{osc} = \sqrt{(2 \times 377)/(2 \times 8)} = 6.85 \quad \text{rad/s}$$
$$f_{osc} = 6.85/2\pi = 1.09 \quad \text{Hz}$$

If MKS units are used, we write

$$f_{osc} = (1/2\pi)\sqrt{\pi f(P_s/S_{B3}H)} \tag{2.39}$$

where f = system frequency in Hz
S_{B3} = three-phase machine rating in MVA
H = inertia constant in s
P_s = synchronizing power coefficient in MW/rad

Next, we should point out that a system of two finite machines can be reduced to a single equivalent finite machine against an infinite bus. The equivalent inertia is $J_1J_2/(J_1 + J_2)$ and the angle is $\delta_{1\Delta} - \delta_{2\Delta}$.

Thus we conclude that each machine oscillates with respect to other machines, each coherent group of machines oscillates with respect to other groups of machines, and so on. The frequencies of oscillations depend on the synchronizing power coefficients and on the inertia constants.

2.7 System of One Machine against an Infinite Bus—The Classical Model

An infinite bus is a source of invariable frequency and voltage (both in magnitude and angle). A major bus of a power system of very large capacity compared to the rating of the machine under consideration is approximately an infinite bus. The inertia of the machines in a large system will make the bus voltage of many high-voltage buses essentially constant for transients occurring outside that system.

Consider a power system consisting of one machine connected to an infinite bus through a transmission line. A schematic representation of this system is shown in Figure 2.7(a).

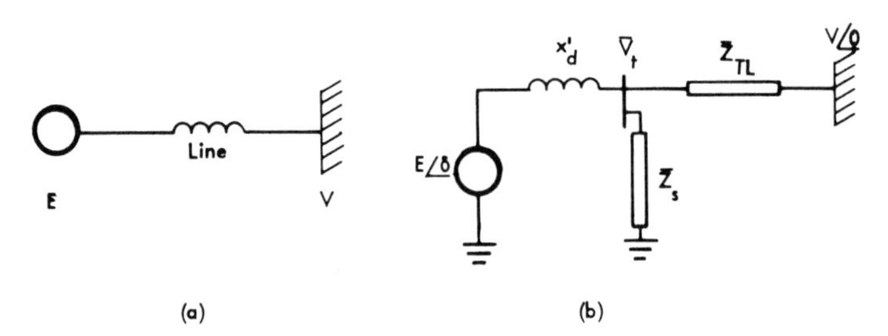

Fig.2.7 One machine connected to an infinite bus through a transmission line: (a) one-line diagram, (b) equivalent circuit.

The equation of motion of the rotor of the finite machine is given by the swing equation (2.7) or (2.10). To obtain a time solution for the rotor angle, we need to develop expressions for the mechanical and the electrical powers. In this section the simplest mathematical model is used. This model, which will be referred to as the classical model, requires the following assumptions:

1. The mechanical power input remains constant during the period of the transient.
2. Damping or asynchronous power is negligible.
3. The synchronous machine can be represented (electrically) by a constant voltage source behind a transient reactance (see Section 2.5.1).
4. The mechanical angle of the synchronous machine rotor coincides with the electrical phase angle of the voltage behind transient reactance.
5. If a local load is fed at the terminal voltage of the machine, it can be represented by a constant impedance (or admittance) to neutral.

The period of interest is the first swing of the rotor angle δ and is usually on the order of one second or less. At the start of the transient, and assuming that the impact initiating the transient creates a positive accelerating power on the machine rotor, the rotor angle increases. If the rotor angle increases indefinitely, the machine loses synchronism and stability is lost. If it reaches a maximum and then starts to decrease, the resulting motion will be oscillatory and with constant amplitude. Thus according to this model and the assumptions used, stability is decided in the first swing. (If damping is present the amplitude will decrease with time, but in the classical model there is very little damping.)

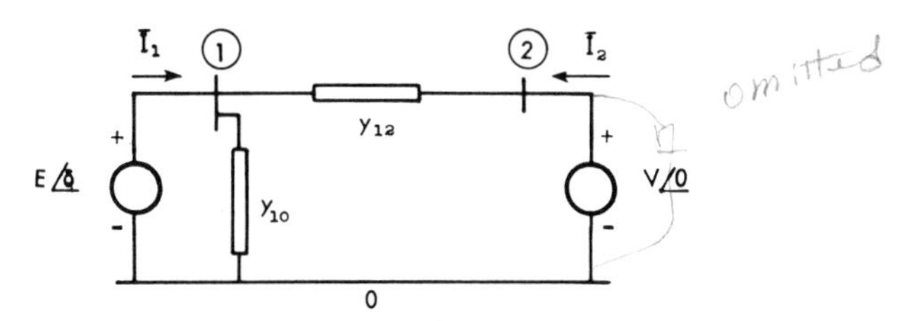

Fig. 2.8 Equivalent circuit for a system of one machine against an infinite bus.

The equivalent electrical circuit for the system is given in Figure 2.7(b). In Figure 2.7 we define

$\overline{V}_t$ = terminal voltage of the synchronous machine
$\overline{V} = V\underline{/0}$ = voltage of the infinite bus, which is used as reference
x_d' = direct axis transient reactance of the machine
$\overline{Z}_{TL}$ = series impedance of the transmission network (including transformers)
$\overline{Z}_s$ = equivalent shunt impedance at the machine terminal, including local loads if any

By using a Y-Δ transformation, the node representing the terminal voltage $\overline{V}_t$ in Figure 2.7 can be eliminated. The nodes to be retained (in addition to the reference node) are the internal voltage behind the transient reactance node and the infinite bus. These are shown in Figure 2.8 as nodes 1 and 2 respectively. Also shown in Figure 2.8 are the admittances obtained by the network reduction. Note that while three admittance elements are obtained (viz., y_{12}, y_{10}, and y_{20}), y_{20} is omitted since it is not needed in the analysis. The two-port network of Figure 2.8 is conveniently described by the equation

$$\begin{bmatrix} \overline{I}_1 \\ \overline{I}_2 \end{bmatrix} = \begin{bmatrix} \overline{Y}_{11} & \overline{Y}_{12} \\ \overline{Y}_{21} & \overline{Y}_{22} \end{bmatrix} \begin{bmatrix} \overline{E} \\ \overline{V} \end{bmatrix} \tag{2.40}$$

The driving point admittance at node 1 is given by $\overline{Y}_{11} = Y_{11}\underline{/\theta_{11}} = \overline{y}_{12} + \overline{y}_{10}$ where we use lower case y's to indicate actual admittances and capital Y's for matrix elements. The negative of the transfer admittance $\overline{y}_{12}$ between nodes 1 and 2 defines the admittance matrix element (1, 2) or $\overline{Y}_{12} = Y_{12}\underline{/\theta_{12}} = -\overline{y}_{12}$.

From elementary network theory we can show that the power at node 1 is given by $P_1 = \Re e\,\overline{E}\overline{I}_1^*$ or

$$P_1 = E^2 Y_{11} \cos\theta_{11} + EVY_{12}\cos(\theta_{12} - \delta)$$

Now define $G_{11} \triangleq Y_{11}\cos\theta_{11}$ and $\gamma = \theta_{12} - \pi/2$, then

$$P_1 = E^2 G_{11} + EVY_{12}\sin(\delta - \gamma) = P_C + P_M \sin(\delta - \gamma) \tag{2.41}$$

The relation between P_1 and δ in (2.41) is shown in Figure 2.9.

Examining Figure 2.9, we note that the power-angle curve of a synchronous machine connected to an infinite bus is a sine curve displaced from the origin vertically by an amount P_C, which represents the power dissipation in the equivalent network, and horizontally by the angle γ, which is determined by the real component of the transfer admittance $\overline{Y}_{12}$. In the special case where the shunt load at the machine terminal $\overline{V}_t$ is open and where the transmission network is reactive, we can easily prove that $P_C = 0$ and $\gamma = 0$. In this case the power-angle curve becomes identical to that given in (2.33).

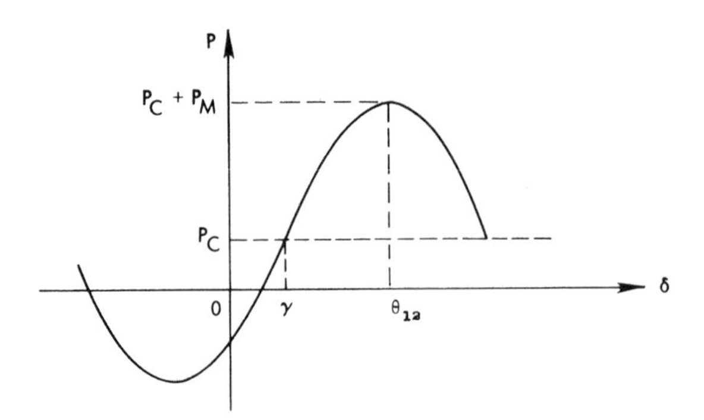

Fig. 2.9 Power output of a synchronous machine connected to an infinite bus.

Example 2.3

A synchronous machine is connected to an infinite bus through a transformer and a double circuit transmission line, as shown in Figure 2.10. The infinite bus voltage $V = 1.0$ pu. The direct axis transient reactance of the machine is 0.20 pu, the transformer reactance is 0.10 pu, and the reactance of each of the transmission lines is 0.40 pu, all to a base of the rating of the synchronous machine. Initially, the machine is delivering 0.8 pu power with a terminal voltage of 1.05 pu. The inertia constant $H = 5$ MJ/MVA. All resistances are neglected. The equation of motion of the machine rotor is to be determined.

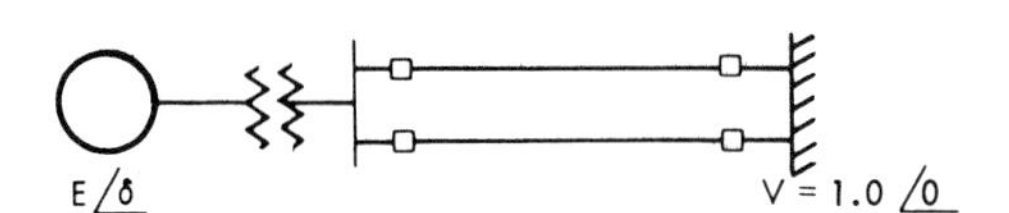

Fig. 2.10 System of Example 2.3.

Solution

The equivalent circuit of the system is shown in Figure 2.11. For this system:

$$\bar{y}_{12} = 1/\mathrm{j}0.5 = -\mathrm{j}2.0 \qquad \overline{Y}_{11} = -\mathrm{j}2.0$$

$$\bar{y}_{10} = 0 \qquad \theta_{11} = -\pi/2$$

$$\overline{Y}_{12} = \mathrm{j}2.0 \qquad \theta_{12} = \pi/2$$

therefore, $P_C = 0$ and $\gamma = 0$.

The electrical power is given by

$$P_e = P_1 = P_C + EVY_{12}\sin(\delta - \gamma) = EVY_{12}\sin\delta = 2E\sin\delta$$

Since the initial power is $P_{e0} = 0.8$ pu, then $E\sin\delta_0 = 0.4$.

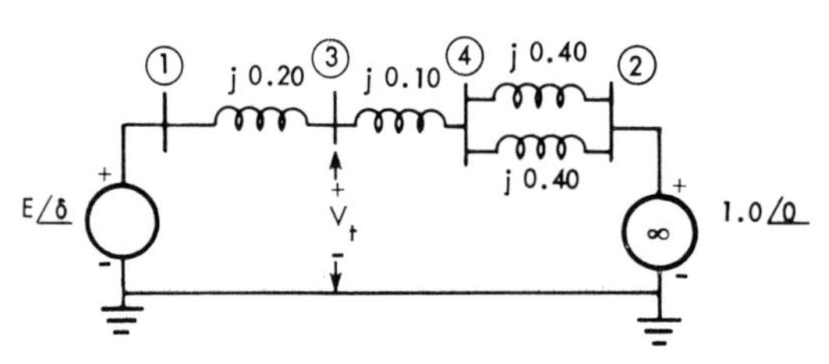

Fig. 2.11 Initial equivalent circuit of the system of Example 2.3.

To find the initial conditions, we solve the network of Figure 2.11. We have the terminal condition

$$\bar{V} = 1.0\,\underline{/0}\ \text{pu} \qquad \bar{V}_t = 1.05\,\underline{/\theta_t}\ \text{pu} \qquad P_e = 0.8\ \text{pu}$$

To find the angle of $\bar{V}_t$, we write, since resistance is zero,

$$\begin{aligned} P_{e0} &= 0.8 = (VV_t/x)\sin\theta_{t0} = (1.05/0.30)\sin\theta_{t0} \\ \sin\theta_{t0} &= 0.8/3.5 = 0.2286 \\ \theta_{t0} &= 13.21^\circ \end{aligned}$$

The current is found from $\bar{V}_t = \bar{Z}\bar{I} + \bar{V}$, or

$$\begin{aligned} \bar{I} &= (\bar{V}_t - \bar{V})/\bar{Z} = (1.05\,\underline{/13.21^\circ} - 1.0\,\underline{/0})/\text{j}0.3 \\ &= (1.022 + \text{j}0.240 - 1.000)/\text{j}0.3 = 0.800 - \text{j}0.074 = 0.803\,\underline{/-5.29^\circ} \end{aligned}$$

Then the internal machine voltage is

$$\begin{aligned} E\,\underline{/\delta} &= 1.05\,\underline{/13.21^\circ} + (0.803\,\underline{/-5.29^\circ})(0.2\,\underline{/90^\circ}) \\ &= 1.022 + \text{j}0.240 + 0.0148 + \text{j}0.160 \\ &= 1.037 + \text{j}0.400 = 1.111\,\underline{/21.09^\circ}\ \text{pu} \end{aligned}$$

Thus $E = 1.111$ is a constant that will be unchanged during the transient, and the initial angel is $\delta_0 = 21.09^\circ = 0.367$ rad. We also may write

$$P_e = [(1.111 \times 1.0)/0.50]\sin\delta = 2.222\sin\delta$$

Then the swing equation is given by

$$\frac{2H}{\omega_R}\frac{d^2\delta}{dt^2} = P_m - P_e$$

or

$$\frac{d^2\delta}{dt^2} = \frac{377}{10}(0.8 - 2.222\sin\delta)\ \text{rad/s}^2$$

From this simple example we observe that the resulting swing equation is nonlinear and will be difficult to solve except by numerical methods. We now extend the example to consider a fault on the system.

Example 2.4

Develop the equation of motion of the system of Figure 2.11 where a fault is applied at the sending end (node 4) of the transmission line. For simplicity we will consider a three-phase fault that presents a balanced impedance of j0.1 to neutral. The network now is as shown in Figure 2.12, where admittances are used for convenience.

Solution

By Y-Δ transformation we compute

$$y_{12} = -\text{j}[(3.333 \times 5)/18.333] = -\text{j}0.909$$

and since $\bar{Y}_{12} = -\bar{y}_{12}$, then $\bar{Y}_{12} = \text{j}0.909$. The electrical power output of the machine is now

$$P_e = (0.909 \times 1.111)\sin\delta = 1.010\sin\delta$$

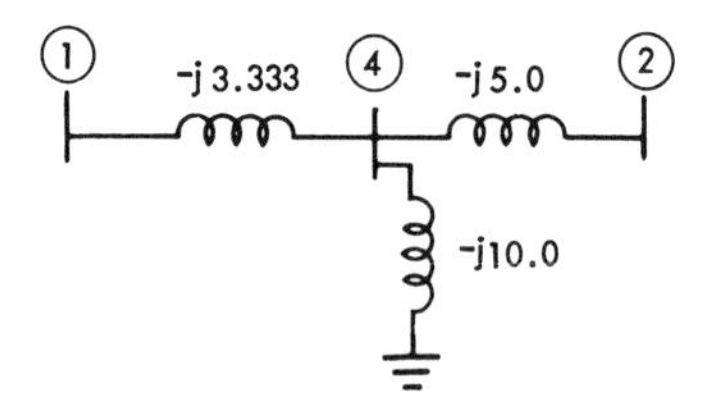

Fig. 2.12 Faulted network for Example 2.4 in terms of admittances.

From Example 2.3 the equation of motion of the rotor is

$$\frac{d^2\delta}{dt^2} = 37.7(0.8 - 1.010\sin\delta)\ \ \text{rad/s}^2$$

At the start of the transient $\sin\delta_0 = 0.36$, and the *initial* rotor acceleration is given by

$$\frac{d^2\delta}{dt^2} = 37.7[0.8 - (1.010 \times 0.368)] = 16.45\ \ \text{rad/s}^2$$

Now let us assume that after some time the circuit breaker at the sending end of the faulted line clears the fault by opening that line. The network now will have a series reactance of j0.70 pu, and the new network (with fault cleared) will have a new value of transfer admittance, $\overline{Y}_{12} = j1.429$ pu. The new swing equation will be

$$\frac{d^2\delta}{dt^2} = 37.7(0.8 - 1.587\sin\delta)\ \ \text{rad/s}^2$$

Example 2.5

Calculate the angle δ as a function of time for the system of Examples 2.3 and 2.4. Assume that the fault is cleared in nine cycles (0.15 s).

Solution

The equations for δ were obtained in Example 2.4 for the faulted network and for the system with the fault cleared. These equations are nonlinear; therefore, time solutions will be obtained by numerical methods. A partial survey of these methods is given in Appendix B.

To illustrate the procedure used in numerical integration, the modified Euler method is used in this example. This method is outlined in Appendix B.

First, the swing equation is replaced by the two first-order differential equations:

$$\dot{\delta} = \omega(t) - \omega_R \qquad \dot{\omega} = (\omega_R/2H)[P_m - P_e(t)] \tag{2.42}$$

The time domain is divided into increments called Δt. With the values of δ and ω and their derivatives known at some time t, an estimate is made of the values of these variables at the end of an interval of time Δt, i.e., at time $t + \Delta t$. These are called the *predicted values* of the variables and are based only on the values of $\delta(t)$, $\omega(t)$, and their derivatives. From the calculated values of $\delta(t + \Delta t)$ and $\omega(t + \Delta t)$, values of the derivatives at $t + \Delta t$ are calculated. A corrected value of $\delta(t + \Delta t)$ and $\omega(t + \Delta t)$ is obtained using the mean derivative over the interval. The process can be repeated until a desired precision is achieved. At the end of this repeated prediction and correction a final value of $\delta(t + \Delta t)$ and $\omega(t + \Delta t)$ is obtained. The process is then repeated for the next interval. The procedure is outlined in detail in Chapter 10 of [8]. From Example 2.4 the initial value of δ is $\sin^{-1}0.368$, and the equation

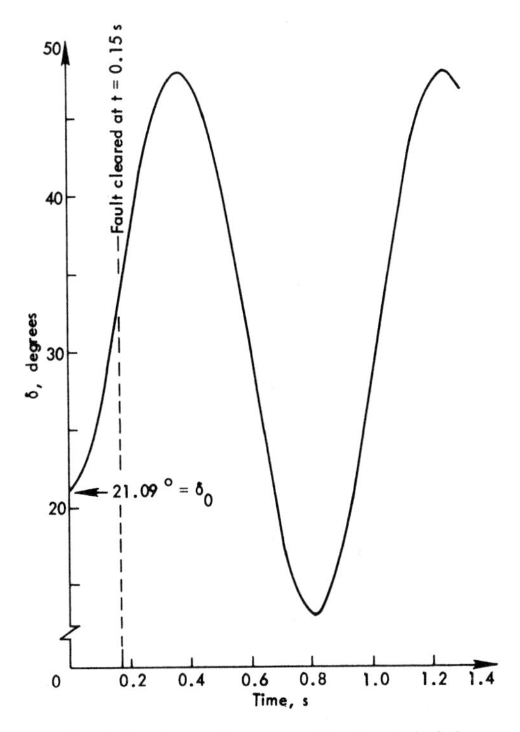

Fig. 2.13 Angle-time curve for Example 2.5.

for ω is given by

$$\begin{aligned} \dot{\omega} &= 37.7(0.800 - 1.010 \sin \delta) \qquad 0 \leqq t < 0.15 \\ &= 37.7(0.800 - 1.587 \sin \delta) \qquad t \geqq 0.15 \end{aligned}$$

The results of the numerical integration of the system equations, performed with the aid of a digital computer, are shown in Figure 2.13. The time solution is carried out for two successive peaks of the angle δ. The first peak of 48.2° is reached at $t = 0.38$ s, after which δ is decreased until it reaches a minimum value of about 13.2° at $t = 0.82$ s, and the oscillation of the rotor angle δ continues.

For the system under study and for the given impact, synchronism is not lost (since the angle δ does not increase indefinitely) and the synchronous machine is stable.

2.8 Equal Area Criterion

Consider the swing equation for a machine connected to an infinite bus derived previously in the form

$$\frac{2H}{\omega_R} \frac{d^2\delta}{dt^2} = P_m - P_e = P_a \text{ pu} \tag{2.43}$$

where P_a is the accelerating power. From (2.43)

$$\frac{d^2\delta}{dt^2} = \frac{\omega_R}{2H} P_a \tag{2.44}$$

Multiplying each side by $2(d\delta/dt)$,

$$2\frac{d\delta}{dt}\frac{d^2\delta}{dt^2} = \left(\frac{\omega_R}{2H}P_a\right)\left(2\frac{d\delta}{dt}\right) \tag{2.45}$$

$$\frac{d}{dt}\left[\left(\frac{d\delta}{dt}\right)^2\right] = \frac{\omega_R}{H}P_a\frac{d\delta}{dt} \tag{2.46}$$

$$d\left[\left(\frac{d\delta}{dt}\right)^2\right] = \frac{\omega_R}{H}P_a d\delta \tag{2.47}$$

Integrating both sides,

$$\left(\frac{d\delta}{dt}\right)^2 = \frac{\omega_R}{H}\int_{\delta_0}^{\delta} P_a d\delta \tag{2.48}$$

or

$$\frac{d\delta}{dt} = \left(\frac{\omega_R}{H}\int_{\delta_0}^{\delta} P_a d\delta\right)^{1/2} \tag{2.49}$$

Equation (2.49) gives the relative speed of the machine with respect to a reference frame moving at constant speed (by the definition of the angle δ). For stability this speed must be zero when the acceleration is either zero or is opposing the rotor motion. Thus for a rotor that is accelerating, the condition of stability is that a value δ_{max} exists such that $P_a(\delta_{max}) \leqq 0$, and

$$\int_{\delta_0}^{\delta_{max}} P_a d\delta = 0 \tag{2.50}$$

If the accelerating power is plotted as a function of δ, equation (2.50) can be interpreted as the area under that curve between δ_0 and δ_{max}. This is shown in Fig-

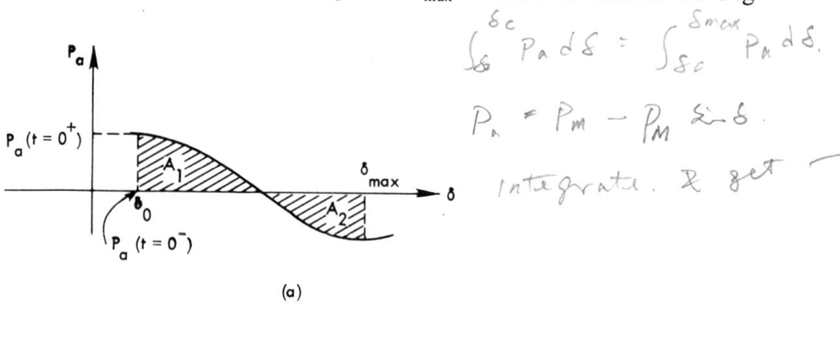

(a)

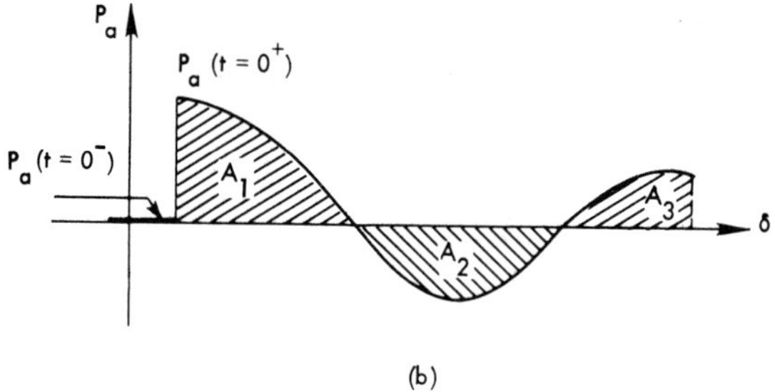

(b)

Fig. 2.14 Equal area criteria: (a) for stability for a stable system, (b) for an unstable system.

ure 2.14(a) where the net area under the P_a
δ_{max} since the two areas A_1 and A_2 are equal
power, and hence the rotor acceleration, is ne
δ_{max} is the maximum rotor angle reached duri

If the accelerating power reverses sign b
synchronism is lost. This situation is shown
than A_1, and as δ increases beyond the val
A_3 is added to A_1. The limit of stability occurs when the angle δ_{max} is such that $P_a(\delta_{max}) = 0$ and the areas A_1 and A_2 are equal. For this case δ_{max} coincides with the angle δ_m on the power-angle curve with the fault cleared such that $P = P_m$ and $\delta > \pi/2$.

Note that the accelerating power need not be plotted as a function of δ. We can obtain the same information if the electrical and mechanical powers are plotted as a function of δ. The former is the power-angle curve discussed in Section 2.7, and in many studies P_m is a constant. The accelerating power curve could have discontinuities due to switching of the network, initiation of faults, and the like.

2.8.1 Critical clearing angle

For a system of one machine connected to an infinite bus and for a given fault and switching arrangement, the critical clearing angle is that switching angle for which the system is at the edge of instability (we will also show that this applies to any two-machine system). The maximum angle δ_{max} corresponds to the angle δ_m on the fault-cleared power-angle curve. Conditions for critical clearing are now obtained (see [1] and [2]).

Let P_m = peak of the prefault power-angle curve
r_1 = ratio of the peak of the power-angle curve of the faulted network to P_M
r_2 = ratio of the peak of the power-angle curve of the network with the fault cleared to P_M
$\delta_0 = \sin^{-1} P_m/P_M < \pi/2$
$\delta_m = \sin^{-1} P_m/r_2 P_M > \pi/2$

Then for $A_1 = A_2$ and for critical clearing,

$$\delta_c = \cos^{-1}[1/(r_2 - r_1)][(P_m/P_M)(\delta_m - \delta_0) + r_2 \cos\delta_m - r_1 \cos\delta_0] \tag{2.51}$$

Note that the corresponding clearing time must be obtained from a time solution of the swing equation.

2.8.2 Application to a one-machine system

The equal area criterion is applied to the power network of Examples 2.4–2.5, and the results are shown in Figure 2.15. The stable system of Examples 2.4–2.5 is illustrated in Figure 2.15. The angle at $t = 0$ is 21.09° and is indicated by the intersection of P_m with the prefault curve. The clearing angle δ_c is obtained from the time solution (see Figure 2.13) and is about 31.6°. The conditions for $A_2 = A_1$ correspond to $\delta_{max} \cong 48°$. This corresponds to the maximum angle obtained in the time solution shown in Figure 2.13.

To illustrate the critical clearing angle, a more severe fault is used with the same system and switching arrangement. A three-phase fault is applied to the same bus with zero impedance. The faulted power-angle curve has zero amplitude. The prefault and

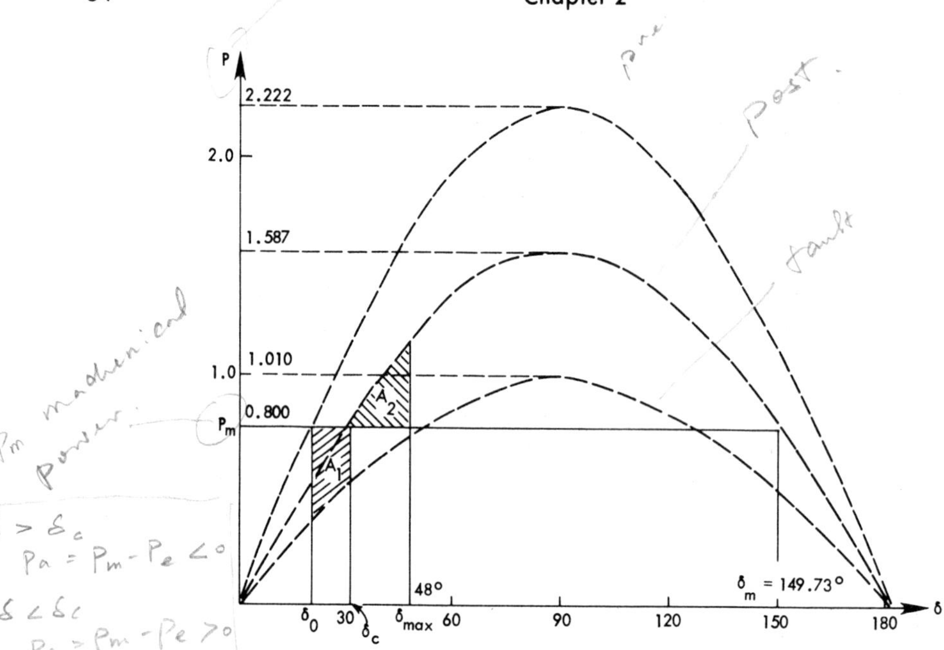

Fig. 2.15 Application of the equal area criterion to a stable system.

postfault networks are the same as before. For this system

$$r_1 = 0 \qquad \delta_0 = 21.09^\circ$$

$$r_2 = 1.587/2.222 = 0.714 \qquad \delta_m = 149.73^\circ$$

Calculation of the critical clearing angle, using (2.51), gives

$$\delta_c = \cos^{-1} 0.26848 = 74.43^\circ$$

This situation is illustrated in Figure 2.16.

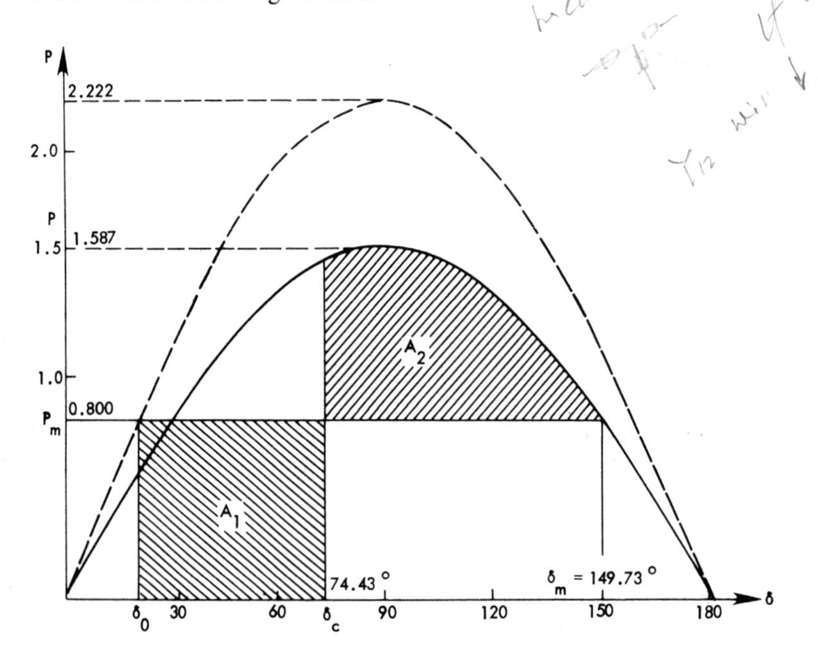

Fig. 2.16 Application of the equal area criterion to a critically cleared system.

2.8.3 Equal area criterion for a two-machine system

It can be shown that the equal area criterion applies to any two-machine system since a two-machine system can be reduced to an equivalent system of one machine connected to an infinite bus (see Problem 2.10). We can show that the expression for the equal area criterion in this case is given by

$$\int_{\delta_{120}}^{\delta_{12}} \left(\frac{P_{a1}}{H_1} - \frac{P_{a2}}{H_2} \right) d\delta_{12} = 0 \tag{2.52}$$

where $\delta_{12} = \delta_1 - \delta_2$.

In the special case where the resistance is neglected, (2.52) becomes

$$\frac{1}{H_0} \int_{\delta_{120}}^{\delta_{12}} P_{a1}\, d\delta_{12} = 0$$

where $H_0 = H_1 H_2/(H_1 + H_2)$.

2.9 Classical Model of a Multimachine System

The same assumptions used for a system of one machine connected to an infinite bus are often assumed valid for a multimachine system:

1. Mechanical power input is constant.
2. Damping or asynchronous power is negligible.
3. Constant-voltage-behind-transient-reactance model for the synchronous machines is valid.
4. The mechanical rotor angle of a machine coincides with the angle of the voltage behind the transient reactance.
5. Loads are represented by passive impedances.

This model is useful for stability analysis but is limited to the study of transients for only the "first swing" or for periods on the order of one second.

Assumption 2 is improved upon somewhat by assuming a linear damping characteristic. A damping torque (or power) $D\omega$ is frequently added to the inertial torque (or power) in the swing equation. The damping coefficient D includes the various damping torque components, both mechanical and electrical. Values of the damping coefficient usually used in stability studies are in the range of 1–3 pu [9, 10, 11, 12]. This represents turbine damping, generator electrical damping, and the damping effect of electrical loads. However, much larger damping coefficients, up to 25 pu, are reported in the literature due to generator damping alone [7, 13].

Assumption 5, suggesting load representation by a constant impedance, is made for convenience in many classical studies. Loads have their own dynamic behavior, which is usually not precisely known and varies from constant impedance to constant MVA. This is a subject of considerable speculation, the major point of agreement being that constant impedance is an inadequate representation. Load representation can have a marked effect on stability results.

The electrical network obtained for an n-machine system is as shown in Figure 2.17. Node 0 is the reference node (neutral). Nodes 1, 2, ..., n are the internal machine buses, or the buses to which the voltages behind transient reactances are applied. Passive impedances connect the various nodes and connect the nodes to the reference at load buses. As in the one-machine system, the initial values of $\bar{E}_1, \bar{E}_2, \ldots, \bar{E}_n$ are determined from the pretransient conditions. Thus a load-flow study for pretransient

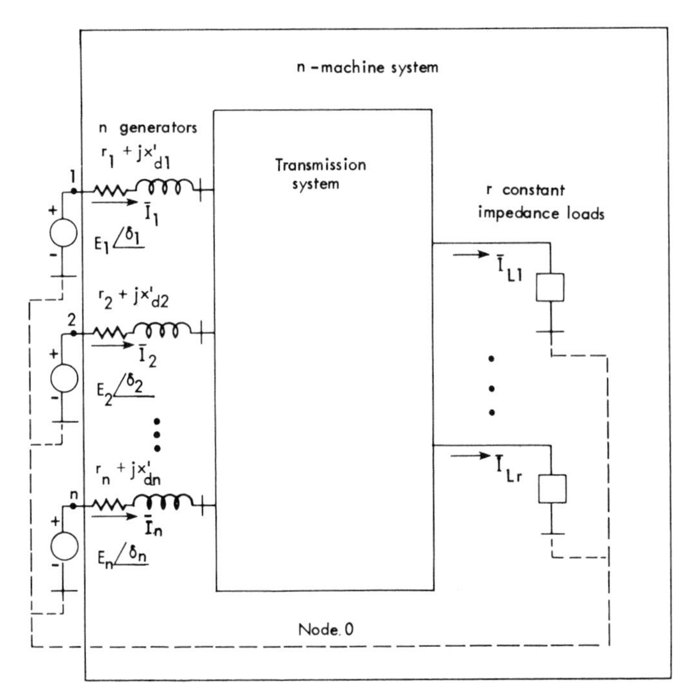

Fig. 2.17 Representation of a multimachine system (classical model).

conditions is needed. The magnitudes E_i, $i = 1, 2, \ldots, n$ are held constant during the transient in classical stability studies.

The passive electrical network described above has n nodes with active sources. The admittance matrix of the n-port network, looking into the network from the terminals of the generators, is defined by

$$\bar{\mathbf{I}} = \bar{\mathbf{Y}}\bar{\mathbf{E}} \tag{2.53}$$

where $\bar{\mathbf{Y}}$ has the diagonal elements $\bar{Y}_{ii}$ and the off-diagonal elements $\bar{Y}_{ij}$. By definition,

$$\begin{aligned}
\bar{Y}_{ii} &= Y_{ii}\underline{/\theta_{ii}} = \text{driving point admittance for node } i\\
&= G_{ii} + \mathrm{j}\,B_{ii}\\
\bar{Y}_{ij} &= Y_{ij}\underline{/\theta_{ij}} = \text{negative of the transfer admittance between nodes } i \text{ and } j\\
&= G_{ij} + \mathrm{j}\,B_{ij}
\end{aligned} \tag{2.54}$$

The power into the network at node i, which is the electrical power output of machine i, is given by $P_i = \mathcal{R}e\ \bar{E}_i\bar{I}_i^*$

$$P_{ei} = E_i^2 G_{ii} + \sum_{\substack{j=1\\ j\neq i}}^{n} E_i E_j Y_{ij}\cos(\theta_{ij} - \delta_i + \delta_j) \qquad i = 1, 2, \ldots, n$$

$$= E_i^2 G_{ii} + \sum_{\substack{j=1\\ j\neq i}}^{n} E_i E_j [B_{ij}\sin(\delta_i - \delta_j) + G_{ij}\cos(\delta_i - \delta_j)] \qquad i = 1, 2, \ldots, n \tag{2.55}$$

The equations of motion are then given by

$$\frac{2H_i}{\omega_R}\frac{d\omega_i}{dt} + D_i\omega_i = P_{mi} - \left[E_i^2 G_{ii} + \sum_{\substack{j=1\\ j\neq i}}^{n} E_i E_j Y_{ij}\cos(\theta_{ij} - \delta_i + \delta_j)\right]$$

$$\frac{d\delta_i}{dt} = \omega_i - \omega_R \qquad i = 1, 2, \ldots, n \tag{2.56}$$

It should be noted that prior to the disturbance ($t = 0^-$) $P_{mi0} = P_{ei0}$

$$P_{mi0} = E_i^2 G_{ii0} + \sum_{\substack{j=1\\ j\neq i}}^{n} E_i E_j Y_{ij0}\cos(\theta_{ij0} - \delta_{i0} + \delta_{j0}) \tag{2.57}$$

The subscript 0 is used to indicate the pretransient conditions. This applies to all machine rotor angles and also to the network parameters, since the network changes due to switching during the fault.

The set of equations (2.56) is a set of n-coupled nonlinear second-order differential equations. These can be written in the form

$$\mathbf{x} = \mathbf{f}(\mathbf{x}, \mathbf{x}_0, t) \tag{2.58}$$

where $\mathbf{x}$ is a vector of dimension ($2n \times 1$),

$$\mathbf{x}^t = [\omega_1, \delta_1, \omega_2, \delta_2, \ldots, \omega_n, \delta_n] \tag{2.59}$$

and $\mathbf{f}$ is a set of nonlinear functions of the elements of the state vector $\mathbf{x}$.

2.10 Classical Stability Study of a Nine-bus System

The classical model of a synchronous machine may be used to study the stability of a power system for a period of time during which the system dynamic response is dependent largely on the stored kinetic energy in the rotating masses. For many power systems this time is on the order of one second or less. The classical model is the simplest model used in studies of power system dynamics and requires a minimum amount of data; hence, such studies can be conducted in a relatively short time and at minimum cost. Furthermore, these studies can provide useful information. For example, they may be used as preliminary studies to identify problem areas that require further study with more detailed modeling. Thus a large number of cases for which the system exhibits a definitely stable dynamic response to the disturbances under study are eliminated from further consideration.

A classical study will be presented here on a small nine-bus power system that has three generators and three loads. A one-line impedance diagram for the system is given in Figure 2.18. The prefault normal load-flow solution is given in Figure 2.19. Generator data for the three machines are given in Table 2.1. This system, while small, is large enough to be nontrivial and thus permits the illustration of a number of stability concepts and results.

2.10.1 Data preparation

In the performance of a transient stability study, the following data are needed:

1. A load-flow study of the pretransient network to determine the mechanical power P_m of the generators and to calculate the values of $E_i\underline{/\delta_{i0}}$ for all the generators. The equivalent impedances of the loads are obtained from the load bus data.

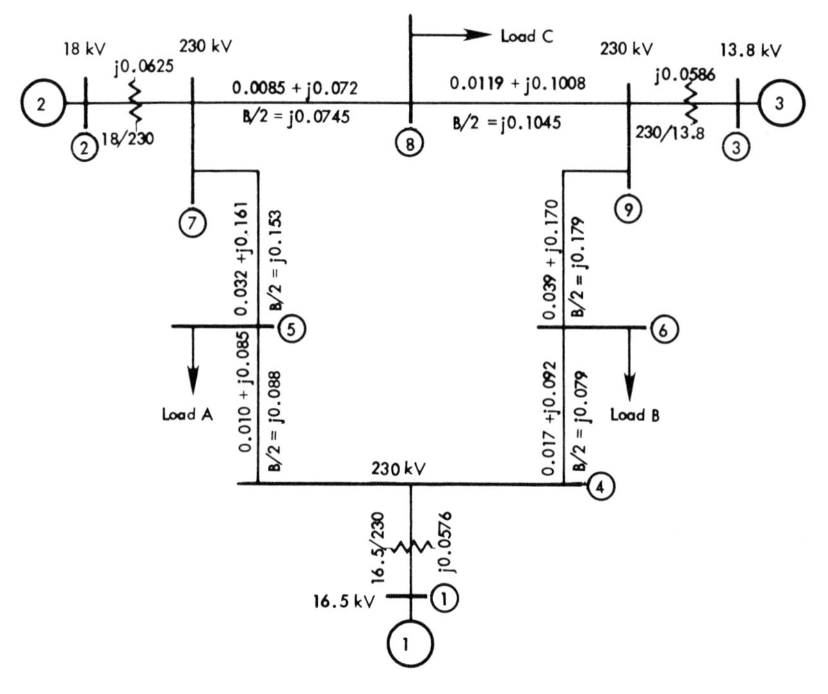

Fig. 2.18 Nine-bus system impedance diagram; all impedances are in pu on a 100-MVA base.

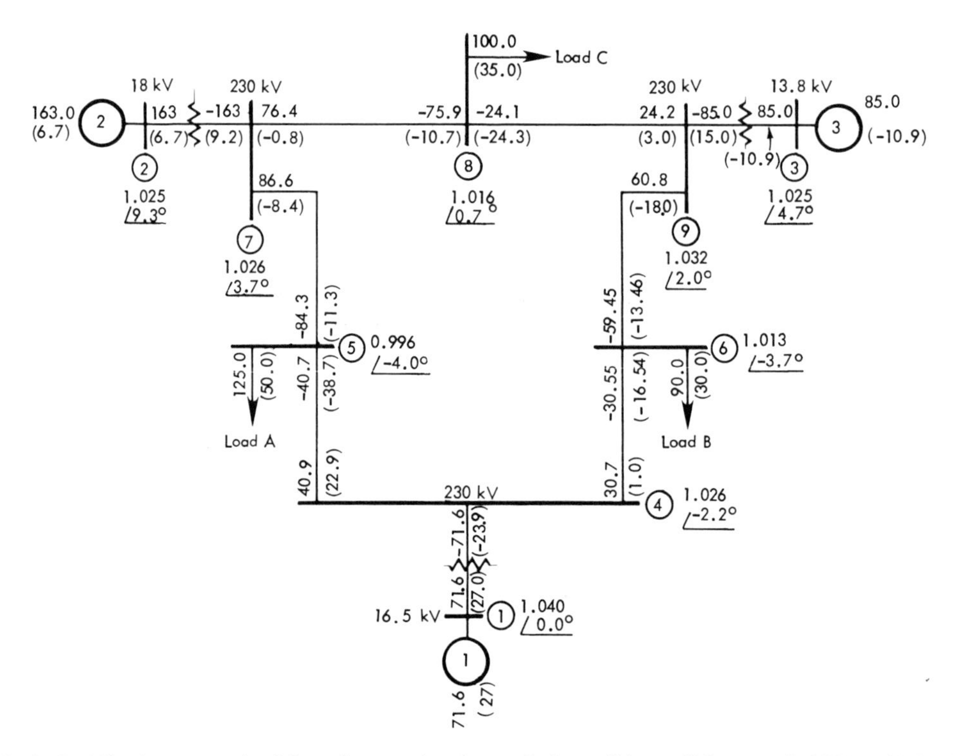

Fig. 2.19 Nine-bus system load-flow diagram showing prefault conditions; all flows are in MW and MVAR.

Table 2.1. Generat

Generator	1	[illegible]	[illegible]
Rated MVA	247.5	[illegible]	[illegible]
kV	16.5	[illegible]	[illegible]
Power factor	1.0	[illegible]	[illegible]
Type	hydro	[illegible]	[illegible]
Speed	180 r/min	3600 r/min	3600 r/min
x_d	0.1460	0.8958	1.3125
x'_d	0.0608	0.1198	0.1813
x_q	0.0969	0.8645	1.2578
x'_q	0.0969	0.1969	0.25
x_ℓ (leakage)	0.0336	0.0521	0.0742
τ'_{d0}	8.96	6.00	5.89
τ'_{q0}	0	0.535	0.600
Stored energy at rated speed	2364 MW·s	640 MW·s	301 MW·s

Note: Reactance values are in pu on a 100-MVA base. All time constants are in s. (Several quantities are tabulated that are as yet undefined in this book. These quantities are derived and justified in Chapter 4 but are given here to provide complete data for the sample system.)

2. System data as follows:
   a. The inertia constant H and direct axis transient reactance x'_d for all generators.
   b. Transmission network impedances for the initial network conditions and the subsequent switchings such as fault clearing and breaker reclosings.
3. The type and location of disturbance, time of switchings, and the maximum time for which a solution is to be obtained.

2.10.2 Preliminary calculations

To prepare the system data for a stability study, the following preliminary calculations are made:

1. All system data are converted to a common base; a system base of 100 MVA is frequently used.
2. The loads are converted to equivalent impedances or admittances. The needed data for this step are obtained from the load-flow study. Thus if a certain load bus has a voltage $\overline{V}_L$, power P_L, reactive power Q_L, and current $\overline{I}_L$ flowing into a load admittance $\overline{Y}_L = G_L + jB_L$, then

$$P_L + jQ_L = \overline{V}_L \overline{I}_L^* = \overline{V}_L[\overline{V}_L^*(G_L - jB_L)] = V_L^2(G_L - jB_L)$$

The equivalent shunt admittance at that bus is given by

$$\overline{Y}_L = P_L/V_L^2 - j(Q_L/V_L^2) \tag{2.60}$$

3. The internal voltages of the generators $E_i\underline{/\delta_{i0}}$ are calculated from the load-flow data. These internal angles may be computed from the pretransient terminal voltages $V\underline{/\alpha}$ as follows. Let the terminal voltage be used temporarily as a reference, as shown in Figure 2.20. If we define $\overline{I} = I_1 + jI_2$, then from the relation $P + jQ = \overline{V}\overline{I}^*$ we have $I_1 + jI_2 = (P - jQ)/V$. But since $E\underline{/\delta'} = \overline{V} + jx'_d\overline{I}$, we compute

$$E\underline{/\delta'} = (V + Qx'_d/V) + j(Px'_d/V) \tag{2.61}$$

The initial generator angle δ_0 is then obtained by adding the pretransient voltage

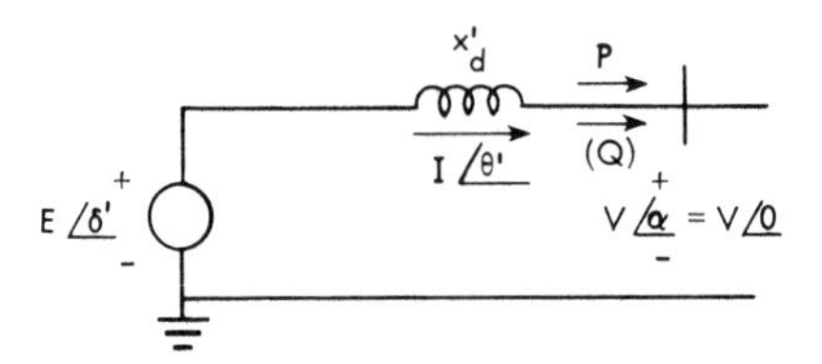

Fig. 2.20 Generator representation for computing δ_0.

angle α to δ', or

$$\delta_0 = \delta' + \alpha \tag{2.62}$$

4. The $\overline{\mathbf{Y}}$ matrix for each network condition is calculated. The following steps are usually needed:
   a. The equivalent load impedances (or admittances) are connected between the load buses and the reference node; additional nodes are provided for the internal generator voltages (nodes 1, 2, ..., n in Figure 2.17) and the appropriate values of x_d' are connected between these nodes and the generator terminal nodes. Also, simulation of the fault impedance is added as required, and the admittance matrix is determined for each switching condition.
   b. All impedance elements are converted to admittances.
   c. Elements of the $\overline{\mathbf{Y}}$ matrix are identified as follows: $\overline{Y}_{ii}$ is the sum of all the admittances connected to node i, and $\overline{Y}_{ij}$ is the negative of the admittance between node i and node j.
5. Finally, we eliminate all the nodes except for the internal generator nodes and obtain the $\overline{\mathbf{Y}}$ matrix for the reduced network. The reduction can be achieved by matrix operation if we recall that all the nodes have zero injection currents except for the internal generator nodes. This property is used to obtain the network reduction as shown below.

   Let

$$\mathbf{I} = \mathbf{YV} \tag{2.63}$$

where

$$\mathbf{I} = \begin{bmatrix} \mathbf{I}_n \\ -- \\ \mathbf{0} \end{bmatrix}$$

Now the matrices $\mathbf{Y}$ and $\mathbf{V}$ are partitioned accordingly to get

$$\begin{bmatrix} \mathbf{I}_n \\ -- \\ \mathbf{0} \end{bmatrix} = \left[\begin{array}{c|c} \mathbf{Y}_{nn} & \mathbf{Y}_{nr} \\ \hline \mathbf{Y}_{rn} & \mathbf{Y}_{rr} \end{array}\right] \begin{bmatrix} \mathbf{V}_n \\ -- \\ \mathbf{V}_r \end{bmatrix} \tag{2.64}$$

where the subscript n is used to denote generator nodes and the subscript r is used for the remaining nodes. Thus for the network in Figure 2.17, $\mathbf{V}_n$ has the dimension $(n \times 1)$ and $\mathbf{V}_r$ has the dimension $(r \times 1)$.

Expanding (2.64),

$$\mathbf{I}_n = \mathbf{Y}_{nn}\mathbf{V}_n + \mathbf{Y}_{nr}\mathbf{V}_r \qquad \mathbf{0} = \mathbf{Y}_{rn}\mathbf{V}_n + \mathbf{Y}_{rr}\mathbf{V}_r$$

from which we eliminate $\mathbf{V}_r$ to find

$$\mathbf{I}_n = (\mathbf{Y}_{nn} - \mathbf{Y}_{nr}\mathbf{Y}_{rr}^{-1}\mathbf{Y}_{rn})\mathbf{V}_n \tag{2.65}$$

The matrix $(\mathbf{Y}_{nn} - \mathbf{Y}_{nr}\mathbf{Y}_{rr}^{-1}\mathbf{Y}_{rn})$ is the desired reduced matrix $\mathbf{Y}$. It has the dimensions $(n \times n)$ where n is the number of the generators.

The network reduction illustrated by (2.63)–(2.65) is a convenient analytical technique that can be used only when the loads are treated as constant impedances. If the loads are not considered to be constant impedances, the identity of the load buses must be retained. Network reduction can be applied only to those nodes that have zero injection current.

Example 2.6

The technique of solving a classical transient stability problem is illustrated by conducting a study of the nine-bus system, the data for which is given in Figures 2.18 and 2.19 and Table 2.1. The disturbance initiating the transient is a three-phase fault occurring near bus 7 at the end of line 5-7. The fault is cleared in five cycles (0.083 s) by opening line 5-7.

For the purpose of this study the generators are to be represented by the classical model and the loads by constant impedances. The damping torques are neglected. The system base is 100 MVA.

Make all the preliminary calculations needed for a transient stability study so that all coefficients in (2.56) are known.

Solution

The objective of the study is to obtain time solutions for the rotor angles of the generators after the transient is introduced. These time solutions are called "swing curves." In the classical model the angles of the generator internal voltages behind transient reactances are assumed to correspond to the rotor angles. Therefore, mathematically, we are to obtain a solution for the set of equations (2.56). The initial conditions, denoted by adding the subscript 0, are given by $\dot{\omega}_{i0} = 0$ and δ_{i0} obtained from (2.57).

Preliminary calculations (following the steps outlined in Section 2.10.2) are:

1. The system base is chosen to be 100 MVA. All impedance data are given to this base.
2. The equivalent shunt admittances for the loads are given in pu as

$$\text{load A: } \bar{y}_{L5} = 1.2610 - j0.5044$$
$$\text{load B: } \bar{y}_{L6} = 0.8777 - j0.2926$$
$$\text{load C: } \bar{y}_{L8} = 0.9690 - j0.3391$$

3. The generator internal voltages and their initial angles are given in pu by

$$E_1\underline{/\delta_{10}} = 1.0566\underline{/2.2717^\circ}$$
$$E_2\underline{/\delta_{20}} = 1.0502\underline{/19.7315^\circ}$$
$$E_3\underline{/\delta_{30}} = 1.0170\underline{/13.1752^\circ}$$

4. The $\overline{\mathbf{Y}}$ matrix is obtained as outlined in Section 2.10.2, step 4. For convenience bus numbers 1, 2, and 3 are used to denote the generator internal buses rather than the generator low-voltage terminal buses. Values for the generator x_d' are added to the reactance of the generator transformers. For example, for generator 2 bus 2 will be the internal bus for the voltage behind transient reactance; the reactance between

Table 2.2. Prefault Network

	Bus no.	Impedance		Admittance	
		R	X	G	B
Generators*					
No. 1	1-4	0	0.1184	0	−8.4459
No. 2	2-7	0	0.1823	0	−5.4855
No. 3	3-9	0	0.2399	0	−4.1684
Transmission lines					
	4-5	0.0100	0.0850	1.3652	−11.6041
	4-6	0.0170	0.0920	1.9422	−10.5107
	5-7	0.0320	0.1610	1.1876	−5.9751
	6-9	0.0390	0.1700	1.2820	−5.5882
	7-8	0.0085	0.0720	1.6171	−13.6980
	8-9	0.0119	0.1008	1.1551	−9.7843
Shunt admittances†					
Load A	5-0			1.2610	−0.2634
Load B	6-0			0.8777	−0.0346
Load C	8-0			0.9690	−0.1601
	4-0				0.1670
	7-0				0.2275
	9-0				0.2835

*For each generator the transformer reactance is added to the generator x'_d.
†The line shunt susceptances are added to the loads.

bus 2 and bus 7 is the sum of the generator and transformer reactances (0.1198 + 0.0625). The prefault network admittances including the load equivalents are given in Table 2.2, and the corresponding $\bar{\mathbf{Y}}$ matrix is given in Table 2.3. The $\bar{\mathbf{Y}}$ matrix for the faulted network and for the network with the fault cleared are similarly obtained. The results are shown in Tables 2.4 and 2.5 respectively.

5. Elimination of the network nodes other than the generator internal nodes by network reduction as outlined in step 5 is done by digital computer. The resulting reduced $\mathbf{Y}$ matrices are shown in Table 2.6 for the prefault network, the faulted network, and the network with the fault cleared respectively.

We now have the values of the constant voltages behind transient reactances for all three generators and the reduced $\mathbf{Y}$ matrix for each network. Thus all coefficients of (2.56) are available.

Example 2.7

For the system and the transient of Example 2.6 calculate the rotor angles versus time. The fault is cleared in five cycles by opening line 5-7 of Figure 2.18. Plot the angles δ_1, δ_2, and δ_3 and their difference versus time.

Solution

The problem is to solve the set of equations (2.56) for $n = 3$ and $D = 0$. All the coefficients for the faulted network and the network with the fault cleared have been determined in Example 2.6. Since the set (2.56) is nonlinear, the desired time solutions for δ_1, δ_2, and δ_3 are obtained by numerical integration. A brief survey of numerical integration of differential equations is given in Appendix B. (For hand calculations see [1] for an excellent discussion of a numerical integration method of the swing equa-

Table 2.3. Y Matrix of Prefault Network

Node	1	2	3	4	5	6	7	8	9
1	−j8.4459			j8.4459					
2		−j5.4855					j5.4855		
3			−j4.1684						j4.1684
4	j8.4459			3.3074 − j30.3937	−1.3652 + j11.6041	−1.9422 + j10.5107			
5				−1.3652 + j11.6041	3.8138 − j17.8426		−1.1876 + j5.9751		
6				−1.9422 + j10.5107		4.1019 − j16.1335			−1.2820 + j5.5882
7		j5.4855			−1.1876 + j5.9751		2.8047 − j24.9311	−1.6171 + j13.6980	
8							−1.6171 + j13.6980	3.7412 − j23.6424	−1.1551 + j9.7843
9			j4.1684			−1.2820 + j5.5882		−1.1551 + j9.7843	2.4371 − j19.2574

Table 2.4. Y Matrix of Faulted Network

Node	1	2	3	4	5	6	7	8	9
1	−j8.4459			j8.4459					
2		−j5.4855							
3			−j4.1684						j4.1684
4	j8.4459			3.3074 − j30.3937	−1.3652 + j11.6041	−1.9422 + j10.5107			
5				−1.3652 + j11.6041	3.8138 − j17.8426				
6				−1.9422 + j10.5107		4.1019 − j16.1335			−1.2820 + j5.5882
7									
8								3.7412 − j23.6424	−1.1551 + j9.7843
9			−j4.1684			−1.2820 + j5.5882		−1.1551 + j9.7843	2.4371 − j19.2574

Table 2.5. Y Matrix of Network with Fault Cleared

Node	1	2	3	4	5	6	7	8	9
1	−j8.4459			j8.4459					
2		−j5.4855					j5.4855		
3			−j4.1684						j4.1684
4	j8.4459			3.3074 − j30.3937	−1.3652 + j11.6041	−1.9422 + j10.5107			
5				−1.3652 + j11.6041	2.6262 − j11.8675				
6				−1.9422 + j10.5107		4.1019 − j16.1335			−1.2820 + j5.5882
7		j5.4855					1.6171 − j18.9559	−1.6171 + j13.6980	
8							−1.6171 + j13.6980	3.7412 − j23.6424	−1.1551 + j9.7843
9			j4.1684			−1.2820 + j5.5882		−1.1551 + j9.7843	2.4371 − j19.2574

Table 2.6. Reduced Y Matrices

Type of network	Node	1	2	3
Prefault	1	0.846 − j2.988	0.287 + j1.513	0.210 + j1.226
	2	0.287 + j1.513	0.420 − j2.724	0.213 + j1.088
	3	0.210 + j1.226	0.213 + j1.088	0.277 − j2.368
Faulted	1	0.657 − j3.816	0.000 + j0.000	0.070 + j0.631
	2	0.000 + j0.000	0.000 − j5.486	0.000 + j0.000
	3	0.070 + j0.631	0.000 + j0.000	0.174 − j2.796
Fault cleared	1	1.181 − j2.229	0.138 + j0.726	0.191 + j1.079
	2	0.138 + j0.726	0.389 − j1.953	0.199 + j1.229
	3	0.191 + j1.079	0.199 + j1.229	0.273 − j2.342

tion. Also see Chapter 10 of [8] for a more detailed discussion of several numerical schemes for solving the swing equation.) The so-called transient stability digital computer programs available at many computer centers include subroutines for solving nonlinear differential equations. Discussion of these programs is beyond the scope of this book.

Numerical integration of the swing equations for the three-generator, nine-bus system is made by digital computer for 2.0 s of simulated real time. Figure 2.21 shows the rotor angles of the three machines. A plot of $\delta_{21} = \delta_2 - \delta_1$ and $\delta_{31} = \delta_3 - \delta_1$ is shown

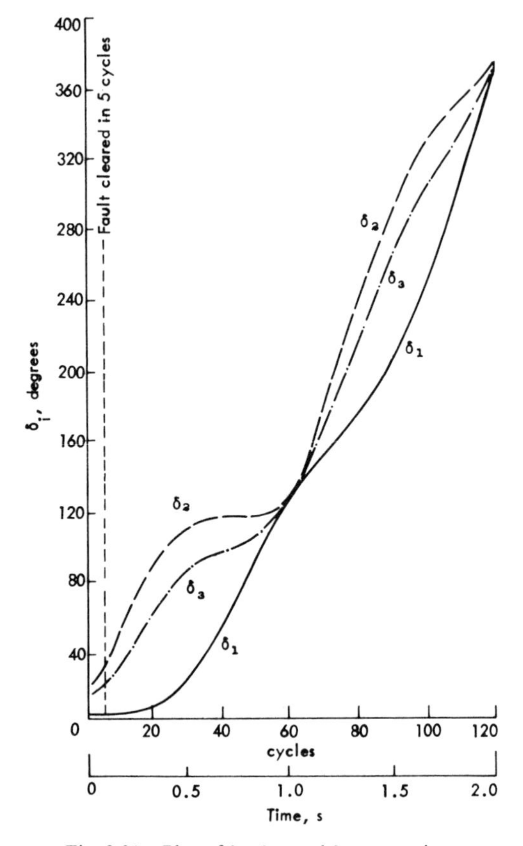

Fig. 2.21 Plot of δ_1, δ_2, and δ_3 versus time.

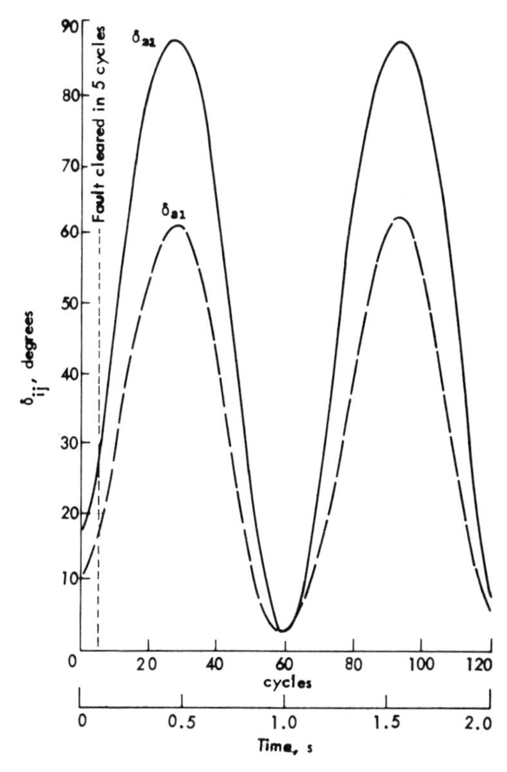

Fig. 2.22 Plot of δ differences versus time.

in Figure 2.22 where we can see that the system is stable. The maximum angle difference is about 85°. This is the value of δ_{21} at $t = 0.43$ s. Note that the solution is carried out for two "swings" to show that the second swing is not greater than the first for either δ_{21} or δ_{31}. To determine whether the system is stable or unstable for the particular transient under study, it is sufficient to carry out the time solution for one swing only. If the rotor angles (or the angle differences) reach maximum values and then decrease, the system is stable. If any of the angle differences increase indefinitely, the system is unstable because at least one machine will lose synchronism.

2.11 Shortcomings of the Classical Model

System stability depends on the characteristics of all the components of the power system. This includes the response characteristics of the control equipment on the turbogenerators, on the dynamic characteristics of the loads, on the supplementary control equipment installed, and on the type and settings of protective equipment used.

The machine dynamic response to any impact in the system is oscillatory. In the past the sizes of the power systems involved were such that the period of these oscillations was not much greater than one second. Furthermore, the equipment used for excitation controls was relatively slow and simple. Thus the classical model was adequate.

Today large system interconnections with the greater system inertias and relatively weaker ties result in longer periods of oscillations during transients. Generator control systems, particularly modern excitation systems, are extremely fast. It is therefore questionable whether the effect of the control equipment can be neglected during these longer periods. Indeed there have been recorded transients caused by large impacts, resulting in loss of synchronism after the system machines had undergone several oscillations. Another aspect is the dynamic instability problem, where growing oscillations have occurred on tie lines connecting different power pools or systems. As this situation has developed, it has also become increasingly important to ensure the security of the bulk power supply. This has made many engineers realize it is time to reexamine the assumptions made in stability studies. This view is well stated by Ray and Shipley [14]:

> We have reached a time when it is appropriate that we appraise the state of the Art of Dynamic Stability Analysis. In conjunction with this we must:
>
> 1. Expand our knowledge of the characteristic time response of our system loads to changes in voltage and frequency—develop new dynamic models of system loads.
> 2. Re-examine old concepts and develop new ideas on changes in system networks to improve system stability.
> 3. Update our knowledge of the response characteristics of the various components of energy systems and their controls (boilers, reactors, turbine governors, generator regulators, field excitation, etc.)
> 4. Reformulate our analytical techniques to adequately simulate the time variation of all of the foregoing factors in system response and accurately determine dynamic system response.

Let us now make a critical appraisal of some of the assumptions made in the classical model:

1. *Transient stability is decided in the first swing.* A large system having many machines will have numerous natural frequencies of oscillations. The capacities of most of the tie lines are comparatively small, with the result that some of these frequencies are quite low (frequencies of periods in the order of 5–6 s are not uncommon). It is quite possible that the worst swing may occur at an instant in time when the peaks of some of these nodes coincide. It is therefore necessary in many cases to study the transient for a period longer than one second.
2. *Constant generator main field-winding flux linkage.* This assumption is suspect on two counts, the longer period that must now be considered and the speed of many modern voltage regulators. The longer period, which may be comparable to the field-winding time constant, means that the change in the main field-winding flux may be appreciable and should be accounted for so that a correct representation of the system voltage is realized. Furthermore, the voltage regulator response could have a significant effect on the field-winding flux. We conclude from this discussion that the constant voltage behind transient reactance could be very inaccurate.
3. *Neglecting the damping powers.* A large system will have relatively weak ties. In the spring-mass analogy used above, this is a rather poorly damped system. It is important to account for the various components of the system damping to obtain a correct model that will accurately predict its dynamic performance, especially in loss of generation studies [8].
4. *Constant mechanical power.* If periods on the order of a few seconds or greater are of interest, it is unrealistic to assume that the mechanical power will not change. The turbine-governor characteristics, and perhaps boiler characteristics should be included in the analysis.
5. *Representing loads by constant passive impedance.* Let us illustrate in a qualitative manner the effect of such representation. Consider a bus having a voltage V to which a load $P_L + jQ_L$ is connected. Let the load be represented by the static ad-

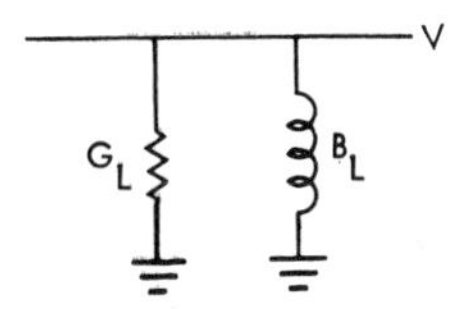

Fig. 2.23 A load represented by passive admittance.

mittances $G_L = P_L/V^2$ and $B_L = Q_L/V^2$ as shown in Figure 2.23. During a transient the voltage magnitude V and the frequency will change. In the model used in Figure 2.17 the change in voltage is reflected in the power and reactive power of the load, while the change in the bus frequency is not reflected at all in the load power. In other words, this model assumes $P_L \propto V^2$, $Q_L \propto V^2$, and that both are frequency independent. This assumption is often on the pessimistic side. (There are situations, however, where this assumption can lead to optimistic results. This discussion is intended to illustrate the errors implied.) To illustrate this, let us assume that the transient has been initiated by a fault in the transmission network. Initially, a fault causes a reduction of the output power of most of the synchronous generators. Some excess generation results, causing the machines to accelerate, and the area frequency tends to increase. At the same time, a transmission network fault usually causes a reduction of the bus voltages near the fault location. In the passive impedance model the load power decreases considerably (since $P_L \propto V^2$), and the increase in frequency does not cause an increase in load power. In real systems the decrease in power is not likely to be proportional to V^2 but rather less than this. An increase in system frequency will result in an increase in the load power. Thus the model used gives a load power lower than expected during the fault and higher than normal after fault removal.

From the foregoing discussion we conclude that the classical model is inadequate for system representation beyond the first swing. Since the first swing is largely an inertial response to a given accelerating torque, the classical model does provide useful information as to system response during this brief period.

2.12 Block Diagram of One Machine

Block diagrams are useful for helping the control engineer visualize a problem. We will be considering the control system for synchronous generators and will do so by analyzing each control function in turn. It may be helpful to present a general block diagram of the entire system without worrying about mathematical details as to what makes up the various blocks. Then as we proceed to analyze each system, we can fill in the blocks with the appropriate equations or transfer functions. Such a block diagram is shown in Figure 2.24 [15].

The basic equation of the dynamic system of Figure 2.24 is (2.18); i.e.,

$$\tau_j \dot{\omega} = P_m - P_e = P_a \text{ pu} \tag{2.66}$$

where $\ddot{\delta}$ has been replaced by $\dot{\omega}$ and J has been replaced by a time constant τ_j, the numerical value of which depends on the rotating inertia and the system of units.

Three separate control systems are associated with the generator of Figure 2.24. The first is the excitation system that controls the terminal voltage. Note that the excitation system also plays an important role in the machine's mechanical oscillations, since it affects the electrical power, P_e. The second control system is the speed control or governor that monitors the shaft speed and controls the mechanical power P_m.

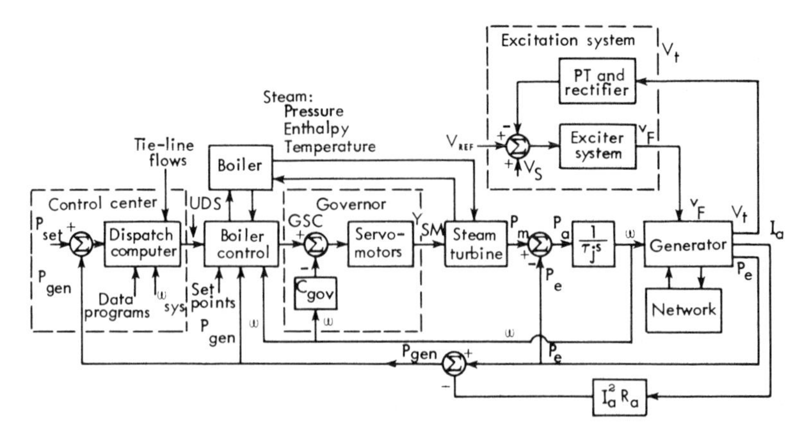

Fig. 2.24 Block diagram of a synchronous generator control system.

Finally, in an interconnected system there is a master controller for each system. This sends a unit dispatch signal (UDS) to each generator and adjusts this signal to meet the load demand or the scheduled tie-line power. It is designed to be quite slow so that it is usually not involved in a consideration of mechanical dynamics of the shaft. Thus in most of our work we can consider the speed reference or governor speed changer (GSC) position to be a constant. In an isolated system the speed reference is the desired system speed and is set mechanically in the governor mechanism, as will be shown later.

In addition to the three control systems, three transfer functions are of vital importance. The first of these is the generator transfer function. The generator equations are nonlinear and the transfer function is a linearized approximation of the behavior of the generator terminal voltage V_t near a quiescent operating point or equilibrium state. The load equations are also nonlinear and reflect changes in the electrical output quantities due to changes in terminal voltage V_t. Finally, the energy source equations are a description of the boiler and steam turbine or of the penstock and hydraulic turbine behavior as the governor output calls for changes in the energy input. These equations are very nonlinear and have several long time constants.

To visualize the stability problem in terms of Figure 2.24, we recognize immediately that the shaft speed ω must be accurately controlled since this machine must operate at precisely the same frequency as all others in the system. If a sudden change in ω occurs, we have two ways of providing controlled responses to this change. One is through the governor that controls the mechanical power P_m, but does so through some rather long time constants. A second controlled response acts through the excitation system to control the electrical power P_e. Time delays are involved here too, but they are smaller than those in the governor loop. Hence much effort has been devoted to refinements in excitation control.

Problems

2.1 Analyze (2.1) dimensionally using a mass, length, time system and specify the units of each quantity (see Kimbark [1]).

2.2 A rotating shaft has zero retarding torque $T_e = 0$ and is supplied a constant full load accelerating torque; i.e., $T_m = T_{FL}$. Let τ_c be the accelerating time constant, i.e., the time required to accelerate the machine from rest to rated speed ω_R. Solve the swing equation to find τ_c in terms of the moment of inertia J, ω_R, and T_{FL}. Then show that τ_c can also be related to H, the pu inertia constant.

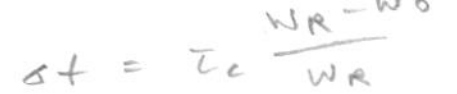

2.3 Solve the swing equation to find the time to reach full load speed ω_R starting from any initial speed ω_0 with constant accelerating torque as in Problem 2.2. Relate this time to τ_c and the slip at speed ω_0.

2.4 Write the equation of motion of the shaft for the following systems:

(a) An electric generator driven by a dc motor, where in the region of interest the generator torque is proportional to the shaft angle and the motor torque decreases linearly with increased speed.

(b) An electric motor driving a fan, where in the region of interest the torques are given by

$$T_{\text{motor}} = a - b\dot{\theta} \qquad T_{\text{fan}} = c\dot{\theta}^2$$

where a, b, and c are constants. State any necessary assumptions. Will this system have a steady-state operating point? Is the system linear?

2.5 In (2.4) assume that T is in N·m, δ is in elec. deg., and J is in lbm·ft^2. What factor must be used to make the units consistent?

2.6 In (2.7) assume that P is in W and M in J·s/rad. What are the units of δ?

2.7 A 500-MVA two-pole machine is to operate in parallel with other U.S. machines. Compute the regulation R of this machine. What are the units of R?

2.8 A 60-MVA two-pole generator and a 600-MVA four-pole generator are to operate in parallel with other U.S. systems and are to share in system governing. Compute the pu constant K that must be used with these machines in their governor simulations if the system base is 100 MVA.

2.9 Repeat problem 2.8 if the constant K is to be computed in MKS units rather than pu.

2.10 In computer simulations it is common to see regulation expressed in two different ways as described below:

(a) $P_m - P_{m0} = s/fR_{su}$

where
- P_m = mechanical power in pu on S_{sB}
- P_{m0} = initial mechanical power in pu on S_{sB}
- f = system base frequency in Hz
- R_{su} = steady-state speed regulation in pu on a system base = $R_u S_{sB}/S_B$
- s = generator slip = $(\omega_R - \omega)/2\pi$ Hz

(b) $P_m - P_{m0} = K_1 \Delta\omega$ pu.

where
- P_m = turbine power in pu on S_{sB}
- P_{m0} = initial turbine power in pu on S_{sB}
- K_1 = $S_B/R_u\omega_R S_{sB}$
- $\Delta\omega$ = speed deviation, rad/s

Verify the expressions in (a) and (b).

2.11 A synchronous machine having inertia constant H = 4.0 MJ/MVA is initially operated in steady state against an infinite bus with angular displacement of 30 elec. deg. and delivering 1.0 pu power. Find the natural frequency of oscillation for this machine, assuming small perturbations from the operating point.

2.12 A solid-rotor synchronous generator is driven by an unregulated turbine with a torque speed characteristic similar to that of Figure 2.3(a). The machine has the same characteristics and operating conditions as given in Problem 2.11 and is connected to an infinite bus. Find the natural frequency of oscillation and the damping coefficient, assuming small perturbations from the operating point.

2.13 Suppose that (2.33) is written for a salient pole machine to include a reluctance torque term; i.e., let $P = P_M \sin\delta + k\sin 2\delta$. For this condition find the expression for P_Δ and for the synchronizing power coefficient.

2.14 Derive an expression similar to that of (2.7) for an interconnection of two finite machines that have inertia constants M_1 and M_2 and angles δ_1 and δ_2. Show that the equations for such a case are exactly equivalent to that of a single finite machine of inertia

$$M = M_1 M_2/(M_1 + M_2)$$

and angle $\delta_{12} = \delta_1 - \delta_2$ connected to an infinite bus.

2.15 Derive linearized expressions (similar to Example 2.2) that describe an interconnection of three finite machines with inertia constants M_1, M_2, and M_3 and angles δ_1, δ_2, and δ_3. Is there a simple expression for the natural frequency of oscillation in this case? Designate synchronizing power between machines 1 and 2 as P_{s12}, etc.

2.16 The system shown in Figure P2.16 has two finite synchronous machines, each represented by a constant voltage behind reactance and connected by a pure reactance. The reactance x includes the transmission line and the machine reactances. Write the swing equation for each machine, and show that this system can be reduced to an equivalent one machine against an infinite bus. Give the inertia constant for the equivalent machine, the mechanical input power, and the amplitude of its power-angle curve. The inertia constants of the two machines are H_1 and H_2 s.

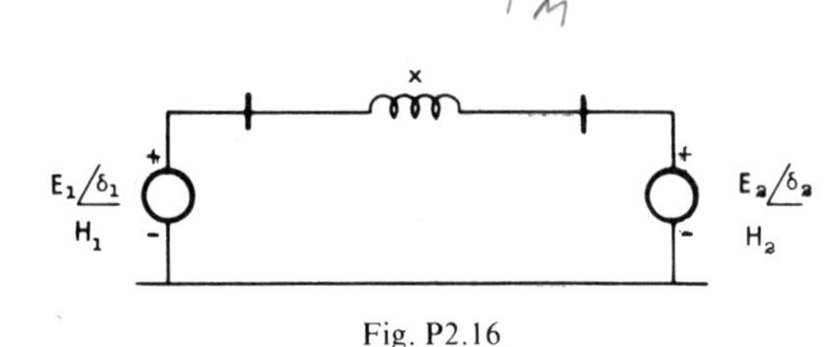

Fig. P2.16

2.17 The system shown in Figure P2.17 comprises four synchronous machines. Machines A and B are 60 Hz, while machines C and D are 50 Hz; B and C are a motor-generator set (frequency changer). Write the equations of motion for this system. Assume that the transmission networks are reactive.

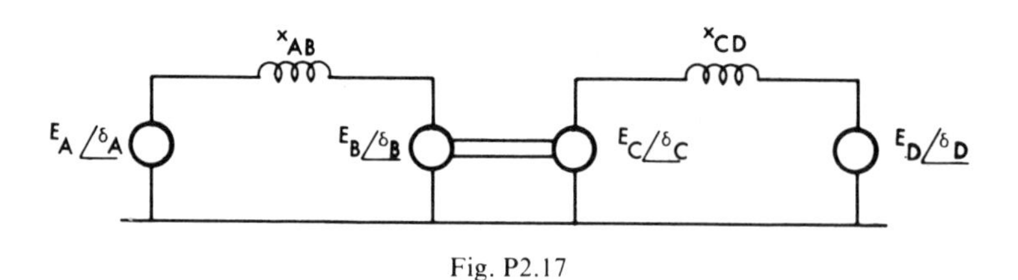

Fig. P2.17

2.18 The system shown in Figure P2.18 has two generators and three nodes. Generator and transmission line data are given below. The result of a load-flow study is also given. A three-phase fault occurs near node 2 and is cleared in 0.1 s by removing line 5.

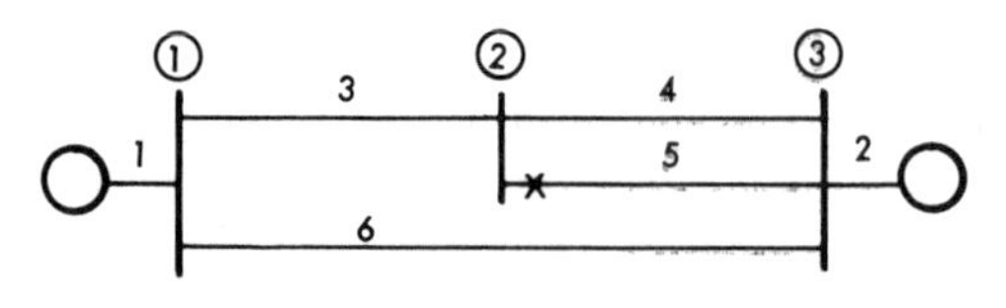

Fig. P2.18

(a) Perform all preliminary calculations for a stability study. Convert the system to a common 100-MVA base, convert the loads to equivalent passive impedances, and calculate the generator internal voltages and initial angles.
(b) Calculate the **Y** matrices for prefault, faulted, and postfault conditions.
(c) Obtain (numerically) time solutions for the internal general angles and determine if the system is stable.

Put together in calculations

Generator Data (in pu to generator MVA base)

Generator number	x'_d (pu)	X_T† (pu)	H (MW·s/MVA)	Rating (MVA)
1	0.28	0.08	5	50
3	0.25	0.07	4	120

†X_T = generator transformer reactance

Transmission Line Data (resistance neglected)

Line number:	3	4	5	6
X pu to 100-MVA base	0.08	0.06	0.06	0.13

Load-Flow Data

Bus	Voltage		Load		Generator	
no.	Magnitude pu	Angle°	MW	MVAR	MW	MVAR
1	1.030	0.0	0.0	0.0	30.0	23.1
2	1.018	−1.0	50.0	20.0	0.0	0.0
3	1.020	−0.5	80.0	40.0	100.0	37.8

2.19 Reduce the system in Problem 2.18 to an equivalent one machine connected to an infinite bus. Write the swing equation for the faulted network and for the network after the fault is cleared. Apply the equal area criterion to the fault discussed in Problem 2.18. What is the critical clearing angle?

2.20 Repeat the calculations of Example 2.4, but with the following changes in the system of Figure 2.11.
(a) Use a fault impedance of $Z_f = 0.01 + j0$ pu. This is more typical of the arcing resistance commonly found in a fault.
(b) Study the damping effect of adding a resistance to the transmission lines of R_L in each line where $R_L = 0.1$ and 0.4 pu. To measure the damping, prepare an analog computer simulation for the system. Implementation will require computation of $\bar{Y}_{11}$, $\bar{Y}_{12}$, the initial conditions, and the potentiometer settings.
(c) Devise a method of introducing additional damping on the analog computer by adding a term $K_d\dot{\delta}$ in the swing equation. Estimate the value of K_d by assuming that a slip of 2.5% gives a damping torque of 50% of full load torque.
(d) Make a parametric study of changes in the analog simulation for various values of H. For example, let $H = 2.5, 5.0, 7.5$ s.

2.21 Repeat Problem 2.20 but with transmission line impedance for each line of $R_L + j0.8$, where $R_L = 0.2, 0.5, 0.8$ pu. Repeat the analog simulation and determine the critical clearing time to the nearest cycle. This will require a means of systematically changing from the fault condition to the postfault (one line open) condition after a measured time lapse. This can be accomplished by logical control on some analog computers or by careful hand switching where logical control is not available. Let $V_\infty = 0.95$.

2.22 Repeat Problem 2.21 using a line impedance of $0.2 + j0.8$. Consider the effect of adding a "local" unity power factor load R_{LD} at bus 3 for the following conditions:

Case 1: $P_{LD} = 0.4$ pu
$P_\infty + jQ_\infty = 0.4 \pm j0.20$ pu
Case 2: P_{LD} = value to give the same generated power as Case 1
$P_\infty + jQ_\infty = 0 + j0$ pu
Case 3: $P_{LD} = 1.2$ pu
$P_\infty + jQ_\infty = -0.4 \mp j0.2$ pu

(a) Compute the values of R_{LD} and E and find the initial condition for δ for each case.

(b) Compute the values of $\overline{Y}_{11}$ and $\overline{Y}_{12}$ for the prefault, faulted, and postfault conditions if the fault impedance is $Z_F = 0.01 + j0$. Use the computer for this, writing the admittance matrices by inspection and reducing to find the two-port admittances.
(c) Compute the analog computer settings for the simulation.
(d) Perform the analog computer simulation and plot the following variables: T_m, T_e, T_a, ω_Δ, δ, $\theta_{12} - \delta$. Also, make a phase-plane plot of ω_Δ versus δ. Compare these results with similar plots with no local load present.
(e) Use the computer simulation to determine the critical clearing angle.

References

1. Kimbark, E. W. *Power System Stability,* Vol. 1. Wiley, New York, 1948.
2. Stevenson, W. D. *Elements of Power System Analysis,* 2nd ed. McGraw-Hill, New York, 1962.
3. Federal Power Commission. *National Power Survey,* Pt. 2. USGPO, Washington, D.C., 1964.
4. Lokay, H. E., and Thoits, P. O. Effects of future turbine-generator characteristics on transient stability. *IEEE Trans.* PAS-90:2427–31, 1971.
5. AIEE Subcommittee on Interconnection and Stability Factors. First report of power system stability. *Electr. Eng.* 56:261–82, 1937.
6. Venikov, V. A. *Transient Phenomena in Electrical Power Systems.* Pergamon Press, Macmillan, New York, 1964.
7. Crary, S. B. *Power System Stability,* Vol. 2. Wiley, New York, 1947.
8. Stagg, G. W., and El-Abiad, A. H. *Computer Methods in Power System Analysis.* McGraw-Hill, New York, 1968.
9. Concordia, C. Effect of steam turbine reheat on speed-governor performance. *ASME J. Eng. Power* 81:201–6, 1959.
10. Kirchmayer, L. K. *Economic Control of Interconnected Systems.* Wiley, New York, 1959.
11. Young, C. C., and Webler, R. M. A new stability program for predicting the dynamic performance of electric power systems. *Proc. Am. Power Conf.* 29:1126–39, 1967.
12. Byerly, R. T., Sherman, D. E., and Shortley, P. B. Stability program data preparation manual. Westinghouse Electric Corp. Rept. 70–736. 1970. (Rev. Dec. 1971.)
13. Concordia, C. Synchronous machine damping and synchronizing torques. *AIEE Trans.* 70:731–37, 1951.
14. Ray, J. J., and Shipley, R. B. Dynamic system performance. Paper 66 CP 709-PWR, presented at the IEEE Winter Power Meeting, New York, 1968.
15. Anderson, P. M., and Nanakorn, S. An analysis and comparison of certain low-order boiler models. *ISA Trans.* 14:17–23, 1975.

chapter 3

System Response to Small Disturbances

3.1 Introduction

This chapter reviews the behavior of an electric power system when subjected to small disturbances. It is assumed the system under study has been perturbed from a steady-state condition that prevailed prior to the application of the disturbance. This small disturbance may be temporary or permanent. If the system is stable, we would expect that for a temporary disturbance the system would return to its initial state, while a permanent disturbance would cause the system to acquire a new operating state after a transient period. In either case synchronism should not be lost. Under normal operating conditions a power system is subjected to small disturbances at random. It is important that synchronism not be lost under these conditions. Thus system behavior is a measure of dynamic stability as the system adjusts to small perturbations.

We now define what is meant by a *small* disturbance. The criterion is simply that the perturbed system can be linearized about a quiescent operating state. An example of this linearization procedure was given in Section 2.5. While the power-angle relationship for a synchronous machine connected to an infinite bus obeys a sine law (2.33), it was shown that for small perturbations the change in power is approximately proportional to the change in angle (2.35). Typical examples of small disturbances are a small change in the scheduled generation of one machine, which results in a small change in its rotor angle δ, or a small load added to the network (say 1/100 of system capacity or less).

In general, the response of a power system to impacts is oscillatory. If the oscillations are damped, so that after sufficient time has elapsed the deviation or the change in the state of the system due to the small impact is small (or less than some prescribed finite amount), the system is stable. If on the other hand the oscillations grow in magnitude or are sustained indefinitely, the system is unstable.

For a linear system, modern linear systems theory provides a means of evaluation of its dynamic response once a good mathematical model is developed. The mathematical models for the various components of a power network will be developed in greater detail in later chapters. Here a brief account is given of the various phenomena experienced in a power system subjected to small impacts, with emphasis on the qualitative description of the system behavior.

3.2 Types of Problems Studied

The method of small changes, sometimes called the perturbation method [1, 2, 3], is very useful in studying two types of problems: system response to small impacts and the distribution of impacts.

3.2.1 System response to small impacts

If the power system is perturbed, it will acquire a new operating state. If the perturbation is small, the new operating state will not be appreciably different from the initial one. In other words, the state variables or the system parameters will usually not change appreciably. Thus the operation is in the neighborhood of a certain quiescent state $\mathbf{x}_0$. In this limited range of operation a nonlinear system can be described mathematically by linearized equations. This is advantageous, since linear systems are more convenient to work with. This procedure is particularly useful if the system contains control elements.

The method of analysis used to linearize the differential equations describing the system behavior is to assume *small* changes in system quantities such as δ_Δ, v_Δ, P_Δ (change in angle, voltage, and power respectively). Equations for these variables are found by making a Taylor series expansion about $\mathbf{x}_0$ and neglecting higher order terms [4, 5, 6]. The behavior or the motion of these changes is then examined. In examining the dynamic performance of the system, it is important to ascertain not only that growing oscillations do not result during normal operations but also that the oscillatory response to small impacts is well damped.

If the stability of the system is being investigated, it is often convenient to assume that the disturbances causing the changes disappear. The motion of the system is then free. Stability is then assured if the system returns to its original state. Such behavior can be determined in a linear system by examining the characteristic equation of the system. If the mathematical description of the system is in state-space form, i.e., if the system is described by a set of first-order differential equations,

$$\dot{\mathbf{x}} = \mathbf{A}\mathbf{x} + \mathbf{B}\mathbf{u} \tag{3.1}$$

the free response of the system can be determined from the eigenvalues of the $\mathbf{A}$ matrix.

3.2.2 Distribution of power impacts

When a power impact occurs at some bus in the network, an unbalance between the power input to the system and the power output takes place, resulting in a transient. When this transient subsides and a steady-state condition is reached, the power impact is "shared" by the various synchronous machines according to their steady-state characteristics, which are determined by the steady-state droop characteristics of the various governors [5, 7]. During the transient period, however, the power impact is shared by the machines according to different criteria. If these criteria differ appreciably among groups of machines, each impact is followed by oscillatory power swings among groups of machines to reflect the transition from the initial sharing of the impact to the final adjustment reached at steady state.

Under normal operating conditions a power system is subjected to numerous random power impacts from sudden application or removal of loads. As explained above, each impact will be followed by power swings among groups of machines that respond to the impact differently at different times. These power swings appear as power oscillations on the tie lines connecting these groups of machines. This gives rise to the term "tie-line oscillations."

In large interconnected power systems tie-line oscillations can become objectionable if their magnitude reaches a significant fraction of the tie-line loading, since they are superimposed upon the normal flow of power in the line. Furthermore, conditions may exist in which these oscillations grow in amplitude, causing instability. This problem is similar to that discussed in Section 3.2.1. It can be analyzed if an adequate mathematical model of the various components of the system is developed and the dynamic response of this model is examined. If we are interested in seeking an approximate answer for the magnitude of the tie-line oscillations, however, such an answer can be reached by a qualitative discussion of the distribution of power impacts. Such a discussion is offered here.

3.3 The Unregulated Synchronous Machine

We start with the simplest model possible, i.e., the constant-voltage-behind-transient-reactance model. The equation of motion of a synchronous machine connected to an infinite bus and the electrical power output are given by (2.18) and (2.41) respectively or

$$\frac{2H}{\omega_R}\frac{d^2\delta}{dt^2} = P_m - P_e$$

$$P_e = P_C + P_M \sin(\delta - \gamma) \tag{3.2}$$

Letting $\delta = \delta_0 + \delta_\Delta$, $P_e = P_{e0} + P_\Delta$, $P_m = P_{m0}$ and using the relationship

$$\sin(\delta - \gamma) = \sin(\delta_0 - \gamma + \delta_\Delta) \cong \sin(\delta_0 - \gamma) + \cos(\delta_0 - \gamma)\delta_\Delta \tag{3.3}$$

the linearized version of (3.2) becomes

$$\frac{2H}{\omega_R}\frac{d^2\delta_\Delta}{dt^2} + P_s\delta_\Delta = 0 \tag{3.4}$$

where

$$P_s = \left.\frac{dP_e}{d\delta}\right]_{\delta_0} = P_M \cos(\delta_0 - \gamma) \tag{3.5}$$

The system described by (3.4) is marginally stable (i.e., oscillatory) for $P_s > 0$. Its response is oscillatory with the frequency of oscillation obtained from the roots of the characteristic equation $(2H/\omega_R)s^2 + P_s = 0$, which has the roots

$$s = \pm j\sqrt{P_s\omega_R/2H} \tag{3.6}$$

If the electrical torque is assumed to have a component proportional to the speed change, a damping term is added to (3.4) and the new characteristic equation becomes

$$(2H/\omega_R)s^2 + (D/\omega_R)s + P_s = 0 \tag{3.7}$$

where D is the damping power coefficient in pu.

The roots of (3.7) are given by

$$s = -\frac{D}{4H} \pm \frac{\omega_R}{4H}\left[\left(\frac{D}{\omega_R}\right)^2 - \frac{8HP_s}{\omega_R}\right]^{1/2} \tag{3.8}$$

Usually $(D/\omega_R)^2 < 8HP_s/\omega_R$, and the roots are complex; i.e., the response is oscillatory with an angular frequency of oscillation essentially the same as that given by (3.6). The system described by (3.7) is stable for $P_s > 0$ and for $D > 0$. If either one of these quantities is negative, the system is unstable.

Venikov [4] reports that a situation may occur where the machine described by (3.4) can be unstable under light load conditions if the network is such that $\delta_0 < \gamma$. This would be the case where there is appreciable series resistance (see [4], Sec. 3.2).

From Chapter 2 we know that the synchronizing power coefficient P_s is negative if the spontaneous change in the angle δ is negative. A negative value of P_s leads to unstable operation.

3.3.1 Demagnetizing effect of armature reaction

The model of constant main field-winding flux linkage neglects some important effects, among them the demagnetizing influence of a change in the rotor angle δ. To account for this effect, another model of the synchronous machine is used. It is not our concern in this introductory discussion to develop the model or even discuss it in detail, as this will be accomplished in Chapter 6. Rather, we will state the assumptions made in such a model and give some of the pertinent results applicable to this discussion. These results are found in de Mello and Concordia [8] and are based on a model previously used by Heffron and Phillips [9]. To account for the field conditions, equations for the direct and quadrature axis quantities are derived (see Chapter 4). Major simplifications are then made by neglecting saturation, stator resistance, and the damper windings. The transformer voltage terms in the stator voltage equations are considered negligible compared to the speed voltage terms. Linearized relations are then obtained between small changes in the electrical power $P_{e\Delta}$, the rotor angle δ_Δ, the field-winding voltage $v_{F\Delta}$, and the voltage proportional to the main field-winding flux E'_Δ.

For a machine connected to an infinite bus through a transmission network, the following s domain relations are obtained,

$$P_{e\Delta} = K_1\delta_\Delta + K_2E'_\Delta \tag{3.9}$$

$$E'_\Delta = \frac{K_3}{1 + K_3\tau'_{d0}s}\, v_{F\Delta} - \frac{K_3K_4}{1 + K_3\tau'_{d0}s}\, \delta_\Delta \tag{3.10}$$

where K_1 is the change in electrical power for a change in rotor angle with constant flux linkage in the direct axis, K_2 is the change in electrical power for a change in the direct axis flux linkages with constant rotor angle, τ'_{d0} is the direct axis open circuit time constant of the machine, K_3 is an impedance factor, and K_4 is the demagnetizing effect of a change in the rotor angle (at steady state). Mathematically, we write

$$K_1 = P_{e\Delta}/\delta_\Delta]_{E'_\Delta=0} \qquad K_2 = P_{e\Delta}/E'_\Delta]_{\delta_\Delta=0}$$

$$K_3 = \text{final value of unit step } v_F \text{ response} = \lim_{t\to\infty} E'_\Delta(t)]_{\delta_\Delta=0}$$

$$K_4 = -\frac{1}{K_3}\lim_{t\to\infty} E'_\Delta(t)\Big]_{\substack{v_{F\Delta}=0\\ \delta_\Delta=u(t)}} \tag{3.11}$$

The constants K_1, K_2, and K_4 depend on the parameters of the machine, the external network, and the initial conditions. Note that K_1 is similar to the synchronizing power coefficient P_s used in the simpler machine model of constant voltage behind

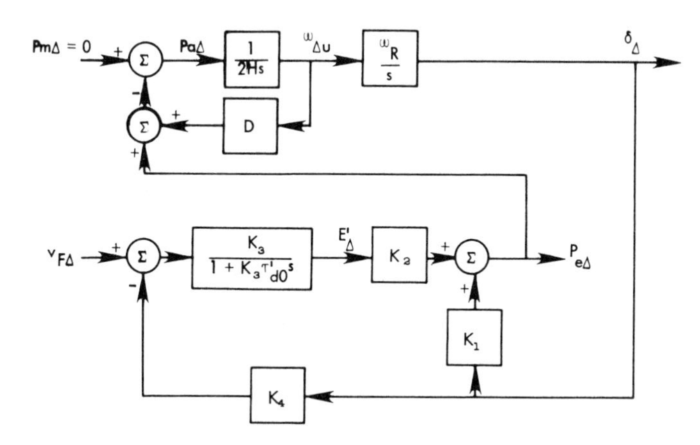

Fig. 3.1 Primitive linearized block diagram representation of a generator model.

transient reactance. Equations (3.9) and (3.10), with the initial equation (3.2), may be represented by the incremental block diagram of Figure 3.1.

$$P_{e\Delta} = \left(K_1 - \frac{K_2K_3K_4}{1 + K_3\tau'_{d0}s}\right)\delta_\Delta + \frac{K_2K_3}{1 + K_3\tau'_{d0}s}v_{F\Delta} \tag{3.12}$$

For the case where $v_{F\Delta} = 0$,

$$P_{e\Delta} = \left(K_1 - \frac{K_2K_3K_4}{1 + K_3\tau'_{d0}s}\right)\delta_\Delta \tag{3.13}$$

where we can clearly identify both the synchronizing and the demagnetizing components.

Substituting in the linearized swing equation (3.4), we obtain the new characteristic equation, (with $D = 0$)

$$\left[\frac{2H}{\omega_R}s^2 + \left(K_1 - \frac{K_2K_3K_4}{1 + K_3\tau'_{d0}s}\right)\right]\delta_\Delta = 0$$

or we have the third-order system

$$s^3 + \frac{1}{K_3\tau'_{d0}}s^2 + \frac{K_1\omega_R}{2H}s + \frac{\omega_R}{2H}\frac{1}{K_3\tau'_{d0}}(K_1 - K_2K_3K_4) = 0 \tag{3.14}$$

Note that all the constants (3.11) are usually positive. Thus from Routh's criterion [10] this system is stable if $K_1 - K_2K_3K_4 > 0$ and $K_2K_3K_4 > 0$.

The first of the above criteria states that the synchronizing power coefficient K_1 must be greater than the demagnetizing component of electrical power. The second criterion is satisfied if the constants K_2, K_3, and K_4 are positive. Venikov [4] points out that if the transmission network has an appreciable series capacitive reactance, it is possible that instability may occur. This would happen because the impedance factor producing the constant K_1 would become negative.

3.3.2 Effect of small changes of speed

In the linearized version of (3.2) we are interested in terms involving changes of power due to changes of the angle δ and its derivative. The change in power due to

δ_Δ was discussed above and was found to include a synchronizing power component and a demagnetizing component due the change in E'_Δ with δ_Δ. The change in speed, $\omega_\Delta = d\delta_\Delta/dt$, causes a change in both electrical and mechanical power. In this case the new differential equation becomes

$$\frac{2H}{\omega_R}\frac{d^2\delta_\Delta}{dt^2} = \frac{\partial P_m}{\partial \omega}\bigg]_{\omega_0}\omega_\Delta - \left(\frac{\partial P_e}{\partial \omega}\bigg]_{\omega_0}\omega_\Delta + \frac{\partial P_e}{\partial \delta}\bigg]_{\delta_0}\delta_\Delta\right) \tag{3.15}$$

As in (3.7) the change in electrical power due to small changes in speed is in the form of

$$P_{e\Delta} = (D/\omega_R)\omega_\Delta \tag{3.16}$$

From Section 2.3 the change in mechanical power due to small changes in speed is also linear

$$P_{m\Delta} = \partial P_m/\partial\omega]_{\omega_0}\omega_\Delta \tag{3.17}$$

where $\partial P_m/\partial\omega]_{\omega_0}$ can be obtained from a relation such as the one given in Figure 2.3. If a transient droop or regulation R is assumed, we may write in pu to the machine base

$$P_{m\Delta} = -(1/R)(\omega_\Delta/\omega_R) \text{ pu} \tag{3.18}$$

which is the equation of an ideal speed droop governor. The system block diagram with speed regulation added is shown in Figure 3.2.

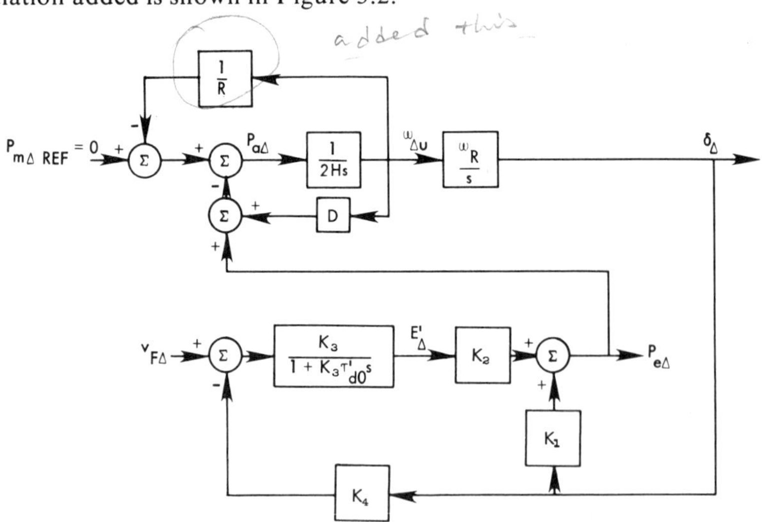

Fig. 3.2 Block diagram representation of the linearized model with speed regulation added.

The characteristic equation of the system now becomes

$$\frac{2H}{\omega_R}s^2 + \frac{1}{\omega_R}\left(D + \frac{1}{R}\right)s + \left(K_1 - \frac{K_2K_3K_4}{1 + K_3\tau'_{d0}s}\right) = 0 \tag{3.19}$$

or

$$\frac{2H}{\omega_R}K_3\tau'_{d0}s^3 + \left[\frac{2H}{\omega_R} + \frac{K_3\tau'_{d0}}{\omega_R}\left(D + \frac{1}{R}\right)\right]s^2 + \left[\frac{1}{\omega_R}\left(D + \frac{1}{R}\right) + K_1K_3\tau'_{d0}\right]s + (K_1 - K_2K_3K_4) = 0 \tag{3.20}$$

Again Routh's criterion may be applied to determine the conditions for stability. This is left as an exercise (see Problem 3.2).

3.4 Modes of Oscillation of an Unregulated Multimachine System

The electrical power output of machine i in an n-machine system is

$$P_{ei} = E_i^2 G_{ii} + \sum_{\substack{j=1 \\ j \neq i}}^{n} E_i E_j Y_{ij} \cos(\theta_{ij} - \delta_{ij})$$

$$= E_i^2 G_{ii} + \sum_{\substack{j=1 \\ j \neq i}}^{n} E_i E_j (B_{ij} \sin \delta_{ij} + G_{ij} \cos \delta_{ij}) \tag{3.21}$$

where $\delta_{ij} = \delta_i - \delta_j$
E_i = constant voltage behind transient reactance for machine i
$\overline{Y}_{ii} = G_{ii} + jB_{ii}$ is a diagonal element of the network short circuit admittance matrix **Y**
$\overline{Y}_{ij} = G_{ij} + jB_{ij}$ is an off-diagonal element of the network short circuit admittance matrix **Y**

Using the incremental model so that $\delta_{ij} = \delta_{ij0} + \delta_{ij\Delta}$, we compute

$$\sin \delta_{ij} = \sin \delta_{ij0} \cos \delta_{ij\Delta} + \cos \delta_{ij0} \sin \delta_{ij\Delta} \cong \sin \delta_{ij0} + \delta_{ij\Delta} \cos \delta_{ij0}$$
$$\cos \delta_{ij} \cong \cos \delta_{ij0} - \delta_{ij\Delta} \sin \delta_{ij0}$$

Finally, for $P_{ei\Delta}$,

$$P_{ei\Delta} = \sum_{\substack{j=1 \\ j \neq i}}^{n} E_i E_j (B_{ij} \cos \delta_{ij0} - G_{ij} \sin \delta_{ij0}) \delta_{ij\Delta} \tag{3.22}$$

For a given initial condition $\sin \delta_{ij0}$ and $\cos \delta_{ij0}$ are known, and the term in parentheses in (3.22) is a constant. Thus we write

$$P_{ei\Delta} = \sum_{\substack{j=1 \\ j \neq i}}^{n} P_{sij} \delta_{ij\Delta} \tag{3.23}$$

where

$$P_{sij} \triangleq \left.\frac{\partial P_{ij}}{\partial \delta_{ij}}\right]_{\delta_{ij0}} = E_i E_j (B_{ij} \cos \delta_{ij0} - G_{ij} \sin \delta_{ij0}) \tag{3.24}$$

is the change in the electrical power of machine i due to a change in the angle between machines i and j, with all other angles held constant. Its units are W/rad or pu power/rad. It is a synchronizing power coefficient between nodes i and j and is identical to the coefficient discussed in Section 2.5.2 for one machine connected to an infinite bus.

We also note that since (3.21) applies to any number of nodes where the voltages are known, the linearized equations (3.22) and (3.23) can be derived for a given machine in terms of the voltages at those nodes and their angles. Thus the concept of the synchronizing power coefficients can be extended to mean "the change in the electrical power of a given machine due to the change in the angle between its internal EMF and any bus, with all other bus angles held constant." (An implied assumption is that the voltage at the remote bus is also held constant.) This expanded definition of the synchronizing power coefficient will be used in Section 3.6.

Using the inertial model of the synchronous machines, we get the set of linearized differential equations,

$$\frac{2H_i}{\omega_R}\frac{d^2\delta_{i\Delta}}{dt^2} + \sum_{\substack{j=1\\ j\neq i}}^{n} E_iE_j(B_{ij}\cos\delta_{ij0} - G_{ij}\sin\delta_{ij0})\delta_{ij\Delta} = 0 \qquad i = 1,2,\ldots,n \tag{3.25}$$

or

$$\frac{2H_i}{\omega_R}\frac{d^2\delta_{i\Delta}}{dt^2} + \sum_{\substack{j=1\\ j\neq i}}^{n} P_{sij}\,\delta_{ij\Delta} = 0 \qquad i = 1,2,\ldots,n \tag{3.26}$$

The set (3.26) is not a set of n-independent second-order equations, since $\Sigma\delta_{ij} = 0$. Thus (3.26) comprises a set of $(n - 1)$-independent equations.

From (3.26) for machine i,

$$\frac{d^2\delta_{i\Delta}}{dt^2} + \frac{\omega_R}{2H_i}\sum_{\substack{j=1\\ j\neq i}}^{n} P_{sij}\,\delta_{ij\Delta} = 0 \qquad i = 1,2,\ldots,n \tag{3.27}$$

Subtracting the nth equation from the ith equation, we compute

$$\frac{d^2\delta_{i\Delta}}{dt^2} - \frac{d^2\delta_{n\Delta}}{dt^2} + \frac{\omega_R}{2H_i}\sum_{\substack{j=1\\ j\neq i}}^{n} P_{sij}\,\delta_{ij\Delta} - \frac{\omega_R}{2H_n}\sum_{j=1}^{n-1} P_{snj}\,\delta_{nj\Delta} = 0 \tag{3.28}$$

Equation (3.28) can be put in the form

$$\frac{d^2}{dt^2}\,\delta_{in\Delta} + \frac{\omega_R}{2H_i}\sum_{\substack{j=1\\ j\neq i}}^{n} P_{sij}\,\delta_{ij\Delta} - \frac{\omega_R}{2H_n}\sum_{j=1}^{n-1} P_{snj}\,\delta_{nj\Delta} = 0 \qquad i = 1,2,\ldots,n-1 \tag{3.29}$$

Since

$$\delta_{ij\Delta} = \delta_{in\Delta} - \delta_{jn\Delta} \tag{3.30}$$

(3.29) can be further modified as

$$\frac{d^2\delta_{in\Delta}}{dt^2} + \sum_{j=1}^{n-1} \alpha_{ij}\,\delta_{jn\Delta} = 0 \qquad i = 1,2,\ldots,n-1 \tag{3.31}$$

where the coefficients α_{ij} depend on the machine inertias and synchronizing power coefficients.

Equation (3.31) represents a set of $n - 1$ linear second-order differential equations or a set of $2(n - 1)$ first-order differential equations. We will use the latter formulation to examine the free response of this system.

Let $x_1, x_2, \ldots, x_{n-1}$ be the angles $\delta_{1n\Delta}, \delta_{2n\Delta}, \ldots, \delta_{(n-1)n\Delta}$ respectively, and let $x_n, \ldots, x_{2n-2}$ be the time derivatives of these angles. The system equations are of the form

$$\begin{bmatrix} \dot{x}_1 \\ \dot{x}_2 \\ \cdots \\ \dot{x}_{n-1} \\ \hline \dot{x}_n \\ \dot{x}_{n+1} \\ \cdots \\ \dot{x}_{2n-2} \end{bmatrix} = \left[\begin{array}{cccc|cccc} & & & & 1 & 0 & \cdots & 0 \\ & \mathbf{0} & & & 0 & 1 & \cdots & 0 \\ & & & & \cdots & \cdots & \cdots & \cdots \\ & & & & 0 & 0 & \cdots & 1 \\ \hline A_{11} & A_{12} & \cdots & A_{1,n-1} & & & & \\ A_{21} & A_{22} & \cdots & A_{2,n-1} & & \mathbf{0} & & \\ \cdots & \cdots & \cdots & \cdots & & & & \\ A_{n-1,1} & A_{n-1,2} & \cdots & A_{n-1,n-1} & & & & \end{array}\right] \begin{bmatrix} x_1 \\ x_2 \\ \cdots \\ x_{n-1} \\ \hline x_n \\ x_{n+1} \\ \cdots \\ x_{2n-2} \end{bmatrix} \tag{3.32}$$

or

$$\begin{bmatrix} \dot{\mathbf{X}}_1 \\ \hline \dot{\mathbf{X}}_2 \end{bmatrix} = \left[\begin{array}{c|c} \mathbf{0} & \mathbf{U} \\ \hline \mathbf{A} & \mathbf{0} \end{array}\right] \begin{bmatrix} \mathbf{X}_1 \\ \hline \mathbf{X}_2 \end{bmatrix} \tag{3.33}$$

where $\mathbf{U}$ = the identity matrix
$\mathbf{X}_1$ = the $n - 1$ vector of the angle changes $\delta_{in\Delta}$
$\mathbf{X}_2$ = the $n - 1$ vector of the speed changes $d\delta_{in\Delta}/dt$

To obtain the free response of the system, we examine the eigenvalues of the characteristic matrix [11, 12]. This is obtained from the characteristic equation derived from equating the determinant of the matrix to zero, as follows:

$$\det \left[\begin{array}{c|c} -\lambda\mathbf{U} & \mathbf{U} \\ \hline \mathbf{A} & -\lambda\mathbf{U} \end{array}\right] = \det \mathbf{M} = 0 \tag{3.34}$$

where λ is the eigenvalue. Since the matrix $-\lambda\mathbf{U}$ is nonsingular, we compute the determinant of $\mathbf{M}$ as

$$\begin{aligned} |\mathbf{M}| &= |-\lambda\mathbf{U}|\,|(-\lambda\mathbf{U}) - \mathbf{A}(-\lambda\mathbf{U})^{-1}\mathbf{U}| \\ &= (-1)^{n-1}\lambda^{n-1}\,|-\lambda\mathbf{U} - (-1/\lambda)^{n-1}\mathbf{A}| = |\lambda^2\mathbf{U} - \mathbf{A}| \end{aligned} \tag{3.35}$$

(See Lefschetz [12], p. 133.) The system described by $|\mathbf{M}| = 0$, or $|\lambda^2\mathbf{U} - \mathbf{A}| = 0$, has $2(n - 1)$ imaginary roots, which occur in $n - 1$ complex conjugate pairs. Thus the system has $n - 1$ frequencies of oscillations.

Example 3.1

Find the modes of oscillation of a three-machine system. The machines are unregulated and classical model representation is used.

Solution

For an unregulated three-machine system, the system equations are given by

$$\frac{2H_1}{\omega_R}\frac{d^2\delta_{\Delta 1}}{dt^2} + P_{s12}\delta_{12\Delta} + P_{s13}\delta_{13\Delta} = 0$$

$$\frac{2H_2}{\omega_R}\frac{d^2\delta_{\Delta 2}}{dt^2} + P_{s21}\delta_{21\Delta} + P_{s23}\delta_{23\Delta} = 0$$

$$\frac{2H_3}{\omega_R}\frac{d^2\delta_{\Delta 3}}{dt^2} + P_{s31}\delta_{31\Delta} + P_{s32}\delta_{32\Delta} = 0$$

Multiplying the above three equations by $\omega_R/2H_i$ and subtracting the third equation from the first two, we get (noting that $\delta_{ij} = -\delta_{ji}$)

$$\frac{d^2\delta_{13\Delta}}{dt^2} + \frac{\omega_R}{2H_1} P_{s12}\delta_{12\Delta} + \left(\frac{\omega_R}{2H_1} P_{s13} + \frac{\omega_R}{2H_3} P_{s31}\right)\delta_{31\Delta} + \frac{\omega_R}{2H_3} P_{s32}\delta_{23\Delta} = 0$$

$$\frac{d^2\delta_{23\Delta}}{dt^2} - \frac{\omega_R}{2H_1} P_{s21}\delta_{12\Delta} + \frac{\omega_R}{2H_3} P_{s31}\delta_{13\Delta} + \left(\frac{\omega_R}{2H_2} P_{s23} + \frac{\omega_R}{2H_3} P_{s32}\right)\delta_{23\Delta} = 0$$

If we eliminate $\delta_{12\Delta}$ by noting that $\delta_{12\Delta} + \delta_{23\Delta} + \delta_{31\Delta} = 0$, the following two equations are obtained:

$$\frac{d^2\delta_{13\Delta}}{dt^2} + \left(\frac{\omega_R}{2H_1} P_{s12} + \frac{\omega_R}{2H_1} P_{s13} + \frac{\omega_R}{2H_3} P_{s31}\right)\delta_{13\Delta} + \left(\frac{\omega_R}{2H_3} P_{s32} - \frac{\omega_R}{2H_1} P_{s12}\right)\delta_{23\Delta} = 0$$

$$\frac{d^2\delta_{23\Delta}}{dt^2} + \left(\frac{\omega_R}{2H_3} P_{s31} - \frac{\omega_R}{2H_1} P_{s21}\right)\delta_{13\Delta} + \left(\frac{\omega_R}{2H_1} P_{s21} + \frac{\omega_R}{2H_2} P_{s23} + \frac{\omega_R}{2H_3} P_{s32}\right)\delta_{23\Delta} = 0$$

or

$$\frac{d^2\delta_{13\Delta}}{dt^2} + \alpha_{11}\delta_{13\Delta} + \alpha_{12}\delta_{23\Delta} = 0 \qquad \frac{d^2\delta_{23\Delta}}{dt^2} + \alpha_{21}\delta_{13\Delta} + \alpha_{22}\delta_{23\Delta} = 0$$

The state-space representation of the above system is

$$\begin{bmatrix} \dot{\delta}_{13\Delta} \\ \dot{\delta}_{23\Delta} \\ \hline \dot{\omega}_{13\Delta} \\ \dot{\omega}_{23\Delta} \end{bmatrix} = \left[\begin{array}{cc|cc} 0 & 0 & 1 & 0 \\ 0 & 0 & 0 & 1 \\ \hline -\alpha_{11} & -\alpha_{12} & 0 & 0 \\ -\alpha_{21} & -\alpha_{22} & 0 & 0 \end{array}\right] \begin{bmatrix} \delta_{13\Delta} \\ \delta_{23\Delta} \\ \hline \omega_{13\Delta} \\ \omega_{23\Delta} \end{bmatrix}$$

To obtain the eigenvalues of this system, the characteristic equation is given by

$$\det \left[\begin{array}{cc|cc} -\lambda & 0 & 1 & 0 \\ 0 & -\lambda & 0 & 1 \\ \hline -\alpha_{11} & -\alpha_{12} & -\lambda & 0 \\ -\alpha_{21} & -\alpha_{22} & 0 & -\lambda \end{array}\right] = 0$$

Now by using (3.35),

$$\det\left\{\begin{bmatrix} \lambda^2 & 0 \\ 0 & \lambda^2 \end{bmatrix} + \begin{bmatrix} \alpha_{11} & \alpha_{12} \\ \alpha_{21} & \alpha_{22} \end{bmatrix}\right\} = 0$$

$$\det\begin{bmatrix} \lambda^2 + \alpha_{11} & \alpha_{12} \\ \alpha_{21} & \lambda^2 + \alpha_{22} \end{bmatrix} = 0$$

$$(\lambda^2 + \alpha_{11})(\lambda^2 + \alpha_{22}) - \alpha_{12}\alpha_{21} = 0$$

$$\lambda^4 + (\alpha_{11} + \alpha_{22})\lambda^2 + (\alpha_{11}\alpha_{22} - \alpha_{12}\alpha_{21}) = 0$$

$$\lambda^2 = (1/2)\{-(\alpha_{11} + \alpha_{22}) \pm [(\alpha_{11} + \alpha_{22})^2 - 4(\alpha_{11}\alpha_{22} - \alpha_{12}\alpha_{21})]^{1/2}\}$$

Examining the coefficients α_{ij}, we can see that both values of λ^2 are negative real quantities. Let these given values be $\lambda = \pm j\beta$, $\lambda = \pm j\gamma$.

The free response will be in the form $\delta_\Delta = C_1 \cos(\beta t + \phi_1) + C_2 \cos(\gamma t + \phi_2)$, where C_1, C_2, ϕ_1, and ϕ_2 are constants.

Example 3.2

Consider the three-machine, nine-bus system of Example 2.6, operating initially in the steady state with system conditions given by Figure 2.18 (load flow) and the computed initial values given in Example 2.6 for $E_i\underline{/\delta_{i0}}$, $i = 1, 2, 3$. A small 10-MW load (about 3% of the total system load of 315 MW) is suddenly added at bus 8 by adding a three-phase fault to the bus through a 10.0 pu impedance. The system base is 100 MVA. Assume that the system load after $t = 0$ is constant and consists of the original load plus the 10 pu shunt resistance at bus 8.

Compute the frequencies of oscillation that will result from this small disturbance. Then compare these computed frequencies against those actually observed in a digital computer solution. Assume there are no governors active on any of the three turbines. Observe the system response for about two seconds.

Solution

First we compute the frequencies of oscillation. From (3.24)

$$P_{sij} = V_i V_j (B_{ij} \cos \delta_{ij0} - G_{ij} \sin \delta_{ij0}) \cong V_i V_j B_{ij} \cos \delta_{ij0}$$

From Example 2.6 we find the data needed to compute P_{sij} with the results shown in Table 3.1.

Table 3.1. Synchronizing Power Coefficients of the Network of Example 2.6

ij	V_i	V_j	B_{ij}	δ_{ij0}	P_{sij}
12	1.0566	1.0502	1.513	−17.4598	1.6015
23	1.0502	1.0170	1.088	6.5563	1.1544
31	1.0170	1.0566	1.226	10.9035	1.2936

Note that the δ_{ij0} are the values of the relative rotor angles at $t = 0^-$. Since these are rotor angles, they will not change at the time of impact, so these are also the correct values for $t = 0^+$. This is also true of angles at load buses to which appreciable inertia is connected. For loads that are essentially constant impedance, however, the voltage angle will exhibit a step change.

Also from Example 2.6 we know $H_i = 23.64, 6.40$, and 3.01 for $i = 1, 2, 3$ respectively. Thus we can compute the values of α_{ij} from Example 3.1 as follows:

$$\alpha_{11} = (\omega_R/2)(P_{s12}/H_1 + P_{s13}/H_1 + P_{s31}/H_3) = 104.096$$
$$\alpha_{12} = (\omega_R/2)(P_{s32}/H_3 - P_{s12}/H_1) = 59.524$$
$$\alpha_{21} = (\omega_R/2)(P_{s31}/H_3 - P_{s21}/H_1) = 68.241$$
$$\alpha_{22} = (\omega_R/2)(P_{s21}/H_2 + P_{s23}/H_2 + P_{s32}/H_3) = 119.065$$

Then

$$\lambda^2 = -(1/2)\left[-(\alpha_{11} + \alpha_{22}) \pm \sqrt{(\alpha_{11} + \alpha_{22})^2 - 4(\alpha_{11}\alpha_{22} - \alpha_{12}\alpha_{21})}\right]$$
$$= -(1/2)\left[-223.61 \pm \sqrt{49800.83 - 33328.85}\right] = -47.409 \text{ or } -175.752$$

Now we can compute the frequencies and periods shown in Table 3.2.

Table 3.2. Frequencies of Oscillation of a Nine-Bus System

Quantity	Eigenvalue 1	Eigenvalue 2
λ	$\pm j6.885$	$\pm j13.257$
ω rad/s	6.885	13.257
f Hz	1.096	2.110
T s	0.912	0.474

Thus two frequencies, about 1.1 Hz and 2.1 Hz, should be observed in the intermachine oscillations of the system. This can be approximately verified by an actual solution of the system by digital computer. The results of such a solution are shown in Figure 3.3, where absolute angles are given in Figure 3.3(a) and angle differences relative to δ_1 are given in Figure 3.3(b). As might be expected, neither of the computed frequencies is clearly observed since the response is a combination of the two frequencies. A rough measurement of the peak-to-peak periods in Figure 3.3(b) gives periods in the neighborhood of 0.7 s.

Methods have been devised [3, 11] by which a system such as the one in Example 3.2 can be transformed to a new frame of reference called the Jordan canonical form. In Jordan form the different frequencies of oscillation are clearly separated. In the form of equations normally used, the variables δ_{12} and δ_{13} (or other angle differences) contain

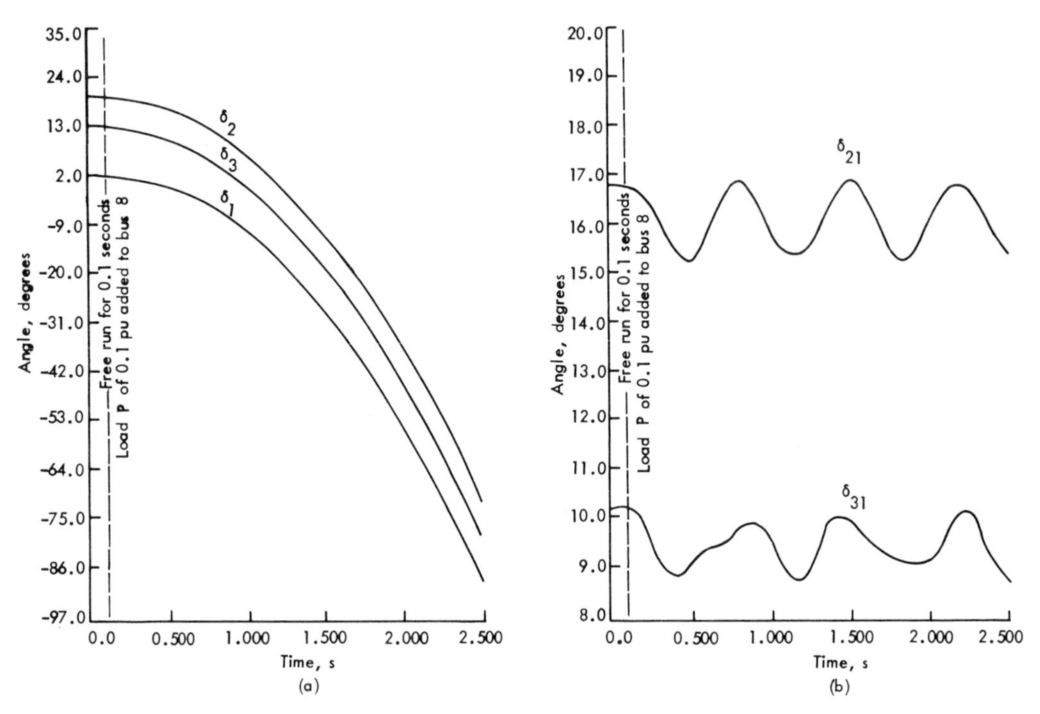

Fig. 3.3 Unregulated response of the nine-bus system to a sudden load application at bus 8: (a) absolute angles, (b) angles relative to δ_1.

"harmonic" terms generally involving all fundamental frequencies of oscillation. Hence we have difficulty observing these frequencies in measured physical variables.

Example 3.3

Transform the system of Example 3.2 into the Jordan canonical form and show that in this form the system frequencies of oscillation are clearly distinguishable.

Solution

The system equations for the three-machine problem are given by

$$\begin{bmatrix} \dot{x}_1 \\ \dot{x}_2 \\ \dot{x}_3 \\ \dot{x}_4 \end{bmatrix} = \left[\begin{array}{cc|cc} \multicolumn{2}{c|}{\multirow{2}{*}{$\mathbf{0}$}} & 1 & 0 \\ & & 0 & 1 \\ \hline -\alpha_{11} & -\alpha_{12} & \multicolumn{2}{c}{\multirow{2}{*}{$\mathbf{0}$}} \\ -\alpha_{21} & -\alpha_{22} & & \end{array}\right] \begin{bmatrix} x_1 \\ x_2 \\ x_3 \\ x_4 \end{bmatrix}$$

or $\dot{\mathbf{x}} = \mathbf{A}\,\mathbf{x}$, where $\mathbf{x}$ is defined by

$$\begin{bmatrix} x_1 \\ x_2 \\ x_3 \\ x_4 \end{bmatrix} = \begin{bmatrix} \delta_{13\Delta} \\ \delta_{23\Delta} \\ \omega_{13\Delta} \\ \omega_{23\Delta} \end{bmatrix}$$

and the α coefficients are computed in Example 3.2.

We now compute the eigenvectors of $\mathbf{A}$, using any method [1, 3, 11] and call these vectors $\mathbf{E}_1$, $\mathbf{E}_2$, $\mathbf{E}_3$, and $\mathbf{E}_4$. We then use these eigenvectors to define a matrix $\mathbf{E}$.

$$\mathbf{E} = [\mathbf{E}_1\ \mathbf{E}_2\ \mathbf{E}_3\ \mathbf{E}_4] = \left[\begin{array}{c|c|c|c} j0.06266 & -j0.06266 & 0.14523 & -0.14523 \\ j0.07543 & -j0.07543 & -0.13831 & 0.13831 \\ 0.83069 & 0.83069 & 1.00000 & 1.00000 \\ 1.00000 & 1.00000 & -0.95234 & -0.95234 \end{array}\right]$$

where the numerical values are found by a suitable computer library routine.

We now define the transformation $\mathbf{x} = \mathbf{E}\,\mathbf{y}$ to compute $\dot{\mathbf{x}} = \mathbf{E}\,\dot{\mathbf{y}} = \mathbf{A}\,\mathbf{x} = \mathbf{A}\,\mathbf{E}\,\mathbf{y}$ or $\dot{\mathbf{y}} = \mathbf{E}^{-1}\mathbf{A}\,\mathbf{E}\,\mathbf{y} = \mathbf{D}\,\mathbf{y}$ where $\mathbf{D} = \text{diag}(\lambda_1, \lambda_2, \lambda_3, \lambda_4)$.

Performing the indicated numerical work, we compute

$$\mathbf{E}^{-1} = \begin{bmatrix} -j3.5245 & -j3.7008 & 0.2659 & 0.2792 \\ j3.5245 & j3.7008 & 0.2659 & 0.2792 \\ -j1.9221 & j1.5967 & 0.2792 & -0.2319 \\ j1.9221 & -j1.5967 & 0.2792 & -0.2319 \end{bmatrix}$$

$$\mathbf{D} = \mathbf{E}^{-1}\mathbf{A}\,\mathbf{E} = \begin{bmatrix} -j13.2571 & 0.0 & 0.0 & 0.0 \\ 0.0 & j13.2571 & 0.0 & 0.0 \\ 0.0 & 0.0 & -j6.8854 & 0.0 \\ 0.0 & 0.0 & 0.0 & j6.8854 \end{bmatrix}$$

Substituting into $\dot{\mathbf{y}} = \mathbf{D}\mathbf{y}$, we can compute the uncoupled solution

$$y_i = C_i e^{\lambda_i t} \qquad i = 1,2,3,4$$

where C_i depends on the initial conditions.

This method of computing the distinct frequencies of oscillation is quite general and may be applied to systems of any size. For very large systems this may not be practical, however, since the eigenvector computation may be too costly.

Finally, we note that the simple model used here assumes that no damping exists. In physical systems damping is usually present; therefore, the oscillatory response given above is usually damped. The magnitude of the damping, however, is such that the frequencies of oscillation given by the above equations are not appreciably affected.

3.5 Regulated Synchronous Machine

In this section we examine the effect of voltage and speed control equipment on the dynamic performance of the synchronous machine. Again we are interested in the free response of the system. We will consider two simple cases of regulation: a simple voltage regulator with one time lag and a simple governor with one time lag.

3.5.1 Voltage regulator with one time lag

Referring to Figure 2.24, we note that a change in the field voltage $v_{F\Delta}$ is produced by changes in either V_{REF} or V_t. If we assume that $V_{REF\Delta} = 0$ and the transducer has no time lags, $v_{F\Delta}$ depends only upon $V_{t\Delta}$, modified by the transfer function of the excitation system. Analysis of such a system is discussed in Chapter 7. To simplify the analysis, a rather simple model of the voltage regulator and excitation system is assumed. This gives the following s domain relation between the change in the exciter voltage $v_{F\Delta}$ and the change in the synchronous machine terminal voltage $V_{t\Delta}$:

$$v_{F\Delta} = -[K_\epsilon/(1 + \tau_\epsilon s)]V_{t\Delta} \tag{3.36}$$

where K_ϵ = regulator gain
τ_ϵ = regulator time constant

To examine the effect of the voltage regulator on the system response, we return to the model discussed in Section 3.3 for a machine connected to an infinite bus through a transmission network. These relations are given in (3.9) and (3.10).

To use (3.36), a relation between $V_{t\Delta}$, δ_Δ, and E'_Δ is needed. Such a relation is developed in reference [8] and is in the form

$$V_{t\Delta} = K_5\delta_\Delta + K_6 E'_\Delta \tag{3.37}$$

where $K_5 = V_{t\Delta}/\delta_\Delta]_{E'_\Delta}$ = change in terminal voltage with change in rotor angle for constant E'
$K_6 = V_{t\Delta}/E'_\Delta]_{\delta_\Delta}$ = change in terminal voltage with change in E' for constant δ

The system block diagram with voltage regulation added is shown in Figure 3.4.

From (3.36) and (3.37)

$$v_{F\Delta} = -[K_\epsilon/(1 + \tau_\epsilon s)](K_5\delta_\Delta + K_6E'_\Delta) \tag{3.38}$$

Substituting in (3.10), we compute

$$E'_\Delta = \frac{K_3}{1 + K_3\tau'_{d0}s}\left[-\frac{K_\epsilon}{1 + \tau_\epsilon s}(K_5\delta_\Delta + K_6E'_\Delta)\right] - \frac{K_3K_4}{1 + K_3\tau'_{d0}s}\delta_\Delta$$

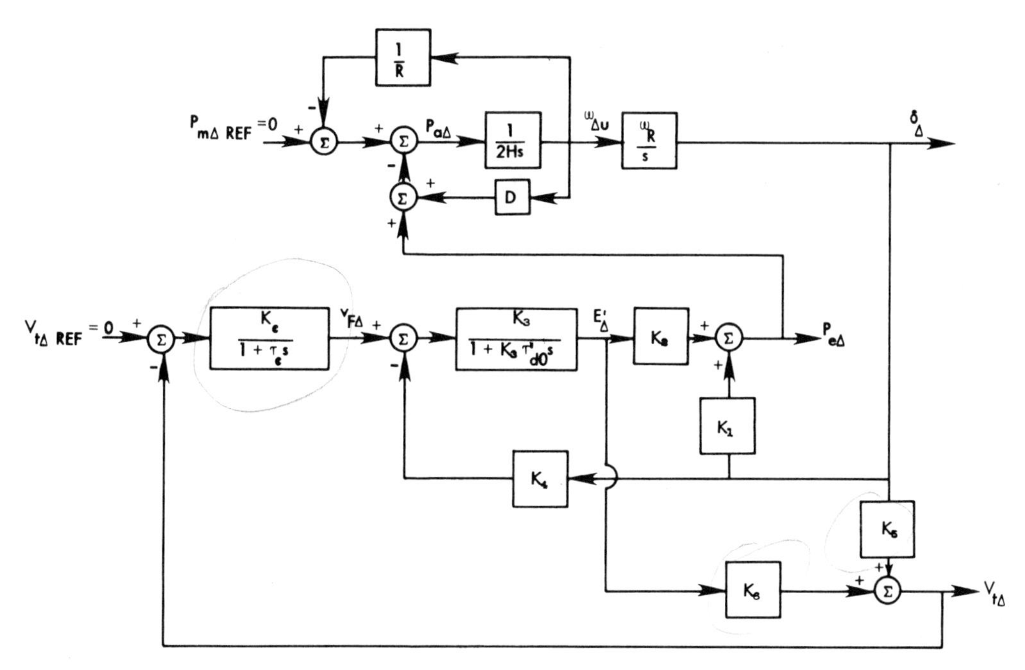

Fig. 3.4 System block diagram with voltage regulation.

or, rearranging,

$$E'_\Delta = \left[-\frac{K_4}{\tau'_{d0}} \frac{s + \left(\dfrac{1}{\tau_\epsilon} + \dfrac{K_5 K_\epsilon}{K_4 \tau_\epsilon}\right)}{s^2 + s\left(\dfrac{1}{\tau_\epsilon} + \dfrac{1}{K_3 \tau'_{d0}}\right) + \dfrac{1 + K_3 K_6 K_\epsilon}{K_3 \tau'_{d0} \tau_\epsilon}} \right] \delta_\Delta \tag{3.39}$$

From (3.39) and (3.9)

$$P_{e\Delta} = \left[K_1 - \frac{K_2 K_4}{\tau'_{d0}} \frac{s + \left(\dfrac{1}{\tau_\epsilon} + \dfrac{K_5 K_\epsilon}{K_4 \tau_\epsilon}\right)}{s^2 + s\left(\dfrac{1}{\tau_\epsilon} + \dfrac{1}{K_3 \tau'_{d0}}\right) + \dfrac{1 + K_3 K_6 K_\epsilon}{K_3 \tau'_{d0} \tau_\epsilon}} \right] \delta_\Delta \tag{3.40}$$

Substituting in the s domain swing equation and rearranging, we obtain the following characteristic equation:

$$s^4 + s^3 \left(\frac{1}{\tau_\epsilon} + \frac{1}{K_3 \tau'_{d0}}\right) + s^2 \left(\frac{1 + K_3 K_6 K_\epsilon}{K_3 \tau'_{d0} \tau_\epsilon} + \frac{K_1 \omega_R}{2H}\right) + s\, \frac{\omega_R}{2H} \left(\frac{K_1}{\tau_\epsilon} + \frac{K_1}{K_3 \tau'_{d0}} - \frac{K_2 K_4}{\tau'_{d0}}\right)$$
$$+ \frac{\omega_R}{2H} \left[\frac{K_1 (1 + K_3 K_6 K_\epsilon)}{K_3 \tau'_{d0} \tau_\epsilon} - \frac{K_2 K_4}{\tau'_{d0}} \left(\frac{1}{\tau_\epsilon} + \frac{K_5 K_\epsilon}{K_4 \tau_\epsilon}\right)\right] = 0 \tag{3.41}$$

Equation (3.41) is of the form

$$s^4 + \alpha_3 s^3 + \alpha_2 s^2 + \alpha_1 s + \alpha_0 = 0 \tag{3.42}$$

Analysis of this fourth-order system for stability is left as an exercise (see Problem 3.3).

3.5.2 Governor with one time lag

Referring to Figure 2.24, we note that a change in the speed ω or in the load or speed reference [governor speed changer (GSC)] produces a change in the mechanical torque T_m. The amount of change in T_m depends upon the speed droop and upon the transfer functions of the governor and the energy source.

For the model under consideration it is assumed that $GSC_\Delta = 0$ and that the combined effect of the turbine and speed governor systems are such that the change in the mechanical power is in the form

$$P_{m\Delta} = -[K_g/(1 + \tau_g s)]\omega_\Delta \tag{3.43}$$

where K_g = gain constant = $1/R$
τ_g = governor time constant

The system block diagram with governor regulation is shown in Figure 3.5.

Then the linearized swing equation in the s domain is in the form

$$(2H/\omega_R)s^2\delta_\Delta(s) = -[K_g/(1 + \tau_g s)]s\delta_\Delta(s) - P_{e\Delta}(s) \tag{3.44}$$

The order of this equation will depend upon the expression used for $P_{e\Delta}(s)$. If we assume the simplest model possible, $P_{e\Delta}(s) = P_s\delta_\Delta(s)$, the characteristic equation of the system is given by

$$(2H/\omega_R)s^2 + [K_g/(1 + \tau_g s)]s + P_s = 0 \tag{3.45}$$

or

$$s^3(2H\tau_g/\omega_R) + s^2(2H/\omega_R) + (K_g + P_s\tau_g)s + P_s = 0 \tag{3.46}$$

The system is now of third order. Applying Routh's criterion, the system is stable if $K_g > 0$ and $P_s > 0$.

If another model is used for $P_{e\Delta}(s)$, such as the model given by (3.9) and (3.10), the system becomes of fourth order, as shown in Figure 3.5. Its dynamic response will change. Information on stability can be obtained from the roots of the characteristic equation or from examining the eigenvalues of its characteristic matrix.

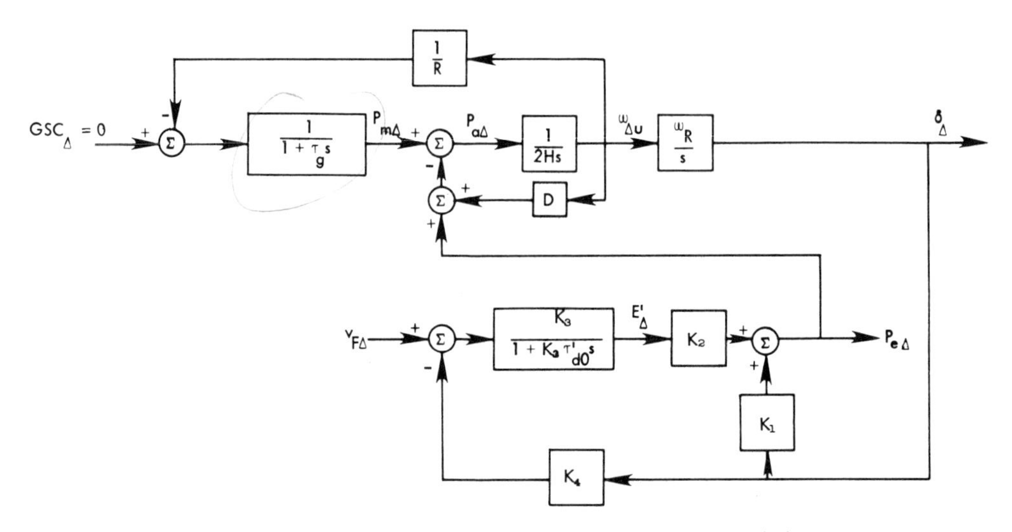

Fig. 3.5 Block diagram of a system with governor speed regulation.

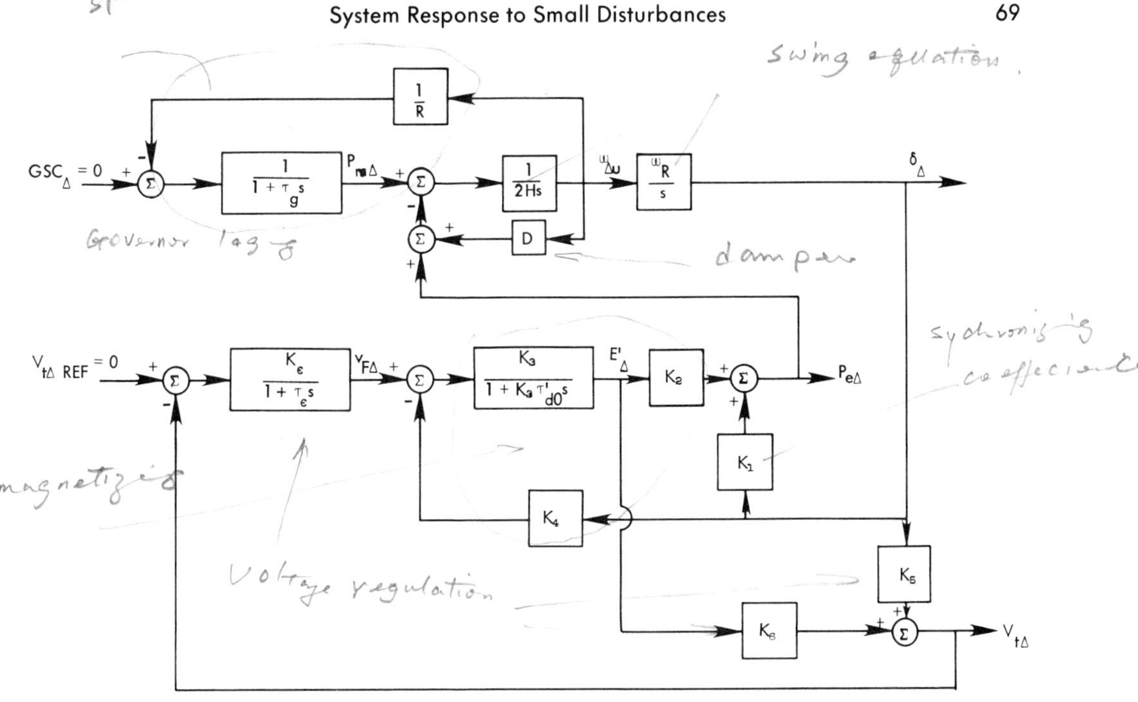

Fig. 3.6 Block diagram of a system with a governor and voltage regulator.

If both speed governor and voltage regulation are added simultaneously, as is usually the case, the system becomes fifth order, as shown in Figure 3.6.

3.6 Distribution of Power Impacts

In this section we consider the effect of the sudden application of a small load $P_{L\Delta}$ at some point in the network. (See also [7, 5].) To simplify the analysis, we also assume that the load has a negligible reactive component. Since the sudden change in load $P_{L\Delta}$ creates an unbalance between generation and load, an oscillatory transient results before the system settles to a new steady-state condition. This kind of impact is continuously occurring during normal operation of power systems. The oscillatory transient is in fact a "spectrum" of oscillations resulting from the random change in loads. These oscillations are reflected in power flow in the tie lines. Thus the scheduled tie-line flows will have "random" power oscillations superimposed upon them. Our concern here is to make an estimate of the magnitude of these power oscillations. Note that the estimates made by the methods outlined below are only approximate, yet they are quite instructive.

We formulate the problem mathematically using the network configuration of Figure 3.7 and the equations of Sections 2.9 and 3.4. Referring to the $(n + 1)$-port network in Figure 3.7, the power into node i is obtained from (3.21) by adding node k.

$$P_i = E_i G_{ii} + \sum_{\substack{j=1 \\ j \neq i,k}}^{n} E_i E_j (B_{ij} \sin \delta_{ij} + G_{ij} \cos \delta_{ij}) + E_i V_k (B_{ik} \sin \delta_{ik} + G_{ik} \cos \delta_{ik})$$

For the case of nearly zero conductance

$$P_i \cong \sum_{\substack{j=1 \\ j \neq i,k}}^{n} E_i E_j B_{ij} \sin \delta_{ij} + E_i V_k B_{ik} \sin \delta_{ik} \tag{3.47}$$

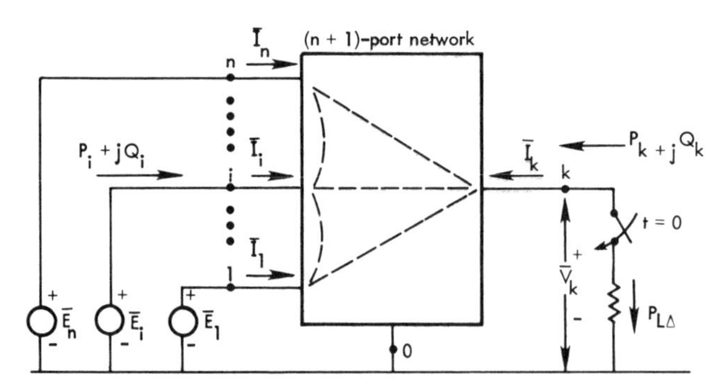

Fig. 3.7 Network with power impact at node k.

and the power into node k (the load bus) is

$$P_k = \sum_{\substack{j=1 \\ j \neq k}}^{n} V_k E_j B_{kj} \sin \delta_{kj} \tag{3.48}$$

Here we assume that the power network has a very high X/R ratio such that the conductances are negligible. The machines are represented by the classical model of constant voltage behind transient reactance. We also assume that the network has been reduced to the internal machine nodes (nodes $1, 2, \ldots, n$ of Figure 2.17) and the node k, where the impact $P_{L\Delta}$ is applied.

The *immediate* effect (assuming the network response to be fast) of the application of $P_{L\Delta}$ is that the angle of bus k is changed while the magnitude of its voltage V_k is unchanged, or $V_k \underline{/\delta_{k0}}$ becomes $V_k \underline{/\delta_{k0} + \delta_{k\Delta}}$. Note also that the internal angles of the machine nodes $\delta_1, \delta_2, \ldots, \delta_n$ do not change instantly because of the rotor inertia.

3.6.1 Linearization

The equations for injected power (3.47) and (3.48) are nonlinear because of the transcendental functions. Since we are concerned only with a small impact $P_{L\Delta}$, we linearize these equations to find

$$P_i = P_{i0} + P_{i\Delta} \qquad P_k = P_{k0} + P_{k\Delta}$$

and determine only the *change* variables $P_{i\Delta}$ and $P_{k\Delta}$.

The transcendental functions are linearized by the relations

$$\begin{aligned} \sin \delta_{kj} &= \sin(\delta_{kj0} + \delta_{kj\Delta}) \cong \sin \delta_{kj0} + (\cos \delta_{kj0})\delta_{kj\Delta} \\ \cos \delta_{kj} &= \cos(\delta_{kj0} + \delta_{kj\Delta}) \cong \cos \delta_{kj0} - (\sin \delta_{kj0})\delta_{kj\Delta} \end{aligned} \tag{3.49}$$

for any k, j. Note that the order kj must be carefully observed since $\delta_{kj} = -\delta_{jk}$. Substituting (3.49) into (3.47) and (3.48) and eliminating the initial values, we compute the linear equations

$$\begin{aligned} P_{i\Delta} &= \sum_{\substack{j=1 \\ j \neq i,k}}^{n} (E_i E_j B_{ij} \cos \delta_{ij0})\delta_{ij\Delta} + (V_k E_i B_{ik} \cos \delta_{ik0})\delta_{ik\Delta} = \sum_{\substack{j=1 \\ j \neq i,k}}^{n} P_{sij}\delta_{ij\Delta} + P_{sik}\delta_{ik\Delta} \\ P_{k\Delta} &= \sum_{j=1}^{n} (V_k E_j B_{kj} \cos \delta_{kj0})\delta_{kj\Delta} = \sum_{j=1}^{n} P_{skj}\delta_{kj\Delta} \end{aligned} \tag{3.50}$$

These equations are valid for any time t following the application of the impact.

3.6.2 A special case: $t = 0^+$

The instant immediately following the impact is of interest. In particular, we would like to determine exactly how much of the impact $P_{L\Delta}$ is supplied by each generator $P_{i\Delta}, i = 1, 2, \ldots, n$.

At the instant $t = 0^+$ we know that $\delta_{i\Delta} = 0$ for all generators because of rotor inertias. Thus we can compute (with both i and j indicating generator subscripts)

$$\delta_{ij\Delta} = 0 \qquad \delta_{ik\Delta} = \delta_{i\Delta} - \delta_{k\Delta} = -\delta_{k\Delta}(0^+) \qquad \delta_{kj\Delta} = \delta_{k\Delta} - \delta_{j\Delta} = \delta_{k\Delta}(0^+)$$

Thus (3.50) becomes

$$P_{i\Delta}(0^+) = -P_{sik}\delta_{k\Delta}(0^+) \qquad P_{k\Delta}(0^+) = \sum_{j=1}^{n} P_{skj}\delta_{k\Delta}(0^+) \tag{3.51}$$

Comparing the above two equations at $t = 0^+$, we note that at node k

$$P_{k\Delta}(0^+) = -\sum_{i=1}^{n} P_{i\Delta}(0^+) \tag{3.52}$$

This is to be expected since we are assuming a nearly reactive network. We also note that at node i $P_{i\Delta}$ depends upon $B_{ik} \cos \delta_{ik0}$. In other words, the higher the transfer susceptance B_{ik} and the lower the initial angle δ_{ik0}, the greater the share of the impact "picked up" by machine i. Note also that $P_{k\Delta} = -P_{L\Delta}$, so the foregoing equations can be written in terms of the load impact as

$$P_{L\Delta}(0^+) = -\sum_{i=1}^{n} P_{ski}\delta_{k\Delta}(0^+) = \sum_{i=1}^{n} P_{i\Delta}(0^+) \tag{3.53}$$

From (3.52) and (3.53) we conclude that

$$\delta_{k\Delta}(0^+) = -P_{L\Delta}(0^+) \Big/ \sum_{i=1}^{n} P_{sik} \tag{3.54}$$

$$P_{i\Delta}(0^+) = \left(P_{sik} \Big/ \sum_{j=1}^{n} P_{sjk}\right) P_{L\Delta}(0^+) \qquad i = 1, 2, \ldots, n \tag{3.55}$$

It is interesting that at the instant of the load impact (i.e., at $t = 0^+$, the source of energy supplied by the generators is the energy stored in their magnetic fields and is distributed according to the synchronizing power coefficients between i and k. Note that the generator rotor angles cannot move instantly; hence the energy supplied by the generators cannot come instantly from the energy stored in the rotating masses. This is also evident from the first equation of (3.51); $P_{i\Delta}$ depends upon P_{sik} or B_{ik}, which depends upon the reactance between generator i and node k. Later on when the rotor angles change, the stored energy in the rotating masses becomes important, as shown below.

Equations (3.52) and (3.55) indicate that the load impact $P_{L\Delta}$ at a network bus k is *immediately* shared by the synchronous generators according to their synchronizing power coefficients with respect to the bus k. Thus the machines electrically close to the point of impact will pick up the greater share of the load *regardless of their size.*

Let us consider next the deceleration of machine i due to the sudden increase in its output power $P_{i\Delta}$. The incremental differential equation governing the motion of machine i is given by

$$\frac{2H_i}{\omega_R}\frac{d\omega_{i\Delta}}{dt} + P_{i\Delta}(t) = 0 \qquad i = 1,2,\ldots,n \tag{3.56}$$

and using (3.55)

$$\frac{2H_i}{\omega_R}\frac{d\omega_{i\Delta}}{dt} + \left(P_{sik} \Big/ \sum_{j=1}^{n} P_{sjk}\right) P_{L\Delta}(0^+) = 0 \qquad i = 1,2,\ldots,n$$

Then if $P_{L\Delta}$ is constant for all t, we compute the acceleration in pu to be

$$\frac{1}{\omega_R}\frac{d\omega_{i\Delta}}{dt} = -\frac{P_{sik}}{2H_i}\left(P_{L\Delta}(0^+) \Big/ \sum_{j=1}^{n} P_{sjk}\right) \qquad i = 1,2,\ldots,n \tag{3.57}$$

Obviously, the shaft decelerates for a positive load $P_{L\Delta}$. The pu deceleration of machine i, given by (3.57), is dependent on the synchronizing power coefficient P_{sik} and inertia H_i. This deceleration will be constant until the governor action begins. Note that after the initial impact the various synchronous machines will be retarded at different rates, each according to its size H_i and its "electrical location" given by P_{sik}.

3.6.3 Average behavior prior to governor action ($t = t_1$)

We now estimate the system behavior during the period $0 < t < t_g$, where t_g is the time at which governor action begins. To designate this period simply, we refer to time as t_1, although there is no specific instant under consideration but a brief time period of no more than a few seconds. Looking at the system as a whole, there will be an overall deceleration of the machines during this period. To obtain the mean deceleration, let us define an "inertial center" that has angle $\bar{\delta}$ and angular velocity $\bar{\omega}$, where by definition,

$$\bar{\delta} \triangleq (1/\textstyle\sum H_i)\sum \delta_i H_i \qquad \bar{\omega} \triangleq (1/\sum H_i)\sum \omega_i H_i \tag{3.58}$$

Summing the set (3.57) for all values of i, we compute

$$\frac{2}{\omega_R}\sum \frac{d}{dt}(H_i\omega_{i\Delta}) = P_{k\Delta} = -P_{L\Delta}(0^+) \tag{3.59}$$

$$\frac{d}{dt}\frac{\bar{\omega}_\Delta}{\omega_R} = -P_{L\Delta}(0^+) \Big/ \sum_{i=1}^{n} 2H_i \tag{3.60}$$

Equation (3.60) gives the *mean* acceleration of all the machines in the system, which is defined here as the acceleration of a fictitious inertial center.

We now investigate the way in which the impact $P_{L\Delta}$ will be shared by the various machines. Note that while the system as a whole is retarding at the rate given by (3.60), the individual machines are retarding at different rates. Each machine follows an oscillatory motion governed by its swing equation. Synchronizing forces tend to pull them toward the mean system retardation, and after the initial transient decays they will acquire the same retardation as given by (3.60). In other words, when the transient decays, $d\omega_{i\Delta}/dt$ will be the same as $d\bar{\omega}_\Delta/dt$ as given by (3.60). Substituting this value of $d\omega_{i\Delta}/dt$ in (3.56), at $t = t_1 > t_0$,

$$P_{i\Delta}(t_1) = \left(H_i \Big/ \sum_{j=1}^{n} H_j\right) P_{L\Delta}(0^+) \tag{3.61}$$

Thus at the end of a brief transient the various machines will share the increase in load as a function *only* of their inertia constants. The time t_1 is chosen large enough so that all the machines will have acquired the mean system retardation. At the same time t_1 is not so large as to allow other effects such as governor action to take place. Equation (3.61) implies that the H constants for all the machines are given to a common base. If they are given for each machine on its own base, the correct powers are obtained if H is replaced by HS_{B3}/S_{sB}, where S_{B3} is the machine rating and S_{sB} is the chosen system base.

Examining (3.56) and (3.61), we note that immediately after the impact $P_{L\Delta}$(i.e., at $t = 0^+$) the machines share the impact according to their electrical proximity to the point of the impact as expressed by the synchronizing power coefficients. After a brief transient period the same machines share the same impact according to entirely different criteria, namely, according to their inertias.

Example 3.4

Consider the nine-bus, three-machine system of Example 2.6 with a small 10-MW resistive load added to bus 8 as in Example 3.2. Solve the system differential equations and plot $P_{i\Delta}$ and $\omega_{i\Delta}$ as functions of time. Compare computed results against theoretical values of Section 3.6.

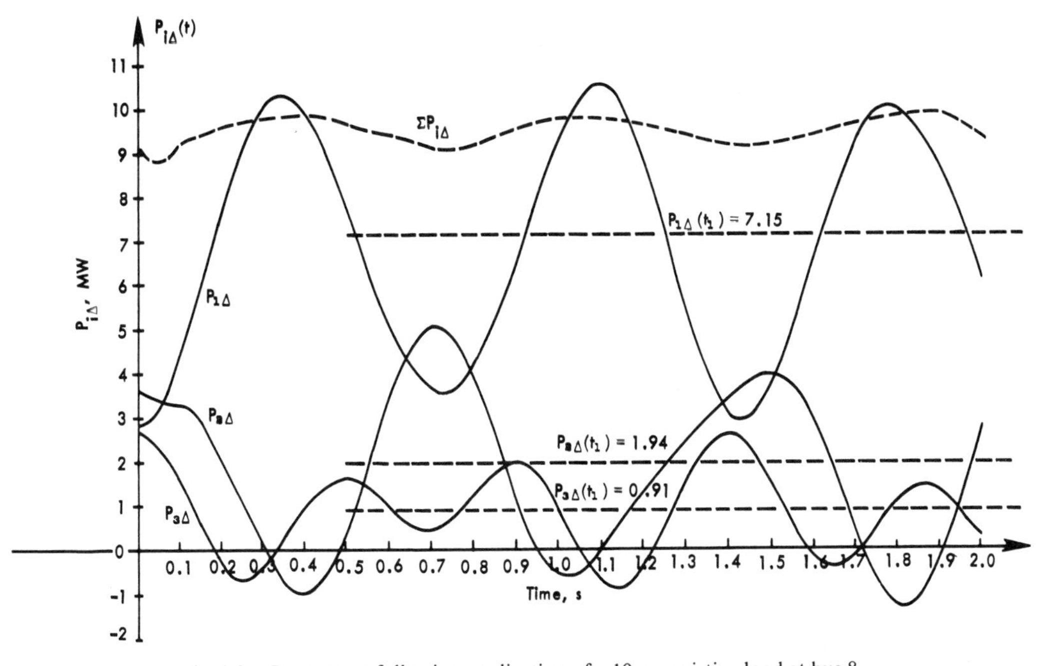

Fig. 3.8 $P_{i\Delta}$ versus t following application of a 10 pu resistive load at bus 8.

Solution

A nominal 10-MW (0.1 pu) load is added to bus 8 by applying a three-phase fault through a 10 pu resistance, using a library transient stability program. The resulting power oscillations $P_{i\Delta}$, $i = 1, 2, 3$, are shown in Figure 3.8 for the system operating without governor action.

The prefault conditions at the generators are given in Table 3.1 and in Example 2.6. From the prefault load flow of Figure 2.19 we determine that $V_{80} = 1.016$ and $\delta_{80} = 0.7°$. A matrix reduction of the nine-bus system, retaining only nodes 1, 2, 3, and 8, gives the system data shown on Table 3.3.

Table 3.3 Transfer Admittances and Initial Angles of a Nine-Bus System

ij	G_{ij}	B_{ij}	δ_{ij0}
1–8	0.01826	2.51242	1.5717
2–8	−0.03530	3.55697	19.0315
3–8	−0.00965	2.61601	12.4752

From (3.24) we compute the synchronizing power coefficients

$$P_{sik} = V_i V_k (B_{ik} \cos \delta_{ik0} - G_{ik} \sin \delta_{ik0})$$

These values are tabulated in Table 3.4. Note that the error in neglecting the G_{ik} term is small.

Table 3.4. Synchronizing Power Coefficients

ik	P_{sik} (neglecting G_{ik})	P_{sik} (with G_{ik} term)
18	2.6961	2.6955
28	3.5878	3.6001
38	2.6392	2.6414
$\sum P_{sik}$	8.9231	8.9370

The values of $P_{i\Delta}(0^+)$ are computed from (3.55) as

$$P_{i\Delta}(0^+) = \left(P_{si8} \bigg/ \sum_{j=1}^{n} P_{sj8} \right) P_{L\Delta}(0^+)$$

where $P_{L\Delta}(0^+) = 10.0$ MW nominally. The results of these calculations and the actual values determined from the stability study are shown in Table 3.5.

Table 3.5. Initial Power Change at Generators Due to 10-MW Load Added to Bus 8

(1) i	(2) $P_{i\Delta}$ (neglecting G_{ik})	(3) $P_{i\Delta}$ (with G_{ik})	(4) $P_{i\Delta}$ (computer study)	(5) $P_{i\Delta}$ [91% of (2)]	(6) $P_{i\Delta}$ [91% of (3)]
1	3.021	3.016	2.8	2.749	2.745
2	4.021	4.028	3.6	3.659	3.665
3	2.958	2.956	2.7	2.692	2.690
$\sum P_{i\Delta}$	10.000	10.000	9.1	9.100	9.100

Note that the actual load pickup is only 9.1 MW instead of the desired 10 MW. This is due in part to the assumption of constant voltage V_k at bus 8 (actually, the voltage drops slightly) and to the assumed linearity of the system. If the computed $P_{i\Delta}$ are scaled down by 0.91, the results agree quite well with values measured from the computer study. These values are also shown on the plot of Figure 3.8 at time $t = 0^+$ and are due only to the synchronizing power coefficients of the generators with respect to bus 8.

The plots of $P_{i\Delta}$ versus time in Figure 3.8 show the oscillatory nature of the power exchange between generators following the impact. These oscillations have frequencies that are combinations of the eigenvalues computed in Example 3.2. The total, labeled $\sum P_{i\Delta}$, averages about 9.5 MW.

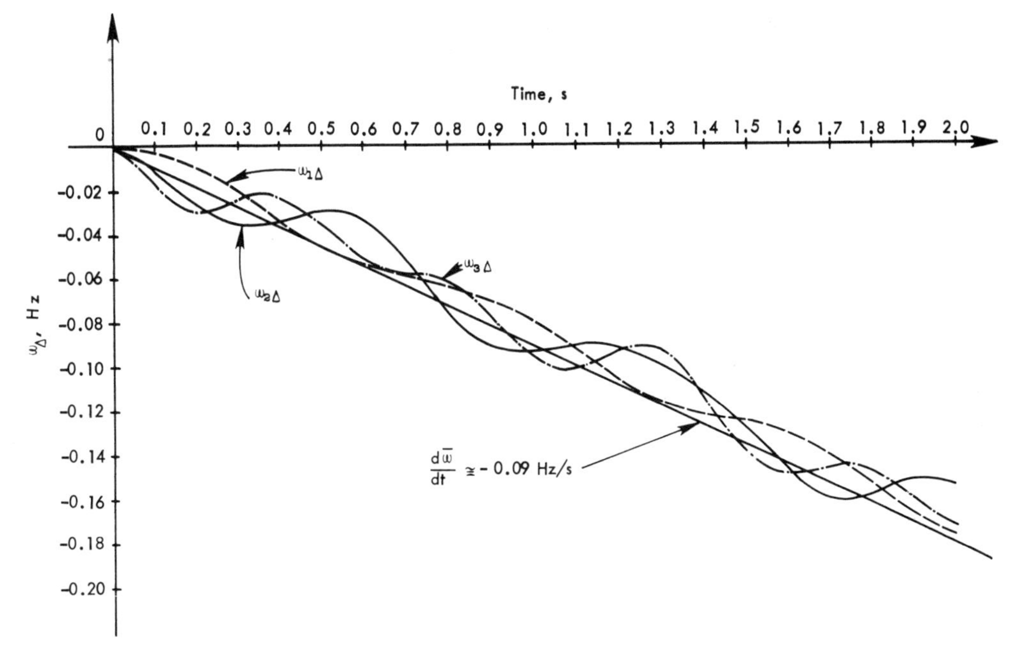

Fig. 3.9 Speed deviation following application of a 10 pu resistive load at bus 8.

Another point of interest in Figure 3.8 is the computed values of $P_{i\Delta}(t_1)$ that depend entirely on the machine inertia. These calculations are made from

$$\begin{aligned} P_{i\Delta}(t_1) &= (H_i/\sum H_i)P_{L\Delta} = 10H_i/(23.64 + 6.40 + 3.01) = 10H_i/33.05 \\ &= 7.15 \text{ MW} \quad i = 1 \\ &= 1.94 \text{ MW} \quad i = 2 \\ &= 0.91 \text{ MW} \quad i = 3 \end{aligned}$$

and the results are plotted in Figure 3.8 as dashed lines. It is fairly obvious that the $P_{i\Delta}(t)$ oscillate about these values of $P_{i\Delta}(t_1)$. It is also apparent that the system has little damping and the oscillations are likely to persist for some time. This is partly due to the inherent nature of this particular system, but the same phenomenon would be present to some extent on any system.

The second plot of interest is the speed deviation or slip as a function of time, shown in Figure 3.9. The computer program provides speed deviation data in Hz and these units are used in Figure 3.9. Note the steady deceleration with all units oscillating about the mean or inertial center. This is computed as

$$\begin{aligned} \frac{d\bar{\omega}}{dt} &= -\frac{P_{L\Delta}}{2\sum H_i} = -\frac{0.10}{2(23.64 + 6.40 + 3.01)} \\ &= -1.513 \times 10^{-3}\,\text{pu/s} = -0.570\,\text{rad/s}^2 = -0.0908\,\text{Hz/s} \end{aligned}$$

The individual machine speed deviations $\omega_{i\Delta}$ are plotted in Figure 3.9 and show graphically the intermachine oscillations that occur as the system slowly retards in frequency. The mean deceleration of about 0.09 Hz/s is plotted in Figure 3.9 as a straight line.

If the governors were active, the speed deviation would level off after a few seconds to a constant value and the oscillations would eventually decay. Since the governors have a drooping characteristic, the speed would then continue at the reduced value as long as the additional load was present. If the speed deviation is great, signifying a substantial load increase on the generators, the governors would need to be readjusted to the new load level so that additional prime-mover torque could be provided.

Example 3.5

Let us examine the effect of the above on the power flow in tie lines. Consider a power network composed of two areas connected with a tie line, as shown in Figure 3.10. The two areas are of comparable size, say 1000 MW each. They are connected with a tie line having a capacity of 100 MW. The tie line is carrying a steady power flow of 80 MW from area 1 to area 2 as shown in Figure 3.10. Now let a load impact $P_{L\Delta} = 10$ MW (1% of the capacity of one area) take place at some point in area 1, and determine the distribution of this added load immediately after its application ($t = 0^+$) and a short time later ($t = t_1$) after the initial transients have subsided. Because of the proximity of the groups of machines in area 1 to the point of impact, their synchronizing power coefficients are larger than those of the groups of machines in area 2. If we define $\sum P_{sik}]_{\text{area 1}} = P_{s1}$, $\sum P_{sik}]_{\text{area 2}} = P_{s2}$, then let us assume that $P_{s1} = 2P_{s2}$.

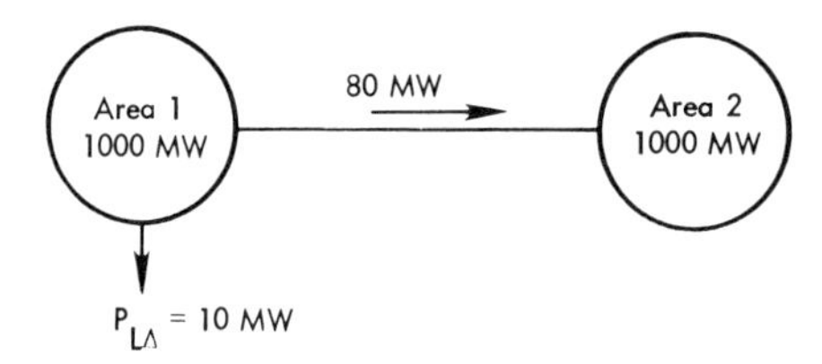

Fig. 3.10 Two areas connected with a tie line.

Solution

Since $P_{s1} = 2P_{s2}$, at the instant of the impact 2/3 of the 10-MW load will be supplied by the groups of machines in area 1, while 1/3 or 3.3 MW will be supplied by the groups of machines in area 2. Thus 3.3 MW will appear as a reduction in tie-line flow. In other words, at that instant the tie-line flow becomes 76.7 MW toward area 2.

At the end of the initial transient the load power impact $P_{L\Delta}$ will be shared by the machines according to their inertias. Let us assume that the machines of area 1 are

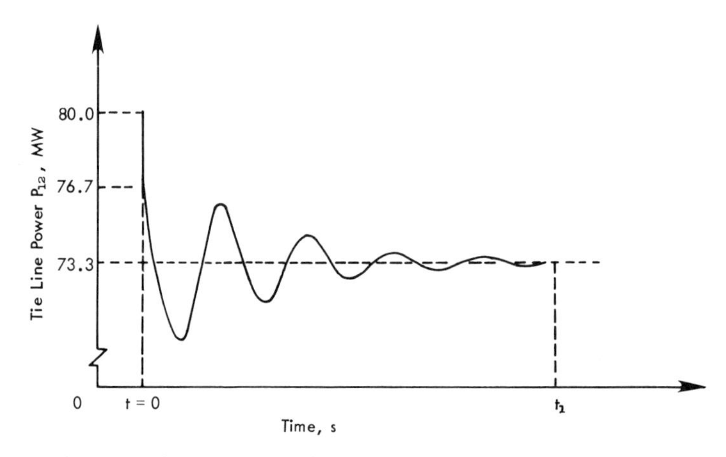

Fig. 3.11 Tie-line power oscillations due to the load impact in area 1.

predominantly hydro units (with relatively small H), while the units of area 2 are of larger inertia constants such that $\sum H_i]_{\text{area2}} = 2\sum H_i]_{\text{area1}}$ where all H's are on a common base. The sharing of the load among the groups of machines will now become 6.7 MW contributed from area 2 and 3.3 MW from area 1. The tie-line flow will now become 73.3 MW (toward area 2).

From the above we can see that in the situation discussed in this example a sudden application of a 10-MW load caused the tie-line flow to drop almost instantly by 3.3 MW, and after a brief transient by 6.7 MW. The transition from 76.7-MW flow to 73.3-MW flow is oscillatory, and power swings of as much as twice the difference between these two values may be encountered. This situation is illustrated in Figure 3.11.

The time t_1 mentioned above is smaller than the time needed by the various controllers to adjust the system generation to match the load and the tie-line flow to meet the scheduled flow.

Example 3.6

We now consider a slightly more complex and more realistic case wherein the area equivalents in Figure 3.10 are represented by their Thevenin equivalents and the tie-line impedance is given. The system data are given in Figure 3.12 in pu on a 1000-MVA base. The capacity of area 1 is 20,000 MW and that of area 2 is 14,000 MW. The inertia constants of the machines in the two areas are about equal.

(a) Find the equations of power for P_1 and P_2.
(b) Find the operating condition when $P_1 = 100$ MW. This would correspond approximately to a 100-MW tie-line flow from area 1 to area 2.
(c) Find the synchronizing power coefficients.
(d) Consider a sudden load addition to area 2, represented by the resistive load $P_{4\Delta}$ at bus 4. If this load is 200 MW (1.43% of the capacity of area 2), find the distribution of this load at $t = 0^+$ and $t = t_1$.

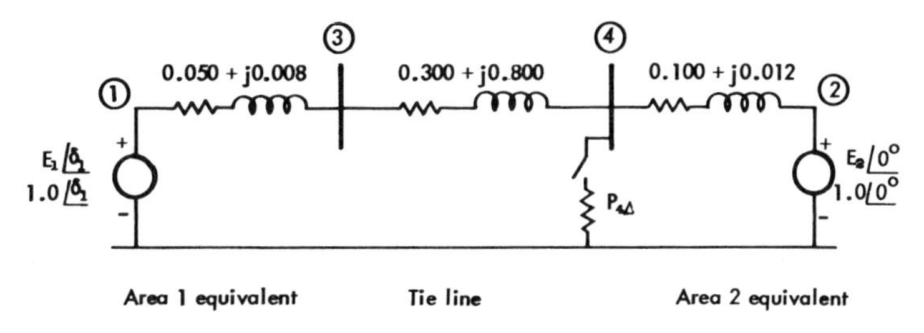

Fig. 3.12 Two areas connected by a tie line.

Solution

Consider the system as a two-port network between nodes 1 and 2. Then we compute

$$\bar{z}_{12} = 0.450 + j1.820 = 1.875\,/76.112^\circ \text{ pu}$$
$$\bar{y}_{12} = 1/\bar{z}_{12} = 0.533\,/-76.112^\circ = 0.128 - j0.518 \text{ pu}$$
$$\bar{Y}_{12} = -\bar{y}_{12} = 0.533\,/103.888^\circ$$
$$G_{11} = 0.128 \qquad g_{10} = g_{20} = 0$$

$$G_{12} = -0.128 \qquad \delta_{12} = \delta_1 - \delta_2 = \delta_1$$
$$B_{12} = 0.518$$

(a)
$$\begin{aligned} P_1 &= V_1^2 g_{10} + V_1 V_2 (G_{12}\cos\delta_{12} + B_{12}\sin\delta_{12}) - V_1^2 G_{12} \\ &= 0 + 1.0(-0.128\cos\delta_1 + 0.518\sin\delta_1) + 0.128 \\ &= 0.128 + 0.533\sin(\delta_1 - 13.796°) \\ P_2 &= V_2^2 g_{20} + V_1 V_2 (G_{12}\cos\delta_{21} + B_{12}\sin\delta_{21}) - V_2^2 G_{21} \\ &= 0 + 1.0(-0.128\cos\delta_1 - 0.518\sin\delta_1) + 0.128 \\ &= 0.128 - 0.533\sin(\delta_1 + 13.796°) \end{aligned}$$

(b) Given that $P_1 = 0.1$ pu

$$0.100 = 0.128 + 0.533\sin(\delta_1 - 13.796°) \qquad \delta_1 = 10.784°$$

(c)
$$\begin{aligned} P_{s12} &= V_1 V_2 (B_{12}\cos\delta_{120} - G_{12}\sin\delta_{120}) \\ &= 1.0(0.518\cos 10.784° + 0.128\sin 10.784°) = 0.533 \\ P_{s21} &= V_1 V_2 (B_{21}\cos\delta_{210} - G_{21}\sin\delta_{210}) \\ &= 1.0[0.518\cos(-10.784°) + 0.128\sin(-10.784°)] = 0.509 \end{aligned}$$

(d) Now add the 200-MW load at bus 4; $P_{4\Delta} = 200/1000 = 0.2$ pu.

To complete the problem, we must know the voltage $\overline{V}_4$ at $t = 0^-$. Thus we compute

$$\begin{aligned} \overline{I}_{12}(0^-) &= (\overline{V}_1 - \overline{V}_2)/\overline{z}_{12} = (1.0\angle 10.784° - 1.0\angle 0)/1.875\angle 76.112° = 0.100\angle 19.280° \\ \overline{V}_4(0^-) &= \overline{E}_2 + (0.100 + j0.012)\overline{I}_{12} = 1.009 + j0.004 = 1.009\angle 0.252° \end{aligned}$$
$$\delta_{40} = 0.252° \qquad \delta_{140} = \delta_{10} - \delta_{40} = 10.532° \qquad \delta_{240} = \delta_{20} - \delta_{40} = -0.252°$$

From the admittance matrix elements

$$\begin{aligned} \overline{Y}_{14} &= -\overline{y}_{14} = -1/\overline{z}_{14} = -0.451 + j1.042 \\ \overline{Y}_{24} &= -\overline{y}_{24} = -1/\overline{z}_{24} = -9.858 + j1.183 \end{aligned}$$

we compute the synchronizing power coefficients

$$\begin{aligned} P_{s14} &= V_1 V_4 (B_{14}\cos\delta_{140} - G_{14}\sin\delta_{140}) \\ &= (1.009)(1.042\cos 10.532° + 0.451\sin 10.532°) = 1.117 \\ P_{s24} &= V_2 V_4 (B_{24}\cos\delta_{240} - G_{24}\sin\delta_{240}) \\ &= 1.009[1.183\cos(-0.252°) + 9.858\sin(-0.252°)] = 1.150 \end{aligned}$$

Then the initial distribution of $P_{4\Delta}$ is

$$\begin{aligned} P_{1\Delta}(0^+) &= P_{s14}(0.2)/(P_{s14} + P_{s24}) = (0.493)(0.2) = 0.0986 \text{ pu} \\ P_{2\Delta}(0^+) &= P_{s24}(0.2)/(P_{s14} + P_{s24}) = (0.507)(0.2) = 0.1014 \text{ pu} \end{aligned}$$

The power distribution according to inertias is computed as

$$\begin{aligned} P_{1\Delta}(t_1) &= 0.2[20{,}000H/(20{,}000H + 14{,}000H)] = 0.11765 \text{ pu} \\ P_{2\Delta}(t_1) &= 0.2[14{,}000H/(20{,}000H + 14{,}000H)] = 0.08235 \text{ pu} \end{aligned}$$

In this example the synchronizing power coefficients P_{s14} and P_{s24} are nearly equal, while the inertias of the two areas are not. Thus while the *initial* distributions of the load $P_{4\Delta}$ are about the same, the distributions at a later time $t = t_1$ are such that area 1 picks up about 59% of the load and area 2 picks up the remaining 41%.

In general, the initial distribution of a load impact depends on the point of impact. Problem 3.10 gives another example where the point of impact is in area 1 (bus 3).

In the above discussion many factors have been neglected, e.g., the effect of the network transfer conductances, the effect of the reactive component of the load impact, the fast primary controllers such as some of the modern exciters, the load frequency and voltage characteristics, and others. Thus the conclusions reached above should be considered qualitative and as rough approximations. Yet these conclusions are basically sound and give a good "feel" for what happens to the machines and to the tie-line flows under the influence of small routine load changes.

If the system is made up of groups of machines separated by tie lines, they share the impacts differently under different conditions. Hence they will oscillate with respect to each other during the transient period following the impact. The power flow in the connecting ties will reflect these oscillations.

The analysis given above could be extended to include governor actions. Following an impact the synchronous machines will share the change first according to their synchronizing power coefficients, then after a brief period according to their inertias. The speed change will be sensed by the prime-mover governors, which will act to make the load sharing according to an entirely different criterion, namely, the speed governor droop characteristic. The transition from the second to the final stage is oscillatory (see Rudenberg [7], Ch. 23). The angular frequency of these oscillations can be estimated as follows. From Section 3.5.2, neglecting $P_{e\Delta}$, the change in the mechanical power $P_{m\Delta}$ is of the form

$$P_{m\Delta} = \frac{-1/R}{1 + \tau_s s} \frac{\omega_\Delta}{\omega_R} \tag{3.62}$$

where R is the regulation and τ_s is the servomotor time constant. The swing equation for machine i becomes, in the s domain,

$$\frac{2H_i s\omega_{i\Delta}}{\omega_R} + \frac{1/R_i}{1 + \tau_{si} s} \frac{\omega_{i\Delta}}{\omega_R} = 0$$

The characteristic equation of the system is given by

$$s^2 + (1/\tau_{si})s + 1/2H_i R_i \tau_{si} = 0 \tag{3.63}$$

from which the natural frequency of oscillation can be estimated.

It is interesting to note the order of magnitude of the frequency of oscillation in the two different transients discussed in this section. For a given machine (or a group of machines) the frequency of oscillation in the first transient is the natural frequency with respect to the point of impact. These frequencies are determined by finding the eigenvalues λ of the $\mathbf{A}$ matrix by solving det $(\mathbf{A} - \lambda\mathbf{U}) = 0$, where $\mathbf{U}$ is the unit matrix and $\mathbf{A}$ is defined by (3.1).

For the second transient, which occurs during the transition from sharing according to inertia to sharing according to governor characteristic, the frequency of oscillation is given by $\nu_{i2}^2 \simeq 1/2H_i R_i \tau_{si}$. Usually these two frequencies are appreciably different.

Problems

3.1 A synchronous machine is connected to a large system (an infinite bus) through a long transmission line. The direct axis transient reactance $x_d' = 0.20$ pu. The infinite bus voltage is 1.0 pu. The transmission line impedance is $Z_{line} = 0.20 + j0.60$ pu. The synchronous machine is to be represented by constant voltage behind transient reactance with $E' = 1.10$ pu. Calculate the minimum and maximum steady-state load delivered at the infinite bus (for stability). Repeat when there is a local load of unity power factor having $R_{load} = 8.0$ pu.

3.2 Use Routh's criterion to determine the conditions of stability for the system where the characteristic equation is given by (3.14).

3.3 Compute the characteristic equation for the system of Figure 3.1, including the damping term, and determine the conditions for stability using Routh's criterion. Compare the results with those of Section 3.3.1.

3.4 Using δ_Δ as the output variable in Figure 3.2, use block diagram algebra to reduce the system block diagram to forward and feedback transfer functions. Then determine the system stability and possible system behavior patterns by sketching an approximate root-locus diagram.

3.5 Use block diagram algebra to reduce the system described by (3.45). Then determine the system behavior by sketching the root loci for variations in K_g.

3.6 Give the conditions for stability of the system described by (3.20).

3.7 A system described by (3.41) has the following data: $H = 4$, $\tau_{d0}' = 5.0$, $\tau_\epsilon = 0.10$, $K_1 = 4.8$, $K_2 = 2.6$, $K_3 = 0.26$, $K_4 = 3.30$, $K_5 = 0.1$, and $K_6 = 0.5$. Find the maximum and minimum values of K_ϵ for stability. Repeat for $K_5 = -0.20$.

3.8 Write the system described by (3.46) in state-space form. Apply Routh's criterion to (3.46).

3.9 The equivalent *prefault* network is given in Table 2.6 for the three-machine system discussed in Section 2.10 and for the given operating conditions. The internal voltages and angles of the generators are given in Example 2.6.
(a) Obtain the synchronizing power coefficients P_{s12}, P_{s13}, P_{s23}, and the corresponding coefficients α_{ij} [see (3.31)] for small perturbations about the given operating point.
(b) Obtain the natural frequencies of oscillation for the angles $\delta_{12\Delta}$ and $\delta_{13\Delta}$. Compare with the periods of the nonlinear oscillations of Example 2.7.

3.10 Repeat Example 3.6 with the impact point shifted to area 1 and let $P_{L\Delta} = 100$ MW as before.

3.11 Repeat Problem 3.10 for an initial condition of $P_{L\Delta} = 300$ MW.

References

1. Korn, G. A., and Korn, T. M. *Mathematical Handbook for Scientists and Engineers.* McGraw-Hill, New York, 1968.
2. Hayashi, C. *Nonlinear Oscillations in Physical Systems.* McGraw-Hill, New York, 1964.
3. Takahashi, Y., Rabins, M. J., and Auslander, D. M. *Control and Dynamic Systems.* Addison-Wesley, Reading, Mass., 1970.
4. Venikov, V. A. *Transient Phenomena in Electric Power Systems.* Trans. by B. Adkins and D. Rutenberg. Pergamon Press, New York, 1964.
5. Hore, R. A. *Advanced Studies in Electrical Power System Design.* Chapman and Hall, London, 1966.
6. Crary, S. B. *Power System Stability,* Vols. 1, 2. Wiley, New York, 1945, 1947.
7. Rudenberg, R. *Transient Performance of Electric Power Systems: Phenomena in Lumped Networks.* McGraw-Hill, New York, 1950. (MIT Press, Cambridge, Mass., 1967.)
8. de Mello, F. P., and Concordia, C. Concepts of synchronous machine stability as affected by excitation control. *IEEE Trans.* PAS-88:316–29, 1969.
9. Heffron, W. G., and Phillips, R. A. Effect of a modern amplidyne voltage regulator on underexcited operation of large turbine generators. *AIEE Trans.* 71 (Pt. 3):692–97, 1952.
10. Routh, E. J. *Dynamics of a System of Rigid Bodies.* Macmillan, London, 1877. (Adams Prize Essay.)
11. Ogata, K. *State-Space Analysis of Control Systems.* Prentice-Hall, Englewood Cliffs, N.J., 1967.
12. Lefschetz, S. *Stability of Nonlinear Control Systems.* Academic Press, New York, London, 1965.

PART II The Electromagnetic Torque

chapter 4

The Synchronous Machine

4.1 Introduction

In this chapter we develop a mathematical model for a synchronous machine for use in stability computations. State-space formulation of the machine equations is used. Two models are developed, one using the currents as state variables and another using the flux linkages. Simplified models, which are often used for stability studies, are discussed. This chapter is not intended to provide an exhaustive treatment of synchronous machine theory. The interested reader should consult one of the many excellent references on this subject (see [1]–[9]).

The synchronous machine under consideration is assumed to have three stator windings, one field winding, and two amortisseur or damper windings. These six windings are magnetically coupled. The magnetic coupling between the windings is a function of the rotor position. Thus the flux linking each winding is also a function of the rotor position. The instantaneous terminal voltage v of any winding is in the form,

$$v = \pm\sum ri \pm \sum\dot{\lambda} \tag{4.1}$$

where λ is the flux linkage, r is the winding resistance, and i is the current, with positive directions of stator currents flowing out of the generator terminals. The notation $\pm\sum$ indicates the summation of all appropriate terms with due regard to signs. The expressions for the winding voltages are complicated because of the variation of λ with the rotor position.

4.2 Park's Transformation

A great simplification in the mathematical description of the synchronous machine is obtained if a certain transformation of variables is performed. The transformation used is usually called Park's transformation [10, 11]. It defines a new set of stator variables such as currents, voltages, or flux linkages in terms of the actual winding variables. The new quantities are obtained from the projection of the actual variables on three axes; one along the direct axis of the rotor field winding, called the direct axis; a second along the neutral axis of the field winding, called the quadrature axis; and the third on a stationary axis. Park's transformation is developed mathematically as follows.[1]

1. The transformation developed and used in this book is not exactly that used by Park [10, 11] but is more nearly that suggested by Lewis [12], with certain other features suggested by Concordia (discussion to [12]) and Krause and Thomas [13].

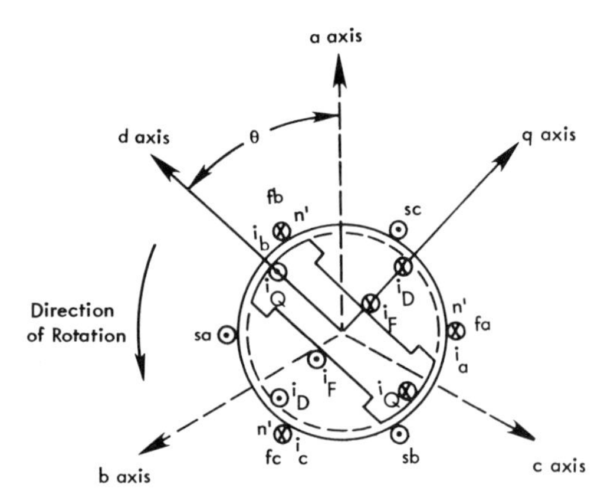

Fig. 4.1 Pictorial representation of a synchronous machine.

We define the d axis of the rotor at some instant of time to be at angle θ rad with respect to a fixed reference position, as shown in Figure 4.1. Let the stator phase currents i_a, i_b, and i_c be the currents *leaving* the generator terminals. If we "project" these currents along the d and q axes of the rotor, we get the relations

$$
\begin{aligned}
i_{q\text{axis}} &= (2/3)[i_a \sin\theta + i_b \sin(\theta - 2\pi/3) + i_c \sin(\theta + 2\pi/3)] \\
i_{d\text{axis}} &= (2/3)[i_a \cos\theta + i_b \cos(\theta - 2\pi/3) + i_c \cos(\theta + 2\pi/3)]
\end{aligned}
\tag{4.2}
$$

We note that for convenience the axis of phase a was chosen to be the reference position, otherwise some angle of displacement between phase a and the arbitrary reference will appear in all the above terms.

The effect of Park's transformation is simply to transform all stator quantities from phases a, b, and c into new variables the frame of reference of which moves with the rotor. We should remember, however, that if we have three variables i_a, i_b, and i_c, we need three new variables. Park's transformation uses two of the new variables as the d and q axis components. The third variable is a stationary current, which is proportional to the zero-sequence current. A multiplier is used to simplify the numerical calculations. Thus *by definition*

$$
\mathbf{i}_{0dq} = \mathbf{P}\mathbf{i}_{abc} \tag{4.3}
$$

where we define the current vectors

$$
\mathbf{i}_{0dq} = \begin{bmatrix} i_0 \\ i_d \\ i_q \end{bmatrix} \qquad \mathbf{i}_{abc} = \begin{bmatrix} i_a \\ i_b \\ i_c \end{bmatrix} \tag{4.4}
$$

and where the Park's transformation $\mathbf{P}$ is defined as

$$
\mathbf{P} = \sqrt{2/3}\begin{bmatrix} 1/\sqrt{2} & 1/\sqrt{2} & 1/\sqrt{2} \\ \cos\theta & \cos(\theta - 2\pi/3) & \cos(\theta + 2\pi/3) \\ \sin\theta & \sin(\theta - 2\pi/3) & \sin(\theta + 2\pi/3) \end{bmatrix} \tag{4.5}
$$

The main field-winding flux is along the direction of the d axis of the rotor. It produces an EMF that *lags* this flux by 90°. Therefore the machine EMF E is primarily along the rotor q axis. Consider a machine having a constant terminal voltage V. For generator action the phasor $\bar{E}$ should be leading the phasor $\bar{V}$. The angle between $\bar{E}$ and $\bar{V}$ is the machine torque angle δ if the phasor $\bar{V}$ is in the direction of the reference phase (phase a).

At $t = 0$ the phasor $\bar{V}$ is located at the axis of phase a, i.e., at the reference axis in Figure 4.1. The q axis is located at an angle δ, and the d axis is located at $\theta = \delta + \pi/2$. At $t > 0$, the reference axis is located at an angle $\omega_R t$ with respect to the axis of phase a. The d axis of the rotor is therefore located at

$$\theta = \omega_R t + \delta + \pi/2 \ \text{rad} \tag{4.6}$$

where ω_R is the rated (synchronous) angular frequency in rad/s and δ is the synchronous torque angle in electrical radians.

Expressions similar to (4.3) may also be written for voltages or flux linkages; e.g.,

$$\mathbf{v}_{0dq} = \mathbf{P}\mathbf{v}_{abc} \qquad \boldsymbol{\lambda}_{0dq} = \mathbf{P}\boldsymbol{\lambda}_{abc} \tag{4.7}$$

If the transformation (4.5) is unique, an inverse transformation also exists wherein we may write

$$\mathbf{i}_{abc} = \mathbf{P}^{-1}\mathbf{i}_{0dq} \tag{4.8}$$

The inverse of (4.5) may be computed to be

$$\mathbf{P}^{-1} = \sqrt{2/3}\begin{bmatrix} 1/\sqrt{2} & \cos\theta & \sin\theta \\ 1/\sqrt{2} & \cos(\theta - 2\pi/3) & \sin(\theta - 2\pi/3) \\ 1/\sqrt{2} & \cos(\theta + 2\pi/3) & \sin(\theta + 2\pi/3) \end{bmatrix} \tag{4.9}$$

and we note that $\mathbf{P}^{-1} = \mathbf{P}^t$, which means that the transformation $\mathbf{P}$ is orthogonal. Having $\mathbf{P}$ orthogonal also means that the transformation $\mathbf{P}$ is power invariant, and we should expect to use the same power expression in either the a-b-c or the 0-d-q frame of reference. Thus

$$\begin{aligned} p &= v_a i_a + v_b i_b + v_c i_c = \mathbf{v}_{abc}^t \mathbf{i}_{abc} = (\mathbf{P}^{-1}\mathbf{v}_{0dq})^t(\mathbf{P}^{-1}\mathbf{i}_{0dq}) \\ &= \mathbf{v}_{0dq}^t(\mathbf{P}^{-1})^t\mathbf{P}^{-1}\mathbf{i}_{0dq} = \mathbf{v}_{0dq}^t\mathbf{P}\mathbf{P}^{-1}\mathbf{i}_{0dq} \\ &= \mathbf{v}_{0dq}^t\mathbf{i}_{0dq} = v_0 i_0 + v_d i_d + v_q i_q \end{aligned} \tag{4.10}$$

4.3 Flux Linkage Equations

The situation depicted in Figure 4.1 is that of a network consisting of six mutually coupled coils. These are the three phase windings sa-fa, sb-fb, and sc-fc; the field winding F-F'; and the two damper windings D-D' and Q-Q'. (The damper windings are often designated by the symbols kd and kq. We prefer the shorter notation used here. Phase-winding designations s and f refer to "start" and "finish" of these coils.) We write the flux linkage equation for these six circuits as

$$\begin{matrix}\text{stator}\\ \\ \\ \text{rotor}\end{matrix}\begin{bmatrix} \lambda_a \\ \lambda_b \\ \lambda_c \\ - \\ \lambda_F \\ \lambda_D \\ \lambda_Q \end{bmatrix} = \left[\begin{array}{ccc|ccc} L_{aa} & L_{ab} & L_{ac} & L_{aF} & L_{aD} & L_{aQ} \\ L_{ba} & L_{bb} & L_{bc} & L_{bF} & L_{bD} & L_{bQ} \\ L_{ca} & L_{cb} & L_{cc} & L_{cF} & L_{cD} & L_{cQ} \\ \hline L_{Fa} & L_{Fb} & L_{Fc} & L_{FF} & L_{FD} & L_{FQ} \\ L_{Da} & L_{Db} & L_{Dc} & L_{DF} & L_{DD} & L_{DQ} \\ L_{Qa} & L_{Qb} & L_{Qc} & L_{QF} & L_{QD} & L_{QQ} \end{array}\right] \begin{bmatrix} i_a \\ i_b \\ i_c \\ - \\ i_F \\ i_D \\ i_Q \end{bmatrix} \ \text{Wb turns} \tag{4.11}$$

$\neq k$

he subscript convention in (4.11) where lower-
ies and uppercase subscripts are used for rotor
of the inductances in (4.11) are functions of
ices may be written as follows

4.3.1 Stator self-inductances

The phase-winding self-inductances are given by

$$
\begin{aligned}
L_{aa} &= L_s + L_m \cos 2\theta \ \text{ H} \\
L_{bb} &= L_s + L_m \cos 2(\theta - 2\pi/3) \ \text{ H} \\
L_{cc} &= L_s + L_m \cos 2(\theta + 2\pi/3) \ \text{ H}
\end{aligned} \tag{4.12}
$$

where $L_s > L_m$ and both L_s and L_m are constants. (All inductance quantities such as L_s or M_s with *single subscripts* are *constants* in our notation.)

4.3.2 Rotor self-inductances

Since saturation and slot effect are neglected, all rotor self-inductances are constants and, according to our subscript convention, we may use a single subscript notation; i.e.,

$$
L_{FF} = L_F \ \text{ H} \qquad L_{DD} = L_D \ \text{ H} \qquad L_{QQ} = L_Q \ \text{ H} \tag{4.13}
$$

4.3.3 Stator mutual inductances

The phase-to-phase mutual inductances are functions of θ but are symmetric,

$$
\begin{aligned}
L_{ab} &= L_{ba} = -M_s - L_m \cos 2(\theta + \pi/6) \ \text{ H} \\
L_{bc} &= L_{cb} = -M_s - L_m \cos 2(\theta - \pi/2) \ \text{ H} \\
L_{ca} &= L_{ac} = -M_s - L_m \cos 2(\theta + 5\pi/6) \ \text{ H}
\end{aligned} \tag{4.14}
$$

where $|M_s| > L_m$. Note that signs of mutual inductance terms depend upon assumed current directions and coil orientations.

4.3.4 Rotor mutual inductances

The mutual inductance between windings F and D is constant and does not vary with θ. The coefficient of coupling between the d and q axes is zero, and all pairs of windings with 90° displacement have zero mutual inductance. Thus

$$
L_{FD} = L_{DF} = M_R \ \text{ H} \qquad L_{FQ} = L_{QF} = 0 \ \text{ H} \qquad L_{DQ} = L_{QD} = 0 \ \text{ H} \tag{4.15}
$$

4.3.5 Stator-to-rotor mutual inductances

Finally, we consider the mutual inductances between stator and rotor windings, all of which are functions of the rotor angle θ. From the phase windings to the field winding we write

$$
\begin{aligned}
L_{aF} &= L_{Fa} = M_F \cos \theta \ \text{ H} \\
L_{bF} &= L_{Fb} = M_F \cos (\theta - 2\pi/3) \ \text{ H} \\
L_{cF} &= L_{Fc} = M_F \cos (\theta + 2\pi/3) \ \text{ H}
\end{aligned} \tag{4.16}
$$

Similarly, from phase windings to damper winding D we have

$$
\begin{aligned}
L_{aD} &= L_{Da} = M_D \cos\theta \ \ \mathrm{H} \\
L_{bD} &= L_{Db} = M_D \cos(\theta - 2\pi/3) \ \ \mathrm{H} \\
L_{cD} &= L_{Dc} = M_D \cos(\theta + 2\pi/3) \ \ \mathrm{H}
\end{aligned} \tag{4.17}
$$

and finally, from phase windings to damper winding Q we have

$$
\begin{aligned}
L_{aQ} &= L_{Qa} = M_Q \sin\theta \ \ \mathrm{H} \\
L_{bQ} &= L_{Qb} = M_Q \sin(\theta - 2\pi/3) \ \ \mathrm{H} \\
L_{cQ} &= L_{Qc} = M_Q \sin(\theta + 2\pi/3) \ \ \mathrm{H}
\end{aligned} \tag{4.18}
$$

The signs on mutual terms depend upon assumed current directions and coil orientation.

4.3.6 Transformation of inductances

Knowing all inductances in the inductance matrix (4.11), we observe that nearly all terms in the matrix are time varying, since θ is a function of time. Only four of the off-diagonal terms vanish, as noted in equation (4.15). Thus in voltage equations such as (4.1) the $\dot{\lambda}$ term is not a simple $L\dot{i}$ but must be computed as $\dot{\lambda} = L\dot{i} + \dot{L}i$.

We now observe that (4.11) with its time-varying inductances can be simplified by referring all quantities to a rotor frame of reference through a Park's transformation (4.5) applied to the a-b-c partition. We compute

$$
\begin{bmatrix} \mathbf{P} & \mathbf{0} \\ \mathbf{0} & \mathbf{U}_3 \end{bmatrix} \begin{bmatrix} \boldsymbol{\lambda}_{abc} \\ \boldsymbol{\lambda}_{FDQ} \end{bmatrix} = \begin{bmatrix} \mathbf{P} & \mathbf{0} \\ \mathbf{0} & \mathbf{U}_3 \end{bmatrix} \begin{bmatrix} \mathbf{L}_{aa} & \mathbf{L}_{aR} \\ \mathbf{L}_{Ra} & \mathbf{L}_{RR} \end{bmatrix} \begin{bmatrix} \mathbf{P}^{-1} & \mathbf{0} \\ \mathbf{0} & \mathbf{U}_3 \end{bmatrix} \begin{bmatrix} \mathbf{P} & \mathbf{0} \\ \mathbf{0} & \mathbf{U}_3 \end{bmatrix} \begin{bmatrix} \mathbf{i}_{abc} \\ \mathbf{i}_{FDQ} \end{bmatrix} \tag{4.19}
$$

where $\mathbf{L}_{aa}$ = stator-stator inductances
$\mathbf{L}_{aR}, \mathbf{L}_{Ra}$ = stator-rotor inductances
$\mathbf{L}_{RR}$ = rotor-rotor inductances

Equation (4.19) is obtained by premultiplying (4.11) by

$$
\begin{bmatrix} \mathbf{P} & \mathbf{0} \\ \mathbf{0} & \mathbf{U}_3 \end{bmatrix}
$$

where $\mathbf{P}$ is Park's transformation and $\mathbf{U}_3$ is the 3×3 unit matrix. Performing the operation indicated in (4.19), we compute

$$
\begin{bmatrix} \lambda_0 \\ \lambda_d \\ \lambda_q \\ - \\ \lambda_F \\ \lambda_D \\ \lambda_Q \end{bmatrix} = \left[\begin{array}{ccc|ccc} L_0 & 0 & 0 & 0 & 0 & 0 \\ 0 & L_d & 0 & kM_F & kM_D & 0 \\ 0 & 0 & L_q & 0 & 0 & kM_Q \\ \hline 0 & kM_F & 0 & L_F & M_R & 0 \\ 0 & kM_D & 0 & M_R & L_D & 0 \\ 0 & 0 & kM_Q & 0 & 0 & L_Q \end{array}\right] \begin{bmatrix} i_0 \\ i_d \\ i_q \\ - \\ i_F \\ i_D \\ i_Q \end{bmatrix} \ \text{Wb turns} \tag{4.20}
$$

where we have defined the following new constants,

$$
\begin{aligned}
L_d &= L_s + M_s + (3/2)L_m \ \ \mathrm{H} \qquad & L_q &= L_s + M_s - (3/2)L_m \ \ \mathrm{H} \\
L_0 &= L_s - 2M_s \ \ \mathrm{H} & k &= \sqrt{3/2}
\end{aligned} \tag{4.21}
$$

In (4.20) λ_d is the flux linkage in a circuit moving with the rotor and centered on the d axis. Similarly, λ_q is centered on the q axis. Flux linkage λ_0 is completely uncoupled from the other circuits, as the first row and column have only a diagonal term.

It is important also to observe that the inductance matrix of (4.20) is a matrix of constants. This is apparent since all quantities have only one subscript, thus conforming with our notation for constant inductances. The power of Park's transformation is that it removes the time-varying coefficients from this equation. This is very important. We also note that the transformed matrix (4.20) is symmetric and therefore is physically realizable by an equivalent circuit. This was not true of the transformation used by Park [10, 11], where he let $\mathbf{v}_{0dq} = \mathbf{Q}\mathbf{v}_{abc}$ with $\mathbf{Q}$ defined as

$$\mathbf{Q} = 2/3 \begin{bmatrix} 1/2 & 1/2 & 1/2 \\ \cos\theta & \cos(\theta - 2\pi/3) & \cos(\theta + 2\pi/3) \\ -\sin\theta & -\sin(\theta - 2\pi/3) & -\sin(\theta + 2\pi/3) \end{bmatrix} \quad (4.22)$$

Other transformations are found in the literature. The transformation (4.22) is not a power-invariant transformation and does not result in a reciprocal (symmetric) inductance matrix. This leads to unnecessary complication when the equations are normalized.

4.4 Voltage Equations

The generator voltage equations are in the form of (4.1). Schematically, the circuits are shown in Figure 4.2, where coils are identified exactly the same as in Figure 4.1 and with coil terminations shown as well. Mutual inductances are omitted from the schematic for clarity but are assumed present with the values given in Section 4.3. Note that the stator currents are assumed to have a positive direction flowing *out* of the machine terminals, since the machine is a generator. For the conditions indicated we may write the matrix equation

$$\mathbf{v} = -\mathbf{r}\mathbf{i} - \dot{\boldsymbol{\lambda}} + \mathbf{v}_n$$

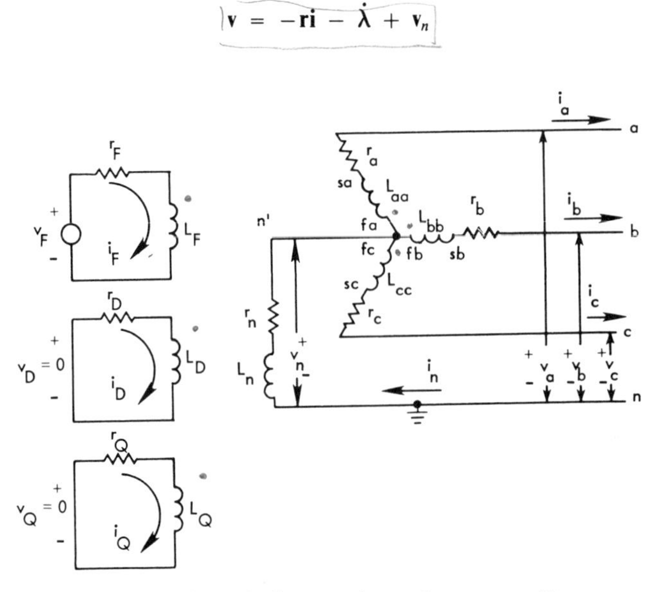

Fig. 4.2 Schematic diagram of a synchronous machine.

or

$$\begin{bmatrix} v_a \\ v_b \\ v_c \\ \hdashline -v_F \\ 0 \\ 0 \end{bmatrix} = - \left[\begin{array}{ccc:ccc} r_a & 0 & 0 & 0 & 0 & 0 \\ 0 & r_b & 0 & 0 & 0 & 0 \\ 0 & 0 & r_c & 0 & 0 & 0 \\ \hdashline 0 & 0 & 0 & r_F & 0 & 0 \\ 0 & 0 & 0 & 0 & r_D & 0 \\ 0 & 0 & 0 & 0 & 0 & r_Q \end{array}\right] \begin{bmatrix} i_a \\ i_b \\ i_c \\ \hdashline i_F \\ i_D \\ i_Q \end{bmatrix} - \begin{bmatrix} \dot{\lambda}_a \\ \dot{\lambda}_b \\ \dot{\lambda}_c \\ \hdashline \dot{\lambda}_F \\ \dot{\lambda}_D \\ \dot{\lambda}_Q \end{bmatrix} + \begin{bmatrix} \mathbf{v}_n \\ \hdashline \mathbf{0} \end{bmatrix} \text{ V} \tag{4.23}$$

where we define the neutral voltage contribution to $\mathbf{v}_{abc}$ as

$$\mathbf{v}_n = -r_n \begin{bmatrix} 1 & 1 & 1 \\ 1 & 1 & 1 \\ 1 & 1 & 1 \end{bmatrix} \begin{bmatrix} i_a \\ i_b \\ i_c \end{bmatrix} - L_n \begin{bmatrix} 1 & 1 & 1 \\ 1 & 1 & 1 \\ 1 & 1 & 1 \end{bmatrix} \begin{bmatrix} \dot{i}_a \\ \dot{i}_b \\ \dot{i}_c \end{bmatrix}$$

$$= -\mathbf{R}_n \mathbf{i}_{abc} - \mathbf{L}_n \dot{\mathbf{i}}_{abc} \text{ V} \tag{4.24}$$

If $r_a = r_b = r_c = r$, as is usually the case, we may also define

$$\mathbf{R}_{abc} = r\mathbf{U}_3 \ \Omega \tag{4.25}$$

where $\mathbf{U}_3$ is the 3×3 unit matrix, and we may rewrite (4.23) in partitioned form as follows:

$$\begin{bmatrix} \mathbf{v}_{abc} \\ \mathbf{v}_{FDQ} \end{bmatrix} = - \begin{bmatrix} \mathbf{R}_{abc} & \mathbf{0} \\ \mathbf{0} & \mathbf{R}_{FDQ} \end{bmatrix} \begin{bmatrix} \mathbf{i}_{abc} \\ \mathbf{i}_{FDQ} \end{bmatrix} - \begin{bmatrix} \dot{\boldsymbol{\lambda}}_{abc} \\ \dot{\boldsymbol{\lambda}}_{FDQ} \end{bmatrix} + \begin{bmatrix} \mathbf{v}_n \\ \mathbf{0} \end{bmatrix} \text{ V} \tag{4.26}$$

where

$$\mathbf{v}_{FDQ} = \begin{bmatrix} -v_F \\ 0 \\ 0 \end{bmatrix} \quad \mathbf{i}_{FDQ} = \begin{bmatrix} i_F \\ i_D \\ i_Q \end{bmatrix} \quad \boldsymbol{\lambda}_{FDQ} = \begin{bmatrix} \lambda_F \\ \lambda_D \\ \lambda_Q \end{bmatrix} \tag{4.27}$$

Thus (4.26) is complicated by the presence of time-varying coefficients in the $\dot{\lambda}$ term, but these terms can be eliminated by applying a Park's transformation to the stator partition. This requires that both sides of (4.26) be premultiplied by

$$\begin{bmatrix} \mathbf{P} & \mathbf{0} \\ \mathbf{0} & \mathbf{U}_3 \end{bmatrix}$$

By definition

$$\begin{bmatrix} \mathbf{P} & \mathbf{0} \\ \mathbf{0} & \mathbf{U}_3 \end{bmatrix} \begin{bmatrix} \mathbf{v}_{abc} \\ \mathbf{v}_{FDQ} \end{bmatrix} = \begin{bmatrix} \mathbf{v}_{0dq} \\ \mathbf{v}_{FDQ} \end{bmatrix} \text{ V} \tag{4.28}$$

for the left side of (4.26). For the resistance voltage drop term we compute

$$\begin{bmatrix} \mathbf{P} & \mathbf{0} \\ \mathbf{0} & \mathbf{U}_3 \end{bmatrix} \begin{bmatrix} \mathbf{R}_{abc} & \mathbf{0} \\ \mathbf{0} & \mathbf{R}_{FDQ} \end{bmatrix} \begin{bmatrix} \mathbf{i}_{abc} \\ \mathbf{i}_{FDQ} \end{bmatrix} = \begin{bmatrix} \mathbf{P} & \mathbf{0} \\ \mathbf{0} & \mathbf{U}_3 \end{bmatrix} \begin{bmatrix} \mathbf{R}_{abc} & \mathbf{0} \\ \mathbf{0} & \mathbf{R}_{FDQ} \end{bmatrix} \begin{bmatrix} \mathbf{P}^{-1} & \mathbf{0} \\ \mathbf{0} & \mathbf{U}_3 \end{bmatrix} \begin{bmatrix} \mathbf{P} & \mathbf{0} \\ \mathbf{0} & \mathbf{U}_3 \end{bmatrix} \begin{bmatrix} \mathbf{i}_{abc} \\ \mathbf{i}_{FDQ} \end{bmatrix}$$

$$= \begin{bmatrix} \mathbf{P}\mathbf{R}_{abc}\mathbf{P}^{-1} & \mathbf{0} \\ \mathbf{0} & \mathbf{R}_{FDQ} \end{bmatrix} \begin{bmatrix} \mathbf{i}_{0dq} \\ \mathbf{i}_{FDQ} \end{bmatrix} = \begin{bmatrix} \mathbf{R}_{abc} & \mathbf{0} \\ \mathbf{0} & \mathbf{R}_{FDQ} \end{bmatrix} \begin{bmatrix} \mathbf{i}_{0dq} \\ \mathbf{i}_{FDQ} \end{bmatrix} \text{ V} \tag{4.29}$$

The second term on the right side of (4.26) is transformed as

$$\begin{bmatrix} \mathbf{P} & \mathbf{0} \\ \mathbf{0} & \mathbf{U}_3 \end{bmatrix} \begin{bmatrix} \dot{\boldsymbol{\lambda}}_{abc} \\ \dot{\boldsymbol{\lambda}}_{FDQ} \end{bmatrix} = \begin{bmatrix} \mathbf{P}\dot{\boldsymbol{\lambda}}_{abc} \\ \dot{\boldsymbol{\lambda}}_{FDQ} \end{bmatrix} \text{ V} \tag{4.30}$$

We evaluate $\mathbf{P}\dot{\boldsymbol{\lambda}}_{abc}$ by recalling the definition (4.7), $\boldsymbol{\lambda}_{0dq} = \mathbf{P}\boldsymbol{\lambda}_{abc}$, from which we compute $\dot{\boldsymbol{\lambda}}_{0dq} = \mathbf{P}\dot{\boldsymbol{\lambda}}_{abc} + \dot{\mathbf{P}}\boldsymbol{\lambda}_{abc}$. Then

$$\mathbf{P}\dot{\boldsymbol{\lambda}}_{abc} = \dot{\boldsymbol{\lambda}}_{0dq} - \dot{\mathbf{P}}\boldsymbol{\lambda}_{abc} = \dot{\boldsymbol{\lambda}}_{0dq} - \dot{\mathbf{P}}\mathbf{P}^{-1}\boldsymbol{\lambda}_{0dq} \text{ V} \tag{4.31}$$

We may show that

$$\dot{\mathbf{P}}\mathbf{P}^{-1}\boldsymbol{\lambda}_{0dq} = \omega \begin{bmatrix} 0 & 0 & 0 \\ 0 & 0 & -1 \\ 0 & 1 & 0 \end{bmatrix} \begin{bmatrix} \lambda_0 \\ \lambda_d \\ \lambda_q \end{bmatrix} = \begin{bmatrix} 0 \\ -\omega\lambda_q \\ \omega\lambda_d \end{bmatrix} \text{ V} \tag{4.32}$$

which is the speed voltage term.

Finally, the third term on the right side of (4.26) transforms as follows:

$$\begin{bmatrix} \mathbf{P} & \mathbf{0} \\ \mathbf{0} & \mathbf{U}_3 \end{bmatrix} \begin{bmatrix} \mathbf{v}_n \\ \mathbf{0} \end{bmatrix} = \begin{bmatrix} \mathbf{P}\mathbf{v}_n \\ \mathbf{0} \end{bmatrix} = \begin{bmatrix} \mathbf{n}_{0dq} \\ \mathbf{0} \end{bmatrix} \text{ V} \tag{4.33}$$

where by definition $\mathbf{n}_{0dq}$ is the voltage drop from neutral to ground in the 0-d-q coordinate system. Using (4.24), we compute

$$\mathbf{n}_{0dq} = \mathbf{P}\mathbf{v}_n = -\mathbf{P}\mathbf{R}_n\mathbf{P}^{-1}\mathbf{P}\mathbf{i}_{abc} - \mathbf{P}\mathbf{L}_n\mathbf{P}^{-1}\mathbf{P}\dot{\mathbf{i}}_{abc} = -\mathbf{P}\mathbf{R}_n\mathbf{P}^{-1}\mathbf{i}_{0dq} - \mathbf{P}\mathbf{L}_n\mathbf{P}^{-1}\dot{\mathbf{i}}_{0dq}$$

$$= -\begin{bmatrix} 3r_n i_0 \\ 0 \\ 0 \end{bmatrix} - \begin{bmatrix} 3L_n \dot{i}_0 \\ 0 \\ 0 \end{bmatrix} \text{ V} \tag{4.34}$$

and observe that this voltage drop occurs only in the zero sequence, as it should.

Summarizing, we substitute (4.28)–(4.31) and (4.33) into (4.26) to write

$$\begin{bmatrix} \mathbf{v}_{0dq} \\ \mathbf{v}_{FDQ} \end{bmatrix} = -\begin{bmatrix} \mathbf{R}_{abc} & \mathbf{0} \\ \mathbf{0} & \mathbf{R}_{FDQ} \end{bmatrix} \begin{bmatrix} \mathbf{i}_{0dq} \\ \mathbf{i}_{FDQ} \end{bmatrix} - \begin{bmatrix} \dot{\boldsymbol{\lambda}}_{0dq} \\ \dot{\boldsymbol{\lambda}}_{FDQ} \end{bmatrix} + \begin{bmatrix} \dot{\mathbf{P}}\mathbf{P}^{-1}\boldsymbol{\lambda}_{0dq} \\ \mathbf{0} \end{bmatrix} + \begin{bmatrix} \mathbf{n}_{0dq} \\ \mathbf{0} \end{bmatrix} \text{ V} \tag{4.35}$$

Note that all terms in this equation are known. The resistance matrix is diagonal.

For balanced conditions the zero-sequence voltage is zero. To simplify the notation, let

$$\mathbf{R} = \begin{bmatrix} r & 0 \\ 0 & r \end{bmatrix} \quad \mathbf{R}_R = \begin{bmatrix} r_F & 0 & 0 \\ 0 & r_D & 0 \\ 0 & 0 & r_Q \end{bmatrix} \quad \mathbf{S} = \begin{bmatrix} -\omega\lambda_q \\ \omega\lambda_d \end{bmatrix}$$

Then for balanced conditions (4.35) may be written without the zero-sequence equation as

$$\begin{bmatrix} \mathbf{v}_{dq} \\ \mathbf{v}_{FDQ} \end{bmatrix} = -\begin{bmatrix} \mathbf{R} & \mathbf{0} \\ \mathbf{0} & \mathbf{R}_R \end{bmatrix}\begin{bmatrix} \mathbf{i}_{dq} \\ \mathbf{i}_{FDQ} \end{bmatrix} + \begin{bmatrix} \mathbf{S} \\ \mathbf{0} \end{bmatrix} - \begin{bmatrix} \dot{\boldsymbol{\lambda}}_{dq} \\ \dot{\boldsymbol{\lambda}}_{FDQ} \end{bmatrix} \quad \text{V} \tag{4.36}$$

4.5 Formulation of State-Space Equations

Recall that our objective is to derive a set of equations describing the synchronous machine in the form

$$\dot{\mathbf{x}} = \mathbf{f}(\mathbf{x}, \mathbf{u}, t) \tag{4.37}$$

where $\mathbf{x}$ = a vector of the state variables
$\mathbf{u}$ = the system driving functions
$\mathbf{f}$ = a set of nonlinear functions

If the equations describing the synchronous machine are linear, the set (4.37) is of the well-known form

$$\dot{\mathbf{x}} = \mathbf{A}\mathbf{x} + \mathbf{B}\mathbf{u} \tag{4.38}$$

Examining (4.35), we can see that it represents a set of first-order differential equations. We may now put this set in the form of (4.37) or (4.38), i.e., in state-space form. Note, however, that (4.35) contains flux linkages and currents as variables. Since these two sets of variables are mutually dependent, we can eliminate one set to express (4.35) in terms of one set of variables only. Actually, numerous possibilities for the choice of the state variables are available. We will mention only two that are common: (1) a set based on the *currents* as state variables; i.e., $\mathbf{x}^t = [i_d\, i_q\, i_F\, i_D\, i_Q]$, which has the advantage of offering simple relations between the voltages v_d and v_q and the state variables (through the power network connected to the machine terminals) and (2) a set based on *flux linkages* as the state variables, where the particular set to be chosen depends upon how conveniently they can be expressed in terms of the machine currents and stator voltages. Here we will use the formulation $\mathbf{x}^t = [\lambda_d\, \lambda_q\, \lambda_F\, \lambda_D\, \lambda_Q]$.

4.6 Current Formulation

Starting with (4.35), we can replace the terms in λ and $\dot{\lambda}$ by terms in i and $\dot{i}$ as follows. The $\dot{\lambda}$ term has been simplified so that we can compute its value from (4.20), which we rearrange in partitioned form. Let

$$\begin{bmatrix} \boldsymbol{\lambda}_{0dq} \\ -- \\ \boldsymbol{\lambda}_{FDQ} \end{bmatrix} = \left[\begin{array}{c|c} \mathbf{L}_{0dq} & \mathbf{L}_m \\ \hline \mathbf{L}_m^t & \mathbf{L}_{FDQ} \end{array}\right]\begin{bmatrix} \mathbf{i}_{0dq} \\ -- \\ \mathbf{i}_{FDQ} \end{bmatrix} \quad \text{Wb turns}$$

where $\mathbf{L}_m^t$ is the transpose of $\mathbf{L}_m$. But the inductance matrix here is a constant matrix, so we may write $\dot{\boldsymbol{\lambda}} = \mathbf{L}\dot{\mathbf{i}}$ V, and the $\dot{\lambda}$ term behaves exactly like that of a passive inductance. Substituting this result into (4.35), expanding to full 6×6 notation, and rearranging,

$$\left[\begin{array}{c} v_0 \\ \hline v_d \\ -v_F \\ v_D = 0 \\ \hline v_q \\ v_Q = 0 \end{array}\right] = -\left[\begin{array}{c|ccc|cc} r + 3r_n & 0 & 0 & 0 & 0 & 0 \\ \hline 0 & r & 0 & 0 & \omega L_q & \omega \mathrm{k} M_Q \\ 0 & 0 & r_F & 0 & 0 & 0 \\ 0 & 0 & 0 & r_D & 0 & 0 \\ \hline 0 & -\omega L_d & -\omega \mathrm{k} M_F & -\omega \mathrm{k} M_D & r & 0 \\ 0 & 0 & 0 & 0 & 0 & r_Q \end{array}\right] \left[\begin{array}{c} i_0 \\ \hline i_d \\ i_F \\ i_D \\ \hline i_q \\ i_Q \end{array}\right]$$

$$- \left[\begin{array}{c|ccc|cc} L_0 + 3L_n & 0 & 0 & 0 & 0 & 0 \\ \hline 0 & L_d & \mathrm{k} M_F & \mathrm{k} M_D & & \\ 0 & \mathrm{k} M_F & L_F & M_R & \multicolumn{2}{c}{\mathbf{0}} \\ 0 & \mathrm{k} M_D & M_R & L_D & & \\ \hline 0 & & \mathbf{0} & & L_q & \mathrm{k} M_Q \\ 0 & & & & \mathrm{k} M_Q & L_Q \end{array}\right] \left[\begin{array}{c} \dot{i}_0 \\ \hline \dot{i}_d \\ \dot{i}_F \\ \dot{i}_D \\ \hline \dot{i}_q \\ \dot{i}_Q \end{array}\right] \quad \mathrm{V} \tag{4.39}$$

where $\mathrm{k} = \sqrt{3/2}$ as before. A great deal of information is contained in (4.39).

First, we note that the zero-sequence voltage is dependent only upon i_0 and $\dot{i}_0$. This equation can be solved separately from the others once the initial conditions on i_0 are given. The remaining five equations are all coupled in a most interesting way. They are similar to those of a passive network *except* for the presence of the speed voltage terms. These terms, consisting of $\omega\lambda$ or ωLi products, appear unsymmetrically and distinguish this equation from that of a passive network. Note that the speed voltage terms in the d axis equation are due only to q axis currents, viz., i_q and i_Q. Similarly, the q axis speed voltages are due to d axis currents, i_d, i_F, and i_D. Also observe that all the terms in the coefficient matrices are constants except ω, the angular velocity. This is a considerable improvement over the description given in (4.23) in the *a-b-c* frame of reference since nearly all inductances in that equation were time varying. The price we have paid to get rid of the time-varying coefficients is the introduction of speed voltage terms in the resistance matrix. Since ω is a variable, this causes (4.39) to be nonlinear. If the speed is assumed constant, which is usually a good approximation, then (4.39) is linear. In any event, the nonlinearity is never great, as ω is usually nearly constant.

4.7 Per Unit Conversion

The voltage equations of the preceding section are not in a convenient form for engineering use. One difficulty is the numerically awkward values with stator voltages in the kilovolt range and field voltage at a much lower level. This problem can be solved by normalizing the equations to a convenient base value and expressing all voltages in pu (or percent) of base. (See Appendix C.)

An examination of the voltage equations reveals the dimensional character shown in Table 4.1, where all dimensions are expressed in terms of a v-i-t (voltage, current, time) system. [These dimensions are convenient here. Other possible systems are

$Lt\mu$ (mass, length, time, permeability).] Ob-
) involve only three dimensions. Thus if we
ll three dimensions, *all* bases are fixed for all
base voltage, base current, and base time, by
olumn 4 of Table 4.1, we may compute base
exactly three base quantities must be chosen
dimensions, v, i, and t.

Table 4.1. Electrical Quantities, Units, and Dimensions

Quantity	Symbol	Units	v-i-t Dimensions	Relationship
Voltage	v	volts (V)	$[v]$	
Current	i	amperes (A)	$[i]$	
Power or voltamperes	p or S	watts (W) voltamperes (VA)	$[vi]$	$p = vi$
Flux linkage	λ	weber turns (Wb turns)	$[vt]$	$v = \dot{\lambda}$
Resistance	r	ohm (Ω)	$[v/i]$	$v = ri$
Inductance	L or M	henry (H)	$[vt/i]$	$v = L\dot{i}$
Time	t	second (s)	$[t]$	
Angular velocity	ω	radians per second (rad/s)	$[1/t]$	
Angle	θ or δ	radian (rad)	dimensionless	

4.7.1 Choosing a base for stator quantities

The variables v_d, v_q, i_d, i_q, λ_d, and λ_q are stator quantities because they relate directly to the a-b-c phase quantities through Park's transformation. (Also see Rankin [15], Lewis [12] and Harris et al. [9] for a discussion of this topic.) Using the subscript B to indicate "base" and R to indicate "rated," we choose the following stator base quantities.

$$
\begin{aligned}
\text{Let}\quad S_B &= S_R = \text{stator rated VA/phase, VA rms}\\
V_B &= V_R = \text{stator rated line-to-neutral voltage, V rms}\\
\omega_B &= \omega_R = \text{generator rated speed, elec rad/s}
\end{aligned}
\tag{4.40}
$$

Before proceeding further, let us examine the effect of this choice on the d and q axis quantities.

First note that the three-phase power in pu is three times the pu power per phase (for balanced conditions). To prove this, let the rms phase quantities be $V\angle\alpha$ V and $I\angle\gamma$ A. The three-phase power is $3\,VI\cos(\alpha-\gamma)$ W. The pu power $P_{3\phi}$ is given by

$$P_{3\phi} = (3VI/V_B I_B)\cos(\alpha-\gamma) = 3V_u I_u \cos(\alpha-\gamma) \tag{4.41}$$

where the subscript u is used to indicate pu quantities. To obtain the d and q axis quantities, we first write the instantaneous phase voltage and currents. To simplify the expression without any loss of generality, we will assume that $v_a(t)$ is in the form,

$$
\begin{aligned}
v_a &= V_m \sin(\theta+\alpha) = \sqrt{2}V\sin(\theta+\alpha)\ \ \text{V}\\
v_b &= \sqrt{2}V\sin(\theta+\alpha-2\pi/3)\ \ \text{V}\\
v_c &= \sqrt{2}V\sin(\theta+\alpha+2\pi/3)\ \ \text{V}
\end{aligned}
\tag{4.42}
$$

Then from (4.5), $\mathbf{v}_{0dq} = \mathbf{P}\mathbf{v}_{abc}$ or

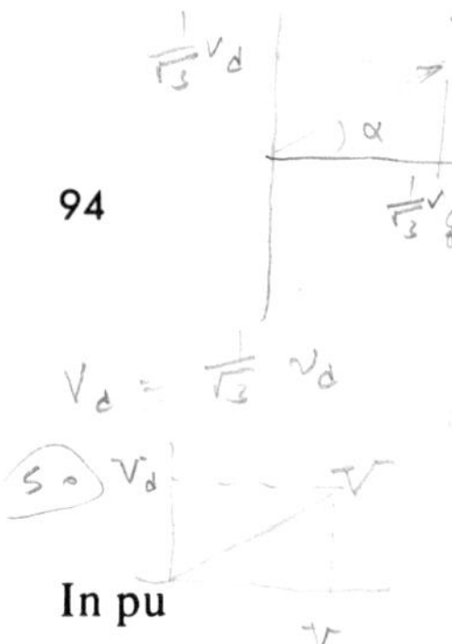

$$\begin{bmatrix} v_0 \\ v_d \\ v_q \end{bmatrix} = \begin{bmatrix} 0 \\ \sqrt{3}V \sin\alpha \\ \sqrt{3}V \cos\alpha \end{bmatrix} \text{ V} \tag{4.43}$$

In pu

$$v_{du} = v_d/V_B = \sqrt{3}(V/V_B)\sin\alpha = \sqrt{3}\,V_u \sin\alpha \tag{4.44}$$

Similarly,

$$v_{qu} = \sqrt{3}V_u \cos\alpha \tag{4.45}$$

Obviously, then

$$v_{du}^2 + v_{qu}^2 = 3V_u^2 \tag{4.46}$$

The above results are significant. They indicate that with this particular choice of the base voltage, the pu d and q axis voltages are numerically equal to $\sqrt{3}$ times the pu phase voltages.

Similarly, we can show that if the *rms phase current* is $I\underline{/\gamma}$ A, the corresponding d and q axis currents are given by,

$$\begin{bmatrix} i_0 \\ i_d \\ i_q \end{bmatrix} = \begin{bmatrix} 0 \\ \sqrt{3}I \sin\gamma \\ \sqrt{3}I \cos\gamma \end{bmatrix} \text{ A} \tag{4.47}$$

and the pu currents are given by

$$i_{du} = \sqrt{3}I_u \sin\gamma \qquad i_{qu} = \sqrt{3}I_u \cos\gamma \tag{4.48}$$

To check the validity of the above, the power in the d and q circuits must be the same as the power in the three stator phases, since **P** is a power-invariant transformation.

$$\begin{aligned} P_{3\phi} &= i_{du}v_{du} + i_{qu}v_{qu} = 3I_u V_u(\sin\alpha\sin\gamma + \cos\alpha\cos\gamma) \\ &= 3I_u V_u \cos(\alpha - \gamma) \text{ pu} \end{aligned} \tag{4.49}$$

We now develop the relations for the various base quantities. From (4.40) and Table 4.1 we compute the following:

$$\begin{aligned} &I_B = S_B/V_B = S_R/V_R \text{ A rms} && t_B = 1/\omega_B = 1/\omega_R \text{ s} \\ &\lambda_B = V_B t_B = V_R/\omega_R = L_B I_B \text{ Wb turn} && \\ &R_B = V_B/I_B = V_R/I_R \ \Omega && L_B = V_B t_B/I_B = V_R/I_R\omega_R \text{ H} \end{aligned} \tag{4.50}$$

Thus by choosing the three base quantities S_B, V_B, and t_B, we can compute base values for all quantities of interest.

To normalize any quantity, it is divided by the base quantity of the same dimension. For example, for currents we write

$$i_u = i(\text{A})/I_B(\text{A}) \text{ pu} \tag{4.51}$$

where we use the subscript u to indicate pu. Later, when there is no danger of ambiguity in the notation, this subscript is omitted.

4.7.2 Choosing a base for rotor quantities

Lewis [12] showed that in circuits coupled electromagnetically, which are to be normalized, it is essential to select the same voltampere and time base in each part of the circuit. (See Appendix C for a more detailed treatment of this subject.) The choice of equal time base throughout all parts of a circuit with mutual coupling is the important constraint. It can be shown that the choice of a common time base t_B forces the VA base to be equal in all circuit parts and also forces the base mutual inductance to be the geometric mean of the base self-inductances if equal pu mutuals are to result; i.e., $M_{12B} = (L_{1B}L_{2B})^{1/2}$. (See Problem 4.18.)

For the synchronous machine the choice of S_B is based on the rating of the stator, and the time base is fixed by the rated radian frequency. These base quantities must be the same for the rotor circuits as well. It should be remembered, however, that the stator VA base is much larger than the VA rating of the rotor (field) circuits. Hence *some* rotor base quantities are bound to be very large, making the corresponding pu rotor quantities appear numerically small. Therefore, care should be exercised in the choice of the remaining free rotor base term, since all other rotor base quantities will then be automatically determined. There is a choice of quantities, but the question is, Which is more convenient?

To illustrate the above, consider a machine having a stator rating of 100×10^6 VA/phase. Assume that its exciter has a rating of 250 V and 1000 A. If, for example, we choose $I_{RB} = 1000$ A, V_{RB} will then be 100,000 V; and if we choose $V_{RB} = 250$ V, then I_{RB} will be 400,000 A.

Is one choice more convenient than the other? Are there other more desirable choices? The answer lies in the nature of the coupling between the rotor and the stator circuits. It would seem desirable to choose some base quantity in the rotor to give the correct base quantity in the stator. For example, we can choose the base rotor current to give, through the magnetic coupling, the correct base stator flux linkage or open circuit voltage. Even then there is some latitude in the choice of the base rotor current, depending on the condition of the magnetic circuit.

The choice made here for the free rotor base quantity is based on the concept of *equal mutual flux linkages*. This means that base field current or base d axis amortisseur current will produce the same space fundamental of air gap flux as produced by base stator current acting in the fictitious d winding.

Referring to the flux linkage equations (4.20) let $i_d = I_B$, $i_F = I_{FB}$, and $i_D = I_{DB}$ be applied one by one with other currents set to zero. If we denote the magnetizing inductances (ℓ = leakage inductances) as

$$
\begin{aligned}
L_{md} &\triangleq L_d - \ell_d \text{ H} \qquad & L_{mq} &\triangleq L_q - \ell_q \text{ H} \\
L_{mF} &\triangleq L_F - \ell_F \text{ H} \qquad & L_{mQ} &\triangleq L_Q - \ell_Q \text{ H} \\
L_{mD} &\triangleq L_D - \ell_D \text{ H}
\end{aligned}
\tag{4.52}
$$

and equate the mutual flux linkages in each winding,

$$
\begin{aligned}
\lambda_{md} &= L_{md} I_B = kM_F I_{FB} = kM_D I_{DB} \text{ Wb} \qquad & \lambda_{mq} &= L_{mq} I_B = kM_Q I_{QB} \text{ Wb} \\
\lambda_{mF} &= kM_F I_B = L_{mF} I_{FB} = M_R I_{DB} \text{ Wb} \qquad & \lambda_{mQ} &= kM_Q I_B = L_{mQ} I_{QB} \text{ Wb} \\
\lambda_{mD} &= kM_D I_B = M_R I_{FB} = L_{mD} I_{DB} \text{ Wb}
\end{aligned}
\tag{4.53}
$$

Then we can show that

$$
\begin{aligned}
L_{md}I_{\mathrm{B}}^2 &= L_{mF}I_{F\mathrm{B}}^2 = L_{mD}I_{D\mathrm{B}}^2 = \mathrm{k}M_F I_{\mathrm{B}} I_{F\mathrm{B}} = \mathrm{k}M_D I_{\mathrm{B}} I_{D\mathrm{B}} = M_{\mathrm{R}} I_{F\mathrm{B}} I_{D\mathrm{B}} \\
L_{mq}I_{\mathrm{B}}^2 &= \mathrm{k}M_Q I_{\mathrm{B}} I_{Q\mathrm{B}} = L_{mQ}I_{Q\mathrm{B}}^2
\end{aligned} \tag{4.54}
$$

and this is the fundamental constraint among base currents.

From (4.54) and the requirement for equal S_{B}, we compute

$$
\begin{aligned}
V_{F\mathrm{B}}/V_{\mathrm{B}} &= I_{\mathrm{B}}/I_{F\mathrm{B}} = (L_{mF}/L_{md})^{1/2} = \mathrm{k}M_F/L_{md} = L_{mF}/\mathrm{k}M_F = M_R/\mathrm{k}M_D \triangleq k_F \\
V_{D\mathrm{B}}/V_{\mathrm{B}} &= I_{\mathrm{B}}/I_{D\mathrm{B}} = (L_{mD}/L_{md})^{1/2} = \mathrm{k}M_D/L_{md} = L_{mD}/\mathrm{k}M_D = M_R/\mathrm{k}M_F \triangleq k_D \\
V_{Q\mathrm{B}}/V_{\mathrm{B}} &= I_{\mathrm{B}}/I_{Q\mathrm{B}} = (L_{mQ}/L_{mq})^{1/2} = \mathrm{k}M_Q/L_{mq} = L_{mQ}/\mathrm{k}M_Q \triangleq k_Q
\end{aligned} \tag{4.55}
$$

These basic constraints permit us to compute

$$
\begin{aligned}
R_{F\mathrm{B}} &= k_F^2 R_{\mathrm{B}}\ \Omega \qquad R_{D\mathrm{B}} = k_D^2 R_{\mathrm{B}}\ \Omega \qquad R_{Q\mathrm{B}} = k_Q^2 R_{\mathrm{B}}\ \Omega \\
L_{F\mathrm{B}} &= k_F^2 L_{\mathrm{B}}\ \mathrm{H} \qquad L_{D\mathrm{B}} = k_D^2 L_{\mathrm{B}}\ \mathrm{H} \qquad L_{Q\mathrm{B}} = k_Q^2 L_{\mathrm{B}}\ \mathrm{H}
\end{aligned} \tag{4.56}
$$

and since the base mutuals must be the geometric mean of the base self-inductances (see Problem 4.18),

$$
M_{F\mathrm{B}} = k_F L_B\ \mathrm{H} \qquad M_{D\mathrm{B}} = k_D L_{\mathrm{B}}\ \mathrm{H} \qquad M_{Q\mathrm{B}} = k_Q L_{\mathrm{B}}\ \mathrm{H} \qquad M_{R\mathrm{B}} = k_F k_D L_{\mathrm{B}}\ \mathrm{H} \tag{4.57}
$$

4.7.3 Comparison with other per unit systems

The subject of the pu system used with synchronous machines has been controversial over the years. While the use of pu quantities is common in the literature, it is not always clear which base quantities are used by the authors. Furthermore, synchronous machine data is usually furnished by the manufacturer in pu. Therefore it is important to understand any major difference in the pu systems adopted. Part of the problem lies in the nature of the original Park's transformation $\mathbf{Q}$ given in (4.22). This transformation is not power invariant; i.e., the three-phase power in watts is given by $p_{abc} = 1.5\,(i_d v_d + i_q v_q)$. Also, the mutual coupling between the field and the stator d axis is not reciprocal. When the $\mathbf{Q}$ transformation is used, the pu system is chosen carefully to overcome this difficulty. Note that the *modified* Park's transformation $\mathbf{P}$ defined by (4.5) was chosen specifically to overcome these problems.

The system most commonly used in the literature is based on the following base quantities:

S_{B} = three-phase rated VA
V_{B} = *peak* rated voltage to neutral
I_{B} = *peak* rated current

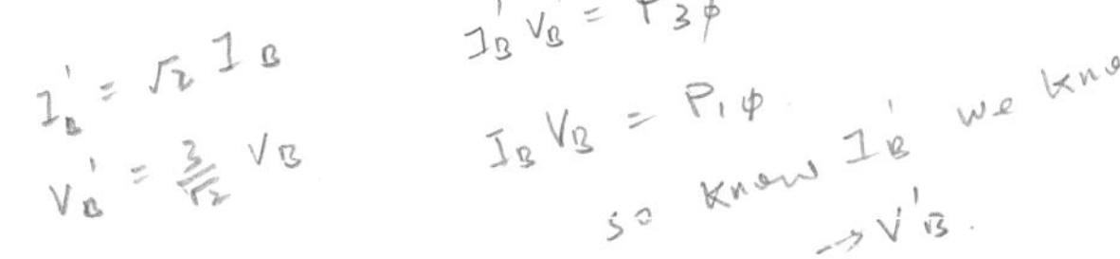

and with rotor base quantities chosen to give equal pu mutual inductances. This leads to the relations

$$
I_{F\mathrm{B}} = \sqrt{2}(L_{md}/M_F)I_{\mathrm{B}} \qquad V_{F\mathrm{B}} = (3/\sqrt{2})(M_F/L_{md})V_{\mathrm{B}}
$$

This choice of base quantities, which is commonly used, gives the *same numerical values* in pu for synchronous machine stator and rotor impedances and self-inductances as the system used in this book. The pu mutual inductances differ by a factor of $\sqrt{3/2}$. Therefore, the terms $\mathrm{k}M_F$ used in this book are numerically equal to M_F in pu as found in the literature. The major differences lie in the following:

1. Since the power in the d and q stator circuits is the three-phase power, one pu current and voltage gives three pu power in the system used here and gives one pu power in the other system.

$= 3V_u^2$, while in the other system $v_{du}^2 + v_{qu}^2 =$
age.

ealing to some engineers than that used by the
the manufacturers' base system is so common,

Find the pu values of the parameters of the synchronous machine for which the following data are given (values are for an actual machine with some quantities, denoted by an asterisk, being estimated for academic study):

Rated MVA	= 160 MVA	L_Q	= 1.423×10^{-3} H*
Rated voltage	= 15 kV, Y connected	$\ell_d = \ell_q$(unsaturated)	= 0.5595×10^{-3} H
Excitation voltage	= 375 V	kM_D	= 5.782×10^{-3} H*
Stator current	= 6158.40 A	kM_Q	= 2.779×10^{-3} H*
Field current	= 926 A	$r(125°C)$	= 1.542×10^{-3} Ω
Power factor	= 0.85	$r_F(125°C)$	= 0.371 Ω
L_d	= 6.341×10^{-3} H	r_D	= 18.421×10^{-3} Ω*
L_F	= 2.189 H	r_Q	= 18.969×10^{-3} Ω*
L_D	= 5.989×10^{-3} H*	Inertia constant	= 1.765 kW·s/hp
L_q	= 6.118×10^{-3} H		

From the no-load magnetization curve, the value of field current corresponding to the rated voltage on *the air gap line* is 365 A.

Solution:

Stator Base Quantities:

$$S_B = 160/3 = 53.3333 \text{ MVA/phase}$$
$$V_B = 15000/\sqrt{3} = 8660.25 \text{ V}$$
$$I_B = 6158.40 \text{ A}$$
$$t_B = 2.6526 \times 10^{-3} \text{ s}$$
$$\lambda_B = 8660 \times 2.65 \times 10^{-3} = 22.972 \text{ Wb turn/phase}$$
$$R_B = 8660.25/6158.40 = 1.406 \ \Omega$$
$$L_B = 8660/(377 \times 6158) = 3.730 \times 10^{-3} \text{ H}$$
$$L_{md} = L_d - \ell_d = (6.341 - 0.5595)10^{-3} = 5.79 \times 10^{-3} \text{ H}$$

To obtain M_F, we use (4.11), (4.16), and (4.23). At open circuit the mutual inductance L_{aF} and the flux linkage in phase a are given by

$$L_{aF} = M_F \cos\theta \qquad \lambda_a = i_F M_F \cos\theta$$

The instantaneous voltage of phase a is $v_a = i_F \omega_R M_F \sin\theta$, where ω_R is the rated synchronous speed. Thus the peak phase voltage corresponds to the product $i_F \omega_R M_F$. From the air gap line of the no-load saturation curve, the value of the field current at rated voltage is 365 A. Therefore,

$$M_F = 8660\sqrt{2}/(377 \times 365) = 89.006 \times 10^{-3} \text{ H}$$
$$kM_F = \sqrt{3/2} \times 89.006 \times 10^{-3} = 109.01 \times 10^{-3} \text{ H}$$

Then $k_F = kM_F/L_{md} = 18.854$.

Then we compute, from (4.55)–(4.57),

$$I_{FB} = 6158.4/18.854 = 326.64 \text{ A}$$
$$M_{FB} = 18.854 \times 3.73 \times 10^{-3} = 70.329 \times 10^{-3} \text{ H}$$

$$V_{FB} = (53.33 \times 10^6)/326.64 = 163280.68 \text{ V}$$
$$R_{FB} = 163280.68/326.64 = 499.89 \ \Omega$$
$$L_{FB} = (18.845)^2 \times 3.73 \times 10^{-3} = 1.326 \text{ H}$$

Amortisseur Base Quantities (estimated for this example):

$$\mathrm{k}M_D/L_{md} = 5.781/5.781 = 1.00 \qquad L_{DB} = L_B \text{ H}$$
$$M_{DB} = L_B \text{ H} \qquad R_{DB} = R_B \ \Omega$$
$$\mathrm{k}M_Q/L_{mq} = 2.779/5.782 = 0.5 \qquad R_{QB} = R_B/4 = 0.352 \ \Omega$$
$$L_{QB} = L_B/4 = 0.933 \times 10^{-3} \text{ H}$$

Inertia Constant:

$$H = 1.765(1.0/0.746) = 2.37 \text{ kW}\cdot\text{s/kVA}$$

The pu parameters are thus given by:

$$L_d = 6.34/3.73 = 1.70$$
$$L_F = 2.189/1.326 = 1.651$$
$$L_D = 5.989/3.730 = 1.605$$
$$\ell_d = \ell_q = 0.5595/3.73 = 0.15$$
$$L_q = 6.118/3.73 = 1.64$$
$$L_Q = 1.423/0.933 = 1.526$$
$$L_{AD} = \mathrm{k}M_D = \mathrm{k}M_F = M_R = 1.70 - 0.15 = 1.55$$
$$L_{AQ} = \mathrm{k}M_Q = 1.64 - 0.15 = 1.49$$
$$r = 0.001542/1.406 = 0.001096$$
$$r_F = 0.371/499.9 = 0.000742$$
$$r_D = 0.018/1.406 = 0.0131$$
$$r_Q = 18.969 \times 10^{-3}/0.351 = 0.0540$$

The quantities L_{AD} and L_{AQ} are defined in Section 4.11.

4.7.4 The correspondence of per unit stator EMF to rotor quantities

We have seen that the particular choice of base quantities used here gives pu values of d and q axis stator currents and voltages that are $\sqrt{3}$ times the rms values. We also note that the coupling between the d axis rotor and stator involves the factor $k = \sqrt{3/2}$, and similarly for the q axis. For example, the contribution to the d axis stator flux linkage λ_d due to the field current i_F is $\mathrm{k}M_F i_F$ and so on. In synchronous machine equations it is often desirable to convert a rotor current, flux linkage, or voltage to an equivalent stator EMF. These expressions are developed in this section.

The basis for converting a field quantity to an equivalent stator EMF is that at open circuit a field current i_F A corresponds to an EMF of $i_F \omega_R M_F$ V peak. If the rms value of this EMF is E, then $i_F \omega_R M_F = \sqrt{2}E$ and $i_F \omega_R \mathrm{k}M_F = \sqrt{3}E$ in MKS units.[2]

---

2. The choice of symbol for the EMF due to i_F is not clearly decided. The American National Standards Institute (ANSI) uses the symbol E_I [16]. A new proposed standard uses E_{af} [17]. The International Electrotechnical Commission (IEC), in a discussion of [17], favors E_q for this voltage. The authors leave this voltage unsubscripted until a new standard is adopted.

Since M_F and ω_R are known constants for a given machine, the field current corresponds to a given EMF by a simple scaling factor. Thus E is the stator air gap rms voltage in pu corresponding to the field current i_F in pu.

We can also convert a field flux linkage λ_F to a corresponding stator EMF. At steady-state open circuit conditions $\lambda_F = L_F i_F$, and this value of field current i_F, when multiplied by $\omega_R M_F$, gives a peak stator voltage the rms value of which is denoted by E'_q. We can show that the d axis stator EMF corresponding to the field flux linkage λ_F is given by

$$\lambda_F(\omega_R k M_F/L_F) = \sqrt{3}\,E'_q \tag{4.58}$$

By the same reasoning a field voltage v_F corresponds (at steady state) to a field current v_F/r_F. This in turn corresponds to a peak stator EMF $(v_F/r_F)\,\omega_R M_F$. If the rms value of this EMF is denoted by E_{FD}, the d axis stator EMF corresponds to a field voltage v_F or

$$(v_F/r_F)\omega_R k M_F = \sqrt{3}\,E_{FD} \tag{4.59}$$

4.8 Normalizing the Voltage Equations

Having chosen appropriate base values, we may normalize the voltage equations (4.39). Having done this, the stator equations should be numerically easier to deal with, as all values of voltage and current will normally be in the neighborhood of unity. For the following computations we add the subscript u to all pu quantities to emphasize their dimensionless character. Later this subscript will be omitted when all values have been normalized.

The normalization process is based on (4.51) and a similar relation for the rotor, which may be substituted into (4.39) to give

$$\begin{bmatrix} v_{0u}V_B \\ v_{du}V_B \\ v_{qu}V_B \\ -v_{Fu}V_{FB} \\ 0 \\ 0 \end{bmatrix} = -\begin{bmatrix} r+3r_n & 0 & 0 & 0 & 0 & 0 \\ 0 & r & \omega L_q & 0 & 0 & \omega k M_Q \\ 0 & -\omega L_d & r & -\omega k M_F & -\omega k M_D & 0 \\ 0 & 0 & 0 & r_F & 0 & 0 \\ 0 & 0 & 0 & 0 & r_D & 0 \\ 0 & 0 & 0 & 0 & 0 & r_Q \end{bmatrix}\begin{bmatrix} i_{0u}I_B \\ i_{du}I_B \\ i_{qu}I_B \\ i_{Fu}I_{FB} \\ i_{Du}I_{DB} \\ i_{Qu}I_{QB} \end{bmatrix}$$

$$-\begin{bmatrix} L_0+3L_n & 0 & 0 & 0 & 0 & 0 \\ 0 & L_d & 0 & kM_F & kM_D & 0 \\ 0 & 0 & L_q & 0 & 0 & kM_Q \\ 0 & kM_F & 0 & L_F & M_R & 0 \\ 0 & kM_D & 0 & M_R & L_D & 0 \\ 0 & 0 & kM_Q & 0 & 0 & L_Q \end{bmatrix}\begin{bmatrix} \dot{i}_{0u}I_B \\ \dot{i}_{du}I_B \\ \dot{i}_{qu}I_B \\ \dot{i}_{Fu}I_{FB} \\ \dot{i}_{Du}I_{DB} \\ \dot{i}_{Qu}I_{QB} \end{bmatrix} \tag{4.60}$$

where the first three equations are on a stator base and the last three are on a rotor base.

Examine the second equation more closely. Dividing through by V_B and setting $\omega = \omega_u\omega_R$, we have

$$v_{du} = -r\frac{I_B}{V_B}i_{du} - \omega_u\omega_R L_q \frac{I_B}{V_B}i_{qu} - \omega_u\omega_R kM_Q\frac{I_{QB}}{V_B}i_{Qu}$$

$$- L_d\frac{I_B}{V_B}\dot{i}_{du} - kM_F\frac{I_{FB}}{V_B}\dot{i}_{Fu} - kM_D\frac{I_{DB}}{V_B}\dot{i}_{Du} \text{ pu} \qquad (4.61)$$

Incorporating base values from (4.50), we rewrite (4.61) as

$$v_{du} = -\frac{r}{R_B}i_{du} - \omega_u\frac{L_q}{L_B}i_{qu} - \omega_u\frac{\omega_R I_{QB}}{V_B}kM_Q i_{Qu} - \frac{L_d}{\omega_R L_B}\dot{i}_{du}$$

$$-\frac{kM_F}{\omega_R}\frac{\omega_R I_{FB}}{V_B}\dot{i}_{Fu} - \frac{kM_D}{\omega_R}\frac{\omega_R I_{DB}}{V_B}\dot{i}_{Du} \text{ pu} \qquad (4.62)$$

We now recognize the following pu quantities.

$$\begin{aligned} r_u &= r/R_B & L_{du} &= L_d/L_B & M_{Fu} &= M_F\omega_R I_{FB}/V_B \\ L_{qu} &= L_q/L_B & M_{Du} &= M_D\omega_R I_{DB}/V_B & M_{Qu} &= M_Q\omega_R I_{QB}/V_B \end{aligned} \qquad (4.63)$$

Incorporating (4.63), the d axis equation (4.62) may be rewritten with all values except time in pu; i.e.,

$$v_{du} = -r_u i_{du} - \omega_u L_{qu}i_{qu} - k\omega_u M_{Qu}i_{Qu} - \frac{L_{du}}{\omega_R}\dot{i}_{du} - k\frac{M_{Fu}}{\omega_R}\dot{i}_{Fu} - k\frac{M_{Du}}{\omega_R}\dot{i}_{Du} \qquad (4.64)$$

The third equation of (4.60) may be analyzed in a similar way to write

$$v_{qu} = \omega_u L_{du}i_{du} - r_u i_{qu} + \omega_u kM_{Fu}i_{Fu} + \omega_u kM_{Du}i_{Du} - \frac{L_{qu}}{\omega_R}\dot{i}_{qu} - k\frac{M_{qu}}{\omega_R}\dot{i}_{Qu} \text{ pu} \qquad (4.65)$$

where all pu coefficients have been previously defined. The first equation is uncoupled from the others and may be written as

$$v_{0u} = -\frac{r + 3r_n}{R_B}i_{0u} - \frac{L_0 + 3L_n}{\omega_R L_B}\dot{i}_{0u}$$

$$= -(r + 3r_n)_u i_{0u} - \frac{1}{\omega_R}(L_0 + 3L_n)_u\dot{i}_{0u} \text{ pu} \qquad (4.66)$$

If the currents are balanced, it is easy to show that this equation vanishes.

The fourth equation is normalized on a rotor basis and may be written from (4.60) as

$$v_{Fu} = r_F\frac{I_{FB}}{V_{FB}}i_{Fu} + k\frac{M_F}{\omega_R}\frac{\omega_R I_B}{V_{FB}}\dot{i}_{du} + \frac{L_F}{\omega_R}\frac{\omega_R I_{FB}}{V_{FB}}\dot{i}_{Fu} + \frac{M_R}{\omega_R}\frac{\omega_R I_{DB}}{V_{FB}}\dot{i}_{Du} \text{ pu} \qquad (4.67)$$

We now incorporate the base rotor inductance to normalize the last two terms as

$$L_{Fu} = L_F/L_{FB} \qquad M_{Ru} = M_R/M_{RB} \qquad (4.68)$$

The normalized field circuit equation becomes

$$v_{Fu} = r_{Fu}i_{Fu} + \frac{kM_{Fu}}{\omega_R}\dot{i}_{du} + \frac{L_{Fu}}{\omega_R}\dot{i}_{Fu} + \frac{M_{Ru}}{\omega_R}\dot{i}_{Du} \qquad (4.69)$$

The damper winding equations can be normalized by a similar procedure. The following equations are then obtained,

$$v_{Du} = 0 = r_{Du}i_{Du} + \frac{kM_{Du}}{\omega_R}\dot{i}_{du} + \frac{M_{Ru}}{\omega_R}\dot{i}_{Fu} + \frac{L_{Du}}{\omega_R}\dot{i}_{Du} \qquad (4.70)$$

$$v_{Qu} = 0 = r_{Qu} i_{Qu} + \frac{\mathrm{k} M_{Qu}}{\omega_{\mathrm{R}}} \dot{i}_{qu} + \frac{L_{Qu}}{\omega_{\mathrm{R}}} \dot{i}_{Qu} \tag{4.71}$$

These normalized equations are in a form suitable for solution in the time domain with time in seconds. However, some engineers prefer to rid the equations of the awkward $1/\omega_{\mathrm{R}}$ that accompanies every term containing a time derivative. This may be done by normalizing time. We do this by setting

$$\frac{1}{\omega_{\mathrm{R}}} \frac{d}{dt} = \frac{d}{d\tau} \tag{4.72}$$

where

$$\tau = \omega_{\mathrm{R}} t \tag{4.73}$$

is the normalized time in rad.

Incorporating all normalized equations in a matrix expression and dropping the subscript u since *all* values are in pu, we write

$$\begin{bmatrix} v_d \\ -v_F \\ 0 \\ \hdashline v_q \\ 0 \end{bmatrix} = - \left[\begin{array}{ccc:cc} r & 0 & 0 & \omega L_q & \omega \mathrm{k} M_Q \\ 0 & r_F & 0 & 0 & 0 \\ 0 & 0 & r_D & 0 & 0 \\ \hdashline -\omega L_d & -\omega \mathrm{k} M_F & -\omega \mathrm{k} M_D & r & 0 \\ 0 & 0 & 0 & 0 & r_Q \end{array}\right] \begin{bmatrix} i_d \\ i_F \\ i_D \\ \hdashline i_q \\ i_Q \end{bmatrix}$$

$$- \left[\begin{array}{ccc:cc} L_d & \mathrm{k} M_F & \mathrm{k} M_D & 0 & 0 \\ \mathrm{k} M_F & L_F & M_R & 0 & 0 \\ \mathrm{k} M_D & M_R & L_D & 0 & 0 \\ \hdashline 0 & 0 & 0 & L_q & \mathrm{k} M_Q \\ 0 & 0 & 0 & \mathrm{k} M_Q & L_Q \end{array}\right] \begin{bmatrix} \dot{i}_d \\ \dot{i}_F \\ \dot{i}_D \\ \hdashline \dot{i}_q \\ \dot{i}_Q \end{bmatrix} \text{ pu} \tag{4.74}$$

where we have omitted the v_0 equation, since we are interested in balanced system conditions in stability studies, and have rearranged the equations to show the d and q coupling more clearly. It is important to notice that (4.74) is *identical* in notation to (4.39). This is always possible if base quantities are carefully chosen and is highly desirable, as the same equation symbolically serves both as a pu and a "system quantity" equation. Using matrix notation, we write (4.74) as

$$\mathbf{v} = -(\mathbf{R} + \omega \mathbf{N})\mathbf{i} - \mathbf{L}\dot{\mathbf{i}} \text{ pu} \tag{4.75}$$

where $\mathbf{R}$ is the resistance matrix and is a diagonal matrix of constants, $\mathbf{N}$ is the matrix of speed voltage inductance coefficients, and $\mathbf{L}$ is a symmetric matrix of constant inductances. If we assume that the inverse of the inductance matrix exists, we may write

$$\dot{\mathbf{i}} = -\mathbf{L}^{-1}(\mathbf{R} + \omega \mathbf{N})\mathbf{i} - \mathbf{L}^{-1}\mathbf{v} \text{ pu} \tag{4.76}$$

This equation has the desired state-space form. It does not express the entire system behavior, however, so we have additional equations to write.

Equation (4.76) may be depicted schematically by the equivalent circuit shown in

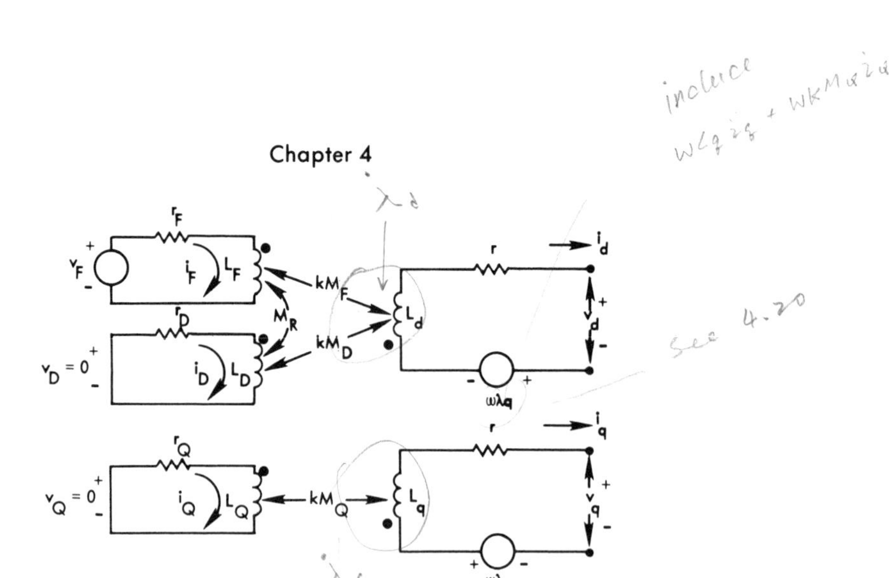

Fig. 4.3 Synchronous generator *d-q* equivalent circuit.

Figure 4.3. Note that all self and mutual inductances in the equivalent circuit are constants, and pu quantities are implied for all quantities, including time. Note also the presence of controlled sources in the equivalent. These are due to speed voltage terms in the equations.

Equation (4.74) and the circuit in Figure 4.3 differ from similar equations found in the literature in two important ways. In this chapter we use the symbols L and M for self and mutual inductances respectively. Some authors and most manufacturers refer to these same quantities by the symbol x or X. This is sometimes confusing to one learning synchronous machine theory because a term XI that appears to be a voltage may be a flux linkage. The use of X for L or M is based on the rationale that ω is nearly constant at 1.0 pu so that, in pu, $X = \omega L \simeq L$. However, as we shall indicate in the sections to follow, ω is certainly not a constant; it is a state variable in our equations, and we must treat it as a variable. Later, in a linearized model we will let ω be approximated as a constant and will simplify other terms in the equations as well.

For convenience of those acquainted with other references we list a comparison of these inductances in Table 4.2. Here the subscript notation kd and kq for D and Q respectively is seen. These symbols are quite common in the literature in reference to the damper windings.

Table 4.2. Comparison of Per Unit Inductance Symbols

Chapter 4	L_d	L_q	L_F	L_D	L_Q	kM_F	M_R	kM_D	kM_Q
Kimbark [2]	L_d	L_q	L_{ff}		L_{gg}	M_F			M_g
Concordia [1]	x_d	x_q	x_{ff}	x_{kdd}	x_{kqq}	x_{af}	x_{fkd}	x_{akd}	x_{akq}

Example 4.2

Consider a 60-Hz synchronous machine with the following pu parameters:

$$L_d = 1.70 \qquad kM_Q = 1.49$$
$$L_q = 1.64 \qquad r = 0.001096$$
$$L_F = 1.65 \qquad r_F = 0.000742$$
$$L_D = 1.605 \qquad r_D = 0.0131$$
$$L_Q = 1.526 \qquad r_Q = 0.0540$$
$$kM_F = M_R = kM_D = 1.55 \qquad H = 2.37 \text{ s}$$
$$\ell_d = \ell_q = 0.15$$

Solution

From (4.75) we have numerically

$$\mathbf{R} + \omega\mathbf{N} = \left[\begin{array}{ccc|cc} 0.0011 & 0 & 0 & 1.64\omega & 1.49\omega \\ 0 & 0.00074 & 0 & 0 & 0 \\ 0 & 0 & 0.0131 & 0 & 0 \\ \hline -1.70\omega & -1.55\omega & -1.55\omega & 0.0011 & 0 \\ 0 & 0 & 0 & 0 & 0.0540 \end{array}\right] \text{pu}$$

$$\mathbf{L} = \left[\begin{array}{ccc|cc} 1.70 & 1.55 & 1.55 & 0 & 0 \\ 1.55 & 1.65 & 1.55 & 0 & 0 \\ 1.55 & 1.55 & 1.605 & 0 & 0 \\ \hline 0 & 0 & 0 & 1.64 & 1.49 \\ 0 & 0 & 0 & 1.49 & 1.526 \end{array}\right] \text{pu}$$

from which we compute by digital computer

$$\mathbf{L}^{-1} = \left[\begin{array}{ccc|cc} 5.405 & -1.869 & -3.414 & 0 & 0 \\ -1.869 & 7.110 & -5.060 & 0 & 0 \\ -3.414 & -5.060 & 8.804 & 0 & 0 \\ \hline 0 & 0 & 0 & 5.406 & -5.280 \\ 0 & 0 & 0 & -5.280 & 5.811 \end{array}\right] \text{pu}$$

Then we may compute

$$-\mathbf{L}^{-1}(\mathbf{R} + \omega\mathbf{N}) = 10^{-3}\left[\begin{array}{ccc|cc} -5.9269 & 1.3878 & 44.7198 & -8864.9\omega & -8504.1\omega \\ 2.0498 & -5.2785 & 66.2818 & 3065.9\omega & 2785.4\omega \\ 3.7433 & 3.7564 & -115.3290 & 5598.9\omega & 5086.8\omega \\ \hline 9190.9\omega & 8379.9\omega & 8379.9\omega & -5.9279 & 284.857 \\ -8975.2\omega & -8183.3\omega & -8183.3\omega & 5.7888 & -313.534 \end{array}\right] \text{pu}$$

and the coefficient matrix is seen to contain ω in 12 of its 25 terms. This gives some idea of the complexity of the equations.

4.9 Normalizing the Torque Equations

In Chapter 2 the swing equation

$$J\ddot{\theta} = (2J/p)\dot{\omega} = T_a \quad \text{N}\cdot\text{m} \tag{4.77}$$

is normalized by dividing both sides of the equation by a shaft torque that corresponds to the rated three-phase power at rated speed (base three-phase torque). The result of this normalization was found to be

$$(2H/\omega_R)\dot{\omega} = T_a \ \text{pu}(3\phi) \tag{4.78}$$

where ω = angular velocity of the revolving magnetic field in elec rad/s

T_a = accelerating torque in pu on a three-phase base

H = W_R/S_{B3} s

and the derivative is with respect to time in seconds. This normalization takes into account the change in angular measurements from mechanical to electrical radians and divides the equations by the base three-phase torque. Equation (4.78) is the swing equation used to determine the speed of the stator revolving MMF wave as a function of time. We need to couple the electromagnetic torque T_e, determined by the generator equations, to the form of (4.78). Since (4.78) is normalized to a three-phase base torque and our chosen generator VA base is a per phase basis, we must use care in combining the pu swing equation and the pu generator torque equation. Rewriting (4.78) as

$$(2H/\omega_B)\dot{\omega} = T_m - T_e \ \text{pu}(3\phi) \tag{4.79}$$

the expression used for T_e must be in pu on a three-phase VA base.

Suppose we define

$$\begin{aligned} T_{e\phi} &= \text{pu generator electromagnetic torque defined on a per phase VA base} \\ &= T_e(\text{N}\cdot\text{m})/(S_B/\omega_B) \ \text{pu} \end{aligned} \tag{4.80}$$

Then

$$T_e = T_{e\phi}/3 \ \text{pu}(3\phi) \tag{4.81}$$

(A similar definition could be used for the mechanical torque; viz., $T_{m\phi} = 3T_m$. Usually, T_m is normalized on a three-phase basis.)

The procedure that must be used is clear. We compute the generator electromagnetic torque in N·m. This torque is normalized along with other generator quantities on a basis of S_B, V_B, I_B, and t_B to give $T_{e\phi}$. Thus for a fully loaded machine at rated speed, we would expect to compute $T_{e\phi} = 3.0$. Equation (4.81) transforms this pu torque to the new value T_e, which is the pu torque on a three-phase basis.

4.9.1 The normalized swing equation

In (4.79), while the torque is normalized, the angular speed ω and the time are given in MKS units. Thus the equation is not completely normalized.

The normalized swing equation is of the form given in (2.66)

$$\tau_j \frac{d}{dt_u}\omega_u = T_{mu} - T_{eu} = T_{au} \ \text{pu} \tag{4.82}$$

where all the terms in the swing equation, including time and angular speed, are in pu.

Beginning with (4.79) and substituting

$$t_u = \omega_B t \qquad \omega_u = \omega/\omega_B \tag{4.83}$$

we have for the normalized swing equation

$$2H\omega_B \frac{d\omega_u}{dt_u} = T_{au} \tag{4.84}$$

thus, when time is in pu,

$$\tau_j = 2H\omega_B \tag{4.85}$$

4.9.2 Forms of the swing equation

There are many forms of the swing equation appearing in the literature of power system dynamics. While the torque is almost always given in pu, it is often not clear which units of ω and t are being used. To avoid confusion, a summary of the different forms of the swing equation is given in this section.

We begin with ω in rad/s and t in s, $(2H/\omega_B)\omega = T_{au}$. If t and T_a are in pu (and ω in rad/s), by substituting $t_u = \omega_B t$ in (4.79),

$$\frac{2H}{\omega_B}\frac{d\omega}{dt} = 2H\frac{d\omega}{dt_u} = T_{au} \quad \text{pu} \tag{4.86}$$

If ω and T_a are in pu (and t in s), by substituting in (4.79),

$$\frac{2H}{\omega_B}\frac{d(\omega_u \omega_B)}{dt} = 2H\frac{d\omega_u}{dt} = T_{au} \quad \text{pu} \tag{4.87}$$

If t, ω, and T_a are all in pu,

$$2H\omega_B\frac{d\omega_u}{dt_u} = \tau_j\frac{d\omega_u}{dt_u} = T_{au} \quad \text{pu} \tag{4.88}$$

If ω is given in elec deg/s, (4.79) and (4.86) are modified as follows:

$$\frac{H}{180 f_B}\frac{d\omega}{dt} = T_{au} \quad \text{pu} \tag{4.89}$$

$$\frac{\pi H}{90}\frac{d\omega}{dt_u} = T_{au} \quad \text{pu} \tag{4.90}$$

It would be tempting to normalize the swing equation on a per phase basis such that all terms in (4.79) are in pu based on S_B rather than S_{B3}. This could indeed be done with the result that all values in the swing equation would be multiplied by three. This is not done here because it is common to express both T_m and T_e in pu on a three-phase base. Therefore, even though S_B is a convenient base to use in normalizing the generator *circuits,* it is considered wise to convert the generator terminal power and torque to a three-phase base S_{B3} to match the basis normally used in computing the machine terminal conditions from the viewpoint of the network (e.g., in load-flow studies). Note there is not a similar problem with the voltage being based on V_B, the phase-to-neutral voltage, since a phase voltage of k pu means that the line-to-line voltage is also k pu on a line-to-line basis.

4.10 Torque and Power

The total three-phase power output of a synchronous machine is given by

$$p_{out} = v_a i_a + v_b i_b + v_c i_c = \mathbf{v}_{abc}^t \mathbf{i}_{abc} \quad \text{pu} \tag{4.91}$$

where the superscript t indicates the transpose of $\mathbf{v}_{abc}$. But from (4.8) we may write $\mathbf{i}_{abc} = \mathbf{P}^{-1}\mathbf{i}_{0dq}$ with a similar expression for the voltage vector. Then (4.91) becomes

$$p_{out} = \mathbf{v}_{0dq}^t (\mathbf{P}^{-1})^t \, \mathbf{P}^{-1}\mathbf{i}_{0dq}$$

Performing the indicated operation and recalling that $\mathbf{P}$ is orthogonal, we find that the power output of a synchronous generator is invariant under the transformation **P**; i.e.,

$$p_{\text{out}} = v_d i_d + v_q i_q + v_0 i_0 \tag{4.92}$$

For simplicity we will assume balanced but not necessarily steady-state conditions. Thus $v_0 = i_0 = 0$ and

$$p_{\text{out}} = v_d i_d + v_q i_q \quad \text{(balanced condition)} \tag{4.93}$$

Substituting for v_d and v_q from (4.36),

$$p_{\text{out}} = (i_d \dot{\lambda}_d + i_q \dot{\lambda}_q) + (i_q \lambda_d - i_d \lambda_q)\omega - r(i_d^2 + i_q^2) \tag{4.94}$$

Concordia [1] observes that the three terms are identifiable as the rate of change of stator magnetic field energy, the power transferred across the air gap, and the stator ohmic losses respectively. The machine torque is obtained from the second term,

$$T_{e\phi} = \partial W_{\text{fld}}/\partial \theta = \partial P_{\text{fld}}/\partial \omega = \partial/\partial \omega \; [(i_q \lambda_d - i_d \lambda_q)\omega] = i_q \lambda_d - i_d \lambda_q \text{ pu} \tag{4.95}$$

The same result can be obtained from a more rigorous derivation. Starting with the three armature circuits and the three rotor circuits, the energy in the field is given by

$$W_{\text{fld}} = \sum_{\substack{k=1 \\ j=1}}^{6} \frac{1}{2} (i_k i_j L_{kj}) \tag{4.96}$$

which is a function of θ. Then using $T = \partial W_{\text{fld}}/\partial \theta$ and simplifying, we can obtain the above relation (see Appendix B of [1]).

Now, recalling that the flux linkages can be expressed in terms of the currents, we write from (4.20), expressed in pu,

$$\lambda_d = L_d i_d + \text{k} M_F i_F + \text{k} M_D i_D \qquad \lambda_q = L_q i_q + \text{k} M_Q i_Q \tag{4.97}$$

Then (4.95) can be written as

$$T_{e\phi} = [L_d i_q \quad \text{k}M_F i_q \quad \text{k}M_D i_q \;\vdots\; -L_q i_d \quad -\text{k}M_Q i_d] \begin{bmatrix} i_d \\ i_F \\ i_D \\ \hline i_q \\ i_Q \end{bmatrix} \text{pu} \tag{4.98}$$

which we recognize to be a bilinear term.

Suppose we express the total accelerating torque in the swing equation as

$$T_a = T_m - T_{e\phi}/3 - T_d = T_m - T_e - T_d \tag{4.99}$$

where T_m is the mechanical torque, T_e is the electrical torque, and T_d is the damping torque. It is often convenient to write the damping torque as

$$T_d = D\omega \text{ pu} \tag{4.100}$$

where D is a damping constant. Then by using (4.81) and (4.98), the swing equation may be written as

$$\dot{\omega} = \frac{T_m}{\tau_j} + \left[-\frac{L_d}{3\tau_j} i_q \quad -\frac{\mathrm{k}M_F}{3\tau_j} i_q \quad -\frac{\mathrm{k}M_D}{3\tau_j} i_q \;\middle|\; \frac{L_q}{3\tau_j} i_d \quad \frac{\mathrm{k}M_Q}{3\tau_j} i_d \;\middle|\; -\frac{D}{3\tau_j} \right] \begin{bmatrix} i_d \\ i_F \\ i_D \\ \hline i_q \\ i_Q \\ \hline \omega \end{bmatrix} \tag{4.101}$$

where τ_j is defined by (4.85) and depends on the units used for ω and t. Finally, the following relation between δ and ω may be derived from (4.6).

$$\dot{\delta} = \omega - 1 \tag{4.102}$$

Incorporating (4.101) and (4.102) into (4.76), we obtain

$$\begin{bmatrix} \dot{i}_d \\ \dot{i}_F \\ \dot{i}_D \\ \dot{i}_q \\ \dot{i}_Q \\ \hline \dot{\omega} \\ \dot{\delta} \end{bmatrix} = \left[\begin{array}{ccccc|cc} \multicolumn{5}{c|}{-\mathbf{L}^{-1}(\mathbf{R} + \omega\mathbf{N})} & \multicolumn{2}{c}{\mathbf{0}} \\ \hline -\frac{L_d i_q}{3\tau_j} & -\frac{\mathrm{k}M_F i_q}{3\tau_j} & -\frac{\mathrm{k}M_D i_q}{3\tau_j} & \frac{L_q i_d}{3\tau_j} & \frac{\mathrm{k}M_Q i_d}{3\tau_j} & -\frac{D}{3\tau_j} & 0 \\ 0 & 0 & 0 & 0 & 0 & 1 & 0 \end{array} \right] \begin{bmatrix} i_d \\ i_F \\ i_D \\ i_q \\ i_Q \\ \hline \omega \\ \delta \end{bmatrix} + \begin{bmatrix} -\mathbf{L}^{-1}\mathbf{v} \\ \\ \\ \\ \\ \hline \frac{T_m}{3\tau_j} \\ -1 \end{bmatrix} \tag{4.103}$$

This matrix equation is in the desired state-space form $\dot{\mathbf{x}} = \mathbf{f}(\mathbf{x}, \mathbf{u}, t)$ as given by (4.37). It is clear from (4.101) that the system is nonlinear. Note that the "inputs" are $\mathbf{v}$ and T_m.

4.11 Equivalent Circuit of a Synchronous Machine

For balanced conditions the normalized flux linkage equations are obtained from (4.20) with the row for λ_0 omitted.

$$\begin{bmatrix} \lambda_d \\ \lambda_q \\ \lambda_F \\ \lambda_D \\ \lambda_Q \end{bmatrix} = \begin{bmatrix} L_d & 0 & \mathrm{k}M_F & \mathrm{k}M_D & 0 \\ 0 & L_q & 0 & 0 & \mathrm{k}M_Q \\ \mathrm{k}M_F & 0 & L_F & M_R & 0 \\ \mathrm{k}M_D & 0 & M_R & L_D & 0 \\ 0 & \mathrm{k}M_Q & 0 & 0 & L_Q \end{bmatrix} \begin{bmatrix} i_d \\ i_q \\ i_F \\ i_D \\ i_Q \end{bmatrix} \tag{4.104}$$

We may rewrite the d axis flux linkages as

$$
\begin{aligned}
\lambda_d &= [(L_d - \ell_d) + \ell_d]i_d + kM_F i_F + kM_D i_D \\
\lambda_F &= kM_F i_d + [(L_F - \ell_F) + \ell_F]i_F + M_R i_D \\
\lambda_D &= kM_D i_d + M_R i_F + [(L_D - \ell_D) + \ell_D]i_D
\end{aligned} \tag{4.105}
$$

where ℓ_d, ℓ_F, and ℓ_D are the leakage inductances of the d, F, and D circuits respectively. Let $i_F = i_D = 0$, and the flux linkage that will be mutually coupled to the other circuits is $\lambda_d - \ell_d i_d$, or $(L_d - \ell_d)i_d$. As stated in Section 4.7.2, $L_d - \ell_d$ is the magnetizing inductance L_{md}. The flux linkage mutually coupled to the other d axis circuits is then $L_{md}i_d$. The flux linkages in the F and D circuits, λ_F and λ_D, are given in this particular case by $\lambda_F = kM_F i_d$, and $\lambda_D = kM_D i_d$. From the choice of the base rotor current, to give equal mutual flux, we can see that the pu values of $L_{md}i_d$, λ_F, and λ_D must be equal. Therefore, the pu values of L_{md}, kM_F, and kM_D are equal. This can be verified by using (4.57) and (4.55),

$$
kM_{Fu} = \frac{kM_F}{M_{FB}} = \frac{kM_F}{(kM_F/L_{md})L_B} = \frac{L_{md}}{L_B} = L_{mdu} \tag{4.106}
$$

In pu, we usually call this quantity L_{AD}; i.e.,

$$
L_{AD} \triangleq L_d - \ell_d = kM_F = kM_D \quad \text{pu} \tag{4.107}
$$

We can also prove that, in pu,

$$
L_{AD} = L_D - \ell_D = L_F - \ell_F = L_d - \ell_d = kM_F = kM_D = M_R \tag{4.108}
$$

Similarly, for the q axis we define

$$
L_{AQ} \triangleq L_q - \ell_q = L_Q - \ell_Q = kM_Q \quad \text{pu} \tag{4.109}
$$

If in each circuit the pu leakage flux linkage is subtracted, the remaining flux linkage is the same as for all other circuits coupled to it. Thus

$$
\lambda_d - \ell_d i_d = \lambda_F - \ell_F i_F = \lambda_D - \ell_D i_D \triangleq \lambda_{AD} \quad \text{pu} \tag{4.110}
$$

where

$$
\lambda_{AD} = i_d(L_d - \ell_d) + kM_F i_F + kM_D i_D = L_{AD}(i_d + i_F + i_D) \quad \text{pu} \tag{4.111}
$$

Similarly, the pu q axis mutual flux linkage is given by

$$
\lambda_{AQ} = (L_q - \ell_q)i_q + kM_Q i_Q = L_{AQ}(i_q + i_Q) \tag{4.112}
$$

Following the procedure used in developing the equivalent circuit of transformers, we can represent the above relations by the circuits shown in Figure 4.4, where we note that the currents add in the mutual branch. To complete the equivalent circuit, we

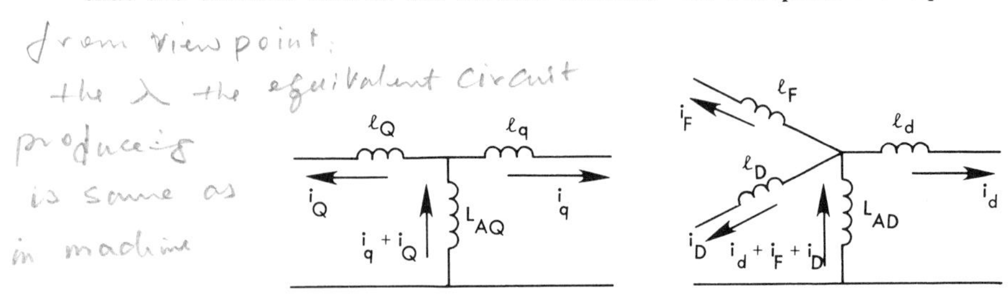

Fig. 4.4 Flux linkage inductances of a synchronous machine.

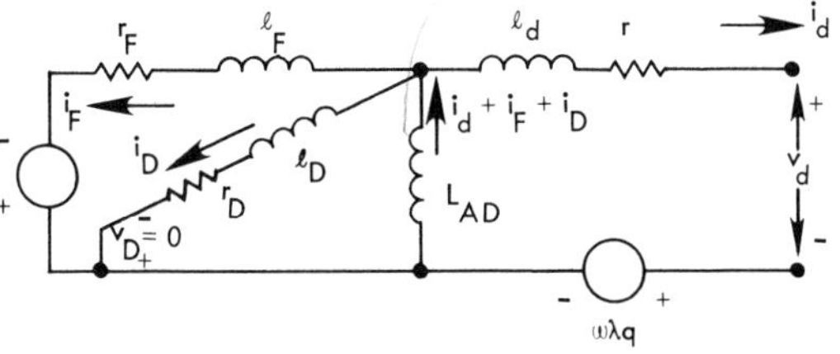

Fig. 4.5 Direct axis equivalent circuit.

consider the voltage equations

$$v_d = -r i_d - \dot{\lambda}_d - \omega \lambda_q$$
$$= -r i_d - \ell_d \dot{i}_d - [(L_d - \ell_d) \dot{i}_d + \mathrm{k} M_F \dot{i}_F + \mathrm{k} M_D \dot{i}_D] - \omega \lambda_q$$

or

$$v_d = -r i_d - \ell_d \dot{i}_d - L_{AD}(\dot{i}_d + \dot{i}_F + \dot{i}_D) - \omega \lambda_q \tag{4.113}$$

Similarly, we can show that

$$-v_F = -r_F i_F - \ell_F \dot{i}_F - L_{AD}(\dot{i}_d + \dot{i}_F + \dot{i}_D) \tag{4.114}$$

$$v_D = 0 = -r_D i_D - \ell_D \dot{i}_D - L_{AD}(\dot{i}_d + \dot{i}_F + \dot{i}_D) \tag{4.115}$$

The above voltage equations are satisfied by the equivalent circuit shown in Figure 4.5. The three d axis circuits (d, F, and D) are coupled through the common magnetizing inductance L_{AD}, which carries the sum of the currents i_d, i_F, and i_D. The d axis circuit contains a controlled voltage source $\omega \lambda_q$ with the polarity as shown.

Similarly, for the q axis circuits

$$v_q = -r i_q - \ell_q \dot{i}_q - L_{AQ}(\dot{i}_q + \dot{i}_Q) + \omega \lambda_d \tag{4.116}$$

$$v_Q = 0 = -r_Q i_Q - \ell_Q \dot{i}_Q - L_{AQ}(\dot{i}_q + \dot{i}_Q) \tag{4.117}$$

These two equations are satisfied by the equivalent circuit shown in Figure 4.6. Note the presence of the controlled source $\omega \lambda_d$ in the stator q circuit.

4.12 The Flux Linkage State-Space Model

We now develop an alternate state-space model where the state variables chosen are λ_d, λ_F, λ_D, λ_q, and λ_Q. From (4.110)

$$i_d = (1/\ell_d)(\lambda_d - \lambda_{AD}) \qquad i_F = (1/\ell_F)(\lambda_F - \lambda_{AD}) \qquad i_D = (1/\ell_D)(\lambda_D - \lambda_{AD}) \tag{4.118}$$

but from (4.111) $\lambda_{AD} = (i_d + i_F + i_D) L_{AD}$, which we can incorporate into (4.118) to get

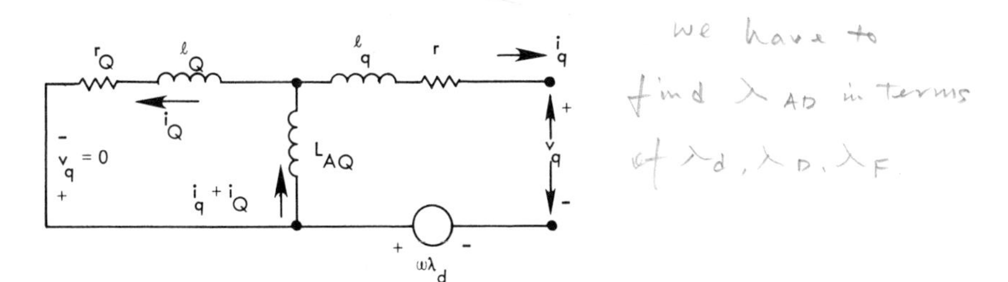

Fig. 4.6 Quadrature axis equivalent circuit.

$$\lambda_{AD}(1/L_{AD} + 1/\ell_d + 1/\ell_F + 1/\ell_D) = \lambda_d/\ell_d + \lambda_F/\ell_F + \lambda_D/\ell_D$$

Now define

$$1/L_{MD} \triangleq 1/L_{AD} + 1/\ell_d + 1/\ell_F + 1/\ell_D \tag{4.119}$$

then

$$\lambda_{AD} = (L_{MD}/\ell_d)\lambda_d + (L_{MD}/\ell_F)\lambda_F + (L_{MD}/\ell_D)\lambda_D \tag{4.120}$$

Similarly, we can show that

$$\lambda_{AQ} = (L_{MQ}/\ell_q)\lambda_q + (L_{MQ}/\ell_Q)\lambda_Q \tag{4.121}$$

where we define

$$1/L_{MQ} \triangleq 1/L_{AQ} + 1/\ell_q + 1/\ell_Q \tag{4.122}$$

and the q axis currents are given by

$$i_q = (1/\ell_q)(\lambda_q - \lambda_{AQ}) \qquad i_Q = (1/\ell_Q)(\lambda_Q - \lambda_{AQ}) \tag{4.123}$$

Writing (4.118) and (4.123) in matrix form,

$$\begin{bmatrix} i_d \\ i_F \\ i_D \\ \hline i_q \\ i_Q \end{bmatrix} = \left[\begin{array}{cccc|ccc} 1/\ell_d & 0 & 0 & -1/\ell_d & & & \\ 0 & 1/\ell_F & 0 & -1/\ell_F & & \mathbf{0} & \\ 0 & 0 & 1/\ell_D & -1/\ell_D & & & \\ \hline & & & & 1/\ell_q & 0 & -1/\ell_q \\ & \mathbf{0} & & & 0 & 1/\ell_Q & -1/\ell_Q \end{array}\right] \begin{bmatrix} \lambda_d \\ \lambda_F \\ \lambda_D \\ \lambda_{AD} \\ \hline \lambda_q \\ \lambda_Q \\ \lambda_{AQ} \end{bmatrix} \tag{4.124}$$

4.12.1 The voltage equations

The voltage equations are derived as follows from (4.36). For the d equation

$$v_d = -r i_d - \dot{\lambda}_d - \omega\lambda_q \tag{4.125}$$

Using (4.124) and rearranging,

$$\dot{\lambda}_d = -r(\lambda_d/\ell_d - \lambda_{AD}/\ell_d) - \omega\lambda_q - v_d$$

or

$$\dot{\lambda}_d = -(r/\ell_d)\lambda_d + (r/\ell_d)\lambda_{AD} - \omega\lambda_q - v_d \tag{4.126}$$

Also from (4.36)

$$-v_F = -r_F i_F - \dot{\lambda}_F \tag{4.127}$$

Substituting for i_F

$$\dot{\lambda}_F = -r_F(\lambda_F/\ell_F - \lambda_{AD}/\ell_F) + v_F$$

or

$$\dot{\lambda}_F = -(r_F/\ell_F)\lambda_F + (r_F/\ell_F)\lambda_{AD} - (-v_F) \tag{4.128}$$

Repeating the procedure for the D circuit,

$$\dot{\lambda}_D = -(r_D/\ell_D)\lambda_D + (r_D/\ell_D)\lambda_{AD} \tag{4.129}$$

The procedure is repeated for the q axis circuits. For the v_q equations we compute

$$\dot{\lambda}_q = -(r/\ell_q)\lambda_q + (r/\ell_q)\lambda_{AQ} + \omega\lambda_d - v_q \tag{4.130}$$

and from the q axis damper-winding equation,

$$\dot{\lambda}_Q = -(r_Q/\ell_Q)\lambda_Q + (r_Q/\ell_Q)\lambda_{AQ} \tag{4.131}$$

Note that λ_{AD} or λ_{AQ} appears in the above equations. This form is convenient if saturation is to be included in the model since the mutual inductances L_{AD} and L_{AQ} are the only inductances that saturate. If saturation can be neglected the λ_{AD} and λ_{AQ} terms can be eliminated (see Section 4.12.3).

4.12.2 The torque equation

From (4.95) $T_{e\phi} = i_q\lambda_d - i_d\lambda_q$. Using (4.124), we substitute for the currents to compute

$$T_{e\phi} = -\lambda_q\left(\frac{\lambda_d - \lambda_{AD}}{\ell_d}\right) + \lambda_d\left(\frac{\lambda_q - \lambda_{AQ}}{\ell_q}\right) = -\frac{1}{\ell_q}\lambda_d\lambda_{AQ} + \frac{1}{\ell_d}\lambda_q\lambda_{AD} + \left(\frac{1}{\ell_q} - \frac{1}{\ell_d}\right)\lambda_d\lambda_q \tag{4.132}$$

We may also take advantage of the relation $\ell_q = \ell_d$ (called ℓ_a in many references). The new electromechanical equation is given by

$$\dot{\omega} = -(\lambda_{AD}/\ell_d 3\tau_j)\lambda_q + (\lambda_{AQ}/\ell_q 3\tau_j)\lambda_d - (D/\tau_j)\omega + T_m/\tau_j \tag{4.133}$$

Finally the equation for δ is given by (4.102). Equations (4.126)–(4.131), (4.133), and (4.102) are in state-space form. The auxiliary equations (4.120) and (4.121) are needed to relate λ_{AD} and λ_{AQ} to the state variables. The state variables are λ_d, λ_F, λ_D, λ_q, λ_Q, ω, and δ. The forcing functions, are v_d, v_q, v_F, and T_m. This form of the equations is particularly convenient for solution where saturation is required, since saturation affects only λ_{AD} and λ_{AQ}.

4.12.3 Machine equations with saturation neglected

If saturation is neglected, L_{AD} and L_{AQ} are constant. Therefore, L_{MD} and L_{MQ} are also constant. The magnetizing flux linkages λ_{AD} and λ_{AQ} will have constant relationships to the state variables as given by (4.120) and (4.121). We can therefore eliminate λ_{AD} and λ_{AQ} from the machine equations.

Substituting for λ_{AD}, as given in (4.120), in (4.118) and rearranging,

$$\begin{aligned}
i_d &= \left(1 - \frac{L_{MD}}{\ell_d}\right)\frac{\lambda_d}{\ell_d} - \frac{L_{MD}}{\ell_d}\frac{\lambda_F}{\ell_F} - \frac{L_{MD}}{\ell_d}\frac{\lambda_D}{\ell_D} \\
i_F &= -\frac{L_{MD}}{\ell_F}\frac{\lambda_d}{\ell_d} + \left(1 - \frac{L_{MD}}{\ell_F}\right)\frac{\lambda_F}{\ell_F} - \frac{L_{MD}}{\ell_F}\frac{\lambda_D}{\ell_D} \\
i_D &= -\frac{L_{MD}}{\ell_D}\frac{\lambda_d}{\ell_d} - \frac{L_{MD}}{\ell_D}\frac{\lambda_F}{\ell_F} + \left(1 - \frac{L_{MD}}{\ell_D}\right)\frac{\lambda_D}{\ell_D}
\end{aligned} \tag{4.134}$$

These currents are substituted in the d axis voltage equations of (4.36) to get

$$
\begin{aligned}
\dot{\lambda}_d &= -r\left(1 - \frac{L_{MD}}{\ell_d}\right)\frac{\lambda_d}{\ell_d} + r\,\frac{L_{MD}}{\ell_d}\frac{\lambda_F}{\ell_F} + r\,\frac{L_{MD}}{\ell_d}\frac{\lambda_D}{\ell_D} - \omega\lambda_q - v_d \\
\dot{\lambda}_F &= r_F\,\frac{L_{MD}}{\ell_F}\frac{\lambda_d}{\ell_d} - r_F\left(1 - \frac{L_{MD}}{\ell_F}\right)\frac{\lambda_F}{\ell_F} + r_F\,\frac{L_{MD}}{\ell_F}\frac{\lambda_D}{\ell_D} + v_F \\
\dot{\lambda}_D &= r_D\,\frac{L_{MD}}{\ell_D}\frac{\lambda_d}{\ell_d} + r_D\,\frac{L_{MD}}{\ell_D}\frac{\lambda_F}{\ell_F} - r_D\left(1 - \frac{L_{MD}}{\ell_D}\right)\frac{\lambda_D}{\ell_D}
\end{aligned}
\tag{4.135}
$$

Similarly, the q axis equations are

$$
\begin{aligned}
\dot{\lambda}_q &= -\,r\left(1 - \frac{L_{MQ}}{\ell_q}\right)\frac{\lambda_q}{\ell_q} + r\,\frac{L_{MQ}}{\ell_q}\frac{\lambda_Q}{\ell_Q} + \omega\lambda_d - v_q \\
\dot{\lambda}_Q &= r_Q\,\frac{L_{MQ}}{\ell_Q}\frac{\lambda_q}{\ell_q} - r_Q\left(1 - \frac{L_{MQ}}{\ell_Q}\right)\frac{\lambda_Q}{\ell_Q}
\end{aligned}
\tag{4.136}
$$

and the equation for the electrical torque is given by

$$
T_{e\phi} = \lambda_d\lambda_q\left(\frac{L_{MD} - L_{MQ}}{\ell_d^2}\right) - \lambda_d\lambda_Q\,\frac{L_{MQ}}{\ell_q\ell_Q} + \lambda_q\lambda_F\,\frac{L_{MD}}{\ell_d\ell_F} + \lambda_q\lambda_D\,\frac{L_{MD}}{\ell_d\ell_D}
\tag{4.137}
$$

The state-space model now becomes

$$
\begin{bmatrix} \dot{\lambda}_d \\ \dot{\lambda}_F \\ \dot{\lambda}_D \\ \hline \dot{\lambda}_q \\ \dot{\lambda}_Q \\ \hline \dot{\omega} \\ \dot{\delta} \end{bmatrix}
=
\left[\begin{array}{ccc|cc|cc}
\overset{\lambda_d}{-\frac{r}{\ell_d}\left(1 - \frac{L_{MD}}{\ell_d}\right)} & \overset{\lambda_F}{\frac{r}{\ell_d}\frac{L_{MD}}{\ell_F}} & \overset{\lambda_D}{\frac{r}{\ell_d}\frac{L_{MD}}{\ell_D}} & \overset{\lambda_q}{-\omega} & \overset{\lambda_Q}{0} & \overset{\omega}{0} & \overset{\delta}{0} \\
\frac{r_F}{\ell_F}\frac{L_{MD}}{\ell_d} & -\frac{r_F}{\ell_F}\left(1 - \frac{L_{MD}}{\ell_F}\right) & \frac{r_F}{\ell_F}\frac{L_{MD}}{\ell_D} & 0 & 0 & 0 & 0 \\
\frac{r_D}{\ell_D}\frac{L_{MD}}{\ell_d} & \frac{r_D}{\ell_D}\frac{L_{MD}}{\ell_F} & -\frac{r_D}{\ell_D}\left(1 - \frac{L_{MD}}{\ell_D}\right) & 0 & 0 & 0 & 0 \\
\hline
\omega & 0 & 0 & -\frac{r}{\ell_q}\left(1 - \frac{L_{MQ}}{\ell_q}\right) & \frac{r}{\ell_q}\frac{L_{MQ}}{\ell_Q} & 0 & 0 \\
0 & 0 & 0 & \frac{r_Q}{\ell_Q}\frac{L_{MQ}}{\ell_q} & -\frac{r_Q}{\ell_Q}\left(1 - \frac{L_{MQ}}{\ell_Q}\right) & 0 & 0 \\
\hline
-\frac{L_{MD}}{3\tau_j\ell_d^2}\lambda_q & -\frac{L_{MD}}{3\tau_j\ell_d\ell_F}\lambda_q & -\frac{L_{MD}}{3\tau_j\ell_d\ell_D}\lambda_q & \frac{L_{MQ}}{3\tau_j\ell_q^2}\lambda_d & \frac{L_{MQ}}{3\tau_j\ell_q\ell_Q}\lambda_d & -\frac{D}{\tau_j} & 0 \\
0 & 0 & 0 & 0 & 0 & 1 & 0
\end{array}\right]
\begin{bmatrix} \lambda_d \\ \lambda_F \\ \lambda_D \\ \hline \lambda_q \\ \lambda_Q \\ \hline \omega \\ \delta \end{bmatrix}
+
\begin{bmatrix} -v_d \\ v_F \\ 0 \\ \hline -v_q \\ 0 \\ \hline \frac{T_m}{\tau_j} \\ -1 \end{bmatrix}
\tag{4.138}
$$

The system described by (4.138) is in the form $\dot{\mathbf{x}} = \mathbf{f}(\mathbf{x}, \mathbf{u}, t)$. Again the description of the system is not complete since v_d and v_q are functions of the currents and will depend on the external load connections. The 7×7 matrix on the right side of (4.138) contains state variables in several terms, and this matrix form of the equation is *not* an appropriate form for solution. It does, however, serve to illustrate the nonlinear nature of the system.

Example 4.3

Repeat Example 4.2 for the flux linkage model.

Solution

From the data of Example 4.1:

$$\ell_d = \ell_q = 0.150 \quad \text{pu}$$
$$\ell_F = 1.651 - 1.550 = 0.101 \quad \text{pu}$$
$$\ell_D = 1.605 - 1.550 = 0.055 \quad \text{pu}$$
$$\ell_Q = 1.526 - 1.490 = 0.036 \quad \text{pu}$$
$$\frac{1}{L_{MD}} = \frac{1}{1.55} + \frac{1}{0.15} + \frac{1}{0.101} + \frac{1}{0.055} = 35.2381 \quad \text{pu}$$
$$L_{MD} = 0.028378 \quad \text{pu}$$
$$\frac{1}{L_{MQ}} = \frac{1}{1.49} + \frac{1}{0.15} + \frac{1}{0.036} = 35.2381 \quad \text{pu}$$
$$L_{MQ} = 0.028378 \quad \text{pu}$$
$$\frac{r}{\ell_d}\left(1 - \frac{L_{MD}}{\ell_d}\right) = 0.005927$$
$$\frac{r}{\ell_d}\frac{L_{MD}}{\ell_F} = 0.002049$$
$$\frac{r}{\ell_d}\frac{L_{MD}}{\ell_D} = 0.003743$$
$$\frac{r_D}{\ell_D}\frac{L_{MD}}{\ell_d} = 0.044720$$
$$\frac{r_D}{\ell_D}\frac{L_{MD}}{\ell_F} = 0.066282$$
$$\frac{r_D}{\ell_D}\left(1 - \frac{L_{MD}}{\ell_D}\right) = 0.115330$$
$$\frac{r}{\ell_q}\left(1 - \frac{L_{MQ}}{\ell_q}\right) = 0.005928$$
$$\frac{r_F}{\ell_F}\frac{L_{MD}}{\ell_d} = 0.001387$$
$$\frac{r_F}{\ell_F}\left(1 - \frac{L_{MD}}{\ell_F}\right) = 0.005278$$
$$\frac{r_F}{\ell_F}\frac{L_{MD}}{\ell_D} = 0.003756$$
$$\frac{r}{\ell_q}\frac{L_{MQ}}{\ell_Q} = 0.005789$$
$$\frac{r_Q}{\ell_Q}\frac{L_{MQ}}{\ell_q} = 0.286058$$
$$\frac{r_Q}{\ell_Q}\left(1 - \frac{L_{MQ}}{\ell_Q}\right) = 0.308485$$
$$\frac{L_{MD}}{3\tau_j \ell_d^2} = 0.000706$$
$$\frac{L_{MD}}{3\tau_j \ell_d \ell_F} = 0.001046$$
$$\frac{L_{MD}}{3\tau_j \ell_d \ell_D} = 0.001910$$
$$\frac{L_{MQ}}{3\tau_j \ell_q \ell_Q} = 0.002954$$
$$\frac{L_{MQ}}{3\tau_j \ell_q^2} = 0.000705$$

and we get for the state-space equation

$$\begin{bmatrix} \dot\lambda_d \\ \dot\lambda_F \\ \dot\lambda_D \\ \dot\lambda_q \\ \dot\lambda_Q \\ \dot\omega \end{bmatrix} = 10^{-3} \begin{bmatrix} -5.927 & 2.050 & 3.743 & -\omega & 0 \\ 1.388 & -5.278 & 3.756 & 0 & 0 \\ 44.720 & 66.282 & -115.330 & 0 & 0 \\ \omega & 0 & 0 & -5.928 & 5.789 \\ 0 & 0 & 0 & 284.854 & -313.530 \\ -0.706\lambda_q & -1.046\lambda_q & -1.910\lambda_q & 0.705\lambda_d & 2.954\lambda_d \end{bmatrix} \begin{bmatrix} \lambda_d \\ \lambda_F \\ \lambda_D \\ \lambda_q \\ \lambda_Q \\ \omega \end{bmatrix} + \begin{bmatrix} -v_d \\ v_F \\ 0 \\ -v_q \\ 0 \\ 0.00056T_m \end{bmatrix}$$

4.12.4 Treatment of saturation

The flux linkage state-space model is convenient for considering the effect of saturation because all the terms in the state equations (4.126)–(4.133) are linear except for the magnetizing flux linkages λ_{AD} and λ_{AQ}. These are affected by saturation of the mutual inductances L_{AD} and L_{AQ}, and only these terms need to be corrected for saturation. In the simulation of the machine, either by digital or analog computer, this can be accom-

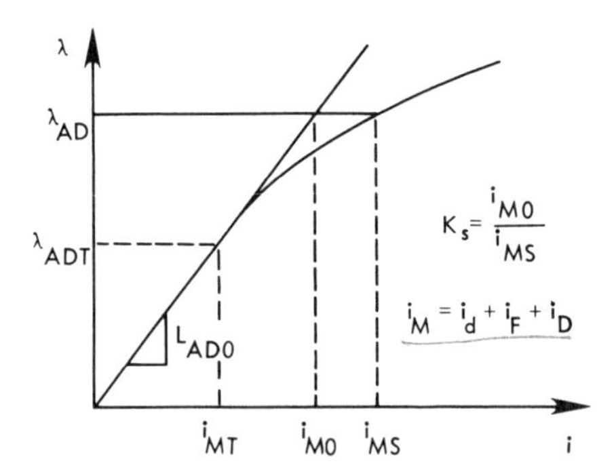

Fig. 4.7 Saturation curve for λ_{AD}.

plished by computing a saturation function to adjust (4.120) and (4.121) at all times to reflect the state of the mutual inductances. As a practical matter, the q axis inductance L_{AQ} seldom saturates, so it is usually necessary to adjust only λ_{AD} for saturation.

The procedure for including the magnetic circuit saturation is given below [18]. Let the unsaturated values of the magnetizing inductances be L_{AD0} and L_{AQ0}. The computations for saturated values of these inductances follow.

For salient pole machines,

$$L_{AD} = K_s L_{AD0} \qquad L_{AQ} = L_{AQ0} \qquad K_s = f(\lambda_{AD}) \tag{4.139}$$

where K_s is a saturation factor determined from the magnetization curve of the machine.

For a round-rotor machine, we compute, according to [16]

$$\begin{aligned} L_{AD} &= K_s L_{AD0} \qquad & L_{AQ} &= K_s L_{AQ0} \\ K_s &= f(\lambda) \qquad & \lambda &= (\lambda_{AD}^2 + \lambda_{AQ}^2)^{1/2} \end{aligned} \tag{4.140}$$

To determine K_s for the d axis in (4.139), the following procedure is suggested. Let the magnetizing current, which is the sum of $i_d + i_F + i_D$, be i_M. The relation between λ_{AD} and i_M is given by the saturation curve shown in Figure 4.7. For a given value of λ_{AD} the unsaturated magnetizing current is i_{M0}, corresponding to L_{AD0}, while the saturated value is i_{MS}. The saturation function K_s is a function of this magnetizing current, which in turn is a function of λ_{AD}.

To calculate the saturated magnetizing current i_{MS}, the current increment needed to satisfy saturation, $i_{M\Delta} = i_{MS} - i_{M0}$, is first calculated. Note that saturation begins at the *threshold* value λ_{ADT} corresponding to a magnetizing current i_{MT}. For flux linkages greater than λ_{ADT} the current $i_{M\Delta}$ increases monotonically in an almost exponential way. Thus we may write approximately

$$i_{M\Delta} = A_s \exp\left[B_s(\lambda_{AD} - \lambda_{ADT})\right] \qquad \lambda_{AD} > \lambda_{ADT} \tag{4.141}$$

where A_s and B_s are constants to be determined from the actual saturation curve.

Knowing $i_{M\Delta}$ for a given value of λ_{AD}, the value of i_{MS} is calculated, and hence K_s is determined. The solution is obtained by an iterative process so that the relation $\lambda_{AD} K_s(\lambda_{AD}) = L_{AD0} i_{MS}$ is satisfied.

4.13 Load Equations

From (4.103) and (4.138) we have a set of equations for each machine in the form

$$\dot{\mathbf{x}} = \mathbf{f}(\mathbf{x}, \mathbf{v}, T_m) \tag{4.142}$$

where $\mathbf{x}$ is a vector of order seven (five currents, ω and δ for the current model, or five flux linkages, ω and δ for the flux linkage model), and $\mathbf{v}$ is a vector of voltages that includes v_d, v_q, and v_F.

Assuming that v_F and T_m are known, the set (4.142) does not completely describe the synchronous machine since there are two additional variables v_d and v_q appearing in the equations. Therefore two additional equations are needed to relate v_d and v_q to the state variables. These are auxiliary equations, which may or may not increase the order of the system depending upon whether the relations obtained are algebraic equations or differential equations and whether new variables are introduced. To obtain equations for v_d and v_q in terms of the state variables, the terminal conditions of the machine must be known. In other words, equations describing the load are required.

There are a number of ways of representing the electrical load on a synchronous generator. For example, we could consider the load to be constant impedance, constant power, constant current, or some composite of all three. For the present we require a load representation that will illustrate the constraints between the generator voltages, currents, and angular velocity. These constraints are found by solving the network, including loads, given the machine terminal voltages. For illustrative purposes here, the load constraint is satisfied by the simple one machine–infinite bus problem illustrated below.

4.13.1 Synchronous machine connected to an infinite bus

Consider the system of Figure 4.8 where a synchronous machine is connected to an infinite bus through a transmission line having resistance R_e and inductance L_e. The voltages and current for phase a only are shown, assuming no mutual coupling between phases. By inspection of Figure 4.8 we can write $v_a = v_{\infty a} + R_e i_a + L_e \dot{i}_a$ or

$$\begin{bmatrix} v_a \\ v_b \\ v_c \end{bmatrix} = \begin{bmatrix} v_{\infty a} \\ v_{\infty b} \\ v_{\infty c} \end{bmatrix} + R_e \mathbf{U} \begin{bmatrix} i_a \\ i_b \\ i_c \end{bmatrix} + L_e \mathbf{U} \begin{bmatrix} \dot{i}_a \\ \dot{i}_b \\ \dot{i}_c \end{bmatrix} \tag{4.143}$$

In matrix notation (4.143) becomes

$$\mathbf{v}_{abc} = \mathbf{v}_{\infty abc} + R_e \mathbf{U} \mathbf{i}_{abc} + L_e \mathbf{U} \dot{\mathbf{i}}_{abc} \tag{4.144}$$

which we transform to the 0-d-q frame of reference by Park's transformation:

$$\mathbf{v}_{0dq} = \mathbf{P}\mathbf{v}_{abc} = \mathbf{P}\mathbf{v}_{\infty abc} + R_e \mathbf{i}_{0dq} + L_e \mathbf{P} \dot{\mathbf{i}}_{abc} \quad \text{V or pu} \tag{4.145}$$

The first term on the right side we may call $\mathbf{v}_{\infty 0dq}$ and may determine its value by assuming that $\mathbf{v}_{\infty abc}$ is a set of *balanced* three-phase voltages, or

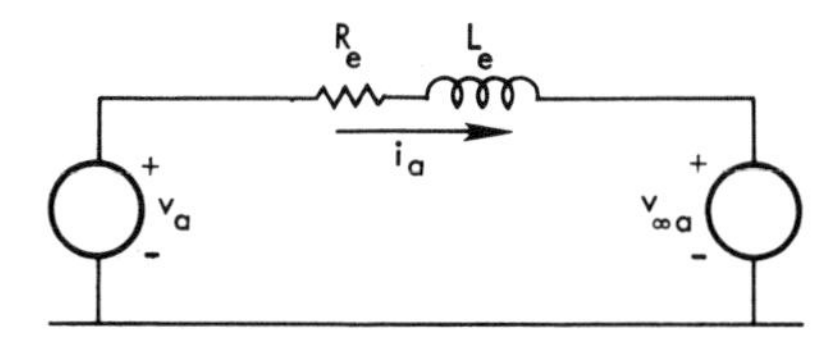

Fig. 4.8 Synchronous generator loaded by an infinite bus.

$$\mathbf{v}_{\infty abc} = \sqrt{2}V_\infty \begin{bmatrix} \cos(\omega_R t + \alpha) \\ \cos(\omega_R t + \alpha - 120°) \\ \cos(\omega_R t + \alpha + 120°) \end{bmatrix} \tag{4.146}$$

where V_∞ is the magnitude of the rms phase voltage. Using the identities in Appendix A and using $\theta = \omega_R t + \delta + \pi/2$, we can show that

$$\mathbf{v}_{\infty 0dq} = \mathbf{P}\mathbf{v}_{\infty abc} = V_\infty\sqrt{3} \begin{bmatrix} 0 \\ -\sin(\delta - \alpha) \\ \cos(\delta - \alpha) \end{bmatrix} \tag{4.147}$$

The last term on the right side of (4.145) may be computed as follows. From the definition of Park's transformation $\mathbf{i}_{0dq} = \mathbf{P}\mathbf{i}_{abc}$, we compute the derivative $\dot{\mathbf{i}}_{0dq} = \mathbf{P}\dot{\mathbf{i}}_{abc} + \dot{\mathbf{P}}\mathbf{i}_{abc}$. Thus

$$\mathbf{P}\dot{\mathbf{i}}_{abc} = \dot{\mathbf{i}}_{0dq} - \dot{\mathbf{P}}\mathbf{i}_{abc} = \dot{\mathbf{i}}_{0dq} - \dot{\mathbf{P}}\mathbf{P}^{-1}\mathbf{i}_{0dq} \tag{4.148}$$

where the quantity $\dot{\mathbf{P}}\mathbf{P}^{-1}$ is known from (4.32). Thus (4.145) may be written as

$$\mathbf{v}_{0dq} = V_\infty\sqrt{3} \begin{bmatrix} 0 \\ -\sin(\delta - \alpha) \\ \cos(\delta - \alpha) \end{bmatrix} + R_e\mathbf{i}_{0dq} + L_e\dot{\mathbf{i}}_{0dq} - \omega L_e \begin{bmatrix} 0 \\ -i_q \\ i_d \end{bmatrix} \text{ V or pu} \tag{4.149}$$

which gives the constraint between the generator terminal voltage $\mathbf{v}_{0dq}$ and the generator current $\mathbf{i}_{0dq}$ for a given torque angle δ. Note that (4.149) is exactly the same whether in MKS units or pu due to our choice of $\mathbf{P}$ and base quantities. Note also that there are two nonlinearities in (4.149). The first is due to the speed voltage term, the $\omega L_e i$ product. There is also a nonlinearity in the trigonometric functions of the first term.

The angle δ is related to the speed by $\dot{\delta} = \omega - 1$ pu or, in radians,

$$\delta = \delta_0 + \int_{t_0}^{t} (\omega - \omega_R)\, dt \tag{4.150}$$

Thus even this simple load representation introduces new nonlinearities, but the order of the system remains at seven.

4.13.2 Current model

Incorporating (4.149) into system (4.75), we may write

$$-\mathbf{L}\dot{\mathbf{i}} = (\mathbf{R} + \omega\mathbf{N})\mathbf{i} + \begin{bmatrix} -K\sin\gamma + R_e i_d + L_e \dot{i}_d + \omega L_e i_q \\ -v_F \\ 0 \\ K\cos\gamma + R_e i_q + L_e \dot{i}_q - \omega L_e i_d \\ 0 \end{bmatrix} \tag{4.151}$$

where $K = \sqrt{3}V_\infty$ and $\gamma = \delta - \alpha$. Now let

$$\hat{R} = r + R_e \qquad \hat{L}_d = L_d + L_e \qquad \hat{L}_q = L_q + L_e \tag{4.152}$$

Using (4.152), we may replace the r, L_d, and L_q terms in $\mathbf{L}$, $\mathbf{R}$, and $\mathbf{N}$ by $\hat{R}$, $\hat{L}_d$, and $\hat{L}_q$ to obtain the new matrices $\hat{\mathbf{L}}$ and $(\hat{\mathbf{R}} + \omega\hat{\mathbf{N}})$. Thus

$$-\hat{\mathbf{L}}\dot{\mathbf{i}} = (\hat{\mathbf{R}} + \omega\hat{\mathbf{N}})\mathbf{i} + \begin{bmatrix} -K\sin\gamma \\ -v_F \\ 0 \\ K\cos\gamma \\ 0 \end{bmatrix} \tag{4.153}$$

Premultiplying by $-\hat{\mathbf{L}}^{-1}$ and adding the equations for $\dot{\omega}$ and $\dot{\delta}$,

$$\begin{bmatrix} \dot{i}_d \\ \dot{i}_F \\ \dot{i}_D \\ \dot{i}_q \\ \dot{i}_Q \\ \hline \dot{\omega} \\ \dot{\delta} \end{bmatrix} = \left[\begin{array}{ccccc|cc} & & -\hat{\mathbf{L}}^{-1}(\hat{\mathbf{R}} + \omega\hat{\mathbf{N}}) & & & \multicolumn{2}{c}{\mathbf{0}} \\ \hline -\dfrac{L_d i_q}{3\tau_j} & -\dfrac{kM_F i_q}{3\tau_j} & -\dfrac{kM_D i_q}{3\tau_j} & \dfrac{L_q i_d}{3\tau_j} & \dfrac{kM_Q i_d}{3\tau_j} & -\dfrac{D}{\tau_j} & 0 \\ 0 & 0 & 0 & 0 & 0 & 1 & 0 \end{array}\right] \begin{bmatrix} i_d \\ i_F \\ i_D \\ i_q \\ i_Q \\ \hline \omega \\ \delta \end{bmatrix} + \left[\begin{array}{c|cc} -\hat{\mathbf{L}}^{-1} & \multicolumn{2}{c}{\mathbf{0}} \\ \hline \mathbf{0} & 1 & 0 \\ & 0 & 1 \end{array}\right] \begin{bmatrix} -K\sin\gamma \\ -v_F \\ 0 \\ K\cos\gamma \\ 0 \\ \hline \dfrac{T_m}{\tau_j} \\ -1 \end{bmatrix} \tag{4.154}$$

The system described by (4.154) is now in the form of (4.37), namely, $\dot{\mathbf{x}} = \mathbf{f}(\mathbf{x}, \mathbf{u}, t)$, where $\mathbf{x}^t = [i_d\, i_F\, i_D\, i_q\, i_Q\, \omega\, \delta]$.

The function $\mathbf{f}$ is a nonlinear function of the state variables and t, and $\mathbf{u}$ contains the system driving functions, which are v_F and T_m. The loading effect of the transmission line is incorporated in the matrices $\hat{\mathbf{R}}$, $\hat{\mathbf{L}}$, and $\hat{\mathbf{N}}$. The infinite bus voltage V_∞ appears in the terms $K \sin \gamma$ and $K \cos \gamma$. Note also that these latter terms are not driving functions, but rather nonlinear functions of the state variable δ.

Because the system (4.154) is nonlinear, determination of its stability depends upon finding a suitable Liapunov function or some equivalent method. This is explored in greater depth in Volume 2.

4.13.3 The flux linkage model

From (4.149) and substituting for i_d and i_q in terms of flux linkages (see Section 4.12.3),

$$\begin{aligned} v_d = &-\sqrt{3}V_\infty \sin(\delta - \alpha) + \frac{R_e}{\ell_d}\left(1 - \frac{L_{MD}}{\ell_d}\right)\lambda_d - \frac{R_e L_{MD}}{\ell_d \ell_F}\lambda_F - \frac{R_e L_{MD}}{\ell_d \ell_D}\lambda_D \\ &+ \frac{\omega L_e}{\ell_q}\left(1 - \frac{L_{MQ}}{\ell_q}\right)\lambda_q - \frac{\omega L_e L_{MQ}}{\ell_q \ell_Q}\lambda_Q + \frac{L_e}{\ell_d}\left(1 - \frac{L_{MD}}{\ell_d}\right)\dot{\lambda}_d - \frac{L_e L_{MD}}{\ell_d \ell_F}\dot{\lambda}_F - \frac{L_e L_{MD}}{\ell_d \ell_D}\dot{\lambda}_D \end{aligned} \tag{4.155}$$

$$
\begin{aligned}
v_q = {} & \sqrt{3}V_\infty \cos(\delta - \alpha) + \frac{R_e}{\ell_q}\left(1 - \frac{L_{MQ}}{\ell_q}\right)\lambda_q - \frac{R_e L_{MQ}}{\ell_q \ell_Q}\lambda_Q - \frac{\omega L_e}{\ell_d}\left(1 - \frac{L_{MD}}{\ell_d}\right)\lambda_d \\
& + \frac{\omega L_e}{\ell_d \ell_F}\lambda_F + \frac{\omega L_e}{\ell_d \ell_D}\lambda_D + \frac{L_e}{\ell_q}\left(1 - \frac{L_{MQ}}{\ell_q}\right)\dot{\lambda}_q - \frac{L_e L_{MQ}}{\ell_q \ell_Q}\dot{\lambda}_Q
\end{aligned}
\tag{4.156}
$$

Combining (4.155) with (4.135),

$$
\begin{aligned}
& \left[1 + \frac{L_e}{\ell_d}\left(1 - \frac{L_{MD}}{\ell_d}\right)\right]\dot{\lambda}_d - \frac{L_e L_{MD}}{\ell_d \ell_F}\dot{\lambda}_F - \frac{L_e L_{MD}}{\ell_d \ell_D}\dot{\lambda}_D = \frac{\hat{R}}{\ell_d}\left(1 - \frac{L_{MD}}{\ell_d}\right)\lambda_d + \frac{\hat{R} L_{MD}}{\ell_d \ell_F}\lambda_F \\
& \quad + \frac{\hat{R} L_{MD}}{\ell_d \ell_D}\lambda_D - \omega\left[1 + \frac{L_e}{\ell_q}\left(1 - \frac{L_{MQ}}{\ell_q}\right)\right]\lambda_q + \frac{\omega L_e L_{MQ}}{\ell_q \ell_Q}\lambda_Q + \sqrt{3}V_\infty \sin(\delta - \alpha)
\end{aligned}
\tag{4.157}
$$

Similarly, we combine (4.156) with (4.136) to get

$$
\begin{aligned}
& \left[1 + \frac{L_e}{\ell_q}\left(1 - \frac{L_{MQ}}{\ell_q}\right)\right]\dot{\lambda}_q - \frac{L_e L_{MQ}}{\ell_q \ell_Q}\dot{\lambda}_Q = -\frac{\hat{R}}{\ell_q}\left(1 - \frac{L_{MQ}}{\ell_q}\right)\lambda_q + \frac{\hat{R} L_{MQ}}{\ell_q \ell_Q}\lambda_Q \\
& \quad + \omega\left[1 + \frac{L_e}{\ell_d}\left(1 - \frac{L_{MD}}{\ell_d}\right)\right]\lambda_d - \frac{\omega L_e L_{MD}}{\ell_d \ell_F}\lambda_F - \frac{\omega L_e L_{MD}}{\ell_d \ell_D}\lambda_D - \sqrt{3}V_\infty \cos(\delta - \alpha)
\end{aligned}
\tag{4.158}
$$

Equations (4.157) and (4.158) replace the first and fourth rows in (4.138) to give the complete state-space model. The resulting equation is of the form

$$
\mathbf{T}\dot{\mathbf{x}} = \mathbf{C}\mathbf{x} + \mathbf{D} \tag{4.159}
$$

where $\mathbf{x}' = [\lambda_d \ \lambda_F \ \lambda_D \ \lambda_q \ \lambda_Q \ \omega \ \delta]$,

$$
\mathbf{T} = \left[\begin{array}{ccc|cc|cc}
1 + \frac{L_e}{\ell_d}\left(1 - \frac{L_{MD}}{\ell_d}\right) & -\frac{L_e L_{MD}}{\ell_d \ell_F} & -\frac{L_e L_{MD}}{\ell_d \ell_D} & & & & \\
0 & 1 & 0 & & \mathbf{0} & & \mathbf{0} \\
0 & 0 & 1 & & & & \\
\hline
 & & & 1 + \frac{L_e}{\ell_q}\left(1 - \frac{L_{MQ}}{\ell_q}\right) & -\frac{L_e L_{MQ}}{\ell_q \ell_Q} & & \\
 & \mathbf{0} & & 0 & 1 & & \mathbf{0} \\
\hline
 & \mathbf{0} & & & & 1 & 0 \\
 & & & & \mathbf{0} & 0 & 1
\end{array}\right]
\tag{4.160}
$$

and the matrix $\mathbf{C}$ is given by

$$\mathbf{C} = \left[\begin{array}{ccc|cc|cc}
-\frac{\hat{R}}{\ell_d}\left(1 - \frac{L_{MD}}{\ell_d}\right) & \frac{\hat{R}L_{MD}}{\ell_d \ell_F} & \frac{\hat{R}L_{MD}}{\ell_d \ell_D} & -\omega\left[1 + \frac{L_e}{\ell_q}\left(1 - \frac{L_{MQ}}{\ell_q}\right)\right] & \frac{\omega L_e L_{MQ}}{\ell_q \ell_Q} & 0 & 0 \\
\frac{r_F L_{MD}}{\ell_F \ell_d} & -\frac{r_F}{\ell_F}\left(1 - \frac{L_{MD}}{\ell_F}\right) & \frac{r_F L_{MD}}{\ell_F \ell_D} & 0 & 0 & 0 & 0 \\
\frac{r_D L_{MD}}{\ell_D \ell_d} & \frac{r_D L_{MD}}{\ell_D \ell_F} & -\frac{r_D}{\ell_D}\left(1 - \frac{L_{MD}}{\ell_D}\right) & 0 & 0 & 0 & 0 \\
\hline
\omega\left[1 + \frac{L_e}{\ell_d}\left(1 - \frac{L_{MD}}{\ell_d}\right)\right] & -\frac{\omega L_e L_{MD}}{\ell_d \ell_F} & -\frac{\omega L_e L_{MD}}{\ell_d \ell_D} & -\frac{\hat{R}}{\ell_q}\left(1 - \frac{L_{MQ}}{\ell_q}\right) & \frac{\hat{R}L_{MQ}}{\ell_q \ell_Q} & 0 & 0 \\
0 & 0 & 0 & \frac{r_Q L_{MQ}}{\ell_Q \ell_q} & -\frac{r_Q}{\ell_Q}\left(1 - \frac{L_{MQ}}{\ell_Q}\right) & 0 & 0 \\
\hline
-\frac{L_{MD}}{3\tau_j \ell_d^2}\lambda_q & -\frac{L_{MD}}{3\tau_j \ell_d \ell_F}\lambda_q & \frac{L_{MD}}{3\tau_j \ell_d \ell_D}\lambda_q & \frac{L_{MQ}}{3\tau_j \ell_d^2}\lambda_d & \frac{L_{MQ}}{3\tau_j \ell_q \ell_Q}\lambda_d & -\frac{D}{\tau_j} & 0 \\
0 & 0 & 0 & 0 & 0 & 1 & 0
\end{array}\right] \tag{4.161}$$

and

$$\mathbf{D} = \begin{bmatrix} \sqrt{3}V_\infty \sin(\delta - \alpha) \\ v_F \\ 0 \\ -\sqrt{3}V_\infty \cos(\delta - \alpha) \\ 0 \\ T_m/\tau_j \\ -1 \end{bmatrix} \tag{4.162}$$

If $\mathbf{T}^{-1}$ exists, premultiply (4.159) by $\mathbf{T}^{-1}$ to get

$$\dot{\mathbf{x}} = \mathbf{T}^{-1}\mathbf{C}\mathbf{x} + \mathbf{T}^{-1}\mathbf{D} \tag{4.163}$$

Equation (4.163) is in the desired form, i.e., in the form of $\dot{\mathbf{x}} = \mathbf{f}(\mathbf{x}, \mathbf{u}, t)$ and completely describes the system. It contains two types of nonlinearities, product nonlinearities and trigonometric functions.

Example 4.4

Extend Examples 4.2 and 4.3 to include the effect of the transmission line and torque equations. The line constants are $R_e = 0$, $L_e = 0.4$ pu, $\tau_j = 2H\omega_R = 1786.94$ rad. The infinite bus voltage constant K and the damping torque coefficient D are left unspecified.

Solution

$$R = r + R_e = 0.001096 \qquad \hat{L}_d = L_d + L_e = 2.10 \qquad \hat{L}_q = L_q + L_e = 2.04$$

Then

$$\hat{\mathbf{R}} + \omega\hat{\mathbf{N}} = \left[\begin{array}{ccc|cc} 0.0011 & 0 & 0 & 2.04\omega & 1.49\omega \\ 0 & 0.00074 & 0 & 0 & 0 \\ 0 & 0 & 0.0131 & 0 & 0 \\ \hline -2.10\omega & -1.55\omega & -1.55\omega & 0.0011 & 0 \\ 0 & 0 & 0 & 0 & 0.0540 \end{array}\right]$$

$$\hat{\mathbf{L}} = \left[\begin{array}{ccc|cc} 2.100 & 1.550 & 1.550 & 0 & 0 \\ 1.550 & 1.651 & 1.550 & 0 & 0 \\ 1.550 & 1.550 & 1.605 & 0 & 0 \\ \hline 0 & 0 & 0 & 2.040 & 1.490 \\ 0 & 0 & 0 & 1.490 & 1.526 \end{array}\right]$$

By digital computer we find

$$\hat{\mathbf{L}}^{-1} = \left[\begin{array}{ccc|cc} 1.709 & -0.591 & -1.080 & & \\ -0.591 & 6.668 & -5.867 & \multicolumn{2}{c}{\mathbf{0}} \\ -1.080 & -5.867 & -7.330 & & \\ \hline & \mathbf{0} & & 1.710 & -1.669 \\ & & & -1.669 & 2.286 \end{array}\right]$$

Then

$$\hat{\mathbf{L}}^{-1}(\hat{\mathbf{R}} + \omega\hat{\mathbf{N}}) = \left[\begin{array}{ccc|cc} 0.00187 & -0.00044 & -0.0141 & 3.487\omega & 2.547\omega \\ -0.00065 & 0.00495 & -0.0769 & -1.206\omega & 0.881\omega \\ -0.00118 & -0.00436 & 0.0960 & -2.202\omega & -1.609\omega \\ \hline -3.590\omega & -2.650\omega & -2.650\omega & 0.00187 & -0.09007 \\ 3.506\omega & 2.588\omega & 2.588\omega & -0.00183 & 0.12332 \end{array}\right]$$

and we compute

$$\hat{\mathbf{L}}^{-1}\begin{bmatrix} -K\sin\gamma \\ -v_F \\ 0 \\ K\cos\gamma \\ 0 \end{bmatrix} = \begin{bmatrix} -1.71K\sin\gamma + 0.589v_F \\ 0.589K\sin\gamma - 6.69v_F \\ 1.08K\sin\gamma + 5.89v_F \\ 1.71K\cos\gamma \\ -1.67K\cos\gamma \end{bmatrix}$$

Therefore the state-space current model is given by

$$
\begin{bmatrix} \dot{i}_d \\ \dot{i}_F \\ \dot{i}_D \\ \dot{i}_q \\ \dot{i}_Q \\ \hline \dot{\omega} \\ \dot{\delta} \end{bmatrix} = \left[\begin{array}{ccccc|cc} 0.00193 & -0.00037 & 0.0143 & -3.49\omega & 2.55\omega & 0 & 0 \\ 0.00067 & -0.00496 & 0.0778 & 1.20\omega & 0.878\omega & 0 & 0 \\ 0.00122 & 0.00437 & -0.0971 & 2.21\omega & 1.61\omega & 0 & 0 \\ 3.59\omega & 2.65\omega & 2.65\omega & -0.0019 & 0.0901 & 0 & 0 \\ -3.50\omega & -2.59\omega & -2.59\omega & 0.0018 & -0.11234 & 0 & 0 \\ \hline -0.00095 i_q & -0.00087 i_q & -0.00087 i_q & 0.00092 i_d & 0.000833 i_d & -0.000559 D & 0 \\ 0 & 0 & 0 & 0 & 0 & 1 & 0 \end{array}\right] \begin{bmatrix} i_d \\ i_F \\ i_D \\ i_q \\ i_Q \\ \hline \omega \\ \delta \end{bmatrix}
$$

$$
+ \begin{bmatrix} 0.171 K \sin\gamma - 0.589 v_F \\ 0.589 K \sin\gamma - 6.69 v_F \\ -1.08 K \sin\gamma - 5.98 v_F \\ 1.71 K \cos\gamma \\ 1.67 K \cos\gamma \\ \hline 0.000559\, T_m \\ -1 \end{bmatrix}
$$

The flux linkage model is of the form $\mathbf{T}\dot{\lambda} = \mathbf{C}\lambda + \mathbf{D}$, where $\mathbf{T}$, $\mathbf{C}$, and $\mathbf{D}$ are given by (4.159)–(4.162). Substituting,

$$
\mathbf{T} = \left[\begin{array}{ccc|cc|cc} 3.1622 & -0.7478 & -1.3656 & 0 & 0 & 0 & 0 \\ 0 & 1.0 & 0 & 0 & 0 & 0 & 0 \\ 0 & 0 & 1.0 & 0 & 0 & 0 & 0 \\ \hline 0 & 0 & 0 & 3.1625 & -2.1118 & 0 & 0 \\ 0 & 0 & 0 & 0 & 1.0 & 0 & 0 \\ \hline 0 & 0 & 0 & 0 & 0 & 1.0 & 0 \\ 0 & 0 & 0 & 0 & 0 & 0 & 1.0 \end{array}\right]
$$

$$
\mathbf{T}^{-1} = \left[\begin{array}{c|c|c} \begin{matrix} 0.3162 & 0.2365 & 0.4319 \\ 0 & 1.0 & 0 \\ 0 & 0 & 1.0 \end{matrix} & \mathbf{0} & \mathbf{0} \\ \hline \mathbf{0} & \begin{matrix} 0.3162 & 0.6678 \\ 0 & 1.0 \end{matrix} & \mathbf{0} \\ \hline \mathbf{0} & \mathbf{0} & \begin{matrix} 1 & 0 \\ 0 & 1 \end{matrix} \end{array}\right]
$$

The matrix $\mathbf{C}$ is mostly the same as that given in Example 4.3 except that the ω terms are modified.

$$
\mathbf{C} = \left[\begin{array}{ccc|cc|cc}
-5.927 & 2.050 & 3.743 & -3162\omega & 2112\omega & & \\
1.388 & -5.278 & 3.756 & 0 & 0 & \mathbf{0} & \\
44.720 & 66.282 & -115.330 & 0 & 0 & & \\
\hline
3162\omega & -747.7\omega & -1366\omega & -5.928 & 5.789 & \mathbf{0} & \\
0 & 0 & 0 & 284.854 & -313.530 & & \\
\hline
-0.7058\lambda_q & 1.046\lambda_q & -1.910\lambda_q & 0.705\lambda_d & 2.954\lambda_d & -0.5596D & 0 \\
0 & 0 & 0 & 0 & 0 & 1 & 0
\end{array}\right] 10^{-3}
$$

$$
\begin{bmatrix} \dot{\lambda}_d \\ \dot{\lambda}_F \\ \dot{\lambda}_D \\ \hline \dot{\lambda}_q \\ \dot{\lambda}_Q \\ \hline \dot{\omega} \\ \dot{\delta} \end{bmatrix}
= 10^{-3} \left[\begin{array}{ccc|cc|cc}
17.766 & 28.024 & -47.733 & 1000\omega & 667.8\omega & & \\
1.388 & -5.278 & 3.756 & 0 & 0 & \mathbf{0} & \\
44.720 & 66.282 & -115.330 & 0 & 0 & & \\
\hline
1000\omega & -236.4\omega & -431.8\omega & 188.337 & -207.529 & \mathbf{0} & \\
0 & 0 & 0 & 284.854 & -313.530 & & \\
\hline
-0.706\lambda_q & -1.046\lambda_q & -1.910\lambda_q & 0.705\lambda_d & 2.954\lambda_d & -0.5596D & 0 \\
0 & 0 & 0 & 0 & 0 & 1 & 0
\end{array}\right]
\begin{bmatrix} \lambda_d \\ \lambda_F \\ \lambda_D \\ \hline \lambda_q \\ \lambda_Q \\ \hline \omega \\ \delta \end{bmatrix}
+ \begin{bmatrix} 0.316\,K\sin\gamma + 0.229\,v_F \\ v_F \\ 0 \\ \hline -0.316\,K\cos\gamma \\ 0 \\ \hline 0.000559\,T_m \\ -1 \end{bmatrix}
$$

4.14 Subtransient and Transient Inductances and Time Constants

If all the rotor circuits are short circuited and balanced three-phase voltages are suddenly impressed upon the stator terminals, the flux linking the d axis circuit will depend *initially* on the subtransient inductances, and after a few cycles on the transient inductances.

Let the phase voltages suddenly applied to the stator be given by

$$
\begin{bmatrix} v_a \\ v_b \\ v_c \end{bmatrix} = \sqrt{2}\,V \begin{bmatrix} \cos\theta \\ \cos(\theta - 120) \\ \cos(\theta + 120) \end{bmatrix} u(t) \tag{4.164}
$$

where $u(t)$ is a unit step function and V is the rms phase voltage. Then from (4.7) we can show that

$$\begin{bmatrix} v_0 \\ v_d \\ v_q \end{bmatrix} = \begin{bmatrix} 0 \\ \sqrt{3}\, V u(t) \\ 0 \end{bmatrix} \tag{4.165}$$

Immediately after the voltage is applied, the flux linkages λ_F and λ_D are still zero, since they cannot change instantly. Thus at $t = 0^+$

$$\lambda_F = 0 = \mathrm{k}M_F i_d + L_F i_F + M_R i_D \qquad \lambda_D = 0 = \mathrm{k}M_D i_d + M_R i_F + L_D i_D \tag{4.166}$$

Therefore

$$i_F = -\frac{\mathrm{k}M_F L_D - \mathrm{k}M_D M_R}{L_F L_D - M_R^2}\, i_d \qquad i_D = -\frac{\mathrm{k}M_D L_F - \mathrm{k}M_F M_R}{L_F L_D - M_R^2}\, i_d \tag{4.167}$$

Substituting in (4.20) for λ_d, we get (at $t = 0^+$)

$$\lambda_d = \left(L_d - \frac{\mathrm{k}^2 M_F^2 L_D + L_F \mathrm{k}^2 M_D^2 - 2\,\mathrm{k}M_F \mathrm{k}M_D M_R}{L_F L_D - M_R^2} \right) i_d \tag{4.168}$$

The subtransient inductance is defined as the *initial* stator flux linkage per unit of stator current, with all the rotor circuits shorted (and previously unenergized). Thus by definition

$$\lambda_d \triangleq L_d'' i_d \tag{4.169}$$

where L_d'' is the d axis subtransient reactance. From (4.168) and (4.169)

$$L_d'' = L_d - \frac{(\mathrm{k}M_F)^2 L_D + (\mathrm{k}M_D)^2 L_F - 2\mathrm{k}M_F \mathrm{k}M_D M_R}{L_F L_D - M_R^2} \tag{4.170}$$

$$= L_d - \frac{L_D + L_F - 2L_{AD}}{(L_F L_D / L_{AD}^2) - 1} \tag{4.171}$$

where L_{AD} is defined in (4.108).

If the balanced voltages described by (4.164) are suddenly applied to a machine with *no* damper winding, the same procedure will yield (at $t = 0^+$)

$$i_F = -(\mathrm{k}M_F / L_F) i_d \tag{4.172}$$

$$\lambda_d = [L_d - (\mathrm{k}M_F)^2 / L_F] i_d = L_d' i_d \tag{4.173}$$

where L_d' is the d axis transient inductance; i.e.,

$$L_d' = L_d - (\mathrm{k}M_F)^2 / L_F = L_d - L_{AD}^2 / L_F \tag{4.174}$$

In a machine *with* damper windings, after a few cycles from the start of the transient described in this section, the damper winding current decays rapidly to zero and the effective stator inductance is the transient inductance.

If the phase of the impressed voltages in (4.164) is changed by 90° ($v_a = \sqrt{2}\, V \sin\theta$), v_d becomes zero and v_q will have a magnitude of $\sqrt{3}\, V$.

Before we examine the q axis inductances, some clarification of the circuits that may exist in the q axis is needed. For a salient pole machine with amortisseur windings a q axis damper circuit exists, but there is no other q axis rotor winding. For such a machine the stator flux linkage after the initial subtransient dies out is determined by essentially the same circuit as that of the steady-state q axis flux linkage. Thus for a salient pole machine it is customary to consider the q axis transient inductance to be the same as the q axis synchronous inductance.

The situation for a round rotor machine is different. Here the solid iron rotor provides multiple paths for circulating eddy currents, which act as equivalent windings during both transient and subtransient periods. Such a machine will have *effective* q axis rotor circuits that will determine the q axis transient and subtransient inductances. Thus for such a machine it is important to recognize that a q axis transient inductance (much smaller in magnitude than L_q) exists.

Repeating the previous procedure for the q axis circuits of a salient pole machine,

$$\lambda_Q = 0 = \mathrm{k}M_Q i_q + L_Q i_Q \tag{4.175}$$

or

$$i_Q = -(\mathrm{k}M_Q/L_Q)i_q \tag{4.176}$$

Substituting in the equation for λ_q,

$$\lambda_q = L_q i_q + \mathrm{k}M_Q i_Q \tag{4.177}$$

or

$$\lambda_q = [L_q - (\mathrm{k}M_Q)^2/L_Q]i_q \triangleq L_q'' i_q \tag{4.178}$$

where L_q'' is the q axis subtransient inductance

$$L_q'' = L_q - (\mathrm{k}M_Q)^2/L_Q = L_q - L_{AQ}^2/L_Q \tag{4.179}$$

We can also see that when i_Q decays to zero after a few cycles, the q axis effective inductance in the "transient period" is the same as L_q. Thus for this type of machine

$$L_q' = L_q \tag{4.180}$$

Since the reactance is the product of the rated angular speed and the inductance and since in pu $\omega_R = 1$, the subtransient and transient reactances are numerically equal to the corresponding values of inductances in pu.

We should again point out that for a round rotor machine $L_q'' < L_q' < L_q$. To identify these inductances would require that two q axis rotor windings be defined. This procedure has not been followed in this book but could be developed in a straightforward way [21, 22].

4.14.1 Time constants

We start with the stator circuits open circuited. Consider a step change in the field voltage; i.e., $v_F = V_F u(t)$. The voltage equations are given by

$$r_F i_F + \dot{\lambda}_F = V_F u(t) \qquad r_D i_D + \dot{\lambda}_D = 0 \tag{4.181}$$

and the flux linkages are given by (note that $i_d = 0$)

$$\lambda_D = L_D i_D + M_R i_F \qquad \lambda_F = L_F i_F + M_R i_D \tag{4.182}$$

Again at $t = 0^+$, $\lambda_D = 0$, which gives for that instant

$$i_F = -(L_D/M_R)i_D \tag{4.183}$$

Substituting for the flux linkages using (4.182) in (4.181),

$$V_F/L_F = (r_F/L_F)i_F + \dot{i}_F + (M_R/L_F)\dot{i}_D \qquad 0 = (r_D/M_R)i_D + \dot{i}_F + (L_D/M_R)\dot{i}_D \tag{4.184}$$

Subtracting and substituting for i_F using (4.183),

$$\dot{i}_D + \frac{r_D L_F + r_F L_D}{L_F L_D - M_R^2} i_D = -V_F \frac{M_R}{L_F L_D - M_R^2} \tag{4.185}$$

Usually in pu $r_D \gg r_F$, while L_D and L_F are of similar magnitude. Therefore we can write, approximately,

$$\dot{i}_D + \frac{r_D}{L_D - M_R^2/L_F} i_D = -V_F \frac{M_R/L_F}{L_D - M_R^2/L_F} \tag{4.186}$$

Equation (4.186) shows that i_D decays with a time constant

$$\tau''_{d0} = \frac{L_D - M_R^2/L_F}{r_D} \tag{4.187}$$

This is the d axis open circuit *subtransient* time constant. It is denoted *open circuit* because by definition the stator circuits are open.

When the damper winding is not available or after the decay of the subtransient current, we can show that the field current is affected only by the parameters of the field circuit; i.e.,

$$r_F i_F + L_F \dot{i}_F = V_F u(t) \tag{4.188}$$

The time constant of this transient is the d axis *transient* open circuit time constant τ'_{d0}, where

$$\tau'_{d0} = L_F/r_F \tag{4.189}$$

Kimbark [2] and Anderson [8] show that when the stator is short circuited, the corresponding d axis time constants are given by

$$\tau''_d = \tau''_{d0} L''_d / L'_d \tag{4.190}$$

$$\tau'_d = \tau'_{d0} L'_d / L_d \tag{4.191}$$

A similar analysis of the transient in the q axis circuits of a salient pole machine shows that the time constants are given by

$$\tau''_{q0} = L_Q/r_Q \tag{4.192}$$

$$\tau''_q = \tau''_{q0} L''_q / L_q \tag{4.193}$$

For a round rotor machine both transient and subtransient time constants are present.

Another time constant is associated with the rate of change of direct current in the stator or with the envelope of alternating currents in the field winding, when the machine is subjected to a three-phase short circuit. This time constant is τ_a and is given by (see [8], Ch. 6)

$$\tau_a = L_2/r \tag{4.194}$$

where L_2 is the negative-sequence inductance, which is given by

$$L_2 = (L'_d + L_q)/2 \tag{4.195}$$

Typical values for the synchronous machine constants are shown in Tables 4.3, 4.4, and 4.5.

Table 4.3. Typical Synchronous Machine Time Constants in Seconds

Time constant	Turbogenerators			Waterwheel generators			Synchronous condensers		
	Low	Avg.	High	Low	Avg.	High	Low	Avg.	High
τ'_{d0}	2.8	5.6	9.2	1.5	5.6	9.5	6.0	9.0	11.5
τ'_d	0.4	1.1	1.8	0.5	1.8	3.3	1.2	2.0	2.8
$\tau''_d = \tau''_q$	0.02	0.035	0.05	0.01	0.035	0.05	0.02	0.035	0.05
τ_a	0.04	0.16	0.35	0.03	0.15	0.25	0.10	0.17	0.30

Source: Reprinted by permission from *Power System Stability*, vol. 3, by E.W. Kimbark. © Wiley, 1956.

Table 4.4. Typical Turbogenerator and Synchronous Condenser Characteristics

Parameter		Generators		Synchronous condensers	
		Range	Recommended average	Range	Recommended average
Nominal rating		300–1000 MW	. . .	50–100 MVA	. . .
Power factor		0.80–0.95	0.90	. . .	. . .
Direct axis synchronous reactance x_d		140–180	160	170–270	220
Transient reactance x'_d		23–35	25	45–65	55
Subtransient reactance x''_d		15–23	20	35–45	40
Quadrature axis synchronous reactance x_q		150–160	155	100–130	115
Negative-sequence reactance x_2		18–20	19	35–45	40
Zero-sequence reactance x_0		12–14	13	15–25	20
Short circuit ratio		0.50–0.72	0.64	0.35–0.65	0.50
Inertia constant H, $\frac{(\text{kW}\cdot\text{s})}{(\text{kVA})}$	3600 r/min	3.0–5.0	4.0	. . .	. . .
	1800 r/min	5.0–8.0	6.0	. . .	. . .

Source: From the 1964 National Power Survey made by the U.S. Federal Power Commission. USGPO.

Note: All reactances in percent on rated voltage and kVA base. kW losses for typical synchronous condensers in the range of sizes shown, excluding losses associated with step-up transformers, are in the order of 1.2–1.5% on rated kVA base. No attempt has been made to show kW losses associated with generators, since generating plants are generally rated on a net power output basis and losses vary widely dependent on the generator plant design.

Table 4.5. Typical Hydrogenerator Characteristics

Parameter	Small units	Large units
Nominal rating (MVA)	0–40	40–200
Power factor	0.80–0.95*	0.80–0.95*
Speed (r/min)	70–350	70–200
Inertia constant H, $\frac{(\text{kW}\cdot\text{s})}{(\text{kVA})}$	1.5–4.0	3.0–5.5
Direct axis synchronous reactance x_d	90–110	80–100
Transient reactance x'_d	25–45	20–40
Subtransient reactance x''_d	20–35	15–30
Quadrature axis synchronous reactance x_q	. . .	. . .
Negative-sequence reactance x_2	20–45	20–35
Zero-sequence reactance x_0	10–35	10–25
Short circuit ratio	1.0–2.0	1.0–2.0

Source: From the 1964 National Power Survey made by the U.S. Federal Power Commission. USGPO.

Note: All reactances in percent on rated voltage and kVA base. No attempt has been made to show kW losses associated with generators, since generating plants are generally rated on a net power output basis and losses vary widely dependent on the generator plant design.

*These power factors cover conditions for generators installed either close to or remote from load centers.

4.15 Simplified Models of the Synchronous Machine

In previous sections we have dealt with a mathematical model of the synchronous machine, taking into account the various effects introduced by different rotor circuits, i.e., both field effects and damper-winding effects. The model includes seven nonlinear differential equations for each machine. In addition to these, other equations describing the load (or network) constraints, the excitation system, and the mechanical torque must be included in the mathematical model. Thus the complete mathematical description of a large power system is exceedingly complex, and simplifications are often used in modeling the system.

In a stability study the response of a large number of synchronous machines to a given disturbance is investigated. The complete mathematical description of the system would therefore be very complicated unless some simplifications were used. Often only a few machines are modeled in detail, usually those nearest the disturbance, while others are described by simpler models. The simplifications adopted depend upon the location of the machine with respect to the disturbance causing the transient and upon the type of disturbance being investigated. Some of the more commonly used simplified models are given in this section. The underlying assumptions as well as the justifications for their use are briefly outlined. In general, they are presented in the order of their complexity.

Some simplified models have already been presented. In Chapter 2 the classical representation was introduced. In this chapter, when the saturation is neglected as tacitly assumed in the current model, the model is also somewhat simplified. An excellent reference on simplified models is Young [19].

4.15.1 Neglecting damper windings—the E_q' model

The mathematical models given in Sections 4.10 and 4.12 assume the presence of three rotor circuits. Situations arise in which some of these circuits or their effects can be neglected.

Machine with solid round rotor [2]. The solid round rotor acts as a q axis damper winding, even with the d axis damper winding omitted. The mathematical model for this type of machine will be the same as given in Sections 4.10 and 4.12 with i_D or λ_D omitted. For example, in (4.103) and (4.138) the third row and column are omitted.

Amortisseur effects neglected. This assumption assumes that the effect of the damper windings on the transient under study is small enough to be negligible. This is particularly true in system studies where the damping between closely coupled machines is not of interest. In this case the effect of the amortisseur windings may be included in the damping torque, i.e., by increasing the damping coefficient D in the torque equation. Neglecting the amortisseur windings can be simulated by omitting i_D and i_Q in (4.103) or λ_D and λ_Q in (4.138). Another model using familiar machine parameters is given below. From (4.118), (4.123), (4.120), and (4.121) with the D and Q circuits omitted,

$$\begin{bmatrix} i_d \\ i_F \\ \hline i_q \end{bmatrix} = \left[\begin{array}{cc|c} (\ell_d - L_{MD})/\ell_d^2 & -L_{MD}/\ell_d\ell_F & 0 \\ -L_{MD}/\ell_d\ell_F & (\ell_F - L_{MD})/\ell_F^2 & 0 \\ \hline 0 & 0 & 1/L_q \end{array}\right] \begin{bmatrix} \lambda_d \\ \lambda_F \\ \hline \lambda_q \end{bmatrix} \tag{4.196}$$

or

$$\begin{bmatrix} \mathbf{i}_{dF} \\ - \\ i_q \end{bmatrix} = \left[\begin{array}{c|c} \mathcal{L}_D^{-1} & \mathbf{0} \\ \hline \mathbf{0} & 1/L_q \end{array}\right] \begin{bmatrix} \boldsymbol{\lambda}_{dF} \\ -- \\ \lambda_q \end{bmatrix} \tag{4.197}$$

We can show that $\mathcal{L}_D^{-1}$ is given by

$$\mathcal{L}_D^{-1} = \begin{bmatrix} 1/L_d' & -L_{AD}/L_d'L_F \\ -L_{AD}/L_d'L_F & L_d/L_d'L_F \end{bmatrix} \tag{4.198}$$

Therefore, the currents are given by

$$\begin{bmatrix} i_d \\ i_F \\ i_q \end{bmatrix} = \begin{bmatrix} 1/L_d' & -L_{AD}/L_d'L_F & 0 \\ -L_{AD}/L_d'L_F & L_d/L_d'L_F & 0 \\ 0 & 0 & 1/L_q \end{bmatrix} \begin{bmatrix} \lambda_d \\ \lambda_F \\ \lambda_q \end{bmatrix} \tag{4.199}$$

The above equations may be in pu or in MKS units. This follows, since the choice of the rotor base quantities is based upon equal flux linkages for base rotor and stator currents. From the stator equation (4.36) and rearranging,

$$\dot{\lambda}_d = -ri_d - \omega\lambda_q - v_d \quad \text{pu} \tag{4.200}$$

or from (4.199) and (4.200)

$$\dot{\lambda}_d = -(r/L_d')\lambda_d + (rL_{AD}/L_d'L_F)\lambda_F - \omega\lambda_q - v_d \quad \text{pu} \tag{4.201}$$

From (4.58) we define

$$\sqrt{3}E_q' \triangleq \omega_R(kM_F/L_F)\lambda_F \quad \text{V} \tag{4.202}$$

and converting to pu

$$\begin{aligned} \sqrt{3}E_{qu}'V_B &= \omega_R(kM_{Fu}M_{FB}/L_{Fu}L_{FB})(\lambda_{Fu}L_{FB}I_{FB}) \\ \sqrt{3}E_{qu}' &= (kM_{Fu}\lambda_{Fu}/L_{Fu})[\omega_R(M_{FB}I_{FB}/V_B)] \end{aligned}$$

or in pu

$$L_{AD}\lambda_F/L_F = \sqrt{3}E_q' \quad \text{pu} \tag{4.203}$$

Now, from (4.201) and (4.203) we compute

$$\dot{\lambda}_d = -(r/L_d')\lambda_d + (r/L_d')\sqrt{3}E_q' - \omega\lambda_q - v_d \quad \text{pu} \tag{4.204}$$

In a similar way we compute λ_q from (4.36), substituting for i_q from (4.199) to write

$$\dot{\lambda}_q = (r/L_q)\lambda_q + \omega\lambda_d - v_q \quad \text{pu} \tag{4.205}$$

Note that in (4.204) and (4.205) all quantities, *including time,* are in pu. For the field voltage, from (4.36) $v_F = r_Fi_F + \dot{\lambda}_F$ pu, and substituting for i_F from (4.199),

$$v_F = r_F[-(L_{AD}/L_d'L_F)\lambda_d + (L_d/L_d'L_F)\lambda_F] + \dot{\lambda}_F \quad \text{pu} \tag{4.206}$$

Now from (4.203)

$$\lambda_F/L_F = \sqrt{3}E_q'/L_{AD} \quad \text{pu} \tag{4.207}$$

Also from (4.59) we define

$$\sqrt{3}E_{FD} \triangleq \omega_R(kM_F/r_F)v_F \quad \text{V} \tag{4.208}$$

and converting to pu

$$\begin{aligned}\sqrt{3}E_{FDu}V_B &= \omega_R[(kM_{Fu}M_{FB}/r_{Fu}R_{FB})v_{Fu}V_{FB}]\\ \sqrt{3}E_{FDu} &= (kM_{Fu}v_{Fu}/r_{Fu})(\omega_R M_{FB}v_{FB}/V_B R_{FB})\\ \sqrt{3}E_{FD} &= L_{AD}v_F/r_F \text{ pu}\end{aligned} \tag{4.209}$$

From (4.207), (4.209), and (4.206) we compute

$$\sqrt{3}\frac{r_F}{L_{AD}}E_{FD} = -\frac{L_{AD}}{L'_d}\frac{r_F}{L_F}\lambda_d + \frac{L_d}{L'_d}\frac{r_F}{L_{AD}}\sqrt{3}E'_q + \sqrt{3}\frac{L_F}{L_{AD}}\dot{E}'_q \text{ pu} \tag{4.210}$$

Rearranging and using $L_{AD}^2/L_F = L_d - L'_d$ and $\tau'_{d0} = L_F/r_F$,

$$\dot{E}'_q = \frac{1}{\tau'_{d0}}\left(E_{FD} - \frac{L_d}{L'_d}E'_q + \frac{L_d - L'_d}{L'_d}\frac{\lambda_d}{\sqrt{3}}\right) \text{ pu} \tag{4.211}$$

We now define rms stator equivalent flux linkages and voltages

$$\Lambda_d = \lambda_d/\sqrt{3} \qquad \Lambda_q = \lambda_q/\sqrt{3} \qquad V_d = v_d/\sqrt{3} \qquad V_q = v_q/\sqrt{3} \tag{4.212}$$

Then (4.204), (4.205), and (4.211) become

$$\dot{\Lambda}_d = -(r/L'_d)\Lambda_d + (r/L'_d)E'_q - \omega\Lambda_q - V_d \text{ pu} \tag{4.213}$$

$$\dot{\Lambda}_q = \omega\Lambda_d - (r/L_q)\Lambda_q - V_q \text{ pu} \tag{4.214}$$

$$\dot{E}'_q = -\frac{L_d}{L'_d\tau'_{d0}}E'_q + \frac{L_d - L'_d}{\tau'_{d0}L'_d}\Lambda_d + \frac{1}{\tau'_{d0}}E_{FD} \text{ pu} \tag{4.215}$$

Note that in the above equations all the variables (including time) and all the parameters are in pu. Thus the time constants must be in radians, or

$$\tau_{pu} = t_{sec}\omega_R \quad \text{rad} \tag{4.216}$$

Now we derive the torque equation. From (4.95) $T_{e\phi} = i_q\lambda_d - i_d\lambda_q$. Substituting for i_d and i_q, from (4.199) we get

$$T_{e\phi} = \lambda_q\lambda_d/L_q - \lambda_d/L'_d - (L_{AD}\lambda_F/L'_dL_F)\lambda_q \text{ pu} \tag{4.217}$$

and by using (4.203) and (4.212),

$$T_e = E'_q\Lambda_q/L'_d - (1/L'_d - 1/L_q)\Lambda_d\Lambda_q \tag{4.218}$$

From the swing equation

$$\tau_j\dot{\omega} = T_m - T_e - D\omega \text{ pu} \tag{4.219}$$

$$\dot{\delta} = \omega - 1 \text{ pu} \tag{4.220}$$

Equations (4.213)–(4.215), (4.219), and (4.220) along with the torque equation (4.218) describe the E'_q model. It is a fifth-order system with "free" inputs E_{FD} and T_m. The signals V_d and V_q depend upon the external network.

Block diagrams of the system equations are found as follows. From (4.213) we write, in the s domain,

$$(r/L'_d)[1 + (L'_d/r)s]\Lambda_d = (r/L'_d)E'_q - \omega\Lambda_q - V_d \text{ pu} \tag{4.221}$$

Similarly, from (4.214)

$$(r/L_q)[1 + (L_q/r)s]\Lambda_q = \omega\Lambda_d - V_q \text{ pu} \tag{4.222}$$

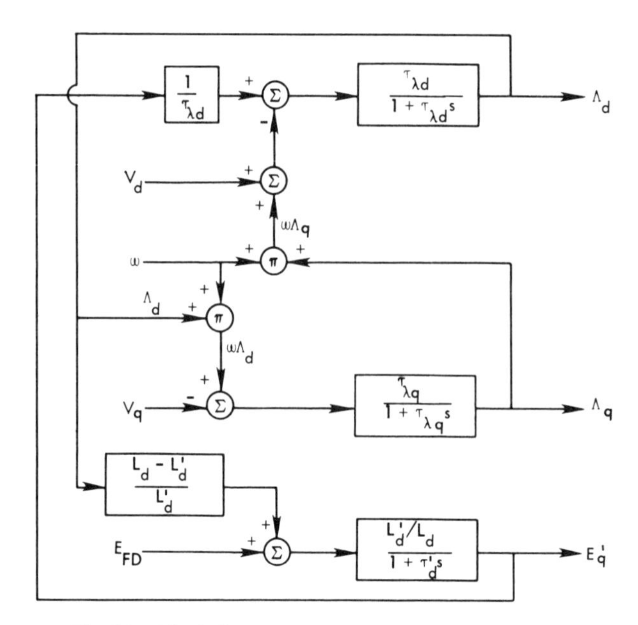

Fig. 4.9 Block diagram representation of the E_q' model.

and from (4.215)

$$(L_d/L_d')\,[1 + \tau_{d0}'(L_d'/L_d)s]E_q' = E_{FD} + [(L_d - L_d')/L_d']\Lambda_d \text{ pu} \tag{4.223}$$

Now define $\tau_{\lambda d} \triangleq L_d'/r$, $\tau_{\lambda q} = L_q/r$, and $\tau_d' = \tau_{d0}'L_d'/L_d$. The above equations are represented by the block diagram shown in Figure 4.9. The remaining system equations can be represented by the block diagrams of Figure 4.10. The block diagrams in Figures 4.9 and 4.10 can be combined to give the block diagram of the complete model. Note that T_m and E_{FD} are assumed to be known and V_d and V_q depend upon the load.

The model developed to this point is for an unsaturated machine. The effect of saturation may be added by computing the additional field current required under saturated operating conditions. From $\lambda_d = L_d i_d + L_{AD} i_F$ and substituting for i_d from (4.199),

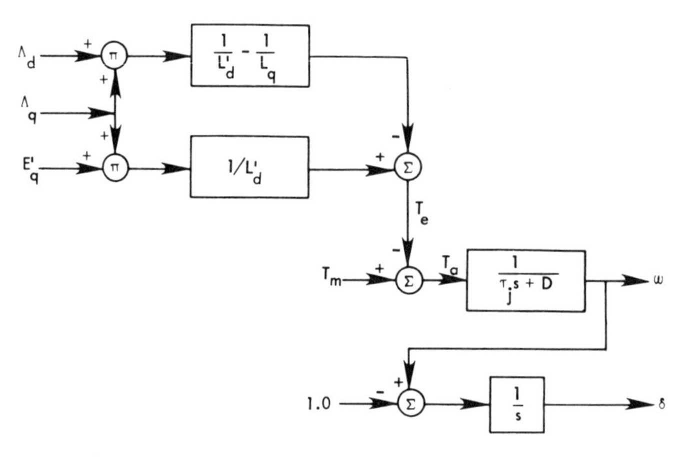

Fig. 4.10 Block diagram representation of (4.218)–(4.220).

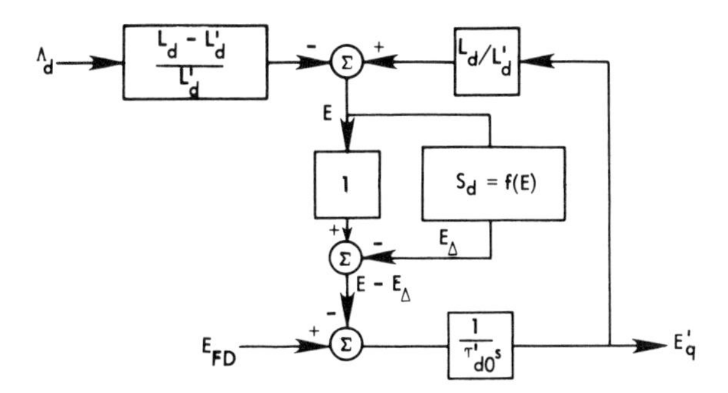

Fig. 4.11 Block diagram for generating E_q' with saturation.

$$\lambda_d = L_d[\lambda_d/L_d' - (L_{AD}/L_d'L_F)\,\lambda_F] + i_F L_{AD} \quad \text{pu} \tag{4.224}$$

then

$$i_F L_{AD} = \lambda_d(1 - L_d/L_d') + [(L_d/L_d')(L_{AD}/L_F)]\lambda_F \quad \text{pu} \tag{4.225}$$

Also, from $\omega_R M_F i_F = \sqrt{2}E$ in Section 4.7.4 we can show that

$$i_F L_{AD} = \sqrt{3}E \quad \text{pu} \tag{4.226}$$

Now from (4.212), (4.203), (4.226), and (4.225)

$$E = (L_d/L_d')E_q' - [(L_d - L_d')/L_d']\Lambda_d \tag{4.227}$$

Substituting (4.227) into (4.215),

$$\tau_{d0}'\dot{E}_q' = E_{FD} - E \tag{4.228}$$

For the treatment of saturation, Young [19] suggests the modification of (4.227) to the form

$$E = (L_d/L_d')E_q' - [(L_d - L_d')/L_d']\Lambda_d + E_\Delta \tag{4.229}$$

where E_Δ corresponds to the additional field current needed to obtain the same EMF on the no-load saturation curve. This additional current is a function of the saturation index and can be determined by a procedure similar to that of Section 4.12.4.

Another method of treating saturation is to consider a saturation function that depends upon E_q'; i.e., let $E_\Delta = f_\Delta(E_q')$. This leads to a solution for E_q' amounting to a negative feedback term and provides a useful insight as to the effect of saturation (see [20] and Problem 4.33).

Equations (4.229) and (4.228) can be represented by the block diagram shown in Figure 4.11. We note that if saturation is to be taken into account, the portion of Figure 4.9 that produces the signal E_q' should be modified according to the Figure 4.11.

Example 4.5

Determine the numerical constants of the E_q' model of Figures 4.9 and 4.10, using the data of Examples 4.1 and 4.2. It is also given that $L_d'' = 0.185$ pu and $L_d' = 0.245$ pu.

Solution

From the given data we compute the time constants required for the model.

$$\tau_{d0}'' = \frac{L_D - M_R^2/L_F}{r_D} = \frac{0.00599 - (0.109)^2/2.189}{0.0184} = 0.03046 \text{ s} = 72.149 \text{ rad}$$

From this we may also compute the short circuit subtransient time constant as

$$\tau_d'' = \tau_{d0}'' L_d''/L_d' = \tau_{d0}''(0.185/0.245) = 0.023 \text{ s} = 8.671 \text{ rad}$$

The fictitious time constants $\tau_{\lambda d}$ and $\tau_{\lambda q}$ are computed as

$$\tau_{\lambda d} = L_d'/r = (0.245)(3.73 \times 10^{-3})/1.542 \times 10^{-3} = 0.593 \text{ s} = 223.446 \text{ rad}$$
$$\tau_{\lambda q} = L_q/r = 6.118 \times 10^{-3}/1.542 \times 10^{-3} = 3.967 \text{ s} = 1495.718 \text{ rad}$$

This large time constant indicates that Λ_q will respond relatively slowly to a change in terminal conditions.

The various gains needed in the model are as follows:

$$L_d'/L_d = 0.245/1.7 = 0.114$$
$$(L_d - L_d')/L_d' = (1.7 - 0.245)/0.245 = 3.939$$
$$1/L_d' - 1/L_q = 1/0.245 - 1/1.64 = 3.473$$
$$1/L_d' = 4.08$$
$$1/\tau_{\lambda d} = 1/0.593\,\omega_R = 0.00447$$

Note the wide range of gain constants required.

4.15.2 Voltage behind subtransient reactance—the E'' model

In this model the transformer voltage terms in the stator voltage equations are neglected compared to the speed voltage terms [19]. In other words, in the equations for v_d and v_q, the terms $\dot{\lambda}_d$ and $\dot{\lambda}_q$ are neglected since they are numerically small compared to the terms $\omega\lambda_q$ and $\omega\lambda_d$ respectively. In addition, it is assumed in the stator voltage equations that $\omega \simeq \omega_R$, and $L_d'' = L_q''$. Note that while some simplifying assumptions are used in this model, the field effects and the effects of the damper circuits are included in the machine representation.

Stator *subtransient* flux linkages are defined by the equations

$$\lambda_d'' = \lambda_d - L_d'' i_d \qquad \lambda_q'' = \lambda_q - L_q'' i_q \tag{4.230}$$

where L_d'' and L_q'' are defined by (4.170) and (4.179) respectively. Note that (4.230) represents the more general case of (4.169), which represents a special case of zero *initial* flux linkage. These flux linkages produce EMF's that lag 90° behind them. These EMF's are defined by

$$e_q'' \triangleq \omega\lambda_d'' = \omega_R \lambda_d'' \qquad e_d'' \triangleq -\omega\lambda_q'' = -\omega_R \lambda_q'' \tag{4.231}$$

(See [8] for a complete derivation.)

From (4.36) the stator voltage equations, under the assumptions stated above, are given by

$$v_d = -ri_d - \omega_R \lambda_q \qquad v_q = -ri_q + \omega_R \lambda_d \tag{4.232}$$

Combining (4.230) and (4.232),

$$v_d = -ri_d - \omega_R i_q L_q'' - \omega_R \lambda_q'' \qquad v_q = -ri_q + \omega_R i_d L_d'' + \omega_R \lambda_d'' \tag{4.233}$$

Now from (4.231) and (4.233),

$$v_d = -ri_d - i_q x'' + e_d'' \qquad v_q = -ri_q + i_d x'' + e_q'' \tag{4.234}$$

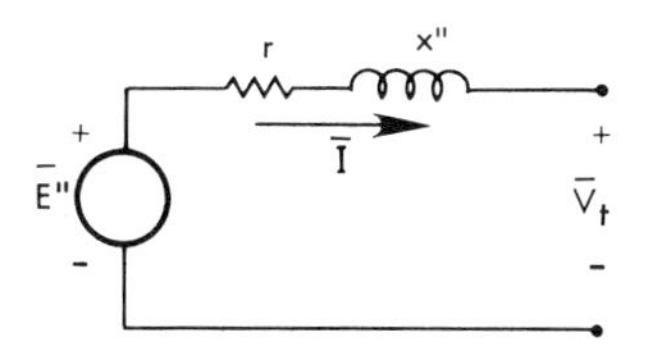

Fig. 4.12 Voltage behind subtransient reactance equivalent.

where, under the assumptions used in this model,

$$x'' \triangleq \omega_R L_d'' = \omega_R L_q'' \tag{4.235}$$

The voltages e_d'' and e_q'' are the d and q axis components of the EMF e'' produced by the subtransient flux linkage, the d and q axis components of which are given by (4.230). This EMF is called the *voltage behind the subtransient reactance.*

Equations (4.234) when transformed to the a-b-c frame of reference may be represented by the equivalent circuit of Figure 4.12. If quasi-steady-state conditions are assumed to apply at any instant, the relations expressed in (4.234) may be represented by the phasor diagram shown in Figure 4.13. In this diagram the q and d axes represent the real and imaginary axes respectively. "Projections" of the different phasors on these axes give the q and d components of these phasors. For example the voltage E'' is represented by the phasor $\bar{E}''$ shown. Its components are E_q'' and E_d'' respectively. From the above we can see that if at any instant the terminal voltage and current of the machine are known, the voltage E'' can be determined. Also if E_d'' and E_q'' are known, E'' can be calculated; and if the current is also known, the terminal voltage can be determined.

We now develop the dynamic model for the subtransient case. Substituting (4.230) into (4.134), we compute

$$\lambda_d'' = \left[1 - \frac{L_d''}{\ell_d}\left(1 - \frac{L_{MD}}{\ell_d}\right)\right]\lambda_d + \frac{L_{MD}L_d''}{\ell_d \ell_F}\lambda_F + \frac{L_{MD}L_d''}{\ell_d \ell_D}\lambda_D \tag{4.236}$$

We can show that

$$\frac{1}{\ell_d}\left(1 - \frac{L_{MD}}{\ell_d}\right) = \frac{\ell_F \ell_D + L_{AD}(\ell_F + \ell_D)}{L_d[\ell_F \ell_D + L_{AD}(\ell_F + \ell_D)] - L_{AD}^2(\ell_F + \ell_D)}$$

$$= \left[L_d - \frac{L_{AD}^2(\ell_F + \ell_D)}{\ell_F \ell_D + L_{AD}(\ell_F + \ell_D)}\right]^{-1} = 1/L_d'' \tag{4.237}$$

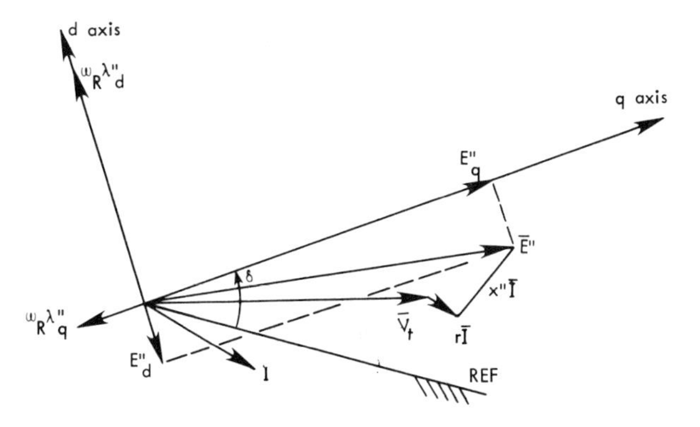

Fig. 4.13 Phasor diagram for the quasi-static subtransient case.

since by definition

$$L_d'' = L_d - \frac{L_{AD}^2(\ell_F + \ell_D)}{\ell_F\ell_D + L_{AD}(\ell_F + \ell_D)} \tag{4.238}$$

Therefore we may write (4.236) as

$$\lambda_d'' = (L_d'' L_{MD}/\ell_d \ell_F)\lambda_F + (L_d'' L_{MD}/\ell_d \ell_D)\lambda_D \tag{4.239}$$

Using (4.203), we can rewrite in terms of E_q' as

$$\lambda_d'' = (L_d'' L_{MD} L_F/\ell_d \ell_F L_{AD}) \sqrt{3}\, E_q' + (L_d'' L_{MD}/\ell_d \ell_D)\lambda_D \tag{4.240}$$

Now we can compute the constants

$$K_1 = \frac{L_d'' L_{MD} L_F}{\ell_d \ell_F L_{AD}} = \frac{L_d'' - \ell_d}{L_d' - \ell_d} = \frac{x_d'' - x_\ell}{x_d' - x_\ell} \tag{4.241}$$

$$K_2 = \frac{L_d'' L_{MD}}{\ell_d \ell_D} = 1 - \frac{L_d'' L_{MD} L_F}{\ell_d \ell_F L_{AD}} = 1 - \frac{x_d'' - x_\ell}{x_d' - x_\ell} = 1 - K_1 \tag{4.242}$$

Substituting in (4.240) and using (4.231), we compute in pu

$$e_q'' = [(x_d'' - x_\ell)/(x_d' - x_\ell)]\,(\sqrt{3}\,E_q' - \lambda_D) + \lambda_D \tag{4.243}$$

Similarly from (4.230) and (4.104),

$$\lambda_q'' = (L_q i_q + L_{AQ} i_Q) - L_q'' i_q = (L_q - L_q'')\, i_q + L_{AQ} i_Q \tag{4.244}$$

which can be substituted into (4.231) to compute

$$e_d'' = -(x_q - x_q'')\, i_q - e_d \tag{4.245}$$

where we define the voltage

$$e_d = \omega_R L_{AQ} i_Q \tag{4.246}$$

We can also show that

$$\lambda_q'' = \lambda_q - L_q'' i_q = (L_{AQ}/L_Q)\lambda_Q \tag{4.247}$$

Now from the field flux linkage equation (4.104) in pu, we incorporate (4.203) and (4.226) to compute

$$E = E_q' - (x_d - x_d')(i_d + i_D)/\sqrt{3} \tag{4.248}$$

From the definition of L_d' (4.174) we can show that

$$L_d - L_d' = L_{AD}^2/L_F \tag{4.249}$$

We can also show that

$$(L_d' - L_d'')/(L_d' - \ell_d)^2 = L_F/(L_F L_D - L_{AD}^2) \tag{4.250}$$

Then from (4.104) in pu

$$\lambda_D = L_{AD}\, i_d + L_{AD}\, i_F + L_D\, i_D \qquad \lambda_F = L_{AD}\, i_d + L_F\, i_F + L_{AD}\, i_D \tag{4.251}$$

Eliminating i_F from (4.251),

$$\frac{L_{AD}}{L_F}\lambda_F - \lambda_D = \left(\frac{L_{AD}^2}{L_F} - L_{AD}\right) i_d + \left(\frac{L_{AD}^2}{L_F} - L_D\right) i_D \tag{4.252}$$

Now substituting (4.203), (4.249), and (4.250) into (4.252),

$$\sqrt{3}E_q' - \lambda_D = (L_d - L_d' - L_{AD})i_d - \frac{(L_d' - \ell_d)^2}{L_d' - L_d''} i_D \tag{4.253}$$

which can be put in the form

$$i_D = \frac{x_d' - x_d''}{(x_d' - x_\ell)^2}[\lambda_D - \sqrt{3}E_q' - (x_d' - x_\ell)i_d] \tag{4.254}$$

In addition to the above auxiliary equations, the following differential equations are obtained. From (4.36) we write

$$r_D i_D + \dot{\lambda}_D = 0 \tag{4.255}$$

Substituting (4.187) and (4.250) in (4.255),

$$\dot{\lambda}_D = -\frac{(x_d' - x_\ell)^2}{(x_d' - x_d'')\tau_{d0}''} i_D \tag{4.256}$$

Similarly, from (4.36) we have

$$r_Q i_Q + \dot{\lambda}_Q = 0$$

which may be written as

$$[\omega_R r_Q(L_{AQ}/L_Q)]i_Q + [(\omega_R L_{AQ})/L_Q]\dot{\lambda}_Q = 0 \tag{4.257}$$

Now from (4.246), (4.247), (4.231), (4.192), and (4.257) we get the differential equation

$$\dot{e}_d'' = e_d/\tau_{q0}'' \tag{4.258}$$

The voltage equation for the field circuit comes from (4.36)

$$v_F = r_F i_F + \dot{\lambda}_F \tag{4.259}$$

which can be put in the same form as (4.228)

$$\tau_{d0}' \dot{E}_q' = E_{FD} - E \tag{4.260}$$

where E is given by (4.248).

Equations (4.256), (4.258), and (4.260) give the time rate of change λ_D, e_d'', and E_q' in terms of i_D, e_d, and E. The auxiliary equations (4.245), (4.248), and (4.254) relate these quantities to i_d and i_q, which in turn depend upon the load configuration. The voltage e_q'' is calculated from (4.243).

To complete the model, the torque equation is needed. From (4.95),

$$T_{e\phi} = i_q \lambda_d - i_d \lambda_q$$

By using (4.230) and recalling that in this model it is assumed that $L_d'' = L_q''$,

$$T_{e\phi} = i_q \lambda_d'' - i_d \lambda_q'' \tag{4.261}$$

and if ω in pu is approximately equal to the synchronous speed, (4.261) becomes

$$T_{e\phi} = e_q'' i_q + e_d'' i_d \tag{4.262}$$

If saturation is neglected, the system equations can be reduced to the following:

$$\dot{e}_d'' = -\frac{1}{\tau_{q0}''} e_d'' - \frac{1}{\tau_{q0}''}(x_q - x_q'')i_q \tag{4.263}$$

$$\dot{\lambda}_D = \frac{1}{\tau''_{d0}}\sqrt{3}E'_q - \frac{1}{\tau''_{d0}}\lambda_D + \frac{1}{\tau''_{d0}}(x'_d - x_\ell)i_d \tag{4.264}$$

$$\sqrt{3}\dot{E}'_q = \frac{\sqrt{3}}{\tau'_{d0}}E_{FD} + \frac{(x_d - x'_d)(x''_d - x_\ell)}{\tau'_{d0}(x'_d - x_\ell)}i_d - \frac{\sqrt{3}}{\tau'_{d0}}\left[1 + \frac{(x_d - x'_d)(x'_d - x''_d)}{(x'_d - x_\ell)^2}\right]E'_q + \frac{(x_d - x'_d)(x'_d - x''_d)}{\tau'_{d0}(x'_d - x_\ell)^2}\lambda_D \tag{4.265}$$

Now from (4.243) and using K_1 and K_2 as defined in (4.241) and (4.242) respectively, we may write

$$e''_q = \sqrt{3}K_1E'_q + K_2\lambda_D \tag{4.266}$$

To complete the description of the system, we add the inertial equations

$$\dot{\omega} = (1/\tau_j)T_m - e''_q i_q/3\tau_j - i_d e''_d/3\tau_j - D\omega/\tau_j \tag{4.267}$$

$$\dot{\delta} = \omega - 1 \tag{4.268}$$

The currents i_d and i_q are determined from the load equations.

The block diagrams for the system may be obtained by rearranging the above equations. In doing so, we eliminate the $\sqrt{3}$ from all equations by using the rms equivalents, similar to (4.212),

$$\Lambda_D = \lambda_D/\sqrt{3} \qquad E'' = e''/\sqrt{3} = E''_q + jE''_d \tag{4.269}$$

Then (4.263)-(4.266) become

$$\begin{aligned}(1 + \tau''_{q0}s)E''_d &= -(x_q - x''_q)I_q\\ (1 + \tau''_{d0}s)\Lambda_D &= E'_q + (x'_d - x_\ell)I_d\\ (1 + \tau'_{d0}s)E'_q &= E_{FD} - K_dE'_q + x_{xd}I_d + K_d\Lambda_D\\ E''_q &= K_1E'_q + K_2\Lambda_D\end{aligned} \tag{4.270}$$

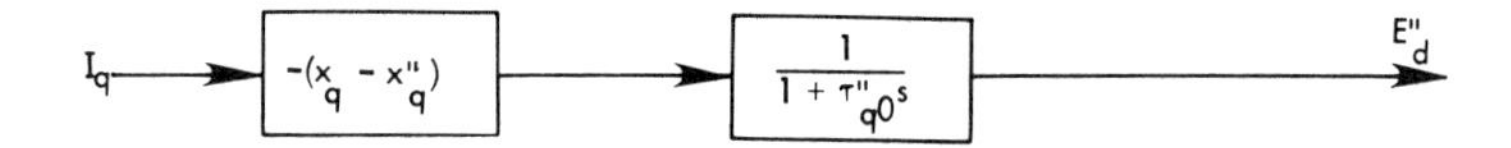

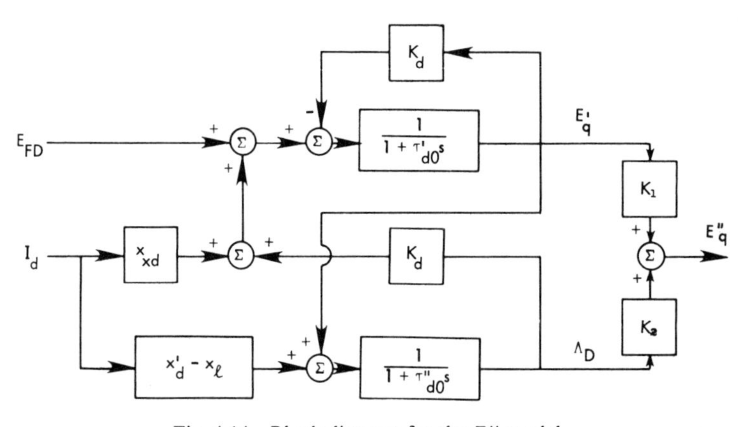

Fig. 4.14 Block diagram for the E'' model.

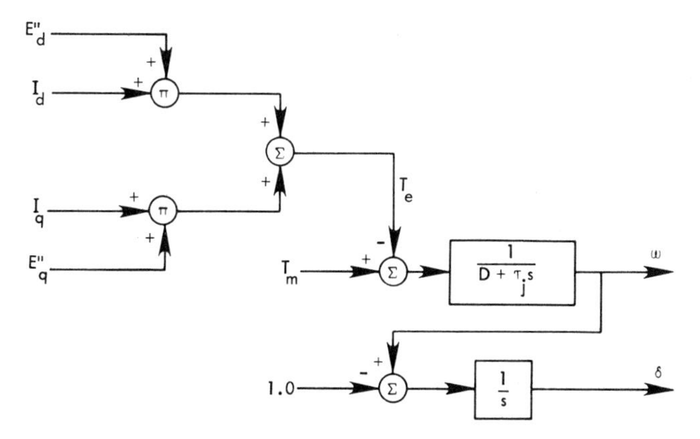

Fig. 4.15 Block diagram for computation of torque and speed in the E'' model.

where we have defined

$$K_1 = \frac{x''_d - x_\ell}{x'_d - x_\ell} \qquad K_d = \frac{(x_d - x'_d)(x'_d - x''_d)}{(x'_d - x_\ell)^2}$$

$$K_2 = 1 - K_1 \qquad x_{xd} = \frac{(x_d - x'_d)(x''_d - x_d)}{x'_d - x_\ell} \tag{4.271}$$

The block diagram for (4.270) is shown in Figure 4.14.

The remaining equations are given by

$$(D + \tau_j s)\omega = T_m - (E''_q I_q + E''_d I_d) \qquad s\delta = \omega - 1 \tag{4.272}$$

The block diagram for equation (4.272) is given in Figure 4.15. Also the block diagram of the complete system can be obtained by combining Figures 4.14 and 4.15.

If saturation is to be included, a voltage increment E_Δ, corresponding to the increase in the field current due to saturation, is to be added to (4.248),

$$E = E'_q + E_\Delta - (x_d - x'_d)(i_d + i_D)/\sqrt{3} \tag{4.273}$$

Example 4.6

Use the machine data from Examples 4.1–4.5 to derive the time constants and gains for the E'' model.

Solution

The time constant $\tau''_{d0} = 0.03046$ s $= 72.149$ rad is already known from Example 4.5. For the E'' model we also need the following additional time constants.

From (4.192) the q axis subtransient open circuit time constant is

$$\tau''_{q0} = L_Q/r_Q = 1.423 \times 10^{-3}/18.969 \times 10^{-3} = 0.075 \text{ s} = 28.279 \quad \text{rad}$$

which is about twice the d axis subtransient open circuit time constant.

We also need the d axis transient open circuit time constant. It is computed from (4.189).

$$\tau'_{d0} = L_F/r_F = 2.189/0.371 = 5.90 \text{ s} = 2224.25 \quad \text{rad}$$

Note that this time constant is about 30 times the subtransient time constant in the d

axis. This means that the integration associated with τ''_{d0} will be accomplished very fast compared to that associated with τ'_{d0}.

To compute the gains, the constant x'_d or L'_d is needed. It is computed from (4.174):

$$L'_d = L_d - L^2_{AD}/L_F = 1.70 - (1.55)^2/1.651 = 0.245 \text{ pu}$$

We can now compute from (4.271)

$$K_1 = \frac{x''_d - x_\ell}{x'_d - x_\ell} = \frac{0.185 - 0.15}{0.245 - 0.15} = 0.368$$

$$K_2 = 1 - K_1 = 0.632$$

$$K_d = \frac{(x_d - x'_d)(x'_d - x''_d)}{(x'_d - x_\ell)^2} = \frac{(1.70 - 0.245)(0.245 - 0.185)}{(0.245 - 0.150)^2} = 9.673$$

$$x_{xd} = \frac{(x_d - x'_d)(x''_d - x_\ell)}{x'_d - x_\ell} = \frac{(1.70 - 0.245)(0.185 - 0.150)}{0.245 - 0.150} = 0.536$$

From (4.179) we compute

$$L''_q = L_q - L^2_{AQ}/L_Q = 1.64 - (1.49)^2/1.526 = 0.185 \text{ pu}$$

Then, from (4.270), we compute the gain, $x_q - x''_q = 1.64 - 0.185 = 1.455$ pu.

4.15.3 Neglecting $\dot{\lambda}_d$ and $\dot{\lambda}_q$ for a cylindrical rotor machine—the two-axis model

In the two-axis model the transient effects are accounted for, while the subtransient effects are neglected [18]. The transient effects are dominated by the rotor circuits, which are the field circuit in the d axis and an equivalent circuit in the q axis formed by the solid rotor. An additional assumption made in this model is that in the stator voltage equations the terms $\dot{\lambda}_d$ and $\dot{\lambda}_q$ are negligible compared to the speed voltage terms and that $\omega \cong \omega_R = 1$ pu.

The machine will thus have two stator circuits and two rotor circuits. However, the number of *differential* equations describing these circuits is reduced by two since $\dot{\lambda}_d$ and $\dot{\lambda}_q$ are neglected in the stator voltage equations (the stator voltage equations are now algebraic equations).

The stator transient flux linkages are defined by

$$\lambda'_d \triangleq \lambda_d - L'_d i_d \qquad \lambda'_q \triangleq \lambda_q - L'_q i_q \tag{4.274}$$

and the corresponding stator voltages are defined by

$$e'_d \triangleq -\omega\lambda'_q = -\omega_R\lambda'_q \qquad e'_q \triangleq \omega\lambda'_d = \omega_R\lambda'_d \tag{4.275}$$

Following a procedure similar to that used in Section 4.15.2,

$$v_d = -r i_d - \omega_R L'_q i_q + e'_d \qquad v_q = -r i_q + \omega_R L'_d i_d + e'_q \tag{4.276}$$

or

$$e'_d = v_d + r i_d + x'_d i_q + (x'_q - x'_d) i_q \tag{4.277}$$

$$e'_q = v_q + r i_q - x'_d i_d \tag{4.278}$$

Since the term $(x'_q - x'_d) i_q$ is usually small, we can write, approximately,

$$e'_d \cong v_d + r i_d + x'_d i_q \tag{4.279}$$

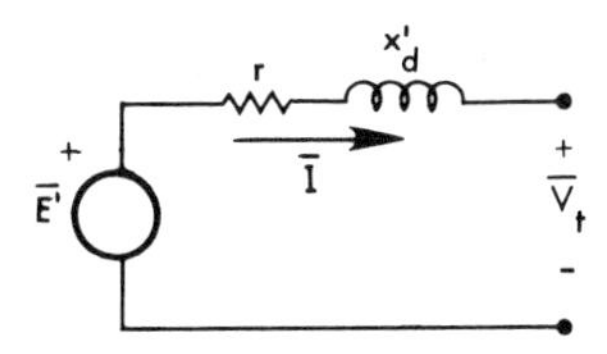

Fig. 4.16 Transient equivalent circuit of a generator.

The voltages e_q' and e_d' are the q and d components of a voltage e' behind transient reactance. Equations (4.279) and (4.278) indicate that during the transient the machine can be represented by the circuit diagram shown in Figure 4.16. It is interesting to note that since e_d' and e_q' are d and q axis stator voltages, they represent $\sqrt{3}$ times the equivalent stator rms voltages. For example, we can verify that $e_q' = \sqrt{3}E_q'$, as given by (4.203). Also, in this model the voltage e', which corresponds to the transient flux linkages in the machine, is not a constant. Rather, it will change due to the changes in the flux linkage of the d and q axis rotor circuits.

We now develop the differential equations for the voltages e_d' and e_q'. The d axis flux linkage equations for this model are

$$\lambda_d = L_d i_d + L_{AD} i_F \text{ pu} \qquad \lambda_F = L_{AD} i_d + L_F i_F \text{ pu} \tag{4.280}$$

By eliminating i_F and using (4.174) and (4.203),

$$\lambda_d - \sqrt{3}E_q' = L_d' i_d \text{ pu}$$

and by using (4.275),

$$e_q' = \sqrt{3}E_q' \text{ pu} \tag{4.281}$$

Similarly, for the q axis

$$\lambda_q = L_q i_q + L_{AQ} i_Q \text{ pu} \qquad \lambda_Q = L_{AQ} i_q + L_Q i_Q \text{ pu} \tag{4.282}$$

Eliminating i_Q, we compute

$$\lambda_q - (L_{AQ}/L_Q)\lambda_Q = (L_q - L_{AQ}^2/L_Q) i_q \text{ pu} \tag{4.283}$$

by defining

$$L_q' = L_q - L_{AQ}^2/L_Q \text{ pu} \tag{4.284}$$

and by using (4.284) and (4.275) we get

$$e_d' \triangleq \sqrt{3}E_d' = -(L_{AQ}/L_Q)\lambda_Q \text{ pu} \tag{4.285}$$

We also define

$$\sqrt{3}E = e_q = L_{AD} i_F \text{ pu} \qquad \sqrt{3}E_d = e_d = L_{AQ} i_Q \text{ pu} \tag{4.286}$$

We can show that [8],

$$E + x_d I_d = E_q' + x_d' I_d \qquad E_d + x_q I_q = E_d' + x_q'' I_q \tag{4.287}$$

From the Q circuit voltage equation $r_Q i_Q + d\lambda_Q/dt = 0$, and by using (4.282) with (4.286),

$$\tau_{q0}' \dot{E}_d' = -E_d' - (x_q - x_q') I_q \tag{4.288}$$

where, for uniformity, we adopt the notation

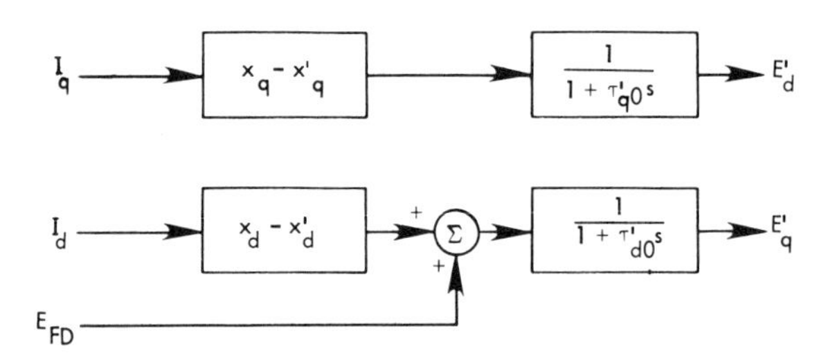

Fig. 4.17 Block diagram representation of the two-axis model.

$$\tau'_{q0} = \tau''_{q0} = L_Q / r_Q \tag{4.289}$$

Similarly, from the field voltage equation we get a relation similar to (4.228)

$$\dot{E}'_q = \frac{1}{\tau'_{d0}} (E_{FD} - E) \tag{4.290}$$

Equations (4.288), (4.290), and (4.287) can be represented by the block diagram shown in Figure 4.17. To complete the description of the system, the electrical torque is obtained from (4.95), $T_{e\phi} = \lambda_d i_q - \lambda_q i_d$, which is combined with (4.274) and (4.275) to compute

$$T_e = E'_d I_d + E'_q I_q - (L'_q - L'_d) I_d I_q \tag{4.291}$$

Example 4.7

Determine the time constants and gains for the two-axis model of Figure 4.17, based on the machine data of Examples 4.1–4.6. In addition we obtain from the manufacturer's data the constant $x'_q = 0.380$ pu.

Solution

Both time constants are known from Example 4.7. The gains are simply the pu reactances

$$x_q - x'_q = 1.64 - 0.380 = 1.260 \text{ pu} \qquad x_d - x'_d = 1.70 - 0.245 = 1.455 \text{ pu}$$

The remaining system equations are given by

$$\begin{aligned} \tau_j \dot{\omega} &= T_m - D\omega - [E'_d I_d + E'_q I_q - (L'_q - L'_d) I_d I_q] \\ \dot{\delta} &= \omega - 1 \end{aligned} \tag{4.292}$$

The block diagram for (4.292) is shown in Figure 4.18.

By combining Figures 4.17 and 4.18, the block diagram for the complete model is obtained. Again saturation can be accounted for by modifying (4.287),

$$E = E'_q - (x_d - x'_d) I_d + E_\Delta \tag{4.293}$$

where E_Δ is a voltage increment that corresponds to the increase in the field current due to saturation (see Young [19]). The procedure for incorporating this modification in the block diagram is similar to that discussed in Section 4.15.2.

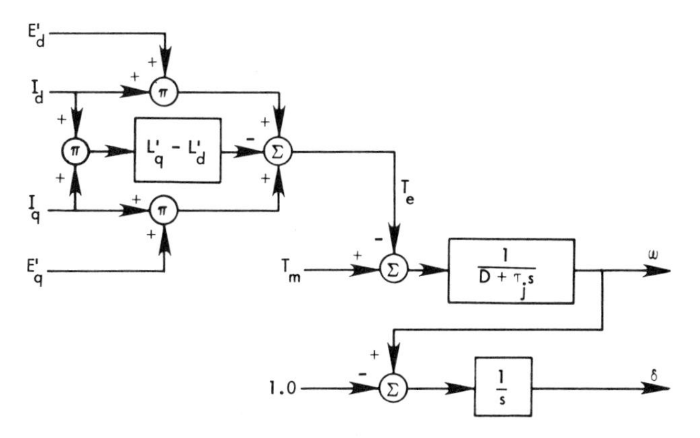

Fig. 4.18 Block diagram representation of (4.292).

4.15.4 Neglecting amortisseur effects and $\dot{\lambda}_d$ and $\dot{\lambda}_q$ terms—the one-axis model

This model is sometimes referred to in the literature as the one-axis model. It is similar to the model presented in the previous section except that the absence of the Q circuit eliminates the differential equation for E_d' or e_d' (which is a function of the current i_Q). The voltage behind transient reactance e' shown in Figure 4.16 has only the component e_q' changing by the field effects according to (4.290) and (4.293). The component e_d' is completely determined from the currents and v_d. Thus, the system equations are

$$\tau_{d0}' \dot{E}_q' = E_{FD} - E \text{ pu} \qquad E = E_q' - (x_d - x_d')I_d \text{ pu} \tag{4.294}$$

The voltage E_d' is obtained from (4.36) with $\dot{\lambda}_d = 0$, and using (4.274) and (4.275),

$$E_d' = V_d + x_q' I_q + r I_d \text{ pu} \tag{4.295}$$

The torque equation is derived from (4.95), $T_{e\phi} = \lambda_d i_q - \lambda_q i_d$. Substituting (4.274) and

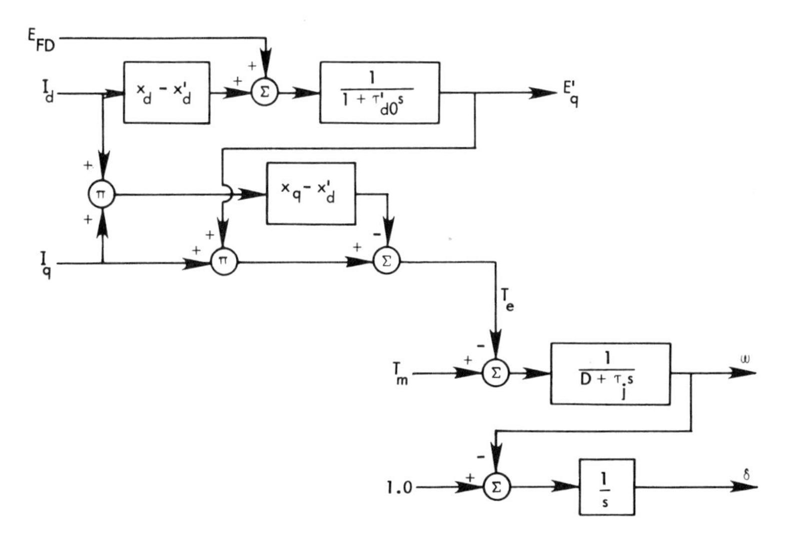

Fig. 4.19 Block diagram representation of the one-axis model.

noting that, in the absence of the Q circuit, $\lambda_q = L_q i_q$,

$$T_e = E'_q I_q - (L_q - L'_d) I_d I_q \quad \text{pu} \tag{4.296}$$

Thus the remaining system equations are

$$\tau_j \dot{\omega} = T_m - D\omega - [E'_q I_q - (L_q - L'_d) I_d I_q] \quad \text{pu} \qquad \dot{\delta} = \omega - 1 \quad \text{pu} \tag{4.297}$$

The block diagram representation of the system is given in Figure 4.19.

4.15.5 Assuming constant flux linkage in the main field winding

From (4.228) we note that the voltage E'_q, which corresponds to the d axis field flux linkage, changes at a rate that depends upon τ'_{d0}. This time constant is on the order of several seconds. The voltage E_{FD} depends on the excitation system characteristics. If E_{FD} does not change very fast and if the impact initiating the transient is short, in some cases the assumption that the voltage E'_q (or e'_q) remains constant during the transient can be justified. Under this assumption the voltage behind transient reactance E' or e' has a q axis component E'_q or e'_q that is always constant. The system equation to be solved is (4.296) with the network constraints (to determine the currents) and the condition that E'_q is constant.

The next step in simplifying the mathematical model of the machine is to assume that E'_q and E' are approximately equal in magnitude and that their angles with respect to the reference voltage are approximately equal (or differ by a small angle that is constant). Under these assumptions E' is considered constant. This is the constant voltage behind transient reactance representation used in the classical model of the synchronous machine.

Example 4.8

The simplified model used in Section 4.15.2 (voltage behind subtransient reactance) is to be used in the system of one machine connected to an infinite bus through a transmission line discussed previously in Section 4.13. The system equations neglecting saturation are to be developed.

Solution

For the case where saturation is neglected, the system equations are given by (4.263)-(4.268). This set of differential equations is a function of the state variables e''_d, λ_D, E'_q, ω, and δ and the currents i_d and i_q. Equation (4.266) expresses e''_q as a linear combination of the variables E'_q and λ_D.

For the mathematical description of the system to be complete, equations for i_d and i_q in terms of the state variables are needed. These equations are obtained from the load constraints.

From the assumptions used in the model, i.e., by neglecting the terms in $\dot{\lambda}_d$ and $\dot{\lambda}_q$ in the stator voltage equations (compared to the speed voltage terms) and also by as-

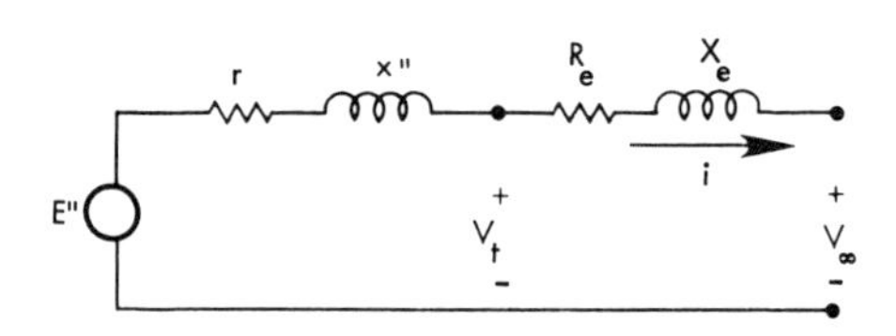

Fig. 4.20 Network representation of the system in Example 4.8.

suming that $\omega \cong \omega_R$, the system reduces to the equivalent network shown in Figure 4.20.

By following a procedure similar to that in Section 4.15.2, equations (4.234) are given by

$$V_{\infty d} = -\hat{R}\, I_d - \hat{X}'' I_q + E_d'' \qquad V_{\infty q} = -\hat{R}\, I_q + \hat{X}'' I_d + E_q'' \tag{4.298}$$

where

$$\hat{R} = r + R_e \qquad \hat{X}'' = x'' + X_e \tag{4.299}$$

and

$$V_{\infty d} = -\sqrt{3}\, V_\infty \sin(\delta - \alpha) \qquad V_{\infty q} = \sqrt{3}\, V \cos(\delta - \alpha) \tag{4.300}$$

From (4.298) I_d and I_q are determined

$$I_d = \frac{1}{(R)^2 + (X'')^2}\left[-\hat{R}(V_{\infty d} - E_d'') + \hat{X}''(V_{\infty q} - E_q'')\right]$$

$$I_q = -\frac{1}{(R)^2 + (X'')^2}\left[\hat{X}''(V_{\infty d} - E_d'') + \hat{R}(V_{\infty q} - E_q'')\right] \tag{4.301}$$

Equations (4.147) and (4.301) along with the set (4.263)–(4.268) complete the mathematical description of the system.

4.16 Turbine Generator Dynamic Models

The synchronous machine models used in this chapter, which are in common use by power system engineers, are based on a classical machine with discrete physical windings on the stator and rotor. As mentioned in Section 4.14, the solid iron rotor used in large steam turbine generators provides multiple paths for circulating eddy currents that act as equivalent damper windings under dynamic conditions. The representation of these paths by one discrete circuit on each axis has been questioned for some time. Another source of concern to the power engineer is that the value of the machine constants (such as L_d', L_d'', etc.) used in dynamic studies are derived from data obtained from ANSI Standard C42.10 [16]. This implicitly assumes two rotor circuits in each axis—the field, one d axis amortisseur, and two q axis amortisseurs. This in turn implies the existence of inductances L_d, L_d', L_d'', L_q, L_q', and L_q'' and time constants τ_{d0}', τ_{d0}'', τ_{q0}', and τ_{q0}'', all of which are intended to define fault current magnitudes and decrements. In some stability studies, discrepancies between computer simulation and field data have been observed. It is now suspected that the reason for these discrepancies is the inadequate definition of machine inductances in the frequency ranges encountered in stability studies.

Studies have been made to ascertain the accuracy of available dynamic models and data for turbine generators [21–25]. These studies show that a detailed representation of the rotor circuits can be more accurately simulated by up to three discrete rotor circuits on the d axis and three on the q axis. Data for these circuits can be obtained from frequency tests conducted with the machine at standstill. To fit the "conventional" view of rotor circuits that influence the so-called subtransient and transient dynamic behavior of the machine, it is found that two rotor circuits (on each axis) are sometimes adequate but the inductances and time constants are not exactly the same as those defined in IEEE Standard No. 115.

The procedure for determining the constants for these circuits is to assume equivalent circuits on each axis made up of a number of circuits in parallel. The transfer function for each is called an operational inductance of the form

$$L(s) = [N(s)/D(s)]L \tag{4.302}$$

where L is the synchronous reactance, and $N(s)$ and $D(s)$ are polynomials in s. Thus for the d axis we write

$$L_d(s) = L_d \frac{(1 + a_1 s)(1 + b_1 s)(1 + c_1 s)}{(1 + a_2 s)(1 + b_2 s)(1 + c_2 s)} \tag{4.303}$$

and the constants L_d, a_1, a_2, b_1, b_2, c_1, and c_2 are determined from the frequency domain response.

If the operational inductance is to be approximated by quadratic polynomials, the constants can be identified approximately with the transient and subtransient parameters. Thus, for the d axis, $L_d(s)$ becomes

$$L_d(s) = L_d \frac{[1 + (L_d'/L_d)\tau_{d0}' s][1 + (L_d''/L_d')\tau_{d0}'' s]}{(1 + \tau_{d0}' s)(1 + \tau_{d0}'' s)} \tag{4.304}$$

The time constants in (4.304) are different from those associated with the exponential decay of d or q axis open circuit voltages, hence the discrepancy with IEEE Standard No. 115.

An example of the data obtained by standstill frequency tests is given in [24] and is reproduced in Figure 4.21. Both third-order and second-order polynomial representations are given. Machine data thus obtained differ from standard data previously obtained by the manufacturer from short circuit tests. Reference [24] gives a comparison between the two sets of data for a 555-MVA turbogenerator. This comparison is given in Table 4.6.

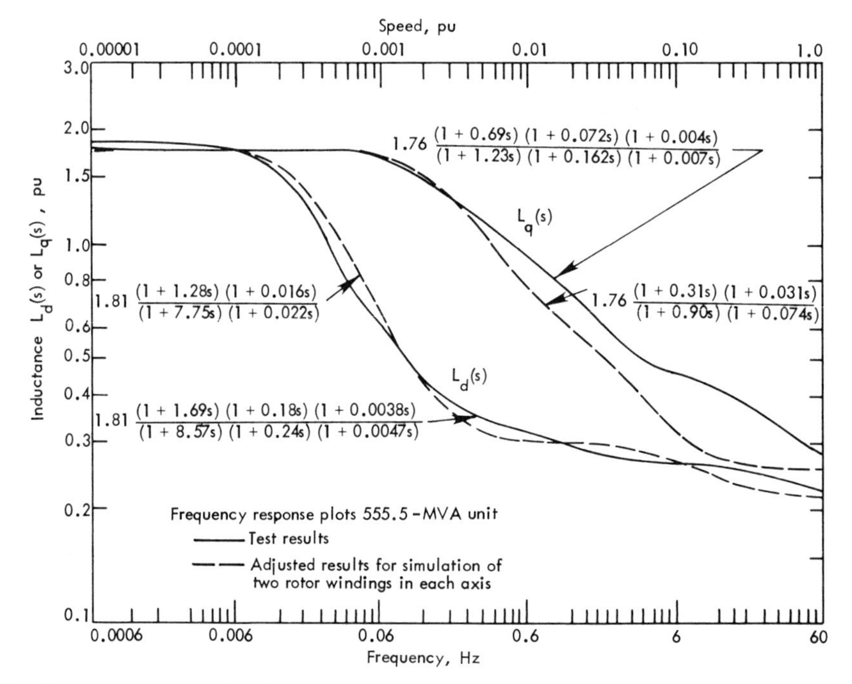

Fig. 4.21 Frequency response plot for a 555-MVA turboalternator. (© IEEE. Reprinted from *IEEE Trans.*, vol. PAS-93, May/June 1974.)

Table 4.6. Comparison of Standard Data with Data Obtained from Frequency Tests for a 555-MVA turboalternator

Constants		Standard data	Adjusted data
L_d	pu	1.97	1.81
L'_d	pu	0.27	0.30
L''_d	pu	0.175	0.217
L_q	pu	1.867	1.76
L'_q	pu	0.473	0.61
L''_q	pu	0.213	0.254
L_ℓ	pu	0.16	0.16
τ'_{d0}	s	4.3	7.8
τ''_{d0}	s	0.031	0.022
τ'_{q0}	s	0.56	0.90
τ''_{q0}	s	0.061	0.074

Source: © IEEE. Reprinted from *IEEE Trans.*, vol. PAS-93, 1974.

The inductance versus frequency plot given in Figure 4.21 is nothing more than the amplitude portion of the familiar Bode plot with the amplitude given in pu rather than in decibels. The transfer functions plotted in Figure 4.21 can be approximated by the superposition of multiple first-order asymptotic approximations. If this is done, the break frequencies should give the constants of (4.304). The machine constants thus obtained are given in the third column of Table 4.6. If, however, the machine constants obtained from the standard data are used to obtain the breakpoints for the straight-line approximation of the amplitude-frequency plots, the approximated curve does not provide a good fit to the experimental data. For example, the *d* axis time constant τ'_{d0} of the machine, as obtained by standard methods, is 4.3 s. If this is used to obtain the first break frequency for log $[1/(1 + \tau'_{d0}s)]$, the computed break frequency is

$$1/\tau'_{d0} = 1/4.3 = 0.2326 \ \text{rad/s} = 0.00062 \ \text{pu} \tag{4.305}$$

The break point that gives a better fit of the experimental data corresponds to a frequency of 0.1282 rad/s or 0.00034 pu. Since the amplitude at this frequency is the reciprocal of the *d* axis transient time constant, this corresponds to an adjusted value, denoted by τ'^*_{d0}, given by

$$\tau'^*_{d0} = 1/0.1282 = 7.8 \ \text{s} \tag{4.306}$$

Reference [24] notes that the proper ajustment of τ'_{d0}, τ'_{q0}, and L'_q are all particularly important in stability studies.

A study conducted by the Northeast Power Coordinating Council [26] concludes that, in general, it is more important in stability studies to use accurate machine data than to use more elaborate machine models. Also, the accuracy of any dynamic machine model is greatly improved when the so-called standard machine data are modified to match the results of a frequency analysis of the solid iron rotor equivalent circuit. At the time of this writing no extensive studies have been reported in the literature to support or dispute these results.

Finally, a comparison of these results and the machine models presented in this chapter are in order. The full model presented here is one of the models investigated in the NPCC study [26] for solid rotor machines. It was found to be inferior to the more elaborate model based on two rotor windings in each axis. This is not surprising since the added detail due to the extra q axis amortisseur should result in an improved simulation. Perhaps more surprising is the fact that the model developed here with F, D, and Q windings provided practically no improvement over a simpler model with only F and Q windings. Furthermore, with the F-Q model based on time constants τ'_{d0} and τ'_{q0}, larger digital integration time steps are possible than with models that use the much shorter time constants τ''_{d0} and τ''_{q0}, as done in this chapter.

As a general conclusion it is apparent that additional studies are needed to identify the best machine data for stability studies and the proper means for testing or estimating these data. This is not to imply that the work of the past is without merit. The traditional models, including those developed in this chapter, are often acceptable. But, as in many technical areas, improvements can and are constantly being made to provide mathematical formulations that better describe the physical apparatus.

Problems

4.1 Park's transformation **P** as defined by (4.5) is an orthogonal transformation. Why? But the transformation **Q** suggested originally by Park [10, 11] is that given by (4.22) and is not orthogonal. Use the transformation **Q** to find voltage equations similar to (4.39).

4.2 Verify (4.9) by finding the inverse of (4.5).

4.3 Verify (4.12) by sketching the stator coils as in Figure 4.1 and observing how the inductance changes with rotor position.

4.4 Verify the following equations:
(a) Equation (4.13). Can you explain why these inductances are constant?
(b) Equation (4.14). Why is the sign of M_s negative? Why is $|M_s| > L_m$?
(c) Explain (4.15) in terms of the coefficient of coupling of these coils.

4.5 Verify (4.16)–(4.18). Explain the signs on these equations by referring to the currents given on Figures 4.1 and 4.2.

4.6 Verify (4.20).

4.7 Explain the signs on all terms of (4.23). Why is the $\dot{\lambda}$ term negative?

4.8 Consider a machine consisting only of the phase winding *sa-fa* shown in Figure 4.1 and the field winding F. Sketch a new physical arrangement where the field flux is stationary and coil *sa-fa* turns clockwise. Are these two physical arrangements equivalent? Explain.

4.9 For the new physical machine proposed in Problem 4.8 we wish to compute the induced EMF in coil *sa-fa*. Do this by two methods and compare your results, *including* the polarity of the induced voltage.
(a) Use the rate of change of flux linkages $\dot{\lambda}$.
(b) Compute the Blv or speed voltage and the transformer-induced voltage.
Do the results agree? They should!

4.10 Verify (4.24) for the neutral voltage drop.

4.11 Check the computation of $\dot{\mathbf{P}}\mathbf{P}^{-1}$ given in (4.32).

4.12 The quantities λ_d and λ_q are given in (4.20). Substitute these quantities into (4.32) and compute the speed voltage terms. Check your result against (4.39).

4.13 Verify (4.34) and explain its meaning.

4.14 Extend Table 4.1 by including the actual dimensions of the voltage equations in an $MLt\mu$ system. Repeat for an $FLtQ$ system.

4.15 Let $v_a(t) = V_m \cos(\omega_R t + \alpha)$
$v_b(t) = V_m \cos(\omega_R t + \alpha - 2\pi/3)$
$v_c(t) = V_m \cos(\omega_R t + \alpha + 2\pi/3)$
(a) For the pu system used in this book find the pu voltages v_d and v_q as related to the rms voltage V.
(b) Repeat part (a) using a pu system based on the following base quantities: S_B = three-phase voltampere and V_B = line-to-line voltage.
(c) For part (b) find the pu power in the d and q circuits and i_d and i_q in pu.

4.16 Using the transformation **Q** of (4.22) (originally used by Park) and the MKS system of units (volt, ampere, etc.), find:
(a) The d and q axis voltages and currents in relation to the rms quantities.
(b) The d and q axis circuit power in relation to the three-phase power.

4.17 Normalize the voltage equations as in Section 4.8 but where the equations are those found from the **Q** transformation of Problem 4.1.

4.18 Show that the choice of a common time base in any coupled circuit automatically forces the equality of VA base in all circuit parts and requires that the base mutual inductance be the geometric mean of the self-inductance bases of the coupled windings; i.e.,

$$S_{1B} = S_{2B} \qquad M_{12B} = (L_{1B}L_{2B})^{1/2}$$

4.19 Show that the constraint among base currents (4.54) based upon equal mutual flux linkages is the same as equal MMF's in each winding.

4.20 Show that the $1/\omega_R$ factors may be eliminated from (4.62) by choosing a pu time $\tau = \omega_R t$ rad.

4.21 Develop the voltage equations for a cylindrical rotor machine, i.e., a machine in which the inductances are not a function of rotor angle except for rotor-stator inductances that are as given in (4.16)–(4.18).

4.22 Consider a synchronous generator for which the following data are given: 2 poles, 2 slots/pole/phase, 3 phases, 6 slots/pole, 12 slots, 5/6 pitch. Sketch the slots and show two coils of the phase a winding, coil 1 beginning in slot 1 (0°) and coil 2 beginning in slot 7 (180°). Label coil 1 sa_1-fa_1 (start a_1 and finish a_1) and coil 2 sa_2-fa_2. Show the position of N and S salient poles and indicate the direction of pole motion.

Now assume the machine is operating at 1.0 PF (internal PF) and note by + and · notation, looking in at the coil ends, the direction of currents at time t_0, where at t_0

$$i_a = I_{max} \qquad i_b = -(1/2)I_{max} \qquad i_c = -(1/2)I_{max}$$

Plot the MMF as positive when radially outward $+i_a$ enters sa_1 and $+i_b$ enters sb_1 but $+i_c$ enters fc_1. Assume the MMF changes abruptly at the center line of the slot. The MMF wave should be a stepwise sine wave. Is it radially outward along d or q?

4.23 Verify (4.138).

4.24 Derive formulas for computing the saturation function parameters A_s and B_s defined in (4.141), given two different values of the variables λ_{AD}, i_{M0}, and i_{MS}.

4.25 Compute the saturation function parameters A_s and B_s given that when

$$\lambda_{AD} = \sqrt{3}, \quad (i_{MS} - i_{M0})/i_{M0} = 0.13$$
$$\lambda_{AD} = 1.2\sqrt{3}, \quad (i'_{MS} - 1.2i_{M0})/1.2i_{M0} = 0.40$$

where i_{MS} and i_{M0} correspond to $\lambda_{AD} = \sqrt{3}$ and i'_{MS} is the saturated current at $\lambda_{AD} = 1.2\sqrt{3}$.

4.26 Compute the saturation function K_s at $\lambda_{AD} = 1.8$, using the data and results of the previous problem. Let $\lambda_{ADT} = 0.8\sqrt{3}$.

4.27 The synchronous machine described in Examples 4.2 and 4.3 is connected to a resistive load of $R_L = 1.0$ pu. Derive the equations for the state-space current model using v_F and T_m as forcing functions. Use the current model.

4.28 Repeat Problem 4.27 using the flux linkage model.

4.29 Derive the state-space model for a synchronous machine connected to an infinite bus with a local load at the machine terminal. The load is to be simulated by a passive resistance.

4.30 Repeat Problem 4.29 for a local load simulated by a passive impedance. The load has a reactive component.

4.31 Obtain the state-space model for a synchronous machine connected to an infinite bus through a series resistance, inductance, and capacitance. **Hint:** Add two state variables related to the voltage (or charge) across the capacitance.

4.32 Incorporate the load equations for the system of one machine against an infinite bus (shown in Figure 4.8) in the simplified models given in Section 4.15:
(a) Neglecting damper effects.

(b) Neglecting $\dot{\lambda}_d$ and $\dot{\lambda}_q$ for a machine with solid round rotor.
(c) Neglecting damper effects and the terms $\dot{\lambda}_d$ and $\dot{\lambda}_q$.

4.33 Show that the voltage-behind-subtransient-reactance model of Figure 4.14 can be rearranged to give the model of Schulz [20] given in Figure P4.33.

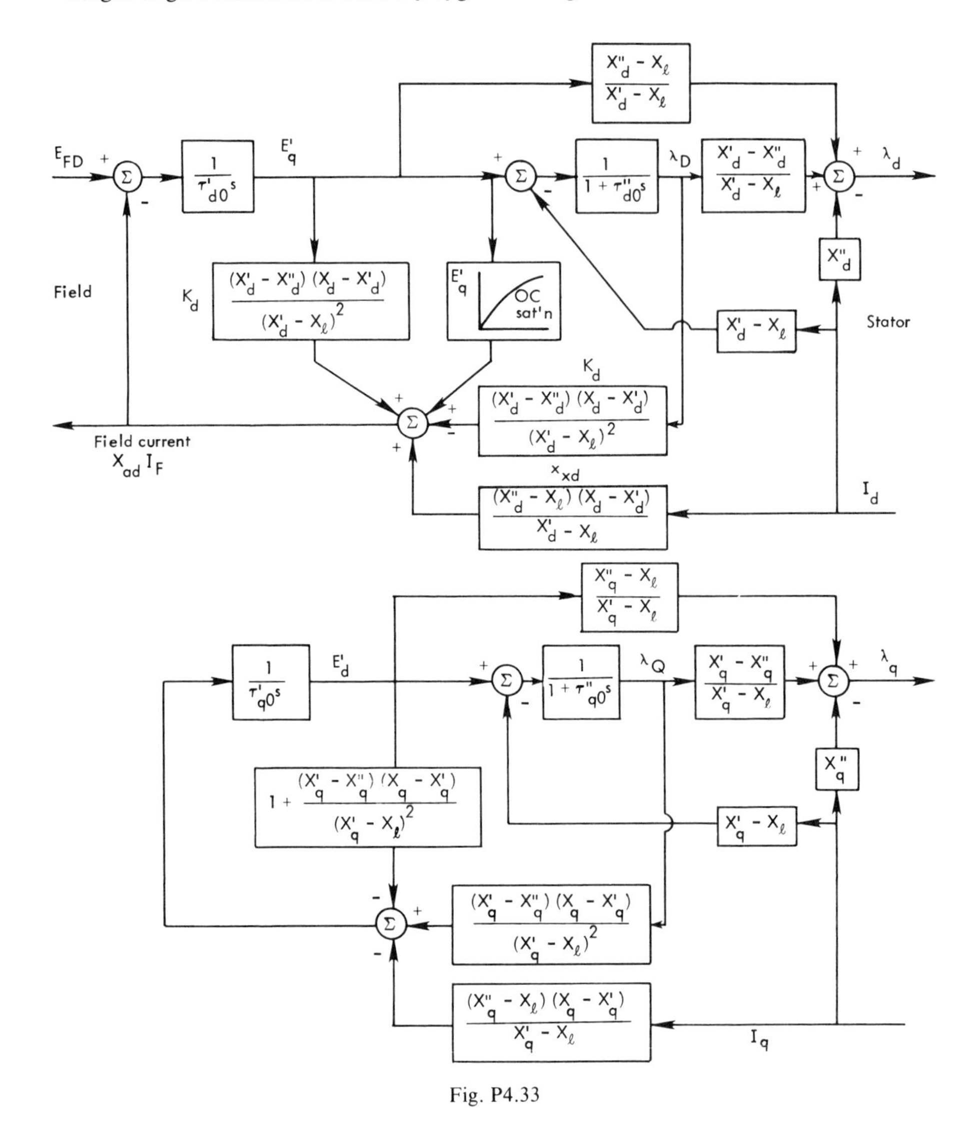

Fig. P4.33

4.34 Using the third-order transfer functions for $L_d(s)$ and $L_q(s)$ given in Figure 4.21, sketch Bode diagrams by making straight-line asymptotic approximations and compare with the given test results.

4.35 Repeat Problem 4.34 using the second-order transfer functions for $L_d(s)$ and $L_q(s)$.

4.36 Repeat Problem 4.35 using the second-order transfer functions of (4.304) and substituting the standard data rather than the adjusted data.

References

1. Concordia, C. *Synchronous Machines*. Wiley, New York, 1951.
2. Kimbark, E. W. *Power System Stability*, Vols. 1, 3. Wiley, New York, 1956.
3. Adkins, B. *The General Theory of Electrical Machines*. Chapman and Hall, London, 1964.

4. Crary, S. B. *Power System Stability,* Vols. 1, 2. Wiley, New York, 1945, 1947.
5. Lynn, T. W., and Walshaw, M. H. *Tensor Analysis of a Synchronous Two-Machine System.* IEE (British) Monograph. Cambridge Univ. Press, London, 1961.
6. Taylor, G. D. *Analysis of Synchronous Machines Connected to Power Network.* IEE (British) Monograph. Cambridge Univ. Press, London, 1962.
7. Westinghouse Electric Corp. *Electrical Transmission and Distribution Reference Book.* Pittsburgh, Pa., 1950.
8. Anderson, P. M. *Analysis of Faulted Power Systems.* Iowa State Univ. Press, Ames, 1973.
9. Harris, M. R., Lawrenson, P. J., and Stephenson, J. M. *Per Unit Systems: With Special Reference to Electrical Machines.* IEE (British) Monograph. Cambridge Univ. Press, London, 1970.
10. Park, R. H. Two reaction theory of synchronous machines, Pt. 1. *AIEE Trans.* 48:716–30, 1929.
11. Park, R. H. Two reaction theory of synchronous machines, Pt. 2. *AIEE Trans.* 52:352–55, 1933.
12. Lewis, W. A. A basic analysis of synchronous machines, Pt. 1. *AIEE Trans.* PAS-77:436–55, 1958.
13. Krause, P. C., and Thomas, C. H. Simulation of symmetrical induction machinery. *IEEE Trans.* PAS-84:1038–52, 1965.
14. Prentice, B. R. Fundamental concepts of synchronous machine reactances. *AIEE Trans.* 56 (Suppl. 1): 716–20, 1929.
15. Rankin, A. W. Per unit impedances of synchronous machines. *AIEE Trans.* 64:569–72, 839–41, 1945.
16. IEEE. Test procedures for synchronous machines. Standard No. 115, March, 1965.
17. IEEE Committee Report. Recommended phasor diagram for synchronous machines. *IEEE Trans.* PAS-88:1593–1610, 1969.
18. Prubhaskankar, K., and Janischewskyj, W. Digital simulation of multimachine power systems for stability studies. *IEEE Trans.* PAS-87:73–80, 1968.
19. Young, C. C. Equipment and system modeling for large-scale stability studies. *IEEE Trans.* PAS-91:99–109, 1972.
20. Schulz, R. P. Synchronous machine modeling. Symposium on Adequacy and Philosophy of Modeling: System Dynamic Performance. IEEE Publ. 75 CH O970-PWR, 1975.
21. Jackson, W. B., and Winchester, R. L. Direct and quadrature axis equivalent circuits for solid-rotor turbine generators. *IEEE Trans.* PAS-88:1121–36, 1969.
22. Schulz, R. P., Jones, W. D., and Ewart, D. N. Dynamic models of turbine generators derived from solid rotor equivalent circuits. *IEEE Trans.* PAS-92:926–33, 1973.
23. Watson, W., and Manchur, G. Synchronous machine operational impedances from low voltage measurements at the stator terminals. *IEEE Trans.* PAS-93:777–84, 1974.
24. Kundur, P., and Dandeno, P. L. Stability performance of 555 MVA turboalternators—Digital comparisons with system operating tests. *IEEE Trans.* PAS-93:767–76, 1974.
25. Dandeno, P. L., Hauth, R. L., and Schulz, R. P. Effects of synchronous machine modeling in large-scale system studies. *IEEE Trans.* PAS-92:574–82, 1973.
26. Northeast Power Coordinating Council. Effects of synchronous machine modeling in large-scale system studies. Final Report, NPCC-10, Task Force on System Studies, System Dynamic Simulation Techniques Working Group, 1971.

chapter 5

The Simulation of Synchronous Machines

5.1 Introduction

This chapter covers some practical considerations in the use of the mathematical models of synchronous machines in stability studies. Among these considerations are the determination of initial conditions, determination of the parameters of the machine from available data, and construction of simulation models for the machine.

In all dynamic studies the initial conditions of the system are required. This includes all the currents, flux linkages, and EMF's for the different machine circuits. The number of these circuits depends upon the model of the machine adopted for the study. The initial position of the rotor with respect to the system reference axis must also be known. These quantities will be determined from the data available at the terminals of the machine.

The machine models used in Chapter 4 require some data not usually supplied by the manufacturer. Here we show how to obtain the required machine parameters from typical manufacturer's data. The remainder of the chapter is devoted to the construction of simulation models for the synchronous machine. Both analog and digital simulations are discussed.

5.2 Steady-State Equations and Phasor Diagrams

The equations of the synchronous machine derived in Chapter 4 are differential equations that describe machine behavior as a function of time. When the machine operates in a steady-state condition, differential equations are not necessary since all variables are either constants or sinusoidal variations with time. For this situation phasor equations are appropriate, and these will be derived. It is common to tacitly assume all machines to be in a steady-state condition prior to a disturbance. The so-called "stability study" examines the system behavior following the disturbance. The phasor equations derived here permit the solution of the initial conditions that exist prior to the application of the disturbance. This is a necessary part of any stability investigation.

From (4.74) at steady state all currents are constant or, mathematically,

$$\dot{i}_d = \dot{i}_F = \dot{i}_D = \dot{i}_q = \dot{i}_Q = 0 \tag{5.1}$$

Then from (4.74)

$$0 = i_D r_D \qquad 0 = i_Q r_Q \tag{5.2}$$

or at steady state

$$i_D = i_Q = 0 \tag{5.3}$$

Using (5.1) we may write the stator voltage equation from (4.74) as

$$v_d = -ri_d - \omega L_q i_q \qquad v_q = -ri_q + \omega L_d i_d + \mathrm{k}M_F\omega i_F \tag{5.4}$$

From (4.5) with balanced conditions, $v_0 = 0$. Therefore, from (4.9) we may compute

$$v_a = \sqrt{2/3}(v_d \cos\theta + v_q \sin\theta) \tag{5.5}$$

where by definition $\theta = \omega_R t + \delta + \pi/2$. Then from (5.4) and (5.5)

$$\begin{aligned} v_a &= \sqrt{2/3}[-(ri_d + \omega L_q i_q)\cos(\omega_R t + \delta + \pi/2) \\ &\quad + (-ri_q + \omega L_d i_d + \mathrm{k}M_F\omega i_F)\sin(\omega_R t + \delta + \pi/2)] \\ &= \sqrt{2/3}[-(ri_d + \omega L_q i_q)\cos(\omega_R t + \delta + \pi/2) \\ &\quad + (-ri_q + \omega L_d i_d + \mathrm{k}M_F\omega i_F)\cos(\omega_R t + \delta)] \end{aligned} \tag{5.6}$$

At steady state the angular speed is constant, $\omega = \omega_R$, and ωL products may be denoted as reactances, or

$$\omega L_q = x_q \qquad \omega L_d = x_d \tag{5.7}$$

From (4.226) we also identify

$$\omega_R M_F i_F = \sqrt{2}E \tag{5.8}$$

where E is the stator equivalent EMF corresponding to i_F. Using phasor notation,[1] the $\sqrt{2}$ multiplier of (5.6) is conveniently used to define the rms voltage phasor

$$\overline{V}_a = -r\left(\frac{i_d}{\sqrt{3}}\angle\delta + \pi/2 + \frac{i_q}{\sqrt{3}}\angle\delta\right) - x_q\frac{i_q}{\sqrt{3}}\angle\delta + \pi/2 + x_d\frac{i_d}{\sqrt{3}}\angle\delta + E\angle\delta \tag{5.9}$$

where the superior bar indicates a total phasor quantity in magnitude and angle (a complex number).

By using the relation $\mathrm{j} = 1\angle\pi/2$ in (5.9),

$$\overline{V}_a = -r\left(\frac{i_q}{\sqrt{3}}\angle\delta + \mathrm{j}\frac{i_d}{\sqrt{3}}\angle\delta\right) - \mathrm{j}x_q\frac{i_q}{\sqrt{3}}\angle\delta + x_d\frac{i_d}{\sqrt{3}}\angle\delta + E\angle\delta \tag{5.10}$$

Note that in this equation V_a and E are stator rms phase voltages in pu, while i_d and i_q are dc currents obtained from the modified Park transformation. The choice of this particular transformation introduced the factor $1/\sqrt{3}$ in the equation. To simplify the notation we define the rms equivalent d and q axis currents as

$$I_d \triangleq i_d/\sqrt{3} \qquad I_q \triangleq i_q/\sqrt{3} \tag{5.11}$$

The stator current i_a expressed as a phasor will have the two rectangular components I_q and I_d. Thus if the phasor reference is the q axis,

$$\overline{I}_a = \overline{I}_q + \mathrm{j}\overline{I}_d \tag{5.12}$$

1. We define the phasor $\overline{A} = Ae^{\mathrm{j}\alpha}$ as a complex number that is related to the corresponding time domain quantity $a(t)$ by the relation $a(t) = \mathrm{Re}\,(\sqrt{2}\overline{A}e^{\mathrm{j}\omega t}) = \sqrt{2}A\cos(\omega t + \alpha)$.

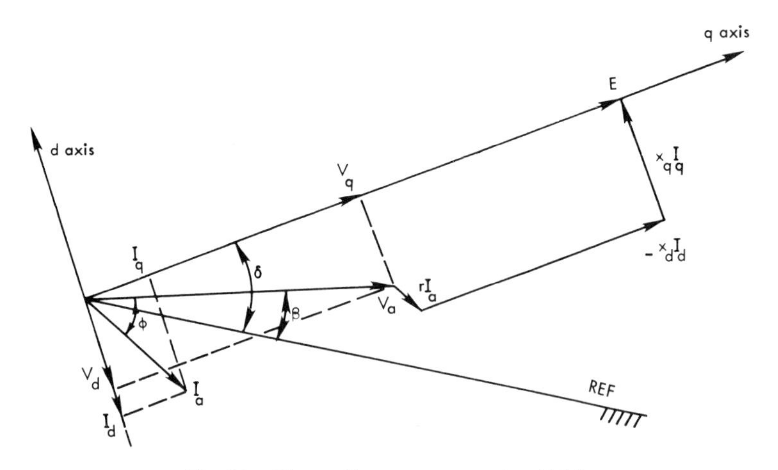

Fig. 5.1 Phasor diagram representing (5.14).

Substituting (5.12) and (5.11) in (5.10) and rearranging,

$$E \underline{/\delta} = \overline{V}_a + r\overline{I}_a + \mathrm{j}x_q I_q \underline{/\delta} - x_d I_d \underline{/\delta} \tag{5.13}$$

and by using $\overline{E} = E\underline{/\delta}$, $\overline{I}_q = I_q\underline{/\delta}$, and $\overline{I}_d = \mathrm{j}I_d\underline{/\delta}$,

$$\overline{E} = \overline{V}_a + r\overline{I}_a + \mathrm{j}x_q\overline{I}_q + \mathrm{j}x_d\overline{I}_d \tag{5.14}$$

The phasor diagram representing (5.14) is shown in Figure 5.1 [1]. Note that the phasor $\mathrm{j}x_q\overline{I}_q$ leads the q axis by 90°. The phasor $\mathrm{j}x_d\overline{I}_d$ makes a 90° angle with the negative d axis since I_d is numerically negative for the case illustrated in Figure 5.1. To obtain v_d and v_q from (5.4), we compute the rms stator equivalent voltages

$$V_d \triangleq v_d/\sqrt{3} = -rI_d - x_qI_q \qquad V_q \triangleq v_q/\sqrt{3} = -rI_q + x_dI_d + E \tag{5.15}$$

Note that V_q and V_d are the projection of V_a along the q and d axes respectively. Also note that in the phasor diagram in Figure 5.1 both V_d and I_d are illustrated as negative quantities. Thus the magnitude of rI_d is subtracted from x_qI_q to obtain the magnitude of V_d. This situation is shown in Figure 5.1 since lagging current (negative I_d) is commonly encountered in practice. Examining Figure 5.1 and (5.15), we note that if the angle δ is known the phasor diagram can be constructed quite readily. If the position of the q axis is not known but the terminal conditions of the machine

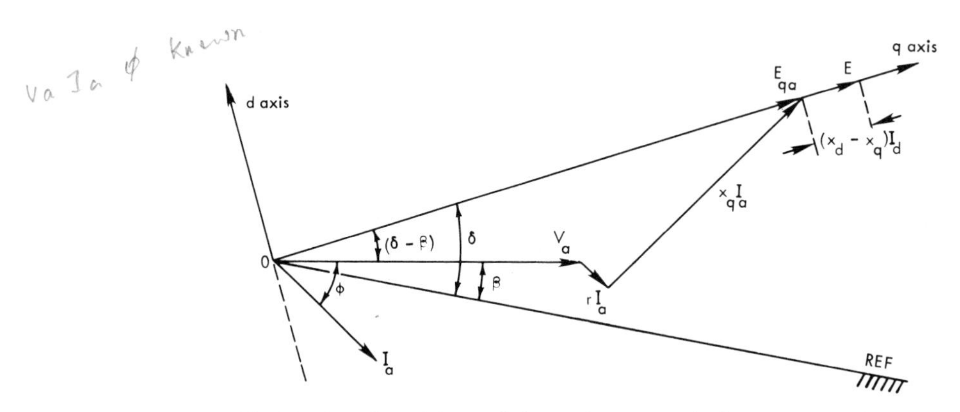

Fig. 5.2 Location of the q axis from a known terminal current and voltage.

are given (i.e., if V_a, I_a, and the angle between them are known), construction of the phasor diagram requires some manipulation of (5.15). However, an alternate procedure for locating the position of the q axis is illustrated in Figure 5.2, where it is assumed that V_a, I_a, and the power factor angle are known. Starting with $\overline{V}_a$ (used here as reference) the voltage drop $r\overline{I}_a$ is drawn parallel to $\overline{I}_a$. Then the voltage drop $jx_q\overline{I}_a$ is added (this is a phasor perpendicular to $\overline{I}_a$). The end of that phasor ($\overline{E}_{qa}$ in Figure 5.2) is located on the q axis. This can be verified by noting that the d axis component of the phasor $jx_q\overline{I}_a$ is $x_q\overline{I}_q$, which is similar to that shown in Figure 5.1. Its q axis component however is $x_q\overline{I}_d$, which is different from that shown in Figure 5.1. Thus to locate the phasor $\overline{E}$ in Figure 5.2, we add the phasor $(x_d - x_q)\overline{I}_d$ to the phasor $\overline{E}_{qa}$.

5.3 Machine Connected to an Infinite Bus through a Transmission Line

To illustrate more fully the procedure for finding the machine steady-state conditions, we solve the simple problem of one machine connected to an infinite bus through a transmission line. Although this one-machine problem is far simpler than actual systems, it serves well to illustrate the procedure of finding initial conditions for any machine. As we shall see later, this simple problem helps us concentrate on concepts without becoming engulfed in details.

The differential equations for one machine connected to an infinite bus through a transmission line with impedance $Z_e = R_e + j\omega_R L_e$ is given by (4.149). Under balanced steady-state conditions with zero derivatives, (4.149) becomes

$$\begin{aligned} v_d &= -\sqrt{3}V_\infty \sin(\delta - \alpha) + R_e i_d + \omega L_e i_q \\ v_q &= \sqrt{3}V_\infty \cos(\delta - \alpha) + R_e i_q - \omega L_e i_d \end{aligned} \tag{5.16}$$

Substituting for v_d and v_q from (5.4) into (5.16),

$$\begin{aligned} -ri_d - \omega L_q i_q &= -\sqrt{3}V_\infty \sin(\delta - \alpha) + R_e i_d + \omega L_e i_q \\ -ri_q + \omega L_d i_d + kM_F \omega i_F &= \sqrt{3}V_\infty \cos(\delta - \alpha) + R_e i_q - \omega L_e i_d \end{aligned}$$

By using (5.7) and (5.11) and rearranging the above equations, we compute

$$\begin{aligned} E &= V_\infty \cos(\delta - \alpha) + (r + R_e)I_q - (x_d + X_e)I_d \\ 0 &= -V_\infty \sin(\delta - \alpha) + (r + R_e)I_d + (x_q + X_e)I_q \end{aligned} \tag{5.17}$$

where $X_e = \omega L_e$. Equations (5.17) represent the components of the voltages along the q and d axes respectively. The phasor diagram described by these equations is shown in Figure 5.3, where the phasor representing the infinite bus voltage $\overline{V}_\infty$, with the q axis as reference, is given by

$$\overline{V}_\infty = V_{\infty q} + jV_{\infty d} = V_\infty \cos(\delta - \alpha) - jV_\infty \sin(\delta - \alpha) \tag{5.18}$$

Note that Figures 5.1 and 5.2 can be combined since the same q and d axes, the same EMF E, and the same current I_a are applicable to both. Thus in Figure 5.3 the machine terminal voltage components V_d and V_q can be obtained using (5.15). An alternate procedure would be to start with the phasor $\overline{V}_\infty$ in Figure 5.3, then add the voltage drop $R_e I_q - X_e I_d$ in the q axis direction and the voltage drop $R_e I_d + X_e I_q$ in the d axis direction to obtain the phasor $\overline{V}_a$.

Again remember that in Figure 5.3 both I_d and $V_{\infty d}$ are shown as negative quantities. The remarks concerning the location of the q axis starting from V_∞ and I_a are also applicable here.

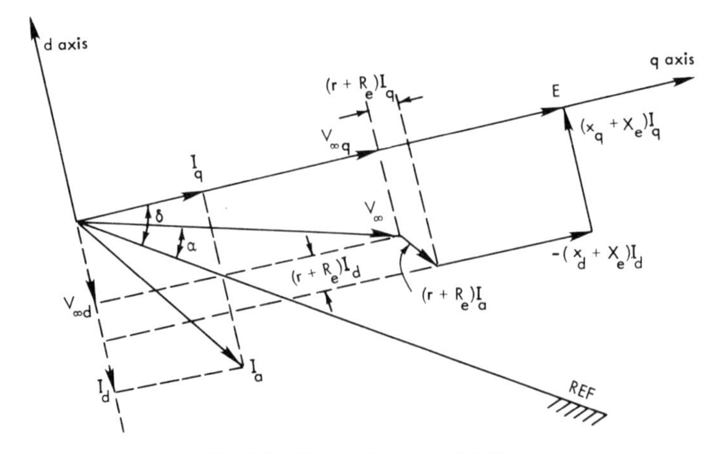

Fig. 5.3 Phasor diagram of (5.17).

5.4 Machine Connected to an Infinite Bus with Local Load at Machine Terminal

The equations that relate the infinite bus voltage V_∞ to the stator equivalent EMF E are given by (5.17). Note that this form of the equations does not give the machine terminal voltage explicitly. Since the terminal voltage is a quantity of considerable interest, we seek a solution in which V_d and V_q are given explicitly. One convenient method is to add a local load at the machine terminals, as shown in Figure 5.4.

For the system shown in Figure 5.4, the steady-state equations for the machine voltages, EMF's, and currents are the same as given by (5.14), (5.15), and (5.12) respectively. Equations (4.149), which at steady-state conditions are the same as (5.16), are still applicable except that the currents i_d and i_q should be replaced by the currents i_{td} and i_{tq}. These are the d and q axis components of the transmission line current i_t. In other words, with the q axis as a reference,

$$\bar{I}_t = I_{tq} + jI_{td} \tag{5.19}$$

where we define

$$I_{tq} = i_{tq}/\sqrt{3} \qquad I_{td} = i_{td}/\sqrt{3} \tag{5.20}$$

The transmission line equations are then given by

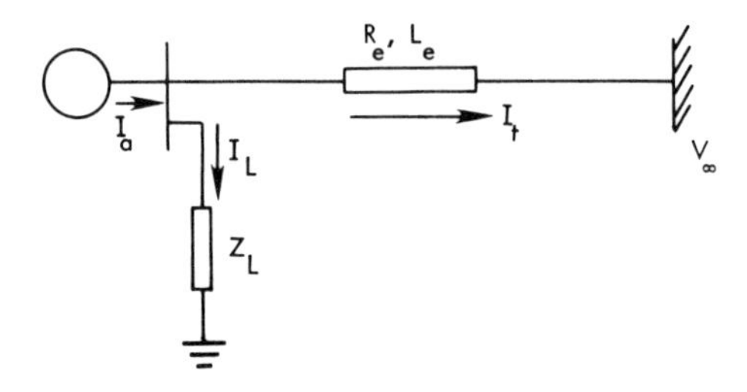

Fig. 5.4 One machine with a local load connected to an infinite bus through a transmission line.

$$\begin{aligned} v_d &= -\sqrt{3}V_\infty \sin(\delta - \alpha) + R_e i_{td} + \omega L_e i_{tq} \\ v_q &= \sqrt{3}V_\infty \cos(\delta - \alpha) + R_e i_{tq} - \omega L_e i_{td} \end{aligned} \tag{5.21}$$

which can be stated in the form

$$\begin{aligned} V_d &= -V_\infty \sin(\delta - \alpha) + R_e I_{td} + X_e I_{tq} \\ V_q &= V_\infty \cos(\delta - \alpha) + R_e I_{tq} - X_e I_{td} \end{aligned} \tag{5.22}$$

To obtain a relation between $\bar{I}_t$ and $\bar{I}_a$, we refer to Figure 5.4. By inspection we can write the phasor relations

$$\bar{I}_L = \bar{I}_a - \bar{I}_t \qquad (\bar{I}_a - \bar{I}_t)(R_L + jX_L) = V_q + jV_d \tag{5.23}$$

where we define $\bar{\bar{Z}}_L = R_L + jX_L$. Separating the real and imaginary components,

$$\begin{aligned} (I_q - I_{tq})R_L - (I_d - I_{td})X_L &= V_q \\ (I_d - I_{td})R_L + (I_q - I_{tq})X_L &= V_d \end{aligned} \tag{5.24}$$

From (5.24) we can solve for I_{tq} and I_{td},

$$I_{td} = I_d + \frac{V_q X_L - V_d R_L}{R_L^2 + X_L^2} \qquad I_{tq} = I_q - \frac{V_q R_L + V_d X_L}{R_L^2 + X_L^2} \tag{5.25}$$

The equations for the q and d axis voltage drops can then be obtained from (5.25), (5.15), and (5.22).

5.4.1 Special case: the resistive load, $\bar{\bar{Z}}_L = R_L + j0$

For this case $X_L = 0$. From (5.25)

$$I_{td} = I_d - V_d/R_L \qquad I_{tq} = I_q - V_q/R_L \tag{5.26}$$

Substituting (5.26) into (5.22),

$$\begin{aligned} V_d &= -V_\infty \sin(\delta - \alpha) + R_e(I_d - V_d/R_L) + X_e(I_q - V_q/R_L) \\ V_q &= V_\infty \cos(\delta - \alpha) + R_e(I_q - V_q/R_L) - X_e(I_d - V_d/R_L) \end{aligned}$$

or

$$\begin{aligned} V_d(1 + R_e/R_L) + V_q(X_e/R_L) &= -V_\infty \sin(\delta - \alpha) + R_e I_d + X_e I_q \\ -V_d(X_e/R_L) + V_q(1 + R_e/R_L) &= V_\infty \cos(\delta - \alpha) + R_e I_q - X_e I_d \end{aligned} \tag{5.27}$$

Substituting (5.15) into (5.27) and rearranging,

$$\begin{aligned} \frac{X_e}{R_L} E &= -V_\infty \sin(\delta - \alpha) + \left(R_e + r\frac{R_e + R_L}{R_L} - \frac{X_e x_d}{R_L}\right) I_d \\ &\quad + \left(x_q \frac{R_L + R_e}{R_L} + r\frac{X_e}{R_L} + X_e\right) I_q \\ \left(1 + \frac{R_e}{R_L}\right) E &= V_\infty \cos(\delta - \alpha) - \left(X_e + r\frac{X_e}{R_L} + x_d \frac{R_L + R_e}{R_L}\right) I_d \\ &\quad + \left(R_e + r\frac{R_e + R_L}{R_L} - \frac{X_e x_q}{R_L}\right) I_q \end{aligned} \tag{5.28}$$

Now define

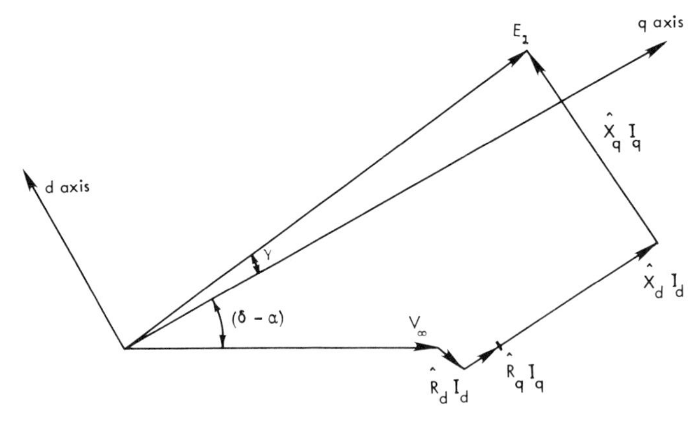

Fig. 5.5 Phasor diagram of a synchronous machine connected to an infinite bus with local resistive load.

$$\hat{R}_d = R_e + r\frac{R_e + R_L}{R_L} - \frac{X_e x_d}{R_L} \qquad \hat{R}_q = R_e + r\frac{R_e + R_L}{R_L} - \frac{X_e x_q}{R_L}$$

$$\hat{X}_q = X_e(1 + r/R_L) + x_q(1 + R_e/R_L) \qquad \hat{X}_d = X_e(1 + r/R_L) + x_d(1 + R_e/R_L)$$

Then (5.28) can be written as

$$\begin{aligned}(X_e/R_L)E &= -V_\infty \sin(\delta - \alpha) + \hat{R}_d I_d + \hat{X}_q I_q \\ (1 + R_e/R_L)E &= V_\infty \cos(\delta - \alpha) - \hat{X}_d I_d + \hat{R}_q I_q\end{aligned} \tag{5.29}$$

Let us define a phasor $\bar{E}_1$:

$$\bar{E}_1 = (1 + R_e/R_L)E + j(X_e/R_L)E \tag{5.30}$$

where the phasor $\bar{E}_1$ makes an angle γ with the q axis

$$\gamma = \arctan[X_e/(R_e + R_L)] \tag{5.31}$$

The phasor diagram for (5.29)-(5.31) is shown in Figure 5.5.

5.4.2 The general case: $\bar{Z}_L$ arbitrary

For $\bar{Z}_L$ arbitrary the equations are more complicated. Substituting (5.25) into (5.22) and rearranging,

$$\begin{aligned}V_d\left(1 + \frac{R_L R_e + X_L X_e}{Z_L^2}\right) + V_q\left(\frac{R_L X_e - X_L R_e}{Z_L^2}\right) &= -V_\infty \sin(\delta - \alpha) + R_e I_d + X_e I_q \\ -V_d\left(\frac{R_L X_e - X_L R_e}{Z_L^2}\right) + V_q\left(1 + \frac{R_L R_e + X_L X_e}{Z_L^2}\right) &= V_\infty \cos(\delta - \alpha) + R_e I_q - X_e I_d\end{aligned} \tag{5.32}$$

or

$$\begin{aligned}V_d(1 + \lambda_1) + V_q\lambda_2 &= -V_\infty \sin(\delta - \alpha) + R_e I_d + X_e I_q \\ -V_d\lambda_2 + V_q(1 + \lambda_1) &= V_\infty \cos(\delta - \alpha) - X_e I_d + R_e I_q\end{aligned} \tag{5.33}$$

where

$$\lambda_1 = (R_L R_e + X_L X_e)/Z_L^2 \qquad \lambda_2 = (R_L X_e - X_L R_e)/Z_L^2 \tag{5.34}$$

Combining (5.33) and (5.15),

$$\begin{aligned} \lambda_2 E &= -V_\infty \sin(\delta - \alpha) + [R_e + r(1 + \lambda_1) - x_d\lambda_2]I_d \\ &\quad + [X_e + x_q(1 + \lambda_1) + r\lambda_2]I_q \\ (1 + \lambda_1)E &= V_\infty \cos(\delta - \alpha) + [-X_e - r\lambda_2 - x_d(1 + \lambda_1)]I_d \\ &\quad + [R_e - x_q\lambda_2 + r(1 + \lambda_1)]I_q \end{aligned} \tag{5.35}$$

Again, by defining $\bar{E}_1 \triangleq (1 + \lambda_1)E + j\lambda_2 E$,

$$\begin{aligned} \hat{R}_d &\triangleq R_e + r(1 + \lambda_1) - x_d\lambda_2 & \hat{R}_q &\triangleq R_e + r(1 + \lambda_1) - x_q\lambda_2 \\ \hat{X}_d &\triangleq X_e + x_d(1 + \lambda_1) + r\lambda_2 & \hat{X}_q &\triangleq X_e + x_q(1 + \lambda_1) + r\lambda_2 \end{aligned} \tag{5.36}$$

we may write (5.35) in the form

$$\begin{aligned} \lambda_2 E &= -V_\infty \sin(\delta - \alpha) + \hat{R}_d I_d + \hat{X}_q I_q \\ (1 + \lambda_1)E &= V_\infty \cos(\delta - \alpha) - \hat{X}_d I_d + \hat{R}_q I_q \end{aligned} \tag{5.37}$$

Since (5.37) is of the same form as (5.29), it can be represented by the same phasor diagram in Figure 5.5.

5.5 Determining Steady-State Conditions

The most common boundary conditions are the terminal voltage V_a and either the current I_a and the power factor F_p or the generated power P and the reactive power Q (per phase). In either case V_a, I_a, and ϕ (the power factor angle) are assumed to be known.

Resolving $\bar{I}_a$ into components with $\bar{V}_a$ as a reference, we write

$$\bar{I}_a = I_r + jI_x \tag{5.38}$$

where I_r is the component of $\bar{I}_a$ in phase with $\bar{V}_a$, and I_x is the quadrature component (which carries its own sign). We also define the power factor F_p as

$$F_p = \cos\phi \tag{5.39}$$

where ϕ is the angle by which I_a *lags* V_a. Then

$$I_r = I_a \cos\phi \qquad I_x = -I_a \sin\phi \tag{5.40}$$

The phasor $\bar{E}_{qa}$ in Figure 5.2 is given by

$$\begin{aligned} \bar{E}_{qa} &\triangleq \bar{V}_a + (r + jx_q)\bar{I}_a = V_a + (I_r + jI_x)(r + jx_q) \\ &= (V_a - x_q I_x + rI_r) + j(x_q I_r + rI_x) \end{aligned} \tag{5.41}$$

The angle between the q axis and the terminal voltage $\bar{V}_a$ (i.e., the angle $\delta - \beta$ in Figures 5.1 and 5.2) is given by

$$\delta - \beta = \tan^{-1}[(x_q I_r + rI_x)/(V_a + rI_r - x_q I_x)] \tag{5.42}$$

The above relations are illustrated in Figure 5.6. Then we compute

$$V_d = -V_a \sin(\delta - \beta) \qquad V_q = V_a \cos(\delta - \beta) \tag{5.43}$$

and v_d and v_q can then be determined from their relationship to V_d and V_q given by (5.15).

The currents are obtained from

$$I_d = -I_a \sin(\delta - \beta + \phi) \qquad I_q = I_a \cos(\delta - \beta + \phi) \tag{5.44}$$

and the rotor quantities i_d and i_q can be determined from (5.11). The remaining

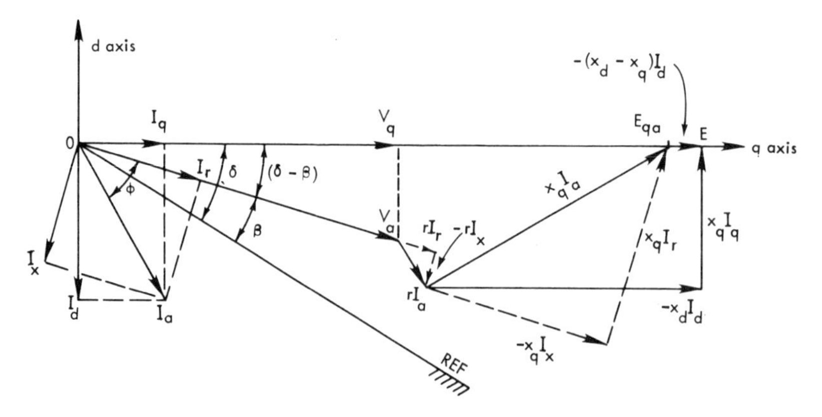

Fig. 5.6 Phasor diagram illustrating (5.41) and (5.42).

currents and flux linkages can readily be determined once these basic quantities are known.

In the case of a synchronous machine connected to an infinite bus the same procedure is followed if the conditions at the machine terminals are given. The voltage of the infinite bus is then determined by subtracting the appropriate voltage drops to the machine terminal voltage $\overline{V}_a$.

If the terminal conditions at the infinite bus are given as the boundary conditions, the position of the q axis is determined by a procedure similar to the above. The machine d and q axis currents and voltages and the machine terminal voltage can then be determined. This is illustrated in Examples 5.1 and 5.2.

5.5.1 Machine connected to an infinite bus with local load

Case 1: V_∞, E, and the machine load angle $\delta - \alpha$ are known.

In this case I_d and I_q can be determined directly from (5.37). Then from (5.15) we can determine V_d and V_q. The three-phase power of the machine can be determined from the relation $P_{3\phi} = 3(V_d I_d + V_q I_q)$. The terminal current I_t is determined from (5.25), and knowing V_∞ we can also determine the power and power factor at the infinite bus.

Case 2: Machine terminal conditions V_a, I_a, and power factor are known.

From I_a, V_a, and the power factor the position of the quadrature axis is determined (see Figure 5.2). From this information I_d, V_d, I_q, and V_q can be found. Also E can be calculated from (5.13). From (5.36) and (5.37) the phasor $\overline{E}_1$ can be constructed. The infinite bus voltage can then be determined by drawing $\hat{R}_d I_d + \hat{X}_q I_q$ parallel to the d axis and $\hat{R}_q I_q - \hat{X}_d I_d$ parallel to the q axis, as shown in Figure 5.7. Thus $\overline{V}_\infty$ and the angle $\delta - \alpha$ are found, from which we can determine $V_{\infty d}$ and $V_{\infty q}$. The current I_t is determined from (5.25), and the power at the infinite bus is given by $3(V_{\infty d} I_{td} + V_{\infty q} I_{tq})$.

Case 3: Conditions at infinite bus are known.

From $\overline{V}_\infty$, $\overline{I}_t$, and Z_e the machine terminal voltage V_a is calculated. Then from $\overline{V}$ and Z_L we can determine $\overline{I}_L$. From $\overline{I}_L$ and $\overline{I}_t$, $\overline{I}_a$ is found. Now the conditions at the terminals of the machine are known and the complete phasor diagram can be constructed.

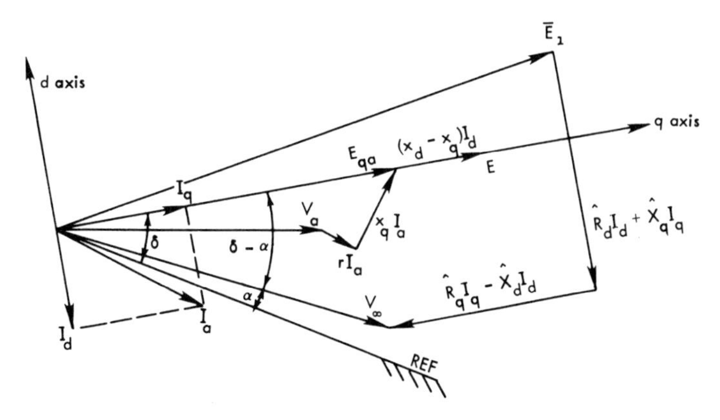

Fig. 5.7 Construction of the phasor diagram for Case 2.

5.6 Examples

The procedures described are illustrated by several examples where different initial conditions are given.

Example 5.1

The machine described in Examples 4.1, 4.2, and 4.3 is to be examined at rated power and 0.85 PF lagging conditions (nameplate loading). The terminal voltage is 1.0 pu. Calculate the steady-state operating conditions. If this machine is connected by a transmission line of 0.02 + j0.40 pu impedance to a large system, find the infinite bus voltage.

Solution

From previous examples and the prescribed boundary conditions the following data are available:

$$x_d = 1.700 \text{ pu} \qquad V_a = 1.000 \text{ pu}$$
$$x_q = 1.640 \text{ pu} \qquad R_e = 0.02 \text{ pu}$$
$$r = 0.001096 \text{ pu} \qquad L_e = 0.4 \text{ pu}$$
$$F_p = \cos\phi = 0.850 \qquad Z_e = 0.4005 \underline{/87.138^\circ}$$

From the given power, power factor, and voltage we compute

$$I_a = 1.0/0.85 = 1.176 \text{ pu}$$

The angle ϕ is computed from F_p as $\phi = \cos^{-1} 0.85 = 31.788^\circ$. Then from (5.40)

$$I_r = I_a \cos\phi = 1.000 \qquad I_x = -I_a \sin\phi = -0.620$$

From (5.42) and Figure 5.7

$$(\delta - \beta) = \arctan \frac{1.00 \times 1.64 - 0.001096 \times 0.620}{1.000 + 0.620 \times 1.64 + 1.00 \times 0.001096}$$
$$= \arctan 0.8126 = 39.096^\circ$$

and $\delta - \beta + \phi = 31.788 + 39.096 = 70.884^\circ$ = angle by which I_a lags the q axis.

Then from (5.44)

0.385 pu $\qquad i_q = 0.667$ pu

1.112 pu $\qquad i_d = -1.925$ pu

$= 0.776$ pu $\qquad v_q = 1.344$ pu

$V_d = -V_a \sin 39.09° = -0.631$ pu $\qquad v_d = -1.092$ pu

From Figure 5.1 by inspection

$$\begin{aligned} E &= V_q + rI_q - x_d I_d \\ &= 0.776 + 0.001096 \times 0.385 + 1.70 \times 1.112 \\ &= 2.666 = E_{FD} \text{ at steady state [from (4.209) and (5.8)]} \end{aligned}$$

Now using (5.8) in pu, $i_F = \sqrt{3}E/L_{AD}$ where, from Example 4.1, $L_{AD} = 1.55$ pu. Then

$$i_F = (\sqrt{3} \times 2.666)/1.55 = 2.979 \text{ pu}$$

The currents i_D and i_Q are both zero. The flux linkages are given in pu by

$$\begin{aligned} \lambda_d &= L_d i_d + \mathrm{k}M_F i_F = 1.70(-1.925) + (1.55)(2.979) = 1.345 \\ \lambda_{AD} &= (i_d + i_F)\mathrm{k}M_F = (2.979 - 1.925)(1.55) = 1.634 \\ \lambda_q &= L_q i_q = 1.64 \times 0.667 = 1.094 \\ \lambda_{AQ} &= \mathrm{k}M_Q i_q = 1.49 \times 0.667 = 0.994 \\ \lambda_F &= \mathrm{k}M_F i_d + L_F i_F = 1.55(-1.925) + (1.651)(2.979) = 1.935 \\ \lambda_D &= \mathrm{k}M_D i_d + M_R i_F = 1.55(2.979 - 1.925) = 1.634 = \lambda_{AD} \\ \lambda_Q &= \mathrm{k}M_Q i_q = 0.994 = \lambda_{AQ} \end{aligned}$$

As a check we calculate the electrical torque T_e, which should be numerically equal to the three-phase power in pu.

$$\begin{aligned} T_{e\phi} &= i_q \lambda_d - i_d \lambda_q \\ &= 0.667 \times 1.345 + 1.925 \times 1.094 = 3.004 \end{aligned}$$

Then $T_e = 1.001$ pu.

If we subtract the three-phase $I^2 r$ losses, we confirm the generated power to be exactly $P = T_e - rI_a^2 = 1.000$. We also calculate the infinite bus voltage for this operating condition. We can write $\overline{V}_\infty = \overline{V}_a - \overline{Z}_e \overline{I}_a$.

Let $\overline{V}_a = V_a \underline{/\beta} = 1.0 \underline{/\beta}$. Then

$$\begin{aligned} \overline{I}_a &= I_a \underline{/\beta - \phi} = 1.176 \underline{/\beta - 31.788°} \\ V_\infty \underline{/\alpha} &= 1.0 \underline{/\beta} - (0.4005 \underline{/87.138°})(1.176 \underline{/\beta - 31.788°}) \end{aligned}$$

or

$$V_\infty \underline{/\alpha - \beta} = 1.0 - 0.4712 \underline{/55.349°} = 0.828 \underline{/-27.899°} \text{ pu}$$

Thus we have $V_\infty = 0.828$ pu, and $\beta - \alpha = 27.899° =$ the angle by which $\overline{V}_a$ leads $\overline{V}_\infty$. The angle between the infinite bus and the q axis is computed as

$$\delta - \alpha = (\delta - \beta) + (\beta - \alpha) = 39.096 + 27.899 = 66.995°$$

Example 5.2

Let the same synchronous machine as in Example 5.1 be connected to an infinite bus through a transmission line having $R_e = 0.02$ pu, and $L_e = X_e = 0.4$ pu. The infinite bus voltage is 1.0 pu. The machine loading remains the same as before ($P = 1.0$ pu at 0.85 PF).

The boundary conditions given in this example are "mixed"; i.e., the voltage is known at one point (the infinite bus), while the power and reactive power are known at a different point (the machine terminal). A slight modification of the procedure of Example 5.1 is needed.

Solution

A good approximation is to assume that the power at the infinite bus is the same as at the machine terminals by neglecting the ohmic power loss in the transmission line (since R_e is small). A better approximation is to assume a power loss in the transmission line based on some estimate of current (say 1.0 pu current).

Let $I_a^2 R_e = (1.00)^2(0.020) = 0.02$ pu. Then the power at the infinite bus is 0.980 pu and the component of the current in phase with V_∞ is $I_r = 0.980$ pu. The angle θ between $\bar{I}_a$ and $\bar{V}_\infty$ is given by

$$\tan\theta = I_x/I_r = 1.020\, I_x$$

The angle β between $\bar{V}_a$ and $\bar{V}_\infty$ is given by an equation similar to (5.32), viz.,

$$\tan\beta = \frac{X_e I_r + R_e I_x}{V_\infty - X_e I_x + R_e I_r} = \frac{0.392 + 0.02 I_x}{1.020 - 0.4 I_x}$$

The power factor angle at the machine terminal ϕ is given by

$$\phi = \beta + \theta = \cos^{-1} 0.85 = 31.788°$$

These angles are shown in Figure 5.8, with V_∞ used as reference; i.e., $\alpha = 0$. Then $\tan\phi = \tan(\cos^{-1} 0.85) = 0.620$. Using the identity

$$\tan\phi = (\tan\beta + \tan\theta)/(1 - \tan\beta\tan\theta)$$

we compute

$$0.620 = \frac{-1.020 I_x + (0.392 + 0.02 I_x)/(1.020 - 0.4 I_x)}{1 + [1.020(0.392 + 0.02 I_x) I_x]/(1.020 - 0.4 I_x)}$$

from which we get $I_x = -0.217$ pu.

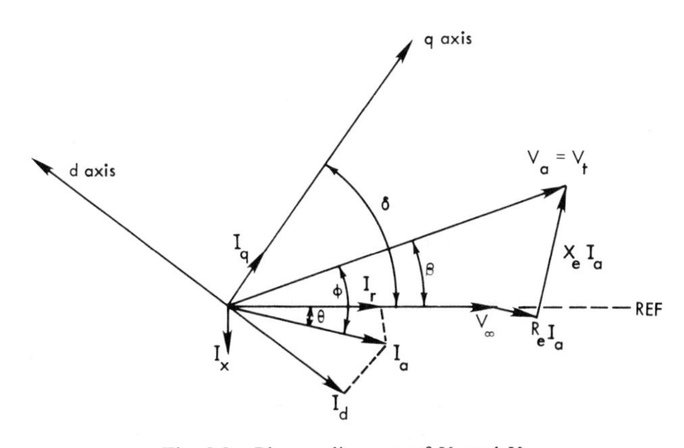

Fig. 5.8 Phasor diagram of V_a and V_∞.

v determine β.

$$- 0.004) \over + 0.082) \Big] = 19.310°$$

$$\phi = 19.310 + 12.483 = 31.793°$$

which is a good ...

The terminal voltage V_a is given by

$$\overline{V}_a = (V_\infty - X_e I_x + R_e I_r) + j(X_e I_r + R_e I_x)$$
$$= 1.106 + j0.388 = 1.172 \underline{/19.31°} \text{ pu}$$

The generator phasor current is

$$\overline{I}_a = 0.980 - j0.217 = 1.003 \underline{/-12.48°} \text{ pu}$$

and $P = V_a I_a \cos\phi = 1.0001$ pu (on a three-phase basis).

The position of the q axis can be determined from an equation similar to (5.41). With $\alpha = 0$,

$$\delta = \tan^{-1} \frac{(x_q + X_e)I_r + (r + R_e)I_x}{V_\infty - (x_q + X_e)I_x + (r + R_e)I_r} = 53.736°$$

The currents, voltages, and flux linkages can then be calculated as in Example 5.1. The results are given below in pu:

$$\begin{aligned} i_d &= -1.591 & \lambda_d &= 1.676 \\ i_q &= 0.701 & \lambda_D &= \lambda_{AD} = 1.914 \\ i_F &= 2.826 & \lambda_q &= 1.150 \\ E &= 2.529 & \lambda_Q &= \lambda_{AQ} = 1.045 \\ v_d &= -1.148 & T_{e\phi} &= 3.004 \\ v_q &= 1.675 & T_e &= 1.001 \end{aligned}$$

In steady-state system studies (often called load-flow studies) it is common to specify the generator boundary conditions in terms of generated power and terminal voltage magnitude, i.e., P and V_t. (Both V_a and V_t are commonly used for the terminal voltage and both are used in this book.) In studies of large systems these boundary conditions are satisfied by iterative techniques, using a digital computer. For the one machine–infinite bus problem the system may be solved explicitly. We now consider the one machine–infinite bus problem with a local load connected to the V_t bus consisting of a shunt resistance R_L and a shunt capacitance C_L, representing the transmission line susceptance.

The system of generator, local load, and line may be conveniently described as a two-port network (Figure 5.9) for which we write, with $\overline{V}_\infty$ as reference ($\alpha = 0$),

$$\begin{bmatrix} \overline{I}_1 \\ \overline{I}_2 \end{bmatrix} = \begin{bmatrix} \overline{Y}_{11} & \overline{Y}_{12} \\ \overline{Y}_{21} & \overline{Y}_{22} \end{bmatrix} \begin{bmatrix} \overline{V}_1 \\ \overline{V}_2 \end{bmatrix} \tag{5.45}$$

The apparent power injected at node 1 may be computed as

$$\overline{S}_1 = P_1 + jQ_1 = \overline{V}_1 \overline{I}_1^* = V_1^2 \overline{Y}_{11}^* + \overline{V}_1 \overline{V}_2^* \overline{Y}_{12}^* \tag{5.46}$$

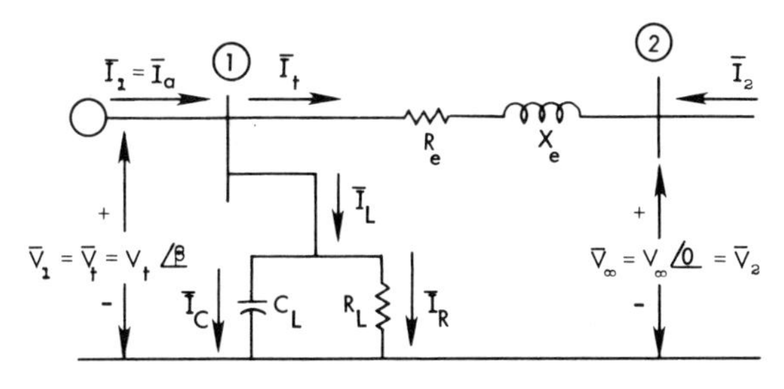

Fig. 5.9 One-machine system as a two-port network.

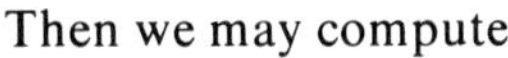

Then we may compute

$$P_1 = G_{11}V_t^2 + V_tV_\infty(G_{12}\cos\beta + B_{12}\sin\beta) \tag{5.47}$$

where we define $\bar{Y}_{km} = G_{km} + jB_{km}$ for all k and m. In (5.47) P_1, V_t, and V_∞ are specified, while G_{11}, G_{12}, and B_{12} are known or computed system parameters. Thus we may solve (5.47) for the angle β. In doing so, it is convenient to define a constant angle γ related to the admittance element $\bar{Y}_{12} = Y_{12}\angle\gamma$. Then from (5.47) we define

$$F = \cos(\gamma - \beta) = (P_1 - G_{11}V_t^2)/Y_{12}V_tV_\infty \tag{5.48}$$

from which β can be found. Obviously, there are limits on the magnitude of P_1 that can be specified in any physical situation, as the cosine function is bounded in (5.48).

Example 5.3

Compute the steady-state conditions for the system of Examples 5.1 and 5.2, where the given boundary conditions in pu are

$$P = 1.0 \text{ (on a three-phase basis)} \qquad V_t = 1.17 \qquad V_\infty = 1.00$$

and where the local load is given in pu as

$$R_L = 100 \qquad B_L = C_L = 0.01$$

Solution

For the numerical data and boundary conditions given, we compute

$$Z_e = R_e + jX_e = 0.02 + j0.4 = 0.4005\angle 87.138^\circ \text{ pu}$$

$$\bar{Y}_{12} = -\bar{y}_{12} = -1/\bar{Z}_e = Y_{12}\angle\gamma$$

$$= -0.1247 + j2.4938 = 2.4969\angle 92.862^\circ \text{ pu}$$

or $\gamma = 92.862^\circ$

We are also given that $R_L = 100$ pu and $B_L = 0.01$ pu. Thus the admittance from node 1 to reference is $\bar{y}_{10} = 0.01 + j0.01$ pu. We then compute

$$\bar{Y}_{11} = \bar{y}_{10} + \bar{y}_{12} = G_{11} + jB_{11} = 0.1347 - j2.4838 \text{ pu}$$

We now compute the quantity F defined in (5.48) as

$$F = (P_1 - G_{11}V_t^2)/Y_{12}V_tV_\infty = 0.2792$$

Then

$$\gamma - \beta = \cos^{-1}F = 73.788^\circ \qquad \beta = 92.862 - 73.788 = 19.074^\circ$$

or $\bar{V}_t = 1.17\angle 19.074^\circ$.

To find the currents, we note from Figure 5.9 that $\bar{I}_a = \bar{I}_t + \bar{I}_L$. Now

$$\bar{I}_L = \bar{I}_R + \bar{I}_C = (V_t/R_L)\underline{/\beta} + (V_t/X_C)\underline{/\beta + 90°}$$
$$= 0.0072 + \mathrm{j}0.0149 \text{ pu}$$

We also write

$$\bar{I}_t = (\bar{V}_t - \bar{V}_\infty)/\bar{Z}_e$$
$$= \frac{[R_e(V_t\cos\beta - V_\infty) + X_eV_t\sin\beta] + \mathrm{j}[R_eV_t\sin\beta - X_e(V_t\cos\beta - V_\infty)]}{Z_e^2}$$
$$= 0.9667 - \mathrm{j}0.2161 \text{ pu}$$

Then, noting that $\bar{I}_a$ lies at an angle θ from V_∞ (Figure 5.8),

$$\bar{I}_a = \bar{I}_t + \bar{I}_L = I_a\underline{/\theta} = 0.9739 - \mathrm{j}0.2012$$
$$= 0.9945\underline{/-11.672°} \text{ pu}$$

We may now compute, as a check,

$$P + \mathrm{j}Q = \bar{V}_t\bar{I}_a^* = 1.000 + \mathrm{j}0.595$$
$$= 1.164\underline{/30.746°} \text{ pu}$$

The power factor is

$$F_p = \tan^{-1}(Q/P) = 0.859$$

The quantity E_{qa} of Figure 5.2 may be computed as a means of finding δ. Thus with $\alpha = 0$ we compute, as in Figure 5.6,

$$\bar{E}_{qa} = E_{qa}\underline{/\delta} = V_t\underline{/\beta} + rI_a\underline{/\theta} + \mathrm{j}x_qI_a\underline{/\theta}$$
$$= 2.446\underline{/54.024°} \text{ pu}$$

and $\delta = 54.024°$. Then we compute

$$\delta - \beta = 34.950° \qquad \phi = \theta + \beta = 30.746° \qquad \delta - \beta + \phi = 65.696°$$

With all the above quantities known, we compute *d-q* currents, voltages, and flux linkages in pu as in Example 5.1, with the result

$$\begin{aligned}
i_d &= -1.570 & \lambda_d &= 1.662\\
i_q &= 0.709 & \lambda_{AD} &= \lambda_D = 1.897\\
v_d &= -1.161 & \lambda_q &= 1.163\\
v_q &= 1.661 & \lambda_{AQ} &= \lambda_Q = 1.056\\
E &= 2.500 & \lambda_F &= 2.180\\
i_F &= 2.794 & T_{e\phi} &= 3.003\\
 & & P_e &= 1.000
\end{aligned}$$

Example 5.4

The same machine at the same loading as in Example 5.1 has a local load of 0.4 pu power at 0.8 PF. It is connected to an infinite bus through a transmission line having $R_e = 0.1$ pu and $X_e = 0.4$ pu. Find the conditions at the infinite bus.

Solution

The internal machine currents, flux linkages, and voltages are the same as in Example 5.1. Thus, in pu,

$$I_d = -1.112 \qquad V_q = 0.776$$
$$I_q = 0.385 \qquad V_d = -0.631$$
$$\delta - \beta = 39.096° \qquad E = 2.666$$

From the local load information

$$|I_L| = 0.4/(1.0 \times 0.8) = 0.5 \text{ pu}$$

Therefore $I_L = 0.4 - j0.3$ pu.

We can also determine that, in pu,

$$R_L = 1.6 \qquad X_L = 1.2 \qquad Z_L = 2.0$$

Thus we compute from (5.34)

$$\lambda_1 = (1.6 \times 0.1 + 1.2 \times 0.4)/(2.0)^2 = 0.16$$
$$\lambda_2 = (1.6 \times 0.4 - 1.2 \times 0.1)/(2.0)^2 = 0.13$$

Then

$$\hat{R}_d = 0.1 + 0.001096 \times 1.16 - 0.13 \times 1.7 = -0.1197$$
$$\hat{R}_q = 0.1 + 0.001096 \times 1.16 - 0.13 \times 1.64 = -0.119$$
$$\hat{X}_d = 0.4 + 1.7 \times 1.16 + 0.001096 \times 0.13 = 2.372$$
$$\hat{X}_q = 0.4 + 1.64 \times 1.16 + 0.001096 \times 0.13 = 2.303$$

From (5.37)

$$V_{\infty d} = V_\infty \sin(\delta - \alpha) = -(-1.112)(-0.1197) - (0.385)(2.303) + (0.13)(2.666)$$
$$= -0.673$$
$$V_{\infty q} = V_\infty \cos(\delta - \alpha) = (-1.112)(2.372) - (0.385)(-0.119) + (1.16)(2.666)$$
$$= 0.501$$
$$V_\infty = [(0.673)^2 + (0.501)^2]^{1/2} = 0.839$$

From (5.25)

$$I_{td} = -1.112 + \frac{0.776 \times 1.2 + 0.631 \times 1.6}{4} = -0.6268$$

$$I_{tq} = 0.385 - \frac{0.776 \times 1.6 - 0.631 \times 1.2}{4} = 0.2639$$

The power delivered to the infinite bus is

$$P_\infty = (-0.673)(-0.6268) + 0.2639 \times 0.501 = 0.554 \text{ pu}$$

The power delivered to the local load is $P_L = 0.4$ pu. Then the transmission losses are 0.14 pu, which is verified by computing $R_l I_t^2$.

5.7 Initial Conditions for a Multimachine System

To initialize the system for a dynamic performance study, the conditions prior to the start of the transient must be known. These are the steady-state conditions that exist before the impact. From the knowledge of these conditions we can assume that the power output, power factor, terminal voltage, and current are known for each machine. If they are not specifically known, a load-flow study is run to determine them.

Assume that a reference frame is adopted for the power system. This reference can be chosen quite arbitrarily. Once it is chosen, however, it should not be changed during the course of the study. In addition, during the study it will be assumed that this reference frame is maintained at synchronous speed.

Consider the ith machine. Let its terminal voltage phasor $\overline{V}_{ai}$ be at an angle β_i with respect to the arbitrary reference frame, and let the q axis be at an angle δ_i with respect to the same reference. Note that β_i is determined from the load-flow study data, while δ_i is the desired initial angle of the machine q axis, which indicates the rotor position. The difference between these two angles $(\delta_i - \beta_i)$ is the load angle or the angle between the q axis and the terminal voltage.

From the load-flow data we can determine for each machine the component I_r of the terminal current in phase with the terminal voltage and the quadrature component I_x. By using an equation similar to (5.42), we can determine the angle $\delta_i - \beta_i$ for this machine. Then by adding the angle β_i, we get the angle δ_i, which is the initial rotor angle of machine i.

From $\overline{V}_{ai}$ and δ_i we can determine I_{qi}, I_{di}, V_{di}, and V_{qi}, which can be used in (5.14) or (5.15) to determine E_i. Then from (5.7) i_{Fi} can be determined. The flux linkages can also be calculated once the d and q components of I_a are known.

5.8 Determination of Machine Parameters from Manufacturers' Data

The machine models given in Chapter 4 are based upon some parameters that are very seldom supplied by the manufacturer. Furthermore, the pu system used here is somewhat different from the manufacturer's pu system. It was noted in Section 4.7.3 that the pu self-inductances of the stator and rotor circuits are numerically equal to the values based on a manufacturer's system, but the mutual inductances between rotor and stator circuits differ by a factor of $\sqrt{3/2}$. We shall attempt to clarify these matters in this section. For a more detailed discussion see Appendix C.

Typical generator data supplied by the manufacturer would include the following.

Ratings:

Three-phase MVA	Stator line current
Frequency and speed	Power factor
Stator line voltage	

Parameters: Of the several reactances supplied, the values of primary interest here are the so-called unsaturated reactances. They are usually given in pu to the base of the machine three-phase rating, *peak*-rated stator voltage to neutral, *peak*-rated stator current, and with the base rotor quantities chosen to force reciprocity in the nonreciprocal Park's transformed equations. This is necessary because of the choice of Park transformation **Q** (4.22) traditionally used by the manufacturers. The following data are commonly supplied.

Reactances (in pu*):*

Synchronous d axis $= x_d$	Subtransient q axis $= x_q''$
Synchronous q axis $= x_q$	Negative-sequence $= x_2$
Transient d axis $= x_d'$	Zero-sequence $= x_0$
Transient q axis $= x_q'$	Armature-leakage $= x_\ell$
Subtransient d axis $= x_d''$	

Time constants (in s):

$$\text{Field open circuit} = \tau'_{d0}$$
$$\text{Subtransient of amortisseur } (d \text{ axis}) = \tau''_d$$
$$\text{Subtransient of amortisseur } (q \text{ axis}) = \tau''_q$$

Resistances (in Ω):

Stator resistance at 25°C
Field circuit resistance at 25°C

Other data:

Moment of inertia in lbm·ft² or WR^2 (sometimes separate data for generator and turbine are given)
No-load saturation curve (at rated speed)
Rated load saturation curve (at rated speed)

Calculations: The base quantities for the stator are readily calculated from the rating data:

$$S_B = \text{VA rating/phase} \quad \text{VA}$$
$$V_B = \text{stator-rated line-to-neutral voltage} \quad \text{V}$$
$$I_B = \text{stator-rated current} \quad \text{A}$$
$$\omega_B = 2\pi \times \text{rated frequency} \quad \text{rad/s}$$

The remaining stator quantities follow:

$$t_B = 1/\omega_B \quad \text{s} \qquad R_B = V_B/I_B \quad \Omega$$
$$\lambda_B = V_B t_B \quad \text{Wb turn} \qquad L_B = V_B t_B/I_B \quad \text{H}$$

Also the stator pu inductances are known from the corresponding reactance values. Thus L_d, L'_d, L''_d, L_q, L'_q, L''_q, L_2, L_0, and ℓ_d are known.

Rotor base quantities: If ℓ_d in pu is known, then L_{AD} in pu is determined from $L_{AD} = L_d - \ell_d$, the corresponding value of L_{AD} in H is then calculated. The mutual field-to-stator inductance M_F in H is determined from the air gap line on the no-load saturation curve as $\sqrt{2}V_B = \omega_B M_F i_F$, where i_F is the field current that gives the rated voltage in the air gap line.

The base rotor quantities are then determined from (4.55) and (4.56); the base mutual inductance M_{FB} is calculated from (4.57).

Rotor per unit quantities: Calculation of the rotor circuit leakage inductances is made with the aid of the equivalent circuits in Figure 5.10. The field-winding leakage inductance ℓ_F is calculated from Figure 5.10(a) by inspection:

$$L'_d = \ell_d + L_{AD}\ell_F/(L_{AD} + \ell_F) \quad \text{pu} \tag{5.49}$$

which can be put in the form

$$\ell_F = L_{AD}[(L'_d - \ell_d)/(L_d - L'_d)] \tag{5.50}$$

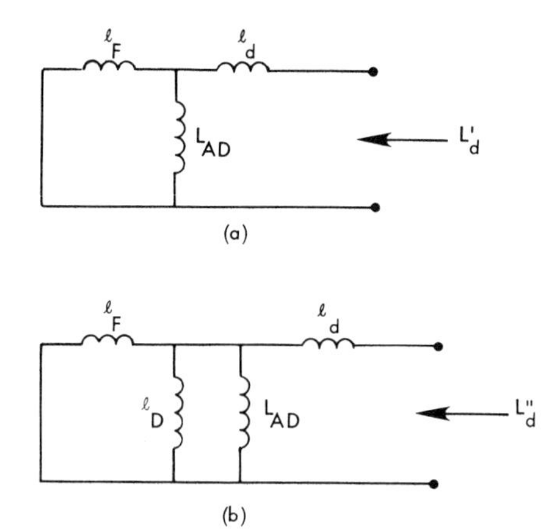

Fig. 5.10 Equivalent circuit for d axis inductances: (a) transient inductance, (b) subtransient inductance.

Similarly, by inspection of Figure 5.10(b),

$$L''_d = \ell_d + \frac{1}{1/L_{AD} + 1/\ell_D + 1/\ell_F} \tag{5.51}$$

from which we can obtain

$$\ell_D = L_{AD}\ell_F(L''_d - \ell_d)/(L_{AD}\ell_F - L_F(L''_d - \ell_d)) \tag{5.52}$$

The self-inductances of the field winding L_F and of the amortisseur L_D are then calculated from

$$L_D = \ell_D + L_{AD} \qquad L_F = \ell_F + L_{AD} \tag{5.53}$$

The same procedure is repeated for the q axis circuits.

$$L_{AQ} = L_q - \ell_q \tag{5.54}$$

where $\ell_q = \ell_d$ and ℓ_Q is determined from Figure 5.11 by inspection:

$$L''_q = \ell_q + \ell_Q L_{AQ}/(\ell_Q + L_{AQ}) \tag{5.55}$$

from which we can obtain

$$\ell_Q = L_{AQ}[(L''_q - \ell_q)/(L_q - L''_q)] \tag{5.56}$$

and the self-inductance of the q axis amortisseur is given by

$$L_Q = L_{AQ} + \ell_Q \tag{5.57}$$

Resistances: The value used for the stator winding resistance should be that which corresponds to the generator operating temperature at the rated load. If this data is not available, a temperature rise of 80–100°C is usually assumed, and the winding resistance

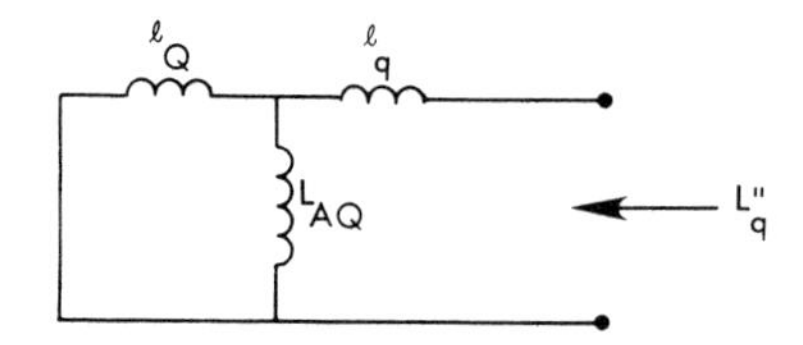

Fig. 5.11 Equivalent circuit of the q axis subtransient inductance.

is calculated accordingly. Thus for copper winding the stator resistance for 100°C temperature rise is given by

$$r_{125} = r_{25}[(234.5 + 125)/(234.5 + 25)] \ \Omega \tag{5.58}$$

The same procedure can be used to estimate the field resistance at an assumed operating temperature. However, other information is available to estimate the field resistance.

From (4.189) we compute

$$r_F = L_F/\tau'_{d0} \ \text{pu} \tag{5.59}$$

where τ'_{d0} is given in pu time. The damper winding resistances may be estimated from the subtransient time constants. From (4.187) and (4.190) the d axis subtransient time constant is given by

$$\tau''_d = [(L_D L_F - L_{AD}^2)/r_D L_F](L''_d/L'_d) \ \text{pu} \tag{5.60}$$

Since all the inductances in (5.60) are known, r_D can be determined. Similarly, from (4.192) and (4.193) r_Q can be found,

$$\tau''_q = (L''_q/L_q)(L_Q/r_Q) \ \text{pu} \tag{5.61}$$

Again note that τ''_d and τ''_q are given in pu.

Finally, data supplied by the manufacturer may not be available in the complete form given in this section. We should also differentiate between data obtained from verified tests and those obtained from manufacturers' quotations. The latter are usually estimated for a machine of given size and type, often long before the machine is fabricated. This may also explain apparent inconsistencies that may be found in a given set of data.

This section illustrates the procedure that can be used to determine the parameters of the machine. When some of the data is not available, the engineer may find it convenient to assign values for this data from typical data available in the literature for machines of the same size and type. We should always ascertain that the parameters thus calculated are self-consistent. Actual values for several existing machines are given in Appendix D.

Example 5.5

The data given by the manufacturer for the machine of Example 4.1 are given below. The machine parameters are to be calculated and compared to those obtained in Example 4.1.

$$x_d = L_d = 1.70 \ \text{pu} \qquad x_\ell = \ell_d = \ell_q = 0.15 \ \text{pu}$$
$$x_q = L_q = 1.64 \ \text{pu} \qquad \tau'_{d0} = 5.9 \ \text{s}$$
$$x'_d = L'_d = 0.245 \ \text{pu} \qquad \tau''_d = 0.023 \ \text{s}$$
$$x'_q = L'_q = 0.380 \ \text{pu} \qquad \tau''_{q0} = 0.075 \ \text{s}$$
$$x''_d = L''_d = 0.185 = L''_q \ \text{pu} \qquad \tau_a = 0.24 \ \text{s}$$

Solution

We begin by calculating the pu d axis mutual inductance

$$L_{AD} = 1.70 - 0.15 = 1.55$$

This is also the same as kM_F, kM_D, and M_R. Similarly,

$$L_{AQ} = kM_Q = 1.64 - 0.15 = 1.49 \text{ pu}$$

Now, from (5.50)

$$\ell_F = 1.55[(0.245 - 0.15)/(1.70 - 0.245)] = 0.101 \text{ pu}$$
$$L_F = 0.101 + 1.55 = 1.651 \text{ pu}$$

From (5.52)

$$\ell_D = \frac{(1.55)(0.101)(0.185 - 0.15)}{(1.55)(0.101) - (1.651)(0.185 - 0.15)} = 0.055 \text{ pu}$$
$$L_D = 1.550 + 0.055 = 1.605 \text{ pu}$$

Also, from (5.56)

$$\ell_Q = 1.49[(0.185 - 0.150)/(1.640 - 0.185)] = 0.036 \text{ pu}$$
$$L_Q = 1.490 + 0.036 = 1.526 \text{ pu}$$

From the open circuit time constant

$$\tau'_{d0} = 5.9 \text{ s} = 2224.25 \text{ rad}$$

We compute from (5.59)

$$r_F = 1.651/2224.25 = 7.423 \times 10^{-4} \text{ pu}$$

and from (5.60)

$$r_D = \frac{(1.605 \times 1.651 - 1.55 \times 1.55)(0.185)}{(1.651)(0.023 \times 377)(0.245)}$$
$$= 0.0131 \text{ pu}$$

From $\tau''_{q0} = 0.075$ s we compute

$$\tau''_q = (L''_q/L_q)\tau''_{q0} = 8.46 \text{ ms} = 3.19 \text{ rad}$$

Then from (5.61)

$$r_Q = (1.526/3.19)(0.185/1.64) = 0.054 \text{ pu}$$

These values are the same as those calculated in Example 4.1.

5.9 Analog Computer Simulation of the Synchronous Machine

The mathematical models describing the dynamic behavior of the synchronous machine were developed in Chapter 4. The remainder of this chapter will be devoted to the simulation of these models by both analog and digital computers. We begin with the analog simulation.

Note that the equations describing the machine are nonlinear. For example (4.154) and (4.163) have two types of nonlinearities, a product nonlinearity of the form $x_i x_j$ (where x_i and x_j are state variables) and the trigonometric nonlinearities $\cos \gamma$ and $\sin \gamma$. These types of nonlinearities can be conveniently represented by special analog computer components. Also, the analog computer can be very useful in representing other nonlinearities such as limiters (in excitation systems) and saturation (in the magnetic circuit). Thus in many ways the analog computer is very well suited for studying synchronous machine problems. A brief description of analog computers is given in Appendix B.

To place the matter in the proper perspective, recall that the state-space model of a synchronous machine connected to an infinite bus is a set of seven first-order, nonlinear differential equations. When the equations for the excitation system (for v_F) and the mechanical torque (for T_m) are also added, the system is typically described by 14 differential equations. Complete representation of only one synchronous machine with its controls would occupy the major part of a large-size analog computer. Thus while the analog computer is well adapted for the study of synchronous machine dynamics, it is usually limited to problems involving one or two machines with full representation or to a small number of machines represented by simplified models [2, 3, 4, 5].

The model most suited for analog computer representation is the flux linkage model. Thus the equations developed in Section 4.12 are used for the analog simulation. The differential equations will be modified, however, to avoid differentiation. For example the state-space equation of the variable x_i is

$$\dot{x}_i = f_i(\mathbf{x}, \mathbf{u}, t) \tag{5.62}$$

where x_j, $j = 1, 2, \ldots, n$, are the state variables, and u_k, $k = 1, 2, \ldots, r$, are the driving functions.

For analog computer simulation (5.62) is written as

$$x_i = \frac{\omega_B}{a} \int_0^t f_i(\mathbf{x}, \mathbf{u}, t)\, dt + x_i(0) \tag{5.63}$$

where a is the computer time scale factor and ω_B is required if time is to be in seconds (see Appendix B).

5.9.1 Direct axis equations

From (4.126)

$$\lambda_d = \frac{\omega_B}{a} \int_0^t \left[\frac{r}{\ell_d} (\lambda_{AD} - \lambda_d) - \omega \lambda_q - v_d \right] dt + \lambda_d(0) \tag{5.64}$$

From (4.128)

$$\lambda_F = \frac{\omega_B}{a} \int_0^t \left[\frac{r_F}{\ell_F} (\lambda_{AD} - \lambda_F) + v_F \right] dt + \lambda_F(0) \tag{5.65}$$

and from (4.129)

$$\lambda_D = \frac{\omega_B}{a} \int_0^t \frac{r_D}{\ell_D} (\lambda_{AD} - \lambda_D)\, dt + \lambda_D(0) \tag{5.66}$$

The mutual flux linkage λ_{AD} is computed from (4.120)

$$\lambda_{AD} = L_{MD}(\lambda_d/\ell_d + \lambda_F/\ell_F + \lambda_D/\ell_D) \tag{5.67}$$

Then from (4.118) the d axis and field currents are given by

$$i_d = (1/\ell_d)(\lambda_d - \lambda_{AD}) \tag{5.68}$$

$$i_F = (1/\ell_F)(\lambda_F - \lambda_{AD}) \tag{5.69}$$

The analog representation of the d axis equations is shown in Figure 5.12. Note that all integrand terms are multiplied by ω_B to compute time in seconds and divided by the time scaling factor a.

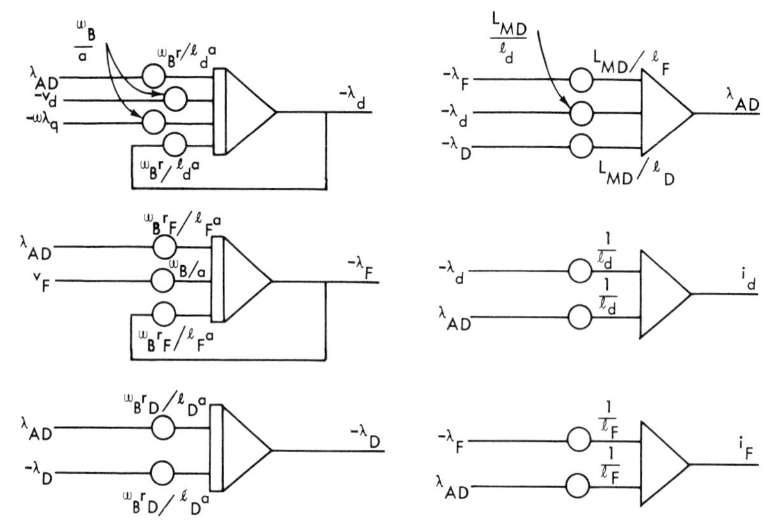

Fig. 5.12 Analog representation of the d axis equations.

5.9.2 Quadrature axis equations

From (4.130)

$$\lambda_q = \frac{\omega_B}{a} \int_0^t \left[\frac{r}{\ell_q} (\lambda_{AQ} - \lambda_q) + \omega \lambda_d - v_q \right] dt + \lambda_q(0) \tag{5.70}$$

and from (4.131)

$$\lambda_Q = \frac{\omega_B}{a} \int_0^t \frac{r_Q}{\ell_Q} (\lambda_{AQ} - \lambda_Q)\, dt + \lambda_Q(0) \tag{5.71}$$

The mutual flux linkage is computed from

$$\lambda_{AQ} = L_{MQ}(\lambda_q/\ell_q + \lambda_Q/\ell_Q) \tag{5.72}$$

Then the q axis current is given by, from (4.123),

$$i_q = (1/\ell_q)(\lambda_q - \lambda_{AQ}) \tag{5.73}$$

The analog simulation of the q axis equations is shown in Figure 5.13.

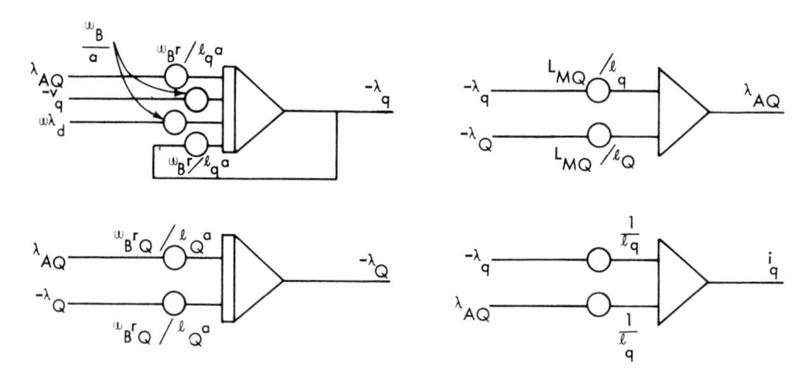

Fig. 5.13 Analog simulation of the q axis equations.

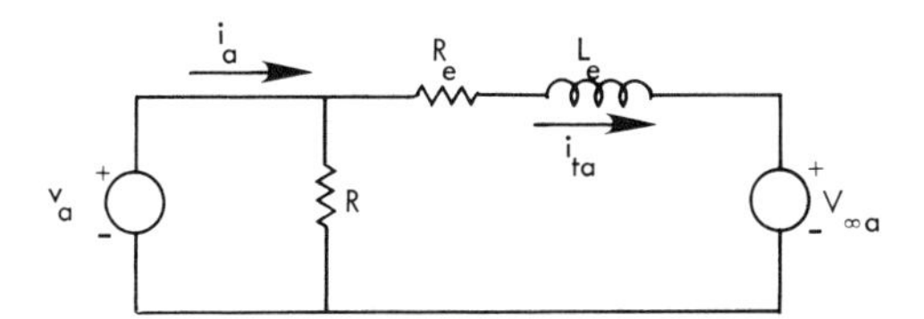

Fig. 5.14 One machine–infinite bus system with local resistive load.

5.9.3 Load equations

In (4.149) $\alpha = 0$ will be used for convenience. Therefore,

$$i_d = \frac{\omega_B}{aL_e}\int_0^t [\sqrt{3}\,V_\infty \sin\delta + v_d - R_e i_d - \omega L_e i_q]\,dt + i_d(0) \tag{5.74}$$

$$i_q = \frac{\omega_B}{aL_e}\int_0^t [-\sqrt{3}\,V_\infty \cos\delta + v_q - R_e i_q + \omega L_e i_d]\,dt + i_q(0) \tag{5.75}$$

Equations (5.74) and (5.75) are useful in generating the voltages v_d and v_q. However, if they are used directly, differentiation of i_d and i_q will be required, which should be avoided in analog computer simulation. To generate v_d and v_q, the following scheme, suggested by Krause [2], is used. The machine is assumed to have a very small resistive load located at its terminal, as shown in Figure 5.14. This load is represented by a large resistance R. From Figure 5.14 the machine terminal voltage and current for phase a are given by

$$v_a = (i_a - i_{ta})R \tag{5.76}$$

where i_{ta} is the phase a current to the infinite bus.

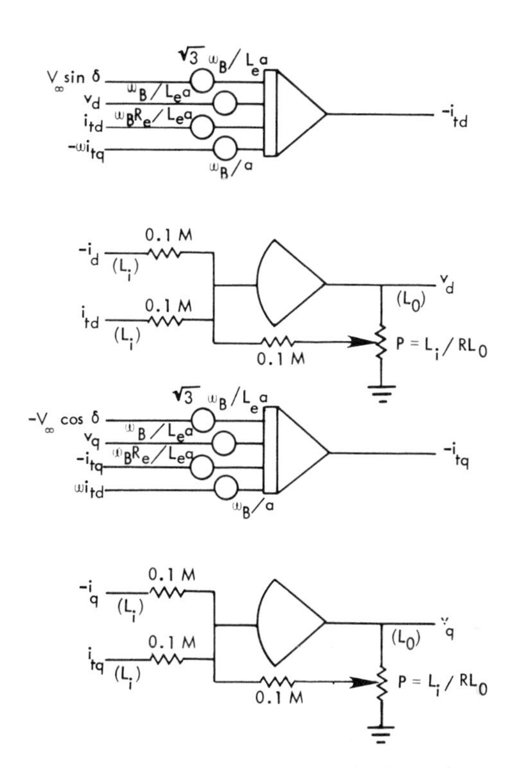

Fig. 5.15 Analog simulation of the load equations.

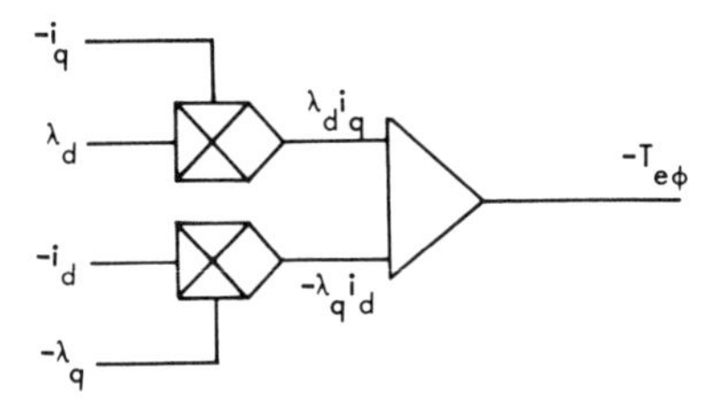

Fig. 5.16 Simulation of the electrical torque $T_{e\phi}$.

Following a procedure similar to that used in Section 5.4, the current i_t can be resolved into d and q axis components i_{dt} and i_{qt} given by (5.74) and (5.75). The currents i_d and i_q are given by (5.68) and (5.73). The v_d and v_q signals are obtained from Figure 5.14 by inspection,

$$v_d = (i_d - i_{td})R \qquad v_q = (i_q - i_{tq})R \tag{5.77}$$

where i_{td} and i_{tq} are obtained from (5.74) and (5.75) respectively, with subscript t added as required by Figure 5.14. The analog computer simulation of the load equations is shown in Figure 5.15.

5.9.4 Equations for ω and δ

From (4.90) and (4.99), with $\dot{\omega}_\Delta = \dot{\omega}$ in pu and $\tau_j = 2H\omega_B$, we can write

$$2H\omega_B \frac{d\omega_{\Delta u}}{dt_u} = 2H \frac{d\omega_{\Delta u}}{dt} = T_m - T_e - D\omega_{\Delta u} \quad \text{pu} \tag{5.78}$$

where $T_e = (i_q\lambda_d - i_d\lambda_q)/3$. Equation (5.78) is integrated with time in seconds to compute, with zero initial conditions and with a time scale factor of a,

$$\omega_{\Delta u} = \frac{1}{2Ha} \int_0^t (T_m - T_e - D\omega_{\Delta u})\, dt \quad \text{pu} \tag{5.79}$$

Note that the load damping signal used is proportional to ω_Δ (pu slip), requiring appropriate values of D.

Most analog computers require that δ be expressed in degrees to find $\sin\delta$ and $\cos\delta$ [6]. Therefore, since $\dot{\delta} = \omega_B(\omega_u - 1) = \omega_B\omega_\Delta$ pu, we compute

$$\delta = \frac{180\,\omega_B}{\pi a} \int_0^t \omega_\Delta\, dt + \frac{180}{\pi}\delta(0) \quad \text{elec deg} \tag{5.80}$$

The analog computer simulation of (5.78)–(5.80) is shown in Figures 5.16 and 5.17. The generation of the signals $-\omega$ and $-\delta$ is shown in Figure 5.17. The analog repre-

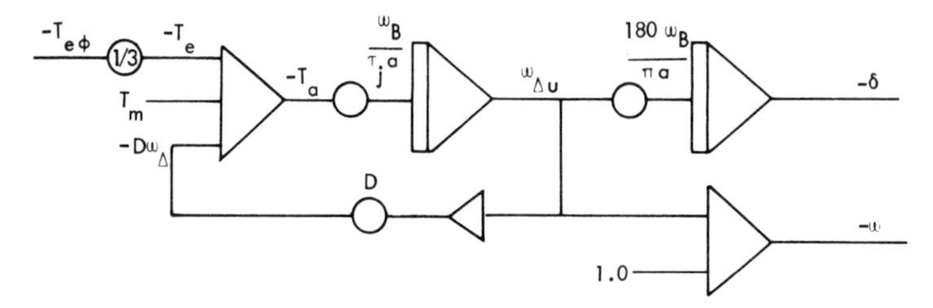

Fig. 5.17 Simulation of ω_Δ, ω, and δ.

sentations shown in Figures 5.12, 5.13, and 5.15–5.17 generate the basic signals needed to simulate a synchronous machine connected to an infinite bus through a transmission line. However, other auxiliary signals are needed. For example to produce the signals $\omega\lambda_q$ and $\omega\lambda_d$ shown in Figures 5.12 and 5.13, additional multipliers are needed. To produce the signals $V_\infty \sin\delta$ and $V_\infty \cos\delta$, an electronic resolver is needed. The complete analog representation of the system is shown in Figure 5.18. It is important to

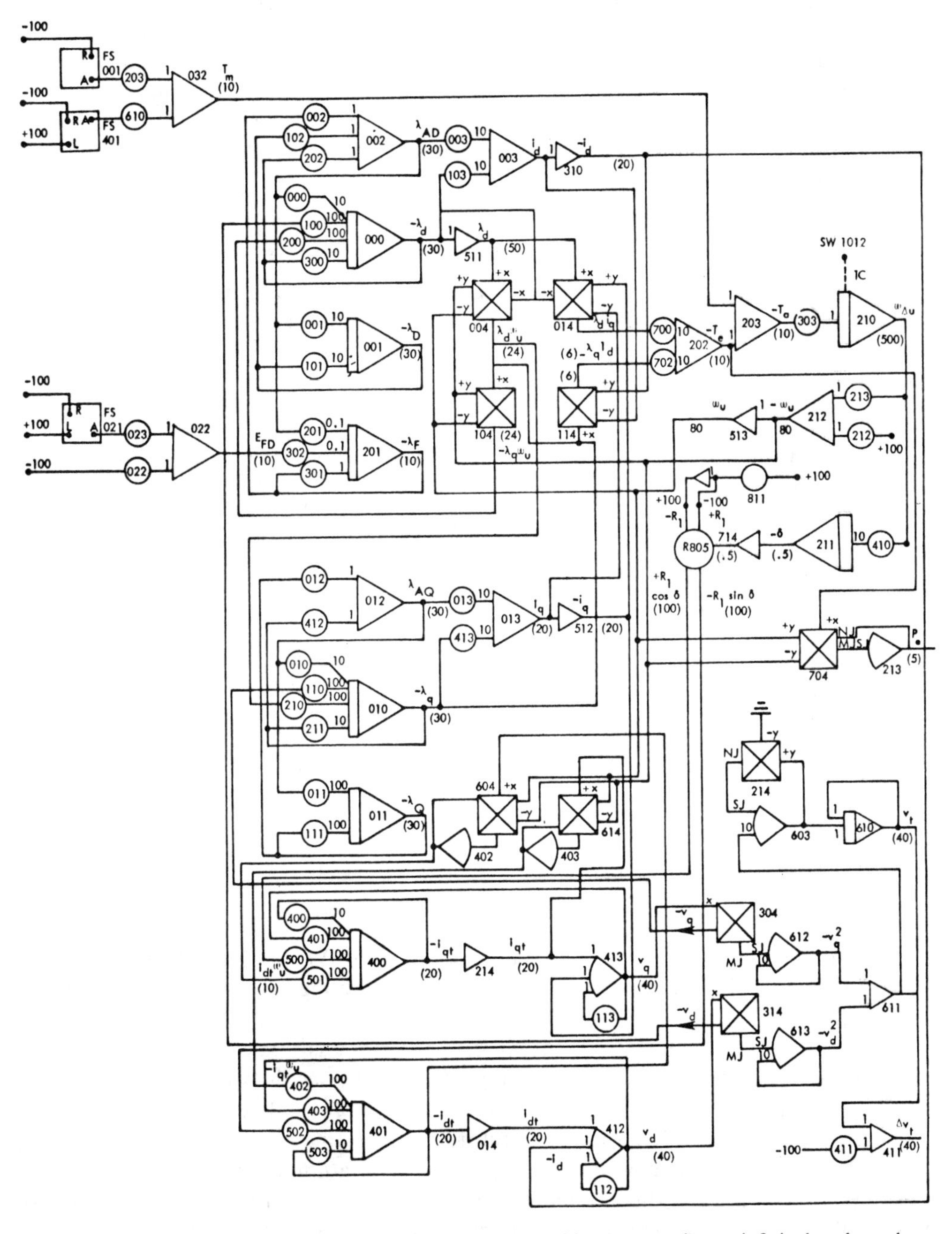

Fig. 5.18 Analog computer patching for a synchronous machine connected to an infinite bus through a transmission line.

note that signals are added by using the appropriate setting for the potentiometers associated with the various amplifiers and integrators scaled to operate within the analog computer rating. This scaling is best illustrated by an example, and in Example 5.6 the scaling is given in detail for the simulation of the synchronous machine.

The initial conditions may be calculated from the steady-state equations (as in Examples 5.1–5.3), and these values may be used to initialize the integrators. However, the analog computer may be used to compute these initial conditions. To initialize the system for analog computation, the following procedure is used. The integrator for the speed is kept at hold position, maintaining the speed constant. The integrators for the flux linkages are allowed to operate with the torque T_m at zero. This builds the flux linkages to values corresponding to the no-load conditions. The load T_m is then applied with the speed integrator in operation. The steady-state conditions thus reached correspond to initialization of the system for transient studies.

Example 5.6

The synchronous machine discussed in Examples 4.1–4.3, 5.1, and 5.2 is to be simulated on an analog computer. The operating conditions as stated in Example 5.1 represent the steady-state conditions. The system response to changes in v_F and T_m is to be examined.

Solution

The data for the synchronous machine and transmission line in pu is given by:

$$L_d = 1.700 \qquad L_{MD} = 0.02838$$
$$L_q = 1.640 \qquad L_{MQ} = 0.02836$$
$$L_D = 1.605 \qquad r = 0.001096$$
$$L_Q = 1.526 \qquad r_F = 0.00074$$
$$L_{AD} = 1.550 \qquad r_D = 0.0131$$
$$L_{AQ} = 1.490 \qquad r_Q = 0.0540$$
$$L_F = 1.651 \qquad R = 100.0$$
$$\ell_d = \ell_q = 0.150 \qquad R_e = 0.02$$
$$\ell_F = 0.101$$
$$\ell_D = 0.055 \qquad H = 2.37 \text{ s}$$
$$\ell_Q = 0.036 \qquad \tau'_{d0} = 5.90 \text{ s}$$
$$L_e = 0.400 \qquad V_\infty = 0.828$$

The additional data needed is $T_m = 1.00$ pu and $E_{FD} = 2.666$. Note that $E_{FD} = E$ in the steady state. This value of E_{FD} with the proper scaling is introduced into the integrator for λ_F.

As explained in Section 5.9.5, the analog computer is made to initialize itself by allowing the integrators to reach the steady-state conditions in two steps. In the first step E_{FD} is applied with $T_m = 0$ and $\omega = \omega_R =$ constant. Then T_m is switched on with all integrators, including the ω integrator, in operation.

The basic connection diagrams for the analog simulation are given in Figures 5.12–5.17. The overall connection diagram is shown in Figure 5.18. In that figure the analog unit numbers and the scaling factors for the various signals are given; e.g., the scaling factor for λ_F is 10, which is given in parentheses. The time scaling used is 20. The settings of the various potentiometers and the scaling are listed in Table 5.1.

Table 5.1. Potentiometer and Gain Settings for Synchronous Machine Simulation by Analog Computer ($a = 20$)

Pot. no.	Amp. no.	L_o	L_i	L_o/L_i	Input	Constant	C (const.)	$(L_o/L_i)C$	Int. cap.	Amp. gain	Pot. set.
000	000	30	30	1.0	λ_{AD}	$\frac{r\omega_B}{\ell_a a} = \frac{(0.0010965)(377)}{(0.15)(20)}$	0.1378	0.1378	0.1	10	0.0138
100	000	30	40	0.75	$-v_d$	$\frac{\omega_B}{a} = \frac{377}{20}$	18.85	14.14	0.1	100	0.1414
200	000	30	24	1.25	$-\omega\lambda_q$	$\frac{\omega_B}{a} = \frac{377}{20}$	18.85	23.56	0.1	100	0.2356
300	000	30	30	1.0	$-\lambda_d$	$\frac{r\omega_B}{\ell_a a}$	0.1378	0.1378	0.1	10	0.0138
001	001	30	30	1.0	λ_{AD}	$\frac{r_D\omega_B}{\ell_D a} = \frac{(0.013099)(377)}{(0.055417)(20)}$	4.456	4.456	1.0	10	0.4456
101	001	30	30	1.0	$-\lambda_d$	$\frac{r_D\omega_B}{\ell_D a} = \frac{(0.013099)(377)}{(0.055417)(20)}$	4.456	4.456	1.0	10	0.4456
201	201	10	30	0.3333	λ_{AD}	$\frac{r_F\omega_B}{\ell_F a} = \frac{(0.74236 \times 10^{-3})(377)}{(0.101203)(20)}$	0.1383	0.04609	10.0	0.1	0.4609
302	201	10	10	1.0	E_{FD}	$\frac{\sqrt{3}r_F\omega_B}{L_{AD}a} = \frac{\sqrt{3}(0.74236 \times 10^{-3})(377)}{(1.55)(20)}$	0.01564	0.01564	10.0	0.1	0.1564
301	201	10	10	1.0	$-\lambda_F$	$\frac{r_F\omega_B}{\ell_F a}$	0.1383	0.1383	10.0	1	0.1383
002	002	30	10	3.0	$-\lambda_F$	$\frac{L_{MD}}{\ell_F} = \frac{0.028378378}{0.101202749}$	0.2804	0.8412	—	1	0.8412
102	002	30	30	1.0	$-\lambda_D$	$\frac{L_{MD}}{\ell_D} = \frac{0.028378378}{0.055416667}$	0.5121	0.5121	—	1	0.5121
202	002	30	30	1.0	$-\lambda_d$	$\frac{L_{MD}}{\ell_d} = \frac{0.028378378}{0.15}$	0.1892	0.1892	—	1	0.1892
003	003	20	30	0.6667	λ_{AD}	$\frac{1}{\ell_d} = \frac{1}{0.15}$	6.667	4.444	—	10	0.4444
103	003	20	30	0.6667	$-\lambda_d$	$\frac{1}{\ell_a}$	6.667	4.444	—	10	0.4444

Table 5.1. *(continued)*

Pot. no.	Amp. no.	L_o	L_i	L_o/L_i	Input	Constant	C (const.)	$(L_o/L_i)C$	Int. cap.	Amp. gain	Pot. set.
010	010	30	30	1.0	λ_{AQ}	$\frac{r\omega_B}{\ell_a a} = \frac{(0.0010965)(377)}{(0.15)(20)}$	0.1378	0.1378	0.1	10	0.0138
110	010	30	40	0.75	$-v_q$	$\frac{\omega_B}{a} = \frac{377}{20}$	18.85	14.14	0.1	100	0.1414
210	010	30	24	1.25	$\omega\lambda_d$	$\frac{\omega_B}{a}$	18.85	23.56	0.1	100	0.2356
211	010	30	30	1.0	$-\lambda_q$	$\frac{r\omega_B}{\ell_a a}$	0.1378	0.1378	0.1	10	0.0138
011	011	30	30	1.0	λ_{AQ}	$\frac{r_Q\omega_B}{\ell_Q a} = \frac{(0.053955)(377)}{(0.035809)(20)}$	28.40	28.40	0.1	100	0.2804
111	011	30	30	1.0	$-\lambda_Q$	$\frac{r_Q\omega_B}{\ell_Q a} = \frac{(0.053955)(377)}{(0.035809)(20)}$	28.40	28.40	0.1	100	0.2804
012	012	30	30	1.0	$-\lambda_Q$	$\frac{L_{MQ}}{\ell_Q} = \frac{0.028357472}{0.035809}$	0.7919	0.7919	—	1	0.7919
412	012	30	30	1.0	$-\lambda_q$	$\frac{L_{MQ}}{\ell_a} = \frac{0.028357472}{0.15}$	0.1890	0.1890	—	1	0.1890
013	013	20	30	0.6667	λ_{AQ}	$\frac{1}{\ell_a} = \frac{1}{0.15}$	6.667	4.444	—	10	0.4444
413	013	20	30	0.6667	$-\lambda_q$	$\frac{1}{\ell_a}$	6.667	4.444	—	10	0.4444
700	202	10	6	1.667	$\lambda_d i_q$	1.0	1.667	1.667	—	10	0.1667
702	202	10	6	1.667	$-\lambda_q i_d$	1.0	1.667	1.667	—	10	0.1667
402	401	20	16	1.25	$-\omega i_{tq}$	$\frac{\omega_B}{a}$	18.85	23.56	0.1	100	0.2356

403	401	20	40	0.5	v_d	$\frac{\omega_B}{L_e a} = \frac{377}{(0.4)(20)}$	47.13	23.56	0.1	100	0.2356
502	401	20	100	0.2	$V_\infty \sin \delta$	$\frac{\sqrt{3}\omega_B}{L_e a} = \frac{\sqrt{3}(377)}{(0.4)(20)}$	81.62	16.32	0.1	100	0.1632
503	401	20	20	1.0	i_{td}	$\frac{R_e \omega_B}{L_e a} = \frac{(0.02)(377)}{(0.4)(20)}$	0.9425	0.9425	0.1	10	0.0943
400	400	20	20	1.0	$-i_{tq}$	$\frac{R_e \omega_B}{L_e a}$	0.9425	0.9425	0.1	10	0.0943
401	400	20	40	0.5	v_q	$\frac{\omega_B}{L_e a}$	47.13	23.56	0.1	100	0.2356
500	400	20	100	0.2	$-V_\infty \cos \delta$	$\frac{\sqrt{3}\omega_B}{L_e a}$	81.62	16.32	0.1	100	0.1632
501	400	20	16	1.25	ωi_{td}	$\frac{\omega_B}{a}$	18.85	23.56	0.1	100	0.2356
112	412 HG	40	20	2.0	$i_{td} - i_d$	$P = L_i/RL_0 = \frac{20}{100(40)}$	0.005	—	—	—	0.005
113	413 HG	40	20	2.0	$i_{tq} - i_q$	$P = L_i/RL_0 = \frac{20}{100(40)}$	0.005	—	—	—	0.005
303	210	500	10	50.0	$-T_a$	$\frac{1}{6Ha} = \frac{1}{6(2.37)(20)}$	3.516×10^{-3}	0.1758	1.0	1	0.1758
410	211	0.5	500	0.001	$\omega_{\Delta u}$	$\frac{180\ \omega_B}{\pi a} = \frac{(180)(377)}{\pi(20)}$	.080	1.080	1.0	10	0.1080
212	212	80	100	0.80	1.0	1.0	1.0	0.8000	—	1	0.8000
213	212	80	500	0.16	$\omega_{\Delta u}$	1.0	1.0	0.1600	—	1	0.1600

Note: In this table ℓ_a is used for either ℓ_d or ℓ_q.

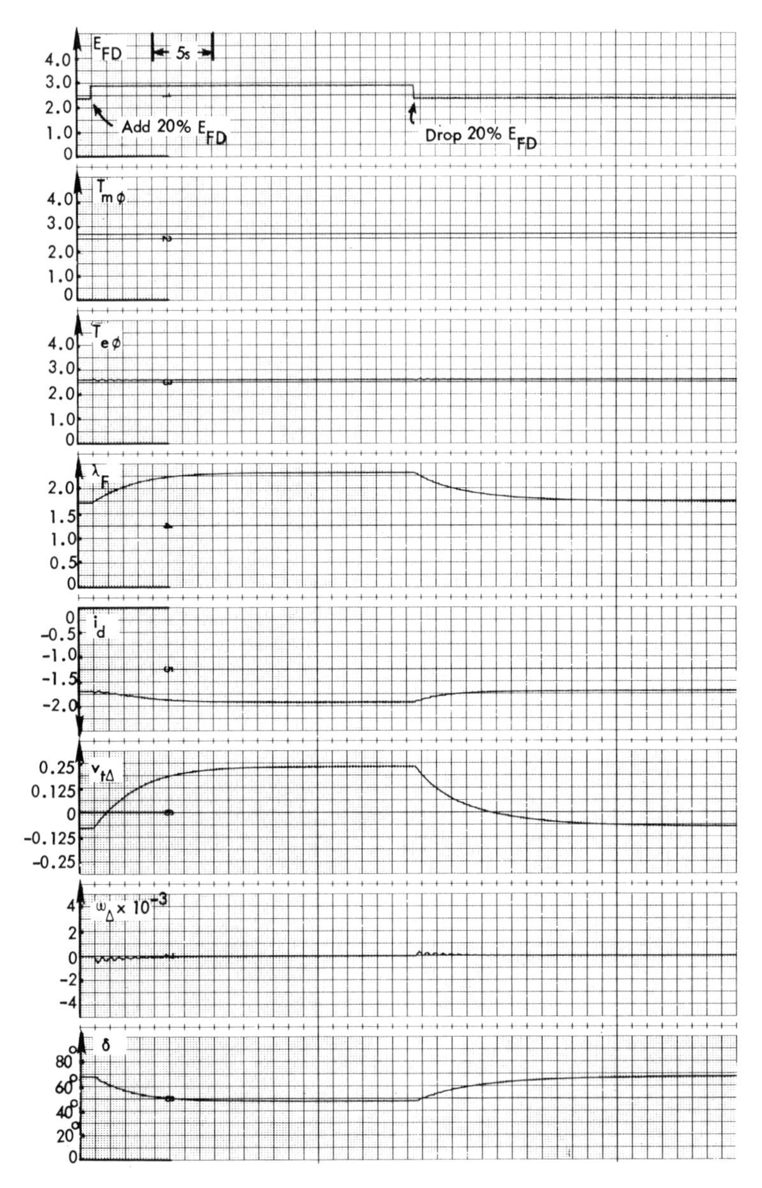

Fig. 5.19 Response of a machine initially at 90% load and 90% excitation to a 20% step change in excitation.

The steady-state conditions reached by the analog computer are listed in Table 5.2. They are compared with the values computed in Example 5.1.

Figures 5.19–5.21 show the following analog computer outputs: the change in the exciter voltage E_{FD}, the mechanical torque $T_{m\phi}$, the electromagnetic torque $T_{e\phi}$, the field flux linkage λ_F, the stator d axis current i_d, the terminal voltage error $V_{t\Delta}$, the angular velocity error ω_Δ, and the rotor angle δ. The results of the simulation are shown in Figures 5.19–5.23, where all plotted quantities are given in pu. Example 5.1 is used as a base for the computer runs. Thus a 10% change in E_{FD} is 0.2666, which is 10% of the nominal value computed in Example 5.1. Similarly, 10% $T_{m\phi}$ is 0.3 pu, and zero $V_{t\Delta}$ corresponds to a terminal voltage V_t of $\sqrt{3}$ pu (or $V_t = 1.0$).

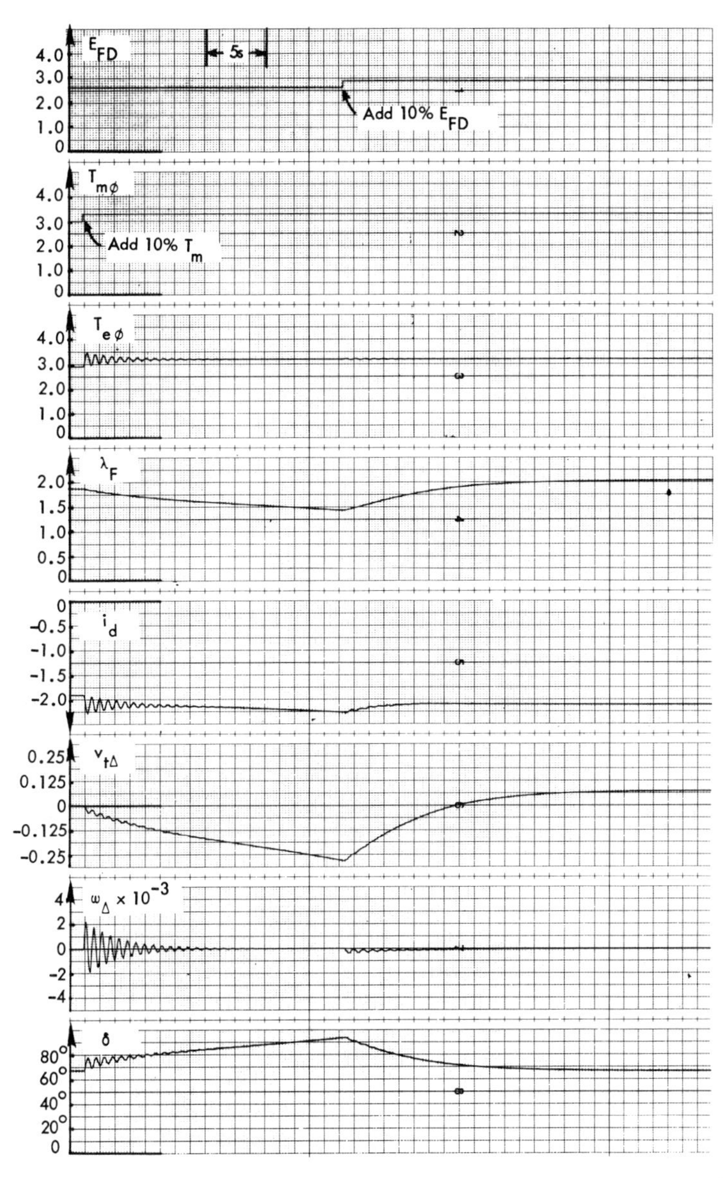

Fig. 5.20 Response of a machine initially at 100% load (Example 5.1 conditions) to a 10% increase in T_m followed by a 10% increase in E_{FD} to assure stable operation.

Figure 5.19 shows the response of the loaded machine to a 20% change in E_{FD}. The generator is initially loaded at 90% of rated load ($T_{m\phi} = 2.7$). Note that the response to this change in E_{FD} does not excite an oscillatory response except for a small, well-damped oscillation in ω_Δ. The terminal voltage responds nearly as a first-order system with a time constant of about 4 s ($\tau'_{d0} = 5.9$ s).

Figure 5.20 shows the system response to 10% step changes in both T_m and E_{FD}. The system is initially in exactly the condition calculated in Example 5.1 with computer voltages given in Table 5.2. A 10% increase in T_m is the first disturbance. This excites a well-damped oscillatory response, particularly in T_e, i_d, V_t, ω, and δ (as well as other variables that are not plotted). A good degree of damping is evident. However, this

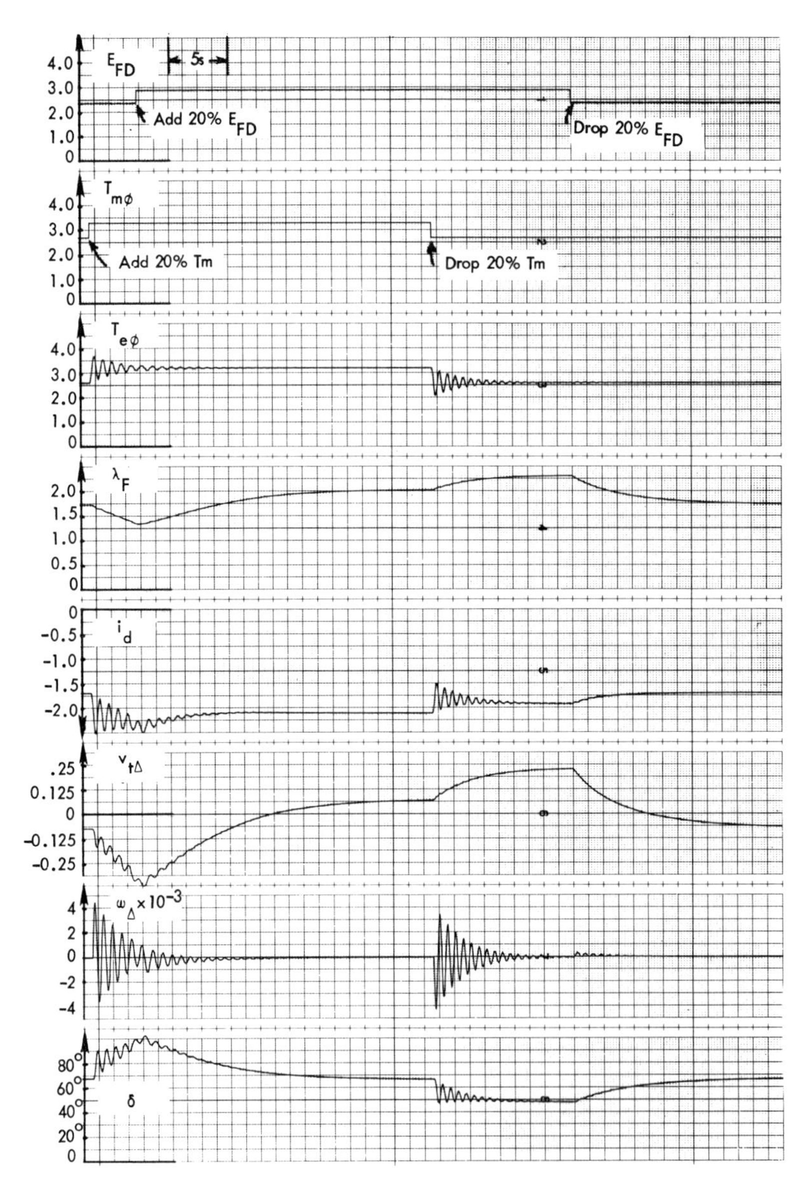

Fig. 5.21 Response of a machine initially at 90% load to a 20% increase in T_m followed by a 20% increase in E_{FD} to restore stability.

overload on the system results in a gradual increase in δ with time, which if not arrested will cause the machine to fall out of step. Repeated runs of the system have indicated that corrective action is required before δ reaches about 95°. The corrective action chosen was a 10% increase in E_{FD}. This quickly restores the system to a stable operating state at about the same angle δ as the initial angle, but at a higher λ_F than the initial value.

Figure 5.21 is similar to 5.20 except that the increments of T_m and E_{FD} are each 20%. The system is initially at 90% load and 90% $E_{FD}(0.9 \times 2.666 = 2.399)$. Then a 20% step increase in T_m is applied. The result is a fast movement toward instability, as evidenced by the rapid increase in δ and the drop in terminal voltage. A 20% increase in E_{FD} is

Table 5.2. Comparison of Digital and Analog Computed Variables

Variable	Computed value pu	Analog computed values		Percent error
		V	pu	
V_t	1.732	68.66	1.717	−0.90
v_d	−1.092	−44.12	−1.103	−1.01
v_q	1.344	52.63	1.316	−2.10
i_d	−1.925	−38.39	−1.920	0.29
i_q	0.667	13.42	0.671	0.60
i_F	2.979			
λ_{AD}	1.634	48.12	1.604	−1.84
λ_{AQ}	0.994	30.10	1.003	0.94
λ_d	1.345	39.49	1.316	−2.13
λ_q	1.094	33.10	1.103	0.85
λ_F	1.935	19.04	1.904	−1.60
T_m	3.004	29.97	2.997	−0.10
δ*	66.995	33.89	67.78	1.17

*Angle between q axis and infinite bus $= \delta - \alpha$.

applied at about the time δ reaches 100°, and the system is quickly restored to a stable operating state. Finally, the excess load and excitation are removed.

Figure 5.22 shows a plot in the phase plane, or ω_Δ versus δ, for exactly the same disturbances as shown in Figure 5.20. The system "spirals" to the right, first very fast and later very slowly, following the 10% increase in T_m. Just prior to loss of synchronism a

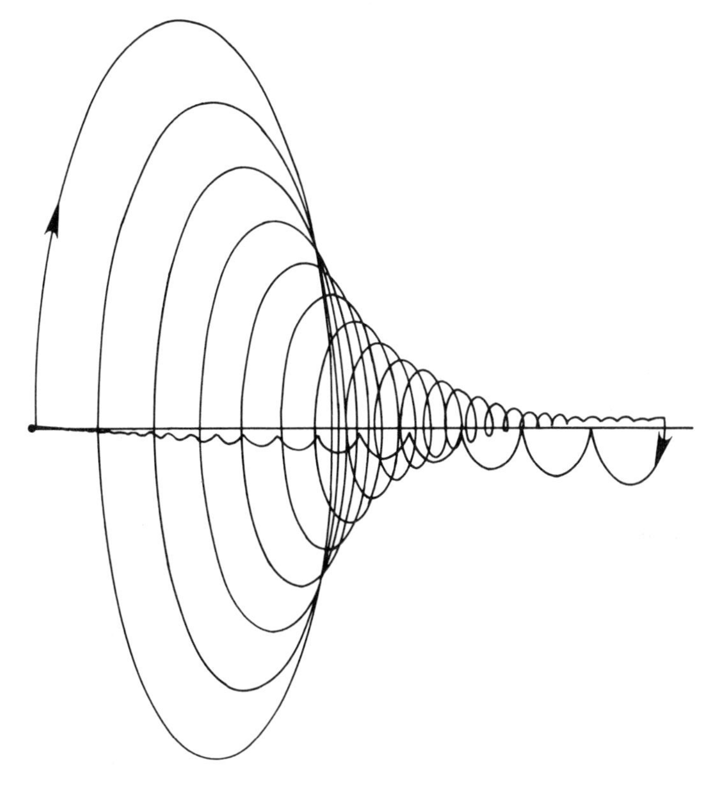

Fig. 5.22 Phase-plane plot ω_Δ versus δ for a 10% step increase in T_m followed by a 10% step increase in E_{FD} (see Figure 5.20). Initial conditions of Example 5.1.

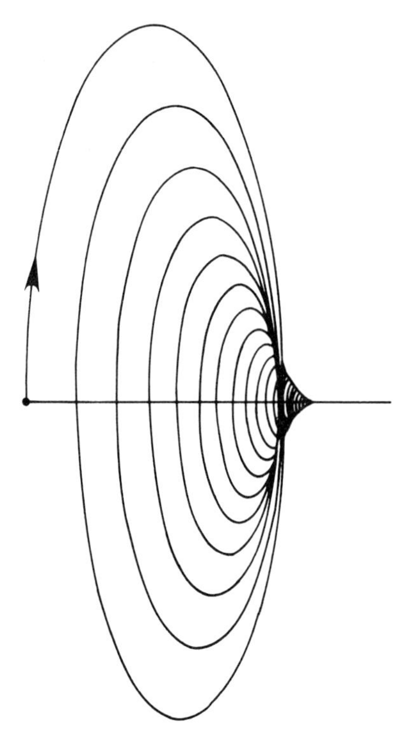

Fig. 5.23 Phase-plane plot ω_Δ versus δ for a 10% step increase in T_m with initial conditions $T_m = 0.9$, $E_{FD} = 2.666$.

10% increase in E_{FD} causes the system to return to about the original δ, following along the lower trajectory.

Figure 5.23 shows an example of a stable phase-plane trajectory. The system is initially at 90% load but with 100% of the Example 5.1 computed value of E_{FD}, or 2.666. A 10% increase in T_m causes the system to oscillate and to seek a new stable value of δ. A comparison of Figures 5.22 and 5.23 shows the more rapid convergence to the target value of δ in the stable case.

5.10 Digital Simulation of Synchronous Machines

Early efforts in solving synchronous machine behavior by digital computer were simply digital applications of the constant-voltage-behind-transient-reactance model, using a step-by-step solution method similar to that of Kimbark [7]. As larger and faster computers became available, engineers quickly realized that the digital computer was a powerful tool for handling very large systems of differential equations. This caused an expansion in power plant modeling to include exciters, governors, and turbines. It also introduced more detailed synchronous machine models into many computer programs, usually in the form of one of the simplified models of Section 4.15. More recent research [8, 9] has been aimed at finding the best machine model for system dynamic studies.

All digital computer simulations must solve the differential equations in a discrete manner; i.e., the time domain is broken up into discrete segments of length t_Δ and the equations solved for each segment. A simple flow chart of the process is shown

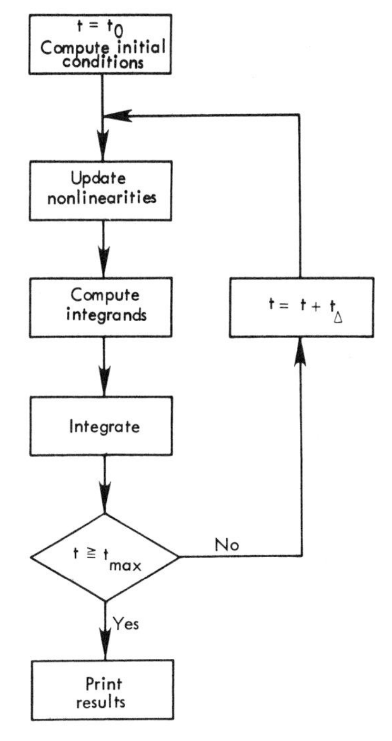

Fig. 5.24 Flow chart of digital integration.

in Figure 5.24. There are several proven methods for performing the actual numerical integration, some of which are presented in Appendix E. Our concern in this book is not with numerical methods, although this is important. Our principal concern is the mathematical model used in the simulation. A number of models are given in Chapter 4. We shall use the flux linkage model of Section 4.12 to illustrate a digital program for calculating synchronous machine behavior in a numerical exercise.

5.10.1 Digital computation of saturation

One of the problems in digital calculation of synchronous machine behavior is the determination of saturation. This is difficult because saturation is an implicit function; i.e., $\lambda_{AD} = f(\lambda_{AD})$. Actually, λ_{AD} is a function of $i_{MD} = i_d + i_F + i_D$, which flows in the magnetizing inductance L_{AD}. But the currents i_d, i_F, and i_D depend upon λ_{AD}, as shown clearly in the analog computer representation of Figure 5.12. Each integration step gives us new λ's by integration. From these λ's we compute i_{MD}. From i_{MD} we estimate saturation, which gives a new λ_{AD}, and this gives new currents, and so on.

The first requirement in computing saturation is to devise some means of determining the amount of saturation corresponding to any given operating point on the saturation curve. For this procedure the saturation curve is represented by a table of data of stator EMF corresponding to given field current, by a polynomial approximation, or by an exponential estimate. The exponential estimate is often used since exponentials are easy to compute. It is based upon computing the offset from the air gap line in pu based on the field current required to produce rated open circuit voltage, shown in Figure 5.25 as i_{F0}. Usually it is assumed there is no saturation at 0.8 pu

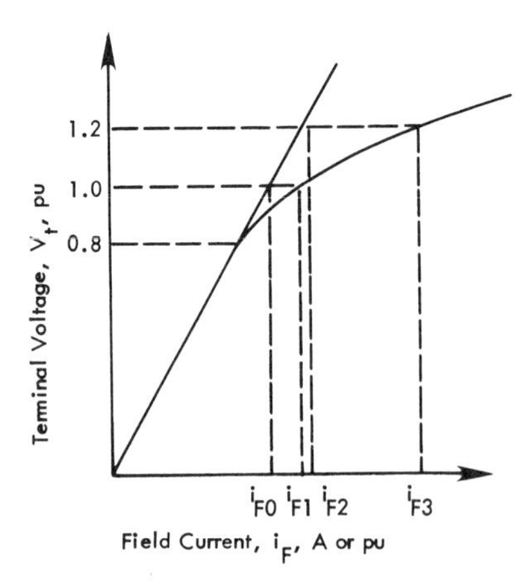

Fig. 5.25 Estimating saturation as an exponential function.

voltage. We then compute the normalized quantities

$$S_{G1} = \frac{i_{F1} - i_{F0}}{i_{F0}} \qquad S_{G2} = \frac{i_{F3} - i_{F2}}{i_{F2}} = \frac{i_{F3} - 1.2i_{F0}}{1.2i_{F0}} \tag{5.81}$$

Then any saturation may be estimated as an exponential function of the form

$$S_G = A_G e^{B_G V_\Delta} \tag{5.82}$$

where $V_\Delta = V_t - 0.8$. Since at open circuit $\lambda_{AD} = \sqrt{3}V_t$, we can also compute saturation in terms of λ_{AD},

$$S_G = A_G \exp[(\lambda_{AD}/\sqrt{3}) - 0.8] \tag{5.83}$$

This is appealing since $\lambda_{AD} = (i_d + i_F + i_D)L_{AD}$ and L_{AD} is the only inductance that saturates appreciably.

If S_{G1} and S_{G2} are given, these values can be substituted into (5.82) to solve for the saturation parameters A_G and B_G. From (5.81) and (5.82) we write

$$S_{G1} = A_G e^{0.2B_G} \qquad 1.2S_{G2} = A_G e^{0.4B_G} \tag{5.84}$$

Rearranging, we compute

$$\ln(S_{G1}/A_G) = 0.2B_G \qquad \ln(1.2S_{G2}/A_G) = 0.4B_G \tag{5.85}$$

Then

$$0.4B_G = \ln(1.2S_{G2}/A_G) = \ln(S_{G1}/A_G)^2$$

or

$$A_G = S_{G1}^2/1.2S_{G2} \tag{5.86}$$

This result may be substituted into (5.85) to compute

$$B_G = 5\ln(1.2S_{G2}/S_{G1}) \tag{5.87}$$

Appendix D shows a plot of S_G as a function of V_t. The function S_G is always positive and satisfies the defined values S_{G1} and S_{G2} at $V_t = 1.0$ and 1.2 respectively. Although we define saturation to be zero for $V_t < 0.8$ pu, actually S_G assumes a very small positive value in this voltage range. The exponential function thus gives a reasonably accurate estimate of saturation for any voltage.

From (5.81) we can write for any voltage level,

$$S_G = (i_F - ki_{F0})/ki_{F0} \tag{5.88}$$

where i_F is the field current required to produce an open circuit voltage V_t, including the effect of saturation. If the air gap line has a slope (resistance) R we have $V_t = Rki_{F0}$. Then, from (5.81)

$$S_G(V_t) = (Ri_F - Rki_{F0})/Rki_{F0} = (Ri_F - V_t)/V_t$$

from which we may write the nonlinear equation

$$V_t = Ri_F - V_tS_G(V_t) \tag{5.89}$$

where Ri_F is the voltage on the air gap line corresponding to field current i_F. Because of saturation, the actual terminal voltage is not Ri_F but is reduced by an amount V_tS_G where S_G is a function of V_t. Equation (5.89) describes only the no-load condition. However, we usually assume that saturation has a similar effect under load; i.e., it reduces the terminal voltage by an amount V_tS_G from the unsaturated value.

Example 5.7

Determine the constants A_G and B_G needed to compute saturation by means of the exponential definition, given the following data from the saturation curve.

$$V_t = 1.0 \text{ pu} \qquad S_{G1} = 30 \text{ A}$$
$$V_t = 1.2 \text{ pu} \qquad S_{G2} = 120 \text{ A}$$

The field current corresponding to $V_t = 1.0$ on the air gap line is $i_{F0} = 365$ A.

Solution

From (5.81) we compute in pu

$$S_{G1} = 30/365 = 0.08219 \qquad S_{G2} = 120/1.2(365) = 0.27397$$

Then from (5.86)

$$A_G = (0.08219)^2/1.2(0.27397) = 0.0205$$

and from (5.87)

$$B_G = 5 \ln [1.2(0.27397/0.08219] = 6.9315$$

5.10.2 Updating the integrands

After computing the new value of saturation for each new time step, we are ready to update the integrands in preparation for numerical integration. This process is illustrated by an example.

Example 5.8

Prepare a FORTRAN computer program to compute the integrands of the flux linkage model for one machine against an infinite bus using the machine data of the Chapter 4 examples. Include in the program a treatment of saturation that can be

```
****CONTINUOUS SYSTEM MODELING PROGRAM****

*** VERSION 1.3 ***

INITIAL
MACRO  SG=GENSAT(WADS)
       KEXP=BGSAT*((WADS/RT3)-0.8)
       SG=AGSAT*EXP(KEXP)
ENDMAC
INITIAL
CONSTANT   PI=3.14159265,RLD=100.0,CLD=0.01
*    MUST SPECIFY GENERATOR POWER,PGEN AND GENERATOR TERMINAL VOLTAGE
*       VT AND INFINITE BUS VOLTAGE, VINF
*
       PGEN=1.00
       VT=1.17
       VINF=1.000
TITLE SATURATED SYNCHRONOUS GENERATOR WITHOUT EXCITER
CONST  NP=2.0,HC=2.37,RMVA=160.0,RKV=15.0,RPF=0.85,XD=1.70,TDOP=5.9
CONST  XQ=1.64,XDP=0.245,XDPP=0.185,XQPP=0.185,XLA=0.15,RAOHM=0.001113
CONST RFOHM=0.267871554,IFLD=365.0,TDPP=.023,TQPP=0.0084603659
PARAM  XE=0.4
PARAM RE=0.02
CONST SATG10=0.0822,SATG12=0.3288
PARAM EXCON=0.01
PARAM TSTART=0.2
PARAM KTM=0.0,KEF=1.0
FIXED KKK
*
*  SATURATION FUNCTION FOR GENERATOR
PROCEDURAL  AGSAT,BGSAT=SATUR(SATG10,SATG12)
       IF(SATG10.EQ.0.0) GO TO 20
       AGSAT=(SATG10**2)/(1.2*SATG12)
       BGSAT=5.0*ALOG((1.2*SATG12)/SATG10)
       GO TO 30
  20   AGSAT=0.0
       BGSAT=0.0
   30  CONTINUE
ENDPRO
*
*
*  COMPUTE INITIAL CONDITIONS
                LA=XLA
               LAD=XD-LA
               LAQ=XQ-LA
                LF=LAD*(XDP-LA)/(LAD-XDP+LA)
               LFF=LAD+LF
              LCON=XDPP-LA
               LKD=LF*LCON*LAD/(LF*LAD-LFF*LCON)
               LKQ=LAQ*(XQPP-LA)/(LAQ-XQPP+LA)
         TBASE=1.0/(2.*PI*60.)
       VBASES=RKV*1000.0/RT3
       SBASE=RMVA*1000000.0/3.0
       IBASES=SBASE/VBASES
       RBASES=VBASES/IBASES
       WBASES=VBASES*TBASE
       LBASES=RBASES*TBASE
       LDB=XD*LBASES
       LAH=XLA*LBASES
       LADH=LDH-LAH
       RT2=SQRT(2.0)
       MFH=RT2*VBASES*TBASE/IFLD
       KMFH=RT3*MFH/RT2
       IFLDB=IBASES*LADH/KMFH
```

Fig. 5.26 CSMP program for computing initial conditions.

```
      VFLDB=SBASE/IFLDB
      RFLDB=VFLDB/IFLDB
      LFLDB=RFLDB*TBASE
      MFBASE=SQRT(LBASES*LFLDB)
      WFLDB=VFLDB*TBASE
      RA=(RAOHM*359.5)/(RBASES*259.5)
            LMD=1.0/((1.0/LAD)+(1.0/LA)+(1.0/LF)+(1.0/LKD))
            LMQ=1.0/((1.0/LAQ)+(1.0/LA)+(1.0/LKQ))
             LE=XE
*
*  PROVIDE LOGIC FOR CASE WHERE TDPP AND TQPP ARE MISSING
PROCEDURAL RKD,RKQ=TFUN(LKD,LKQ,OMB,TDPP,TQPP,LAD,LAQ,LA,LF)
      IF(TDPP .EQ. 0.0) GO TO 104
        RKD=LKD/(OMB*TDPP)+(LAD*LA*LF)/(OMB*TDPP*(LAD*LA+LAD*LF+LA*LF))
      GO TO 105
  104 RKD=1.0E+8
  105 IF(TQPP .EQ. 0.0) GO TO 106
            RKQ=LKQ/(OMB*TQPP)+(LAQ*LA)/(OMB*TQPP*(LAQ+LA))
      GO TO 110
  106 RKQ=1.0E+8
  110       OMU=1.0
ENDPRO
*  PROVIDE LOGIC FOR CALCULATING BEST POSSIBLE RF
PROCEDURAL RF=TRFD(TDOP,LFF,LFLDB,RFLDB,RFOHM)
      IF(TDOP.EQ.0.0) GO TO 120
      RF=(LFF*LFLDB)/(TDOP*RFLDB)
      GO TO 125
  120 RF=(RFOHM*359.5)/(RFLDB*259.5)
  125 CONTINUE
ENDPRO
            RT3=SQRT(3.0)
            DPR=180.0/PI
            OMB=120.0*PI
            ZES=RE**2+XE**2
            G12=-RE/ZES
            B12=XE/ZES
         GAMMA=PI+ATAN(B12/G12)
            G1G=1.0/RLD
            G11=G1G-G12
            Y12=SQRT(G12**2+B12**2)
            NUM=PGEN-G11*VT**2
            DEN=Y12*VT*VINF
            FAC=NUM/DEN
            DUM=SQRT(1.-FAC**2)
          ZETA=ATAN(DUM/FAC)
          BETA=GAMMA-ZETA
         COSAL=COS(BETA)
         SINAL=SIN(BETA)
          ILRE=VT*((COSAL/RLD)-(SINAL*CLD))
          ILIM=VT*((SINAL/RLD)+(COSAL*CLD))
          ITRE=(RE*(VT*COSAL-VINF)+XE*VT*SINAL)/ZES
          ITIM=(RE*VT*SINAL-XE*(VT*COSAL-VINF))/ZES
          IARE=ILRE+ITRE
          IAIM=ILIM+ITIM
         THETA=ATAN(IAIM/IARE)
             IA=SQRT(IARE**2+IAIM**2)
          EQRE=VT*COSAL+RA*IA*COS(THETA)-XQ*IA*SIN(THETA)
          EQIM=VT*SINAL+RA*IA*SIN(THETA)+XQ*IA*COS(THETA)
           EQ0=SQRT(EQRE**2+EQIM**2)
      TECHK=3*PGEN+3*RA*IA**2
      DL=ATAN(EQIM/EQRE)
PROCEDURE VD,VQ,ID,IQ,WD,WQ,WADS,WAQS,SGD,SGQ,IFUN,IMU,TEDEL,DELDL,...
      TE,NN=FUNC(DL,VT,IA,BETA,THETA,RT3,RA,LAD,LA,LF,LAQ,TECHK)
      NN=0
```

Fig. 5.26 *(continued)*

```
  100 VD=-RT3*VT*SIN(DL-BETA)
      VQ=RT3*VT*COS(DL-BETA)
      ID=-RT3*IA*SIN(DL-THETA)
      IQ=RT3*IA*COS(DL-THETA)
*  UNSATURATED FIELD CURRENT = IFUN
      IFUN=((VQ+RA*IQ)/LAD)-((LA+LAD)*ID)/LAD
      IMU=ID+IFUN
      WADS=LAD*IMU
*  SATURATED D-AXIS FLUX LINKAGES
      SGD=GENSAT(WADS)
      WD=LA*ID+WADS
*  SATURATED Q-AXIS FLUX LINKAGES
      WAQZ=LAQ*IQ
      WAQS=IMPL(WAQZ,0.00001,GAQS)
      SGQ=GENSAT(WAQS)
      GAQS=LAQ*IQ/(1.0+SGQ)
      WQ=LA*IQ+WAQS
      TE=WD*IQ-WQ*ID
      TEDEL=TE-TECHK
      DELDL=0.1*(TEDEL/TECHK)*TAN(DL)
      IF(DELDL.GT.0.00001) GO TO 200
      IF(DELDL.LT.-0.00001) GO TO 200
      GO TO 400
 200  DL=DL+DELDL
      NN=NN+1
      IF(NN.GT.50) GOTO 400
      GO TO 100
  400 CONTINUE
ENDPRO
*  FINAL INITIAL COMPUTATIONS
      IFF=IFUN+SGD*IMU
      WF=LF*IFF+WADS
      WKD=WADS
      WKQ=WAQS
             VF=RF*IFF
            DLD=DPR*DL
             OM=OMB
           DOMU=0.0
          ITMAG=SQRT(ITRE**2+ITIM**2)
            PSI=ATAN(ITIM/ITRE)
            IDT=-RT3*ITMAG*SIN(DL-PSI)
            IQT=RT3*ITMAG*COS(DL-PSI)
            ILDD=ID-IDT
            ILDQ=IQ-IQT
              TM=TE
              TA=0.0
           VTCHK=(1./RT3)*SQRT(VD**2+VQ**2)
             WDZ=WD
             WFZ=WF
            WKDZ=WKD
             WQZ=WQ
            WKQZ=WKQ
            DOMZ=DOMU
             DLZ=DL
            IDTZ=IDT
            IQTZ=IQT
             TMZ=TM
             VDZ=VD
             VQZ=VQ
      KFD=RT3*RF/LAD
      EFD=VF/KFD
      EFDZ=EFD
*
 NOSORT
```

Fig. 5.26 *(continued)*

executed prior to integration at each time step. Include a local load on the generator bus in the computation. Use the Continuous System Modeling Program (CSMP) [10] for solving the equations and plotting the results.

Solution

An essential part of the computer program is a routine to compute the initial conditions. As noted in Examples 5.1–5.3, this computation depends upon the boundary conditions that are specified. The boundary conditions chosen for this example are those of Example 5.3, viz., P and V_t at the generator terminals. The FORTRAN coding for this section of the program is included in the portion of the program listing in Figure 5.26 called INITIAL. Note that the statement of the problem does not give any explicit numerical boundary condition. This is one of the advantages of a computer program; once it is written and verified, problems with different boundary conditions but of the same type can be solved with ease. The boundary conditions specified in Figure 5.26 give $P = 1.00$ (PGEN), $V_t = 1.17$ (VT), and $V_\infty = 1.00$ (VINF).

1. Make a preliminary estimate of λ_{AD} (λ_{AD} is named WADS in the program; W being used for λ and S meaning "saturated").

$$\lambda_{AD} = \lambda_{AD}]_{\text{last}\,\Delta t} = \text{WADS}]_{\text{last}\,\Delta t} \tag{5.90}$$

2. Compute the new currents. From the equations

$$\begin{aligned} i_d &= (\lambda_d - \lambda_{AD})/\ell_d \qquad & i_F &= (\lambda_F - \lambda_{AD})/\ell_F \\ i_D &= (\lambda_D - \lambda_{AD})/\ell_D \qquad & i_{MD} &= i_d + i_F + i_D \end{aligned} \tag{5.91}$$

we compute an estimate of the new currents. This estimate is not exact because the value of λ_{AD} used in (5.91) is the value computed at the start of the *last* Δt, whereas the flux linkages λ_d, λ_F, and λ_D are the integrated new values. Thus i_{MD} computed by (5.91) does not correspond to point A of Figure 5.27, but to some new point *B*. Since λ_{AD} is a function of the currents and of saturation, we must find the correct new λ_{AD} iteratively. We do this by changing our estimated λ_{AD} slightly until i_{MD} agrees with λ_{AD} on the saturation curve, or until points *A* and *B* of Figure 5.27 coincide.

3. To estimate the new λ_{AD}, we compute the saturation function $S_{GD} = f(\lambda_{AD})$ in the

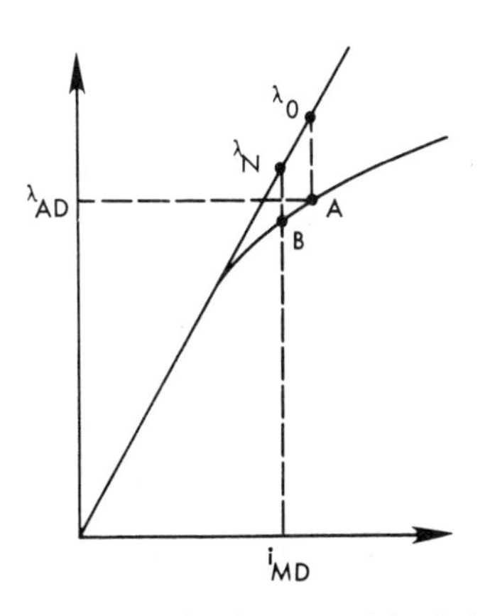

Fig. 5.27 Saturation curve for the magnetizing inductance L_{AD}.

```
DYNAMIC
NOSORT
*
*     GENERATOR CURRENTS
           WADSO=WADS
            WADS=IMPL(WADSO,0.0001,FWAD)
              ID=(WD-WADS)/LA
             IFF=(WF-WADS)/LF
             IKD=(WKD-WADS)/LKD
             IMD=ID+IFF+IKD
       SGD=GENSAT(WADS)
            GADS=LAD*IMD/(1.0+SGD)
            FWAD=WADS+(GADS-WADS)*EXCON
           WAQSO=WAQS
            WAQS=IMPL(WAQSO,0.0001,FWAQ)
              IQ=(WQ-WAQS)/LA
             IKQ=(WKQ-WAQS)/LKQ
             IMQ=IQ+IKQ
       SGQ=GENSAT(WAQS)
            GAQS=LAQ*IMQ/(1.0+SGQ)
            FWAQ=WAQS+(GAQS-WAQS)*EXCON
SORT
*
*     TORQUE
              TE=WD*IQ-WQ*ID
              TA=TM-TE
*
*     SPEED
           IGDOM=TA/(6.0*HC)
            DOMU=INTGRL(DOMZ,IGDOM)
             OMU=DOMU+1.0
              OM=OMU*OMB
*
*     ANGLE
            IGDL=OMB*DOMU
              DL=INTGRL(DLZ,IGDL)
             DLD=DPR*DL
*
*     LOCAL LOAD
            IGVD=(OMB/CLD)*(ID-IDT-(VD/RLD)-(OMU*CLD*VQ))
              VD=INTGRL(VDZ,IGVD)
            IGVQ=(OMB/CLD)*(IQ-IQT-(VQ/RLD)+(OMU*CLD*VD))
              VQ=INTGRL(VQZ,IGVQ)
*
*     TRANSMISSION LINE
           IGIDT=(OMB/LE)*(VD+RT3*SIN(DL)-RE*IDT-OMU*LE*IQT)
             IDT=INTGRL(IDTZ,IGIDT)
*
           IGIQT=(OMB/LE)*(VQ-RT3*COS(DL)-RE*IQT+OMU*LE*IDT)
             IQT=INTGRL(IQTZ,IGIQT)
*
*     D-AXIS FLUX LINKAGES
             IGD=OMB*(-VD-(RA*ID)-OMU*WQ)
              WD=INTGRL(WDZ,IGD)
       VF=KFD*EFD
*
             IGF=OMB*(VF-RF*IFF)
              WF=INTGRL(WFZ,IGF)
*
            IGKD=OMB*(-RKD*IKD)
             WKD=INTGRL(WKDZ,IGKD)
*
*     Q-AXIS FLUX LINKAGES
             IGQ=OMB*(-VQ-(RA*IQ)+OMU*WD)
              WQ=INTGRL(WQZ,IGQ)
*
            IGKQ=OMB*(-RKQ*IKQ)
             WKQ=INTGRL(WKQZ,IGKQ)
*
*
*     TERMINAL VOLTAGE
              VT=((VD**2+VQ**2)**0.5)/RT3
              IA=((ID**2+IQ**2)**0.5)/RT3
              PE=OMU*TE/3.0
*
*     DRIVING FUNCTIONS
       TM=TMZ+KTM*TMZ/10.0*STEP(TSTART)
       EFD=EFDZ+KEF*EFDZ/20.0*STEP(TSTART)
*
NOSORT
```

Fig. 5.28 CSMP program for updating integrands.

```
TERMINAL
RANGE DELT
OUTPUT WD
PAGE  GROUP=(1.55,1.75)
OUTPUT WF
PAGE  GROUP=(2.1,2.3)
OUTPUT WKD
PAGE  GROUP=(1.8,2.0)
OUTPUT WADS
PAGE  GROUP=(1.8,2.0)
OUTPUT SGD
PAGE  GROUP=(0.12,0.24)
OUTPUT IA
PAGE  GROUP=(0.98,1.18)
OUTPUT IFF
PAGE  GROUP=(3.0,3.24)
OUTPUT IKD
PAGE GROUP=(-0.02,0.04)
OUTPUT VT
PAGE  GROUP=(1.15,1.19)
OUTPUT DLD
PAGE  GROUP=(50,62)
OUTPUT DOMU
PAGE  GROUP=(-0.0016,0.002)
OUTPUT TE
PAGE GROUP=(2.8,3.6)
TIMER FINTIM=2.5,OUTDEL=0.05,DELT=0.0001
END
STOP
```

Fig. 5.28 *(continued)*

usual way, using (5.83). Then we compute λ_0 and λ_N, defined in Figure 5.27,

$$\lambda_0 = \lambda_{AD}(1 + S_{GD}) \qquad \lambda_N = L_{AD} i_{MD}$$

Then the error measured on the air gap line is $\lambda_E = \lambda_N - \lambda_0$, and the error measured on the saturation curve is approximately

$$\lambda_\Delta = \lambda_E/(1 + S_{GD})$$

Now define a new λ_{AD} to be G_{AD}, defined as $G_{AD} = \lambda_{AD} + \lambda_\Delta$. Then we compute

$$G_{AD} = \lambda_{AD} + (\lambda_N - \lambda_0)/(1 + S_{GD}) = L_{AD} i_{MD}/(1 + S_{GD})$$

4. Now we test G_{AD} to see if it is significantly different from λ_{AD}; i.e., we compute

$$| G_{AD} - \lambda_{AD} | \overset{?}{<} \epsilon$$

where ϵ is any convenient precision index, such as 10^{-4}. If the test fails, we estimate a new λ_{AD} from

$$\text{new } \lambda_{AD} \triangleq F_{AD} = \lambda_{AD} - h(G_{AD} - \lambda_{AD})$$

where h is chosen to be a number small enough to prevent overshoot; typically, $h = 0.01$. Now the entire procedure is repeated, returning to step 1 with the $\lambda_{AD} = F_{AD}$, finding new currents, etc. As the process converges, we will know both the new current and the new saturated value of λ_{AD}.

B. The second part of the program computes the integrands of all equations in preparation for integration (integration is indicated in the program by the macro INTGTL). The computer program for updating the integrands is shown in Figure 5.28.

The computed output of several variables to a step change in T_m and E_{FD} is shown in Figures 5.29–5.40. Computer mnemonics are given in Table 5.3. In both cases, the step input is applied at t = TSTART = 0.2 s.

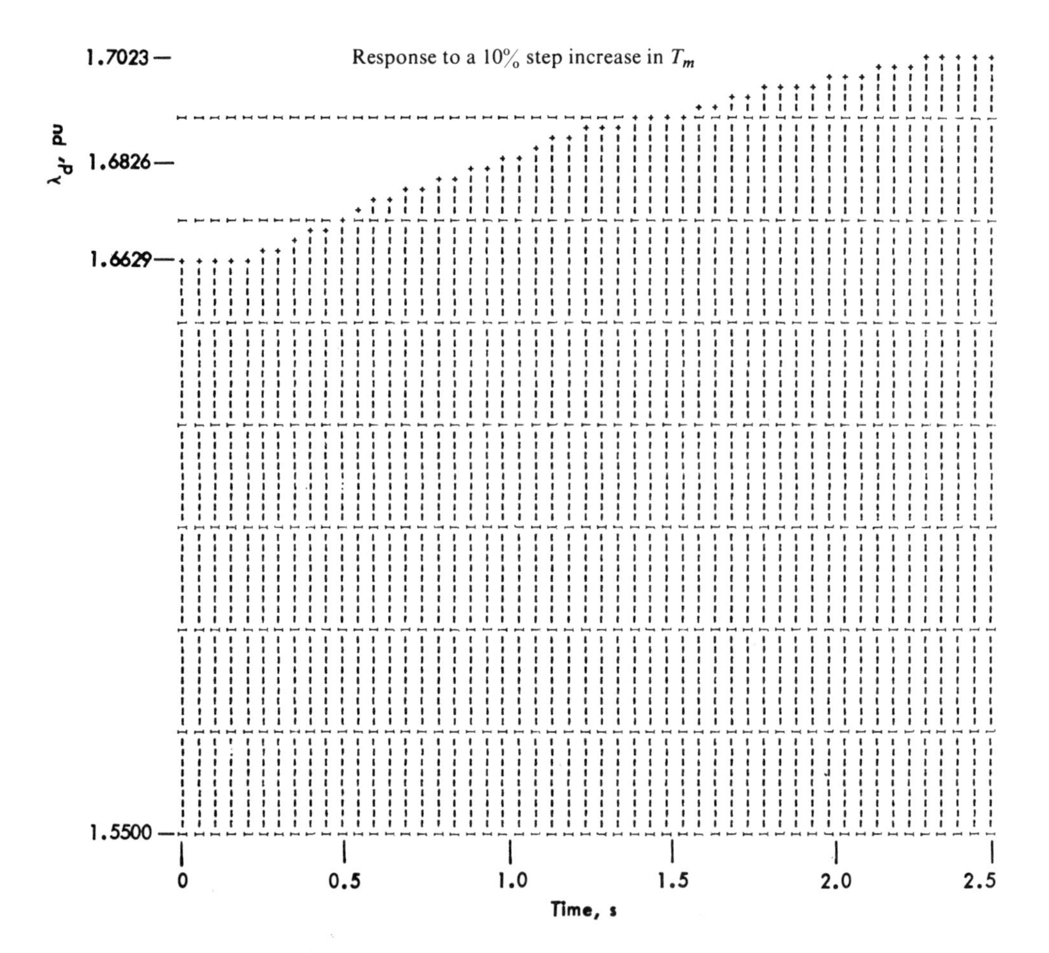

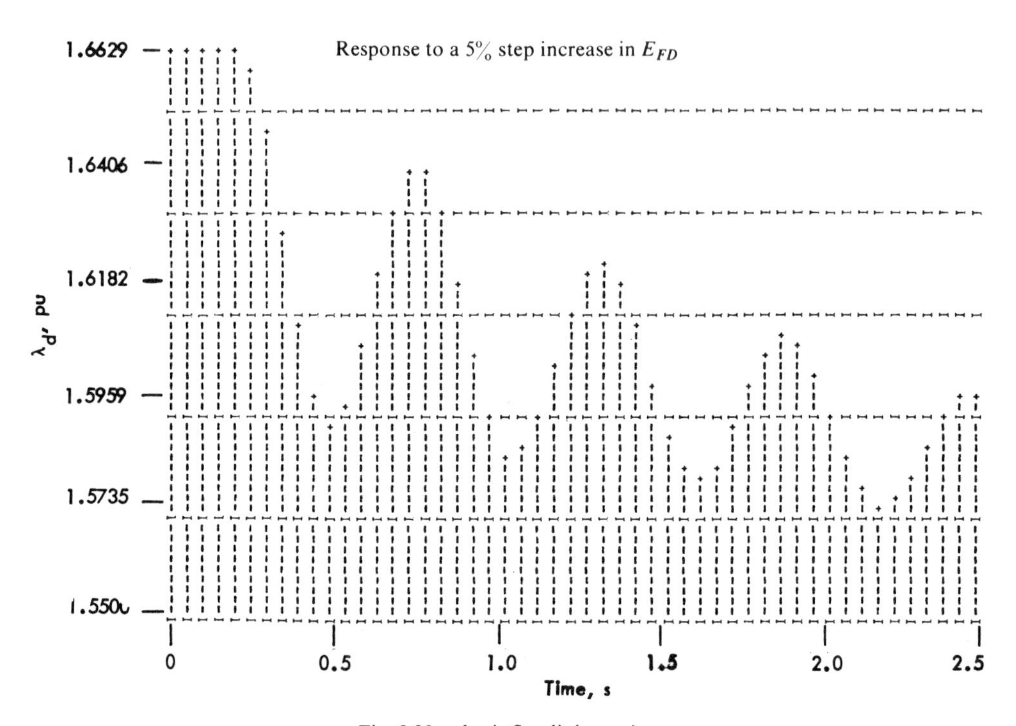

Fig. 5.29 d axis flux linkages λ_d.

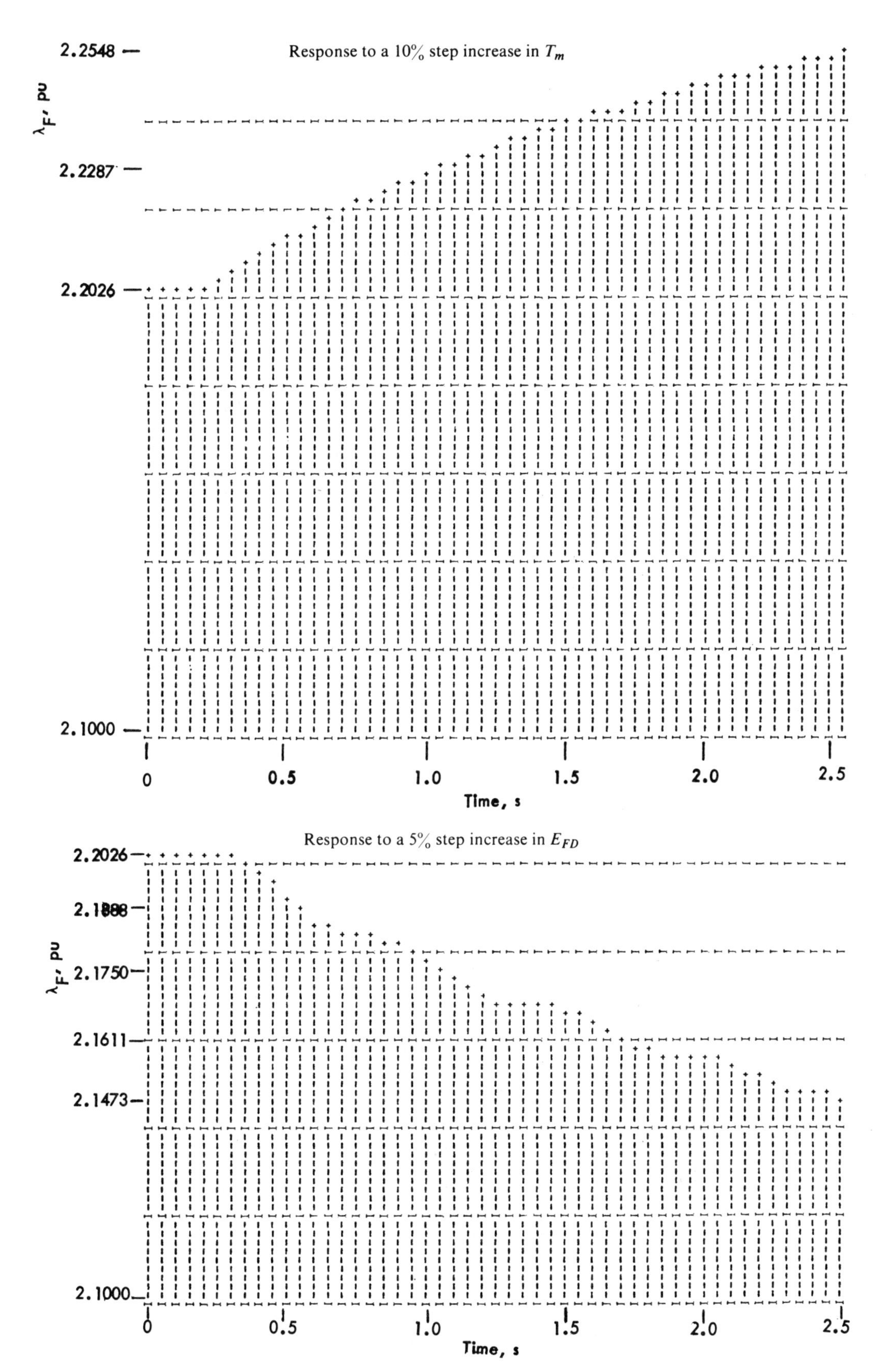

Fig. 5.30 Field flux linkages λ_F.

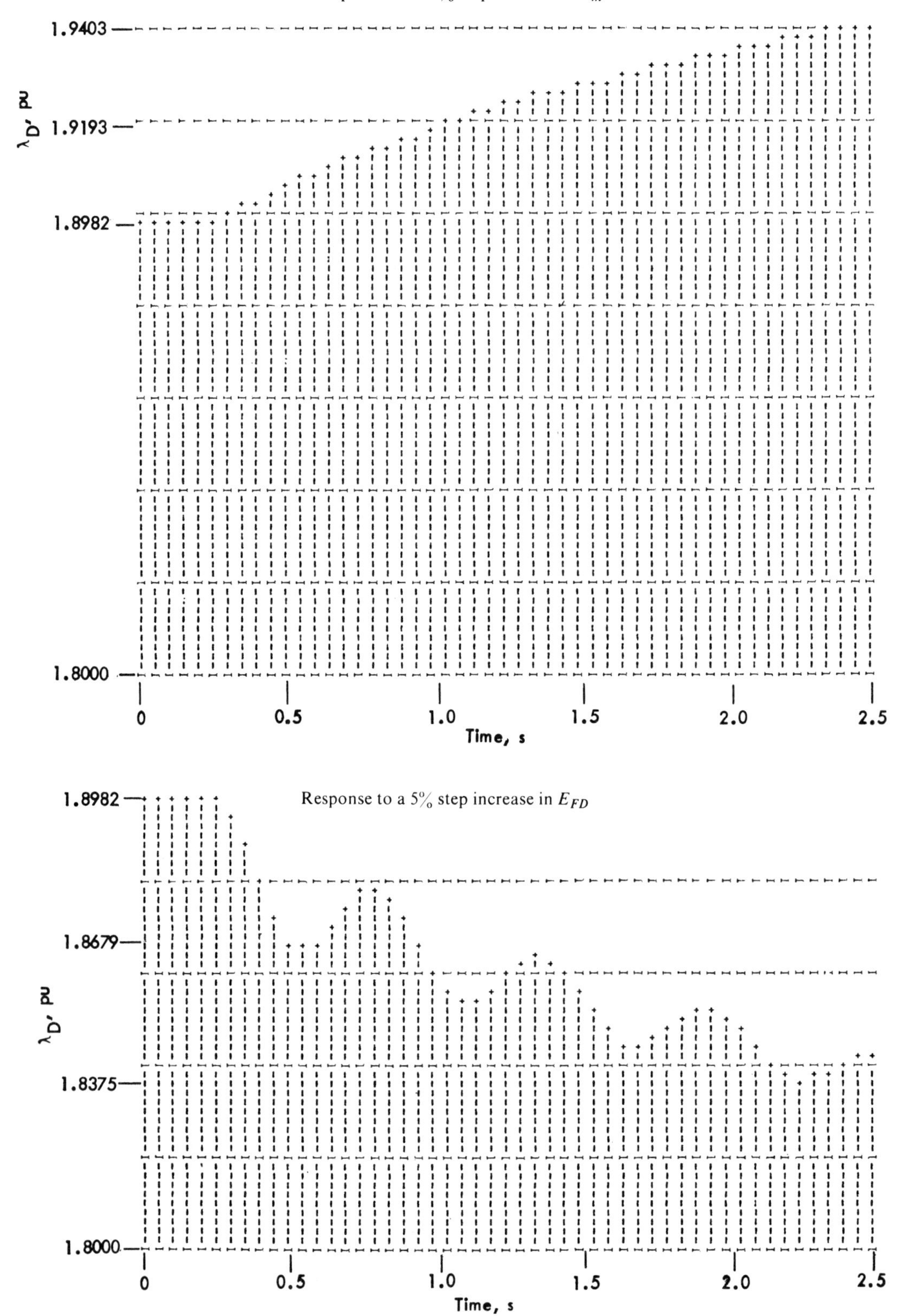

Fig. 5.31 d axis amortisseur flux linkages λ_D.

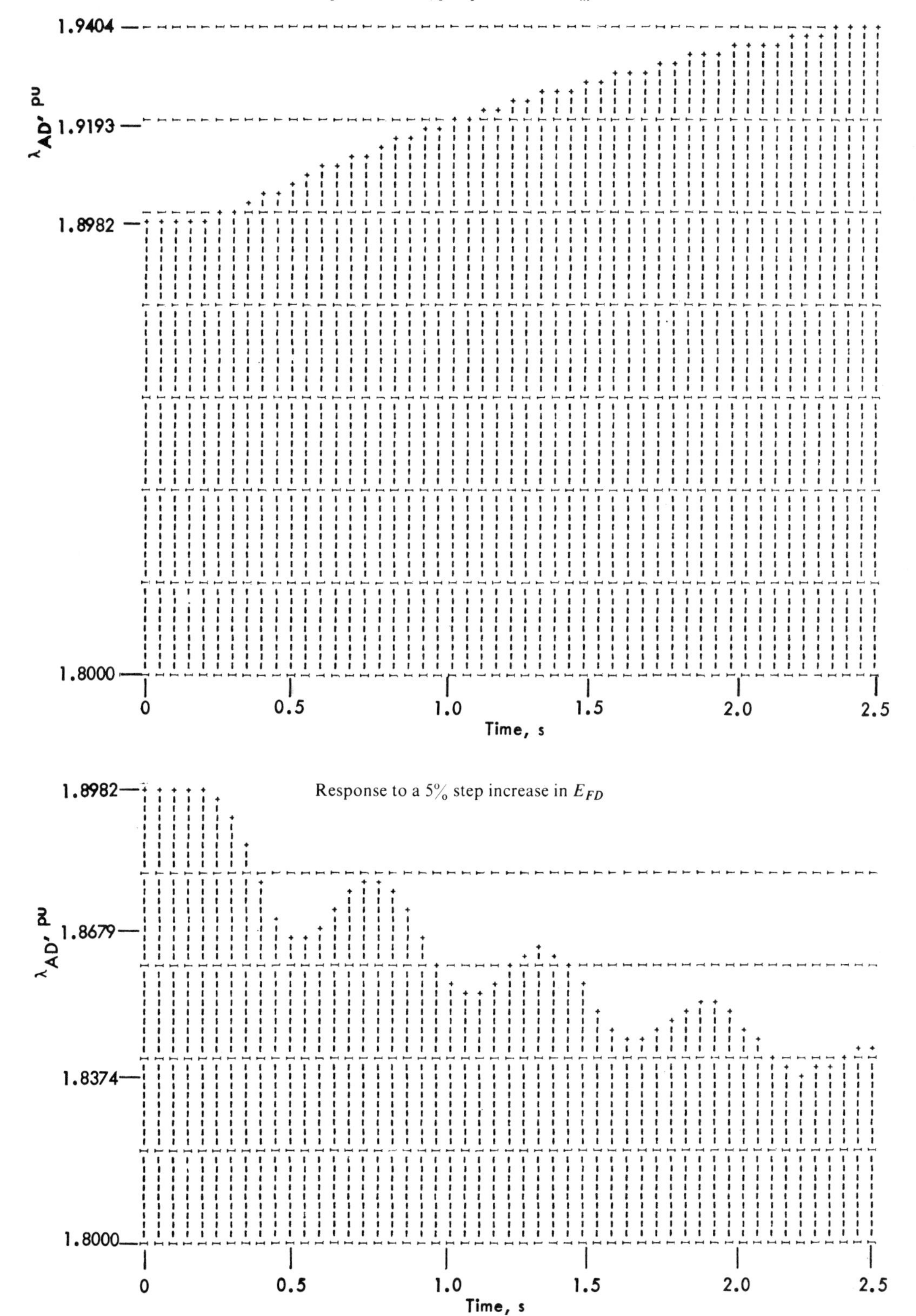

Fig. 5.32 Saturated d axis mutual flux linkages λ_{ADS}.

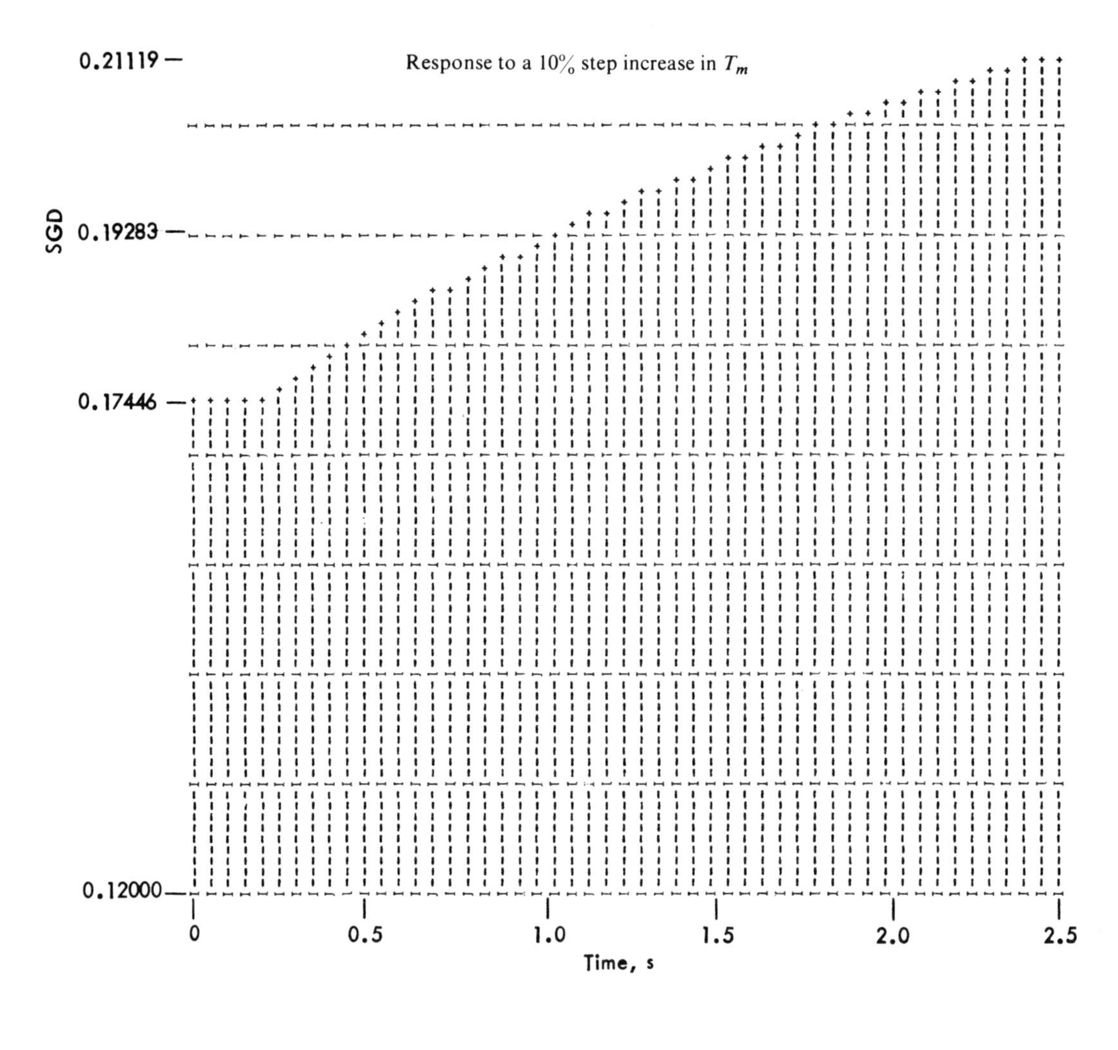

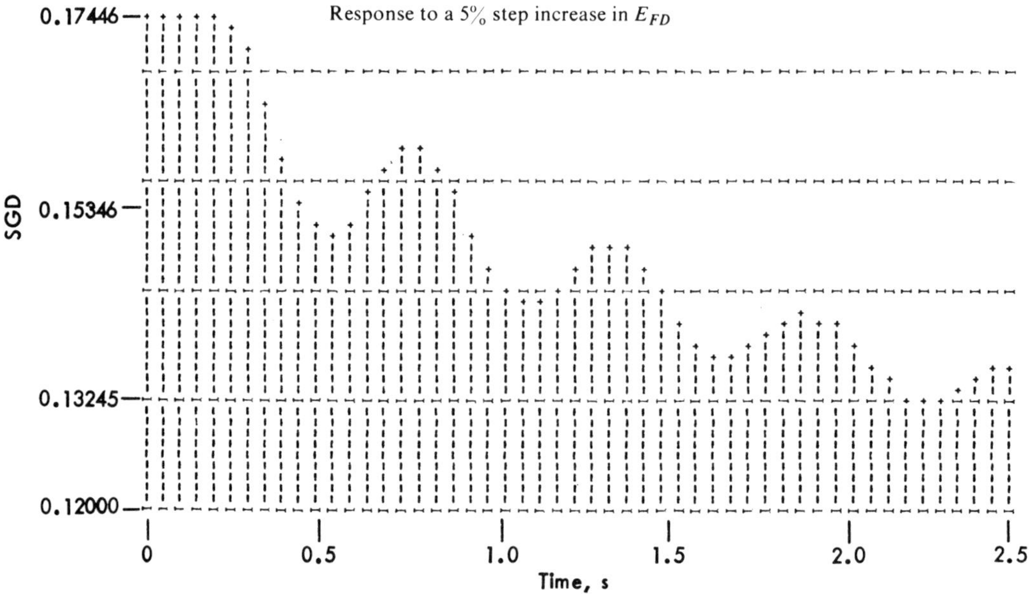

Fig. 5.33 d axis saturation function S_{GD}.

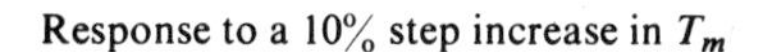

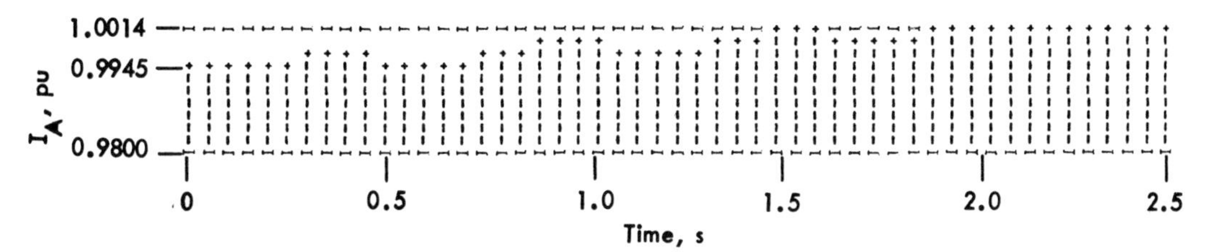

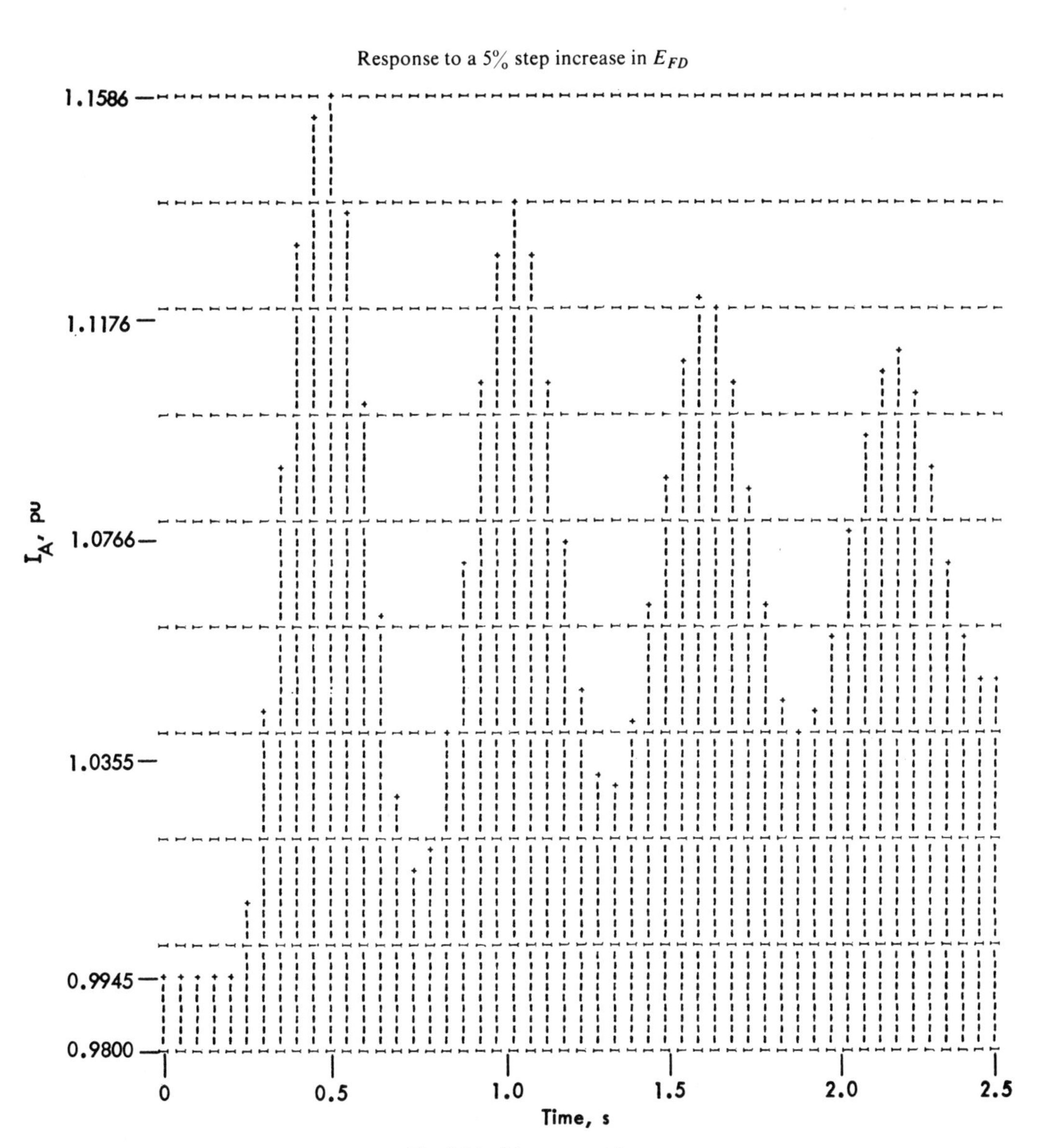

Fig. 5.34 Line current i_a.

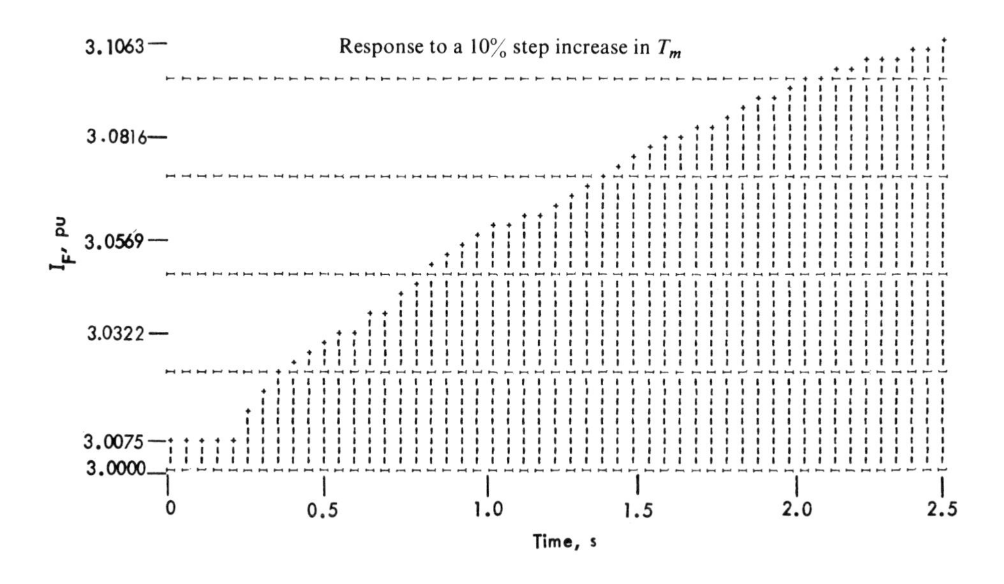

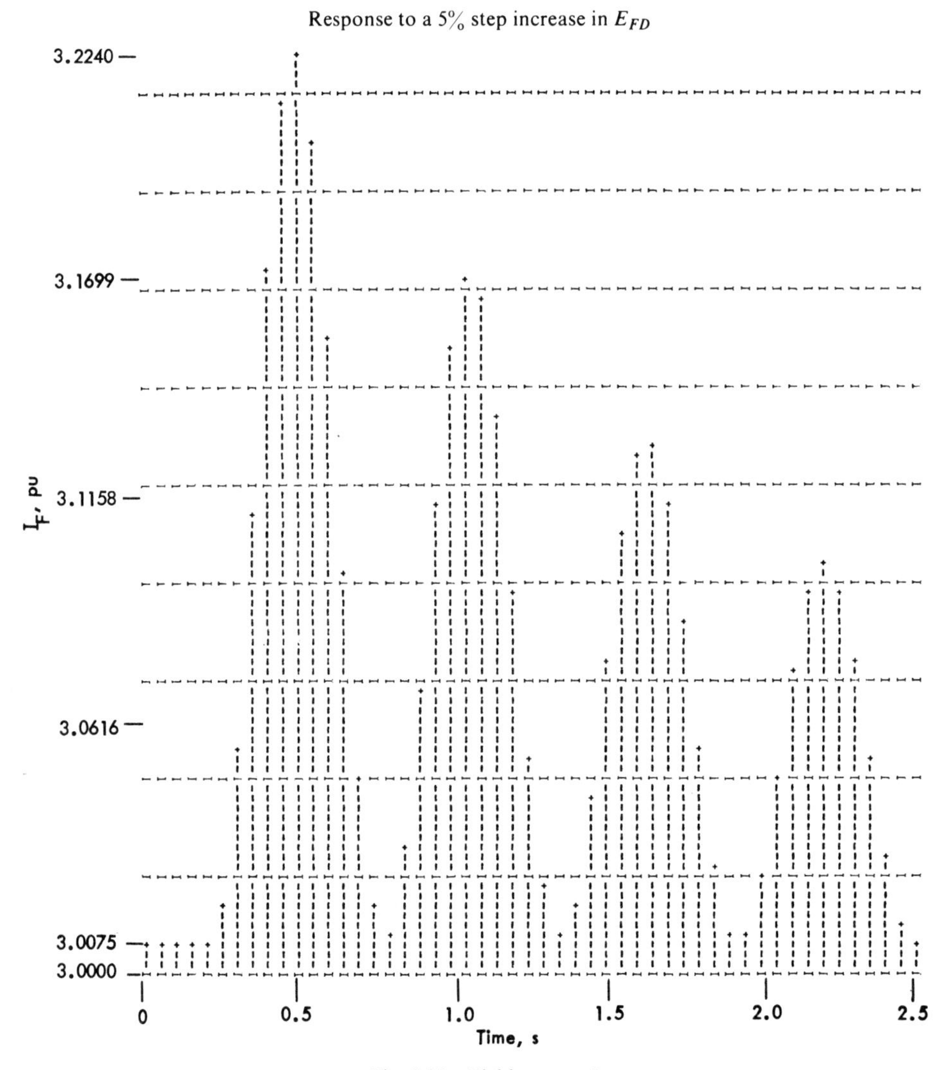

Fig. 5.35 Field current i_F.

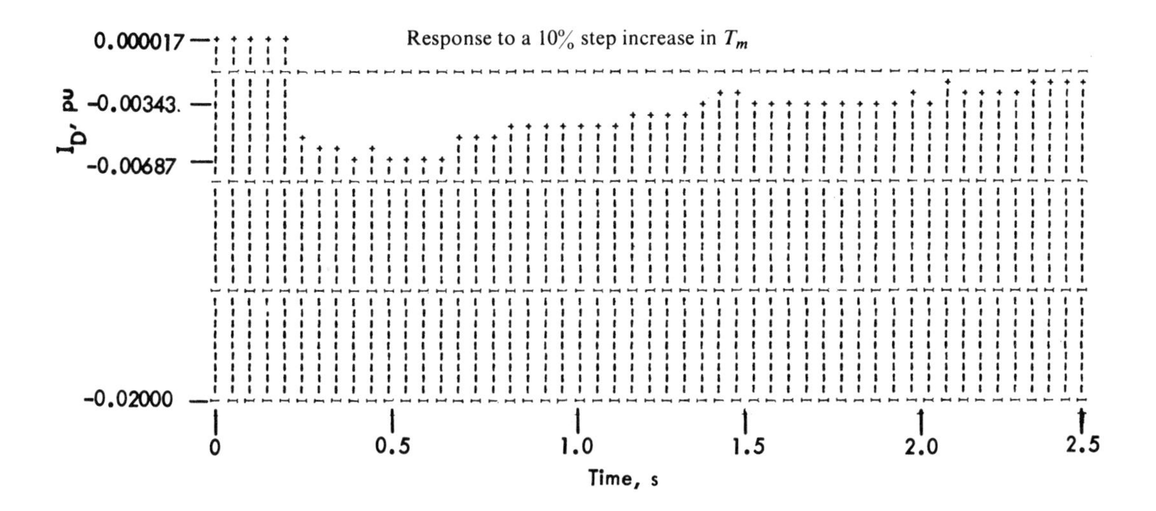

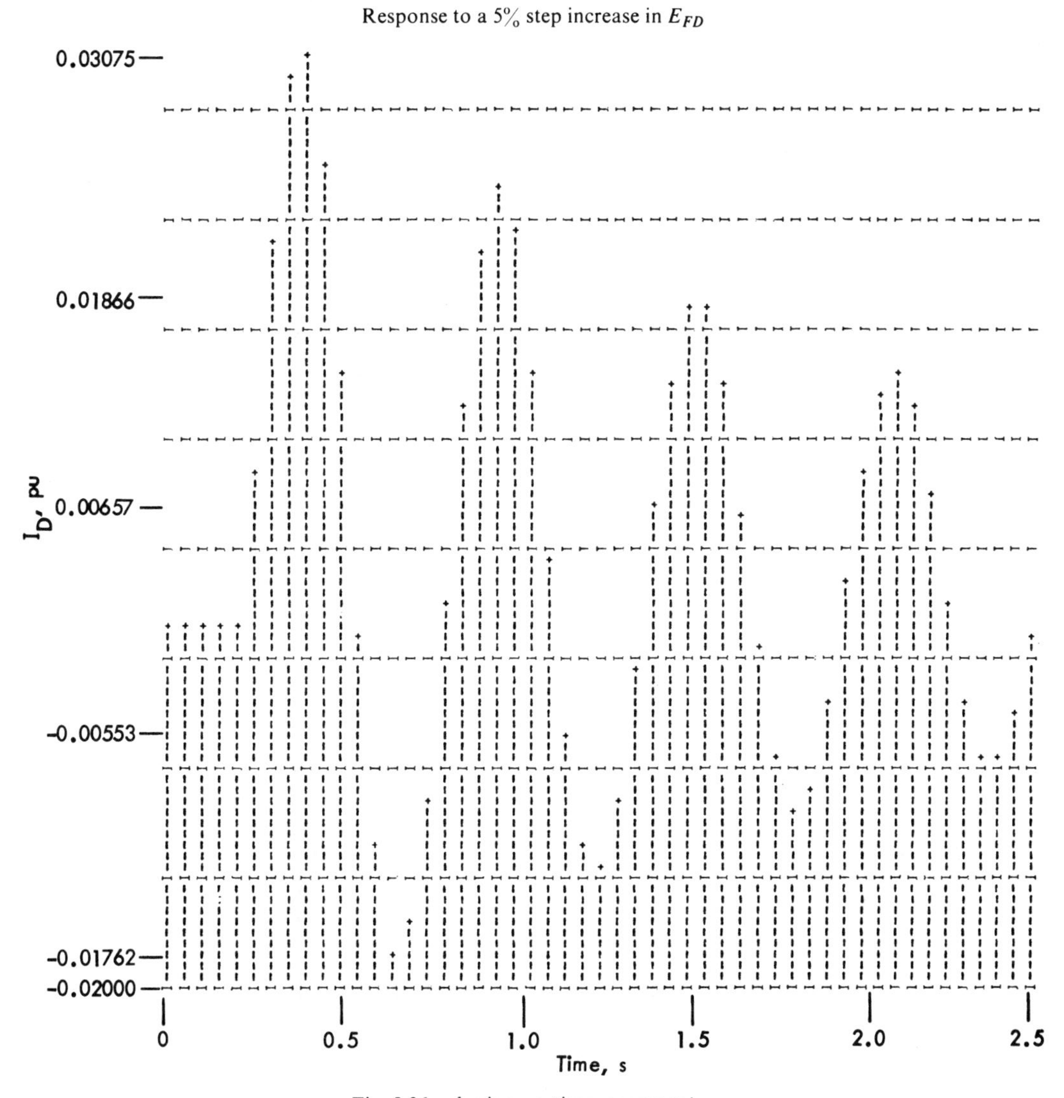

Fig. 5.36 d axis amortisseur current i_D.

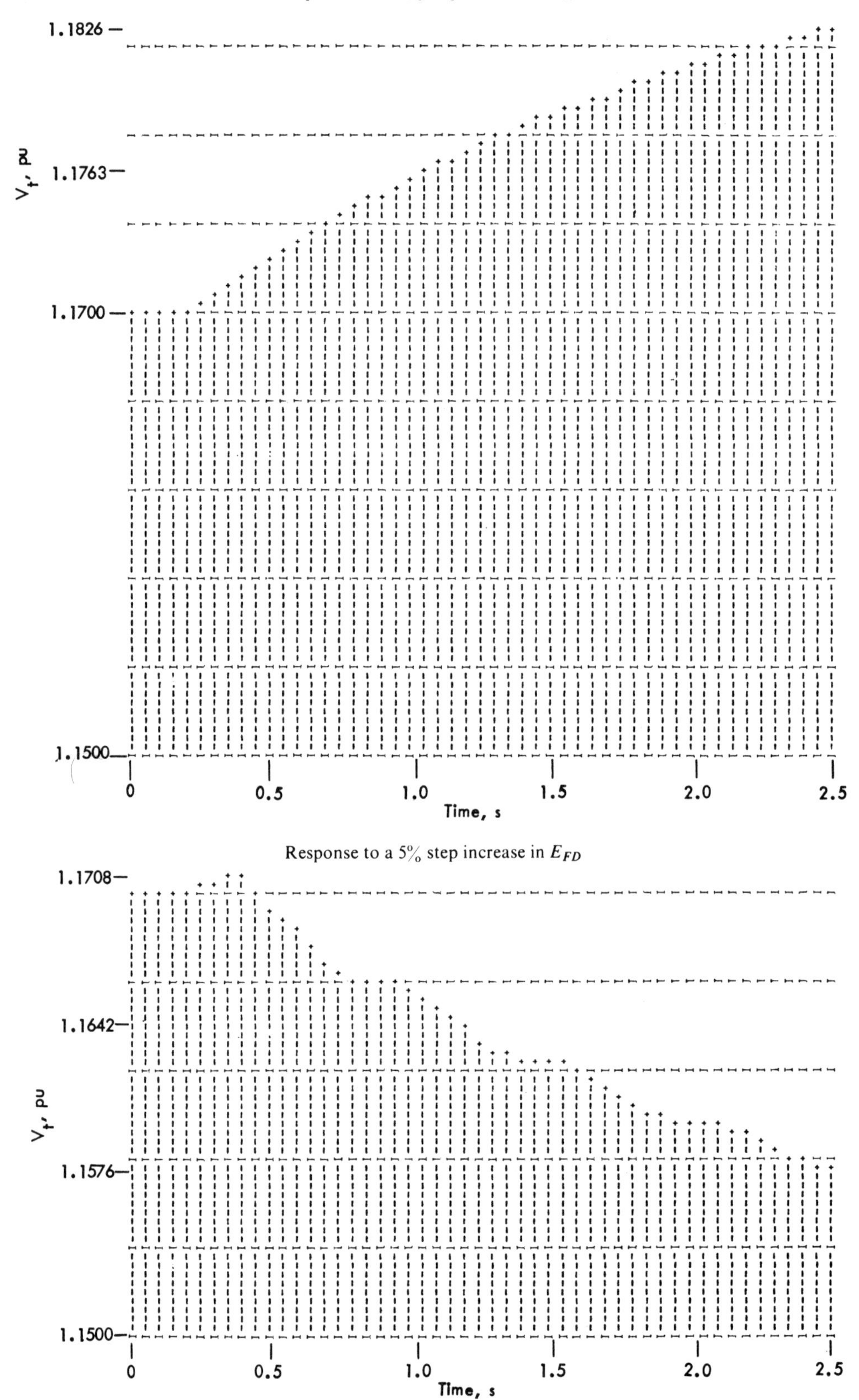

Fig. 5.37 Terminal voltage V_t.

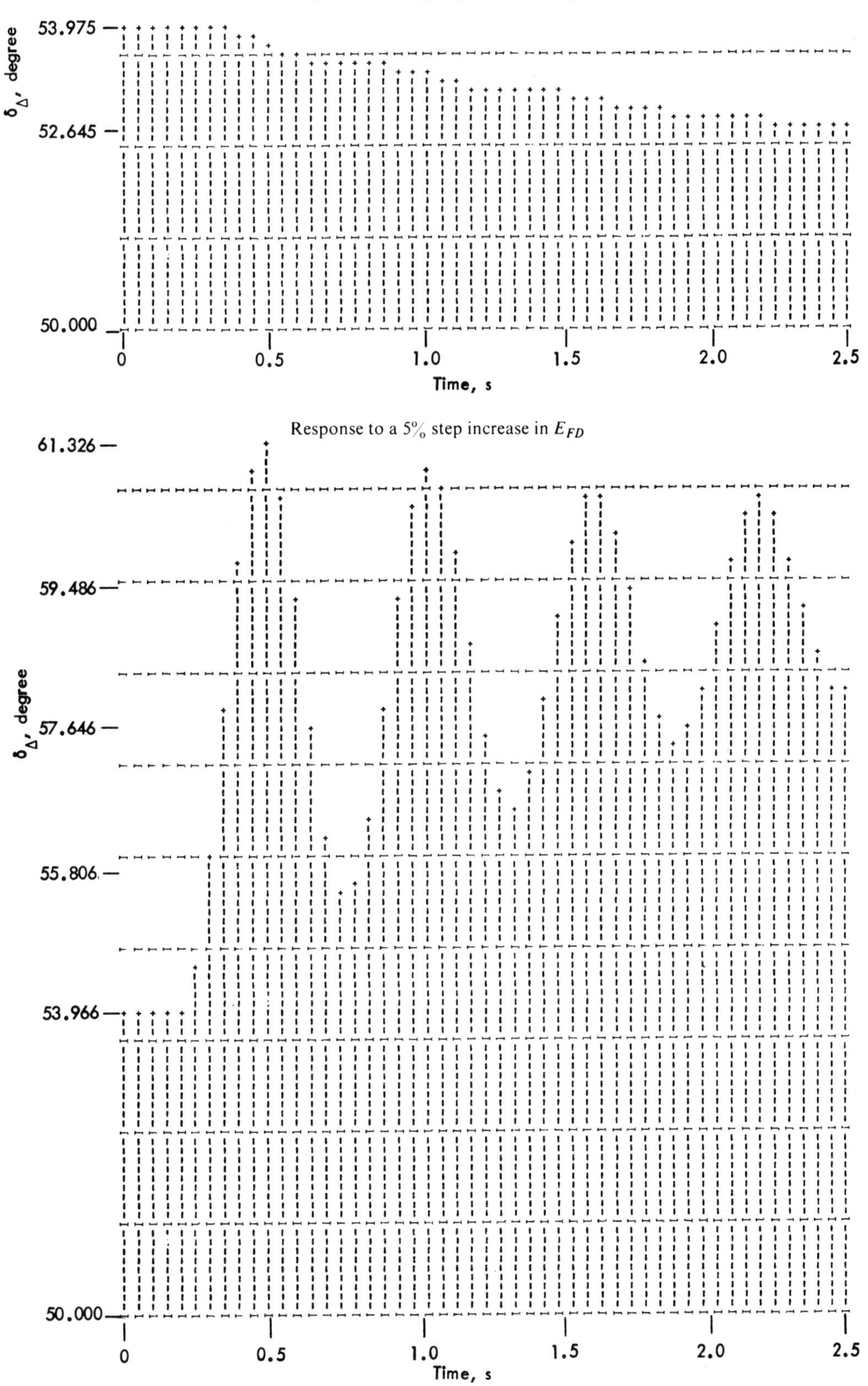

Fig. 5.38 Torque angle δ in degrees.

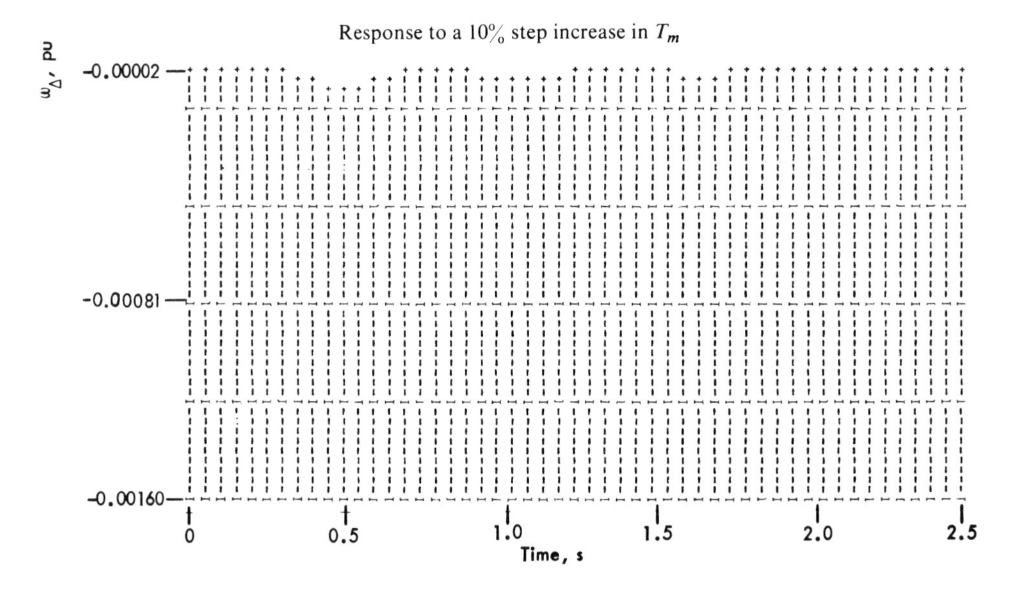

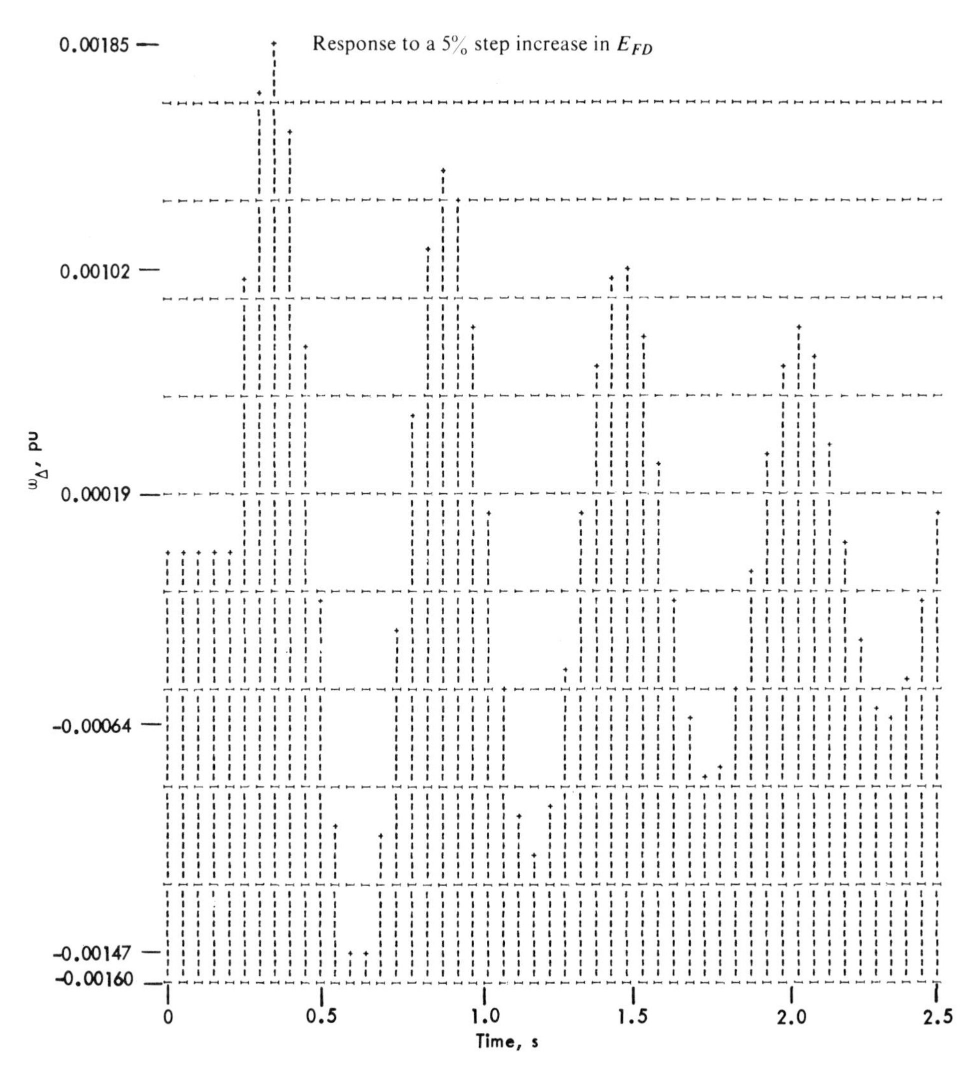

Fig. 5.39 Speed deviation ω_Δ in pu.

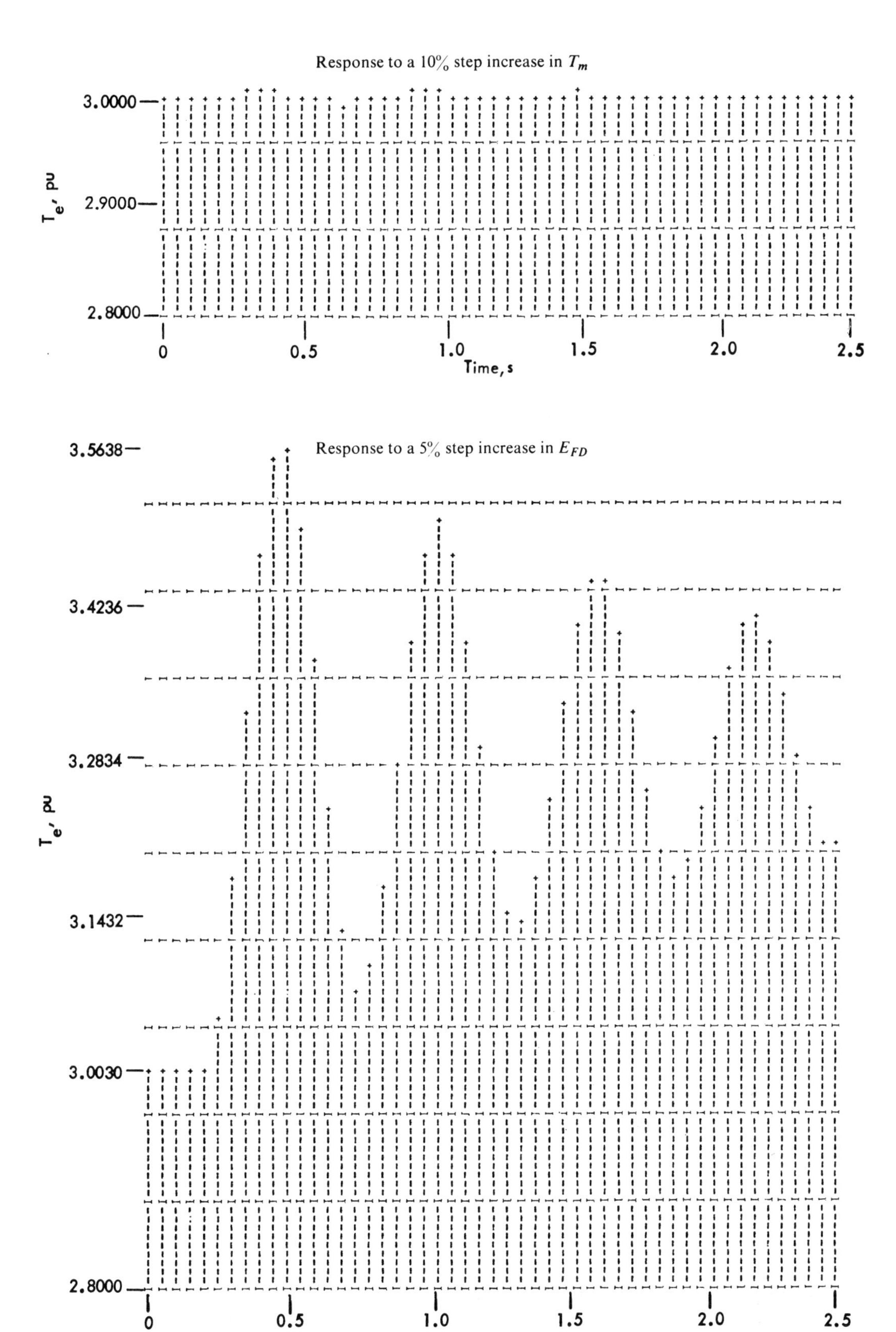

Fig. 5.40 Electromagnetic torque $T_{e\phi}$.

Table 5.3. Computer Mnenomics of Output Variables

Figure	Variable	Computer mnemonic
5.29	λ_d	WD
5.30	λ_F	WF
5.31	λ_D	WKD
5.32	λ_{ADS}	WADS
5.33	S_{GD}	SGD
5.34	i_a	IA
5.35	i_F	IFF
5.36	i_D	IKD
5.37	V_t	VT
5.38	δ (in degrees)	DLD
5.39	ω_Δ (in pu)	DOMU
5.40	$T_{e\phi}$	TE

Problems

5.1 The synchronous machine discussed in Examples 5.1 and 5.2 is operating at rated terminal voltage, and its output power is 0.80 pu. The angle between the q axis and the terminal voltage is 45°. Find the steady-state operating condition: the d and q axis voltages, currents, flux linkages, and the angle δ.

5.2 The same synchronous machine connected to the same transmission line, as in Examples 5.1 and 5.2, has a local load of unity power factor, which is represented by a resistance $R = 10$ pu. The infinite bus voltage is 1.0 pu. The power at the infinite bus is 0.9 pu at 0.9 PF lagging. Find the operating condition of the machine.

5.3 Repeat Problem 5.2 with the machine output power being 0.9 pu at 0.9 PF lagging.

5.4 In the system of one synchronous machine connected to an infinite bus through a transmission line (discussed in Examples 5.1, 5.2, and 5.6) the synchronous machine is to be represented by the simplified model known as the one-axis model given in Section 4.15. Prepare a complete analog computer simulation of this system. Indicate the signal levels for the operating conditions of Example 5.1, the amplitude and time scaling, the potentiometer settings, and the amplifier gains. **Note:** In the load equations, assume that $L_e i_d = L_e i_q \cong 0$.

5.5 Repeat Problem 5.4 using the two-axis model of Section 4.15.

5.6 Repeat Problem 5.4 using the voltage-behind-subtransient-reactance model of Section 4.15.

5.7 In the analog computer simulation shown in Figure 5.13 and Table 5.1, the time scaling is (20). If the time scaling is changed to (10), identify the amplifiers and potentiometers in Table 5.1 that will be affected.

5.8 In Figure 5.13 the signal to the resolver represents the infinite bus voltage. If the level of this signal is reduced by a factor of 2 while the level of all the other signals are maintained, identify the potentiometer and amplifier settings that need adjustment.

References

1. IEEE Committee Report. Recommended phasor diagram for synchronous machines. *IEEE Trans.* PAS-88:1593–1610, 1969.
2. Krause, P. C. Simulation of a single machine–infinite bus system. Mimeo notes, Electr. Eng. Dept., Purdue Univ., West Lafayette, Ind., 1967.
3. Buckley, D. F. Analog computer representation of a synchronous machine. Unpubl. M.S. thesis, Iowa State Univ., Ames, 1968.
4. Riaz, M. Analogue computer representations of synchronous generators in voltage regulator studies. *AIEE Trans.* PAS-75:1178–84, 1956.
5. Schroder, D. C., and Anderson, P. M. Compensation of synchronous machines for stability. Paper C 73 313-4, presented at the IEEE Summer Power Meeting, Vancouver, B.C., Canada, 1973.
6. Electronic Associates, Inc. *Handbook of Analog Computation*, 2nd ed. Publ. 00800.0001-3. Princeton, N.J., 1967.

7. Kimbark, E. W. *Power System Stability,* Vol. 1. Wiley, New York, 1948.
8. Dandeno, P. L., Hauth, R. L., and Schulz, R. P. Effects of synchronous machine modeling in large-scale system studies. *IEEE Trans.* PAS-92:574–82, 1973.
9. Schulz, R. P., Jones, W. D., and Ewart, D. N. Dynamic models of turbine generators derived from solid rotor equivalent circuits. *IEEE Trans.* PAS-92:926–33, 1973.
10. International Business Machines. System/360 Continuous System Modeling Program Users Manual, GH20-0367-4. IBM Corp., 1967.

chapter 6

Linear Models of the Synchronous Machine

6.1 Introduction

A brief review of the response of a power system to small impacts is given in Chapter 3. It is shown that when the system is subjected to a small load change, it tends to acquire a new operating state. During the transition between the initial state and the new state the system behavior is oscillatory. If the two states are such that all the state variables change only slightly (i.e., the variable x_i changes from x_{i0} to $x_{i0} + x_{i\Delta}$ where $x_{i\Delta}$ is a small change in x_i), the system is operating near the initial state. The initial state may be considered as a quiescent operating condition for the system.

To examine the behavior of the system when it is perturbed such that the new and old equilibrium states are nearly equal, the system equations are *linearized* about the quiescent operating condition. By this we mean that first-order approximations are made for the system equations. The new linear equations thus derived are assumed to be valid in a region near the quiescent condition.

The dynamic response of a linear system is determined by its characteristic equation (or equivalent information). Both the forced response and the free response are decided by the roots of this equation. From a point of view of stability the free response gives the needed information. If it is stable, any bounded input will give a bounded and therefore a stable output.

The synchronous machine models developed in Chapter 4 have two types of nonlinearities: product nonlinearities and trigonometric functions. The first-order approximations for these have been illustrated in previous chapters and are outlined below.

As an example of product nonlinearities, consider the product $x_i x_j$. Let the state variables x_i and x_j have the initial values x_{i0} and x_{j0}. Let the changes in these variables be $x_{i\Delta}$ and $x_{j\Delta}$. Initially their product is given by $x_{i0}x_{j0}$. The new value becomes

$$(x_{i0} + x_{i\Delta})(x_{j0} + x_{j\Delta}) = x_{i0}x_{j0} + x_{i0}x_{j\Delta} + x_{j0}x_{i\Delta} + x_{i\Delta}x_{j\Delta}$$

The last term is a second-order term, which is assumed to be negligibly small. Thus for a first-order approximation, the *change* in the product $x_i x_j$ is given by

$$(x_{i0} + x_{i\Delta})(x_{j0} + x_{j\Delta}) - x_{i0}x_{j0} = x_{j0}x_{i\Delta} + x_{i0}x_{j\Delta} \tag{6.1}$$

We note that x_{j0} and x_{i0} are known quantities and are treated here as coefficients, while $x_{i\Delta}$ and $x_{j\Delta}$ are "incremental" variables.

The trigonometric nonlinearities are treated in a similar manner as

$$\cos(\delta_0 + \delta_\Delta) = \cos\delta_0 \cos\delta_\Delta - \sin\delta_0 \sin\delta_\Delta$$

with $\cos\delta_\Delta \cong 1$ and $\sin\delta_\Delta \cong \delta_\Delta$. Therefore,

$$\cos(\delta_0 + \delta_\Delta) - \cos\delta_0 \cong (-\sin\delta_0)\delta_\Delta \tag{6.2}$$

The incremental *change* in $\cos\delta$ is then $(-\sin\delta_0)\delta_\Delta$; the incremental variable is δ_Δ and its coefficient is $-\sin\delta_0$. Similarly, we can show that the incremental change in the term $\sin\delta$ is given by

$$\sin(\delta_0 + \delta_\Delta) - \sin\delta_0 \cong (\cos\delta_0)\delta_\Delta \tag{6.3}$$

6.2 Linearization of the Generator State-Space Current Model

Let the state-space vector $\mathbf{x}$ have an initial state $\mathbf{x}_0$ at time $t = t_0$; e.g., if the current model is used,

$$\mathbf{x}_0^t = [i_{d0}\ i_{F0}\ i_{D0}\ i_{q0}\ i_{Q0}\ \omega_0\ \delta_0] \tag{6.4}$$

At the occurrence of a small disturbance, i.e., after $t = t_0^+$, the states will change slightly from their previous positions or values. Thus

$$\mathbf{x} = \mathbf{x}_0 + \mathbf{x}_\Delta \tag{6.5}$$

Note that $\mathbf{x}_0$ need not be constant, but we do require that it be known.

The state-space model is in the form

$$\dot{\mathbf{x}} = \mathbf{f}(\mathbf{x}, t) \tag{6.6}$$

which, by using (6.5), reduces to

$$\dot{\mathbf{x}}_0 + \dot{\mathbf{x}}_\Delta = \mathbf{f}(\mathbf{x}_0 + \mathbf{x}_\Delta, t) \tag{6.7}$$

In expanding (6.7) all second-order terms are neglected; i.e., terms of the form $x_{i\Delta}x_{j\Delta}$ are assumed to be negligibly small. The system (6.7) becomes

$$\dot{\mathbf{x}}_0 + \dot{\mathbf{x}}_\Delta \cong \mathbf{f}(\mathbf{x}_0, t) + \mathbf{A}(\mathbf{x}_0)\mathbf{x}_\Delta + \mathbf{B}(\mathbf{x}_0)\mathbf{u} \tag{6.8}$$

from which we obtain the linearized state-space equation

$$\dot{\mathbf{x}}_\Delta = \mathbf{A}(\mathbf{x}_0)\mathbf{x}_\Delta + \mathbf{B}(\mathbf{x}_0)\mathbf{u} \tag{6.9}$$

The elements of the $\mathbf{A}$ matrix depend upon the initial values of the state vector $\mathbf{x}_0$. For a specific dynamic study it is considered constant. The dynamic properties of the system described by (6.9) are determined from the nature of the eigenvalues of the $\mathbf{A}$ matrix.

The state space may be thought of as an n-dimensional space, and the operating conditions constrain the operation to a particular surface in this n space. Being nonlinear, the surface is not flat, although we would expect it to be continuous and relatively smooth. The quiescent operating point $\mathbf{x}_0$ and the functions $\mathbf{A}(\mathbf{x}_0)$ and $\mathbf{B}(\mathbf{x}_0)$ are different for every new initial condition.

We may also compute the $\mathbf{A}(\mathbf{x}_0)$ by finding the total differential $d\mathbf{x}$ at $\mathbf{x}_0$ with respect to all variables; i.e., with $dx \cong x_\Delta$

$$\mathbf{x} = \mathbf{x}_0 + \mathbf{x}_\Delta$$

$$\dot{\mathbf{x}}_\Delta = \frac{\partial \mathbf{f}}{\partial x_1}\bigg]_{\mathbf{x}_0} x_{1\Delta} + \frac{\partial \mathbf{f}}{\partial x_2}\bigg]_{\mathbf{x}_0} x_{2\Delta} + \cdots + \frac{\partial \mathbf{f}}{\partial x_n}\bigg]_{\mathbf{x}_0} x_{n\Delta}$$

$$= \left[\frac{\partial \mathbf{f}}{\partial x_1}\,\frac{\partial \mathbf{f}}{\partial x_2}\cdots\frac{\partial \mathbf{f}}{\partial x_n}\right]_{\mathbf{x}_0} \mathbf{x}_\Delta = \mathbf{A}(\mathbf{x}_0)\mathbf{x}_\Delta$$

where the quantity in brackets defines $\mathbf{A}(\mathbf{x}_0)$.

We begin by linearizing (4.74), proceeding one row at a time. For the first equation (of the d circuit) we write

$$v_{d0} + v_{d\Delta} = -r(i_{d0} + i_{d\Delta}) - (\omega_0 + \omega_\Delta)L_q(i_{q0} + i_{q\Delta}) - (\omega_0 + \omega_\Delta)\mathrm{k}M_Q(i_{Q0} + i_{Q\Delta})$$
$$- L_d(\dot{i}_{d0} + \dot{i}_{d\Delta}) - \mathrm{k}M_F(\dot{i}_{F0} + \dot{i}_{F\Delta}) - \mathrm{k}M_D(\dot{i}_{D0} + \dot{i}_{D\Delta})$$

Expanding the product terms and dropping the second-order terms,

$$v_{d0} + v_{d\Delta} = (-ri_{d0} - \omega_0 L_q i_{q0} - \omega_0 \mathrm{k}M_Q i_{Q0} - L_d\dot{i}_{d0} - \mathrm{k}M_F\dot{i}_{F0} - \mathrm{k}M_D\dot{i}_{D0})$$
$$- ri_{d\Delta} - \omega_0 L_q i_{q\Delta} - i_{q0}L_q\omega_\Delta - \omega_0\mathrm{k}M_Q i_{Q\Delta} - i_{Q0}\mathrm{k}M_Q\omega_\Delta$$
$$- L_d\dot{i}_{d\Delta} - \mathrm{k}M_F\dot{i}_{F\Delta} - \mathrm{k}M_D\dot{i}_{D\Delta}$$

The quantity in parenthesis on the right side is exactly equal to v_{d0}. Rearranging the remaining quantities,

$$v_{d\Delta} = -ri_{d\Delta} - \omega_0 L_q i_{q\Delta} - \omega_0\mathrm{k}M_Q i_{Q\Delta} - (i_{q0}L_q + \mathrm{k}M_Q i_{Q0})\omega_\Delta$$
$$- L_d\dot{i}_{d\Delta} - \mathrm{k}M_F\dot{i}_{F\Delta} - \mathrm{k}M_D\dot{i}_{D\Delta} \tag{6.10}$$

which is equal to

$$v_{d\Delta} = -ri_{d\Delta} - \omega_0 L_q i_{q\Delta} - \omega_0\mathrm{k}M_Q i_{Q\Delta} - \lambda_{q0}\omega_\Delta - L_d\dot{i}_{d\Delta} - \mathrm{k}M_F\dot{i}_{F\Delta} - \mathrm{k}M_D\dot{i}_{D\Delta} \tag{6.11}$$

Similarly, for the q axis voltage change we write

$$v_{q\Delta} = \omega_0 L_d i_{d\Delta} + \omega_0\mathrm{k}M_F i_{F\Delta} + \omega_0\mathrm{k}M_D i_{D\Delta} + (i_{d0}L_d + i_{F0}\mathrm{k}M_F + i_{D0}\mathrm{k}M_D)\omega_\Delta$$
$$- ri_{q\Delta} - L_q\dot{i}_{q\Delta} - \mathrm{k}M_Q\dot{i}_{Q\Delta} \tag{6.12}$$

which is equal to

$$v_{q\Delta} = \omega_0 L_d i_{d\Delta} + \omega_0\mathrm{k}M_F i_{F\Delta} + \omega_0\mathrm{k}M_D i_{d\Delta} + \lambda_{d0}\omega_\Delta - ri_{q\Delta} - L_q\dot{i}_{q\Delta} - \mathrm{k}M_Q\dot{i}_{Q\Delta} \tag{6.13}$$

For the field winding we compute

$$-v_{F\Delta} = -r_F i_{F\Delta} - \mathrm{k}M_F\dot{i}_{d\Delta} - L_F\dot{i}_{F\Delta} - M_R\dot{i}_{D\Delta} \tag{6.14}$$

The linearized damper-winding equations are given by

$$0 = -r_D i_{D\Delta} - \mathrm{k}M_D\dot{i}_{d\Delta} - M_R\dot{i}_{F\Delta} - L_D\dot{i}_{D\Delta} \tag{6.15}$$
$$0 = -r_Q i_{Q\Delta} - \mathrm{k}M_Q\dot{i}_{q\Delta} - L_Q\dot{i}_{Q\Delta} \tag{6.16}$$

From (4.101) the linearized torque equation may be established as

$$\tau_j\dot{\omega}_\Delta = (1/3)(-L_d i_{q0} i_{d\Delta} - L_d i_{d0} i_{q\Delta} - \mathrm{k}M_F i_{q0} i_{F\Delta} - \mathrm{k}M_F i_{F0} i_{q\Delta}$$
$$- \mathrm{k}M_D i_{q0} i_{D\Delta} - \mathrm{k}M_D i_{D0} i_{q\Delta} + L_q i_{d0} i_{q\Delta} + L_q i_{q0} i_{d\Delta}$$
$$+ \mathrm{k}M_Q i_{d0} i_{Q\Delta} + \mathrm{k}M_Q i_{Q0} i_{d\Delta}) - D\omega_\Delta + T_{m\Delta} \tag{6.17}$$

which can be put in the form

$$\tau_j\dot{\omega}_\Delta = T_{m\Delta} - (1/3)[(L_d i_{q0} - \lambda_{q0})i_{d\Delta} - (\lambda_{d0} - L_q i_{d0})i_{q\Delta} - kM_F i_{q0} i_{F\Delta} - kM_D i_{q0} i_{D\Delta} + kM_Q i_{d0} i_{Q\Delta}] - D\omega_\Delta \tag{6.18}$$

Finally, the torque angle equation given by (4.102) may be written as

$$\dot{\delta}_\Delta = \omega_\Delta \tag{6.19}$$

Equations (6.11)–(6.19) are the linearized system equations for a synchronous machine (not including the load equation). If we drop the Δ subscript, since all variables are now small displacements, we may write these equations in the following matrix form:

$$\begin{bmatrix} v_d \\ -v_F \\ 0 \\ \hline v_q \\ 0 \\ \hline T_m \\ 0 \end{bmatrix} = - \left[\begin{array}{ccc|cc|cc} r & 0 & 0 & \omega_0 L_q & \omega_0 kM_Q & \lambda_{q0} & 0 \\ 0 & r_F & 0 & 0 & 0 & 0 & 0 \\ 0 & 0 & r_D & 0 & 0 & 0 & 0 \\ \hline -\omega_0 L_d & -\omega_0 kM_F & -\omega_0 kM_D & r & 0 & -\lambda_{d0} & 0 \\ 0 & 0 & 0 & 0 & r_Q & 0 & 0 \\ \hline \dfrac{\lambda_{q0} - L_d i_{q0}}{3} & \dfrac{-kM_F i_{q0}}{3} & \dfrac{-kM_D i_{q0}}{3} & \dfrac{-\lambda_{d0} + L_q i_{d0}}{3} & \dfrac{kM_Q i_{d0}}{3} & -D & 0 \\ 0 & 0 & 0 & 0 & 0 & -1 & 0 \end{array}\right] \begin{bmatrix} i_d \\ i_F \\ i_D \\ \hline i_q \\ i_Q \\ \hline \omega \\ \delta \end{bmatrix}$$

$$- \left[\begin{array}{ccc|cc|cc} L_d & kM_F & kM_D & 0 & 0 & 0 & 0 \\ kM_F & L_F & M_R & 0 & 0 & 0 & 0 \\ kM_D & M_R & L_D & 0 & 0 & 0 & 0 \\ \hline 0 & 0 & 0 & L_q & kM_Q & 0 & 0 \\ 0 & 0 & 0 & kM_Q & L_Q & 0 & 0 \\ \hline 0 & 0 & 0 & 0 & 0 & -\tau_j & 0 \\ 0 & 0 & 0 & 0 & 0 & 0 & 1 \end{array}\right] \begin{bmatrix} \dot{i}_d \\ \dot{i}_F \\ \dot{i}_D \\ \hline \dot{i}_q \\ \dot{i}_Q \\ \hline \dot{\omega} \\ \dot{\delta} \end{bmatrix} \tag{6.20}$$

or in matrix form

$$\mathbf{v} = -\mathbf{Kx} - \mathbf{M\dot{x}} \quad \text{pu} \tag{6.21}$$

Note that the matrix **M** is related to the matrix **L** of equation (4.74) by

$$\mathbf{M} = \left[\begin{array}{c|cc} \mathbf{L} & \multicolumn{2}{c}{\mathbf{0}} \\ \hline \mathbf{0} & -\tau_j & 0 \\ & 0 & 1 \end{array}\right]$$

Assuming that $\mathbf{M}^{-1}$ exists, the state equation for the synchronous generator, not including the load equations, is

$$\mathbf{\dot{x}} = \mathbf{M}^{-1}\mathbf{Kx} - \mathbf{M}^{-1}\mathbf{v} \quad \text{pu} \tag{6.22}$$

which is the same form as

$$\dot{\mathbf{x}} = \mathbf{A}\mathbf{x} + \mathbf{B}\mathbf{u} \tag{6.23}$$

Example 6.1

As a preparation for later examples involving a loaded machine, determine the matrices **M** and **K** for the generator described in Examples 4.1–4.3. Let $\tau_j = 2H\omega_R = 1786.94$ rad.

Solution

The matrix **M** is related to the matrix **L** of Example 4.2 as follows

$$\mathbf{M} = \left[\begin{array}{c|cc} \mathbf{L} & 0 & 0 \\ \hline 0 & -\tau_j & 0 \\ 0 & 0 & 1 \end{array}\right]$$

Then we write

$$\mathbf{M} = \left[\begin{array}{ccc|cc|cc} 1.700 & 1.550 & 1.550 & & & & \\ 1.550 & 1.651 & 1.550 & & \mathbf{0} & & \mathbf{0} \\ 1.550 & 1.550 & 1.506 & & & & \\ \hline & & & 1.640 & 1.490 & & \\ & \mathbf{0} & & & & & \mathbf{0} \\ & & & 1.490 & 1.526 & & \\ \hline & & & & & -1786.94 & 0 \\ & \mathbf{0} & & & \mathbf{0} & & \\ & & & & & 0 & 1 \end{array}\right]$$

The matrix **K** is defined by (6.20)

$$\mathbf{K} = -\left[\begin{array}{ccc|cc|cc} 0.0011 & 0 & 0 & 1.64 & 1.49 & \lambda_{q0} & 0 \\ 0 & 0.0007 & 0 & 0 & 0 & 0 & 0 \\ 0 & 0 & 0.0131 & 0 & 0 & 0 & 0 \\ \hline -1.700 & -1.550 & -1.550 & 0.0011 & 0 & -\lambda_{d0} & 0 \\ 0 & 0 & 0 & 0 & 0.0540 & 0 & 0 \\ \hline \dfrac{\lambda_{q0} - L_d i_{q0}}{3} & \dfrac{-L_{AD} i_{q0}}{3} & \dfrac{-L_{AD} i_{q0}}{3} & \dfrac{-\lambda_{d0} + L_q i_{d0}}{3} & \dfrac{L_{AQ} i_{d0}}{3} & -D & 0 \\ 0 & 0 & 0 & 0 & 0 & -1 & 0 \end{array}\right]$$

When the machine is loaded, certain terms in these matrices change from the numeric values given to reflect the impedance of the connecting system. For example, when loaded through a transmission line to a large system, r, L_d, and L_q change to $\hat{R}$, $\hat{L}_d$, and $\hat{L}_q$ as noted in Section 4.13. Other terms are load dependent (such as the currents and flux linkages) and must be determined from the initial conditions.

6.3 Linearization of the Load Equation for the One-Machine Problem

Equation (4.149) is repeated here for convenience:

$$\begin{aligned} v_d &= -K\sin(\delta - \alpha) + R_e i_d + L_e \dot{i}_d + \omega L_e i_q \\ v_q &= K\cos(\delta - \alpha) + R_e i_q + L_e \dot{i}_q - \omega L_e i_d \end{aligned} \tag{6.24}$$

where $K = \sqrt{3}V_\infty$ and α is the angle of $\overline{V}_\infty$.

The same procedure followed previously is used to linearize this equation, with the result

$$\begin{aligned} v_{d\Delta} &= -K\cos(\delta_0 - \alpha)\delta_\Delta + R_e i_{d\Delta} + \omega_0 L_e i_{q\Delta} + i_{q0} L_e \omega_\Delta + L_e \dot{i}_{d\Delta} \\ v_{q\Delta} &= -K\sin(\delta_0 - \alpha)\delta_\Delta + R_e i_{q\Delta} + L_e \dot{i}_{q\Delta} - \omega_0 L_e i_{d\Delta} - i_{d0} L_e \omega_\Delta \end{aligned} \tag{6.25}$$

Substituting (6.25) into (6.11) and (6.12),

$$\begin{aligned} &-K\cos(\delta_0 - \alpha)\delta_\Delta + R_e i_{d\Delta} + L_e \dot{i}_{d\Delta} + \omega_0 L_e i_{q\Delta} + i_{q0} L_e \omega_\Delta \\ &\quad = -r i_{d\Delta} - \omega_0 L_q i_{q\Delta} - \omega_0 \mathrm{k} M_Q i_{Q\Delta} - \lambda_{q0}\omega_\Delta - L_d \dot{i}_{d\Delta} - \mathrm{k} M_F \dot{i}_{F\Delta} - \mathrm{k} M_D \dot{i}_{D\Delta} \\ &-K\sin(\delta_0 - \alpha)\delta_\Delta + R_e i_{q\Delta} + L_e \dot{i}_{q\Delta} - \omega_0 L_e i_{d\Delta} - i_{d0} L_e \omega_\Delta \\ &\quad = \omega_0 L_d i_{d\Delta} + \omega_0 \mathrm{k} M_F i_{F\Delta} + \omega_0 \mathrm{k} M_D i_{D\Delta} + \lambda_{d0}\omega_\Delta - r i_{q\Delta} - L_q \dot{i}_{q\Delta} - \mathrm{k} M_Q \dot{i}_{Q\Delta} \end{aligned} \tag{6.26}$$

Rearranging (6.26) and making the substitution

$$\begin{aligned} \hat{\lambda}_d &= \lambda_d + L_e i_d & \hat{L}_q &= L_q + L_e \\ \hat{\lambda}_q &= \lambda_q + L_e i_q & R &= r + R_e \\ \hat{L}_d &= L_d + L_e \end{aligned} \tag{6.27}$$

we get, after dropping the subscript Δ,

$$\begin{aligned} 0 &= -\hat{R} i_d - \omega_0 \hat{L}_q i_q - \omega_0 \mathrm{k} M_Q i_Q - \hat{\lambda}_{q0}\omega + K\cos(\delta_0 - \alpha)\delta \\ &\quad - \hat{L}_d \dot{i}_d - \mathrm{k} M_F \dot{i}_F - \mathrm{k} M_D \dot{i}_D \ \text{pu} \\ 0 &= -\hat{R} i_q + \omega_0 \hat{L}_d i_d + \omega_0 \mathrm{k} M_F i_F + \omega_0 \mathrm{k} M_D i_D + \hat{\lambda}_{d0}\omega \\ &\quad + K\sin(\delta_0 - \alpha)\delta - L_q \dot{i}_q - \mathrm{k} M_Q \dot{i}_Q \ \text{pu} \end{aligned} \tag{6.28}$$

Combining (6.28) with (6.14)–(6.16), (6.18), and (6.19), we get for the linearized system equations

$$\begin{bmatrix} 0 \\ -v_F \\ 0 \\ \hline 0 \\ 0 \\ \hline T_m \\ 0 \end{bmatrix} = - \left[\begin{array}{ccc|cc|cc} \hat{R} & 0 & 0 & \omega_0 \hat{L}_q & \omega_0 \mathrm{k} M_Q & \hat{\lambda}_{q0} & -K\cos(\delta_0 - \alpha) \\ 0 & r_F & 0 & 0 & 0 & 0 & 0 \\ 0 & 0 & r_D & 0 & 0 & 0 & 0 \\ \hline -\omega_0 \hat{L}_d & -\omega_0 \mathrm{k} M_F & -\omega_0 \mathrm{k} M_D & \hat{R} & 0 & -\hat{\lambda}_{d0} & -K\sin(\delta_0 - \alpha) \\ 0 & 0 & 0 & 0 & r_Q & 0 & 0 \\ \hline \dfrac{\lambda_{q0} - L_d i_{q0}}{3} & \dfrac{-\mathrm{k} M_F i_{q0}}{3} & \dfrac{-\mathrm{k} M_D i_{q0}}{3} & \dfrac{-\lambda_{d0} + L_q i_{d0}}{3} & \dfrac{\mathrm{k} M_Q i_{d0}}{3} & -D & 0 \\ 0 & 0 & 0 & 0 & 0 & -1 & 0 \end{array} \right] \begin{bmatrix} i_d \\ i_F \\ i_D \\ \hline i_q \\ i_Q \\ \hline \omega \\ \delta \end{bmatrix}$$

$$\begin{bmatrix} \hat{L}_d & kM_F & kM_D & & & & \\ kM_F & L_F & M_R & \mathbf{0} & & \mathbf{0} & \\ kM_D & M_R & L_D & & & & \\ & \mathbf{0} & & \hat{L}_q & kM_Q & \mathbf{0} & \\ & & & kM_Q & L_Q & & \\ & & & & & -\tau_j & 0 \\ & \mathbf{0} & & \mathbf{0} & & 0 & 1 \end{bmatrix} \begin{bmatrix} \dot{i}_d \\ \dot{i}_F \\ \dot{i}_D \\ \dot{i}_q \\ \dot{i}_Q \\ \dot{\omega} \\ \dot{\delta} \end{bmatrix} \tag{6.29}$$

Equation (6.29) is a linearized set of seven first-order differential equations with constant coefficients. In matrix form (6.29) becomes $\mathbf{v} = -\mathbf{Kx} - \mathbf{M\dot{x}}$, and assuming that $\mathbf{M}^{-1}$ exists,

$$\dot{\mathbf{x}} = -\mathbf{M}^{-1}\mathbf{Kx} - \mathbf{M}^{-1}\mathbf{v} = \mathbf{Ax} + \mathbf{Bu} \tag{6.30}$$

where $\mathbf{A} = -\mathbf{M}^{-1}\mathbf{K}$. Note that the new matrices $\mathbf{M}$ and $\mathbf{K}$ are now expanded to include the transmission line constants and the infinite bus voltage.

It is convenient to compute $\mathbf{A}$ as follows. Let

$$\mathbf{M} = \begin{bmatrix} \mathbf{M}_1 & 0 & 0 \\ 0 & \mathbf{M}_2 & 0 \\ 0 & 0 & \mathbf{M}_3 \end{bmatrix} \qquad \mathbf{K} = \begin{bmatrix} \mathbf{K}_{11} & \mathbf{K}_{12} & \mathbf{K}_{13} \\ \mathbf{K}_{21} & \mathbf{K}_{22} & \mathbf{K}_{23} \\ \mathbf{K}_{31} & \mathbf{K}_{32} & \mathbf{K}_{33} \end{bmatrix}$$

Then

$$\mathbf{M}^{-1} = \begin{bmatrix} \mathbf{M}_1^{-1} & \mathbf{0} & \mathbf{0} \\ \mathbf{0} & \mathbf{M}_2^{-1} & \mathbf{0} \\ \mathbf{0} & \mathbf{0} & \mathbf{M}_3^{-1} \end{bmatrix}$$

$$\mathbf{A} = -\begin{bmatrix} \mathbf{M}_1^{-1}\mathbf{K}_{11} & \mathbf{M}_1^{-1}\mathbf{K}_{12} & \mathbf{M}_1^{-1}\mathbf{K}_{13} \\ \mathbf{M}_2^{-1}\mathbf{K}_{21} & \mathbf{M}_2^{-1}\mathbf{K}_{22} & \mathbf{M}_2^{-1}\mathbf{K}_{23} \\ \mathbf{M}_3^{-1}\mathbf{K}_{31} & \mathbf{M}_3^{-1}\mathbf{K}_{32} & \mathbf{M}_3^{-1}\mathbf{K}_{33} \end{bmatrix} \tag{6.31}$$

Note that the only driving functions in the system (6.29) are the field voltage $v_{F\Delta}$ and the mechanical torque $T_{m\Delta}$. Initially, the machine is spinning at synchronous speed and is delivering some known power to the infinite bus. A change in either v_F or T_m will cause the system to seek a new operating point, and this change is usually accompanied by damped oscillations of the variables.

Example 6.2

Complete Example 6.1 for the operating conditions described in Example 5.2, taking into account the load equation. Find the new expanded $\mathbf{A}$ matrix. Assume $D = 0$.

Solution

From Example 5.2 we compute

$$\hat{R} = 0.0011 + 0.020 = 0.0211$$
$$\hat{L}_d = 1.700 + 0.400 = 2.100$$
$$\hat{L}_q = 1.640 + 0.400 = 2.040$$

The matrix **M** is given by

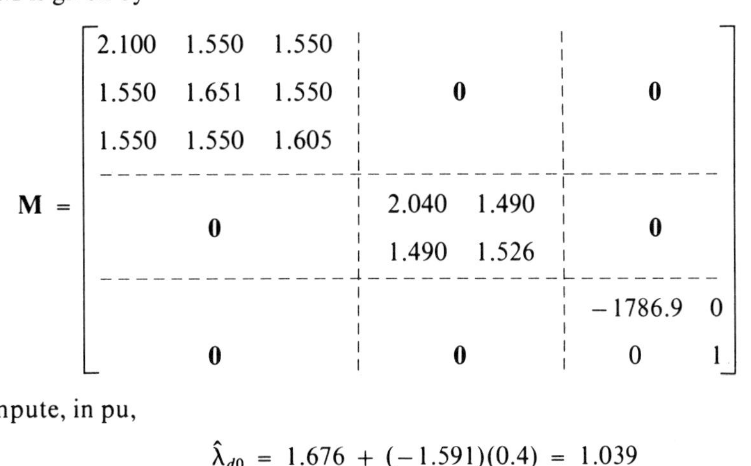

$$\mathbf{M} = \left[\begin{array}{ccc|cc|cc} 2.100 & 1.550 & 1.550 & & & & \\ 1.550 & 1.651 & 1.550 & \multicolumn{2}{c|}{\mathbf{0}} & \multicolumn{2}{c}{\mathbf{0}} \\ 1.550 & 1.550 & 1.605 & & & & \\ \hline & \mathbf{0} & & 2.040 & 1.490 & \multicolumn{2}{c}{\mathbf{0}} \\ & & & 1.490 & 1.526 & & \\ \hline & & & & & -1786.9 & 0 \\ & \mathbf{0} & & \multicolumn{2}{c|}{\mathbf{0}} & 0 & 1 \end{array}\right]$$

We also compute, in pu,

$$\hat{\lambda}_{d0} = 1.676 + (-1.591)(0.4) = 1.039$$
$$\hat{\lambda}_{q0} = 1.150 + (0.701)(0.4) = 1.430$$
$$K\cos(\delta_0 - \alpha) = \sqrt{3}(\cos 53.735^\circ) = 1.025$$
$$K\sin(\delta_0 - \alpha) = \sqrt{3}(\sin 53.735^\circ) = 1.397$$
$$\frac{1}{3}(\lambda_{q0} - L_d i_{q0}) = \frac{1.150 - 1.70 \times 0.701}{3} = -0.014$$
$$\frac{1}{3}(-\mathrm{k}M_F i_{q0}) = \frac{-1.55 \times 0.701}{3} = -0.362$$
$$\frac{1}{3}(-\mathrm{k}M_D i_{q0}) = \frac{-(1.676 + 1.64 \times 1.591)}{3} = -1.428$$
$$\frac{1}{3}(\mathrm{k}M_Q i_{d0}) = \frac{1.490(-1.591)}{3} = -0.790$$

The matrix **K** is given by

$$\mathbf{K} = \left[\begin{array}{ccc|cc|cc} 0.0211 & 0 & 0 & 2.040 & 1.490 & 1.430 & -1.025 \\ 0 & 0.0007 & 0 & 0 & 0 & 0 & 0 \\ 0 & 0 & 0.0131 & 0 & 0 & 0 & 0 \\ \hline -2.100 & -1.550 & -1.550 & 0.0211 & 0 & -1.039 & -1.397 \\ 0 & 0 & 0 & 0 & 0.0540 & 0 & 0 \\ \hline -0.014 & -0.362 & -0.362 & -1.428 & -0.790 & -D & 0 \\ 0 & 0 & 0 & 0 & 0 & -1 & 0 \end{array}\right]$$

The new **A** matrix is given by $\mathbf{A} = -\mathbf{M}^{-1}\mathbf{K}$, or with $D = 0$,

$$
\mathbf{A} = \left[\begin{array}{ccc|cc|cc}
-36.062 & 0.439 & 14.142 & -3487.18 & -2547.01 & -2444.63 & 1751.33 \\
12.472 & -4.950 & 76.857 & 1206.01 & 880.86 & 845.46 & -605.68 \\
22.776 & 4.356 & -96.017 & 2202.43 & 1608.63 & 1543.98 & -1106.10 \\
\hline
3589.95 & 2649.72 & 2649.72 & -36.064 & 90.072 & 1776.71 & 2387.40 \\
-3505.70 & -2587.54 & -2587.54 & 35.218 & -123.320 & -1735.01 & -2331.37 \\
\hline
-0.0078 & -0.2027 & -0.2027 & -0.7993 & -0.4422 & 0.0 & 0.0 \\
0.0 & 0.0 & 0.0 & 0.0 & 0.0 & 1.0 & 0.0
\end{array}\right] 10^{-3}
$$

Example 6.3

Find the eigenvalues of the **A** matrix of the linearized system of Example 6.2. Examine the stability of the system. Generator loading is that of Example 5.2.

Solution

To perform the computation of the eigenvalues for the **A** matrix obtained in Example 6.2, a digital computer program is used. The results are given below.

$$
\begin{array}{ll}
\lambda_1 = -0.0359 + j0.9983 & \lambda_5 = -0.0016 + j0.0289 \\
\lambda_2 = -0.0359 - j0.9983 & \lambda_6 = -0.0016 - j0.0289 \\
\lambda_3 = -0.0991 & \lambda_7 = -0.0007 \\
\lambda_4 = -0.1217 &
\end{array}
$$

All the eigenvalues are given in rad/rad. Note that there are two pairs of complex eigenvalues. The pair λ_5 and λ_6 correspond to frequencies of approximately 1.73 Hz; they are damped with a time constant of $1/(0.0016 \times 377)$ or 1.66 s. This complex pair and the real pole due to λ_7 dominate the transient response of the system. The other complex pair corresponds to a very fast transient of about 60 Hz, which is damped at a much faster rate. This is the 60-Hz component injected into the rotor circuits to balance the MMF caused by the stator dc currents. Note also that the real parts of all the eigenvalues are negative, which means that the system is stable under the conditions assumed in the development of this model, namely small perturbation about a quiescent operating condition.

Example 6.4

Repeat the above example for the system conditions stated in Example 5.1.

Solution

A procedure similar to that followed in Examples 6.2 and 6.3 gives the following results:

$$
\mathbf{A} = \left[\begin{array}{ccc|cc|cc}
-36.062 & 0.439 & 14.142 & -3487.18 & -2547.01 & -2327.01 & 958.54 \\
12.472 & -4.950 & 76.857 & 1206.01 & 880.86 & 804.78 & -331.50 \\
22.776 & 4.356 & -96.017 & 2202.43 & 1608.63 & 1469.69 & -605.39 \\
\hline
3589.95 & 2649.72 & 2649.72 & -36.064 & 90.071 & 982.66 & 2257.70 \\
-3505.70 & -2587.54 & -2587.54 & 35.218 & -1233.20 & -959.60 & -2204.72 \\
\hline
-0.0075 & -0.1929 & -0.1929 & -0.8399 & -0.5351 & 0.0 & 0.0 \\
0.0 & 0.0 & 0.0 & 0.0 & 0.0 & 1.0 & 0.0
\end{array}\right] 10^{-3}
$$

and the eigenvalues are given by

$$\begin{aligned}
\lambda_1 &= -0.0359 + j0.9983 & \lambda_5 &= -0.0009 + j0.0248\\
\lambda_2 &= -0.0359 - j0.9983 & \lambda_6 &= -0.0009 - j0.0248\\
\lambda_3 &= -0.0991 & \lambda_7 &= -0.0005\\
\lambda_4 &= -0.1230
\end{aligned}$$

Note that this new operating condition has a slightly reduced natural frequency (1.49 Hz) and a greatly increased time constant (2.95 s) compared to the previous example. Thus damping is substantially reduced by the change in operating point.

6.4 Linearization of the Flux Linkage Model

We now linearize the flux linkage model of a synchronous machine, following a procedure similar to that used above for the current model. From (4.135) we can compute the linear equations

$$\begin{aligned}
\dot{\lambda}_{d\Delta} = &-\frac{r}{\ell_d}\left(1 - \frac{L_{MD}}{\ell_d}\right)\lambda_{d\Delta} + r\frac{L_{MD}}{\ell_d\ell_F}\lambda_{F\Delta} + r\frac{L_{MD}}{\ell_d\ell_D}\lambda_{D\Delta}\\
&- \omega_0\lambda_{q\Delta} - \lambda_{q0}\omega_\Delta - v_{d\Delta}
\end{aligned} \tag{6.32}$$

$$\dot{\lambda}_{F\Delta} = r_F\frac{L_{MD}}{\ell_F\ell_d}\lambda_{d\Delta} - \frac{r_F}{\ell_F}\left(1 - \frac{L_{MD}}{\ell_F}\right)\lambda_{F\Delta} + r_F\frac{L_{MD}}{\ell_F\ell_D}\lambda_{D\Delta} + v_{F\Delta} \tag{6.33}$$

$$\dot{\lambda}_{D\Delta} = r_D\frac{L_{MD}}{\ell_D\ell_d}\lambda_{d\Delta} + r_D\frac{L_{MD}}{\ell_D\ell_F}\lambda_{F\Delta} - \frac{r_D}{\ell_D}\left(1 - \frac{L_{MD}}{\ell_D}\right)\lambda_{D\Delta} \tag{6.34}$$

Similarly the q axis equation (4.136) can be linearized to give

$$\dot{\lambda}_{q\Delta} = \omega_0\lambda_{d\Delta} - \frac{r}{\ell_q}\left(1 - \frac{L_{MQ}}{\ell_q}\right)\lambda_{q\Delta} + r\frac{L_{MQ}}{\ell_q\ell_Q}\lambda_{Q\Delta} + \lambda_{d0}\omega_\Delta - v_{q\Delta} \tag{6.35}$$

$$\dot{\lambda}_{Q\Delta} = r_Q\frac{L_{MQ}}{\ell_Q\ell_q}\lambda_{q\Delta} - \frac{r_Q}{\ell_Q}\left(1 - \frac{L_{MQ}}{\ell_Q}\right)\lambda_{Q\Delta} \tag{6.36}$$

The torque equation (4.137) becomes

$$\begin{aligned}
T_{e\phi\Delta} = &\left(\frac{L_{MD} - L_{MQ}}{\ell_d^2}\lambda_{q0} - \frac{L_{MQ}}{\ell_q\ell_Q}\lambda_{Q0}\right)\lambda_{d\Delta} + \frac{L_{MD}}{\ell_d\ell_F}\lambda_{q0}\lambda_{F\Delta} + \frac{L_{MD}}{\ell_d\ell_D}\lambda_{q0}\lambda_{D\Delta}\\
&+ \left(\frac{L_{MD} - L_{MQ}}{\ell_d^2}\lambda_{d0} + \frac{L_{MD}}{\ell_d\ell_F}\lambda_{F0} + \frac{L_{MD}}{\ell_d\ell_D}\lambda_{D0}\right)\lambda_{q\Delta} - \frac{L_{MQ}}{\ell_q\ell_Q}\lambda_{d0}\lambda_{Q\Delta}
\end{aligned} \tag{6.37}$$

Similarly, the swing equation becomes

$$\begin{aligned}
\dot{\omega}_\Delta = &\frac{1}{3\tau_j}\left(\frac{L_{MQ}}{\ell_q\ell_Q}\lambda_{Q0} - \frac{L_{MD} - L_{MQ}}{\ell_d^2}\lambda_{q0}\right)\lambda_{d\Delta} - \frac{1}{3\tau_j}\left(\frac{L_{MD}}{\ell_d\ell_F}\lambda_{q0}\right)\lambda_{F\Delta}\\
&- \frac{1}{3\tau_j}\left(\frac{L_{MD}}{\ell_d\ell_D}\lambda_{q0}\right)\lambda_{D\Delta} - \frac{1}{3\tau_j}\left(\frac{L_{MD} - L_{MQ}}{\ell_d^2}\lambda_{d0} + \frac{L_{MD}}{\ell_d\ell_F}\lambda_{F0} + \frac{L_{MD}}{\ell_d\ell_D}\lambda_{D0}\right)\lambda_{q\Delta}\\
&+ \frac{1}{3\tau_j}\left(\frac{L_{MQ}}{\ell_q\ell_Q}\lambda_{d0}\right)\lambda_{Q\Delta} - \frac{D}{\tau_j}\omega_\Delta + \frac{1}{\tau_j}T_{m\Delta}
\end{aligned}$$

and finally

$$\dot{\delta}_\Delta = \omega_\Delta . \tag{6.38}$$

For a system of one machine connected to an infinite bus through a transmission line, the load equations are given by (4.157) and (4.158). These are then linearized to give

$$\begin{aligned}
&\left[1 + \frac{L_e}{\ell_d}\left(1 - \frac{L_{MD}}{\ell_d}\right)\right] \dot{\lambda}_{d\Delta} - \frac{L_e L_{MD}}{\ell_q \ell_F} \dot{\lambda}_{F\Delta} - \frac{L_e L_{MD}}{\ell_d \ell_D} \dot{\lambda}_{D\Delta} \\
&\qquad = -\frac{\hat{R}}{\ell_d}\left(1 - \frac{L_{MD}}{\ell_d}\right) \lambda_{d\Delta} + \hat{R}\frac{L_{MD}}{\ell_d \ell_F} \lambda_{F\Delta} + \hat{R}\frac{L_{MD}}{\ell_d \ell_D} \lambda_{D\Delta} \\
&\qquad\quad - \omega_0 \left[1 + \frac{L_e}{\ell_q}\left(1 - \frac{L_{MQ}}{\ell_q}\right)\right] \lambda_{q\Delta} + \omega_0 \frac{L_e L_{MQ}}{\ell_q \ell_Q} \lambda_{Q\Delta} \\
&\qquad\quad - \hat{\lambda}_{q0}\,\omega_\Delta + [K \cos(\delta_0 - \alpha)]\delta_\Delta
\end{aligned} \tag{6.39}$$

$$\begin{aligned}
&\left[1 + \frac{L_e}{\ell_q}\left(1 - \frac{L_{MQ}}{\ell_q}\right)\right] \dot{\lambda}_{q\Delta} - \frac{L_e L_{MQ}}{\ell_q \ell_Q} \dot{\lambda}_{Q\Delta} \\
&\qquad = \omega_0 \left[1 + \frac{L_e}{\ell_d}\left(1 - \frac{L_{MD}}{\ell_d}\right)\right] \lambda_{d\Delta} \\
&\qquad\quad - \omega_0 \frac{L_e L_{MD}}{\ell_d \ell_F} \lambda_{F\Delta} - \omega_0 \frac{L_e L_{MD}}{\ell_d \ell_D} \lambda_{D\Delta} - \frac{\hat{R}}{\ell_q}\left(1 - \frac{L_{MQ}}{\ell_q}\right) \lambda_{q\Delta} \\
&\qquad\quad + \hat{R}\frac{L_{MQ}}{\ell_q \ell_Q} \lambda_{Q\Delta} + \hat{\lambda}_{d0}\,\omega_\Delta + [K \sin(\delta_0 - \alpha)]\delta_\Delta
\end{aligned} \tag{6.40}$$

where

$$\hat{\lambda}_{q0} = \left[1 + \frac{L_e}{\ell_q}\left(1 - \frac{L_{MQ}}{\ell_q}\right)\right] \lambda_{q0} - \frac{L_e L_{MQ}}{\ell_q \ell_Q} \lambda_{Q0}$$

$$\hat{\lambda}_{d0} = \left[1 + \frac{L_e}{\ell_d}\left(1 - \frac{L_{MD}}{\ell_d}\right)\right] \lambda_{d0} - \frac{L_e L_{MD}}{\ell_d \ell_F} \lambda_{F0} - \frac{L_e L_{MD}}{\ell_d \ell_D} \lambda_{D0}$$

and $\hat{R} = r + R_e$ and $K = \sqrt{3}V_\infty$. The linearized equations of the system are (6.33), (6.34), (6.36), and (6.37)–(6.40) and $\dot{\delta}_\Delta = \omega_\Delta$. In matrix form we write

$$\mathbf{T}\dot{\boldsymbol{\lambda}} = \mathbf{C}\boldsymbol{\lambda} + \mathbf{D} \tag{6.41}$$

where the matrices **T**, **C**, and **D** are similar to those defined in Section 4.13.3 for the nonlinear model.

If the state equations are written out in the form of (6.41) and compared with the nonlinear equations (4.159)–(4.162), several interesting observations can be made. First, we can show that the matrix **T** is exactly the same as (4.160). The matrix **C** is similar, but not exactly the same as (4.161). If we write **C** as

$$\mathbf{C} = \begin{array}{c} \begin{matrix} d\,F\,D & q\,Q & \omega\,\delta \end{matrix} \\ \left[\begin{array}{c|c|c} \mathbf{C}_1 & \mathbf{C}_2 & \mathbf{C}_3 \\ \hline \mathbf{C}_4 & \mathbf{C}_5 & \mathbf{C}_6 \\ \hline \mathbf{C}_7 & \mathbf{C}_8 & \mathbf{C}_9 \end{array}\right] \end{array} \tag{6.42}$$

with partitioning as in (4.161), we can observe that $\mathbf{C}_1$, $\mathbf{C}_5$, and $\mathbf{C}_9$ are exactly the same as in the nonlinear equation. Submatrices $\mathbf{C}_2$ and $\mathbf{C}_4$ are exactly as in (4.161) if ω is replaced by ω_0. Submatrices $\mathbf{C}_3$, $\mathbf{C}_6$, $\mathbf{C}_7$, and $\mathbf{C}_8$ are considerably changed, however, and $\mathbf{C}_3$ and $\mathbf{C}_6$, which were formerly zero matrices, now become

$$\mathbf{C}_3 = \begin{bmatrix} -\hat{\lambda}_{q0} & \sqrt{3}\,V_\infty \cos(\delta_0 - \alpha) \\ 0 & 0 \\ 0 & 0 \end{bmatrix}$$

$$\mathbf{C}_6 = \begin{bmatrix} \hat{\lambda}_{d0} & \sqrt{3}\,V_\infty \sin(\delta_0 - \alpha) \\ 0 & 0 \end{bmatrix} \tag{6.43}$$

where α is the angle of $\overline{V}_\infty$ and δ_0 is the initial angle of the q axis, each measured from the arbitrary reference.

We may write matrices $\mathbf{C}_7$ and $\mathbf{C}_8$ as

$$\mathbf{C}_7 = \left[\begin{array}{c|c|c} \dfrac{1}{3\tau_j \ell_d}\left(\lambda_{AQ0} - \dfrac{L_{MD}\lambda_{q0}}{\ell_d}\right) & -\dfrac{L_{MD}\lambda_{q0}}{3\tau_j \ell_d \ell_F} & -\dfrac{L_{MD}\lambda_{q0}}{3\tau_j \ell_d \ell_D} \\ \hline 0 & 0 & 0 \end{array}\right]$$

$$\mathbf{C}_8 = \left[\begin{array}{c|c} \dfrac{-1}{3\tau_j \ell_d}\left(\lambda_{AD0} - \dfrac{L_{MQ}\lambda_{d0}}{\ell_d}\right) & \dfrac{L_{MQ}\lambda_{d0}}{3\tau_j \ell_q \ell_Q} \\ \hline 0 & 0 \end{array}\right] \tag{6.44}$$

where λ_{AD0} and λ_{AQ0} are the initial values of λ_{AD} and λ_{AQ} respectively. Finally, we note the new $\mathbf{D}$ matrix to be

$$\mathbf{D} = [0 \quad v_{F\Delta} \quad 0 \mid 0 \quad 0 \mid T_{m\Delta}/\tau_j \quad 0]^t \tag{6.45}$$

Assuming that the inverse of $\mathbf{T}$ exists, we can premultiply both sides of (6.42) by $\mathbf{T}^{-1}$ to obtain

$$\dot{\boldsymbol{\lambda}} = \mathbf{T}^{-1}\mathbf{C}\boldsymbol{\lambda} + \mathbf{T}^{-1}\mathbf{D} \tag{6.46}$$

which is of the form

$$\dot{\mathbf{x}} = \mathbf{A}\mathbf{x} + \mathbf{B}\mathbf{u} \tag{6.47}$$

The matrices $\mathbf{A}$ and $\mathbf{B}$ will have constant coefficients, which are dependent upon the quiescent operating conditions.

Note that the matrices $\mathbf{A}$ and $\mathbf{B}$ will not be the same here as in the current model. Since the choice of the state variables is arbitrary, there are many other equations that could be written. The *order* of the system does not change, however, and there are still seven degrees of freedom in the solution.

Example 6.5

Obtain the matrices $\mathbf{T}$, $\mathbf{C}$, and $\mathbf{A}$ of the flux linkage model for the operating conditions discussed in the previous examples.

Solution

Machine and line data are taken from previous examples in pu as:

$$L_d = 1.700 \qquad \ell_Q = 0.036$$
$$L_q = 1.640 \qquad r = 0.0011$$
$$\ell_d = \ell_q = 0.150 \qquad r_F = 0.00074$$
$$L_{AD} = kM_F = kM_D = 1.550 \qquad r_D = 0.0131$$
$$L_{AQ} = kM_Q = 1.490 \qquad r_Q = 0.0540$$
$$L_F = 1.651 \qquad L_{MD} = 0.02838$$
$$\ell_F = 0.101 \qquad L_{MQ} = 0.02836$$
$$L_D = 1.605 \qquad R_e = 0.020$$
$$\ell_D = 0.055 \qquad L_e = 0.400$$
$$L_Q = 1.526 \qquad \tau_j = 1786.94 \text{ rad}$$

The matrix **T** is independent of load and is given by

$$\mathbf{T} = \left[\begin{array}{ccc|cc|cc} 3.1622 & -0.7478 & -1.3656 & 0 & 0 & 0 & 0 \\ 0 & 1.0 & 0 & 0 & 0 & 0 & 0 \\ 0 & 0 & 1.0 & 0 & 0 & 0 & 0 \\ \hline 0 & 0 & 0 & 3.1625 & -2.1118 & 0 & 0 \\ 0 & 0 & 0 & 0 & 1.0 & 0 & 0 \\ \hline 0 & 0 & 0 & 0 & 0 & 1.0 & 0 \\ 0 & 0 & 0 & 0 & 0 & 0 & 1.0 \end{array}\right]$$

and $\mathbf{T}^{-1}$ is computed as

$$\mathbf{T}^{-1} = \left[\begin{array}{ccc|cc|cc} 0.3162 & 0.2364 & 0.4318 & & & & \\ 0 & 1.0 & 0 & & \mathbf{0} & & \mathbf{0} \\ 0 & 0 & 1.0 & & & & \\ \hline & & & 0.3162 & 0.6678 & & \\ & \mathbf{0} & & 0 & 1.0 & & \mathbf{0} \\ \hline & & & & & 1 & 0 \\ & \mathbf{0} & & & \mathbf{0} & 0 & 1 \end{array}\right]$$

To calculate the matrix **C**, the following data is obtained from the initial operating conditions as given in Example 5.2:

$$\lambda_{q0} = 1.150 \qquad \sqrt{3}\,V_\infty \cos(\delta_0 - \alpha) = 1.025$$
$$\lambda_{Q0} = 1.045 \qquad \sqrt{3}\,V_\infty \sin(\delta_0 - \alpha) = 1.397$$
$$\lambda_{d0} = 1.676$$
$$\lambda_{F0} = 2.200$$
$$\lambda_{D0} = 1.914$$

The matrix **C** corresponding to Example 5.2 loading is then calculated to be

$$\mathbf{C} = \left[\begin{array}{ccc|cc|cc}
-114.035 & 39.438 & 72.022 & -3162.53 & 2111.78 & -1430.11 & 1024.53 \\
1.388 & -5.278 & 3.756 & 0 & 0 & 0 & 0 \\
44.720 & 66.282 & -115.330 & 0 & 0 & 0 & 0 \\
\hline
3162.16 & -747.76 & -1365.58 & -114.055 & 111.378 & 1039.32 & 1396.55 \\
0 & 0 & 0 & 284.854 & -313.530 & 0 & 0 \\
\hline
-1.0285 & -0.4009 & -0.7322 & -1.9867 & 1.6503 & 0 & 0 \\
0 & 0 & 0 & 0 & 0 & 1 & 0
\end{array}\right] 10^{-3}$$

Note that some of the elements of the matrices $\mathbf{C}_1$ and $\mathbf{C}_5$ in this example are somewhat different from those in Example 4.4 since the resistance $\hat{R}$ is not the same in both examples.

The $\mathbf{A}$ matrix is given by

$$\mathbf{A} = \left[\begin{array}{ccc|cc|cc}
-16.422 & 39.848 & -26.141 & -1000.12 & 667.83 & -452.26 & 324.00 \\
1.388 & -5.278 & 3.756 & 0 & 0 & 0 & 0 \\
44.720 & 66.282 & -115.330 & 0 & 0 & 0 & 0 \\
\hline
999.88 & -236.44 & -431.80 & 154.147 & -174.142 & 328.63 & 441.59 \\
0 & 0 & 0 & 284.854 & -313.530 & 0 & 0 \\
\hline
1.0285 & -0.4009 & -0.7322 & -1.9867 & 1.6503 & 0 & 0 \\
0 & 0 & 0 & 0 & 0 & 1 & 0
\end{array}\right] 10^{-3}$$

The eigenvalues of this matrix are the same as those obtained in Example 6.3 and correspond to the loading condition of Example 5.2.

For the operating condition of Example 5.1 we obtain the same matrix $\mathbf{T}$. For this operating condition the initial conditions in pu are given by $\lambda_{d0} = 1.345$, $\lambda_{F0} = 1.935$, $\lambda_{D0} = 1.634$, $\lambda_{q0} = 1.094$, $\lambda_{Q0} = 0.994$, $K\cos(\delta_0 - \alpha) = 0.5607$, and $K\sin(\delta_0 - \alpha) = 1.3207$.

The matrix $\mathbf{C}$ for the operating conditions of Example 5.1 is given by

$$\mathbf{C} = \left[\begin{array}{ccc|cc|cc}
-114.035 & 39.437 & 72.022 & -3162.53 & 2111.78 & -1361.30 & 560.75 \\
1.388 & -5.278 & 3.756 & 0 & 0 & 0 & 0 \\
44.720 & 66.282 & -115.330 & 0 & 0 & 0 & 0 \\
\hline
3162.16 & -747.76 & -1365.58 & -114.055 & 111.378 & 574.48 & 1320.68 \\
0 & 0 & 0 & 284.854 & -313.530 & 0 & 0 \\
\hline
-0.9790 & -0.3816 & -0.6969 & -1.7155 & 1.3246 & 0 & 0 \\
0 & 0 & 0 & 0 & 0 & 1.0 & 0
\end{array}\right] 10^{-3}$$

and the matrix $\mathbf{A}$ is given by

$$
\mathbf{A} = \left[\begin{array}{ccc|cc|cc}
-16.422 & 39.848 & -26.141 & -1000.12 & 667.83 & -430.50 & 177.33 \\
1.388 & -5.278 & 3.756 & 0 & 0 & 0 & 0 \\
44.720 & 66.282 & -115.330 & 0 & 0 & 0 & 0 \\
\hline
999.88 & -236.44 & -431.80 & 154.15 & -174.14 & 181.76 & 417.60 \\
0 & 0 & 0 & 284.85 & -313.53 & 0 & 0 \\
\hline
0.9790 & -0.3816 & -0.6969 & -1.7155 & 1.3246 & 0 & 0 \\
0 & 0 & 0 & 0 & 0 & 1.0 & 0
\end{array}\right] 10^{-3}
$$

The eigenvalues obtained are the same as those given in Example 6.4 and correspond to the loading condition of Example 5.1.

6.5 Simplified Linear Model

A simplified linear model for a synchronous machine connected to an infinite bus through a transmission line having resistance R_e and inductance L_e (or a reactance X_e) can be developed (see references [1] and [2]). Let the following assumptions be made:

1. Amortisseur effects are neglected.
2. Stator winding resistance is neglected.
3. The $\dot{\lambda}_d$ and $\dot{\lambda}_q$ terms in the stator and load voltage equations are neglected compared to the speed voltage terms $\omega\lambda_q$ and $\omega\lambda_d$.
4. The terms $\omega\lambda$ in the stator and load voltage equations are assumed to be approximately equal to $\omega_R\lambda$.
5. Balanced conditions are assumed and saturation effects are neglected.

Under the assumptions stated above the equations describing the system are given below in pu.

6.5.1 The E' equation

From (4.74) and (4.104) the field equations are given by

$$v_F = r_F i_F + \dot{\lambda}_F \qquad \lambda_F = L_F i_F + kM_F i_d \tag{6.48}$$

Eliminating i_F, we get

$$v_F = (r_F/L_F)\lambda_F + \dot{\lambda}_F - (r_F/L_F)kM_F i_d \tag{6.49}$$

Now let $e_q' = \sqrt{3}E_q'$ be the stator EMF proportional to the main winding flux linking the stator; i.e., $\sqrt{3}E_q' = \omega_R kM_F\lambda_F/L_F$. Also let E_{FD} be the stator EMF that is produced by the field current and corresponds to the field voltage v_F; i.e.,

$$\sqrt{3}E_{FD} = \omega_R kM_F v_F/r_F$$

Using the above definitions and τ_{d0}' defined by (4.189), we get from (6.49) in the s domain

$$E_{FD} = (1 + \tau_{d0}' s)E_q' - (x_d - x_d')I_d \tag{6.50}$$

where $I_d = i_d/\sqrt{3}$ and s is the Laplace transform variable. Also using the above definition for E_q', we can arrange the second equation in (6.48) to give

$$E_q' = \omega_R kM_F i_F/\sqrt{3} + (x_d - x_d')I_d = E + (x_d - x_d')I_d \tag{6.51}$$

where E is as defined in Section 4.7.4. Note that (6.50) and (6.51) are linear.

From (4.149) and (4.74) and from the assumptions made in the simplified model, we compute v_d and v_q for infinite bus loading to be

$$\begin{aligned} v_d &= -\omega_R L_q i_q = -\sqrt{3}V_\infty \sin(\delta - \alpha) + R_e i_d + \omega_R L_e i_q \\ v_q &= \omega_R L_d i_d + \omega_R k M_F i_F = \sqrt{3}V_\infty \cos(\delta - \alpha) + R_e i_q - \omega_R L_e i_d \end{aligned} \quad (6.52)$$

Linearizing (6.52),

$$\begin{aligned} 0 &= -R_e i_{q\Delta} + (x_d + X_e) i_{d\Delta} + \omega_R k M_F i_{F\Delta} + [K \sin(\delta_0 - \alpha)]\delta_\Delta \\ 0 &= -R_e i_{d\Delta} - (x_q + X_e) i_{q\Delta} + [K \cos(\delta_0 - \alpha)]\delta_\Delta \end{aligned} \quad (6.53)$$

where $K = \sqrt{3}V_\infty$ and V_∞ is the infinite bus voltage to neutral.

Rearranging (6.51) and (6.53),

$$\begin{aligned} -(x_d' + X_e) I_{d\Delta} + R_e I_{q\Delta} &= E'_{q\Delta} + [V_\infty \sin(\delta_0 - \alpha)]\delta_\Delta \\ R_e I_{d\Delta} + (x_q + X_e) I_{q\Delta} &= [V_\infty \cos(\delta_0 - \alpha)]\delta_\Delta \end{aligned} \quad (6.54)$$

Solving (6.54) for $I_{d\Delta}$ and $I_{q\Delta}$, we compute

$$\begin{bmatrix} I_{d\Delta} \\ I_{q\Delta} \end{bmatrix} = K_I \begin{bmatrix} -(x_q + X_e) & R_e \cos(\delta_0 - \alpha) - (x_q + X_e)\sin(\delta_0 - \alpha) \\ R_e & (x_d' + X_e)\cos(\delta_0 - \alpha) + R_e \sin(\delta_0 - \alpha) \end{bmatrix} \begin{bmatrix} E'_{q\Delta} \\ V_\infty \delta_\Delta \end{bmatrix} \quad (6.55)$$

where

$$K_I = 1/[R_e^2 + (x_q + X_e)(x_d' + X_e)] \quad (6.56)$$

We now substitute I_d into an incremental version of (6.50) to compute

$$E_{FD} = (1/K_3 + \tau_{d0}' s) E'_{q\Delta} + K_4 \delta_\Delta \quad (6.57)$$

where we define (in agreement with [2])

$$\begin{aligned} 1/K_3 &= 1 + K_I(x_d - x_d')(x_q + X_e) \\ K_4 &= V_\infty K_I (x_d - x_d')[(x_q + X_e)\sin(\delta_0 - \alpha) - R_e \cos(\delta_0 - \alpha)] \end{aligned} \quad (6.58)$$

Then from (6.58) and (6.57) we get the following s domain relation

$$E'_{q\Delta} = \frac{K_3}{1 + K_3 \tau_{d0}' s} E_{FD\Delta} - \frac{K_3 K_4}{1 + K_3 \tau_{d0}' s} \delta_\Delta \quad (6.59)$$

[Note that (6.59) differs from (3.10) because of the introduction here of E_{FD} rather than v_F.] From (6.59) we can identify that K_3 is an impedance factor that takes into account the loading effect of the external impedance, and K_4 is related to the demagnetizing effect of a change in the rotor angle; i.e.,

$$K_4 = \frac{1}{K_3} \frac{E_q'}{\delta_\Delta}\bigg]_{E_{FD} = \text{constant}} \quad (6.60)$$

6.5.2 Electrical torque equation

The pu electrical torque T_e is numerically equal to the three-phase power. Therefore,

$$T_e = (1/3)(v_d i_d + v_q i_q) = (V_d I_d + V_q I_q) \text{ pu} \quad (6.61)$$

where under the assumptions used in this model,

$$V_d = -x_q I_q \qquad V_q = x_d I_d + \omega_R k M_F i_F/\sqrt{3} \tag{6.62}$$

Using (6.51) in the second equation of (6.62),

$$V_d = -x_q I_q \qquad V_q = x_d' I_d + E_q' \tag{6.63}$$

From (6.63) and (6.61)

$$T_e = [E_q' - (x_q - x_d')I_d]I_q \tag{6.64}$$

Linearizing (6.64), we compute

$$\begin{aligned} T_{e\Delta} &= I_{q0}E'_{q\Delta} + [E'_{q0} - (x_q - x_d')I_{d0}]I_{q\Delta} - (x_q - x_d')I_{q0}I_{d\Delta} \\ &= I_{q0}E'_{q\Delta} + E_{qa0}I_{q\Delta} - (x_q - x_d')I_{q0}I_{d\Delta} \end{aligned} \tag{6.65}$$

where we have used the q axis voltage E_{qa} defined in Figure 5.2 as $E_{qa} = E + (x_d - x_q)I_d$ with E taken from (6.51) to write the initial condition

$$\begin{aligned} E_{qa0} &= E_0 + (x_d - x_q)I_{d0} = E'_{q0} - (x_d - x_d')I_{d0} + (x_d - x_q)I_{d0} \\ &= E'_{q0} - (x_q - x_d')I_{d0} \end{aligned} \tag{6.66}$$

Substituting (6.55) and (6.56) into (6.65), we compute the incremental torque to be

$$\begin{aligned} T_{e\Delta} &= K_I V_\infty \{E_{qa0}[R_e \sin(\delta_0 - \alpha) + (x_d' + X_e)\cos(\delta_0 - \alpha)] \\ &\quad + I_{q0}(x_q - x_d')[(x_q + X_e)\sin(\delta_0 - \alpha) - R_e\cos(\delta_0 - \alpha)]\}\delta_\Delta \\ &\quad + K_I\{I_{q0}[R_e^2 + (x_q + X_e)^2] + E_{qa0}R_e\}E'_{q\Delta} \\ &\triangleq K_1\delta_\Delta + K_2 E'_{q\Delta} \end{aligned} \tag{6.67}$$

Where K_1 is the change in electrical torque for a small change in rotor angle at constant d axis flux linkage; i.e., the synchronizing torque coefficient

$$\begin{aligned} K_1 &= \left.\frac{T_{e\Delta}}{\delta_\Delta}\right]_{E_q' = E_{q0}'} \\ &= K_I V_\infty \{E_{qa0}[R_e \sin(\delta_0 - \alpha) + (x_d' + X_e)\cos(\delta_0 - \alpha)] \\ &\quad + I_{q0}(x_q - x_d')[(X_e + x_q)\sin(\delta_0 - \alpha) - R_e\cos(\delta_0 - \alpha)]\} \end{aligned}$$

K_2 is the change in electrical torque for small change in the d axis flux linkage at constant rotor angle

$$K_2 = \left.\frac{T_e}{E'_{q\Delta}}\right]_{\delta = \delta_0} = K_I\{R_e E_{qa0} + I_{q0}[R_e^2 + (x_q + X_e)^2]\}$$

We should point out the similarity between the constant K_1 in (6.67) and the synchronizing power coefficient discussed in Chapter 2 and given by (2.36). If the field flux linkage is constant, E_q' will also be constant and $K_2 = 0$. The model is reduced to the classical model of Chapter 2.

6.5.3 Terminal voltage equation

From (4.41) the synchronous machine terminal voltage V_t is given by

$$V_t^2 = (1/3)(v_d^2 + v_q^2)$$

or in rms equivalent variables

$$V_t^2 = V_d^2 + V_q^2 \tag{6.68}$$

This equation is linearized to obtain

$$V_{t\Delta} = (V_{d0}/V_{t0})\,V_{d\Delta} + (V_{q0}/V_{t0})\,V_{q\Delta} \tag{6.69}$$

Substituting (6.63) in (6.69),

$$V_{t\Delta} = -(V_{d0}/V_{t0})x_q I_{q\Delta} + (V_{q0}/V_{t0})(x_d' I_{d\Delta} + E_{q\Delta}') \tag{6.70}$$

Substituting for $I_{q\Delta}$ and $I_{d\Delta}$ from (6.55),

$$\begin{aligned} V_{t\Delta} &= \{(K_l V_\infty x_d' V_{q0}/V_{t0})[R_e \cos(\delta_0 - \alpha) - (x_q + X_e)\sin(\delta_0 - \alpha)] \\ &\quad - (K_l V_\infty x_q V_{d0}/V_{t0})[(x_d' + X_e)\cos(\delta_0 - \alpha) + R_e \sin(\delta_0 - \alpha)]\}\,\delta_\Delta \\ &\quad + \{(V_{q0}/V_{t0})[1 - K_l x_d'(x_q + X_e)] - (V_{d0}/V_{t0})K_l x_q R_e\}\,E_{q\Delta}' \\ &\triangleq K_5\delta_\Delta + K_6 E_{q\Delta}' \end{aligned} \tag{6.71}$$

where K_5 is the change in the terminal voltage V_t for a small change in rotor angle at constant d axis flux linkage, or

$$K_5 = \left.\frac{V_{t\Delta}}{\delta_\Delta}\right]_{E_q' = E_{q0}'}$$

and K_6 is the change in the terminal voltage V_t for a small change in the d axis flux linkage at constant rotor angle, or

$$K_6 = \left.\frac{V_{t\Delta}}{E_{q\Delta}'}\right]_{\delta = \delta_0}$$

6.5.4 Summary of equations

Equations (6.59), (6.67), and (6.71) are the basic equations for the simplified linear model, i.e.,

$$\begin{aligned} E_{q\Delta}' &= \frac{K_3}{1 + K_3\tau_{d0}'s}E_{FD\Delta} - \frac{K_3K_4}{1 + K_3\tau_{d0}'s}\delta_\Delta \\ T_{e\Delta} &= K_1\delta_\Delta + K_2E_{q\Delta}' \\ V_{t\Delta} &= K_5\delta_\Delta + K_6E_{q\Delta}' \end{aligned} \tag{6.72}$$

We note that the constants K_1, K_2, K_3, K_4, K_5, and K_6 depend upon the network parameters, the quiescent operating conditions, and the infinite bus voltage.

To complete the model, the linearized swing equation from (4.90) is used.

$$\tau_j\dot{\omega}_\Delta = T_{m\Delta} - T_{e\Delta} \tag{6.73}$$

The angle δ_Δ in radians is obtained by integrating on $\dot{\omega}_\Delta$ twice.

In the above equations the time is in pu to a base quantity of 1/377 s, T is the total torque to a base quantity of the three-phase machine power, and $\tau_j = 2H\omega_R$.

Example 6.6

Find the constants K_1 through K_6 of the simplified model for the system and conditions stated in Example 5.1.

Solution

We can tabulate the data from Example 5.1 as follows.

Transmission line data:

$$R_e = 0.02 \qquad X_e = 0.40 \text{ pu}$$

Infinite bus voltage:

$$V_\infty = 0.828$$

Synchronous machine data:

$$x_d = 1.700 \text{ pu} \qquad x_q = 1.640 \text{ pu}$$
$$x_d' = 1.700 - [(1.55)^2/1.651] = 0.245 \text{ pu}$$

Also, from Example 5.1

$$i_{F0} = 2.979 \qquad I_{q0} = 0.385$$
$$I_{d0} = -1.112 \qquad V_{q0} = 0.776$$
$$V_{d0} = -0.631$$
$$V_a = 1.000$$

We can calculate the angle between the infinite bus and the q axis to be $\delta_0 - \alpha = 66.995°$. Then $\sin(\delta_0 - \alpha) = 0.9205$, $\cos(\delta_0 - \alpha) = 0.3908$. From (6.66) we compute

$$E_{qa0} = 1.55 \times 2.979/\sqrt{3} - 1.112(1.70 - 1.64) = 2.5995$$

Also,

$$1/K_I = R_e^2 + (x_q + X_e)(x_d' + X_e) = 1.3162$$
$$K_I = 0.7598$$

Then we compute from (6.58)

$$K_3 = [1 + (1/1.3162)(1.455)(2.04)]^{-1} = 0.3072$$
$$K_4 = 0.828 \times 0.7598 \times 1.455(2.04 \times 0.9205 - 0.02 \times 0.3908) = 1.7124$$

We then calculate K_1 and K_2 from (6.67).

$$\begin{aligned} K_1 &= K_I V_\infty \{E_{qa0}[R_e \sin(\delta_0 - \alpha) + (x_d' + X_e)\cos(\delta_0 - \alpha)] \\ &\quad + I_{q0}(x_q - X_d')[(x_q + X_e)\sin(\delta_0 - \alpha) - R_e \cos(\delta_0 - \alpha)]\} \\ &= 0.7598 \times 0.828[2.5995(0.02 \times 0.9205 + 0.645 \times 0.3908) \\ &\quad + 0.3853 \times 1.395(2.04 \times 0.9205 - 0.02 \times 0.3908)] \\ &= 1.0755 \end{aligned}$$

$$\begin{aligned} K_2 &= K_I\{I_{q0}[R_e^2 + (x_q + X_e)^2] + E_{qa0}R_e\} \\ &= 0.7598\{0.385[(0.02)^2 + (2.04)^2] + 2.5995 \times 0.02\} \\ &= 1.2578 \end{aligned}$$

K_5 and K_6 are calculated from (6.71):

$$\begin{aligned} K_5 &= (K_I V_\infty x_d' V_{q0}/V_{t0})[R_e \cos(\delta_0 - \alpha) - (x_q + X_e)\sin(\delta_0 - \alpha)] \\ &\quad - (K_I V_\infty x_q V_{d0}/V_{t0})[(x_d' + X_e)\cos(\delta_0 - \alpha) + R_e \sin(\delta_0 - \alpha)] \\ &= [(0.7598)(0.828)(0.245)(0.776/1.0)][(0.02)(0.3908) - (2.04)(0.9205)] \\ &\quad - (0.7598)(0.828)(1.64)(-0.631/1.0)[(0.645)(0.3908) + (0.02)(0.9205)] \\ &= -0.0409 \end{aligned}$$

$$K_6 = (V_{q0}/V_{t0})[1 - K_I x_d'(x_q + X_e)] - (V_{d0}/V_{t0})K_I x_q R_e$$
$$= 0.776[1 - (0.7598)(0.245)(2.04)]$$
$$+ (0.631)(0.7598)(1.64)(0.02) = 0.4971$$

Therefore at this operating condition the linearized model of the system is given by

$$E'_{q\Delta} = [0.3072/(1 + 1.813\,s)]E_{FD\Delta} - [0.5261/(1 + 1.813\,s)]\delta_\Delta$$
$$T_{e\Delta} = 1.0755\,\delta_\Delta + 1.2578\,E'_{q\Delta}$$
$$V_{t\Delta} = -0.0409\,\delta_\Delta + 0.4971\,E'_{q\Delta}$$

Example 6.7

Repeat Example 6.6 for the operating conditions given in Example 5.2.

Solution

From Example 5.2

$$i_{F0} = 2.8259 \qquad I_{q0} = 0.4047 \text{ pu}$$
$$I_{d0} = -0.9185 \qquad V_{q0} = 0.9670 \text{ pu}$$
$$V_{d0} = -0.6628 \qquad V_\infty = 1.000 \text{ pu}$$
$$V_{t0} = 1.172 \qquad \delta_0 - \alpha = 53.736°$$

and $\sin(\delta_0 - \alpha) = 0.8063$, $\cos(\delta_0 - \alpha) = 0.5915$.

From this data we calculate E'_{q0} and E_{qa0}

$$E'_{q0} = 1.55 \times 2.826/\sqrt{3} - 1.455 \times 0.9185 = 1.1925$$
$$E_{qa0} = 1.1925 - 1.395(-0.9185) = 2.4738$$
$$1/K_I = R_e^2 + (x_q + X_e)(x_d' + X_e) = 1.3162$$
$$K_I = 0.7598$$

Then

$$K_3 = \left(1 + \frac{2.04 \times 1.455}{1.316}\right)^{-1} = 0.3072$$

$$K_4 = \frac{1.0 \times 1.455}{1.3162}(2.04 \times 0.8063 - 0.02 \times 0.5915) = 1.805$$

$$\tau'_{d0} = 5.90 \text{ s}$$

The effective field-winding time constant under this loading is given by

$$K_3\tau'_{d0} = 0.3072 \times 5.9 = 1.8125 \text{ s}$$
$$K_1 = (0.7598)(1.0)\{(2.474)[(0.02)(0.8063) + (0.645)(0.5915)]$$
$$+ (0.4047)(1.395)[(2.04)(0.8063) - (0.02)(0.5915)]\} = 1.4479$$

We note that for this example the constant K_1 is greater in magnitude than in Example 6.6. The constant K_1 corresponds to the synchronizing power coefficient discussed in Chapter 2. The greater value in this example is indicative of a lower loading condition or a greater ability in this case to transmit synchronizing power.

$$K_2 = 0.7598\{0.4047[(0.02)^2 + (2.04)^2] + (2.474)(0.02)\} = 1.3174$$

$$K_5 = (0.7598)(1.0)(0.245)\left(\frac{0.9670}{1.172}\right)[(0.02)(0.5915) - (2.04)(0.8063)]$$

$$- (0.7598)(1.0)(1.64)\left(\frac{-0.6628}{1.172}\right)[(0.645)(0.5915) + (0.02)(0.8063)] = 0.0294$$

$$K_6 = \left(\frac{0.9670}{1.172}\right)[1 - (0.7598)(0.245)(2.041)]$$

$$- \left(\frac{-0.6628}{1.172}\right)(0.7598)(1.64)(0.02) = 0.5257$$

The linearized model of the system at the given operating point is given in pu by

$$E'_{q\Delta} = [0.3072/(1 + 1.813\,s)]E_{FD\Delta} - [0.5546/(1 + 1.813\,s)]\delta_\Delta$$
$$T_{e\Delta} = 1.4479\,\delta_\Delta + 1.3174\,E'_{q\Delta}$$
$$V_{t\Delta} = 0.0294\,\delta_\Delta + 0.5257\,E'_{q\Delta}$$

6.5.5 Effect of loading

Examining the values of the constants K_1 through K_6 for the loading conditions of Examples 6.6 and 6.7, we note the following:

1. The constant K_3 is the same in both cases. From (6.57) and (6.58) we note that K_3 is an impedance factor and hence is independent of the machine loading.
2. The constants K_1, K_2, K_4, and K_6 are comparable in magnitude in both cases, while K_5 has reversed sign. From (6.58), (6.67), and (6.71) we note that these constants depend on the initial machine loading.

The cases studied in the above examples represent heavy load conditions. Certain effects are clearly demonstrated. In the heavier loading condition of Example 6.6, K_5 has a value of -0.0409, and in the less severe loading condition of Example 6.7 its value is 0.0294. This is rather significant, and in Chapter 8 it will be pointed out that in machines with voltage regulators, the system damping is affected by the constant K_5. If this constant is negative, the voltage regulator decreases the natural damping of the system (at that operating condition). This is usually compensated for by the use of supplementary signals to produce artificial damping.

From Examples 6.6 and 6.7 we note that the demagnetizing effect of the armature reaction as manifested by the $E'_{q\Delta}$ dependence is quite significant. This effect is more pronounced in relation to the change in the terminal voltage.

To illustrate the demagnetizing effect of the armature reaction, let $E_{FD\Delta} = 0$; then

$$E'_{q\Delta} = [K_3K_4/(1 + K_3\,\tau'_{d0}\,s)]\,\delta_\Delta \tag{6.74}$$

and substituting in the expression for $T_{e\Delta}$ we get,

$$T_{e\Delta} = [K_1 - K_2K_3K_4/(1 + K_3\,\tau'_{d0}\,s)]\,\delta_\Delta \tag{6.75}$$

The bracketed term is the synchronizing torque coefficient taking into account the effect of the armature reaction. *Initially,* the coefficient K_1 is reduced by a factor K_2K_4/τ'_{d0}.

Similarly, substituting in the expression for $V_{t\Delta}$,

$$V_{t\Delta} = [K_5 - K_3K_4K_6/(1 + K_3\,\tau'_{d0}\,s)]\,\delta_\Delta \tag{6.76}$$

The second term is usually much larger in magnitude than K_5, and *initially* the change in the terminal voltage is given by

$$V_{t\Delta}\big]_{t=0} = -(K_4K_6/\tau'_{d0})\,\delta_\Delta \tag{6.77}$$

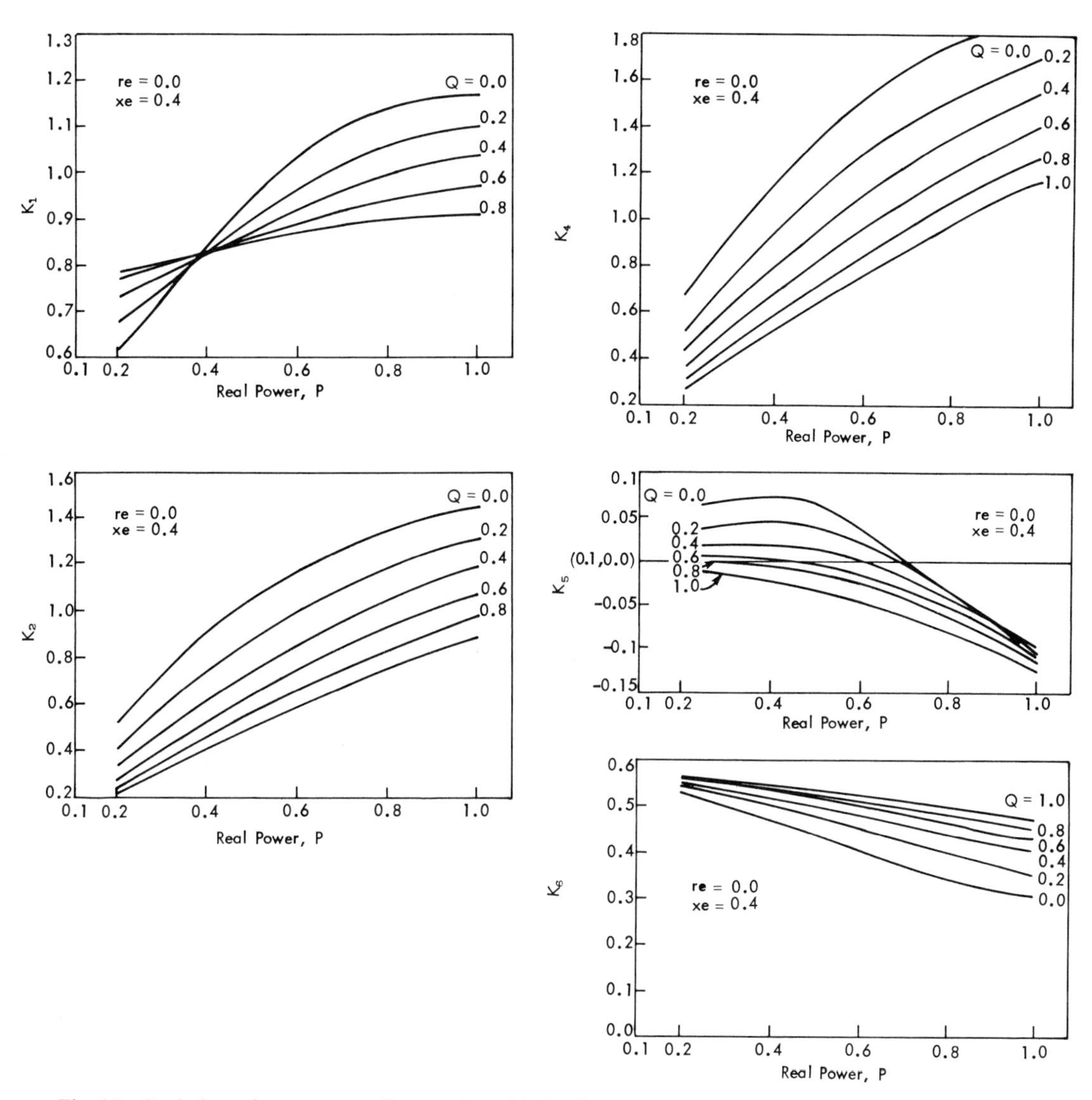

Fig. 6.1 Variation of parameters $K_1, \ldots, K_6$ with loading: (a) K_1 versus P (real power) and Q (reactive power) as parameter, (b) K_2 versus P and Q, (c) K_4 versus P and Q, (d) K_5 versus P and Q, (e) K_6 versus P and Q. (© IEEE. Reprinted from *IEEE Trans.*, vol. PAS-92, Sept./Oct. 1973.)

The effects of the machine loading on the constants K_1, K_2, K_4, K_5, and K_6 are studied in reference [3] for a one machine–infinite bus system very similar to the system in the above examples except for zero external resistance. The results are shown in Figure 6.1.

6.5.6 Comparison with classical model

The machine model discussed in this section is almost as simple as the classical model discussed in Chapter 2, except for the variation in the main field-winding flux. It is interesting to compare the two models.

The classical model does not account for the demagnetizing effect of the armature reaction, manifested as a change in E'_q. Thus (6.67) in the classical model would have $K_2 = 0$. Also in (6.59) the effective time constant is assumed to be very large so that $E'_q \simeq$ constant. In (6.72) the classical model will have $K_6 = 0$.

To illustrate the difference between the two models, the same system in Example 6.7 is solved by the classical model.

Example 6.8

Using the classical model discussed in Chapter 2, solve the system of Example 6.7.

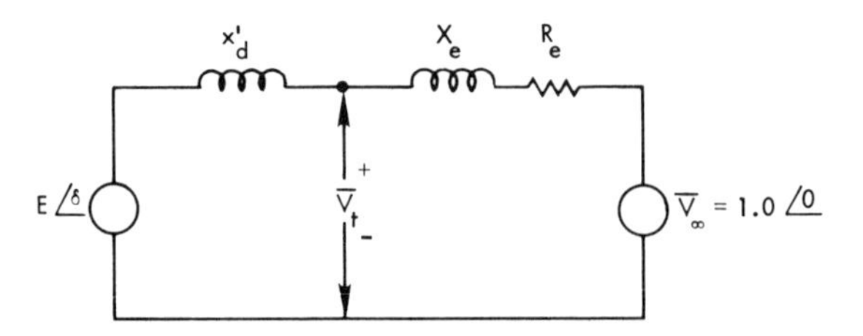

Fig. 6.2 Network of Example 6.7.

Solution

The network used in the classical model is shown in Figure 6.2. The phasor $\overline{E} = E\,/\underline{\delta}$ is the constant voltage behind transient reactance. Note that the angle δ here is *not* the same as the rotor angle δ discussed previously; it is the angle of the fictitious voltage $\overline{E}$. The phasors $\overline{V}_t$ and $\overline{V}_\infty$ are the machine terminal voltage and the infinite bus voltage respectively.

For convenience we will use the pu system used (or implied) in Chapter 2, i.e., based on the three-phase power. Therefore,

$$\begin{aligned}\overline{E} &= E\,/\underline{\delta} = 1 + \text{j}0.0 + (0.020 + \text{j}0.645)(0.980 - \text{j}0.217)\\ &= 1.3186\,/\underline{28.43^\circ}\end{aligned}$$

The synchronizing power coefficient is given by

$$P_s = \left.\frac{P_\Delta}{\delta_\Delta}\right]_{\delta=\delta_0} = EV_\infty(B_{12}\cos\delta_0 - G_{12}\sin\delta_0) = (EV_\infty/Z^2)[(x_d' + X_e)\cos\delta_0 + R_e\sin\delta_0)]$$

$$= \frac{1.3164 \times 1.0}{0.4162}(0.645 \times 0.8794 - 0.02 \times 0.4761) = 1.826$$

To compare with the value of K_1 in Example 6.7 we note the difference in the pu system, $K_1 = 1.448$. Thus the classical model gives a larger value of the synchronizing power coefficient than that obtained when the demagnetizing effect of the armature reaction is taken into account.

To obtain the linearized equation for V_t, neglecting R_e we get

$$\begin{aligned}\overline{I}_a &= [(1.3186\cos\delta - 1.00) + \text{j}1.3186\sin\delta]/\text{j}0.645\\ \overline{V}_t &= 1.000 + \text{j}0.0 + \text{j}0.40\,\overline{I}_a\end{aligned}$$

Substituting, we get for the magnitude of V_t

$$\begin{aligned}V_t^2 &= (0.3798 + 0.8177\cos\delta)^2 + (0.8177)^2\sin^2\delta\\ 2V_{t0}V_{t\Delta} &= -(0.62\sin\delta_0)\,\delta_\Delta\end{aligned}$$

or

$$V_{t\Delta} = -0.1261\,\delta_\Delta$$

The corresponding *initial* value in Example 6.7 is given by

$$V_{t\Delta}\Big]_{t=0^+} = -(K_4 K_6/\tau'_{d0})\,\delta_\Delta = -0.1252\,\delta_\Delta$$

6.6 Block Diagrams

The block diagram representation of (6.73) and the equation for $\dot{\delta}_\Delta$ is shown in Figure 6.3. This block diagram "generates" the rotor angle δ_Δ. When combined with (6.59), (6.67), and (6.72) the resulting block diagram is shown in Figure 6.4. In both diagrams the subscript Δ is omitted for convenience. Note that Figure 6.4 is similar to Figure 3.1.

Figure 6.4 has two inputs or forcing functions, namely, E_{FD} and T_m. The output is the terminal voltage change V_t. Other significant quantities are identified in the diagram, such as E'_q, T_e, ω, and δ. The diagram and its equations show that the simplified model of the synchronous machine is a third-order system.

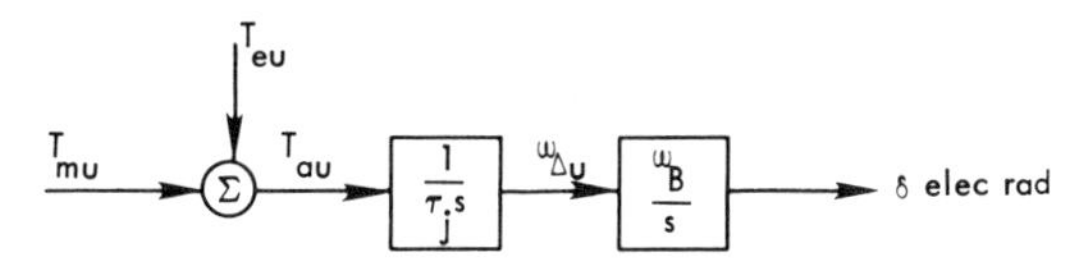

Fig. 6.3 Block diagram of (6.73).

6.7 State-Space Representation of Simplified Model

From Section 6.5 the system equations are given by

$$\begin{aligned}
K_3\tau'_{d0}\dot{E}'_{q\Delta} + E'_{q\Delta} &= K_3 E_{FD\Delta} - K_3 K_4\,\delta_\Delta \\
T_{e\Delta} &= K_1\,\delta_\Delta + K_2 E'_{q\Delta} \\
V_{t\Delta} &= K_5\delta_\Delta + K_6 E'_{q\Delta} \\
\tau_j\dot{\omega}_\Delta &= T_{m\Delta} - T_{e\Delta} \\
\dot{\delta}_\Delta &= \omega_\Delta
\end{aligned} \tag{6.78}$$

Eliminating $V_{t\Delta}$ and $T_{e\Delta}$ from the above equations,

$$\begin{aligned}
\dot{E}'_{q\Delta} &= -(1/K_3\tau'_{d0})E'_{q\Delta} - (K_4/\tau'_{d0})\,\delta_\Delta + (1/\tau'_{d0})E_{FD\Delta} \\
\dot{\omega}_\Delta &= -(K_2/\tau_j)E'_{q\Delta} - (K_1/\tau_j)\,\delta_\Delta + (1/\tau_j)T_{m\Delta} \\
\dot{\delta}_\Delta &= \omega_\Delta
\end{aligned} \tag{6.79}$$

By designating the state variables as $E'_{q\Delta}$, ω_Δ, and δ_Δ and the input signals as $E_{FD\Delta}$ and

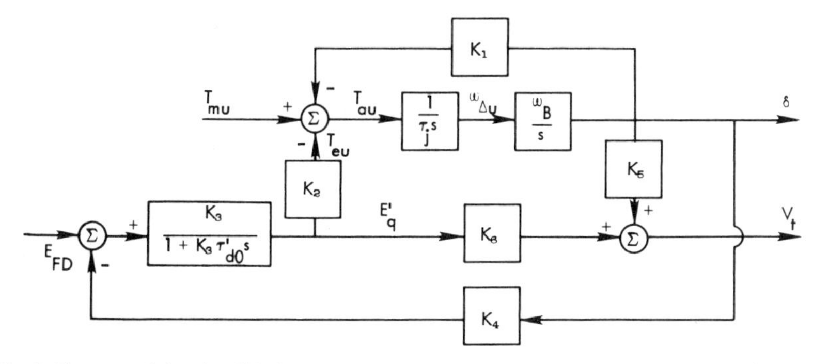

Fig 6.4 Block diagram of the simplified linear model of a synchronous machine connected to an infinite bus.

$T_{m\Delta}$, the above equation is in the desired state-space form

$$\dot{\mathbf{x}} = \mathbf{Ax} + \mathbf{Bu}$$

where

$$\mathbf{x}^t = [E'_{q\Delta} \quad \omega_\Delta \quad \delta_\Delta] \qquad \mathbf{u} = \begin{bmatrix} E_{FD\Delta} \\ T_{m\Delta} \end{bmatrix} \qquad \mathbf{B} = \begin{bmatrix} 1/\tau'_{d0} & 0 \\ 0 & 1/\tau_j \\ 0 & 0 \end{bmatrix} \tag{6.80}$$

$$\mathbf{A} = \begin{bmatrix} -1/K_3\tau'_{d0} & 0 & -K_4/\tau'_{d0} \\ -K_2/\tau_j & 0 & -K_1/\tau_j \\ 0 & 1 & 0 \end{bmatrix} \tag{6.81}$$

In the above equations the driving functions $E_{FD\Delta}$ and $T_{m\Delta}$ are determined from the detailed description of the voltage regulator–excitation systems and the mechanical turbine–speed governor systems respectively. The former will be discussed in Chapter 7 while the latter is discussed in Volume 2.

Problems

6.1 The generator of Example 5.2 is loaded to 75% of nameplate rating at rated terminal voltage and with constant turbine output. The excitation is then varied from 90% PF lagging to unity and finally to 90% leading. Compute the current model **A** matrix for these three power factors. How many elements of the **A** matrix vary as the power factor is changed? How sensitive are these elements to change in power factor?

6.2 Use a digital computer to compute the eigenvalues of the three **A** matrices determined in Problem 6.1. What conclusions, if any, can you draw from the results? Let $D = 0$.

6.3 Using the data of Problem 6.1 at 90% PF lagging, compute the eigenvalues of the **A** matrix with the damping $D = 1, 2$, and 3. Find the sensitivity of the eigenvalues to this parameter.

6.4 Repeat Problem 6.1 using the flux linkage model

6.5 Repeat Problem 6.2 using the flux linkage model.

6.6 Repeat Problem 6.3 using the flux linkage model.

6.7 Make an analog computer study using the linearized model summarized in Section 6.5.4. Note in particular the system damping as compared to the analog computer results of Chapter 5. Determine a value of D that will make the linear model respond with damping similar to the nonlinear model.

6.8 Examine the linear system (6.79) and write the equation for the eigenvalues of this system. Find the characteristic equation and see if you can identify any system constraints for stability using Routh's criterion.

6.9 For the generator and loading conditions of Problem 6.1, calculate the constants K_1 through K_6 for the simplified linear model.

6.10 Repeat Example 6.8 for the system of Example 6.6. Find the synchronizing power coefficient and $V_{t\Delta}$ as a function of δ_Δ for the classical model and compare with the corresponding values obtained by the simplified linear model.

References

1. Hefron, W. G., and Phillips, R. A. Effect of a modern voltage regulator on underexcited operation of large turbine generators. *AIEE Trans.* 71:692–97, 1952.
2. de Mello, F. P., and Concordia, C. Concepts of synchronous machine stability as affected by excitation control. *IEEE Trans.* PAS-88:316–29, 1969.
3. El-Sherbiny, M. K., and Mehta, D. M. Dynamic system stability. Pt. 1. *IEEE Trans.* PAS-92:1538–46, 1973.

chapter 7

Excitation Systems

Three principal control systems directly affect a synchronous generator: the boiler control, governor, and exciter. This simplified view is expressed diagramatically in Figure 7.1, which serves to orient our thinking from the problems of *representation* of the machine to the problems of *control*. In this chapter we shall deal exclusively with the excitation system, leaving the consideration of governors and boiler control for Volume 2.

7.1 Simplified View of Excitation Control

Referring again to Figure 7.1, let us examine briefly the function of each control element. Assume that the generating unit is lossless. This is not a bad assumption when total losses of turbine and generator are compared to total output. Under this assumption all power received as steam must leave the generator terminals as electric power. Thus the unit pictured in Figure 7.1 is nothing more than an energy conversion device that changes heat energy of steam into electrical energy at the machine terminals. The amount of steam power admitted to the turbine is controlled by the governor. The excitation system controls the generated EMF of the generator and therefore controls not only the output voltage but the power factor and current magnitude as well. An example will illustrate this point further.

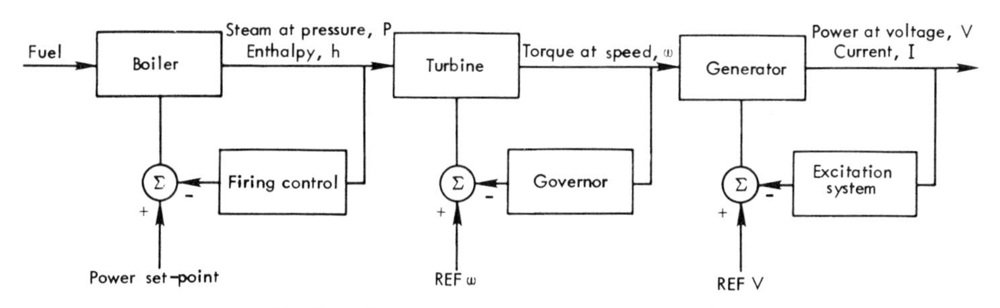

Fig. 7.1 Principal controls of a generating unit.

Refer to the schematic representation of a synchronous machine shown in Figure 7.2 where, for convenience, the stator is represented in its simplest form, namely, by an EMF behind a synchronous reactance as for round rotor machines at steady state. Here

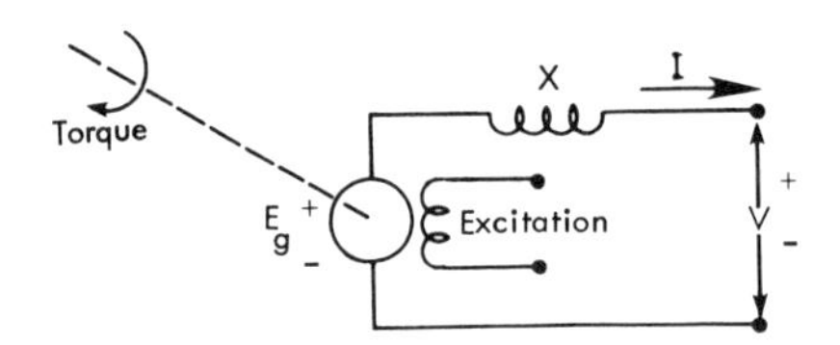

Fig. 7.2 Equivalent circuit of a synchronous machine.

the governor controls the torque or the shaft power input and the excitation system controls E_g, the internally generated EMF.

Example 7.1

Consider the generator of Figure 7.2 to be operating at a lagging power factor with a current I, internal voltage E_g, and terminal voltage V. Assume that the input power is held constant by the governor. Having established this initial operating condition, assume that the excitation is increased to a new value E_g'. Assume that the bus voltage is held constant by other machines operating in parallel with this machine, and find the new value of current I', the new power factor $\cos \theta'$, and the new torque angle δ'.

Solution

This problem without numbers may be solved by sketching a phasor diagram. Indeed, considerable insight into learning how the control system functions is gained by this experience.

The initial operating condition is shown in the phasor diagram of Figure 7.3. Under the operating conditions specified, the output power per phase may be expressed in two ways: first in terms of the generator terminal conditions

$$P = VI \cos \theta \tag{7.1}$$

and second in terms of the power angle, with saliency effects and stator resistance neglected,

$$P = (E_g V/X) \sin \delta \tag{7.2}$$

In our problem P and V are constants. Therefore, from (7.1)

$$I \cos \theta = k_1 \tag{7.3}$$

where k_1 is a constant. Also from (7.2)

$$E_g \sin \delta = k_2 \tag{7.4}$$

where k_2 is a constant.

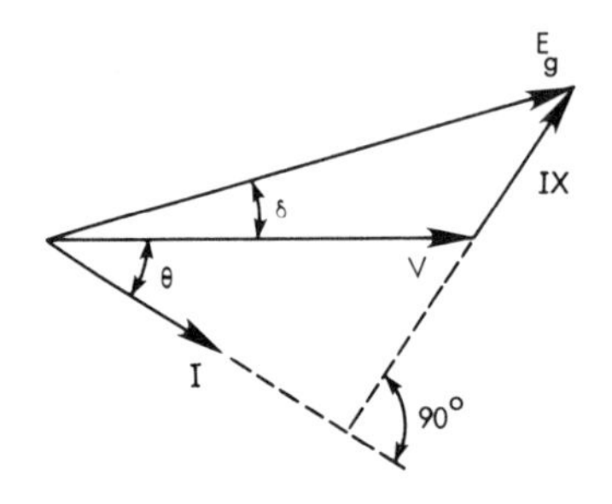

Fig. 7.3 Phasor diagram of the initial condition.

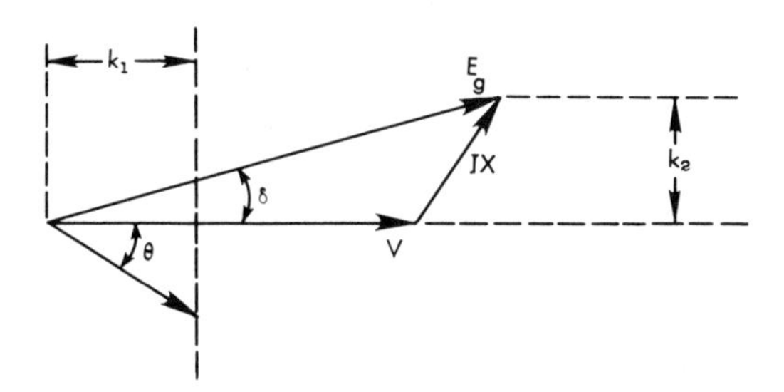

Fig. 7.4 Phasor diagram showing control constraints.

Figure 7.4 shows the phasor diagram of Figure 7.3, but with k_1 and k_2 shown graphically. Thus as the excitation is increased, the tip of E_g is constrained to follow the dashed line of Figure 7.4, and the tip of I is similarly constrained to follow the vertical dashed line. We also must observe the physical law that requires that phasor $\overline{IX}$ and phasor $\overline{I}$ lie at right angles. Thus we construct the phasor diagram of Figure 7.5, which shows the "before and after" situation. We observe that the new equilibrium condition requires that (1) the torque angle is decreased, (2) the current is increased, and (3) the power factor is more lagging; but the output power and voltage are the same.

By similar reasoning we can evaluate the results of decreasing the excitation and of changing the governor setting. These mental exercises are recommended to the student as both interesting and enlightening.

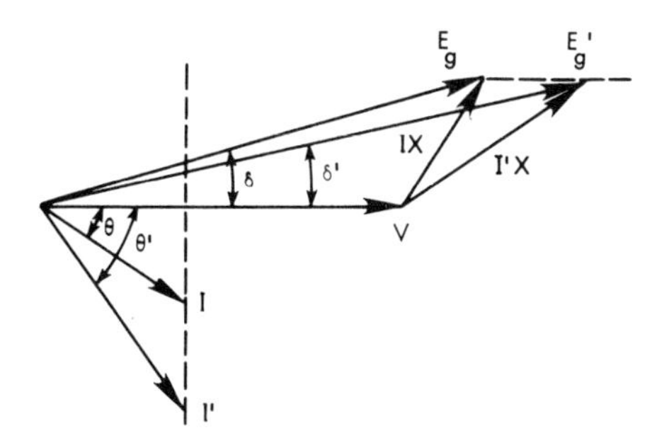

Fig. 7.5 Solution for increasing E_g at constant P and V.

Note that in Example 7.1 we have studied the effect of going from one stable operating condition to another. We have ignored the transient period necessary to accomplish this change, with its associated problems—the speed of response, the nature of the transient (overdamped, underdamped, or critically damped), and the possibility of saturation at the higher value of E_g. These will be topics of concern in this chapter.

7.2 Control Configurations

We now consider the physical configuration of components used for excitation systems. Figure 7.6 shows in block form the arrangement of the physical components in

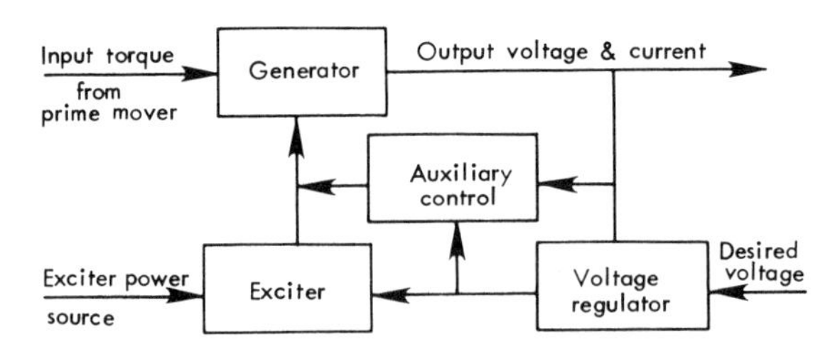

Fig. 7.6 Arrangement of excitation components.

any system. In many present-day systems the exciter is a dc generator driven by either the steam turbine (on the same shaft as the generator) or an induction motor. An increasing number are solid-state systems consisting of some form of rectifier or thyristor system supplied from the ac bus or from an alternator-exciter.

The voltage regulator is the intelligence of the system and controls the output of the exciter so that the generated voltage and reactive power change in the desired way. In earlier systems the "voltage regulator" was entirely manual. Thus the operator observed the terminal voltage and adjusted the field rheostat (the voltage regulator) until the desired output conditions were observed. In most modern systems the voltage regulator is a controller that senses the generator output voltage (and sometimes the current) then initiates corrective action by changing the exciter control in the desired direction. The *speed* of this device is of great interest in studying stability. Because of the high inductance in the generator field winding, it is difficult to make rapid changes in field current. This introduces considerable "lag" in the control function and is one of the major obstacles to be overcome in designing a regulating system.

The auxiliary control illustrated in Figure 7.6 may include several added features. For example, damping is sometimes introduced to prevent overshoot. A comparator may be used to set a lower limit on excitation, especially at leading power factor operation, for prevention of instability due to very weak coupling across the air gap. Other auxiliary controls are sometimes desirable for feedback of speed, frequency, acceleration, or other data [1].

7.3 Typical Excitation Configurations

To further clarify the arrangement of components in typical excitation systems, we consider here several possible designs without detailed discussion.

7.3.1 Primitive systems

First we consider systems that can be classified in a general way as "slow response" systems. Figure 7.7 shows one arrangement consisting of a main exciter with manual or automatic control of the field. The "regulator" in this case detects the voltage level and includes a mechanical device to change the control rheostat resistance. One such direct-acting rheostatic device (the "Silverstat" regulator) is described in reference [2] and consists of a regulating coil that operates a plunger, which in turn acts on a row of spaced silver buttons to systematically short out sections of the rheostat. In application, the device is installed as shown in Figure 7.8. In operation, an increase in generator output voltage will cause an increase in dc voltage from the rectifier. This will cause an increase in current through the regulator coil that mechanically operates a solenoid to insert exciter field resistance elements. This reduces excitation field flux and voltage, thereby lowering the field current in the generator field, hence lowering the generator

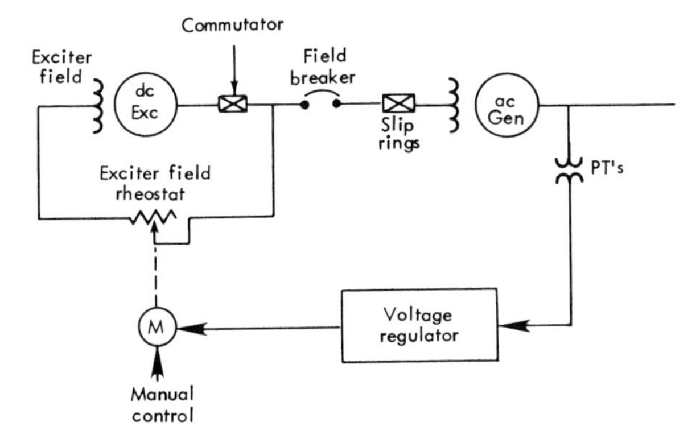

Fig. 7.7 Main exciter with rheostat control.

voltage. Two additional features of the system in Figure 7.8 are the damping transformer and current compensator. The damping transformer is an electrical "dashpot" or antihunting device to damp out excessive action of the moving plunger. The current compensator feature is used to control the division of reactive power among parallel generators operating under this type of control. The current transformer and compensator resistance introduce a voltage drop in the potential circuit proportional to the line current. The phase relationship is such that for lagging current (positive generated reactive power) the voltage drop across the compensating resistance adds to the voltage from the potential transformer. This causes the regulator to lower the excitation voltage for an increase in lagging current (increase in reactive power output) and provides a drooping characteristic to assure that the load reactive power is equally divided among the parallel machines.

The next level of complication in excitation systems is the main exciter and pilot

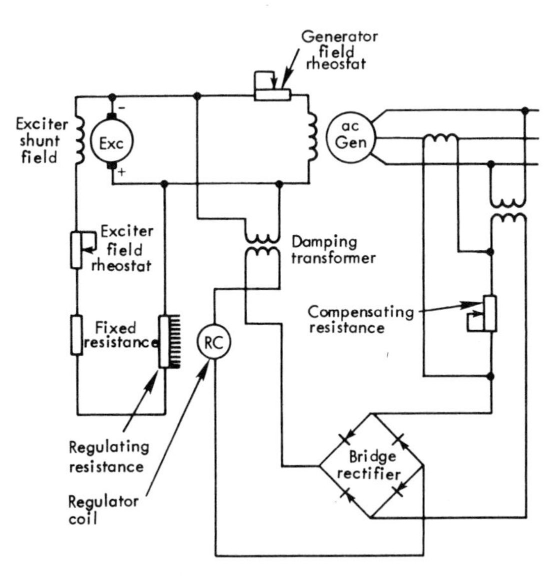

Fig. 7.8 Self-excited main exciter with Silverstat regulator. (Used with permission from *Electrical Transmission and Distribution Reference Book*, Westinghouse Electric Corp., 1950.)

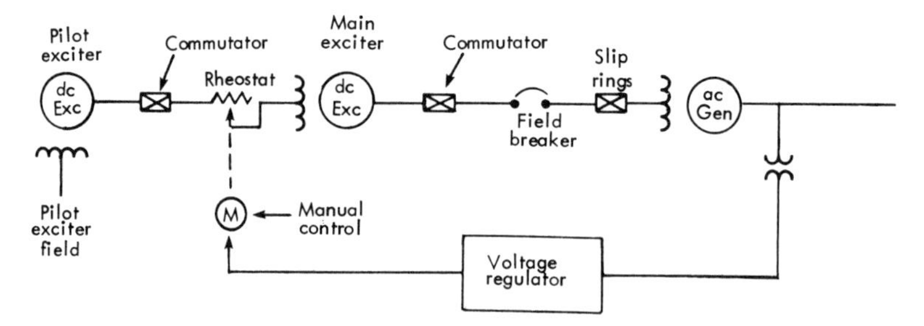

Fig. 7.9 Main exciter and pilot exciter system.

exciter system shown in Figure 7.9. This system has a much faster response than the self-excited main exciter, since the exciter field control is independent of the exciter output voltage. Control is achieved in much the same way as for the self-excited case. Because the rheostat positioner is electromechanical, the response may be slow compared to more modern systems, although it is faster than the self-excited arrangement.

The two systems just described are examples of older systems and represent direct, straightforward means of effecting excitation control. In terms of present technology in control systems they are primitive and offer little promise for really fast system response because of inherent friction, backlash, and lack of sensitivity.

The first step in sophistication of the primitive systems was to include in the feedback path an amplifier that would be fast acting and could magnify the voltage error and induce faster excitation changes. Gradually, as generators have become larger and interconnected system operation more common, the excitation control systems have become more and more complex. The following sections group these modern systems according to the type of exciter [3].

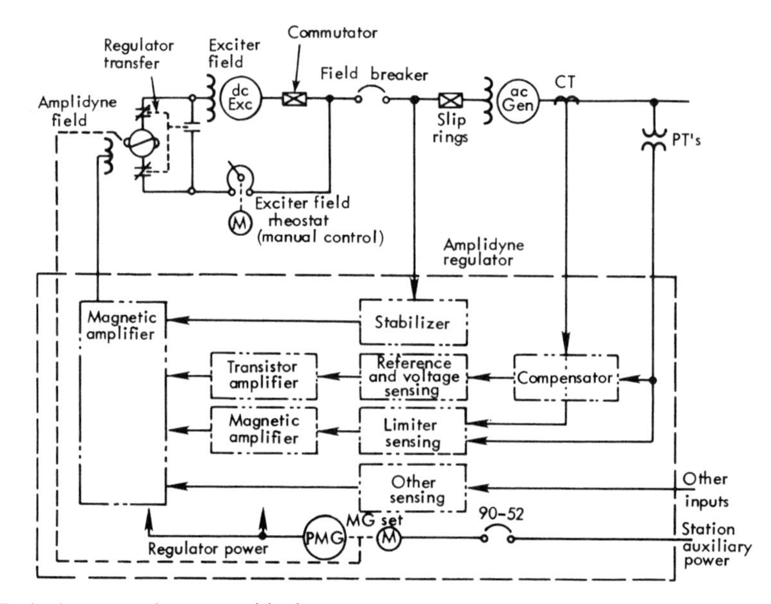

Fig. 7.10 Excitation control system with dc generator-commutator exciter. (© IEEE. Reprinted from *IEEE Trans.*, vol. PAS-88, Aug. 1969.) Example: General Electric type NA143 amplidyne system [4].

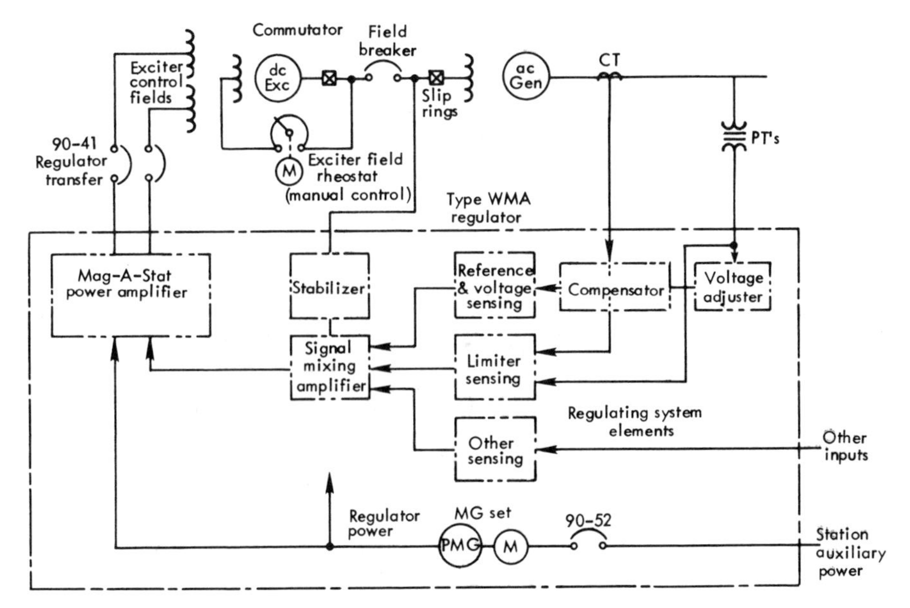

Fig. 7.11 Excitation control system with dc generator-commutator exciter. (© IEEE. Reprinted from *IEEE Trans.*, vol. PAS-88, Aug. 1969.) Example: Westinghouse type WMA Mag-A-Stat system [6].

7.3.2 Excitation control systems with dc generator-commutator exciters

Two systems of U.S. manufacture have dc generator-commutator exciters. Both have amplifiers in the feedback path; one a rotating amplifier, the other a magnetic amplifier.

Figure 7.10 [3] shows one such system that incorporates a rotating amplifier or amplidyne [5] in the exciter field circuit. This amplifier is used to force the exciter field in the desired direction and results in much faster response than with a self-excited machine acting unassisted.

Another system with a similar exciter is that of Figure 7.11 where the amplifier is a static magnetic amplifier deriving its power supply from a permanent-magnet generator-motor set. Often the frequency of this supply is increased to 420 Hz to increase the amplifier response. Note that the exciter in this system has two control fields, one for boost and one for buck corrections. A third field provides for self-excited manual operation when the amplifier is out of service.

7.3.3 Excitation control systems with alternator-rectifier exciters

With the advent of solid-state technology and availability of reliable high-current rectifiers, another type of system became feasible. In this system the exciter is an ac generator, the output of which is rectified to provide the dc current required by the generator field. The control circuitry for these units is also solid-state in most cases, and the overall response is quite fast [3].

An example of alternator-rectifier systems is shown in Figure 7.12. In this system the alternator output is rectified and connected to the generator field by means of slip rings. The alternator-exciter itself is shunt excited and is controlled by electronically adjusting the firing angle of thyristors (SCR's). This means of control can be very fast

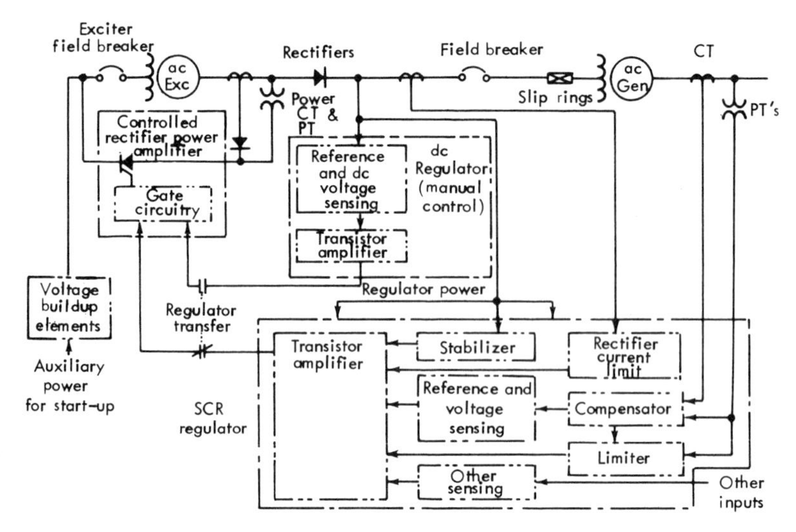

Fig. 7.12 Excitation control system with alternator-rectifier exciter using stationary noncontrolled rectifiers. (© IEEE. Reprinted from *IEEE Trans.*, vol. PAS-88, Aug. 1969.) Example: General Electric Alterrex excitation system [7].

since the firing angle can be adjusted very quickly compared to the other time constants involved.

Another example of an alternator-rectifier system is shown in Figure 7.13. This system is unique in that it is brushless; i.e., there is no need for slip rings since the alternator-exciter and diode rectifiers are rotating with the shaft. The system incorporates a pilot permanent magnet generator (labeled PMG in Figure 7.13) with a permanent magnet field to supply the (stationary) field for the (rotating) alternator-exciter. Thus all coupling between stationary and rotating components is electromagnetic. Note, however, that it is impossible to meter any of the generator field quantities directly since these components are all moving with the rotor and no slip rings are used.

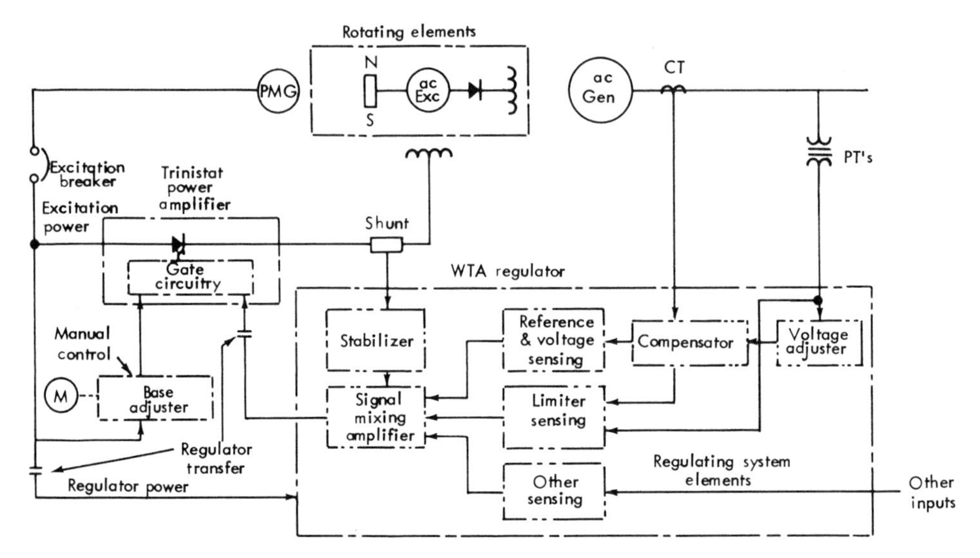

Fig. 7.13 Excitation control system with alternator-rectifier exciter employing rotating rectifiers. (© IEEE. Reprinted from *IEEE Trans.*, vol. PAS-88, Aug. 1969.) Example: Westinghouse type WTA Brushless excitation system [8, 9].

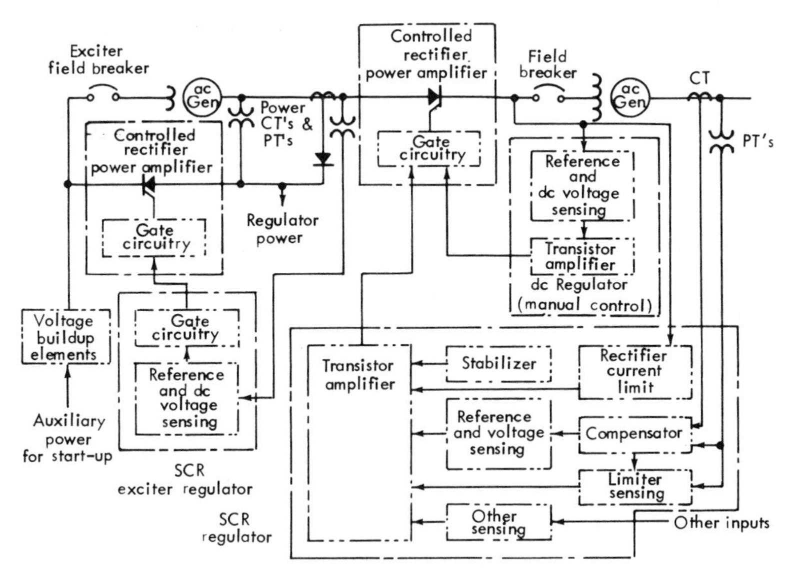

Fig. 7.14 Excitation control system with alternator-SCR exciter system. (© IEEE. Reprinted from *IEEE Trans.*, vol. PAS-88, Aug. 1969.) Example: General Electric Althyrex excitation system [11].

The response of systems with alternator-rectifier exciters is improved by designing the alternator for operation at frequencies higher than that of the main generator. Recent systems have used 420-Hz and 300-Hz alternators for this reason and report excellent response characteristics [8, 10].

7.3.4 Excitation control systems with alternator-SCR exciter systems

Another important development in excitation systems has been the alternator-SCR design shown in Figure 7.14 [3]. In this system the alternator excitation is supplied di-

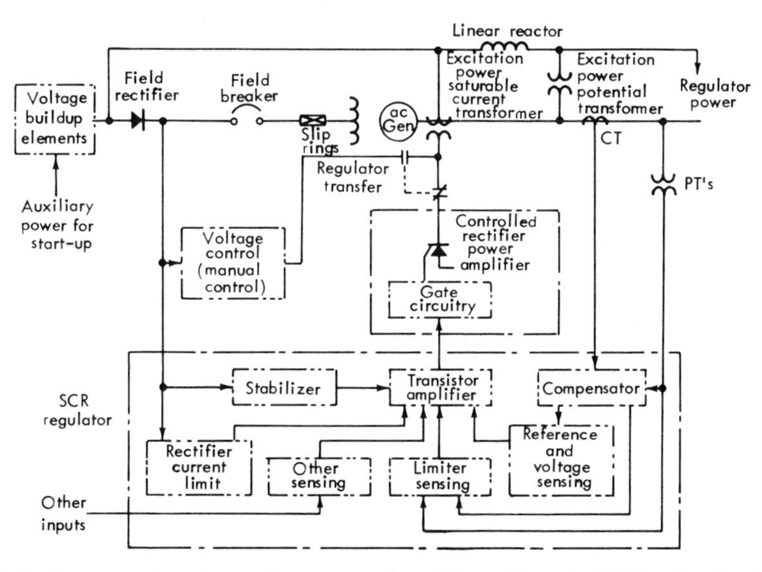

Fig. 7.15 Excitation control system with compound-rectifier exciter. (© IEEE. Reprinted from *IEEE Trans.*, vol. PAS-88, Aug. 1969.) Example: General Electric SCTP static excitation system [12, 13].

rectly from an SCR system with an alternator source. Hence it is only necessary to adjust the SCR firing angle to change the excitation level, and this involves essentially no time delay. This requires a somewhat larger alternator-exciter than would otherwise be necessary since it must have a rating capable of continuous operation at ceiling voltage. In slower systems, ceiling voltage is reached after a delay, and sustained operation at that level is unlikely.

7.3.5 Excitation control systems with compound-rectifier exciter systems

The next classification of exciter systems is referred to as a "compound-rectifier" exciter, of which the system shown in Figure 7.15 is an example [3].

This system can be viewed as a form of self-excitation of the main ac generator. Note that the exciter input comes from the generator electrical output terminals, not from the shaft as in previous examples. This electrical feedback is controlled by saturable reactors, the control for which is arranged to use both ac output and exciter values as intelligence sources. The system is entirely static, and this feature is important. Although originally designed for use on smaller units [12, 13], this same principle may be applied to large units as well.

Self-excited units have the inherent disadvantage that the ac output voltage is low at the same time the exciter is attempting to correct the low voltage. This may be partially compensated for by using output current as well as voltage in the control scheme so that (during faults, for example) feedback is still sufficient to effect adequate control. Such is the case in the unit shown in Figure 7.15.

7.3.6 Excitation control system with compound-rectifier exciter plus potential-source-rectifier exciter

A variation of the compound-rectifier scheme is one in which a second rectified output is added to the self-excited feedback to achieve additional control of excitation.

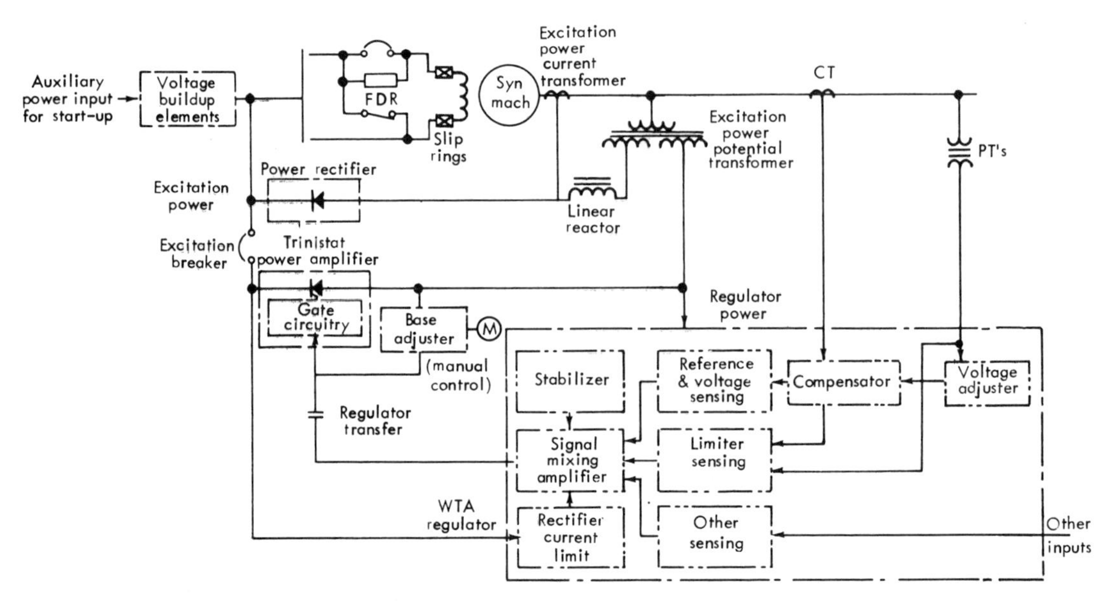

Fig. 7.16 Excitation control system with compound-rectifier exciter plus potential-source-rectifier exciter. (© IEEE. Reprinted from *IEEE Trans.*, vol. PAS-88, Aug. 1969.) Example: Westinghouse type WTA-PCV static excitation system [14].

This scheme is depicted in Figure 7.16 [3]. Again the basic self-excited main generator scheme is evident. Here, however, the voltage regulator controls a second rectifier system (called the "Trinistat power amplifier" in Figure 7.16) to achieve the desired excitation control. Note that the system is entirely static and can be inherently very fast, the only time constants being those of the reactor and the regulator.

7.3.7 Excitation control systems with potential-source-rectifier exciter

The final category of excitation systems is the self-excited main generator where the rectification is done by means of SCR's rather than diodes. Two such systems are shown in Figure 7.17 and Figure 7.18 [3]. Both circuits have static voltage regulators that use potential, current, and excitation levels to generate a control signal by which the SCR gating may be controlled. This type of control is very fast since there is no time delay in shifting the firing angle of the SCR's.

7.4 Excitation Control System Definitions

Most of the foregoing excitation system configurations are described in reference [3], which also gives definitions of the control system quantities of interest in this application. Only the most important of these are reviewed here. Other definitions, including those referred to by number here, are stated in Appendix E.

All excitation control systems may be visualized as automatic control systems with feed forward and feedback elements as shown in Figure 7.19. Viewed in this way, the excitation control systems discussed in the preceding section may be arranged in a general way, as indicated in Figure 7.20 and further described in Table 7.1. Note that the synchronous machine is considered a part of the "excitation control system," but the control elements themselves are referred to simply as the "excitation system."

The type of transfer function belonging in each block of Figure 7.20 is discussed in reference [15]. The reference to systems of Type 1, Type 3, etc., in the last column of Table 7.1 also refers to system types defined in that reference. This will be discussed in greater detail in Section 7.9. Our present concern is to learn the general configuration

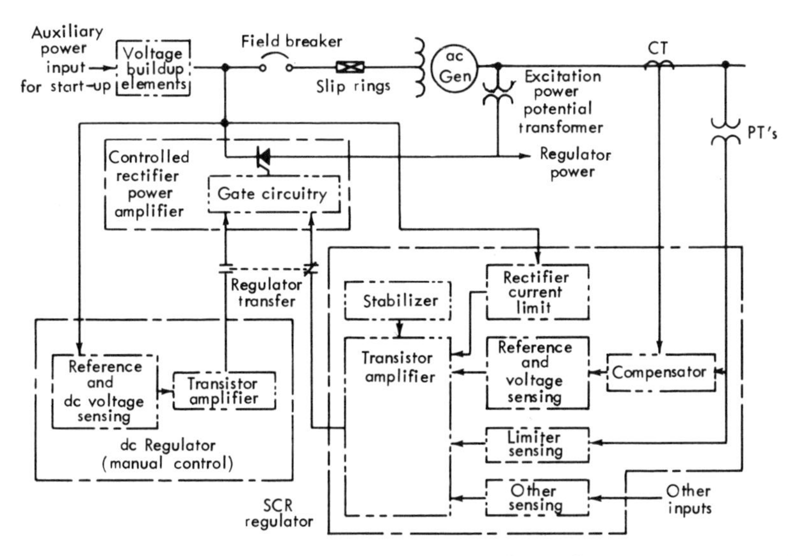

Fig. 7.17 Excitation control system with potential-source-rectifier exciter. (© IEEE. Reprinted from *IEEE Trans.*, vol. PAS-88, Aug. 1969.) Example: General Electric SCR static excitation system [14].

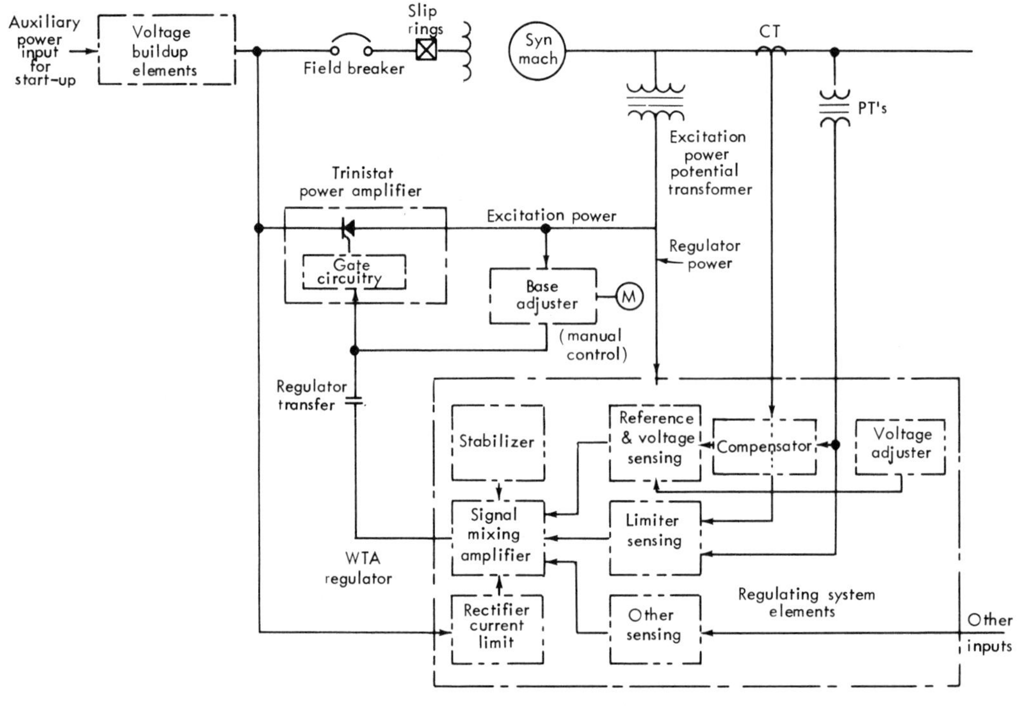

Fig. 7.18 Excitation control system with potential-source-rectifier exciter. (© IEEE. Reprinted from *IEEE Trans.*, vol. PAS-88, Aug. 1969.) Example: Westinghouse type WTA-Trinistat excitation system.

of modern excitation control systems and to become familiar with the language used in describing them.

7.4.1 Voltage response ratio

An important definition used in describing excitation control systems is that of the *response ratio* defined in Appendix E, Def. 3.15–3.19. This is a rough measure of how fast the exciter open circuit voltage will rise in 0.5 s if the excitation control is adjusted suddenly in the maximum increase direction. In other words, the voltage reference in Figure 7.20 is a step input of sufficient magnitude to drive the exciter voltage to its ceiling (Def. 3.03) value with the exciter operating under no-load conditions. Figure 7.21 shows a typical response of such a system where the voltage v_F starts at the rated load field voltage (Def. 3.21) that is the value of v_F, which will produce rated

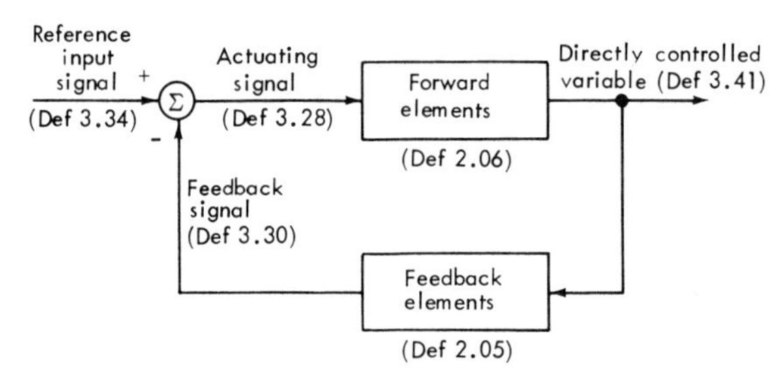

Fig. 7.19 Essential elements of an automatic feedback control system (Def. 1.02). (© IEEE. Reprinted from *IEEE Trans.*, vol. PAS-88, Aug. 1969.) Note: In excitation control system usage the actuating signal is commonly called the error signal (Def. 3.29). (See Appendix E for definitions.)

Table 7.1. Components Commonly Used in Excitation Control Systems

1	2	3	4	5	6	7		8		
Type of exciter	Primary detecting element & reference input	Pre-amplifier	Power amplifier	Manual control	Signal modifiers	Power sources		Excitation system stabilizers	System diagram reference	Type of computer representation*
						Regulator	Exciter			
dc Generator-commutator exciter	See note 2	See note 3	Rotating, magnetic, thyristor	Self-excited or separately excited exciter	See note 6	MG set	MG set, synchronous machine shaft	See note 8	Figs. 7.10 & 7.11	1
Alternator-rectifier exciter			Rotating, thyristor	Compensated input to power amplifier. Self-excited field voltage regulator		Synchronous machine shaft, MG set, alternator output	Synchronous machine shaft		Figs. 7.12 & 7.13	1
Alternator-rectifier (controlled) exciter			Thyristor	Exciter output voltage regulator		Alternator output	Synchronous machine shaft		Fig. 7.14	1
Compound-rectifier exciter			Magnetic, thyristor	Self-excited		Synchronous machine terminals	Synchronous machine terminals		Fig. 7.15	3
Compound-rectifier exciter plus potential-source rectifier exciter			Thyristor	Compensated input to power amplifier		Synchronous machine terminals	Synchronous machine terminals		Fig. 7.16	Similar to 3
Potential-source rectifier (controlled) exciter			Thyristor	Exciter output voltage regulator. Compensated input to power amplifier		Synchronous machine terminals	Synchronous machine terminals		Figs. 7.17 & 7.18	1S

Source: © IEEE. Reprinted from *IEEE Trans.*, vol. PAS-88, Aug. 1969.

2. Primary detecting element and reference input: can consist of many types of circuits on any system including differential amplifier, amplifier-turn comparison, intersecting impedance, and bridge circuits.

3. Preamplifier: Consists of all types but on newer systems is usually a solid-state amplifier.

6. Signal modifiers: (A) Auxiliary inputs—reactive and active current compensators; system stabilizing signals proportional to power, frequency, speed, etc. (B) Limiters—maximum excitation, minimum excitation, maximum V/Hz.

8. Excitation control system stabilizers: can consist of all types from series lead-lag to rate feedback around any element or group of elements of the system.

*IEEE committee report [15].

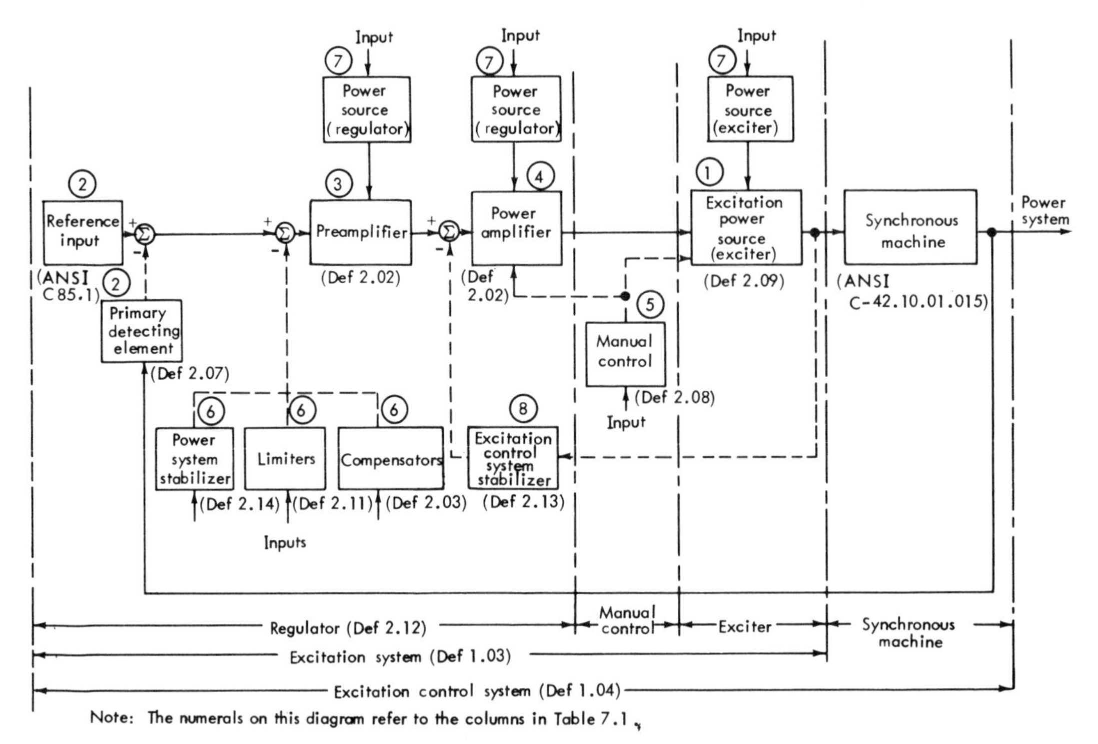

Fig. 7.20 Excitation control systems. (© IEEE. Reprinted from *IEEE Trans.*, vol. PAS-88, Aug. 1969.) Note: The numerals on this diagram refer to the columns in Table 7.1. (See Appendix E for definitions.)

generator voltage under nameplate loading. Then, responding to a step change in the reference, the open-circuited field is forced at the maximum rate to ceiling along the curve *ab*. Since the response is nonlinear, the response ratio is defined in terms of the area under the curve *ab* for exactly 0.5 s. We can easily approximate this area by a straight line *ac* and compute

$$\text{Response ratio} = cd/(0a)(0.5) \quad \text{pu V/s} \tag{7.5}$$

Kimbark [16] points out that since the exciter feeds a highly inductive load (the generator field), the voltage across the load is approximately $v = k\,d\phi/dt$. Then in a short time Δt the total flux change is

$$\Delta\phi = \frac{1}{k}\int_0^{\Delta t} v\,dt = \text{area under buildup curve} \tag{7.6}$$

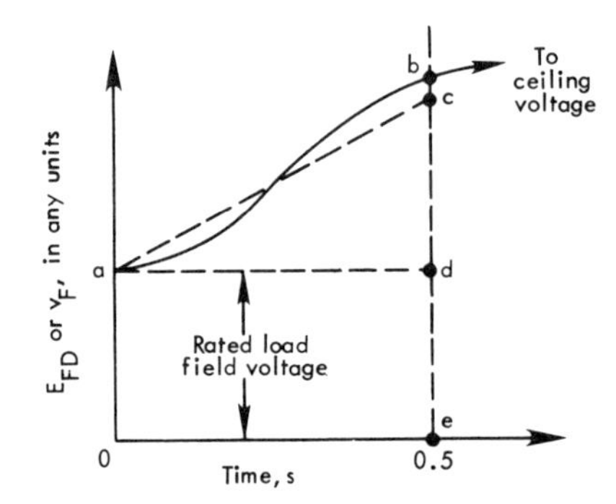

Fig. 7.21 Definition of a voltage response ratio.

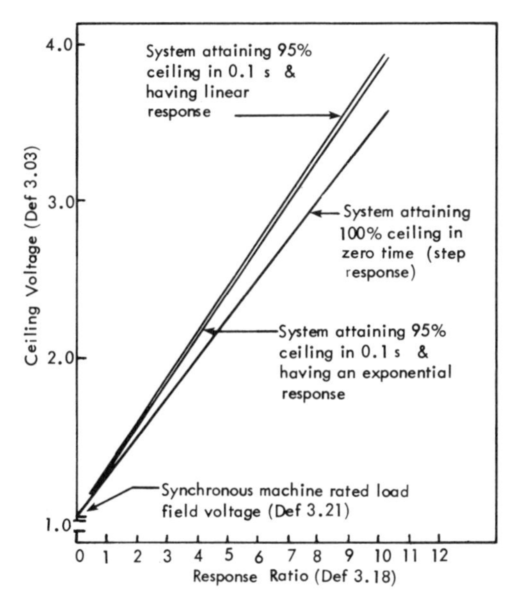

Fig. 7.22 Exciter ceiling voltage as a function of response ratio for a high initial response excitation system. (© IEEE. Reprinted from *IEEE Trans.*, vol. PAS-88, Aug. 1969.)

The time $\Delta t = 0.5$ was chosen because this is about the time interval of older "quick-response" regulators between the recognition of a step change in the output voltage and the shorting of field rheostat elements. Buildup rather than build-down is used because there is usually more interest in the response to a drop in terminal voltage, such as a fault condition. In dynamic operation where the interest is in small, fast changes, build-down may be equally important.

Equation (7.5) is an adequate definition if the voltage response is rather slow, such as the one shown in Figure 7.21. It has been recognized for some time, however, that modern fast systems may reach ceiling in 0.1 s or less, and extending the triangle *acd* out to 0.5 s is almost meaningless. This is discussed in reference [3], and a new definition is introduced (Def. 1.05) that replaces the 0.5-s interval $0e$ in Figure 7.21 by an interval $0e = 0.1$ s for "systems having an excitation voltage response time of 0.1 s or less" [the voltage response time (Def. 3.16) is the time required to reach 95% of ceiling]. A comparison of three systems, each attaining 95% ceiling voltage in 0.1 s, is given in Figure 7.22 [3] and shows how close the 0.1-s response is to the ideal system, a step function.

7.4.2 Exciter voltage ratings

Some additional comments are in order concerning certain of the excitation voltage definitions. First, it may be helpful to state certain numerical values of exciter ratings offered by the manufacturers (see [2] for a discussion of exciter ratings). Briefly, exciters are usually rated at 125 V for small generators, say 10 MVA and below. Larger units usually have 250-V exciters, say up to 100 MVA; with still larger machines being equipped with 350-V, 375-V, or 500-V exciters.

The voltage rating and the ceiling voltage are both important in considering the speed of response [1, 17]. Reference [1] tabulates the pattern of ceiling voltages for various response characteristics in Table 7.2, which shows the improved response for higher ceiling voltage ratings (and the lower ceiling voltage for solid-state exciters). It is reasonable that an exciter with a high ceiling voltage will build up to a particular voltage level faster than a similar exciter with a lower ceiling voltage simply because it saturates at a higher value. This is an important consideration in comparing types and ratings of both conventional and solid state exciters as shown in Table 7.2.

Table 7.2. Typical Ceiling Voltages for Various Exciter Response Ratios

Response ratio	Per unit ceiling voltage conventional exciters*	SCR exciters
0.5	1.25–1.35	1.20
1.0	1.40–1.50	1.20–1.25
1.5	1.55–1.65	1.30–1.40
2.0	1.70–1.80	1.45–1.55
4.0	...	2.00–2.10

*Based on rated exciter voltage.

In adopting a pu system for the exciter, there is no obvious choice as to what base voltage to use. Some possibilities are (also see [2]): (A) exciter rated voltage, (B) rated load field voltage, (C) rated air-gap voltage (the voltage necessary to produce rated voltage on the air gap line of the main machine in the case of a dc generator exciter), and (D) no-load field voltage. The IEEE [3] recommends the use of system B, the rated load field voltage. Consider, as an example, an exciter rated at 250 V. For this rating some typical values of other defined voltages are given in Figure 7.23. The pu system A

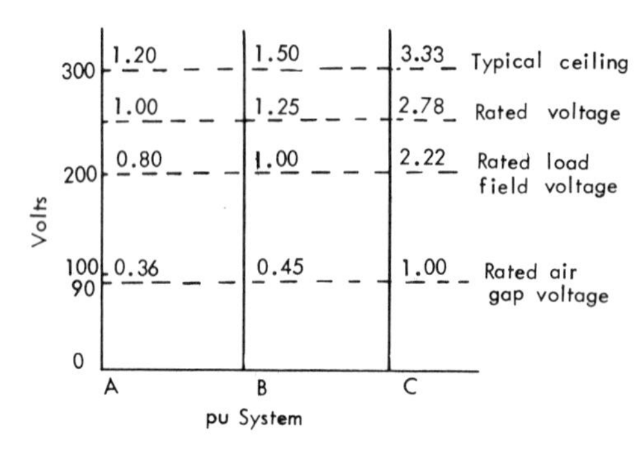

Fig. 7.23 Per unit voltages for a 250-V exciter.

of Figure 7.23 has little merit and is seldom used. System B is often used. System C is often convenient since, with rated air gap voltage as a base, pu exciter voltage, pu field current, and pu synchronous internal voltage are all equal under steady-state conditions with no saturation. System D is not illustrated in Figure 7.23 and is seldom used.

7.4.3 Other specifications

Excitation control system response should be compared against a suitable criterion of performance if the system is to be judged or graded. System performance could be measured under any number of forcing conditions. It is generally agreed that the quantity of primary interest is the exciter voltage-time characteristic in response to a step change in the generated voltage of from 10 to 20% [18, 19]. The problem is how to state in words the various possible slopes, delays, overshoots, damping, and the like. One useful description, often used in control system specification, is that based on the

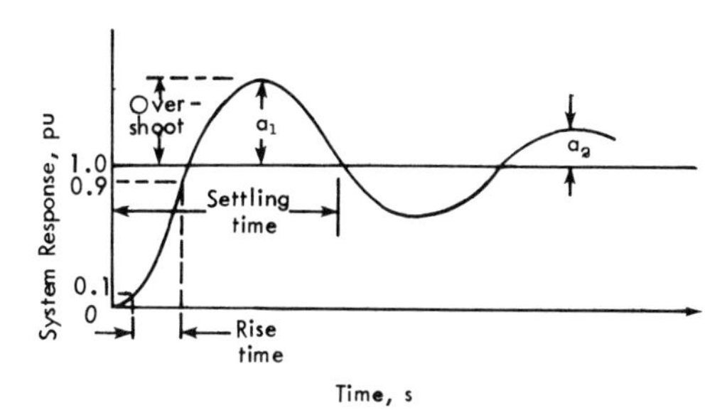

Fig. 7.24 Time domain specifications [22].

curve shown in Figure 7.24. Here the curve is the response to a step change in one of the system variables, such as the terminal voltage. This response, based on that of a second-order system, is a reasonable one on which to base time domain specifications since many systems tend to exhibit two "least-damped" poles that give a response of this general shape at some value of gain [20, 21]. Three quantities describe this response: the overshoot, the rise time, and the settling time.

The *overshoot* is the amount that the response exceeds the steady-state response—in Figure 7.24, a_1 pu.

The *rise time* is the time for the response to rise from 10 to 90% of the steady-state response.

The *settling time* is the time required for the response to a step function to stay within a certain percentage of its final value. Sometimes it is given as the time required to arrive at the final value after first overshooting this value. The first definition is preferred.

The *damping ratio* is that value for a second-order system defined by ζ in the expression

$$G(s) = K/(s^2 + 2\zeta\omega_n s + \omega_n^2) \tag{7.7}$$

and is related to the values a_1 and a_2 of two successive overshoots [23]. The *natural resonant frequency* ω_n is also of interest and may be given as a specification.

In the case of the second-order system (7.7), the response to a step change of a driving variable is

$$c(t) = 1 - e^{-\zeta\omega_n t}\{\cos\omega_r t + [\zeta/(1 - \zeta^2)]\sin\omega_r t\} \tag{7.8}$$

where

$$\omega_r = \omega_n(1 - \zeta^2)^{1/2} \tag{7.9}$$

When $\zeta = 0$, the system is oscillatory; when $\zeta = 0.7$, it has very little overshoot (about 5%). Critical damping is said to occur when $\zeta = 1.0$.

In dealing with an exciter being forced to ceiling due to a step change in the voltage regulator control, the system is often "overdamped"; i.e., $\zeta > 1$. In this case the voltage rise is more "sluggish," as shown in Figure 7.25. Here the overshoot is zero, the settling time is T_s (i.e., the time for the response to settle within k of its final value), and the rise time is T_R. Reference [19] suggests testing an excitation system to determine the response, such as in Figure 7.25. Then determine the area under this curve for 0.5 s and use this as a specification of response in the time domain. For newer, fast systems reference [3] suggests simulation of the excitation as preferable to actual testing since on some systems certain parameters are unavailable for measurement [8, 9].

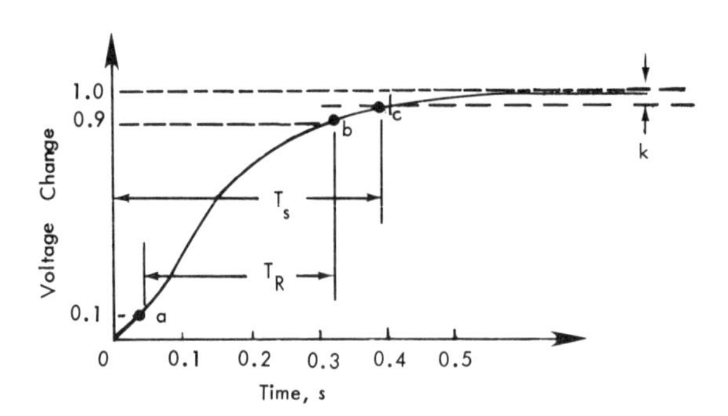

Fig. 7.25 Response of an excitation system.

7.5 Voltage Regulator

In several respects the heart of the excitation system is the voltage regulator (Def. 2.12). This is the device that senses changes in the output voltage (and current) and causes corrective action to take place. No matter what the exciter speed of the response, it will not alter its response until instructed to do so by the voltage regulator. If the regulator is slow, has deadband or backlash, or is otherwise insensitive, the system will be a poor one. Thus we need to be very critical of this important system component.

In addition to high reliability and availability for maintenance, it is necessary that the voltage regulator be a continuously acting proportional system. This means that any corrective action should be proportional to the deviation in ac terminal voltage from the desired value, no matter how small the deviation. Thus no deadband is to be tolerated, and large errors are to receive stronger corrective measures than small errors. In the late 1930s and early 1940s several types of regulators, electronic and static, were developed and tested extensively [24, 25]. These tests indicated that continuously acting proportional control "increased the generator steady-state stability limits well beyond the limits offered by the rheostatic regulator" [24, 26]. This type of system was therefore studied intensively and widely applied during the 1940s and 1950s, beginning with application to synchronous condensers; then to turbine generators; and finally, in the early 1950s, to hydrogenerators. (Reference [24] gives an interesting tabulation of the progress of these developments.)

7.5.1 Electromechanical regulators

The rather primitive direct-acting regulator shown in Figure 7.8 is an example of an electromechanical regulator. In such a system the voltage reference is the spring tension against which the solenoid must react. It is reliable and independent of auxiliaries of any kind. The response, however, is sluggish and includes deadband and backlash due to mechanical friction, stiction, and loosely fitting parts.

Two types of electromechanical regulators are often recognized; the direct-acting and the indirect-acting. *Direct-acting* regulators, such as the Silverstat [2] and the Tirrell [24], have been in use for many years, some dating back to about 1900. Such devices were widely used and steadily improved, while maintaining essentially the same form. As machines of larger size became more common in the 1930s the *indirect-acting* rheostatic regulators began to appear. These devices use a relay as the voltage-sensitive element [24]; thus the reference is essentially a spring, as in the direct-acting device. This relay operates to control a motor-operated rheostat, usually connected between the pilot exciter and the main exciter, as in Figure 7.9. This regulator is limited in its speed of response by various mechanical delays. Once the relay closes, to short out a rheostat section, the response is quite fast. In some cases, high-speed relays are used to permit faster excitation changes. These devices were considered quite successful, and nearly all large units installed between about 1930 and 1945 had this type of control. Many are still in service.

Another type of indirect-acting regulator that has seen considerable use employs a polyphase torque motor as a voltage-sensitive element [27]. In such a device the output torque is proportional to the average three-phase voltage. This torque is balanced against a spring in torsion so that each value of voltage corresponds to a different angular position of the rotor. A contact assembly attached to the rotor responds by closing contacts in the rheostat as the shaft position changes. A special set of contacts closes very fast with rapid rotor accelerations that permit faster than normal response due to sudden system voltage changes. The response of this type of regulator is fairly fast, and much larger field currents can be controlled than with the direct-acting regulator. This is due to the additional current "gain" introduced by the pilot exciter–main exciter scheme. The contact type of control, however, has inherent deadband and this, coupled with mechanical backlash, constitutes a serious handicap.

7.5.2 Early electronic regulators

About 1930 work was begun on electronic voltage regulators, electronic exciters, and electronic pilot exciters used in conjunction with a conventional main exciter [24, 25]. In general, these early electronic devices provided "better voltage regulation as well as smoother and faster generation excitation control" [24] than the competitive indirect-acting systems. They never gained wide acceptance because of anticipated high maintenance cost due to limited tube life and reliability, and this was at least partly justified in later analyses [25]. Generally speaking, electronic voltage regulators were of two types and used either to control electronic pilot exciters or electronic main exciters [25]. The electronic exciters or pilot exciters were high-power dc sources usually employing thyratron or ignitron tubes as rectifying elements.

7.5.3 Rotating amplifier regulators

In systems using a rotating amplifier to change the field of a main exciter, as in Figure 7.10, it is not altogether clear whether the rotating amplifier is a part of the "voltage regulator" or is a kind of pilot exciter. Here we take the view that the rotating amplifier is the final, high-gain stage in the voltage regulator.

The development of rotating amplifiers in the late 1930s and the application of these devices to generator excitation systems [28, 29] have been accompanied by the development of entirely "static" voltage sensing circuitry to replace the electromechanical devices used earlier. Usually, such static circuits were designed to exclude any electronic active components so that the reliability of the device would be more independent of component aging. For example, devices employing saturable reactors and selenium rectifiers showed considerable promise. Such circuits supplied the field windings of the rotating amplifiers, which were connected in series with the main exciter field, as in Figure 7.10. This scheme has the feature that the rotating amplifier can be bypassed for maintenance and the generator can continue to operate normally by manual regulation through a field rheostat. This connection is often called a "boost-buck" connection since, depending on polarity, the rotating amplifier is in a position to aid or oppose the exciter field.

The operation of a typical rotating amplifier regulating system can be analyzed by reference to Figure 7.10. The generator is excited by a self-excited shunt exciter. The

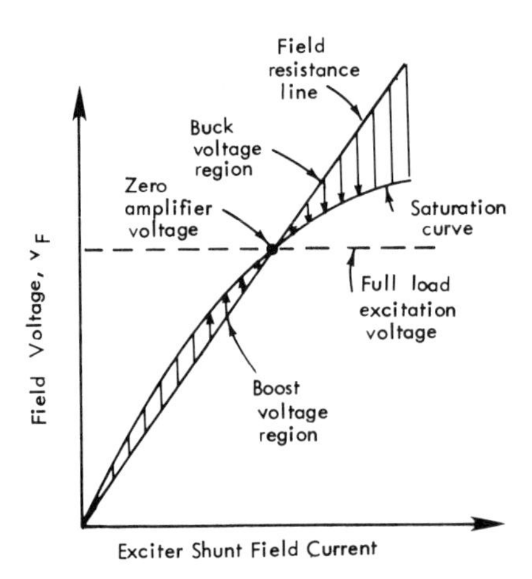

Fig. 7.26 V-I characteristic defining boost and buck regions.

field circuit can be controlled either manually by energizing a relay whose contacts bypass the rotating amplifier or automatically, with the amplifier providing a feedback of the error voltage to increase or decrease the field current.

The control characteristic may be better understood by examining Figure 7.26. The field rheostat is set to intersect the saturation curve at a point corresponding to rated terminal voltage, i.e., the exciter voltage required to hold the generated voltage at rated value with full load. Under this condition the rotating amplifier voltage is zero.

Now suppose the generator load is reduced and the generator terminal voltage begins to rise. The voltage sensing circuit (described later) detects this rise and causes the rotating amplifier to reduce the field current in the exciter field. This reduces the exciter voltage, which in turn reduces i_F, the generator field current. Thus the shaded area above the set point in Figure 7.26 is called the *buck* voltage region. A similar reasoning defines the area below the set point to be the *boost* voltage region.

Rotating amplifier systems have a moderate response ratio, often quoted as about 0.5 (e.g., see Appendix D). The speed of response is due largely to the main exciter time constant, which is much greater than the amplidyne time constant. The ceiling voltage is an important factor too, exciters with higher ceilings having much faster response than exciters of similar design but with lower ceiling voltage (see [17] for a discussion of this topic). The voltage rating of the rotating amplifier in systems of this type is often comparable to the main exciter voltage rating, and the voltage swings of the amplifier change rapidly in attempting to regulate the system [24].

7.5.4 Magnetic amplifier regulators

Another regulator-amplifier scheme capable of zero deadband proportional control is the magnetic amplifier system [6, 30, 31]. (We use the generic term "magnetic amplifier" although those accustomed to equipment of a particular manufacturer use trade names, e.g., Magamp of the Westinghouse Electric Corporation and Amplistat of the General Electric Company.) In this system a magnetic amplifier, i.e., a static amplifying device [32, 33], replaces the rotating amplifier. Usually, the magnetic amplifier consists of a saturable core reactor and a rectifier. It is essentially an amplifying device with the advantages of no rotating parts, zero warm-up time, long life, and sturdy construction. It is restricted to low or moderate frequencies, but this is no drawback in power applications.

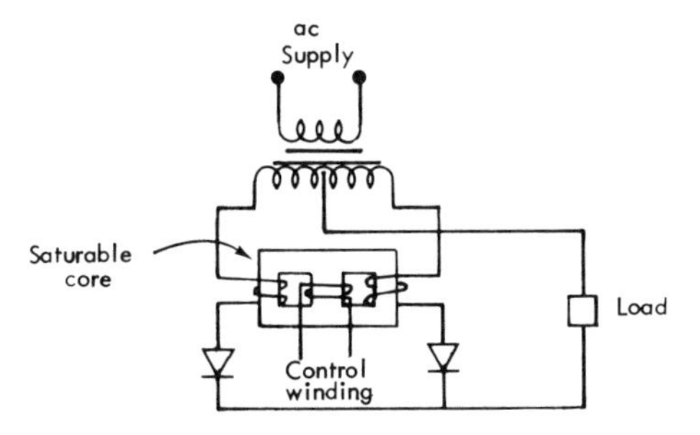

Fig. 7.27 Magnetic amplifier.

Basically, the magnetic amplifier is similar to that shown in Figure 7.27 [33]. The current flowing through the load is basically limited by the very large inductance in the saturable core main windings. As the core becomes saturated, however, the current jumps to a large value limited only by the load resistance. By applying a small (low-power) signal to the control winding, we control the firing point on each voltage (or current) cycle, and hence the average load current. This feature, of controlling a large output current by means of a small control current, is the essence of any amplifier. The fact that this amplifier is very nonlinear is of little concern.

One type of regulator that uses a magnetic amplifier is shown in block diagram form in Figure 7.10 [4]. Here the magnetic amplifier is used to amplify a voltage error signal to a power level satisfactory for supplying the field of a rotating amplifier. The rotating amplifier is located in series with the exciter field in the usual boost-buck connection. One important feature of this system is that the magnetic amplifier is relatively insensitive to variations in line voltage and frequency, making this type of regulator favorable to remote (especially hydro) locations.

Another application of magnetic amplifiers in voltage regulating systems, shown in Figure 7.11 [6], has several features to distinguish it from the previous example. First, the magnetic amplifiers and reference are usually supplied from a 420-Hz system supplied by a permanent-magnet motor-generator set for maximum security and reliability. The power amplifier supplies the main exciter directly in this system. Note, however, that the exciter must have two field windings for boost or buck corrections since magnetic amplifiers are not reversible in polarity. The main exciter also has a self-excited, rheostat-controlled field and can continue to operate with the magnetic amplifiers out of service.

The magnetic amplifier in the system of Figure 7.11 consists of a two-stage push-pull input amplifier that, with 1-mW input signals, can respond to maximum output in three cycles of the 420-Hz supply. The second stage is driven to maximum output when the input stage is at half-maximum, and its transient response is also about three cycles. The figures of merit [34] are about 200/cycle for the input stage and 500/cycle for the output stage. This compares with about 500/s for a conventional pilot exciter. The power amplifier has a figure of merit of 1500/cycle with an overall delay of less than 0.01 s. (The figure of merit of an amplifier has been defined as the ratio of the power amplification to the time constant. It is shown in [34] that for static magnetic amplifiers it is equal to one-half the ratio of power output to stored magnetic energy.)

Reference [6] reviews the operating experience of a magnetic amplifier regulator installation on one 50-MW machine in a plant consisting of seven units totaling over 300 MW, only two units of which are regulated. The experience indicates that, since the magnetic amplifier regulator is so much faster than the primitive rheostatic regulator, it causes the machine on which it is installed to absorb much of the swing in load, particularly reactive load. In fact, close observation of operating oscillograms, when operating with an arc furnace load, reveals that both exciter voltage and line currents undergo rapid fluctuations when regulated but are nearly constant when unregulated. This is to be expected since the regulation of machine terminal voltage to a nearly constant level makes this machine appear to have a lower reactance, hence it absorbs changes faster than its neighbors. In the case under study, the machine terminal voltage was regulated to $\pm 0.25\%$, whereas a $\pm 1\%$ variation was observed with the regulator disconnected [6].

7.5.5 Solid-state regulators

Some of the amplification and comparison functions in modern regulators consist of solid-state active circuits [3]. Various configurations are used depending on the manufacturer, but all have generally fast operation with no appreciable time delay compared to other system time constants. The future will undoubtedly bring more applications of solid-state technology in these systems because of the inherent reliability, ease of maintenance, and low initial cost of these devices.

7.6 Exciter Buildup

Exciter response has been defined as the rate of increase or decrease of exciter voltage when a change is demanded (see Appendix E, Def. 3.15). Usually we interpret this demand to be the greatest possible control effort, such as the complete shorting of the field resistance. Since the exciter response ratio is defined in terms of an unloaded exciter (Def. 3.19), we compute the response under no-load conditions. This serves to satisfy the terms of the response ratio definition and also simplifies the computation or test procedure.

The best way to determine the exciter response is by actual test where this is possible. The exciter is operated at rated speed (assuming it is a rotating machine) and with no load. Then a step change in a reference variable is made, driving the exciter voltage to ceiling while the voltage is recorded as a function of time. This is called a "buildup curve." In a similar way, a "build-down" curve can also be recorded. Curves thus recorded do not differ a great deal from those obtained under loaded conditions. If it is impractical to stage a test on the exciter, the voltage buildup must be computed. We now turn our attention to this problem.

7.6.1 The dc generator exciter

In dealing with conventional dc exciters three configurations (i.e., separately excited, self-excited, and boost-buck) are of interest. They must be analyzed independently, however, because the equations describing them are different. (Portions of this analysis parallel that of Kimbark [16], Rudenberg [20], and Dahl [35] to which the reader is referred for additional study.)

Consider the separately excited exciter shown in Figure 7.28. Summing voltage drops around the pilot exciter terminal connection, we have

$$\dot{\lambda}_E + Ri = v_p \tag{7.10}$$

where λ_E = flux linkages of the main exciter field, Wb turns

R = main exciter field resistance, Ω

i = current, A

v_p = pilot exciter voltage, V

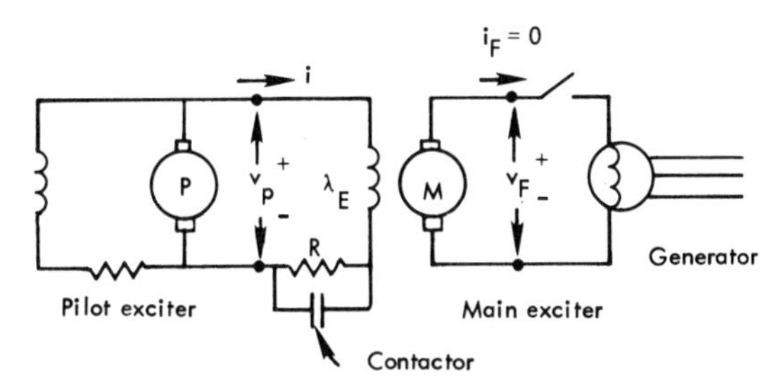

Fig. 7.28 Separately excited exciter.

It is helpful to think in terms of the field flux ϕ_E rather than the field flux linkages. If we assume the field flux links N turns, we have

$$N\dot{\phi}_E + Ri = v_p \tag{7.11}$$

The voltage of the pilot exciter v_p may be treated as a constant [16]. Thus we have an equation in terms of i and ϕ_E with all other terms constant. The problem is that i depends on the exact location of the operating point on the saturation curve and is not linearly related to v_F. Furthermore, the flux ϕ_E has two components, leakage flux and armature flux, with relative magnitudes also depending on saturation. Therefore, (7.11) is nonlinear.

Since magnetization curves are plotted in terms of v_F versus i, we replace ϕ_E in (7.11) by a term involving the voltage ordinate v_F. Assuming the main exciter to be running at constant speed, its voltage v_F is proportional to the air gap flux ϕ_a; i.e.,

$$v_F = k\phi_a \tag{7.12}$$

The problem is to determine how ϕ_a compares with ϕ_E. The field flux has two components, as shown in Figure 7.29. The leakage component, comprising 10–20% of the total, traverses a high-reluctance path through the air space between poles. It does not link all N turns of the pole on the average and is usually treated either as proportional to ϕ_a or proportional to i. Let us assume that ϕ_ℓ is proportional to ϕ_a (see [16] for a more detailed discussion), then

$$\phi_\ell = C\phi_a \tag{7.13}$$

where C is a constant. Also, since

$$\phi_E = \phi_a + \phi_\ell \tag{7.14}$$

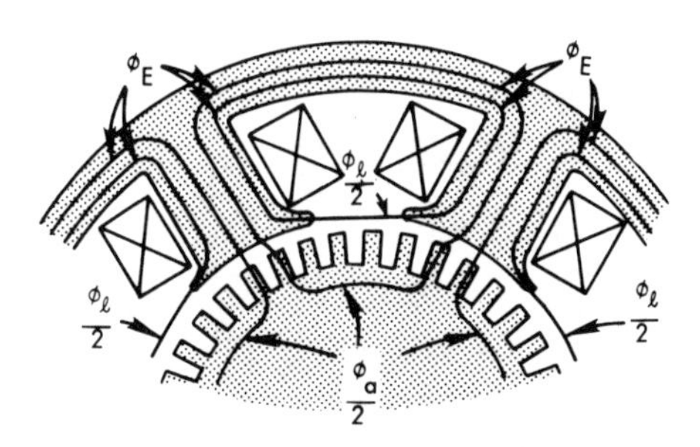

Fig. 7.29 Armature of air gap flux ϕ_a, leakage flux ϕ_ℓ, and field flux $\phi_E = \phi_a + \phi_\ell$. (Reprinted by permission from *Power System Stability,* vol. 3, by E. W. Kimbark. © Wiley, 1956.)

we have

$$\phi_E = (1 + C)\phi_a = \sigma \phi_a \tag{7.15}$$

where σ is called the coefficient of dispersion and takes on values of about 1.1 to 1.2. Substituting (7.15) into (7.11),

$$\tau_E \dot{v}_F + Ri = v_p \tag{7.16}$$

where $\tau_E = (N\sigma/k)$ s, and where we usually assume σ to be a constant. This equation is still nonlinear, however, as v_F is not a linear function of i. We usually assume v_p to be a constant.

In a similar way we may develop the differential equation for the self-excited exciter shown in Figure 7.30, where we have $\dot{\lambda}_E + Ri = v_F$ or

$$N\dot{\phi}_E + Ri = v_F \tag{7.17}$$

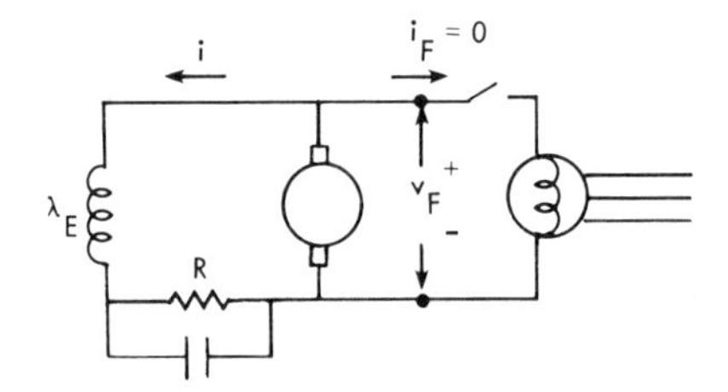

Fig. 7.30 Self-excited exciter.

Following the same logic regarding the fluxes as before, we may write the nonlinear equation

$$\tau_E \dot{v}_F + Ri = v_F \tag{7.18}$$

for the self-excited case where τ_E is the same as in (7.16).

In a similar way we establish the equation for the self-excited exciter with boost-buck rotating amplification as shown in Figure 7.31. Writing the voltage equation with the usual assumptions,

$$\tau_E \dot{v}_F + Ri = v_F + v_R \tag{7.19}$$

Kimbark [16] suggests four methods of solution for (7.16)–(7.19). These are (1) formal integration, (2) graphical integration (area summation), (3) step-by-step integration (manual), and (4) analog or digital computer solution. Formal integration requires that the relationship between v_F and i, usually expressed graphically by means of the magnetization curve, be known explicitly. An empirical relation, the Frohlich equation [35]

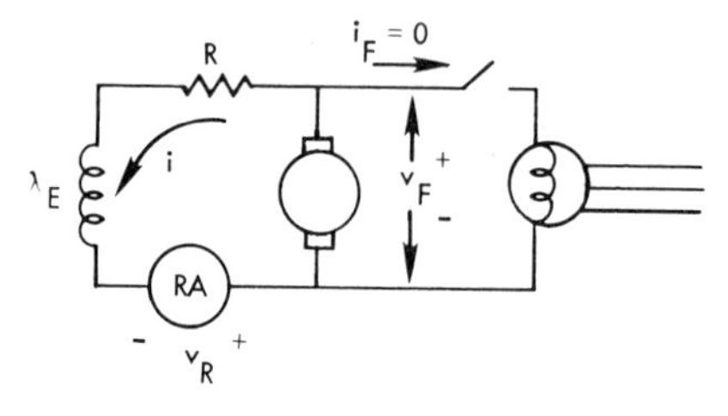

Fig. 7.31 Self-excited exciter with a rotating amplifier (boost-buck).

$$v_F = ai/(b + i) \tag{7.20}$$

may be used, or the so-called modified Frohlich equation

$$v_F = ai/(b + i) + ci \tag{7.21}$$

can be tried. In either case the constants a, b, and c must be found by cut-and-try techniques. If this is reasonably successful, the equations can be integrated by separation of variables.

Method 2, graphical integration, makes use of the saturation curve to integrate the equations. This method, although somewhat cumbersome, is quite instructive. It is unlikely, however, that anyone except the most intensely interested engineer would choose to work many of these problems because of the labor involved. (See Kimbark [16], Rudenberg [20], and Dahl [35] for a discussion of this method.)

Method 3, the step-by-step method (called the point-by-point method by some authors [16, 35]), is a manual method similar to the familiar solution of the swing equation by a stepwise procedure [36]. In this method, the time derivatives are assumed constant over a small interval of time, with the value during the interval being dependent on the value at the middle of the interval.

Method 4 is probably the method of greatest interest because digital and analog computers are readily available, easy to use, and accurate. The actual methods of computation are many but, in general, nonlinear functions can be handled with relative ease and with considerable speed compared to methods 2 and 3.

In this chapter the buildup of a dc generator will be computed by the formal integration method only. However, an analog computer solution and a digital computer technique are outlined in Appendix B.

To use formal integration, a nonlinear equation is necessary to represent the saturation curve. For convenience we shall use the Frohlich equation (7.20), which may be solved for i to write

$$i = bv_F/(a - v_F) \tag{7.22}$$

We illustrate the application of (7.22) by an example.

Example 7.2

A typical saturation curve for a separately excited generator is given in Figure 7.32. Approximate this curve by the Frohlich equation (7.22).

Solution

By examination of Figure 7.32 we make the several voltage and current observations given in Table 7.3.

Table 7.3. Exciter Generated Voltages and Field Currents

i	A	0	1	2	3	4	5	6	7	8	9	10
v_F	V	0	30	60	90	116	134	147	156	164	172	179

Since there are two unknowns in the Frohlich equation, we select two known points on the saturation curve, substitute into (7.20) or (7.22), and solve for a and b. One experienced in the selection process may be quite successful in obtaining a good match. To illustrate this, we will select two pairs of points and obtain two different solutions.

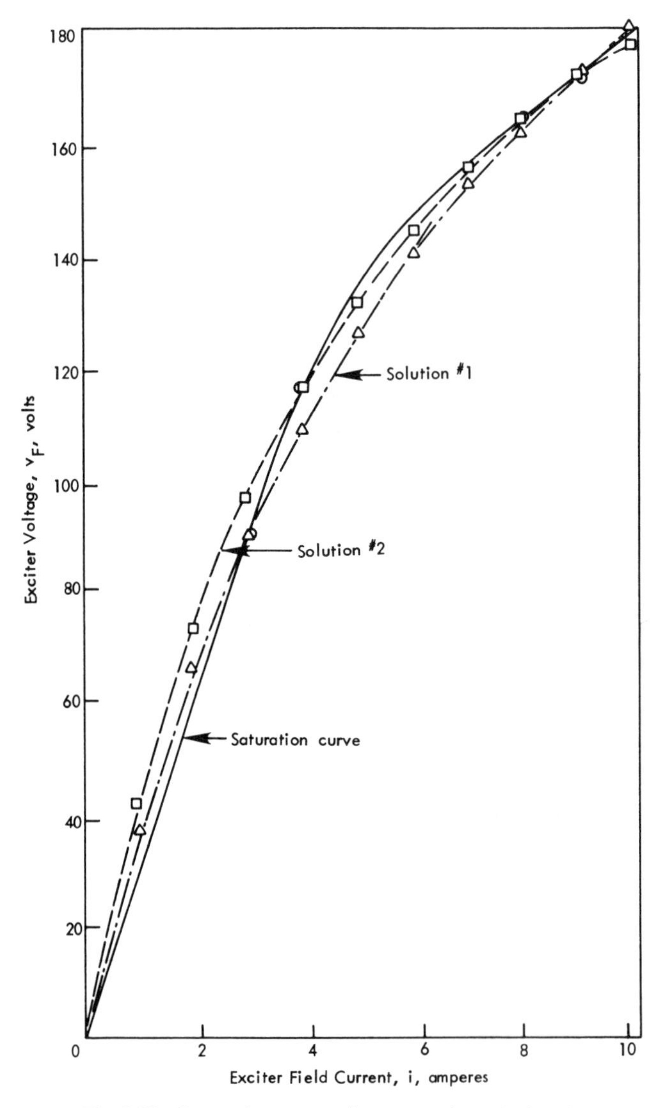

Fig. 7.32 Saturation curve of a separately excited exciter.

	Solution #1	*Solution #2*
Select		
	$i = 3, v_F = 90$	$i = 4, v_F = 116$
	$i = 9, v_F = 172$	$i = 8, v_F = 164$
Then the equations to solve are		
	$90 = 3a/(3 + b)$	$116 = 4a/(4 + b)$
	$172 = 9a/(9 + b)$	$164 = 8a/(8 + b)$
for which the solutions are		
	$a_1 = 315.9$ V	$a_2 = 279.9$ V
	$b_1 = 7.53$ A	$b_2 = 5.65$ A

Both solutions are plotted on Figure 7.32. For solution 1

$$v_F = 315.9i/(7.53 + i) \quad \text{or} \quad i = 7.53\,v_F/(315.9 - v_F) \tag{7.23}$$

and for solution 2

$$v_F = 279.9i/(5.65 + i) \quad \text{or} \quad i = 5.65\,v_F/(279.9 - v_F) \tag{7.24}$$

Example 7.3

Approximate the saturation curve of Figure 7.32 by a modified Frohlich equation. Select values of $i = 2$, 5, and 10.

Solution

$$\begin{aligned} i &= 2 & 60 &= 2a/(2 + b) + 2c \\ i &= 5 & 134 &= 5a/(5 + b) + 5c \\ i &= 10 & 179 &= 10a/(10 + b) + 10c \end{aligned}$$

Solving simultaneously for a, b, and c,

$$a = 359 \qquad b = -21.95 \qquad c = 48.0$$

This gives us the modified formula

$$v_F = 359i/(i - 21.95) + 48i \tag{7.25}$$

Equation (7.25) is not plotted on Figure 7.32 but is a better fit than either of the other two solutions.

Separately excited buildup by integration. For simplicity, let the saturation curve be represented by the Frohlich equation (7.22). Then, substituting for the current in (7.16),

$$\tau_E \dot{v}_F + bR\,v_F/(a - v_F) = v_p \tag{7.26}$$

This equation may be solved by separation of variables. Rearranging algebraically, we write

$$dt = [\tau_E(a - v_F)/(av_p - hv_F)]\,dv_F \tag{7.27}$$

where we have defined for convenience, $h = v_p + bR$. Integrating (7.27),

$$(t - t_0)/\tau_E = (1/h)(v_F - v_{F0}) - (abR/h^2)\ln[(av_p - hv_F)/(av_p - hv_{F0})] \tag{7.28}$$

This equation cannot be solved explicitly for v_F, so we leave it in this form.

Example 7.4

Using the result of formal integration for the separately excited case (7.28), compute the v_F versus t relationship for values of t from 0 to 1 s and find the voltage response ratio by graphical integration of the area under the curve. Assume that the following constants apply and that the saturation curve is the one found in Example 7.2, solution 2.

$$\begin{aligned} N &= 2500 \text{ turns} & v_p &= 125 \text{ V} & R &= 34\ \Omega \\ \sigma &= 1.2 & k &= 12{,}000 & v_{F0} &= 90 \text{ V} \end{aligned}$$

Solution

First we compute the various constants involved. From (7.16)

$$\tau_E = N\sigma/k = (2500)(1.2)/12{,}000 = 0.25 \text{ s}$$

Also, from Example 7.2

$$a = 279.9 \simeq 280 \qquad b = 5.65$$

Now, from the given data, the initial voltage v_{F0} is 90 V. Then from the Frohlich equation (7.22) we compute

$$i_0 = 5.65(90)/(280 - 90) = 2.675 \text{ A}$$

This means that there is initially a total resistance of

$$R_T = 125/2.675 = 46.7 \ \Omega$$

of which all but 34 Ω is in the field rheostat. Assume that we completely short out the field rheostat, changing the resistance from 46.7 to 34 Ω at $t = 0$.

Since v_p is 125 V, we compute the final values of the system variables. From the field circuit,

$$i_\infty = v_p/R = 125/34 = 3.675 \text{ A}$$

Then, from the Frohlich equation the ceiling voltage is

$$v_{F\infty} = a i_\infty/(b + i_\infty) = 280(3.675)/(5.65 + 3.675) = 110.3 \text{ V}$$

Using the above constants we compute the v_F versus t relationship shown in Table 7.4 and illustrated in Figure 7.33.

Table 7.4. Buildup of Separately Excited v_F for Example 7.4

t	v_F	t	v_F
0.00	90.00	0.45	109.55
0.05	95.85	0.50	109.79
0.10	100.12	0.55	109.96
0.15	103.18	0.60	110.08
0.20	105.35	0.65	110.16
0.25	106.87	0.70	110.21
0.30	107.94	0.75	110.25
0.35	108.68	0.80	110.28
0.40	109.19	0.85	110.30

From Figure 7.33, by graphical construction we find the triangle *acd*, which has the same area as that under the v_F curve *abd*. Then from (7.5) with $cd = 27.9$ V, as shown in the figure, the response ratio $= 27.9/90(0.5) = 0.62$.

Self-excited buildup by integration. For a self-excited machine whose saturation curve is represented by the Frohlich approximation (7.22), we have

$$\tau_E \dot{v}_F + bR\, v_F/(a - v_F) = v_F \tag{7.29}$$

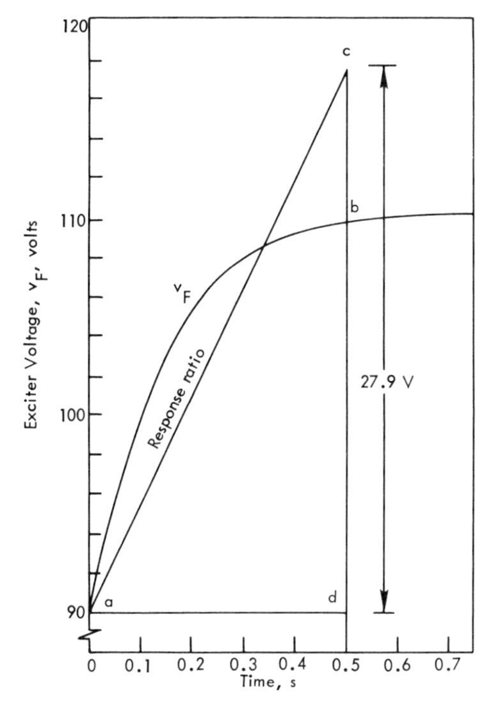

Fig. 7.33 Buildup of the separately excited exciter for Example 7.4.

This is recognized to be identical to the previous case except that the term on the right side is v_F instead of v_p. Again we rearrange the equation to separate the variables as

$$dt = \frac{\tau_E(a - v_F)dv_F}{(a - bR)v_F - v_F^2} \tag{7.30}$$

This equation can be integrated from t_0 to t with the result

$$\frac{t - t_0}{\tau_E} = \frac{a}{K}\ln\frac{v_F}{v_{F0}} - \frac{bR}{K}\ln\frac{v_F - K}{v_{F0} - K} \tag{7.31}$$

where $K = a - bR$.

Example 7.5

Compute the self-excited buildup for the same exciter studied in Example 7.4. Change the final resistance (field resistance) so that the self-excited machine will achieve the same ceiling voltage as the separately excited machine. Compare the two buildup curves by plotting the results on the same graph and by comparing the computed response ratios.

Solution

The ceiling voltage is to be 110.3 V, at which point the current in the field is 3.68 A (from the Frohlich equations). Then the resistance must be $R = 110.3/3.68 = 30\,\Omega$. Solving (7.31) with this value of R and using Frohlich parameters from Example 7.4, we have the results in Table 7.5 and the solution curve of Figure 7.34. The response ratio $= 15.4/90(0.5) = 0.342$ for the self-excited case.

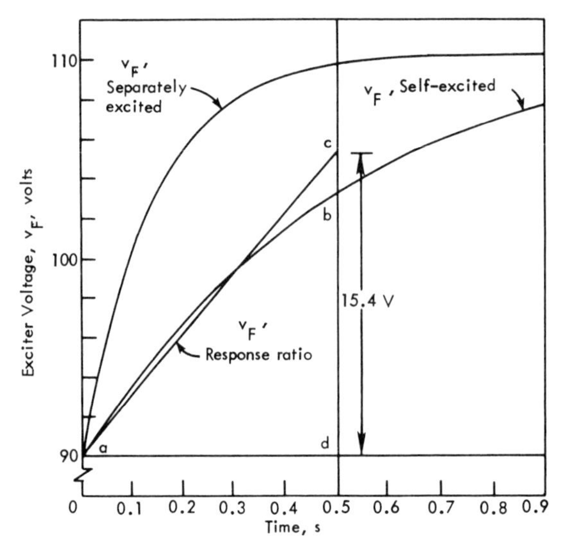

Fig. 7.34 Buildup of the self-excited exciter for Example 7.5.

Table 7.5. Buildup of Self-excited v_F for Example 7.5

t	v_F	t	v_F
0.00	90.00	0.50	103.38
0.05	91.87	0.55	104.15
0.10	93.61	0.60	104.85
0.15	95.23	0.65	105.47
0.20	96.73	0.70	106.03
0.25	98.10	0.75	106.52
0.30	99.37	0.80	106.96
0.35	100.52	0.85	107.36
0.40	101.57	0.90	107.71
0.45	102.52	...	...

Boost-buck buildup by integration. The equation for the boost-buck case is the same as the self-excited case except the amplifier voltage is added to the right side, or

$$\tau_E \dot{v}_F + bRv_F/(a - v_F) = v_F + v_R \tag{7.32}$$

Rearranging, we may separate variables to write

$$dt = \tau_E(a - v_F)dv_F/(A + Mv_F - v_F^2) \tag{7.33}$$

where $A = av_R$ and $M = a - v_R - bR$.

Integrating (7.33), we compute

$$\frac{t - t_0}{\tau_E} = \frac{2a - M}{Q} \ln \frac{(M - Q - 2v_F)(M + Q - 2v_{F0})}{(M - Q - 2v_{F0})(M + Q - 2v_F)} + \frac{1}{2} \ln \frac{(A + Mv_F - v_F^2)}{(A + Mv_{F0} - v_{F0}^2)} \tag{7.34}$$

where $Q = \sqrt{4A + M^2}$.

Example 7.6

Compute the boost-buck buildup for the exciter of Example 7.4 where the amplifier voltage is assumed to be a step function at $t = t_0$ with a magnitude of 50 V. Compare with previous results by adjusting the resistance until the ceiling voltage is again 110.3 V. Repeat for an amplifier voltage of 100 V.

Solution

With a ceiling voltage of 110.3 V and an amplifier voltage of 50 V, we compute with $\dot{v}_F = 0$, $Ri_\infty = v_F + v_R = 160.3$. This equation applies as long as v_R maintains its value of 50 V. This requires that i_∞ again be equal to 3.68 A so that R may be computed as $R = 160.3/3.68 = 43.6\ \Omega$. This value of R will insure that the ceiling voltage will again be 110.3 V. Using this R in (7.34) results in the tabulated values given in Table 7.6. Repeating with $v_R = 100$ V gives a second set of data, also tabulated, in which $R = 57.2\ \Omega$.

Table 7.6. Buildup of Boost-Buck v_F for Example 7.6

t	v_F for $v_R = 50$	v_F for $v_R = 100$
0.00	90.00	90.00
0.05	94.23	96.32
0.10	97.70	100.84
0.15	100.50	103.98
0.20	102.72	106.12
0.25	104.47	107.56
0.30	105.84	108.51
0.35	106.90	109.14
0.40	107.72	109.56
0.45	108.34	109.83
0.50	108.82	110.00
0.55	109.19	110.12
0.60	109.47	110.20
0.65	109.68	110.24
0.70	109.84	110.27
0.75	109.96	110.30
0.80	110.05	110.31
0.85	110.12	110.32
0.90	110.17	110.33

These results are plotted in Figure 7.35. Note that increasing the amplifier voltage has the effect of increasing the response ratio. In this case changing v_R from 50 to 100 V gives a result that closely resembles the separately excited case. In each case the response ratio (RR) may be calculated as follows:

$$\mathrm{RR}\ (v_R = 50) = 2cd/0a = 2(24.15)/90 = 0.537$$

$$\mathrm{RR}\ (v_R = 100) = 2c'd/0a = 2(29)/90 = 0.645$$

7.6.2 Linear approximations for dc generator exciters

Since the Frohlich approximation fails to provide a simple v_F versus t relationship, other possibilities may be worth investigating. One method that looks attractive because of its simplicity is to assume a *linear* magnetization curve as shown in Figure 7.36, where

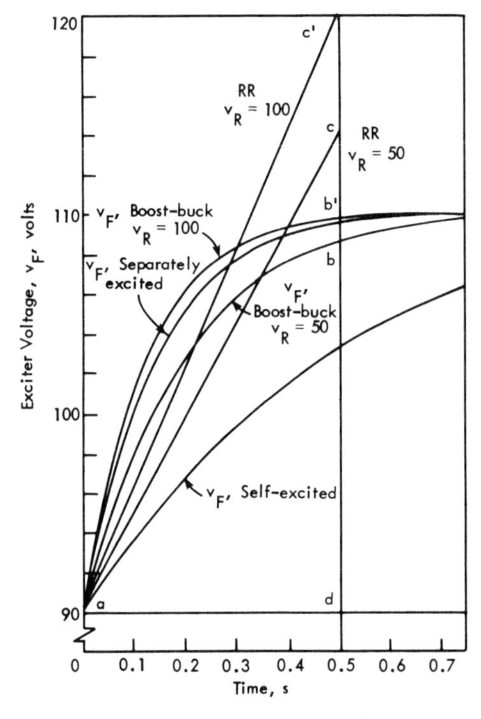

Fig. 7.35 Buildup of boost-buck exciters for Example 7.6.

$$v_F = mi + n \tag{7.35}$$

Substituting (7.35) into the excitation equation we have the linear ordinary differential equation

$$\tau_E \dot{v}_F = v - (R/m)(v_F - n) \tag{7.36}$$

where v = v_p separately excited
= v_F self-excited
= $v_F + v_R$ boost-buck excited

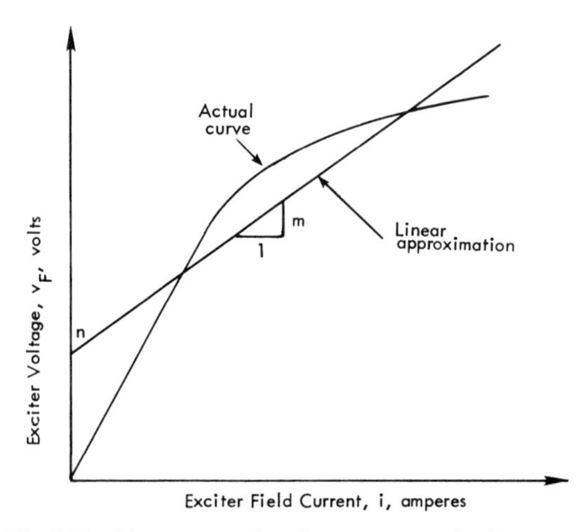

Fig. 7.36 Linear approximation to a magnetization curve.

This equation may be solved by conventional techniques. The question of interest is, What values of m and n, if any, will give solutions close to the actual nonlinear solutions? This can be resolved by solving (7.36) for each case and then systematically trying various values of m and n to find the best "fit." This extremely laborious process becomes much less painful, or even fun, if the comparison is made by analog computer. In this process, both the linear and nonlinear problems are solved simultaneously and the solutions compared on an oscilloscope. A simple manipulation of two potentiometers, one controlling the slope and one controlling the intercept, will quickly and easily permit an optimum choice of these parameters. The procedure will be illustrated for the separately excited case.

Linear approximation of the separately excited case. In the separately excited case we set $v = v_p$ so that (7.36) becomes $\dot{v}_F = k_1 - k_2 v_F$ where

$$k_1 = (1/\tau_E)(v_p + nR/m) \qquad k_2 = R/\tau_E m \tag{7.37}$$

Solution of (7.37) gives

$$v_F(t) = (k_1/k_2)(1 - e^{-k_2 t})u(t) + v_{F0}e^{-k_2 t}u(t) \tag{7.38}$$

Equation (7.38) is solved by the analog computer connection shown in Figure 7.37 and compared with the solution of (7.76) given in Appendix B, shown in Figure B.9.

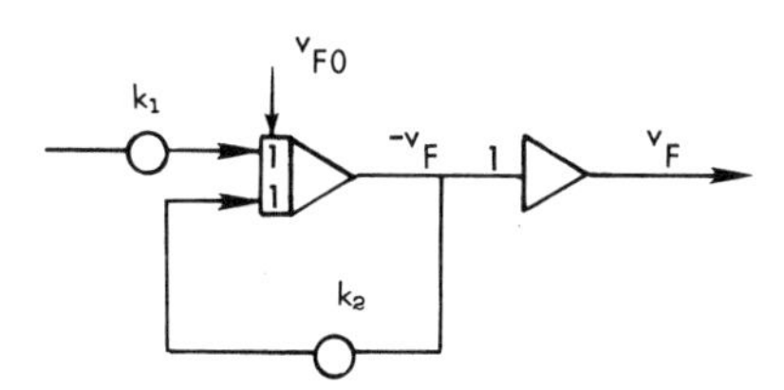

Fig. 7.37 Solution of the linear equation.

Adjusting potentiometers k_1 and k_2 quickly provides the "best fit" solution shown in Figure 7.38, which is a graph made directly by the computer. Having adjusted k_1 and k_2 for the best fit, the potentiometer settings are read and the factors m and n computed. In a similar way linear approximations can be found for the self-excited and boost-buck connections.

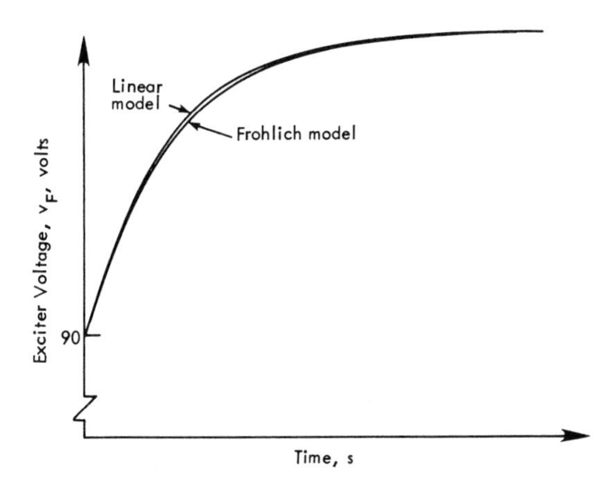

Fig. 7.38 Analog computer comparison of linear and Frohlich models of the separately excited buildup.

7.6.3 The ac generator exciters

As we observed in Chapter 4, there is no simple relationship between the terminal voltage and the field voltage of a synchronous generator. Including all the detail of Chapter 4 in the analysis of the exciter would be extremely tedious and would not be warranted in most cases. We therefore seek a reasonable approximation for the ac exciter voltage, taking into account the major time constants and ignoring other effects.

Kimbark [16] has observed that the current in the dc field winding changes much more slowly than the corresponding change in the ac stator winding. Therefore, since the terminal voltage is proportional to i_F (neglecting saturation), the ac exciter voltage will change approximately as fast as its field current changes. The rate of change of field current depends a great deal on the external impedance of the stator circuit or on the load impedance. But, using the response ratio definition (see Def. 3.19, Appendix E) we may assume that the ac exciter is open circuited. In this case the field current in the exciter changes according to the "direct-axis transient open circuit time constant" τ'_{d0} where

$$\tau'_{d0} = L_F/r_F \text{ s} \tag{7.39}$$

This will give the most conservative (pessimistic) result since, with a load impedance connected to the stator, the effective inductance seen by the field current is smaller and the time constant is smaller.

Using relation (7.39) we write, in the Laplace domain,

$$v_F(s) = [K/(1 + \tau'_{d0}s)]\, v_R(s) \tag{7.40}$$

where $v_F(s)$ is the Laplace transform of the open circuit field voltage and $v_R(s)$ is the transform of the regulator voltage. If the regulator output experiences a step change of magnitude D at $t = t_0$, the field voltage may be computed from (7.40) to be

$$v_F = v_{F0} + KD(1 - e^{-(t-t_0)/\tau'_{d0}})\, u(t - t_0) \tag{7.41}$$

This linearized result does not include saturation or other nonlinearities, but does include the major time delay in the system. An ac exciter designed for operation at a few hundred Hz could have a very respectable τ'_{d0}, much lower than that of the large 60-Hz generator that is being controlled.

7.6.4 Solid-state exciters

Modern solid-state exciters, such as the SCR exciter of Figure 7.14, can go to ceiling without any appreciable delay. In systems of this type a small delay may be required for the amplifiers and other circuits involved. The field voltage may then be assumed to depend only on this delay.

One way to solve this system is to assume that v_F changes linearly to ceiling in a given time delay of t_d s, where t_d may be very small. This is nearly the same as permitting a step change in v_F. For such fast systems the time constants are so much smaller than others involved in the system that assuming a step change in v_F should be fairly accurate.

7.6.5 Buildup of a loaded dc exciter

Up to this point we have considered the response characteristics of unloaded exciters, i.e., with $i_F = 0$. If the exciter is loaded, the load current will affect the terminal voltage of the exciter v_F by an amount depending upon the internal impedance of the exciter. In modern solid-state circuits this effect will usually be small, amounting to essentially a small series $i_F R$ drop. In rotating dc machines the effect is greater, since in addition to the $i_F R$ drop there is also the brush drop, the drop due to armature reaction, and the drop due to armature inductance. (Dahl [35] provides an exhaustive treatment of this subject and Kimbark [16] also has an excellent analysis.)

We can analyze the effect of load current in a dc machine as follows. First, we recognize that the armature inductance is small, and at the relatively slow rate of buildup to be experienced this voltage drop is negligible. Furthermore, if the machine has interpoles, we may neglect demagnetizing armature reaction. However, we do have to estimate the effect of cross-magnetizing armature reaction, which causes a net decrease in the air gap flux. Thus, the net effect of load is in the resistance drop (including brush drop) and in the decrease in flux due to cross-magnetizing armature reaction.

To facilitate analysis, we assume the load current i_F has a constant value. This means the $i_F R$ drop is constant, and the armature reaction effect depends on the value of current in the field, designated i in our notation. The combined effect is determined most easily by test, a typical result of which is shown in Figure 7.39. To the load

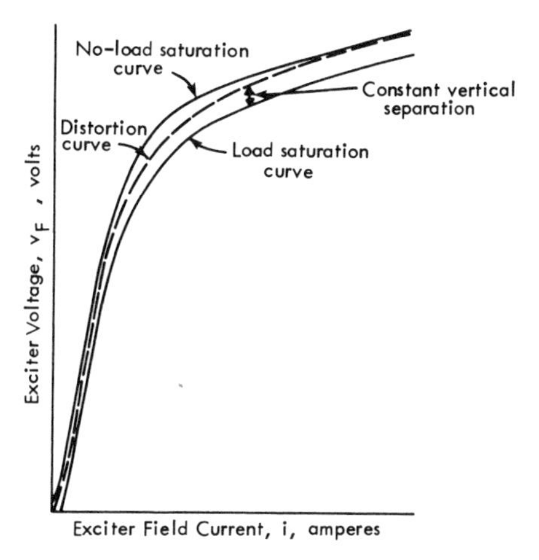

Fig. 7.39 No-load and load saturation curves. (Reprinted by permission from *Power System Stability*, vol. 3, by E. W. Kimbark. © Wiley, 1956.)

saturation curve is added the resistance drop to obtain a fictitious curve designated "distortion curve." This curve shows the voltage generated by air gap flux at this value of i_F as a function of i, and it differs from the no-load saturation curve by an amount due to armature reaction. The magnitude of this difference is greatest near the knee of the curve.

Kimbark [16] treats this subject thoroughly and is recommended to the interested reader. We will ignore the loading effect in our analysis in the interest of finding a reasonable solution that is a fair representation of the physical device. As in all engineering problems, certain complications must be ignored if the solution is to be manageable.

7.6.6 Normalization of Exciter Equations

The exciter equations in this book are normalized on the basis of rated air gap voltage, i.e., exciter voltage that produces rated no-load terminal voltage with no saturation. This is the pu system designated as C in Figure 7.23. Thus at no load and with no saturation, $E_{FD} = 1.0$ pu corresponds to $V_t = 1.0$ pu.

The slip ring voltage corresponding to 1.0 pu E_{FD} is not the same base voltage as that chosen for the field circuit in normalizing the synchronous machine. From (4.55) we have

$$V_{FB} = V_B I_B / I_{FB} = S_B / I_{FB} \quad \text{V}$$

This base voltage is usually a very large number (163 kV in Example 4.1, for example). The base voltage for E_{FD}, on the other hand, would be on the order of 100 V or so. Simply stated, the exciter base voltage and the synchronous machine base for the field voltage differ, and a change of base between the two quantities is required. The required relationship is given by (4.59), which can be written as

$$E_{FD} = (L_{AD}/\sqrt{3}r_F)\, v_F \ \text{pu} \qquad E_{FD} = (\omega_B k M_F/\sqrt{3}r_F)\, v_F \ \text{V} \tag{7.42}$$

Thus any exciter equation may be divided through by V_{FB} to obtain an equation in v_{Fu} and then multiplied by $L_{AD}/\sqrt{3}r_F$ to convert to an equation in E_{FDu}. For example, for the dc generator exciter we have an equation of the form $\tau_E \dot{v}_F = f(v_F)$ V. Dividing through by V_{FB} we have the pu equation $\tau_E \dot{v}_{Fu} = f(v_{Fu})$. Multiplying by $L_{AD}/\sqrt{3}r_F$, we write the exciter equation $\tau_E \dot{E}_{FDu} = f(E_{FDu})$.

It is necessary, of course, to always maintain the "gain constant" $\sqrt{3}r_F/L_{AD}$ between the exciter E_{FD} output and the v_F input to the synchronous machine. This constant is the change of base needed to connect the pu equations of the two machines.

7.7 Excitation System Response

The response of the exciter alone does not determine the overall excitation system response. As noted in Figure 7.20, the excitation system includes not only the exciter but the voltage regulator as well. The purpose of this section is to compute the response of typical systems, including the voltage regulators. This will give us a feel for the equations that describe these systems and will illustrate the way a mathematical model is constructed.

7.7.1 Noncontinuously regulated systems

Early designs of voltage regulating schemes, many of which are still in service, used an electromechanical means of changing the exciter field rheostat to cause the desired change in excitation. A typical scheme is shown in Figure 7.40, which may be explained as follows. Any given level of terminal voltage will, after rectification, result in a given voltage v_c across the regulating coil and a given coil current i_c. This current flowing in the regulating coil exerts a pull on the plunger that works against the spring K and dashpot B. Thus, depending on the reference screw setting, the arm attached to the plunger will find a new position x for each voltage V_t. High values of V_t will increase the coil voltage v_c and pull the arm to the right, reducing x, etc. Note that the reference is the mechanical setting of the reference screw.

Now imagine a gradual increase in V_t that pulls the arm slowly to the right, reducing x until the lower contact L is made. This causes current to flow in the coil L, closing the rheostat motor contact and moving the rheostat in the direction to increase R_H. This, as we have seen, will reduce V_t. Note that there is no corrective action at all until a contact is closed. This constitutes an intentional dead zone in which no control action is taken. Once control action is begun, the rheostat setting will change at an assumed constant rate until the maximum or minimum setting is reached.

Mathematically, we can describe this action as follows. From (7.16) we have, for the separately excited arrangement,

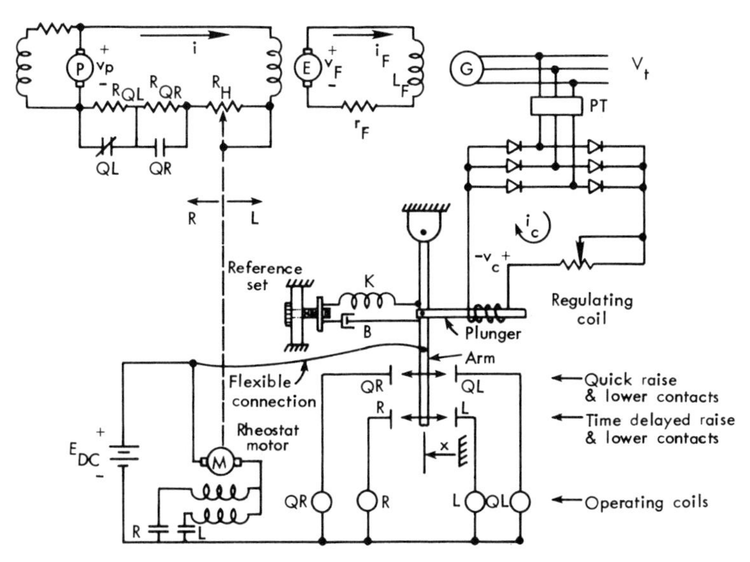

Fig. 7.40 A noncontinuous regulator for a separately excited system. The scheme illustrated is a simplified sketch similar to the Westinghouse type BJ system [2].

$$\tau_E \dot{v}_F = v_p - Ri \tag{7.43}$$

and in this case the regulating is accomplished by a change in R. But R changes as a function of time whenever the arm position x is greater than some threshold value K_x. This condition is shown in Figure 7.41 where the choice of curve depends on the magnitude of x being greater than the dead zone $\pm K_x$. Note that any change in x from the equilibrium position is a measure of the error in the terminal voltage magnitude. This control action is designated the "raise-lower mode" of operation. It results in a slow excitation change, responding to a change in V_t large enough to exceed the threshold K_x, where the rheostat motor steadily changes the rheostat setting. A block diagram of this control action is shown in Figure 7.42.

The balanced beam responds to an accelerating force

$$F_a = K(x_0 + \ell) - F_c = M\ddot{x} + B\dot{x} + Kx \tag{7.44}$$

where x_0 is the reference position; ℓ is the unstressed length of the spring; F_c is the plunger force; and M, B, and K are the mass, damping, and spring constants respectively. If the beam mass is negligible, the right side of (7.44) can be simplified.

In operation the beam position x is changed continuously in response to variations

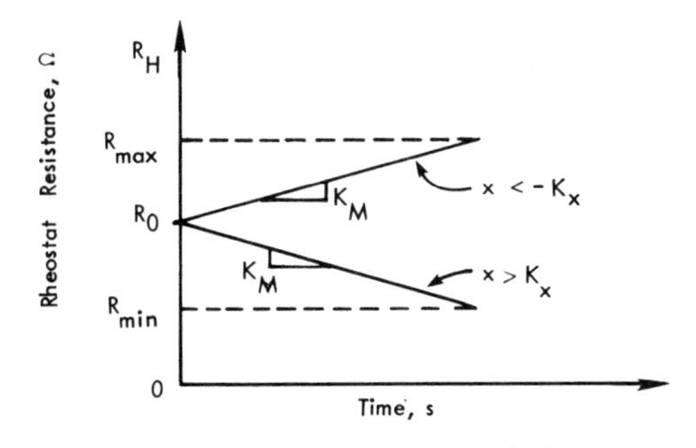

Fig. 7.41 R_H versus t for the condition $|x| > K_x > 0$.

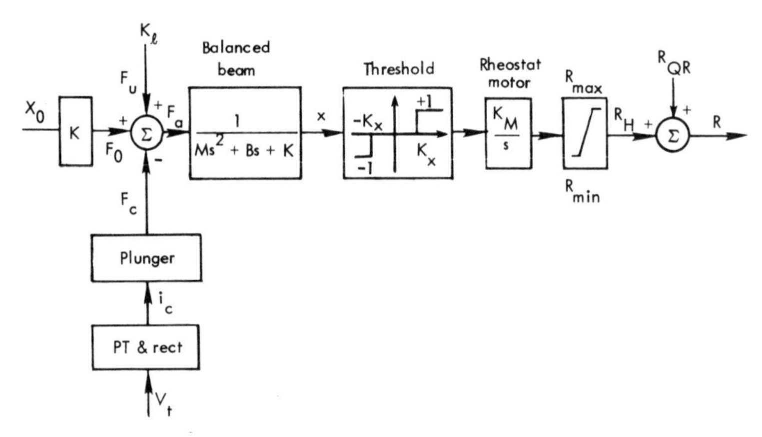

Fig. 7.42 Block diagram of the raise-lower control mode.

in V_t. Any change in V_t large enough to cause $|x| \geqq K_x$ results in the rheostat motor changing the setting of R_H. As the rheostat is reset, the position x returns to the threshold region $|x| < K_x$ and the motor stops, leaving R_H at the value finally reached. At any instant the total resistance R is given by

$$R = R_{QR} + R_H = R_{QR} + R_0 - K_M t \quad \text{(raise)}$$
$$= R_{QR} + R_0 + K_M t \quad \text{(lower)} \qquad (7.45)$$

Thus the exact R depends on the integration time and on the direction of rotation of the rheostat motor. In (7.45) and Figure 7.41, R_0 is the value of R_H retained following the last integration. This value is constrained by the physical size of the rheostat so that for any time t, $R_{\min} < (R_0 \pm K_M t) < R_{\max}$.

The foregoing discussion pertains to the raise-lower mode only. Referring again to Figure 7.40, a second possible mode of operation is recognized. If the x deflection is large enough to make the QL or QR contacts, the fixed field resistors R_{QL} or R_{QR} are switched into or out of the field respectively, initiating a quick response in the exciter. This control scheme is shown in Figure 7.43 as an added quick control mode to the original controller. The quick raise-lower mode is initiated whenever $|x| > K_v$, with the resulting action described by

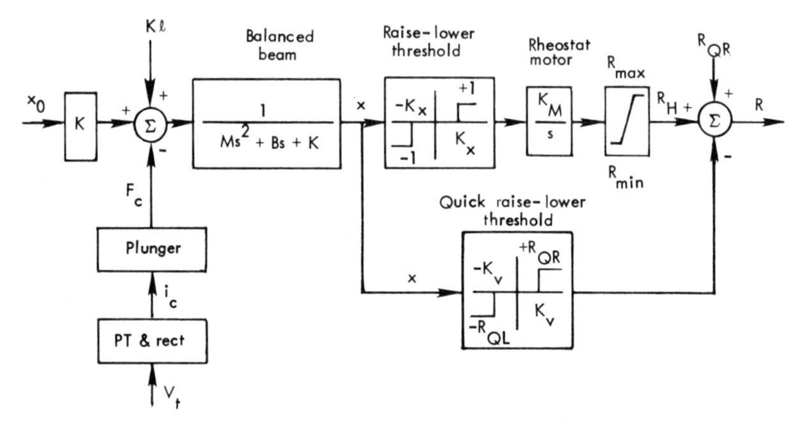

Fig. 7.43 Block diagram of the combined raise-lower and quick-raise-lower control modes.

$$R = R_H \qquad x > K_v \quad \text{(quick raise)}$$
$$= R_H + R_{QL} \qquad x < K_v \quad \text{(quick lower)} \tag{7.46}$$

If we set $K_v > K_x$, this control mode will be initiated only for large changes in V_t and will provide a fast response. Thus, although the raise-lower mode will also be operational when $|x| > K_v$, it will probably not have time to move appreciably before x returns to the deadband.

The controller of Figure 7.43 operates to adjust the total field resistance R to the desired value. Mathematically, we can describe the complete control action by combining (7.45)–(7.46). The resulting change in R affects the solution for v_F in the exciter equation (7.43). If saturation is added, a more realistic solution results. Saturation is often treated as shown in Figure 7.44, where we define the saturation function

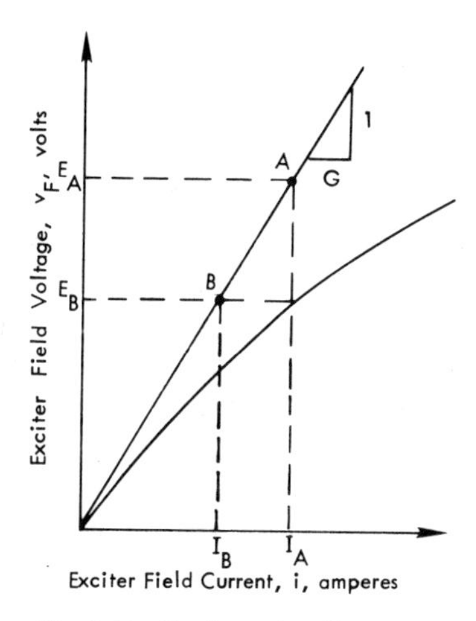

Fig. 7.44 Exciter saturation curve.

$$S_E = (I_A - I_B)/I_B \tag{7.47}$$

Then we can show that

$$I_A = (1 + S_E)I_B \qquad E_A = (1 + S_E)E_B \tag{7.48}$$

The function S_E is nonlinear and can be approximated by any convenient nonlinear function throughout the operating range (See Appendix D). If the air gap line has slope $1/G$, we can write the total (saturated) current as

$$i = Gv_F(1 + S_E) = Gv_F + Gv_FS_E \tag{7.49}$$

Substituting (7.49) into (7.43) the exciter equation is

$$\tau_E\dot{v}_F = v_p - Ri = v_p - RGv_F - RGv_FS_E \tag{7.50}$$

A block diagram for use in computer simulation of this equation is shown in Figure 7.45, where the exciter voltage is converted to the normalized exciter voltage E_{FD}. The complete excitation system is the combination of Figures 7.43 and 7.45.

7.7.2 Continuously regulated systems

Usually it is preferable for a control system to be a continuously acting, proportional system, i.e., the control signal is always present and exerts an effort proportional

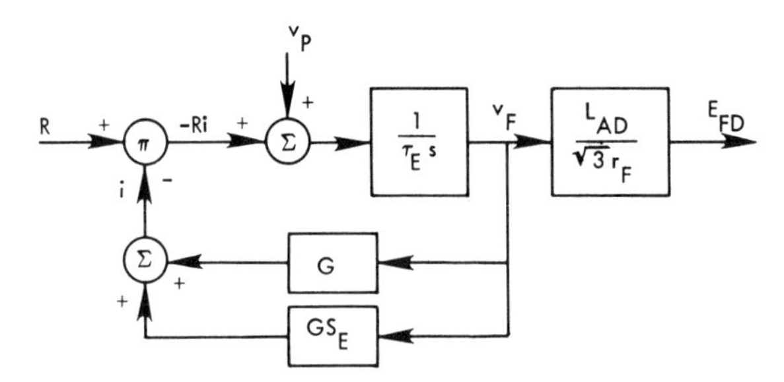

Fig. 7.45 Exciter block diagram.

to the system error (see Def. 2.12.1). Most of the excitation control systems in use today are of this type. Here we shall analyze one system, the familiar boost-buck system, since it is typical of this kind of excitation system.

Consider the system shown in Figure 7.10 where the feedback signal is applied to the rotating amplifier in the exciter field circuit. Reduced to its fundamental components, this is shown in Figure 7.46. We analyze each block separately.

Potential transformer and rectifier. One possible connection for this block is that shown in Figure 7.47, where the potential transformer secondaries are connected to bridged rectifiers connected in series. Thus the output voltage V_{dc} is proportional to the sum or average of the rms values of the three phase voltages. If we let the average rms voltage be represented by the symbol V_t, we may write

$$V_{dc} = K_R V_t/(1 + \tau_R s) \tag{7.51}$$

where K_R is a proportionality constant and τ_R is the time constant due to the filtering or first-order smoothing in the transformer-rectifier assembly. The actual delay in this system is small, and we may assume that $0 < \tau_R < 0.06$ s.

Voltage regulator and reference (comparator). The second block compares the voltage V_{dc} against a fixed reference and supplies an output voltage V_e, called the error voltage, which is proportional to the difference; i.e.,

$$V_e = k(V_{\text{REF}} - V_{dc}) \tag{7.52}$$

This can be accomplished in several ways. One way is to provide an electronic difference amplifier as shown in Figure 7.48, where the time constant of the electronic amplifier is usually negligible compared to other time delays in the system. There is often an objection, however, to using active circuits containing vacuum tubes, transistors, and the associated electronic power supplies because of reliability and the need for replace-

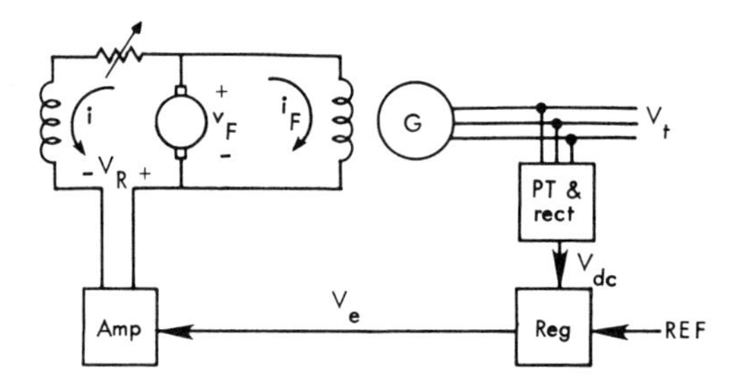

Fig. 7.46 Simplified diagram of a boost-buck system.

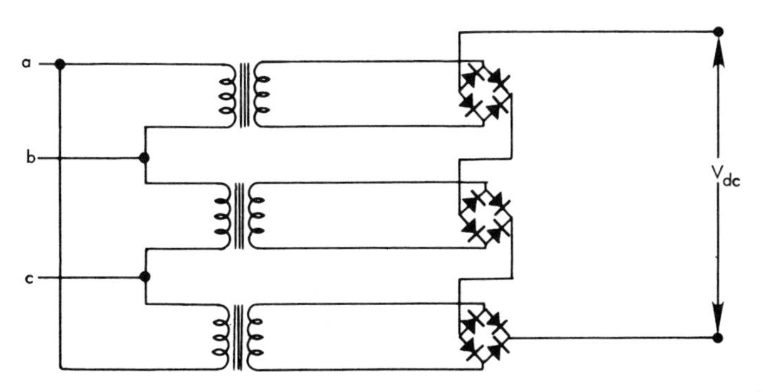

Fig. 7.47 Potential transformer and rectifier connection.

ment of aging components. This difficulty could be overcome by having a spare amplifier with automatic switching upon the detection of faulty operation.

Another solution to the problem is to make the error comparison by an entirely *passive* network such as the nonlinear bridge circuit in Figure 7.49. Here the input current i_{dc} sees parallel paths i_a and i_b or $i_{dc} = i_a + i_b$. But since the output is connected to an amplifier, we assume that the voltage gain is large and that the input current is negligible, or $i_e = 0$. Under this condition the currents i_a and i_b are equal. Then the output voltage V_e is

$$V_e = v_N - v_L \tag{7.53}$$

The operation of the bridge is better understood by examination of Figure 7.50 where the v-i characteristics of each resistance are given and the characteristic for the total resistance $R_L + R_N$ seen by i_a and i_b is also given. Since $i_a = i_b$, the sum of voltage drops v_L and v_N is always equal to V_{dc}, the applied voltage. If we choose the nonlinear elements carefully, the operation in the neighborhood of V_{REF} is essentially linear; i.e., a deviation v_Δ above or below V_{REF} results in a change i_Δ in the total current, which is also displaced equally above and below i_{REF}. Note that the nonlinear resistance shown is quite linear in this critical region. Thus we may write for a voltage deviation v_Δ,

$$v_N = v_x + k_N v_\Delta \qquad v_L = v_x + k_L v_\Delta \tag{7.54}$$

where $k_L > k_N$. Combining (7.54) and (7.53), we compute

$$V_e = -(k_L - k_N)v_\Delta = -k\, v_\Delta \tag{7.55}$$

But for a deviation v_Δ, $V_{dc} = V_{\text{REF}} + v_\Delta$, which may be incorporated into (7.55) to write

$$V_e = k(V_{\text{REF}} - V_{dc}) \tag{7.56}$$

We note that (7.56) has the same block diagram representation as the difference amplifier shown in Figure 7.48(b), where we set $\tau = 0$ for the passive circuit.

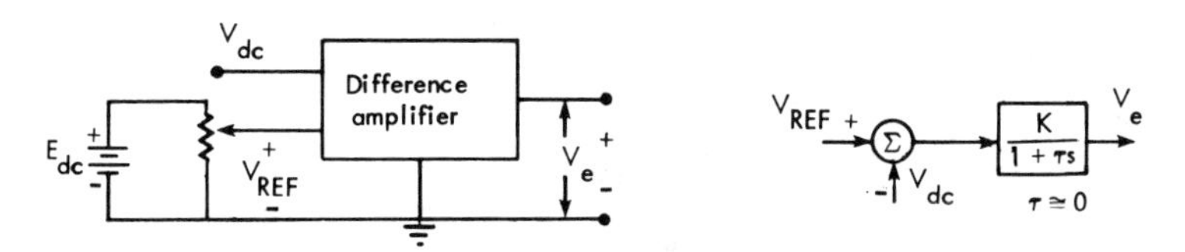

Fig. 7.48 Electronic difference amplifier as a comparator: (a) circuit connection, (b) block diagram.

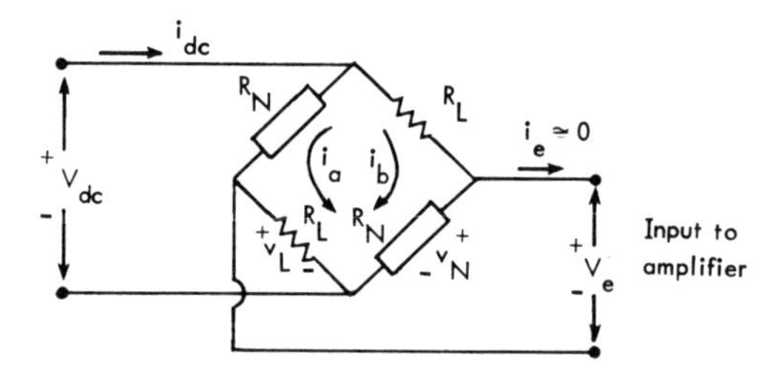

Fig. 7.49 Nonlinear bridge comparison circuit.

A natural question to ask at this point is, What circuit element constitutes the voltage reference? Note that no external reference voltage is applied. A closer study of Figure 7.50 will reveal that the linear resistance R_L is a convenient reference and that two identical gang-operated potentiometers in the bridge circuit would provide a convenient means of setting the reference voltage.

The nonlinear bridge circuit has the obvious advantage of being simple and entirely passive. If nonlinear resistances of appropriate curvature are readily available, this circuit makes an inexpensive comparator that should have long life without component aging.

The amplifier. The amplifier portion of the excitation system may be a rotating amplifier, a magnetic amplifier, or conceivably an electronic amplifier. In any case we will assume linear voltage amplification K_A with time constant τ_A, or

$$V_R = K_A V_e/(1 + \tau_A s) \tag{7.57}$$

As with any amplifier a saturation value must be specified, such as $V_{R\min} < V_R < V_{R\max}$. These conditions are both shown in the block diagram of Figure 7.51.

The exciter. The exciter output voltage is a function of the regulator voltage as derived in (7.48) and with block diagram representation as shown in Figure 7.44. The major difference between that case and this is in the definition of the constant K_E. Since the exciter is a boost-buck system, we can write the normalized equation

$$E_{FD} = (V_R - E_{FD}S_E)/(K_E + \tau_E s) \tag{7.58}$$

where

$$K_E = RG - 1 \tag{7.59}$$

The generator. The generator voltage response to a change in v_F was examined in

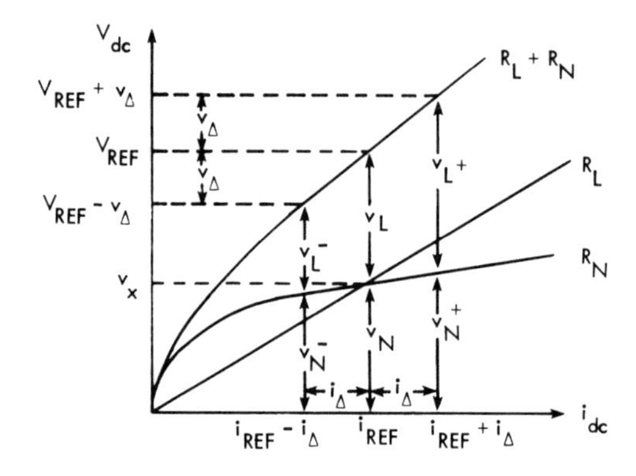

Fig. 7.50 The v versus i characteristics for the nonlinear bridge.

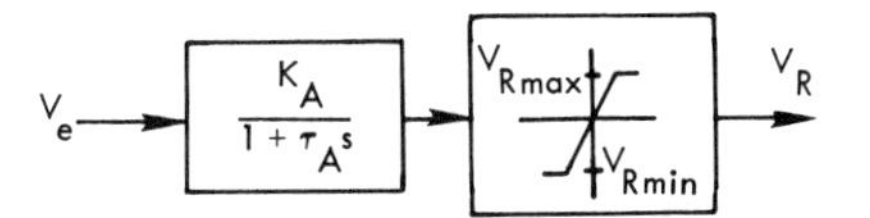

Fig. 7.51 Block diagram of the regulator amplifier.

Chapter 5. Looking at the problem heuristically, we would expect the generator to respond nearly as a linear amplifier with time constant τ'_{d0} when unloaded and τ'_d when shorted, with the actual time constant being load dependent and between these two extremes. Let us designate this value as τ_G and the gain as K_G to write, neglecting saturation,

$$V_t = K_G E_{FD}/(1 + \tau_G s) \tag{7.60}$$

In the region where linear operation may be assumed, there is no need to consider saturation of the generator since its output is not undergoing large changes. If saturation must be included, it could be done by employing the same technique as used for the exciter, where a saturation function S_G would be defined as in Figure 7.44.

Example 7.7

1. Construct the block diagram of the system described in Section 7.7.1 and compute the system transfer function.
2. Find the open-loop transfer function for the case where

$$\tau_A = 0.1 \quad \tau_G = 1.0 \quad K_E = -0.05$$
$$\tau_E = 0.5 \quad \tau_R = 0.05 \quad K_A = 40$$
$$K_G = 1.0$$

3. Sketch a root locus for this system and discuss the problem of making the system stable.

Solution 1

The block diagram for the system is shown in Figure 7.52. If we designate the feed-forward gain and transfer function as KG and the feedback transfer function as H, the system transfer function is [23]

$$V_t/V_{\text{REF}} = KG(s)/[1 + KG(s)H(s)]$$

where, neglecting saturation and limiting, we have

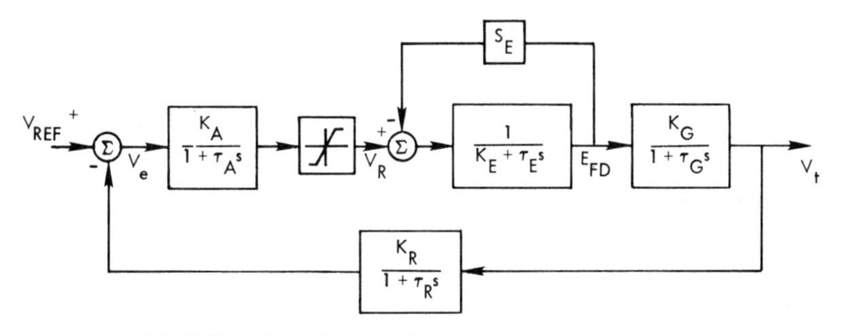

Fig. 7.52 Block diagram of the excitation control system.

$$KG(s) = \frac{K_A K_G}{(1 + \tau_A s)(K_E + \tau_E s)(1 + \tau_G s)} \qquad H(s) = \frac{K_R}{1 + \tau_R s}$$

or

$$\frac{V_t}{V_{\text{REF}}} = \frac{K_A K_G(1 + \tau_R s)}{(1 + \tau_A s)(K_E + \tau_E s)(1 + \tau_G s)(1 + \tau_R s) + K_A K_G K_R}$$

and the system is observed to be fourth order.

Solution 2

The open-loop transfer function is KGH, or

$$KGH = \frac{K_A K_G K_R}{(1 + \tau_A s)(K_E + \tau_E s)(1 + \tau_G s)(1 + \tau_R s)}$$

Using the values specified and setting $K = 400\, K_A K_R K_G$, we have

$$KGH = \frac{K}{\underset{\text{(amp)}}{(s + 10)}\underset{\text{(exc)}}{(s - 0.1)}\underset{\text{(gen)}}{(s + 1)}\underset{\text{(reg)}}{(s + 20)}}$$

Solution 3

Using the open-loop transfer function computed in Solution 2, we have the root-locus plot shown in Figure 7.53, where we compute [22]

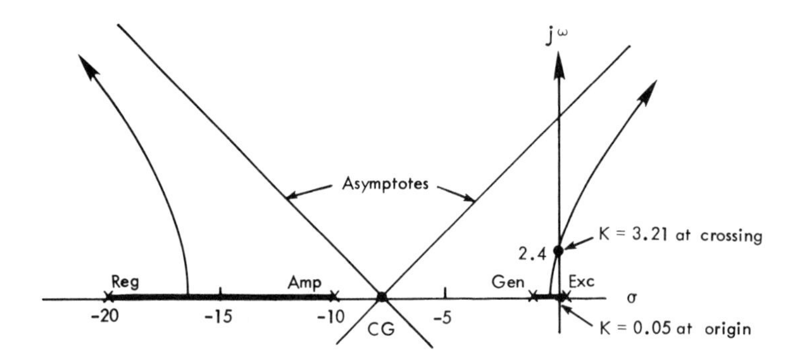

Fig. 7.53 Root locus for the system of Figure 7.52.

(1) Center of gravity $= (\sum P - \sum Z)/(\#P - \#Z) = -(30.9 - 0.0)/4 = -7.75$

(2) Breakaway points (by trial and error):

left breakaway at -16.4: $1/3.6 = 1/6.4 + 1/15.4 + 1/16.5$

$0.278 \simeq 0.281$

right breakaway at -0.43: $1/19.57 + 1/9.57 + 1/0.57 = 1/0.53$

$1.91 \simeq 1.89$

(3) Gain at $j\omega$ axis crossing:

From the closed-loop transfer function we compute the characteristic equation

$$\phi(s) = s^4 + 30.9s^3 + 226.9s^2 + 177s + K'$$

where $K' = 400K - 20$ and $K = K_A K_R K_G = 40K_R$.

Then by Routh's criterion we have

s^4	1	226.9	K'
s^3	30.9	177	
s^2	221.2	K'	
s^1	$177 - 0.14K'$	0	
s^0	K'		

For the first column we have:

From row s^0

$$K' = 400K - 20 > 0 \qquad K > 0.05$$

From row s^1

$$K' = 400K - 20 < (177/0.14) = 1266 \qquad K < 3.21$$

We may also compute the point of $j\omega$ axis crossing from the auxiliary polynomial in s^2 with $K' = 1266$, or

$$221.2s^2 + 1266 = 0 \qquad s^2 = -5.73 \qquad s = \pm j2.4$$

An examination of the root locus reveals several important system characteristics. We note that for any reasonable gain the roots due to the regulator and amplifier excite response modes that die out very fast and will probably be overdamped. Thus the response is governed largely by the generator and exciter poles that are very close to the origin. Even modest values of gain are likely to excite unstable modes in the solution. This can be improved by (a) moving the exciter pole into the left half of the s plane, which requires that R in (7.59) have a greater value; (b) moving the generator pole to the left, which would need to be done as part of the generator design rather than afterwards; and (c) adding some kind of compensation that will bend the locus to a more favorable shape in the neighborhood of the $j\omega$ axis. Of these options only (c) is of practical interest.

Excitation system compensation. Example 7.7 illustrates the need for compensation in the excitation control system. This can take many forms but usually involves some sort of rate or derivative feedback and lead or lead-lag compensation. (It is

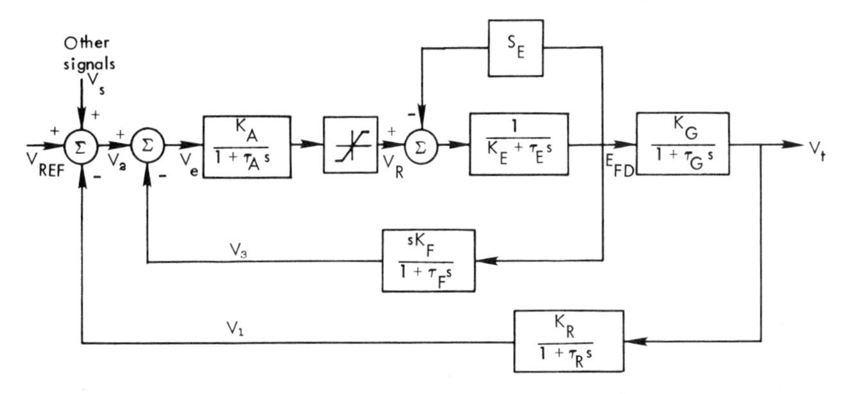

Fig. 7.54 Block diagram of a typical compensated system.

interesting to note that Gabriel Kron recognized the need for this kind of compensation as early as 1954 when he patented an excitation system incorporating these features [37].) This can be accomplished by adding the rate feedback loop shown in Figure 7.54, where time constant τ_F and gain K_F are introduced. Such a compensation scheme can be adapted to bend the root locus near the $j\omega$ axis crossing to improve stability substantially. Also notice that provision is made for the introduction of other compensating signals if they should be necessary or desirable. The effect of compensation will be demonstrated by an example.

Example 7.8

1. Repeat Example 7.7 for the system shown in Figure 7.54.
2. Use a digital computer solution to obtain the "best" values for τ_F and K_F to minimize the rise time and settling time with minimum overshoot.
3. Repeat part 2 using an analog computer solution.

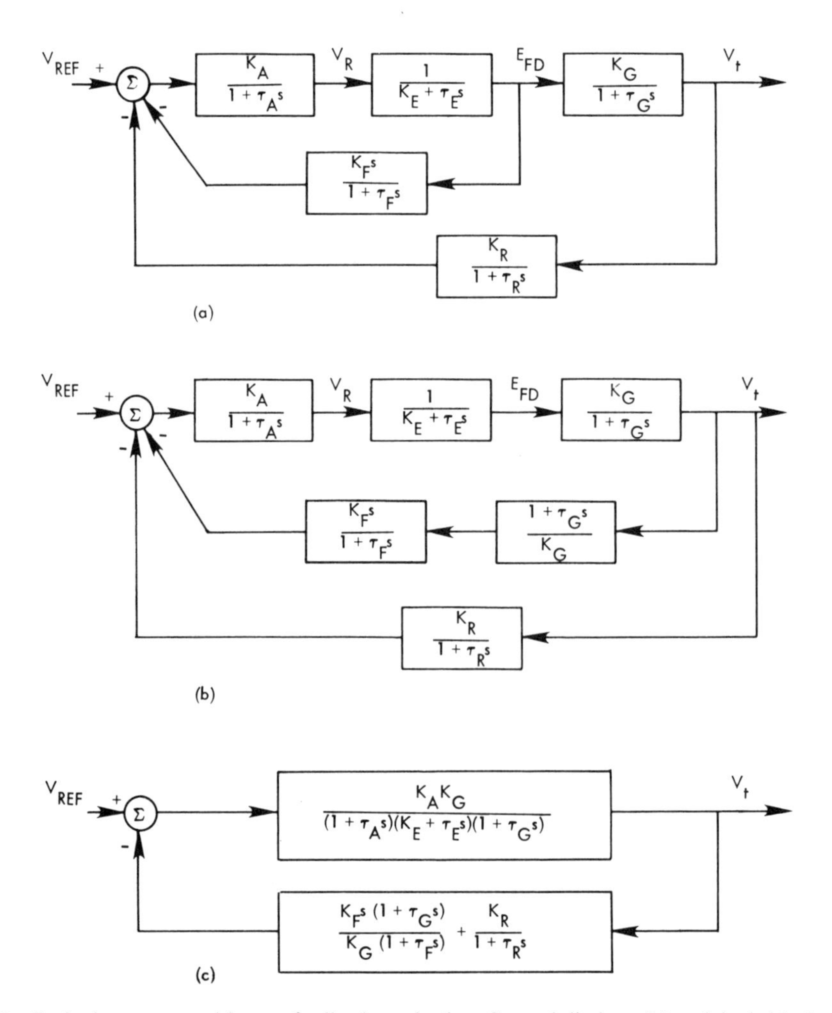

Fig. 7.55 Excitation system with rate feedback neglecting S_E and limiter: (a) original block diagram, (b) with rate feedback take-off point moved to V_t, (c) with combined feedback.

Solution 1

The system transfer function can be easily computed for $S_E = 0$ and with limiting ignored. Figure 7.55(a) shows a block diagram of the system with $S_E = 0$ and without the limiter. By using block diagram reduction, the takeoff point for the rate feedback signal is moved to V_t, as shown in Figure 7.55(b), then the two feedback signals are combined in Figure 7.55(c). The forward loop has a transfer function $KG(s)$ given by

$$KG(s) = \frac{K_A K_G}{\tau_A \tau_E \tau_G} \frac{1}{(s + 1/\tau_A)(s + K_E/\tau_E)(s + 1/\tau_G)}$$

and the feedback transfer function $H(s)$ is given by

$$H(s) = \frac{(K_F \tau_G / K_G \tau_F) s(s + 1/\tau_G)(s + 1/\tau_R) + (K_R/\tau_R)(s + 1/\tau_F)}{(s + 1/\tau_F)(s + 1/\tau_R)}$$

The open loop transfer function is thus given by

$$KGH = \frac{K_A K_F}{\tau_A \tau_E \tau_F} \frac{s(s + 1/\tau_G)(s + 1/\tau_R) + (K_R K_G \tau_F / \tau_R \tau_G K_F)(s + 1/\tau_F)}{(s + 1/\tau_A)(s + K_E/\tau_E)(s + 1/\tau_G)(s + 1/\tau_F)(s + 1/\tau_R)}$$

Substituting the values $\tau_A = 0.1$, $\tau_E = 0.5$, $\tau_R = 0.05$, $\tau_G = 1.0$ $K_E = -0.05$, $K_G = 1.0$, and $K_R = 1.0$,

$$KGH = 20\, K_A \frac{K_F}{\tau_F} \frac{s(s + 1)(s + 20) + 20(\tau_F/K_F)(s + 1/\tau_F)}{(s + 10)(s - 0.1)(s + 1)(s + 1/\tau_F)(s + 20)} \tag{7.61}$$

A given τ_F fixes all poles of (7.61). Then the shape of the locus depends on the location of the zeros. Thus we examine the zeros of (7.61). From the numerator we write

$$s(s + 1)(s + 20) + 20(\tau_F/K_F)(s + 1/\tau_F) = 0$$

$$0 = 1 + \frac{20(\tau_F/K_F)(s + 1/\tau_F)}{s(s + 1)(s + 20)} = 1 + \frac{K(s + a)}{s(s + 1)(s + 20)} \tag{7.62}$$

where we let $K = 20(\tau_F/K_F)$ and $a = 1/\tau_F$.

The locus of the roots of (7.62), which gives the zeros of (7.61), depends upon the value of $a = 1/\tau_F$. There are three cases of interest (note that $a > 0$): Case I, $0 < a < 1$; Case II, $1 < a < 20$; and Case III, $a > 20$. These cases are shown in Figure 7.56 where $-m$ is the location of the asymptote.

Case I is sketched in Figure 7.56(a), where a zero falls on the negative real axis at $-a$, which is between the origin and -1. The *locus* therefore falls between the origin and $-a$. This means that (7.61) would have a zero on the real axis near the origin. Thus the open loop transfer function of (7.61) will have a pole at 0.1 and a zero on the real axis at $-a$. The locus of the roots for this system will have a branch on the real

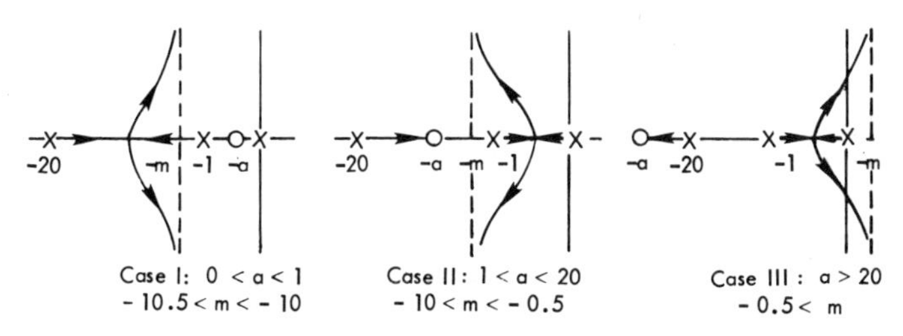

Fig. 7.56 Locus of zeros for the open loop transfer function of (7.62).

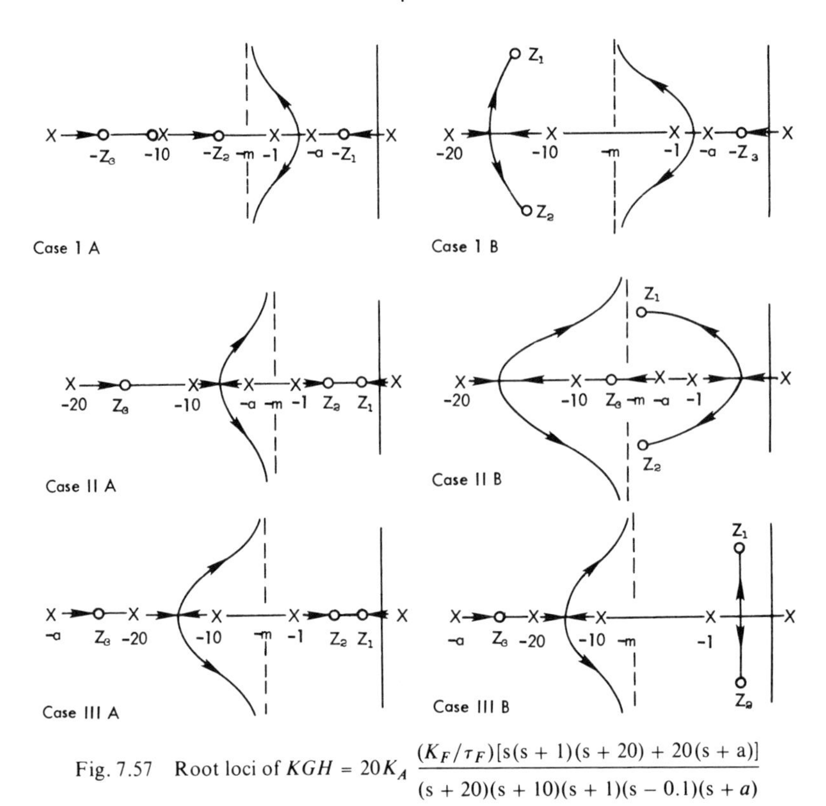

Fig. 7.57 Root loci of $KGH = 20K_A \dfrac{(K_F/\tau_F)[s(s+1)(s+20) + 20(s+a)]}{(s+20)(s+10)(s+1)(s-0.1)(s+a)}$

axis near the origin, and the system dynamic performance will be dominated by this root. Its dynamic response will be sluggish. Cases II and III are shown in Figures 7.56(b) and (c). In both cases, the root-locus plots of (7.62) have branches that, with the proper choice of the ratio K, give a pair of complex roots near the imaginary axis. Again, these are the zeros for the system described by (7.61). However, in Case II the loci approach the asymptotes to the left of the imaginary axis, while for Case III the loci approach the asymptotes to the right of the origin. The position of the roots of (7.62) and hence the zeros of (7.61), are more likely to be located further to the left of the imaginary axis in Case II than in Case III.

A further examination of the possible loci of zeros in Figure 7.56 reveals that for the three zeros, two may appear as a complex pair. Thus there are two situations of interest: (A) all zeros real and (B) one real zero and a complex pair of zeros. Furthermore, both conditions can appear in all cases. Figure 7.57 provides a pictorial summary of all six possibilities. In all but two cases the system response is dominated by a root very near the origin. Only in Cases IIB and IIIB is there any hope of pulling this dominant root away from the origin; and of these two, Case IIB is clearly the better choice. Thus we will concentrate on Case IIB for further study. (Also see [38] for a further study of this subject.)

From (7.61) the open loop transfer function is given by

$$KGH = 20\,K_A\,\frac{K_F}{\tau_F}\,\frac{s^3 + 21s^2 + 20(1 + \tau_F/K_F)s + 20/K_F}{(s+10)(s-0.1)(s+1)(s+20)(s+1/\tau_F)} \tag{7.63}$$

where $1 < 1/\tau_F < 20$.

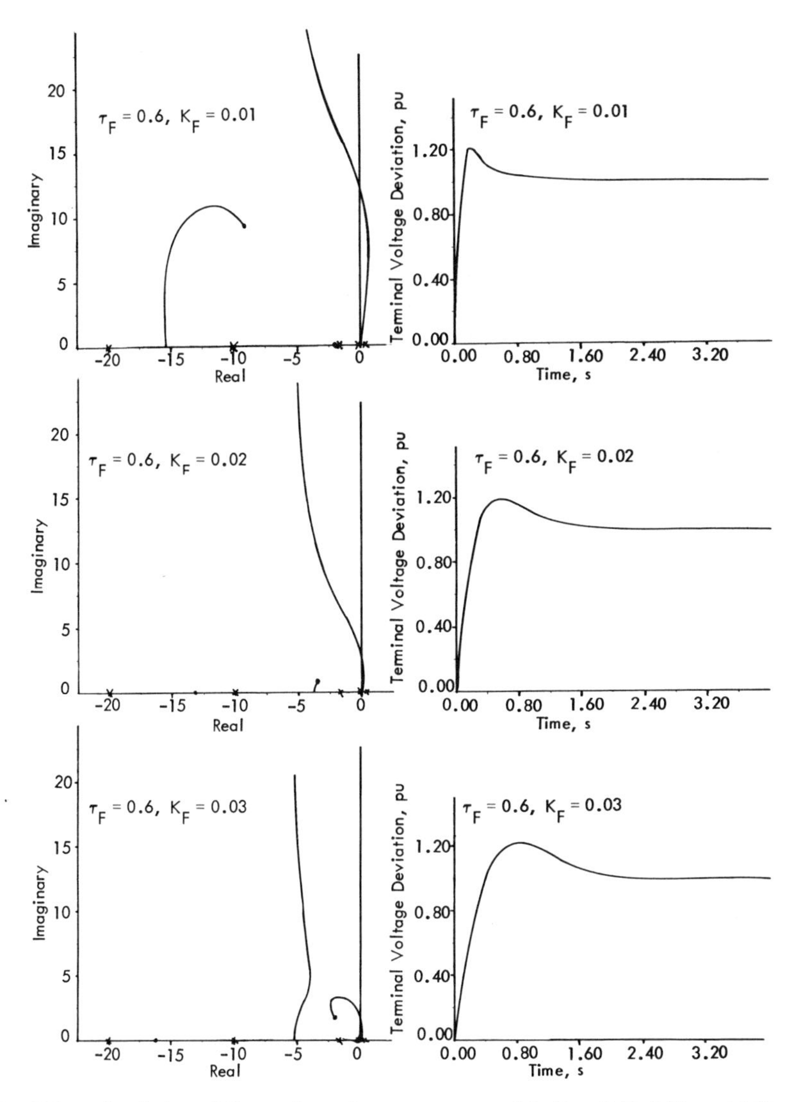

Fig. 7.58(a) Effect of variation of K_F on dynamic response: $\tau_F = 0.6$, $K_F = 0.01$, 0.02, and 0.03 respectively. Type 1 excitation system.

Solution 2

The above system is studied for different values of τ_F and K_F with the aid of special digital computer programs. The programs used are a root-finding subroutine for polynomials to obtain the zeros of equation (7.63), a root-locus program, and a time-response program. Two sample runs to illustrate the effect of τ_F and K_F are shown in Figure 7.58.

In Figure 7.58(a) τ_F is held constant at 0.6 while K_F is varied between 0.01 and 0.03. Plots of the loci of the roots are shown for the three cases, along with the time-response for the "rated" value of K_A. The most obvious effect of reducing K_F is to reduce the settling time.

In Figure 7.58(b), K_F is held constant at 0.02 while τ_F is varied between 0.5 and 0.7. The root-locus plots and the time-response for the system are repeated. The effect of increasing τ_F is to reduce the overshoot.

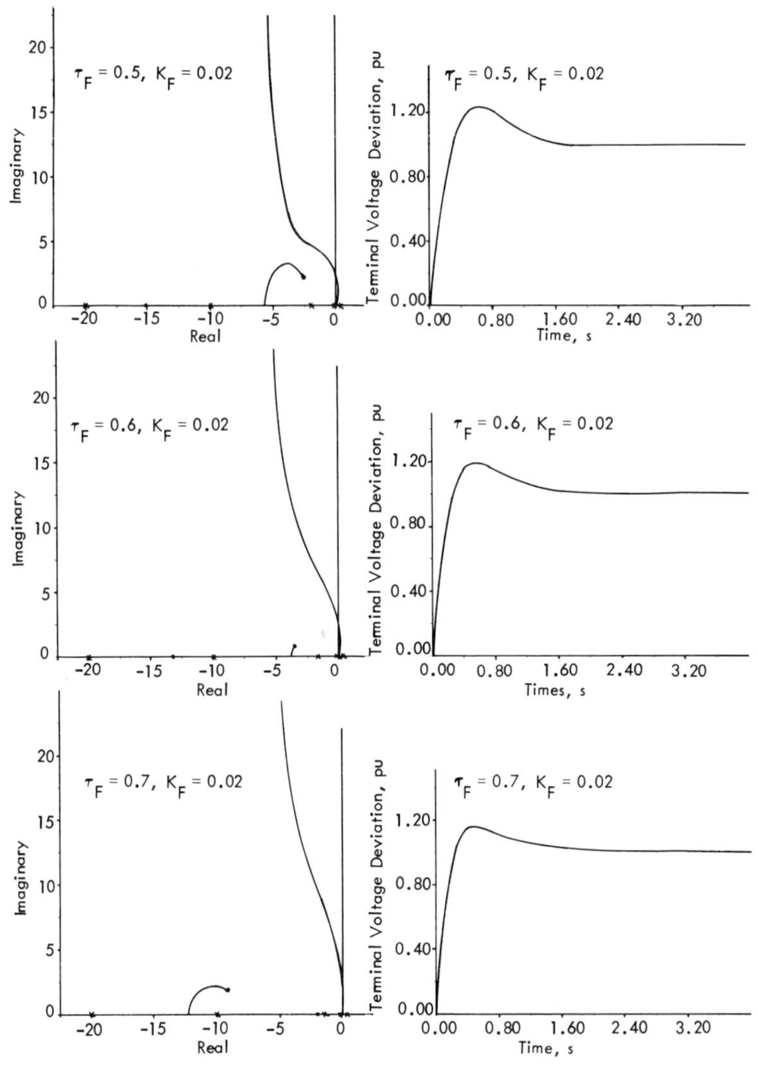

Fig. 7.58(b) Effect of variation of τ_F on dynamic response: $K_F = 0.02$, $\tau_F = 0.5$, 0.6, and 0.7 respectively. Type 1 excitation system.

From Figures 7.58(a) and 7.58(b) we can see that the values of τ_F and K_F significantly influence the dynamic performance of the system. There is, however, a variety of choices of K_F and τ_F, which gives a reasonably good dynamic response. For this particular system, $\tau_F = 0.6$ and $K_F = 0.02$ seem to give the best results.

Solution 3

An engineer with experience in s plane design may be able to guess a workable location for the zero and estimate the value of K_F that will give satisfactory results. For most engineers, the *analog computer* can be a great help in speeding up the design procedure, and we shall consider this technique as an alternate design procedure.

From Figure 7.54 we write, with $V_s = 0$,

$$V_{\text{REF}} - V_1 = V_2 \qquad V_2 - V_3 = V_e$$

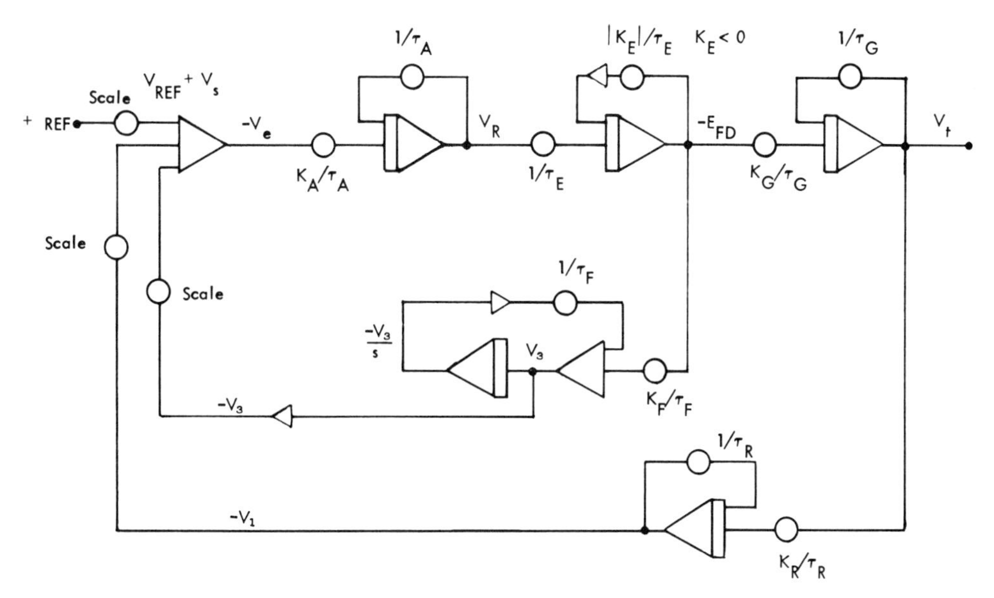

Fig. 7.59 Analog computer diagram for a linear excitation system with derivative feedback.

or

$$V_e = V_{\text{REF}} - V_1 - V_3 \tag{7.64}$$

For the amplifier block of Figure 7.54 we have $V_R = K_A V_e/(1 + \tau_A s)$, which may be rearranged as

$$V_R = (1/s)[(K_A/\tau_A) V_e - (1/\tau_A) V_R] \tag{7.65}$$

Equation (7.64) may be represented on the analog computer by a summer and (7.65) by an integrator with feedback. All other blocks except the derivative feedback term are similar to (7.65). For the derivative feedback we have $V_3 = sK_F E_{FD}/(1 + \tau_F s)$, which can be rewritten as

$$V_3 = (K_F/\tau_F) E_{FD} - (1/\tau_F s) V_3 \tag{7.66}$$

Using (7.64)–(7.66), we may construct the analog computer diagram shown in Figure 7.59. Then we may systematically move the zero from $s = 0$ to the left and check the response. In each case both the forward loop gain and feedback gains may be optimized.

Table 7.7 shows the results of several typical runs of this kind. In all cases K_R has been adjusted to unity, and other gains have been chosen to optimize V_t in a qualitative sense. The constants in these studies may be used to compute the cubic coefficients (7.62), and the equation may then be factored. If the roots are known, a root locus

Table 7.7. Summary of Analog Computer Studies for Example 7.8

Run	$v_0 = \frac{1}{\tau_F}$	K_F	K_A	Settling time, s	Percent overshoot	0–90% rise time, s
1	1.75	0.16	50	1.35	9.2	0.37
2	1.50	0.16	50	1.05	8.0	0.30
3	1.25	0.16	50	1.05	22.8	0.25
4	1.00	0.16	50	2.05	42.0	0.215
5	0.75	0.16	50	very long	70.0	0.20

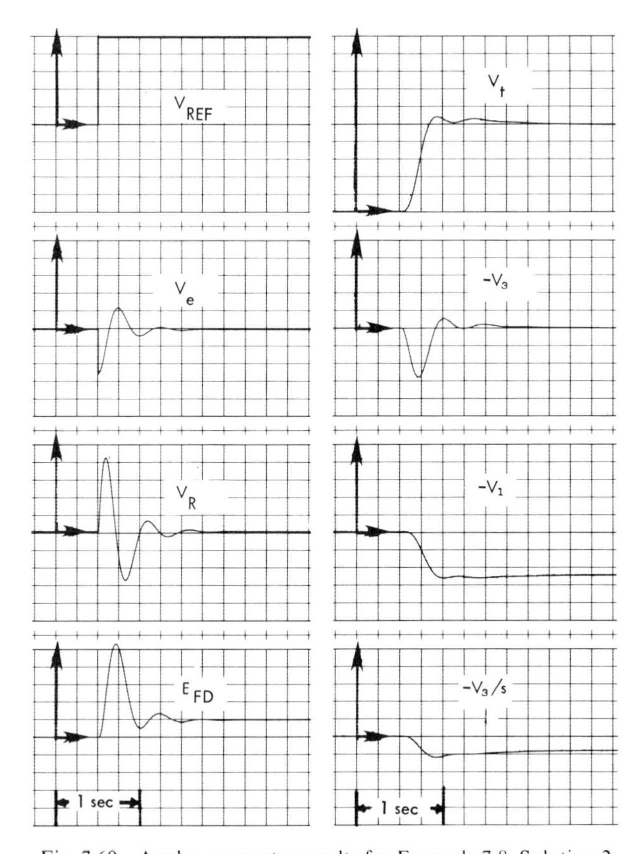

Fig. 7.60 Analog computer results for Example 7.8, Solution 2.

may be plotted and a comparison made between this and the previous uncompensated solution.

The actual analog computer outputs for run 2 are shown in Figure 7.60. One-second timing pulses are shown on the chart. The plot is made so that 20 such pulses correspond to 1 s of real time. This system is tuned to optimize the output V_t, which responds with little overshoot and displays good damping. Note, however, that this requires excessive overshoot of E_{FD} and V_R, which in physical systems would both be limited by saturation. Inclusion of saturation is a practical necessity, even in linear simulation.

Examples 7.7 and 7.8 are intended to give us some feeling for the derivative feedback of Figure 7.54. A study of the eigenvalues of a synchronous machine indicates that a first-order approximation to the generator voltage response is only approximately true. Nevertheless, making this simplification helps us to concentrate on the characteristics of the excitation system without becoming confused by the added complexity of the generator. Visualizing the root locus of the control is helpful and shows clearly how the compensated system can be operated at much greater gain while still holding a suitable damping ratio. These studies also suggest how further improvements could be realized by adding series compensation, but this is left as an exercise for the interested reader.

7.8 State-Space Description of the Excitation System

Refer again to the analog computer diagram of Figure 7.59. By inspection we write the following equations (including saturation).

$$\begin{aligned}
\dot{V}_1 &= (K_R/\tau_R)V_t - (1/\tau_R)V_1 \\
\dot{V}_3 &= (K_F/\tau_F)E_{FD} - (1/\tau_F)V_3 \\
\dot{V}_R &= (K_A/\tau_A)V_e - (1/\tau_A)V_R \qquad V_R < V_{R\max},\ V_R > V_{R\min} \\
\dot{E}_{FD} &= (1/\tau_E)V_R - [(S_E + K_F)/\tau_E]E_{FD} \\
V_e &= V_{\text{REF}} + V_s - V_1 - V_3
\end{aligned} \tag{7.67}$$

Since $S_E = S_E(E_{FD})$ is a nonlinear function of E_{FD}, we linearize at the operating point to write

$$S_{E\Delta} = \left.\frac{\partial S_E}{\partial E_{FD}}\right]_0 E_{FD\Delta} \triangleq S'_E E_{FD\Delta}$$

where we define the coefficient S'_E to describe saturation in the vicinity of the initial operating point.

Suppose we arbitrarily assign a state to each integrator associated with the excitation. Arbitrarily, we set

$$\begin{bmatrix} x_8 \\ x_9 \\ x_{10} \\ x_{11} \end{bmatrix} = \begin{bmatrix} V_1 \\ V_3 \\ V_R \\ E_{FD} \end{bmatrix}$$

and rewrite (7.67) as

$$\begin{bmatrix} \dot{V}_1 \\ \dot{V}_3 \\ \dot{V}_R \\ \dot{E}_{FD} \end{bmatrix} = \begin{bmatrix} \dot{x}_8 \\ \dot{x}_9 \\ \dot{x}_{10} \\ \dot{x}_{11} \end{bmatrix} = \begin{bmatrix} -\dfrac{1}{\tau_R} & 0 & 0 & 0 \\ 0 & -\dfrac{1}{\tau_F} & \dfrac{K_F}{\tau_F\tau_E} & -\dfrac{K_F(S'_E + K_E)}{\tau_F\tau_E} \\ -\dfrac{K_A}{\tau_A} & -\dfrac{K_A}{\tau_A} & -\dfrac{1}{\tau_A} & 0 \\ 0 & 0 & \dfrac{1}{\tau_E} & -\dfrac{(S'_E + K_E)}{\tau_E} \end{bmatrix} \begin{bmatrix} x_8 \\ x_9 \\ x_{10} \\ x_{11} \end{bmatrix} + \begin{bmatrix} \dfrac{K_R}{\tau_R} V_t \\ 0 \\ \dfrac{K_A}{\tau_A}(V_{\text{REF}} + V_s) \\ 0 \end{bmatrix} \tag{7.68}$$

In equation (7.68) the term $(K_R/\tau_R)V_t$ is a function of the state variables. From (4.46) or (6.69)

$$V_t^2 = (1/3)(v_d^2 + v_q^2) \tag{7.69}$$

where v_d and v_q are functions of the state variables; thus (7.69) is nonlinear. If the system equations are linearized about a quiescent operating state, a linear relation between the change in the terminal voltage $V_{t\Delta}$ and the change in the d and q axis voltages $v_{d\Delta}$ and $v_{q\Delta}$ is obtained. Such a relation is given in (6.69) and repeated here:

$$V_{t\Delta} = (1/3)\left(\frac{v_{d0}}{V_{t0}}\, v_{d\Delta} + \frac{v_{q0}}{V_{t0}}\, v_{q\Delta}\right) = d_0 v_{d\Delta} + q_0 v_{q\Delta} \tag{7.70}$$

The linear model is completed by substituting for $v_{d\Delta}$ and $v_{q\Delta}$ in terms of the state variables and from (6.20) and by setting $v_F = (\sqrt{3}\, r_F/L_{AD})E_{FD}$.

7.8.1 Simplified linear model

A simplified linear model can be constructed based on the linear model discussed in Section 6.5. The linearized equations for the synchronous machine are given by (the Δ subscripts are dropped for convenience)

$$E'_q = \frac{K_3}{1 + K_3\tau'_{d0}s}\, E_{FD} - \frac{K_3K_4}{1 + K_3\tau'_{d0}s}\,\delta \tag{7.71}$$

$$T_e = K_1\,\delta + K_2E'_q \tag{7.72}$$

$$V_t = K_5\,\delta + K_6E'_q \tag{7.73}$$

From (7.71)

$$\dot{E}'_q = -(1/K_3\tau'_{d0})\, E'_q - (K_4/\tau'_{d0})\,\delta + (1/\tau'_{d0})\, E_{FD} \tag{7.74}$$

From the torque equation (6.73) and (7.72)

$$\dot{\omega} = T_m/\tau_j - (K_1/\tau_j)\,\delta - (K_2/\tau_j)\, E'_q - (D/\tau_j)\,\omega \tag{7.75}$$

and from the definition of ω_Δ

$$\dot{\delta} = \omega \tag{7.76}$$

The system is now described by (7.68) and (7.72)–(7.76). The state variables are $\mathbf{x}^t = [E'_q\, \omega\, \delta\, V_1\, V_3\, V_R\, E_{FD}]$. The driving functions are V_{REF} and T_m assuming that V_s in (7.68) is zero. The complete state-space description of the system is given by

$$\begin{bmatrix} \dot{E}'_q \\ \dot{\omega} \\ \dot{\delta} \\ \dot{V}_1 \\ \dot{V}_3 \\ \dot{V}_R \\ \dot{E}_{FD} \end{bmatrix} = \begin{array}{c} \begin{array}{ccccccc} \quad E'_q & \quad\omega & \quad\delta & \quad V_1 & \quad V_3 & \quad V_R & \qquad E_{FD} \end{array} \\ \begin{bmatrix} -\dfrac{1}{K_3\tau'_{d0}} & 0 & -\dfrac{K_4}{\tau'_{d0}} & 0 & 0 & 0 & \dfrac{1}{\tau'_{d0}} \\ -\dfrac{K_2}{\tau_j} & 0 & -\dfrac{K_1}{\tau_j} & 0 & 0 & 0 & 0 \\ 0 & 1 & 0 & 0 & 0 & 0 & 0 \\ \dfrac{K_6K_R}{\tau_R} & 0 & \dfrac{K_5K_R}{\tau_R} & -\dfrac{1}{\tau_R} & 0 & 0 & 0 \\ 0 & 0 & 0 & 0 & -\dfrac{1}{\tau_F} & \dfrac{K_F}{\tau_F\tau_E} & -\dfrac{K_F(S'_E + K_E)}{\tau_F\tau_E} \\ 0 & 0 & 0 & -\dfrac{K_A}{\tau_A} & -\dfrac{K_A}{\tau_A} & -\dfrac{1}{\tau_A} & 0 \\ 0 & 0 & 0 & 0 & 0 & \dfrac{1}{\tau_E} & -\dfrac{(S'_E + K_E)}{\tau_E} \end{bmatrix} \end{array} \begin{bmatrix} E'_q \\ \omega \\ \delta \\ V_1 \\ V_3 \\ V_R \\ E_{FD} \end{bmatrix} + \begin{bmatrix} 0 \\ \dfrac{T_m}{\tau_j} \\ 0 \\ 0 \\ 0 \\ \dfrac{K_A}{\tau_A}V_{\text{REF}} \\ 0 \end{bmatrix} \tag{7.77}$$

7.8.2 Complete linear model

By using the linearized model for a synchronous machine connected to an infinite bus developed in Chapter 6, the excitation system equations are added to the system of (6.20). Before this is done, V_t must be expressed in terms of the state variables, using (6.25) and (7.70). These are repeated here (with the Δ subscript omitted),

$$
\begin{aligned}
V_t &= d_0 v_q + q_0 v_q \\
v_d &= -K\cos(\delta_0 - \alpha)\delta + R_e i_d + L_e \dot{i}_d + \omega_0 L_e i_q + i_{q0} L_e \omega \\
v_q &= -K\sin(\delta_0 - \alpha)\delta + R_e i_q + L_e \dot{i}_q - \omega_0 L_e i_d - i_{d0} L_e \omega
\end{aligned} \tag{7.78}
$$

From (7.78) and using

$$
\sqrt{3}\, V_{\infty q0} \triangleq K\cos(\delta_0 - \alpha) \qquad \sqrt{3}\, V_{\infty d0} \triangleq -K\sin(\delta_0 - \alpha)
$$

we get

$$
\begin{aligned}
V_t = -\sqrt{3}(d_0 V_{\infty q0} - q_0 V_{\infty d0})\delta &+ (d_0 R_e - q_0 \omega_0 L_e) i_d + (d_0 \omega_0 L_e + q_0 R_e) i_q \\
&+ (d_0 i_{q0} L_e - q_0 i_{d0} L_e)\omega + d_0 L_e \dot{i}_d + q_0 L_e \dot{i}_q
\end{aligned} \tag{7.79}
$$

Substituting in the first equation in (7.68),

$$
\begin{aligned}
\dot{V}_1 = &-(1/\tau_R) V_1 - (K_R/\tau_R)\sqrt{3}(d_0 V_{\infty q0} - q_0 V_{\infty d0})\,\delta + (K_R/\tau_R)(d_0 R_e - q_0 \omega_0 L_e)\, i_d \\
&+ (K_R/\tau_R)(d_0 \omega_0 L_e + q_0 R_e) i_q + (K_R/\tau_R)(d_0 i_{q0} - q_0 i_{d0}) L_e \omega + (K_R/\tau_R) d_0 L_e \dot{i}_d \\
&+ (K_R/\tau_R)\, q_0 L_e \dot{i}_q
\end{aligned}
$$

The remaining equations in (7.68) will be unchanged. The equations introduced by the exciter (for $V_s = 0$) will thus become

$$
\begin{aligned}
\dot{V}_1 - (K_R/\tau_R)\, d_0 L_e \dot{i}_d - (K_R/\tau_R)\, q_0 L_e \dot{i}_q &= (K_R/\tau_R)(d_0 R_e - q_0 \omega_0 L_e)\, i_d \\
&\quad + (K_R/\tau_R)(d_0 \omega_0 L_e + q_0 R_e)\, i_q + (K_R/\tau_R)(d_0 i_{q0} - q_0 i_{d0}) L_e \omega \\
&\quad - (K_R/\tau_R)\sqrt{3}(d_0 V_{\infty q0} - q_0 V_{\infty d0})\,\delta - (1/\tau_R)\, V_1 \\
\dot{V}_3 &= -(1/\tau_F)\, V_3 + (K_F/\tau_F \tau_E)\, V_R - [K_F(S_E' + K_E)/\tau_F \tau_E]\, E_{FD} \\
\dot{V}_R &= -(K_A/\tau_A)\, V_1 - (K_A/\tau_A)\, V_3 - (1/\tau_A)\, V_R + (K_A/\tau_A)\, V_{\text{REF}} \\
\dot{E}_{FD} &= (1/\tau_E)\, V_R - [(S_E' + K_E)/\tau_E]\, E_{FD}
\end{aligned} \tag{7.80}
$$

This set of equations is incorporated in the set (6.20) to obtain the complete mathematical description. The new **A** matrix for the system is given by $\mathbf{A} = -\mathbf{M}^{-1}\mathbf{K}$.

Note that in (7.80) the state variable for the field voltage is E_{FD} and not v_F. Therefore, the equation for the field current is adjusted accordingly. In this equation the term v_F is changed to $(\sqrt{3} r_F / L_{AD})\, E_{FD}$.

The matrices **M** and **K** are thus given by the defining equation $\mathbf{v} = -\mathbf{K}\mathbf{x} - \mathbf{M}\dot{\mathbf{x}}$, where

$$
\begin{array}{cccccccccccc}
 & i_d & i_F & i_D & i_q & i_Q & \omega & \delta & V_1 & V_3 & V_R & E_{FD} \\
\mathbf{v}^t = & [\,0 & 0 & 0 \;\vdots & 0 & 0 \;\vdots & T_m & 0 \;\vdots & 0 & 0 & -\dfrac{K_A V_{\text{REF}}}{\tau_A} & 0\,]
\end{array}
$$

M is given by

$$
\mathbf{M} = \begin{array}{c|ccc:cc:cc:cccc|}
 & i_d & i_F & i_D & i_q & i_Q & \omega & \delta & V_1 & V_3 & V_R & E_{FD} \\
i_d & \hat{L}_d & \mathrm{k}M_F & \mathrm{k}M_D & & & & & & & & \\
i_F & \mathrm{k}M_F & L_F & M_R & \mathbf{0} & & \mathbf{0} & & & \mathbf{0} & & \\
i_D & \mathrm{k}M_D & M_R & L_D & & & & & & & & \\
\hdashline
i_q & & \mathbf{0} & & \hat{L}_q & \mathrm{k}M_Q & \mathbf{0} & & & \mathbf{0} & & \\
i_Q & & & & \mathrm{k}M_Q & L_Q & & & & & & \\
\hdashline
\omega & & \mathbf{0} & & \mathbf{0} & & -\tau_j & 0 & & \mathbf{0} & & \\
\delta & & & & & & 0 & 1 & & & & \\
\hdashline
V_1 & -\dfrac{K_R}{\tau_R} d_0 L_e & 0 & 0 & -\dfrac{K_R}{\tau_R} q_0 L_e & 0 & & & 1 & 0 & 0 & 0 \\
V_3 & 0 & 0 & 0 & 0 & 0 & \mathbf{0} & & 0 & 1 & 0 & 0 \\
V_R & 0 & 0 & 0 & 0 & 0 & & & 0 & 0 & 1 & 0 \\
E_{FD} & 0 & 0 & 0 & 0 & 0 & & & 0 & 0 & 0 & 1
\end{array}
\tag{7.81}
$$

And the matrix **K** becomes

$$
\mathbf{K} = \begin{array}{c|ccc:cc:cc:cccc|}
 & i_d & i_F & i_D & i_q & i_Q & \omega & \delta & V_1 & V_3 & V_R & E_{FD} \\
i_d & \hat{R} & 0 & 0 & \omega_0 \hat{L}_q & \omega_0 \mathrm{k}M_Q & \hat{\lambda}_{q0} & -\sqrt{3}V_{\infty d0} & 0 & 0 & 0 & 0 \\
i_F & 0 & r_F & 0 & 0 & 0 & 0 & 0 & 0 & 0 & 0 & -\dfrac{\sqrt{3}r_F}{\omega_R \mathrm{k}M_F} \\
i_D & 0 & 0 & r_D & 0 & 0 & 0 & 0 & 0 & 0 & 0 & 0 \\
\hdashline
i_q & -\omega_0 \hat{L}_d & -\omega_0 \mathrm{k}M_F & -\omega_0 \mathrm{k}M_D & \hat{R} & 0 & -\hat{\lambda}_{d0} & -\sqrt{3}V_{\infty q0} & 0 & 0 & 0 & 0 \\
i_Q & 0 & 0 & 0 & 0 & r_Q & 0 & 0 & 0 & 0 & 0 & 0 \\
\hdashline
\omega & \frac{1}{3}(\lambda_{q0} - L_d i_{q0}) & -\frac{1}{3}\mathrm{k}M_F i_{q0} & -\frac{1}{3}\mathrm{k}M_D i_{q0} & \frac{1}{3}(-\lambda_{d0} + L_q i_{d0}) & \frac{1}{3}\mathrm{k}M_Q i_{d0} & -D & 0 & 0 & 0 & 0 & 0 \\
\delta & 0 & 0 & 0 & 0 & 0 & -1 & 0 & 0 & 0 & 0 & 0 \\
\hdashline
V_1 & K_{81} & 0 & 0 & K_{84} & 0 & K_{86} & K_{87} & \dfrac{1}{\tau_R} & 0 & 0 & 0 \\
V_3 & 0 & 0 & 0 & 0 & 0 & 0 & 0 & 0 & \dfrac{1}{\tau_F} & -\dfrac{K_F}{\tau_F \tau_E} & \dfrac{K_F(S_E' + K_E)}{\tau_F \tau_E} \\
V_R & 0 & 0 & 0 & 0 & 0 & 0 & 0 & \dfrac{K_A}{\tau_A} & \dfrac{K_A}{\tau_A} & \dfrac{1}{\tau_A} & 0 \\
E_{FD} & 0 & 0 & 0 & 0 & 0 & 0 & 0 & 0 & 0 & -\dfrac{1}{\tau_E} & \dfrac{S_E' + K_E}{\tau_E}
\end{array}
\tag{7.82}
$$

where

$$
\begin{aligned}
K_{81} &= -(K_R/\tau_R)(d_0 R_e - q_0 \omega_0 L_e) \\
K_{84} &= -(K_R/\tau_R)(q_0 R_e + d_0 \omega_0 L_e) \\
K_{86} &= -(K_R/\tau_R)(d_0 i_{q0} - q_0 i_{d0})\, L_e \\
K_{87} &= -(\sqrt{3}\, K_R/\tau_R)(q_0 V_{\infty d0} - d_0 V_{\infty q0})
\end{aligned}
$$

Example 7.9

Expand Example 6.2 to include the excitation system using the mathematical description of (7.80). Assume that the machine is operating initially at the load specified in Example 6.2. The excitation system parameters are given by

$$\tau_R = 0.01 \text{ s} = 3.77 \text{ pu} \qquad \tau_E = 0.5 \text{ s} = 188.5 \text{ pu}$$
$$K_R = 1.0 \qquad K_E = -0.05$$
$$\tau_A = 0.05 \text{ s} = 18.85 \text{ pu} \qquad \tau_F = 0.715 = 269.55 \text{ pu}$$
$$K_A = 40 \qquad K_F = 0.04$$

Let the exciter saturation be represented by the nonlinear function

$$S_E = A_{EX} \exp(B_{EX} E_{FD}) = 0.0039 \exp(1.555\, E_{FD})$$

Solution

From the initial conditions

$$v_{d0} = -1.148 \qquad i_{d0} = -1.59 \qquad \sqrt{3}\, V_{\infty d0} = -1.397$$
$$v_{q0} = 1.675 \qquad i_{q0} = 0.70 \qquad \sqrt{3}\, V_{\infty q0} = 1.025$$
$$V_{t0} = 1.172 \qquad E_{FD0} = 2.529$$
$$d_0 = (1/3)(v_{d0}/V_{t0}) = -(1/3)(1.148/1.172) = -0.3264$$
$$q_0 = (1/3)(v_{q0}/V_{t0}) = (1/3)(1.675/1.172) = 0.4762$$

The linear saturation coefficient at the initial operating point is

$$S_E' = \frac{\partial S_E}{\partial E_{FD}} = 1.555\,[0.0039 \exp(1.555 \times 2.529)] = 0.3095$$

The exciter time constants should be given in pu time (radians). The new terms in the **K** matrix are

$$K_{81} = -(1.0/3.77)(-0.326 \times 0.02 - 0.476 \times 0.4) = 0.0523$$
$$K_{84} = -(1.0/3.77)(-0.326 \times 0.4 + 0.476 \times 0.02) = 0.0321$$
$$K_{86} = -(1.0/3.77)(-0.326 \times 0.70 + 0.476 \times 1.59)0.4 = -0.0561$$
$$K_{87} = (1.0/3.77)(-0.326 \times 1.025 + 0.476 \times 1.397) = 0.0751$$
$$K_{88} = 1/\tau_R = 0.265$$
$$K_{99} = 1/\tau_F = 0.0037$$
$$K_{9\text{-}10} = K_F/\tau_F\tau_E = 0.04/(269.5 \times 188.5) = 0.787 \times 10^{-6}$$
$$K_{9\text{-}11} = K_F(S_E' + K_E)/\tau_F\tau_E = 0.787 \times 10^{-6} \times 0.15 = 0.118 \times 10^{-6}$$
$$K_{10\text{-}8} = K_A/\tau_A = 40/18.85 = 2.122 = K_{10\text{-}9}$$
$$K_{10\text{-}10} = 1/\tau_A = 1/18.85 = 0.053$$
$$K_{11\text{-}10} = 1/\tau_E = 0.0053$$
$$K_{11\text{-}11} = (S_E' + K_E)/\tau_E = 0.15 \times 0.0053 = 0.000796$$
$$K_{2\text{-}11} = \sqrt{3} r_F/\omega_R \mathrm{k} M_F = \sqrt{3}(0.000742)/1.55 = -0.000829$$

The new **K** matrix is given by

$$
\mathbf{K} = \left[\begin{array}{ccc|cc|cc|cccc}
0.021 & 0 & 0 & 2.040 & 1.490 & 1.430 & -1.025 & 0 & 0 & 0 & 0 \\
0 & 0.00074 & 0 & 0 & 0 & 0 & 0 & 0 & 0 & 0 & -0.0008 \\
0 & 0 & 0.013 & 0 & 0 & 0 & 0 & 0 & 0 & 0 & 0 \\
\hline
-2.100 & -1.550 & -1.550 & 0.021 & 0 & -1.039 & -1.397 & 0 & 0 & 0 & 0 \\
0 & 0 & 0 & 0 & 0.054 & 0 & 0 & 0 & 0 & 0 & 0 \\
\hline
-0.042 & -1.087 & -1.087 & -4.285 & -2.370 & 0 & 0 & 0 & 0 & 0 & 0 \\
0 & 0 & 0 & 0 & 0 & -1 & 0 & 0 & 0 & 0 & 0 \\
\hline
0.0523 & 0 & 0 & 0.032 & 0 & 0.056 & 0.075 & 0.265 & 0 & 0 & 0 \\
0 & 0 & 0 & 0 & 0 & 0 & 0 & 0 & 0.0037 & 0.79 \times 10^{-6} & 0.118 \times 10^{-6} \\
0 & 0 & 0 & 0 & 0 & 0 & 0 & 2.122 & 2.122 & 0.053 & 0 \\
0 & 0 & 0 & 0 & 0 & 0 & 0 & 0 & 0 & -0.0053 & 0.0008
\end{array}\right]
$$

The new **M** matrix is given by

$$M_{81} = K_R d_0 L_e / \tau_R = -0.0479$$
$$M_{84} = K_R q_0 L_e / \tau_R = 0.0211$$

$$
\mathbf{M} = \left[\begin{array}{ccc|cc|cc|cccc}
2.100 & 1.550 & 1.550 & & & & & & & & \\
1.550 & 1.651 & 1.550 & & \mathbf{0} & & \mathbf{0} & & \mathbf{0} & & \\
1.550 & 1.550 & 1.605 & & & & & & & & \\
\hline
 & & & 2.040 & 1.490 & & & & & & \\
 & \mathbf{0} & & 1.490 & 1.526 & & \mathbf{0} & & \mathbf{0} & & \\
\hline
 & & & & & -5360.8 & 0 & & & & \\
 & \mathbf{0} & & & \mathbf{0} & 0 & 1 & & \mathbf{0} & & \\
\hline
0.048 & 0 & 0 & -0.021 & 0 & & & 1 & 0 & 0 & 0 \\
0 & 0 & 0 & 0 & 0 & & \mathbf{0} & 0 & 1 & 0 & 0 \\
0 & 0 & 0 & 0 & 0 & & & 0 & 0 & 1 & 0 \\
0 & 0 & 0 & 0 & 0 & & & 0 & 0 & 0 & 1
\end{array}\right]
$$

The **A** matrix is given by

$$
\mathbf{A} = \left[\begin{array}{ccc|cc|cc|cccc}
-36.062 & 0.4388 & 14.142 & -3487.2 & -2547.0 & -2444.6 & 1751.3 & 0 & 0 & 0 & -0.4904 \\
12.472 & -4.9503 & 76.857 & 1206.0 & 880.86 & 845.46 & -0.6057 & 0 & 0 & 0 & 5.5317 \\
22.776 & 4.3557 & -96.017 & 2202.4 & 1608.6 & 1544.0 & -1106.1 & 0 & 0 & 0 & -4.8673 \\
\hline
3590.0 & 2649.7 & 2649.7 & -36.064 & 90.072 & 1776.7 & 2387.4 & 0 & 0 & 0 & 0 \\
-3505.7 & -2587.5 & -2587.5 & 35.218 & -123.32 & -1735.0 & -2331.4 & 0 & 0 & 0 & 0 \\
\hline
-0.0078 & -0.2027 & -0.2027 & -0.7993 & -0.4422 & 0 & 0 & 0 & 0 & 0 & 0 \\
0 & 0 & 0 & 0 & 0 & 1.00 & 0 & 0 & 0 & 0 & 0 \\
\hline
25.394 & 56.019 & 55.361 & 134.50 & 124.15 & 211.02 & -108.65 & -265.26 & 0 & 0 & 0.0235 \\
0 & 0 & 0 & 0 & 0 & 0 & 0 & 0 & -3.7099 & 0.00078 & 0.00012 \\
0 & 0 & 0 & 0 & 0 & 0 & 0 & -2122.1 & -2122.1 & -53.052 & 0 \\
0 & 0 & 0 & 0 & 0 & 0 & 0 & 0 & 0 & 53.052 & -0.7958
\end{array}\right] 10^{-3}
$$

The eigenvalues obtained are

$\lambda_1 = -0.0359 + j0.9983$ $\lambda_7 = -0.0015 + j0.0290$

$\lambda_2 = -0.0359 - j0.9983$ $\lambda_8 = -0.0015 - j0.0290$

$\lambda_3 = -0.2653$ $\lambda_9 = -0.0002 + j0.0064$

$\lambda_4 = -0.0986$ $\lambda_{10} = -0.0002 - j0.0064$

$\lambda_5 = -0.1217$ $\lambda_{11} = -0.0037$

$\lambda_6 = -0.0548$

Example 7.10

Repeat Example 7.9 for different exciters. Use the same machine loading. Tabulate the data used and the eigenvalues obtained.

Solution

For this example we will use the same machine loading of Example 5.1 and three exciters made by the same manufacturer: W TRA, W Brushless, and W Low τ_E Brushless. Data for the exciters and the appropriate M and K constants are given in Table 7.8. The eigenvalues obtained are tabulated in Table 7.9.

Table 7.8. Exciter Data and Elements of Matrices **M** and **K** (Loading of Example 5.1)

Constants and matrix elements	IEEE type 1 exciter		
	W TRA	W Brushless	W low τ_E Brushless
K_A	400	400	400
τ_A	0.05	0.02	0.02
K_E	−0.17	1.0	1.0
τ_E	0.95	0.80	0.015
K_F	0.04	0.03	0.04
τ_F	1.0	1.0	0.50
K_R	1.0	1.0	1.0
τ_R	0.0*	0.0*	0.0*
A_{EX}	0.0027	0.098	0.0761
B_{EX}	1.304	0.553	0.4475
S_{E0}	0.0874	0.4282	0.2510
S'_E	0.1140	0.2368	0.1123
$V_{R\max}$	3.5	7.3	6.96
$V_{R\min}$	−3.5	−7.3	−6.96
$M_{8\text{-}1}$	3.862069	3.862069	3.862069
$M_{8\text{-}4}$	−4.753316	−4.753316	−4.753316
$K_{8\text{-}1}$	4.9464	4.9464	4.9464
$K_{8\text{-}4}$	3.6244	3.6244	3.6244
$K_{8\text{-}6}$	−6.5741	−6.5741	−6.5741
$K_{8\text{-}7}$	10.2754	10.2754	10.2754
$K_{8\text{-}8}$	26.5252	26.5252	26.5252
$K_{9\text{-}9}$	0.002653	0.002653	0.005305
$K_{9\text{-}10}$	−0.000112	−0.000099	−0.014147
$K_{9\text{-}11}$	−0.000006	0.000123	0.015735
$K_{10\text{-}8} = K_{10\text{-}9}$	21.220159	53.050398	53.050398
$K_{10\text{-}10}$	0.053050	0.132626	0.132626
$K_{11\text{-}10}$	−0.002792	−0.003316	−0.176835
$K_{11\text{-}11}$	−0.000156	0.004101	0.196693

*Where $\tau_R = 0.0$ take $\tau_R = 10^{-4}$.

Table 7.9. Eigenvalues for System of Example 7.10 (Loading of Example 5.1)

Exciter type		
W TRA	W Brushless	W low τ_E Brushless
$-0.03594 + j0.99826$	$-0.03594 + j0.99826$	$-0.03594 + j0.99827$
$-0.03594 - j0.99826$	$-0.03594 - j0.99826$	$-0.03594 - j0.99827$
-0.265×10^2	-0.265×10^2	-0.26525×10^2
-0.09804	-0.07300	-0.09763
-0.12299	-0.12315	-0.12302
$-0.02536 + j0.03912$	$-0.07870 + j0.02139$	$-0.16664 + j0.86637$
$-0.02536 - j0.03912$	$-0.07870 - j0.02139$	$-0.16664 - j0.86637$
$-0.00076 + j0.02444$	$-0.00071 + j0.02444$	$-0.00082 + j0.02468$
$-0.00076 - j0.02444$	$-0.00071 - j0.02444$	$-0.00082 - j0.02468$
$-0.00340 + j0.00249$	$-0.00447 + j0.00185$	$-0.00177 + j0.00353$
$-0.00340 - j0.00249$	$-0.00447 - j0.00185$	$-0.00177 - j0.00353$

The results tabulated in Table 7.9 are for the same machine and loading condition as used in Example 6.3 except for the addition of the exciter models. Comparing the results of Examples 6.3 and 7.10, we note that two pairs of complex eigenvalues and two real eigenvalues are essentially present in all the results. We can conclude that these eigenvalues are identified with the parameters of the machine and are not dependent on the exciter parameters. The additional eigenvalues obtained in Example 7.10 and not previously present are comparable in magnitude except for one complex pair associated with the W Low τ_E Brushless exciter. For this exciter a frequency of approximately 50 Hz is obtained, which might be introduced by the extremely low exciter time constant.

The same example was repeated for the loading of Example 5.2 and for the same exciters. The results obtained indicate that only one pair of complex eigenvalues change with the machine loading. This pair is one of the two complex pairs associated with the machine parameters. The eigenvalues associated with the exciter parameters did not change significantly with the machine loading.

7.9 Computer Representation of Excitation Systems

Most of the problems in which the transient behavior of the excitation system is being studied will require the use of computers. It is therefore recognized that the solution of systems can be greatly simplified if a standard set of mathematical models can be chosen. Then each manufacturer can specify the constants for the model that will best represent his systems, and the data acquisition problem will be simplified for the user.

As the use of computers has increased and programs have been developed that represent excitation systems, several models have evolved for such systems. Actually, the differences in these representations was more in the form of the data than in the accuracy of the representation. Recognizing this fact, the *IEEE* formed a working group in the early 1960s to study standardization. This group, which presented its final report in 1967 [15], standardized the representation of excitation systems in four different types and identified specific commercial systems with each type. These models allow for several degrees of complexity, depending upon the available data or importance of a particular exciter in a large system problem. Thus, anything from a very simple linear model to a more complex nonlinear model may be formulated by following these generalized descriptions. We describe the four *IEEE* models below.

The excitation system models described use a pu system wherein 1.0 pu generator voltage is the rated generator voltage and 1.0 pu exciter voltage is that voltage required to produce rated generator voltage on the generator air gap line (see Def. 3.20 in Appendix E). This means that at no load and neglecting saturation, $E_{FD} = 1.0$ pu gives exactly $V_t = 1.0$ pu. Table 7.10 gives a list of symbols used in the four *IEEE* models, changed slightly to conform to the notation used throughout this chapter.

Table 7.10. Excitation System Model Symbols

Symbol	Description	Symbol	Description
E_{FD}	= exciter output voltage	τ_A	= regulator amplifier time constant
I_F	= generator field current	τ_E	= exciter time constant
V_t	= generator terminal voltage		
I_t	= generator terminal current	τ_F	= regulator stabilizing circuit time constant
K_A	= regulator gain	τ_{F1}, τ_{F2}	= same as τ_F for rotating rectifier system
K_E	= exciter constant related to self-excited field	τ_R	= regulator input filter time constant
K_F	= regulator stabilizing circuit gain	τ_{RH}	= rheostat time constant, Type 4
K_I	= current circuit gain in Type 3 system	V_R	= regulator output voltage
K_P	= potential circuit gain in Type 1S or Type 3 system	$V_{R\max}$	= maximum value of V_R
K_V	= fast raise/lower constant setting, Type 4 system	$V_{R\min}$	= minimum value of V_R
S_E	= exciter saturation function	V_{REF}	= regulator reference voltage setting
V_s	= auxiliary (stabilizing) input signal	V_{RH}	= field rheostat setting

Note: Voltages and currents are s domain quantities.

7.9.1 Type 1 system—continuously acting regulator and exciter

The block diagram for the Type 1 system is shown in Figure 7.61. Note that provision is made for first-order smoothing or filtering of the terminal voltage V_t with a filter time constant of τ_R. Usually τ_R is very small and is often approximated as zero.

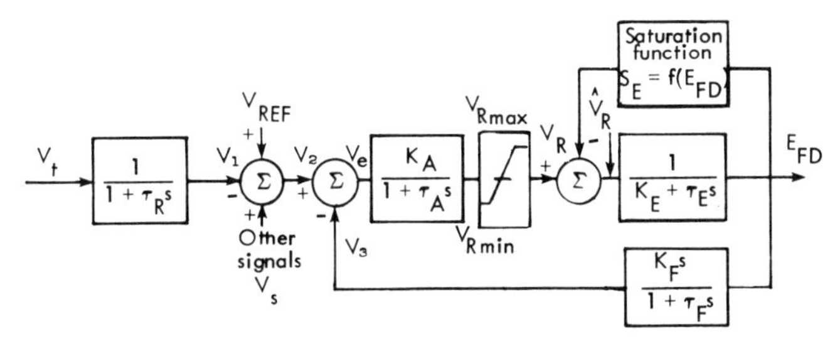

Fig. 7.61 Type 1 excitation system representation for a continuously acting regulator and exciter. (© IEEE. Reprinted from *IEEE Trans.*, vol. PAS-87, 1968.)

The amplifier has time constant τ_A and gain K_A, and its output is limited by $V_{R\max}$ and $V_{R\min}$. Note that if we have no filter and the rate feedback is zero ($K_F = 0$), the input to the rotating amplifier is the error voltage

$$V_e = V_{\mathrm{REF}} - V_t \tag{7.83}$$

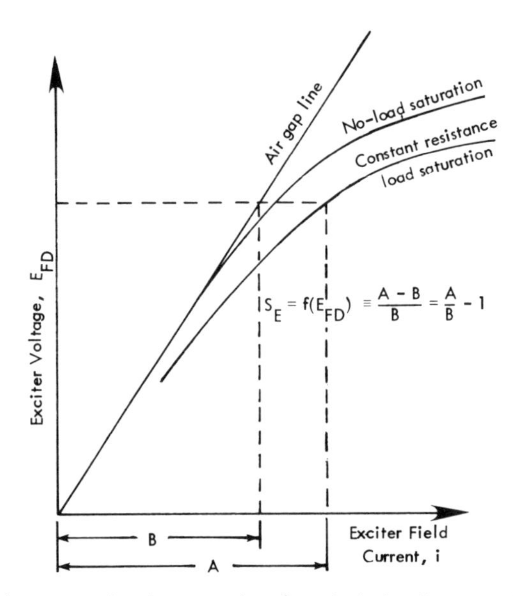

Fig. 7.62 Exciter saturation curves showing procedure for calculating the saturation function S_E. (© IEEE. Reprinted from *IEEE Trans.*, vol. PAS-87, 1968.)

and this voltage is small, but finite in the steady state. The exciter itself is represented as a first-order linear system with time constant τ_E. However, a provision is made to include the effect of saturation in the exciter by the saturation function S_E. The saturation function is defined as shown in Figure 7.62 by the relation

$$S_E = (A - B)/B \tag{7.84}$$

and is thus a function of E_{FD} that is nonlinear. This alters the amplifier voltage V_R by an amount $S_E E_{FD}$ to give a new effective value of $\hat{V}_R$, viz.,

$$\hat{V}_R = V_R - S_E E_{FD} \tag{7.85}$$

This altered value $\hat{V}_R$ is operated upon linearly by the exciter transfer function. Note that for sufficiently small E_{FD} the system is nearly linear ($S_E = 0$). Note also that the exciter transfer function contains a constant K_E. This transfer function

$$G(s) = 1/(K_E + \tau_E s) \tag{7.86}$$

is not in the usual form for a linear transfer function for a first-order system (usually stated as $1/(1 + \tau s)$. From the block diagram we write $E_{FD} = \hat{V}_R/(K_E + \tau_E s)$, and substituting (7.85) for $\hat{V}_R$ we have

$$\tau_E s E_{FD} = -K_E E_{FD} + V_R - S_E E_{FD} \tag{7.87}$$

which includes the nonlinear function $S_E E_{FD}$. Equation (7.87) corresponds in the time domain to

$$\tau_E \dot{E}_{FD} = -K_E E_{FD} + V_R - S_E E_{FD} \tag{7.88}$$

Comparing with (7.32), for example, where we computed

$$\tau_E \dot{v}_F = v_F + v_R - bRv_F/(a - v_F)$$

with the nonlinearity approximated by a Frohlich equation, we can observe the obvious similarity. Reference [15] suggests taking

$$K_E = S_E]_{E_{FD}(0)} = f[E_{FD}(0)] \tag{7.89}$$

which corresponds to the resistance in the exciter field circuit at $t = 0$.

Some engineers approximate the saturation function by an exponential function, i.e.,

$$S_E = f(E_{FD}) = A_{EX}\exp(B_{EX}E_{FD}) \tag{7.90}$$

The coefficients A_{EX} and B_{EX} are computed from saturation data, where S_E and E_{FD} are specified at two points, usually the exciter ceiling voltage and 75% of ceiling. The function (7.90) is easy to compute and provides a simple way to represent exciter saturation with reasonable accuracy. See Appendix D.

Finally we examine the feedback transfer function of Figure 7.61

$$H(s) = K_F s/(1 + \tau_F s) \tag{7.91}$$

where K_F and τ_F are respectively the gain constant and the time constant of the regulator stabilizing circuit. This time constant introduces a zero on the negative real axis. Note that (7.91) introduces both a derivative feedback and a first-order lag.

Reference [15] points out that the regulator ceiling $V_{R\max}$ and the exciter ceiling $E_{FD\max}$ are interrelated through S_E and K_E. Under steady-state conditions we compute

$$V_R = K_E E_{FD} + S_E E_{FD} \tag{7.92}$$

with the constraint $V_{R\min} < V_R < V_{R\max}$, then

$$V_{R\max} = (K_E + S_{E\max})E_{FD\max} \tag{7.93}$$

Thus there exists a constraint between the maximum (or minimum) values of $E_{FD\max}$ and $V_{R\max}$ ($E_{FD\min}$ and $V_{R\min}$).

7.9.2 Type 1S system—controlled rectifier system with terminal potential supply only

This is a special case of continuously acting systems where excitation is obtained through rectification of the terminal voltage as in Figures 7.17 and 7.18. In this case the maximum regulator voltage is not a constant but is proportional to V_t, i.e.,

$$V_{R\max} = K_P V_t \tag{7.94}$$

Such systems have almost instantaneous response of their main excitation components such that in Figure 7.61 $K_E = 1$, $\tau_E = 0$, and $S_E = 0$. This system is shown in Figure 7.63.

A state-space representation of the Type 1S system can be derived by referring to (7.67) (written for the Type 1 system), setting $V_R = E_{FD}$ and eliminating (7.65), with

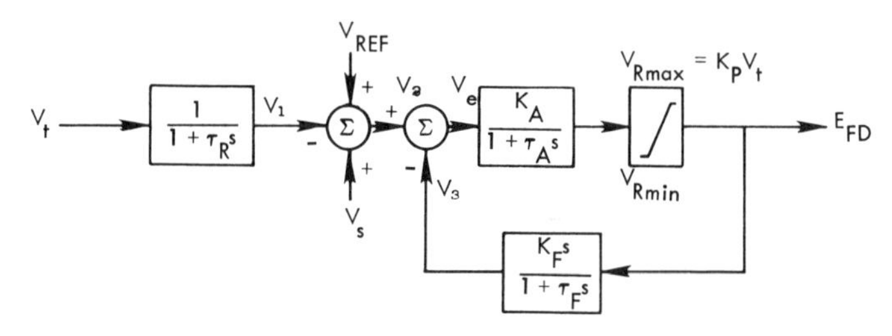

Fig. 7.63 Type 1S system. (© IEEE. Reprinted from *IEEE Trans.*, vol. PAS-87, 1968.)

the result

$$\begin{aligned} \dot{V}_1 &= (K_R/\tau_R)\,V_t - (1/\tau_R)\,V_1 & \dot{V}_3 &= (K_F/\tau_F)\,\dot{E}_{FD} - (1/\tau_F)\,V_3 \\ \dot{E}_{FD} &= (K_A/\tau_A)\,V_e - (1/\tau_A)\,E_{FD} & E_{FD} &< V_{R\max},\ E_{FD} > V_{R\min} \\ V_e &= V_{\text{REF}} + V_s - V_1 - V_3 & & \end{aligned} \tag{7.95}$$

By using (7.79) and substituting for i_d and i_q, we can express V_t as a function of the state variables. For the linearized system discussed in Chapter 6 where the state variables

$$\mathbf{x}^t = [i_d\, i_F\, i_D\, i_q\, i_Q\, \omega\, \delta] = [x_1\, x_2\, x_3\, x_4\, x_5\, x_6\, x_7]$$

we can show that

$$V_t = f_f E_{FD} + \sum_{k=1}^{7} f_k x_k \tag{7.96}$$

where the f coefficients are constants. Rearranging, we write

$$\begin{bmatrix} \dot{V}_1 \\ \dot{V}_3 \\ \dot{E}_{FD} \end{bmatrix} = \begin{bmatrix} \dot{x}_8 \\ \dot{x}_9 \\ \dot{x}_{10} \end{bmatrix} = \begin{bmatrix} -\dfrac{1}{\tau_R} & 0 & f_f \\ -\dfrac{K_F K_A}{\tau_F \tau_A} & -\dfrac{(K_F K_A + \tau_A)}{\tau_F \tau_A} & -\dfrac{K_F}{\tau_F \tau_A} \\ -\dfrac{K_A}{\tau_A} & -\dfrac{K_A}{\tau_A} & -\dfrac{1}{\tau_A} \end{bmatrix} \begin{bmatrix} x_8 \\ x_9 \\ x_{10} \end{bmatrix} + \begin{bmatrix} \dfrac{1}{\tau_R} & 0 & 0 \\ 0 & 0 & \dfrac{K_F}{\tau_F} \\ 0 & 0 & 1 \end{bmatrix} \begin{bmatrix} \sum_{k=1}^{7} f_k x_k \\ 0 \\ \dfrac{K_A}{\tau_A}(V_{\text{REF}} + V_s) \end{bmatrix} \tag{7.97}$$

where

$$V_t = f_f x_{10} + \sum_{k=1}^{7} f_x x_k$$

Note that only three states are needed in this case.

7.9.3 Type 2 system—rotating rectifier system

Another type of system, the rotating rectifier system of Figure 7.13, incorporates damping loops that originate from the regulator output rather than from the excitation voltage [39] since, being brushless, the excitation voltage is not available to feed back. The *IEEE* description of this system is shown in Figure 7.64, where the damping feedback loop is seen to be different from that of Figure 7.61. Note that two time constants appear in the damping loop of this new system, τ_{F1} and τ_{F2}, one of which approximates

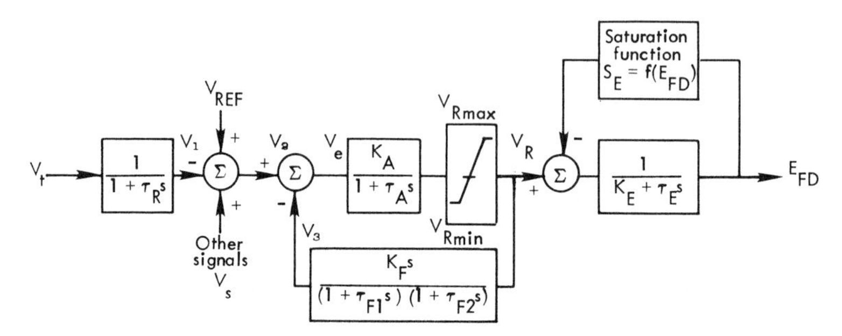

Fig. 7.64 Type 2 excitation system representation—rotating rectifier system before 1967. (© IEEE. Reprinted from *IEEE Trans.*, vol. PAS-87, 1968.)

the exciter time delay [39] and is considered "major damping," with the second or "minor damping" being present to damp higher frequencies.

A state-space representation of this system may be derived from the following equations:

$$
\begin{aligned}
\dot{V}_1 &= (K_R/\tau_R)\, V_t - (1/\tau_R)\, V_1 \\
\dot{V}_R &= (K_A/\tau_A)\, V_e - (1/\tau_A)\, V_R \\
\dot{E}_{FD} &= (1/\tau_E)\, V_R - [(K_E + S_E)/\tau_E]\, E_{FD} \\
\ddot{V}_3 &= \frac{K_F K_A}{\tau_{F1}\tau_{F2}\tau_A} V_e - \frac{K_F}{\tau_{F1}\tau_{F2}\tau_A} V_R - \frac{(\tau_{F1} + \tau_{F2})}{\tau_{F1}\tau_{F2}} \dot{V}_3 - \frac{1}{\tau_{F1}\tau_{F2}} V_3 \\
V_e &= V_{\mathrm{REF}} + V_s - V_1 - V_3
\end{aligned}
\tag{7.98}
$$

Rearranging, we may write as

$$
\begin{bmatrix} \dot{V}_1 \\ \dot{V}_3 \\ \ddot{V}_3 \\ \dot{V}_R \\ \dot{E}_{FD} \end{bmatrix}
=
\begin{bmatrix} \dot{x}_8 \\ \dot{x}_9 \\ \dot{x}_{10} \\ \dot{x}_{11} \\ \dot{x}_{12} \end{bmatrix}
=
\begin{bmatrix}
-\frac{1}{\tau_R} & 0 & 0 & 0 & \frac{K_R}{\tau_R} f_f \\
0 & 0 & 1 & 0 & 0 \\
-\frac{K_A K_F}{\tau_{F1}\tau_{F2}\tau_A} & -\frac{\tau_A + K_A K_F}{\tau_{F1}\tau_{F2}\tau_A} & -\frac{\tau_{F1} + \tau_{F2}}{\tau_{F1}\tau_{F2}} & -\frac{K_F}{\tau_{F1}\tau_{F2}\tau_A} & 0 \\
-\frac{K_A}{\tau_A} & -\frac{K_A}{\tau_A} & 0 & -\frac{1}{\tau_A} & 0 \\
0 & 0 & 0 & \frac{1}{\tau_R} & -\frac{K_E + S_E}{\tau_E}
\end{bmatrix}
\begin{bmatrix} x_8 \\ x_9 \\ x_{10} \\ x_{11} \\ x_{12} \end{bmatrix}
+
\begin{bmatrix}
1 & 0 & 0 & 0 & 0 \\
0 & 0 & 0 & 0 & 0 \\
0 & 0 & 0 & \frac{K_F}{\tau_{F1}\tau_{F2}} & 0 \\
0 & 0 & 0 & 1 & 0 \\
0 & 0 & 0 & 0 & 0
\end{bmatrix}
\begin{bmatrix} \sum_{k=1}^{7} f_k x_k \\ 0 \\ 0 \\ \frac{K_A}{\tau_A}(V_{\mathrm{REF}} + V_s) \\ 0 \end{bmatrix}
\tag{7.99}
$$

The Type 2 excitation system representation is intended for use in simulating the Westinghouse Brushless excitation system. An alternate representation developed by the manufacturer is reported to represent the physical equipment more accurately. This revised Type 2 representation is shown in Figure 7.65 [40].

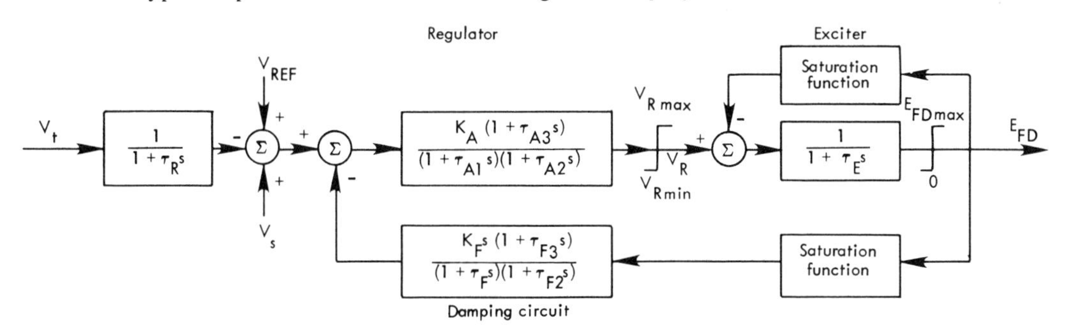

Fig. 7.65 Revised Type 2 excitation system representation. (Used with permission from *Stability Program Data Preparation Manual*, Advanced Systems Technology Rept. 70–736, Westinghouse Electric Corp., Dec. 1972.)

7.9.4 Type 3 system—static with terminal potential and current supplies

Some systems use a combination of current and voltage intelligence as a feedback signal to be compared against the reference, e.g., the systems of Figures 7.15 and 7.16. These systems are not properly represented by Type 1 or 1S and require special treatment, as shown in Figure 7.66. (The reactance x_L is the commutating reactance of the transformer and is discussed in [41].) Here the regulator and input smoothing are similar to the Type 1 system. However, the signal denoted V_B incorporates information fed forward from V_t with added information concerning both I_t and I_F. Thus

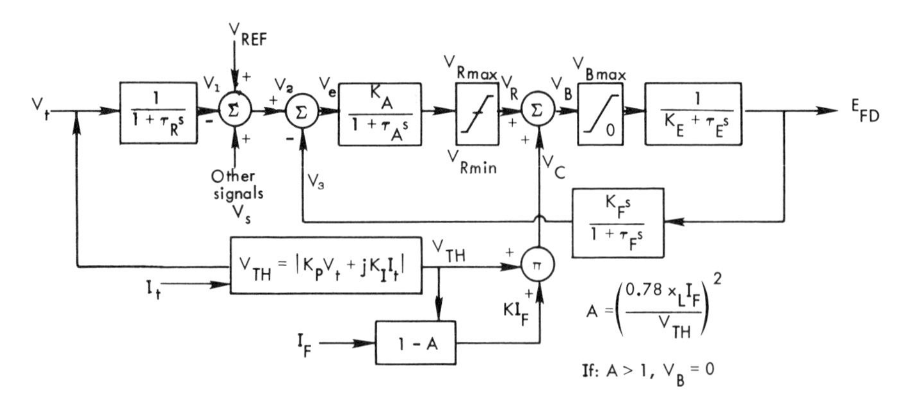

Fig. 7.66 Type 3 excitation system representation—static with terminal potential and current supplies. (© IEEE. Reprinted from *IEEE Trans.*, vol. PAS-87, 1968.)

V_C represents the self-excitation from the generator terminals. Constants K_P and K_I are proportionality factors indicating the proportion of the "Thevenin voltage," V_{TH} due to potential and current information. Multiplying V_{TH} is a signal proportional to I_F, which accounts for variation of self-excitation with change in the angular relation of field current (I_F) and self-excitation voltage (V_{TH}) [15].

Obviously, systems of this type are nonlinear. To formulate a linearized state-space representation, we may write the self-excitation components as

$$V_C = K_1 V_t + K_2 I_t + K_3 I_F \tag{7.100}$$

Then we write for the entire system

$$\begin{aligned} V_B &= V_R + V_C \qquad & E_{FD} &= V_B/(K_E + \tau_E s) \qquad & V_R &= [K_A/(1 + \tau_A s)]\, V_e \\ V_3 &= K_F E_{FD} s/(1 + \tau_F s) \qquad & V_1 &= K_R V_t/(1 + \tau_R s) \end{aligned} \tag{7.101}$$

But we may write the terminal voltage in the time domain as

$$V_t = f_f E_{FD} + \sum_{k=1}^{7} f_k x_k = f_f E_{FD} + v_x \tag{7.102}$$

where for brevity we let v_x be the term on the right. Also, for the terminal current we may write

$$i_t = M_d i_d + M_q i_q = M_d x_1 + M_q x_4 \tag{7.103}$$

If we define the states as in (7.68), we reduce (7.101)–(7.103) to the following form:

$$\begin{bmatrix} \dot{V}_1 \\ \dot{V}_3 \\ \dot{V}_R \\ \dot{E}_{FD} \end{bmatrix} = \begin{bmatrix} \dot{x}_8 \\ \dot{x}_9 \\ \dot{x}_{10} \\ \dot{x}_{11} \end{bmatrix} = \begin{bmatrix} -\dfrac{1}{\tau_R} & 0 & 0 & \dfrac{K_R}{\tau_R} f_f \\ 0 & -\dfrac{1}{\tau_F} & \dfrac{K_F}{\tau_E \tau_F} & \dfrac{K_F}{\tau_E \tau_F}(f_f K_1 - K_E) \\ -\dfrac{K_A}{\tau_A} & -\dfrac{K_A}{\tau_A} & -\dfrac{1}{\tau_A} & 0 \\ 0 & 0 & \dfrac{1}{\tau_E} & \dfrac{f_f K_1 - K_E}{\tau_E} \end{bmatrix} \begin{bmatrix} x_8 \\ x_9 \\ x_{10} \\ x_{11} \end{bmatrix} + \begin{bmatrix} \dfrac{K_R}{\tau_R} v_x \\ \dfrac{K_F K_1}{\tau_E \tau_F} v_x + \dfrac{K_F K_2}{\tau_E \tau_F} i_t + \dfrac{K_F K_3}{\tau_E \tau_F} i_F \\ \dfrac{K_A}{\tau_A}(V_{\mathrm{REF}} + V_s) \\ \dfrac{K_1}{\tau_E} v_x + \dfrac{K_2}{\tau_E} i_t + \dfrac{K_3}{\tau_E} i_F \end{bmatrix} \tag{7.104}$$

Note that v_x, i_t, and i_F are all linear functions of x_1–x_7.

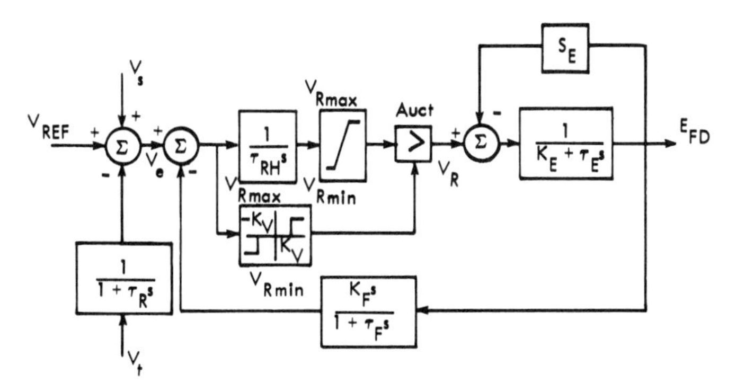

Fig. 7.67 Type 4 excitation system representation—noncontinuously setting regulator. Note: V_{RH} limited between $V_{R\min}$ and $V_{R\max}$; time constant of rheostat travel = τ_{RH}.

7.9.5 Type 4 system—noncontinuous acting

The previous systems are similar in the sense that they are all continuous acting with relatively high gain and are usually fast acting. However, a great many systems are of an earlier design similar to the rheostatic system of Section 7.7.1 and are noncontinuous acting; i.e., they have dead zones in which the system operates essentially open loop. In addition to this, they are generally characterized as slow due to friction and inertia of moving parts.

Type 4 systems (e.g., Westinghouse BJ30 or General Electric GFA4 regulated systems) often have two speeds of operation depending upon the magnitude of the voltage error. Thus a large-error voltage may cause several rheostat segments to be shorted out, while a small-error voltage will cause the segments to be shorted one at a time. The computer representation of a system is illustrated in Figure 7.67, where K_V is the raise-lower contact setting, typically set at 5%, that controls the fast-change mechanism on the rheostat. If V_e is below this limiting value of K_V, the rheostat setting is changed by motor action with an integrating time constant of τ_{RH}. An "auctioneer" circuit sets the output V_R to the higher of the two input quantities.

Because the Type 4 system is so nonlinear, there is no advantage in representing it in state variable form. The equations for the Type 4 system are similar to those derived for the electromechanical system of Section 7.7.1. A comparison of these two systems is recommended.

7.10 Typical System Constants

Reference [15] gives, in addition to the system representations, a table of typical constants of physical systems. These data are given in Table 7.11 and, although typical, do not necessarily represent any physical system accurately. For any real system all quantities should be obtained from the manufacturer.

Also note that the values in Table 7.11 are for a system with a response ratio of 0.5 which, although common, is certainly not fast by today's standards. The RR of modern fast systems are often in the range of 2.0–3.5.

Note that the values of $V_{R\max}$ and $V_{R\min}$ given in Table 7.11 are unity in column 1 and higher values in columns 2 and 3. This difference is due to the different choice of base voltage for V_R by the different exciter manufacturers and does not necessarily imply any marked difference in the regulator ceilings or performance. Changing the base voltage of V_R to $V_{R\max}$ affects all the other constants in the forward loop. There-

Table 7.11. Typical Constants of Excitation Systems in Operation on 3600 r/min Steam Turbine Generators (excitation system voltage response ratio = 0.5)

Symbol	Self-excited exciters, commutator, or silicon diode with amplidyne voltage regulators (1)	Self-excited commutator exciter with Mag-A-Stat voltage regulator (2)	Rotating rectifier exciter with static voltage regulator (3)
τ_R	0.0–0.06	0.0	0.0
K_A	25–50*	400	400
τ_A	0.06–0.20	0.05	0.02
$V_{R\max}$	1.0	3.5	7.3
$V_{R\min}$	−1.0	−3.5	−7.3
K_F	0.01–0.08	0.04	0.03
τ_F	0.35–1.0	1.0	1.0
K_E	−0.05	−0.17	1.0
τ_E	0.5	0.95	0.80
$S_{E\max}$	0.267	0.95	0.86
$S_{E.75\max}$	0.074	0.22	0.50

*For generators with open circuit field time constants greater than 4 s.

fore, caution must be used in comparing gains, time constants, and limits for systems of different manufacture.

As experience has accumulated in excitation system modeling, the manufacturer and utility engineers have determined excitation system parameters for many existing units. Since these constants are specified on a normalized basis, they can often be used with reasonable confidence on other simulations where data is unavailable. Tables 7.12–7.15 give examples of excitation system parameters that can be used for estimating new systems or for cases where exact data is unavailable.

Since the formation of the National Electric Reliability Council (NERC) a set of de-

Table 7.12. Westinghouse Excitation System Constants for System Studies (excitation system voltage response ratio = 0.5)

Symbol	Mag-A-Stat	Rotating-rectifier		BJ30	Rototrol	Silverstat	TRA
Excitation system type	1	1		4	1	1	1
τ_R (s)	0.0	0.0		...	0.05	0.02	0.05
K_A	400	400		...	200	200	400
τ_A (s)	0.05	0.02		...	0.25	0.1	0.0
$E_{FD\max}$ (pu)*	4.5	3.9		4.28	4.5	4.5	4.5
$E_{FD\min}$ (pu)*	−4.5	0		1.70	−4.5	0.3	0.2
K_E	−0.17	1.0		1.0	−0.17	−0.17	−0.17
K_F	0.04	0.03		...	0.105	0.028	0.028
τ_F (s)	1.0	1.0		...	1.25	0.5	0.5
K_V	...	...		0.05	...	...	...
τ_{RH}	...	...		20	...	...	...
		3600 r/min	1800 r/min				
$V_{R\max}$ (pu)*	3.5	7.3	8.2	8.3	3.5	3.5	3.5
$V_{R\min}$ (pu)*	−3.5	−7.3	−8.2	1.7	−3.5	0.3	0.2
$S_{E\max}$	0.95	0.86	1.10	0.95	0.95	0.95	0.95
$S_{E.75\max}$	0.22	0.50	0.50	0.22	0.22	0.22	0.22
τ_E (s)	0.95	0.8	1.30	0.76	0.85	0.50	0.50

Source: Used with permission from *Stability Program Data Preparation Manual*, Advanced Systems Technology Rept. 70-736, Westinghouse Electric Corp., Dec. 1972.

*Values given assume v_F (full load) = 3.0 pu. If not, multiply * values by $v_F/3.0$.

Table 7.13. Typical Excitation System Constants

Type of regulator	τ_R	K_A	τ_A	$V_{R\max}$	$V_{R\min}$	K_F/τ_F	τ_F
Mag-A-Stat (Type 1)	0	400	0.05†	3.5	−3.5	0.04	1.0
SCPT (Type 3)	0	120	0.15	1.2	−1.2	$0.2/\tau'_{d0}$	$\tau'_{d0}/10.0$
BJ30 (Type 4)	20.0	0.05	0	8.3	1.8	0	0
Rototrol (Type 1)	0.05	200	0.25	3.5	−3.5	0.084	1.25
Silverstat (Type 1)	0	200	0.10	3.5	−0.05	0.056	0.5
TRA (Type 1)	0	400	0.05†	3.5	−0.04	0.056	0.45
GFA4 (Type 4)	0.05	20	0	1.0	0	0	1.0
NA101 (Type 1) Amplidyne	0.06	*	0.2	1.0	−1.0	$11.5\tau_E/K_A$	0.35
NA108 (Type 1) Amplidyne	0	*	0.2	1.0	−1.0	$4\tau_E/K_A$	1.0
NA143 (Type 1) Amplidyne < 5 kW	0	*	0.2	1.0	−1.0	$4\tau_E/K_A$	1.0
NA143 (Type 1) Amplidyne > 5 kW	0	*	0.06	1.0	−1.0	$8\tau_E/K_A$	1.0
Brushless (Type 2) 3600 r/min	0	400	0.02	7.3	−7.8	0.03	1.0
Brushless (Type 2) 1800 r/min	0	400	0.02	8.2	−8.2	0.03	1.0

Source: Used by permission from *Power System Stability Program User's Guide*, Philadelphia Electric Co., 1971.

*Data obtained from curves supplied by manufacturer. For typical values see Appendix D and Table 7.15.

†High-speed contact setting, if known.

sign criteria has been established specifying the conditions under which power systems must be proven stable. This has caused an enlarged interest and concern in the accuracy of modeling all system components, particularly the generators, governors, exciters, and loads. Thus it is becoming common for the manufacturer to specify the exciter model to be used in system studies and to provide accurate gains and time constants for the system purchased.

Table 7.14. Typical Excitation System Constants

Type of regulator	K_E	τ_E	A_{EX}	B_{EX}
Mag-A-Stat (Type 1)	−0.17	0.95	0.0039	1.555
SCPT* (Type 3)	1.0	0.05	0	0
BJ30 (Type 4)	1.0	0.76	0.0052	1.555
Rototrol (Type 1)	−0.17	0.85	0.0039	1.555
Silverstat (Type 1)	−0.17	0.5	0.0039	1.555
TRA (Type 1)	−0.17	0.5	0.0039	1.555
GFA4 (Type 4)	0.051†	0.5	0.00105	1.465
Brushless (Type 2) 3600 r/min	1.0	0.8	0.12	0.855
Brushless (Type 2) 1800 r/min	1.0	1.3	0.059	1.1

Source: Used by permission from *Power System Stability Program User's Guide*, Philadelphia Electric Co., 1971.

*$K_p = 1.19$

$$K_I = 1.19\left[-\sin(\cos^{-1}F_p) + \sqrt{E_{FDFL}^2 - F_p}\right]\left[\frac{\text{study MVA base}}{\text{generator MVA base}}\right]$$

$V_{B\max} = 1.4E_{FDFL}$

†High-speed contact setting, if known.

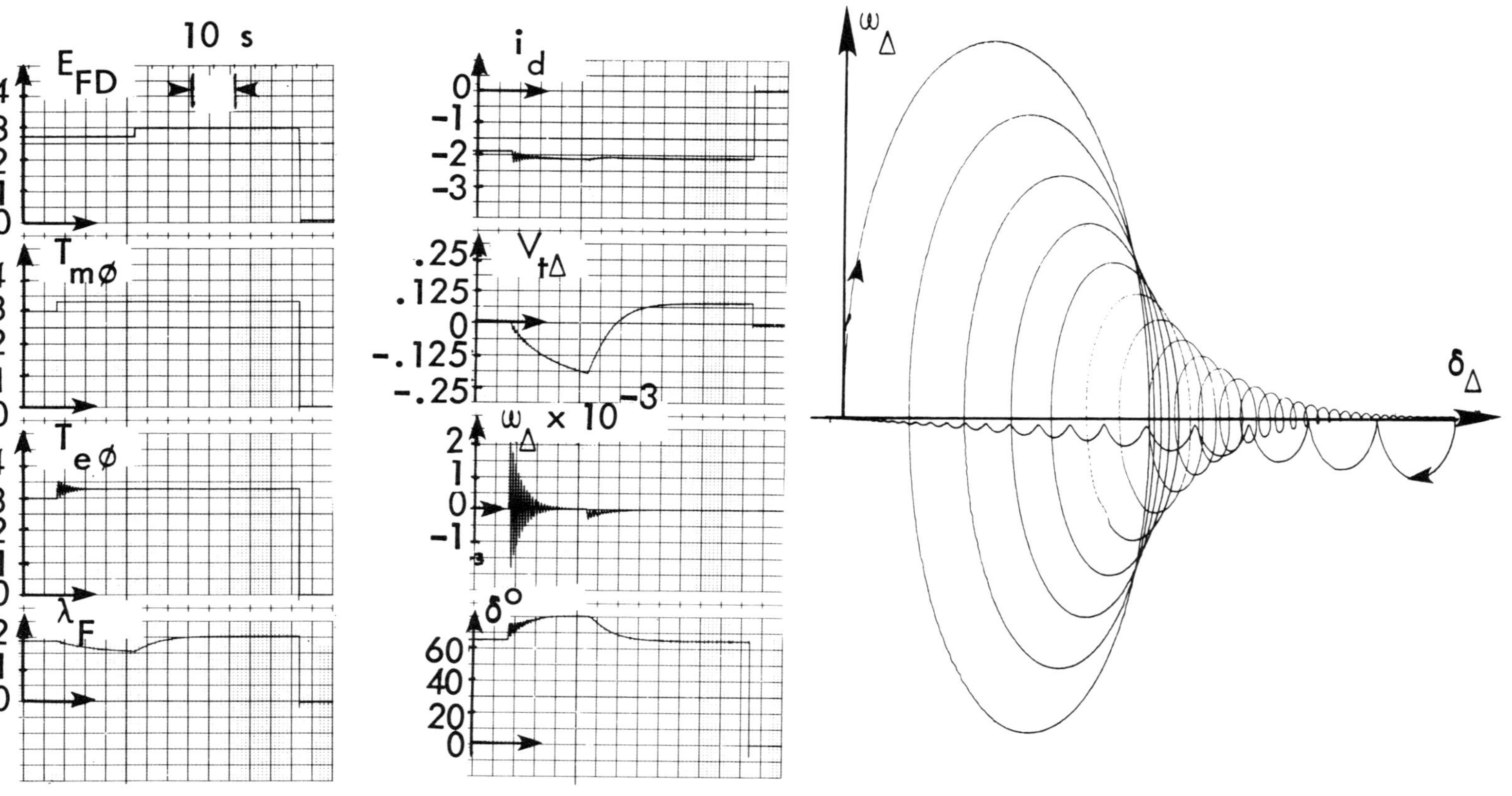

Fig. 7.68 Full model generator response of 10% step increase in T_m and E_{FD}. Initial loading of Example 5.1, with no exciter and no generator saturation.

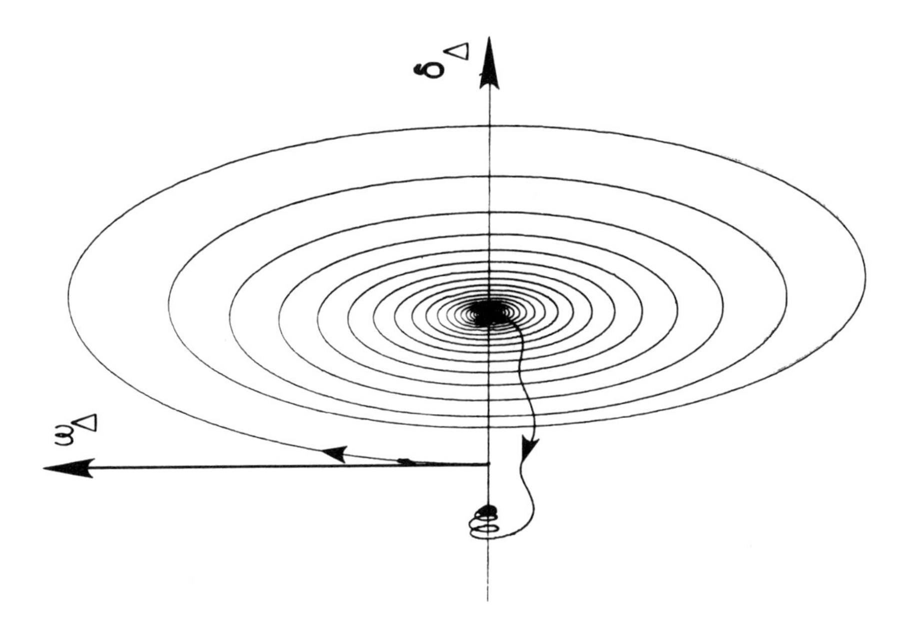

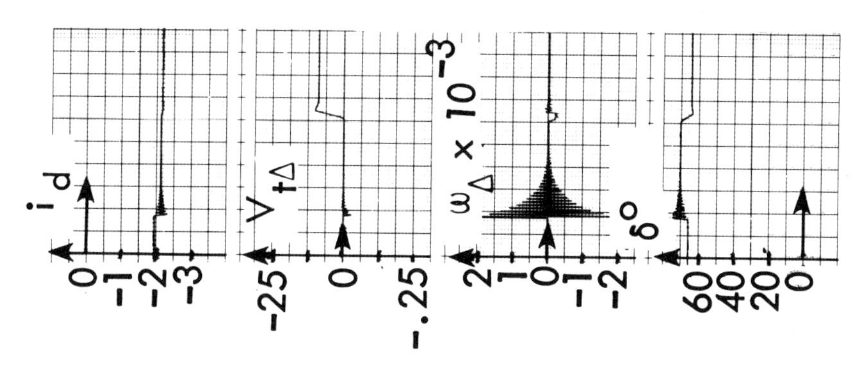

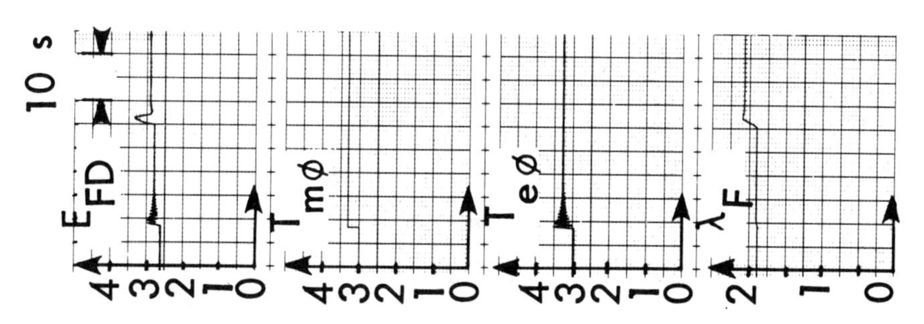

Fig. 7.69 Full model generator response to 10% step increase in T_m and 5% step increase in V_{REF}, with initial loading of Example 5.1. Exciter parameters (Westinghouse Brushless): $K_A = 400$, $\tau_A = 0.02$, $K_E = 1.0$, $\tau_E = 0.8$, $K_F = 0.03$, $\tau_F = 1.0$, $K_R = 1.0$, $\tau_R = 0.0$, $V_{R\max} = 7.3$, $V_{R\min} = -7.3$, $E_{FD\max} = 3.93$; no generator or exciter saturation.

Table 7.15. Typical Excitation System Constants for Exciters with Amplidyne Voltage Regulators (NA101, NA108, NA143)

Exciter nominal response	K_E	τ_E	K_A*	$K_{A\max}$*	K_A†	$K_{A\max}$†	A_{EX}‡	B_{EX}‡
0.5	−0.0445	0.5	$20\tau'_{d0}/3$	25	$20\tau'_{d0}$	50	0.0016	1.465
1.0	−0.0333	0.25	$10\tau'_{d0}/3$	25	$10\tau'_{d0}$	50	0.0058	1.06
1.5	−0.0240	0.1428	$25\tau'_{d0}/13$	25	$17\tau'_{d0}/3$	50	0.0093	0.898
2.0	−0.0171	0.0833	$25\tau'_{d0}/22$	25	$10\tau'_{d0}/3$	50	0.0108	0.79

Source: Used by permission from *Power System Stability Program User's Guide*, Philadelphia Electric Co., 1971.

*For all NA101, NA108, and NA143 5 kW or less.

†For NA143 over 5 kW.

‡See (7.90).

7.11 The Effect of Excitation on Generator Performance

Using the models of excitation systems presented in this chapter and the full model of the generator developed in Chapters 4 and 5, we can construct a computer simulation of a generator with an excitation system. The results of this simulation are interesting and instructive and demonstrate clearly the effect of excitation on system performance.

For the purpose of illustration, a Type 1 excitation system similar to Figure 7.61, has been added to the generator analog simulation of Figure 5.18. Appropriate switching is arranged so the simulation can be operated with the exciter active or with constant E_{FD}. The results are shown in Figure 7.68 for constant E_{FD} and Figure 7.69 with the exciter operative. The exciter modeled for this illustration is similar to the Westinghouse Brushless exciter.

Both Figures 7.68 and 7.69 show the response of the system to a 10% step increase in T_m, beginning with the full-load condition of Example 5.1. For the generator with no exciter, this torque increase causes a monotone decay in both λ_F and V_t and an increase in δ that will eventually cause the generator to pull out of step. This increase in δ is most clearly shown in the phase plane plot.

Adding the excitation system, as shown in Figure 7.69, improves the system response dramatically. Note that the exciter holds λ_F and V_t nearly constant when T_m is changed. As a result, δ is increased to its new operating level in a damped oscillatory manner. The phase plane plot shows a stable focus at the new δ.

Following the increase in torque the system is subjected to an increase in E_{FD}. This is accomplished by switching the unregulated machine E_{FD} from 100% to 110% of the Example 5.1 level. In the regulated machine a 5% step increase in V_{REF} is made. The results are roughly the same with increases noted in λ_F and V_t, and with a decrease in δ to just below the initial value.

We conclude that for the load change observed, the exciter has a stabilizing influence due to its ability to hold the flux linkages and voltage nearly constant. This causes the change in δ to be more stable. In Chapter 8 we will consider further the effects of excitation on stability, both in the transient and dynamic modes of operation.

Problems

7.1 Consider the generator of Figure 7.2 as analyzed in Example 7.1. Repeat Example 7.1 but assume that the machine is located at a remote location so that the terminal voltage V_t increases roughly in proportion to E_g. Assume, however, that the output power is held constant by the governor.

7.2 Consider the generator of Example 7.1 connected in parallel with an infinite bus and operating with constant excitation. By means of a phasor diagram analyze the change in δ, I, and θ when the governor setting is changed to increase the power output by 20%. Note particularly the change in δ in both direction and magnitude.

7.3 Following the change described in Problem 7.2, what action would be required, and in what amount, to restore the power factor to its original value?

7.4 Repeat Example 7.1 except that instead of increasing the excitation, decrease E_g to a magnitude less than that of V_t. Observe the new values of δ and θ and, in particular, the change in δ and θ.

7.5 Comparing results of Example 7.1 and Problems 7.1–7.4, can you make any general statement regarding the sensitivity of δ and θ to changes in P and E_g?

7.6 Establish a line of reasoning to show that a heavily cumulative compounded exciter is not desirable. Assume linear variations where necessary to establish your arguments.

7.7 Consider the separately excited exciter E shown in Figure P7.7. The initial current in the generator field is ρ when the exciter voltage $v_F = k_0$. At time $t = a$ a step function in the voltage v_F is introduced; i.e., $v_F = k_0 + k_1 u(t - a)$.

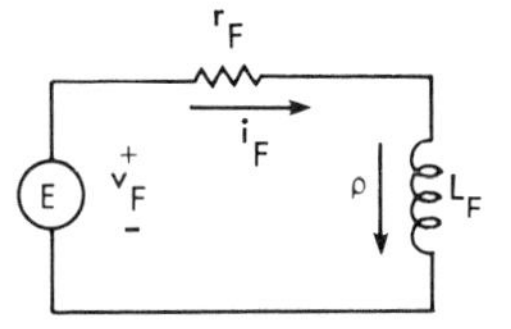

Fig. P7.7

Compute the current i_F. Sketch this result for the cases where the time constant L_F/r_F is both very large and very small. Plot the current function in the s plane.

7.8 Consider the exciter shown in Figure P7.8, where the main exciter M is excited by a pilot exciter P such that the relation $v_F = k'\omega_e \simeq ki_1$ holds. What assumptions must be made for the above relation to be approximately valid? Compute the current i_2 due to a step change in the pilot exciter voltage, i.e., for $v_P = u(t)$.

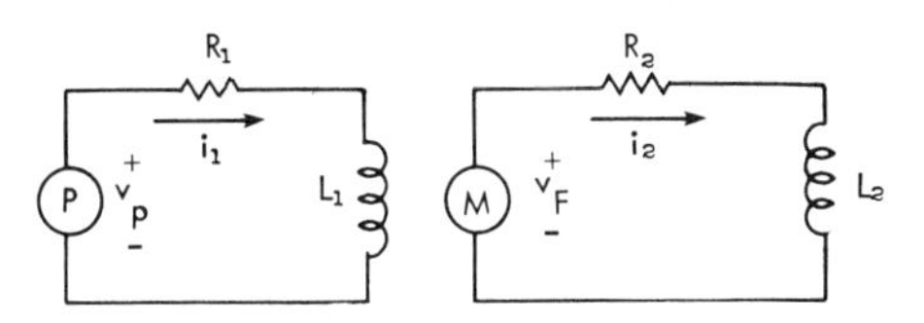

Fig. P7.8

7.9 A solenoid is to be used as the sensing and amplification mechanism for a crude voltage regulator. The system is shown in Figure P7.9. Discuss the operation of this device and comment on the feasibility of the proposed design. Write the differential equations that describe the system.

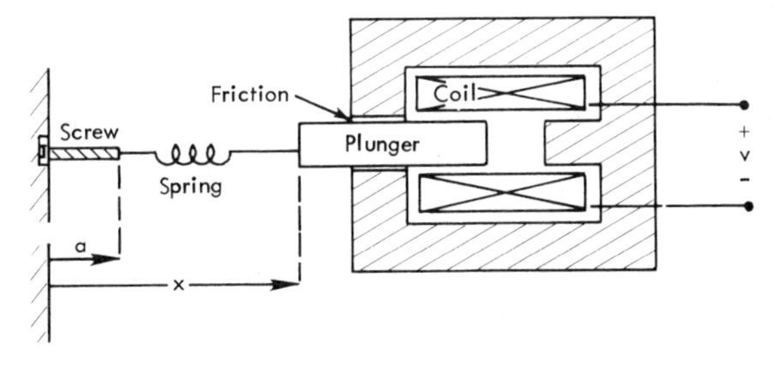

Fig. P7.9

7.10 An exciter for an ac generator, instead of being driven from the turbine-generator shaft, is driven by a separate motor with a large flywheel. Consider the motor to have a constant output torque and write the equations for this system.

7.11 Analyze the system given in Figure P7.11 to determine the effectiveness of the damping transformer in stabilizing the system to sudden changes. Write the equations for this system and show that, with parameters carefully selected, a degree of stabilization is achieved, particularly for large values of R_P. Assume no load on the exciter.

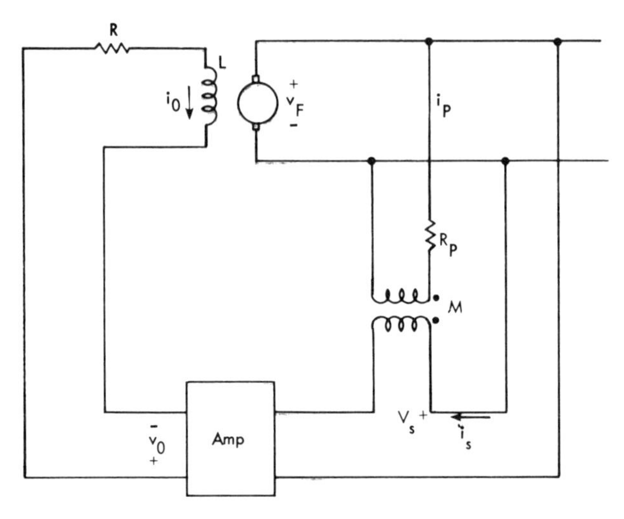

Fig. P7.11

7.12 The separately excited exciter shown in Figure P7.12 has a magnetization curve as given in Table 7.3. Other constants of interest are

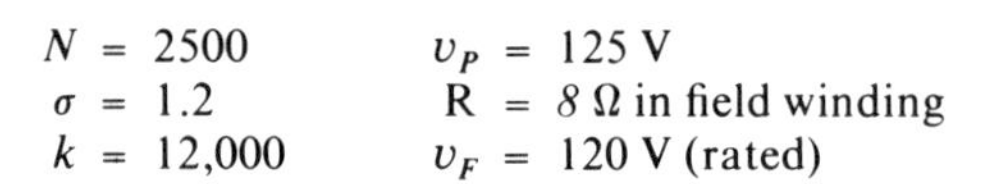

$N = 2500$ $\quad v_P = 125$ V
$\sigma = 1.2$ $\quad$ R $= 8\ \Omega$ in field winding
$k = 12{,}000$ $\quad v_F = 120$ V (rated)

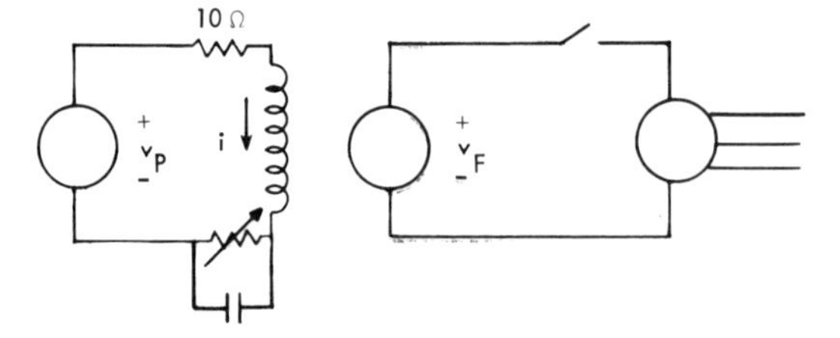

Fig. P7.12

(a) Determine the buildup curve beginning at rated voltage; i.e., $v_{F1} = 120$ V. What are the initial and final values of resistance in the field circuit?

(b) What is the main exciter response ratio?

7.13 Given the same exciter of Problem 7.12, consider a self-excited connection with an amplidyne boost-buck regulation system that quickly goes to its saturation voltage of +100 V following a command from the voltage regulator. If this forcing voltage is held constant, compute the buildup. Assume $v_{F1} = 40$ V, $v_{F2} = 180$ V.

7.14 Assume that the constants τ_A, τ_E, τ_G, K_E, K_G, and K_A are the same as in Example 7.7. Let τ_R take the values of 0.001, 0.01, and 0.1. Find the effect of τ_R on the branch of the root locus near the imaginary axis.

7.15 Repeat Problem 7.14 with $\tau_R = 0.05$ and for values of $\tau_A = 0.05$ and 0.2.

7.16 Obtain the loci of the roots for the polynomial of (7.63) for $\tau_F = 0.3$ and for values of K_F between 0.02 and 0.10.

7.17 Obtain (or sketch) a root-locus plot for the system of Example 7.8 for $K_F = 0.05$ and $\tau_F = 0.3$.

7.18 Complete the analog computer simulation of the system of one machine connected to an infinite bus (given in Chapter 5) by adding the simulation of the excitation system. Use a Type 1 exciter. Also include the effect of saturation in the simulation.

7.19 For the excitation system described in Example 7.9 and for the machine model and operating conditions described in Example 6.6, obtain the **A** matrix of the system and find the eigenvalues.

7.20 Repeat Problem 7.19 for the conditions of Example 6.7.

7.21 Repeat Example 7.9 for the operating condition of Example 6.1.

7.22 Repeat Example 7.9 (with the same operating condition) using a Type 2 excitation system. Data for the excitation system is given in Table 7.11.

7.23 Show how the choice of base voltage for the voltage regulator output V_R affects other constants in the forward loop. Assume the usual bases for V_t and E_{FD}.

References

1. Concordia, C., and Temoshok, M. Generator excitation systems and power system performance. Paper 31 CP 67-536, presented at the IEEE Summer Power Meeting, Portland, Oreg., 1967.
2. Westinghouse Electric Corp. *Electrical Transmission and Distribution Reference Book.* Pittsburgh, Pa., 1950.
3. IEEE Committee Report. Proposed excitation system definitions for synchronous machines. *IEEE Trans.* PAS-88:1248–58, 1969.
4. Chambers, G. S., Rubenstein, A. S., and Temoshok, M. Recent developments in amplidyne regulator excitation systems for large generators. *AIEE Trans.* PAS-80:1066–72, 1961.
5. Alexanderson, E. F. W., Edwards, M. A., and Bowman, K. K. The amplidyne generator—A dynamoelectric amplifier for power control. *General Electric Rev.* 43:104–6, 1940.
6. Bobo, P. O., Carlson, J. T., and Horton, J. F. A new regulator and excitation system. *IEEE Trans.* PAS-72:175–83, 1953.
7. Barnes, H. C., Oliver, J. A., Rubenstein, A. S., and Temoshok, M. Alternator-rectifier exciter for Cardinal Plant. *IEEE Trans.* PAS-87:1189–98, 1968.
8. Whitney, E. C., Hoover, D. B., and Bobo, P. O. An electric utility brushless excitation system. *AIEE Trans.* PAS-78:1821–24, 1959.
9. Myers, E. H., and Bobo, P. O. Brushless excitation system. *Proc. Southwest IEEE Conf.* (SWIEEECO), 1966.
10. Myers, E. H. Rotating rectifier exciters for large turbine-driven ac generators. *Proc. Am. Power Conf.*, Vol. 27, Chicago, 1965.
11. Rubenstein, A. S., and Temoshok, M. Excitation systems—Designs and practices in the United States. Presented at Association des Ingénieurs Électriciens de l'Institute Électrotechnique Montefiore, A.I.M., Liege, Belgium, 1966.
12. Domeratzky, L. M., Rubenstein, A. S., and Temoshok, M. A static excitation system for industrial and utility steam turbine-generators. *AIEE Trans.* PAS-80:1072–77, 1961
13. Lane, L. J., Rogers, D. F., and Vance, P. A. Design and tests of a static excitation system for industrial and utility steam turbine-generators. *AIEE Trans.* PAS-80:1077–85, 1961.
14. Lee, C. H., and Keay, F. W. A new excitation system and a method of analyzing voltage response. *IEEE Int. Conv. Rec.* 12:5–14, 1964.
15. IEEE Committee Report. Computer representation of excitation systems. *IEEE Trans.* PAS-87: 1460–64, 1968.
16. Kimbark, E. W. *Power System Stability,* Vol. 3. Wiley, New York, 1956.
17. Cornelius, H. A., Cawson, W. F., and Cory, H. W. Experience with automatic voltage regulation on a 115-megawatt turbogenerator. *AIEE Trans.* PAS-71:184–87, 1952.
18. Dandeno, P. L., and McClymont, K. R. Excitation system response: A utility viewpoint. *AIEE Trans.* PAS-76:1497–1501, 1957.
19. Temoshok, M., and Rothe, F. S. Excitation voltage response definitions and significance in power systems. *AIEE Trans.* PAS-76:1491–96, 1957.
20. Rudenberg, R. *Transient Performance of Electric Power Systems: Phenomena in Lumped Networks.* McGraw-Hill, New York, 1950. (MIT Press, Cambridge, Mass., 1967).
21. Takahashi, J., Rabins, M. J., and Auslander, D. M. *Control and Dynamic Systems.* Addison-Wesley, Reading, Mass., 1970.

22. Brown, R. G., and Nilsson, J. W. *Introduction to Linear Systems Analysis.* Wiley, New York, 1962.
23. Savant, C. J., Jr. *Basic Feedback Control System Design.* McGraw-Hill, New York, 1958.
24. Hunter, W. A., and Temoshok, M. Development of a modern amplidyne voltage regulator for large turbine generators. *AIEE Trans.* PAS-71:894–900, 1952.
25. Porter, F. M., and Kinghorn, J. H. The development of modern excitation systems for synchronous condensers and generators. *AIEE Trans.* PAS-65:1070–27, 1946.
26. Concordia, C. Effect of boost-buck voltage regulator on steady-state power limit. *AIEE Trans.* PAS-69:380–84, 1950.
27. McClure, J. B., Whittlesley, S. I., and Hartman, M. E. Modern excitation systems for large synchronous machines. *AIEE Trans.* PAS-65:939–45, 1946.
28. General Electric Co. Amplidyne regulator excitation systems for large generators. Bull. GET-2980, 1966.
29. Harder, E. L., and Valentine, C. E. Static voltage regulator for Rototrol exciter. *Electr. Eng.* 64: 601, 1945.
30. Kallenback, G. K., Rothe, F. S., Storm, H. F., and Dandeno, P. L. Performance of new magnetic amplifier type voltage regulator for large hydroelectric generators. *AIEE Trans.* PAS-71:201–6, 1952.
31. Hand, E. W., McClure, F. N., Bobo, P. O., and Carleton, J. T. Magamp regulator tests and operating experience on West Penn Power System. *AIEE Trans.* PAS-73:486–91, 1954.
32. Carleton, J. T., and Horton, W. F. The figure of merit of magnetic amplifiers. *AIEE Trans.* PAS-71:239–45, 1952.
33. Ogle, H. M. The amplistat and its applications. *General Electric Rev.* Pt. 1, Feb.; Pt. 2, Aug.; Pt. 3, Oct., 1950.
34. Hanna, C. R., Oplinger, K. A., and Valentine, C. E. Recent developments in generator voltage regulation. *AIEE Trans.* 58:838–44, 1939.
35. Dahl, O. G. C. *Electric Power Circuits, Theory and Application,* Vol. 2. McGraw-Hill, New York, 1938.
36. Kimbark, E. W. *Power System Stability,* Vol. 1. *Elements of Stability Calculations.* Wiley, New York, 1948.
37. Kron, G. Regulating system for dynamoelectric machines. Patent No. 2,692,967, U.S. Patent Office, 1954.
38. Oyetunji, A. A. Effects of system nonlinearities on synchronous machine control. Unpubl. Ph.D. thesis, Research Rept. ERI-71130. Iowa State Univ., Ames, 1971.
39. Ferguson, R. W., Herbst, R., and Miller, R. W. Analytical studies of the brushless excitation system. *AIEE Trans.* PAS-78:1815–21, 1959.
40. Westinghouse Electric Corp. Stability program data preparation manual. Advanced Systems Technology Rept. 70-736, 1972.
41. Lane, L. J., Mendel, J. E., Ewart, D. N., Crenshaw, M. L., and Todd, J. M. A static excitation system for steam turbine generators. Paper CP 65-208, presented at the IEEE Winter Power Meeting, New York, 1965.
42. Philadelphia Electric Co. Power system stability program. Power System Planning Div., Users Guide U6004-2, 1971.

chapter 8

The Effect of Excitation on Stability

8.1 Introduction

Considerable attention has been given in the literature to the excitation system and its role in improving power system stability. Early investigators realized that the so-called "steady-state" power limits of power networks could be increased by using the then available high-gain continuous-acting voltage regulators [1]. It was also recognized that the voltage regulator gain requirement was different at no-load conditions from that needed for good performance under load. In the early 1950s engineers became aware of the instabilities introduced by the (then) modern voltage regulators, and stabilizing feedback circuits came into common use [2]. In the 1960s large interconnected systems experienced growing oscillations that disrupted parallel operation of large systems [3–12]. It was discovered that the inherently weak natural damping of large and weakly coupled systems was the main cause and that situations of negative damping were further aggravated by the regulator gain [13]. Engineers learned that the system damping could be enhanced by artificial signals introduced through the excitation system. This scheme has been very successful in combating growing oscillation problems experienced in the power systems of North America.

The success of excitation control in improving power system dynamic performance in certain situations has lead to greater expectations among power system engineers as to the capability of such control. Because of the small effective time constants in the excitation system control loop, it was assumed that a large control effort could be expended through excitation control with a relatively small input of control energy. While basically sound, this control is limited in its effectiveness. A part of the engineer's job, then, is to determine this limit, i.e., to find the exciter design and control parameters that can provide good performance at reasonable cost [14].

The subject of excitation control is further complicated by a conflict in control requirements in the period following the initiation of a transient. In the first few cycles these requirements may be significantly different from those needed over a few seconds. Furthermore, it has been shown that the best control effort in the shorter period may tend to cause instability later. This suggests the separation of the excitation control studies into two distinct problems, the *transient* (short-term) problem and the *dynamic* (long-term) problem. It should be noted that this terminology is not universally used. Some authors call the dynamic stability problem by the ambiguous name of "steady-state stability." Other variations are found in the literature, but usually the two problems are treated separately as noted.

8.1.1 Transient stability and dynamic stability considerations

In transient stability the machine is subjected to a large impact, usually a fault, which is maintained for a short time and causes a significant reduction in the machine terminal voltage and the ability to transfer synchronizing power. If we consider the one machine–infinite bus problem, the usual approximation for the power transfer is given by

$$P = (V_t V_\infty / x) \sin \delta \tag{8.1}$$

where V_t is the machine terminal voltage and V_∞ is the infinite bus voltage. Note that if V_t is reduced, P is reduced by a corresponding amount. Prevention of this reduction in P requires very fast action by the excitation system in forcing the field to ceiling and thereby holding V_t at a reasonable value. Indeed, the most beneficial attributes the voltage regulator can have for this situation is speed and a high ceiling voltage, thus improving the chances of holding V_t at the needed level. Also, when the fault is removed and the reactance x of (8.1) is increased due to switching, another fast change in excitation is required. These violent changes affect the machine's ability to release the power it is receiving from the turbine. These changes are effectively controlled by very fast excitation changes.

The dynamic stability problem is different from the transient problem in several ways, and the requirements on the excitation system are also different. By dynamic stability we mean the ability of all machines in the system to adjust to small load changes or impacts. Consider a multimachine system feeding a constant load (a condition never met in practice). Let us assume that at a given instant the load is changed by a small amount, say by the energizing of a very large motor somewhere in the system. Assume further that this change in load is just large enough to be recognized as such by a certain group of machines we will call the control group. The machines nearest the load electrically will see the largest change, and those farther away will experience smaller and smaller changes until the change is not perceptible at all beyond the boundary of the control group.

Now how will this load change manifest itself at the several machines in the control group? Since it is a load increase, there is an immediate increase in the output power requirements from each of the machines. Since step changes in power to turbines are not possible, this increased power requirement will come first from stored energy in the control group of machines. Thus energy stored in the magnetic field of the machines is released, then somewhat later, rotating energy $[(1/2)mv^2]$ is used to supply the load requirements until the governors have a chance to adjust the power input to the various generators. Let us examine the behavior of the machines in the time interval prior to the governor action. This interval may be on the order of 1 s. In this time period the changes in machine voltages, currents, and speeds will be different for each machine in the control group because of differences in unit size, design, and electrical location with respect to the load. Thus each unit responds by contributing its share of the load increase, with its share being dictated by the impedance it sees at its terminals (its Thevenin impedance) and the size of the unit. Each unit has its own natural frequency of response and will oscillate for a time until damping forces can decay these oscillations. Thus the one change in load, a step change, sets up all kinds of oscillatory responses and the system "rings" for a time with many frequencies present, these induced changes causing their own interaction with neighboring machines (see Section 3.6).

Now visualize the excitation system in this situation. In the older electromechanical systems there was a substantial deadband in the voltage regulator, and unless the generator was relatively close to the load change, the excitation of these machines would remain unchanged. The machines closer to the load change would recognize a need for increased excitation and this would be accomplished, although somewhat slowly. Newer excitation systems present a different kind of problem. These systems recognize the change in load immediately, either as a perceptible change in terminal voltage, terminal current, or both. Thus each oscillation of the unit causes the excitation system to try to correct accordingly, since as the speed voltage changes, the terminal voltage also changes. Moreover, the oscillating control group machines react with one another, and each action or reaction is accompanied by an excitation change.

The excitation system has one major handicap to overcome in following these system oscillations; this is the effective time constant of the main exciter field which is on the order of a few seconds or so. Thus from the time of recognition of a desired excitation change until its partial fulfillment, there is an unavoidable delay. During this delay time the state of the oscillating system will change, causing a new excitation adjustment to be made. This system lag then is a detriment to stable operation, and several investigators have shown examples wherein systems are less oscillatory with the voltage regulators turned off than with them operating [7, 12].

Our approach to this problem must obviously depend upon the type of impact under consideration. For the large impact, such as a fault, we are concerned with maximum forcing of the field, and we examine the response in building up from normal excitation to ceiling excitation. This is a nonlinear problem, as we have seen, and the shape of the magnetization curve cannot be neglected. The small impact or dynamic stability problem is different. Here we are concerned with small excursions from normal operation, and linearization about this normal or "quiescent" point is possible and desirable. Having done this, we may study the response using the tools of linear systems analysis; in this way not only can we analyze but possibly compensate the system for better damping and perhaps faster response.

8.2 Effect of Excitation on Generator Power Limits

We begin with a simple example, the purpose of which is to show that the excitation system can have an effect upon stability.

Example 8.1

Consider the two-machine system of Figure 8.1, where we consider one machine against an infinite bus. (This problem was introduced and analyzed by Concordia [1].) The power output of the machine is given by

$$P = [E_1 E_2/(X_1 + X_2)] \sin \delta$$
$$\delta = \delta_1 + \delta_2 \tag{8.2}$$

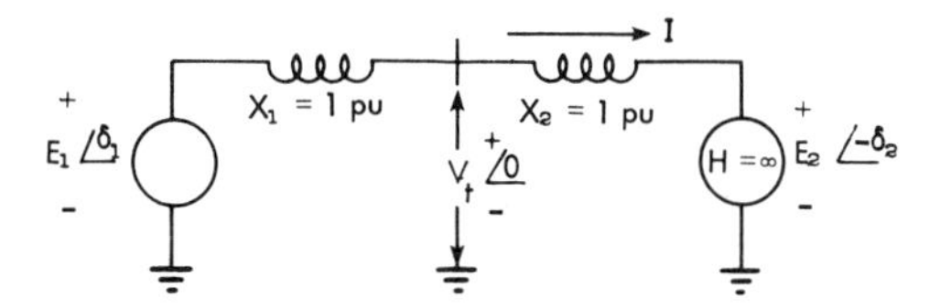

Fig. 8.1 One machine–infinite bus system.

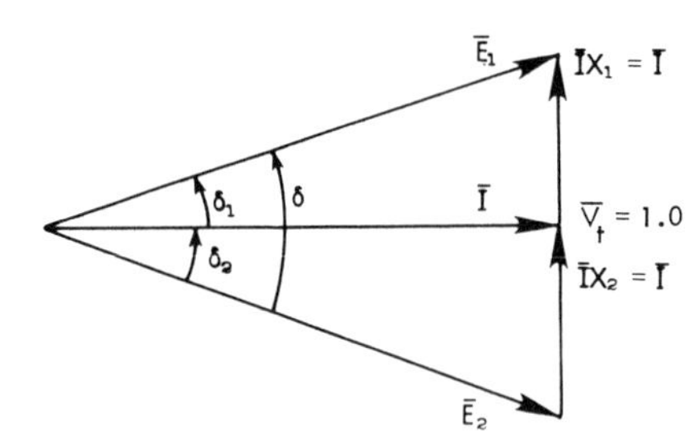

Fig. 8.2 Phasor diagram for Example 8.1.

This equation applies whether or not there is a voltage regulator. Determine the effect of excitation on this equation.

Solution

We now establish the boundary conditions for the problem. First we assume that $X_1 = X_2 = 1.0$ pu and that $V_t = 1.0$ pu. Then for any given load the voltages $\overline{E}_1$ and $\overline{E}_2$ must assume a certain value to hold $\overline{V}_t$ at 1.0 pu. If the power factor is unity, $\overline{E}_1$ and $\overline{E}_2$ have the same magnitude as shown in the phasor diagram of Figure 8.2. If $\overline{E}_1$ and $\overline{E}_2$ are held constant at these values, the power transferred to the infinite bus varies sinusoidally according to (8.2) and has a maximum when δ is 90°.

Now assume that $\overline{E}_1$ and $\overline{E}_2$ are both subject to perfect regulator action and that the key to this action is that V_t is to be held at 1.0 pu and the power factor is to be held at unity. We write in phasor notation

$$\overline{E}_1 = 1 + j\overline{I} = \sqrt{1 + I^2}\, e^{j\delta/2} \qquad \overline{E}_2 = 1 - j\overline{I} = \sqrt{1 + I^2}\, e^{-j\delta/2}$$

Adding these equations we have

$$\overline{E}_1 + \overline{E}_2 = 2 = 2\sqrt{1 + I^2}\cos\delta/2$$

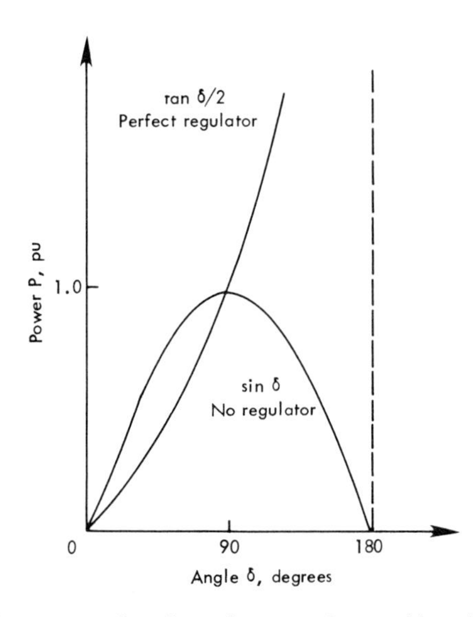

Fig. 8.3 Comparison of power transferred at unity power factor with and without excitation control.

or

$$E_1 = E_2 = \frac{1}{\cos \delta/2} \tag{8.3}$$

Substituting (8.3) into (8.2) and simplifying, we have for the perfect regulator, at unity power factor,

$$P = \tan \delta/2 \tag{8.4}$$

The result is plotted in Figure 8.3 along with the same result for the case of constant (unregulated) E_1 and E_2.

In deriving (8.4), we have tacitly assumed that the regulators acting upon E_1 and E_2 do so instantaneously and continuously. The result is interesting for several reasons. First, we observe that with this ideal regulation there is no stability limit. Second, it is indicated that operation in the region where $\delta > 90°$ is possible. We should comment that the assumed physical system is not realizable since there is always a lag in the excitation response even if the voltage regulator is ideal. Also, excitation control of the infinite bus voltage is not a practical consideration, as this remote bus is probably not infinite and may not be closely regulated.

Example 8.2

Consider the more practical problem of holding the voltage E_2 constant at 1.0 pu and letting the power factor vary, other things being the same.

Solution

Under this condition we have the phasor diagram of Figure 8.4 where we note that the locus of E_2 is the dashed circular arc of length 1.0. Note that the power factor is constrained by the relation

$$\theta_1 = \delta_2/2 \tag{8.5}$$

where $\theta_1 = 2\pi - \theta$ and $\delta = \delta_1 - \delta_2$.

Writing phasor equations for the voltages, we have

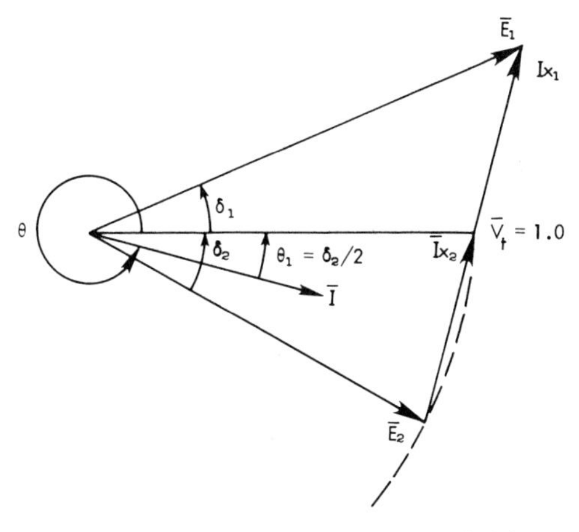

Fig. 8.4 Phasor diagram for Example 8.2.

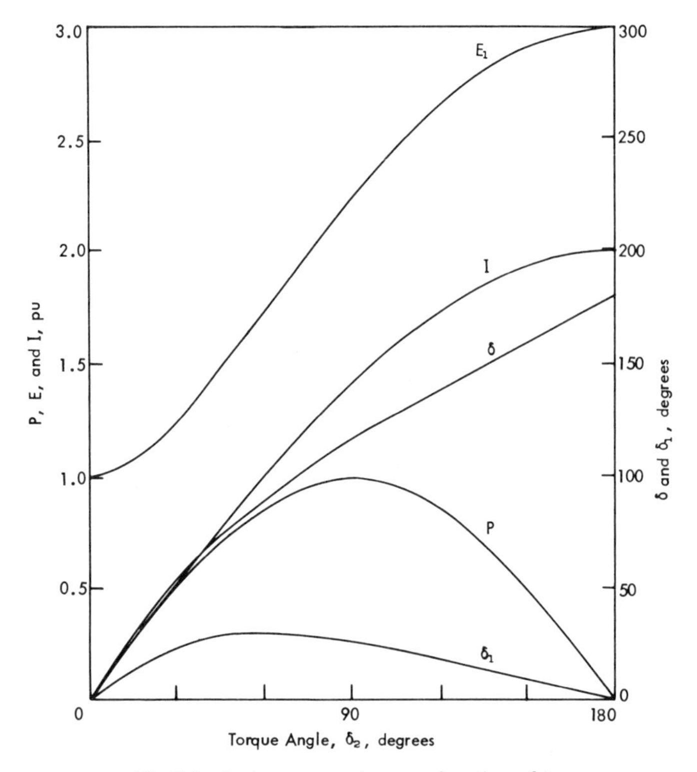

Fig. 8.5 System parameters as a function of δ_2.

$$\bar{E}_1 = 1 + j\bar{I} = 1 - I\sin\theta + jI\cos\theta = E_1 e^{j\delta_1}$$
$$\bar{E}_2 = 1 - j\bar{I} = 1 + I\sin\theta - jI\cos\theta = E_2 e^{-j\delta_2} \tag{8.6}$$

where θ, θ_1, δ_1, and δ_2 are all measured positive as counterclockwise. Noting that $E_2 =$ 1, we can establish that

$$I = 2\sin\theta_1 \qquad E_1 \sin\delta = 2\sin\delta_2$$
$$\sin\delta = 2\sin\delta_1 \qquad \tan\delta_1 = \sin\delta_2/(2 - \cos\delta_2) \tag{8.7}$$

Thus once we establish δ_2, we also fix θ, I, δ, and δ_1, although the relationships among these variables are nonlinear. These results are plotted in Figure 8.5 where equations (8.7) are used to determine the plotted values. We also note that

$$P = V_t I \cos\theta \tag{8.8}$$

but from the second of equations (8.6) we can establish that $I\cos\theta = \sin\delta_2$ or

$$P = \sin\delta_2 \tag{8.9}$$

so δ_2 also establishes P. Thus P does have a maximum in this case, and this occurs when $\delta_2 = 90°$ ($\bar{E}_2$ pointing straight down in Figure 8.4). In this case we have at maximum power

$$\bar{E}_1 = 2 + j1 = 2.235\,\underline{/26.6°} \qquad I = 1.414$$
$$\theta = -45° \qquad \delta = 116.6°$$

The important thing to note is that P is again limited, but we see that δ may go

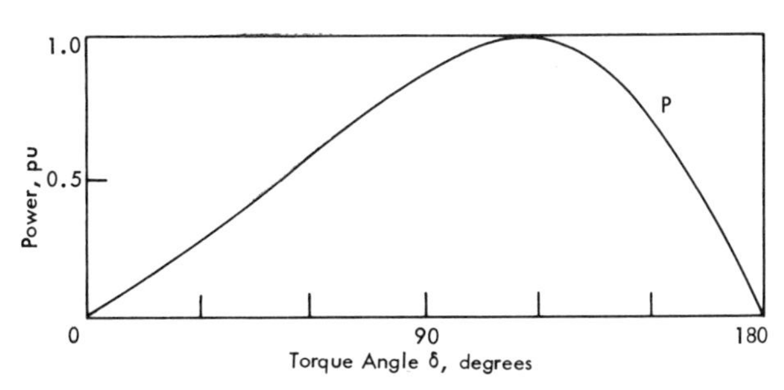

Fig. 8.6 Variation of P with δ.

beyond 90° to achieve maximum power and that this requires over 2 pu E_1. The variation of P with δ is shown in Figure 8.6.

These simple examples show the effect of excitation under certain ideal situations. Obviously, these ideal conditions will not be realized in practice. However, they provide limiting values of the effect of excitation on changing the effective system parameters. A power system is nearly a constant voltage system and is made so because of system component design and close voltage control. This means that the Thevenin impedance seen looking into the source is very small. Fast excitation helps keep this impedance small during disturbances and contributes to system stability by allowing the required transfer of power even during disturbances. Finally, it should be stated that while the ability of exciters to accomplish this task is limited, other considerations make it undesirable to achieve perfect control and *zero* Thevenin impedance. Among these is the fault-interrupting capability.

8.3 Effect of the Excitation System on Transient Stability

In the transient stability problem the performance of the power system when subjected to severe impacts is studied. The concern is whether the system is able to maintain synchronism during and following these disturbances. The period of interest is relatively short (at most a few seconds), with the first swing being of primary importance. In this period the generator is suddenly subjected to an appreciable change in its output power causing its rotor to accelerate (or decelerate) at a rate large enough to threaten loss of synchronism. The important factors influencing the outcome are the machine behavior and the power network dynamic relations. For the sake of this discussion it is assumed that the power supplied by the prime movers does not change in the period of interest. Therefore the effect of excitation control on this type of transient depends upon its ability to help the generator maintain its output power in the period of interest.

To place the problem in the proper perspective, we should review the main factors that affect the performance during severe transients. These are:

1. The disturbing influence of the impact. This includes the type of disturbance, its location, and its duration.
2. The ability of the transmission system to maintain strong synchronizing forces during the transient initiated by a disturbance.
3. The turbine-generator parameters.

The above have traditionally been the main factors affecting the so-called first-swing transients. The system parameters influencing these factors are:

1. The synchronous machine parameters. Of these the most important are: (a) the inertia constant, (b) the direct axis transient reactance, (c) the direct axis open circuit time constant, and (d) the ability of the excitation system to hold the flux level of the synchronous machine and increase the output power during the transient.
2. The transmission system impedances under normal, faulted, and postfault conditions. Here the flexibility of switching out faulted sections is important so that large transfer admittances between synchronous machines are maintained when the fault is isolated.
3. The protective relaying scheme and equipment. The objective is to detect faults and isolate faulted sections of the transmission network very quickly with minimum disruption.

8.3.1 The role of the excitation system in classical model studies

In the classical model it is assumed that the flux linking the main field winding remains constant during the transient. If the transient is initiated by a fault, the armature reaction tends to decrease this flux linkage [15]. This is particularly true for the generators electrically close to the location of the fault. The voltage regulator tends to force the excitation system to boost the flux level. Thus while the fault is on, the effect of the armature reaction and the action of the voltage regulator tend to counteract each other. These effects, along with the relatively long effective time constant of the main field winding, result in an almost constant flux linkage during the first swing of 1 s or less. (For the examples in Chapter 6 this time constant $K_3 \tau'_{d0}$ is about 2.0 s.)

It is important to recognize what the above reasoning implies. First, it implies the presence of a voltage regulator that tends to hold the flux linkage level constant. Second, it is significant to note that the armature reaction effects are particularly pronounced during a fault since the reactive power output of the generator is large. Therefore the duration of the fault is important in determining whether a particular type of voltage regulator would be adequate to maintain constant flux linkage.

A study reported by Crary [2] and discussed by Young [15] illustrates the above. The system studied consists of one machine connected to a larger system through a 200-mile double circuit transmission line. The excitation system for the generator is Type 1 (see Chapter 7) with provision to change the parameters such that the response ratio (RR) varies from 0.10 to 3.0 pu. The former corresponds to a nearly constant field voltage condition. The latter would approximate the response of a modern fast excitation system. Data of the system used in the study are shown in Figure 8.7. A transient stability study was made for a three-phase fault near the generator. The sending end power limits versus the fault clearing time are shown in Figure 8.8 for different exciter responses (curves 1–5) and for the classical model (curve 6).

From Figure 8.8 it appears that the classical model corresponds to a very slow and weak excitation system for very short fault clearing times, while for longer clearing times it approximates a rather fast excitation system. If the nature of the stability study is such that the fault clearing time is large, as in "stuck breaker" studies [15], the actual power limits may be lower than those indicated when using the classical model.

In another study of excitation system representation [16] the authors report (in a certain stability study they conducted) that a classical representation showed a certain generator to be stable, while detailed representation of the generator indicated that loss of synchronism resulted. The authors conclude that the dominant factor affecting loss

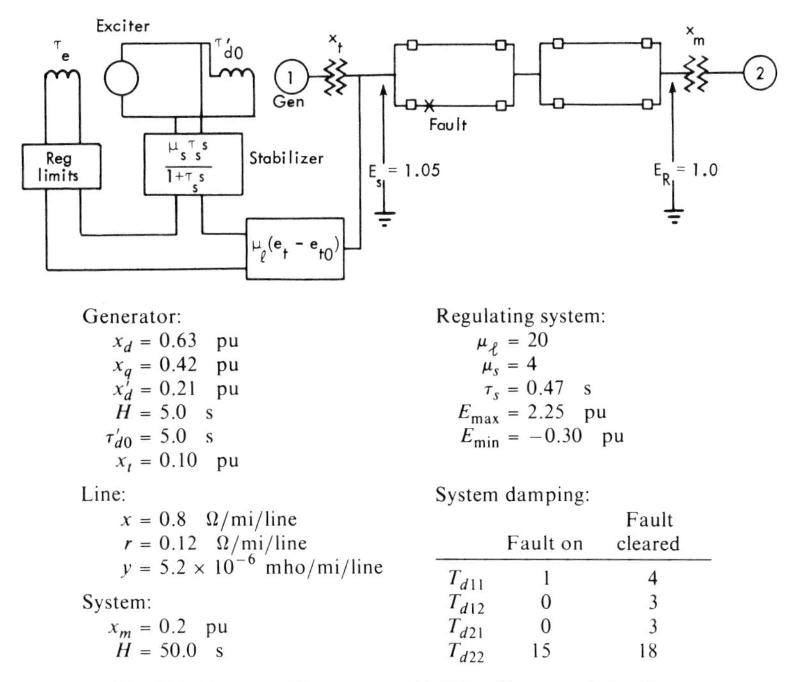

Generator:
$x_d = 0.63$ pu
$x_q = 0.42$ pu
$x'_d = 0.21$ pu
$H = 5.0$ s
$\tau'_{d0} = 5.0$ s
$x_t = 0.10$ pu

Line:
$x = 0.8$ Ω/mi/line
$r = 0.12$ Ω/mi/line
$y = 5.2 \times 10^{-6}$ mho/mi/line

System:
$x_m = 0.2$ pu
$H = 50.0$ s

Regulating system:
$\mu_\ell = 20$
$\mu_s = 4$
$\tau_s = 0.47$ s
$E_{max} = 2.25$ pu
$E_{min} = -0.30$ pu

System damping:

	Fault on	Fault cleared
T_{d11}	1	4
T_{d12}	0	3
T_{d21}	0	3
T_{d22}	15	18

Fig. 8.7 Two-machine system with 200-mile transmission lines.

of synchronism is the inability of the excitation system of that generator, with response ratio of 0.5, to offset the effects of armature reaction.

8.3.2 Increased reliance on excitation control to improve stability

Trends in the design of power system components have resulted in lower stability margins. Contributing to this trend are the following:

1. Increased rating of generating units with lower inertia constants and higher pu reactances.
2. Large interconnected system operating practices with increased dependence on the transmission system to carry greater loading.

These trends have led to the increased reliance on the use of excitation control as a

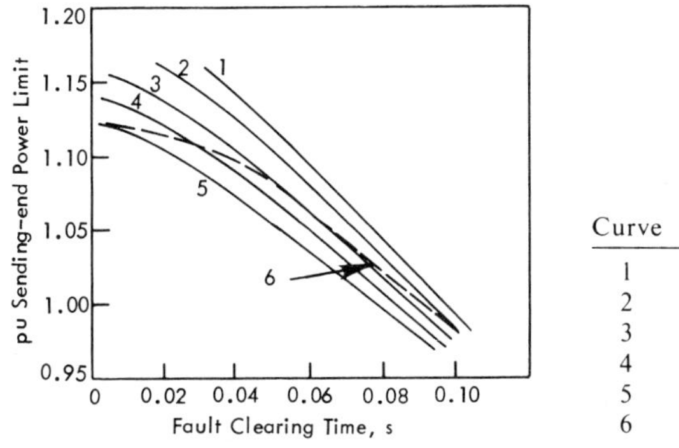

Curve	Clearing Time	RR
1	0.042 s	3.0
2	0.17 s	2.0
3	0.68 s	1.0
4	2.70 s	0.25
5	11.0 s	0.10
6	Classical model	

Fig. 8.8 Sending-end power versus fault clearing time for different excitation system responses.

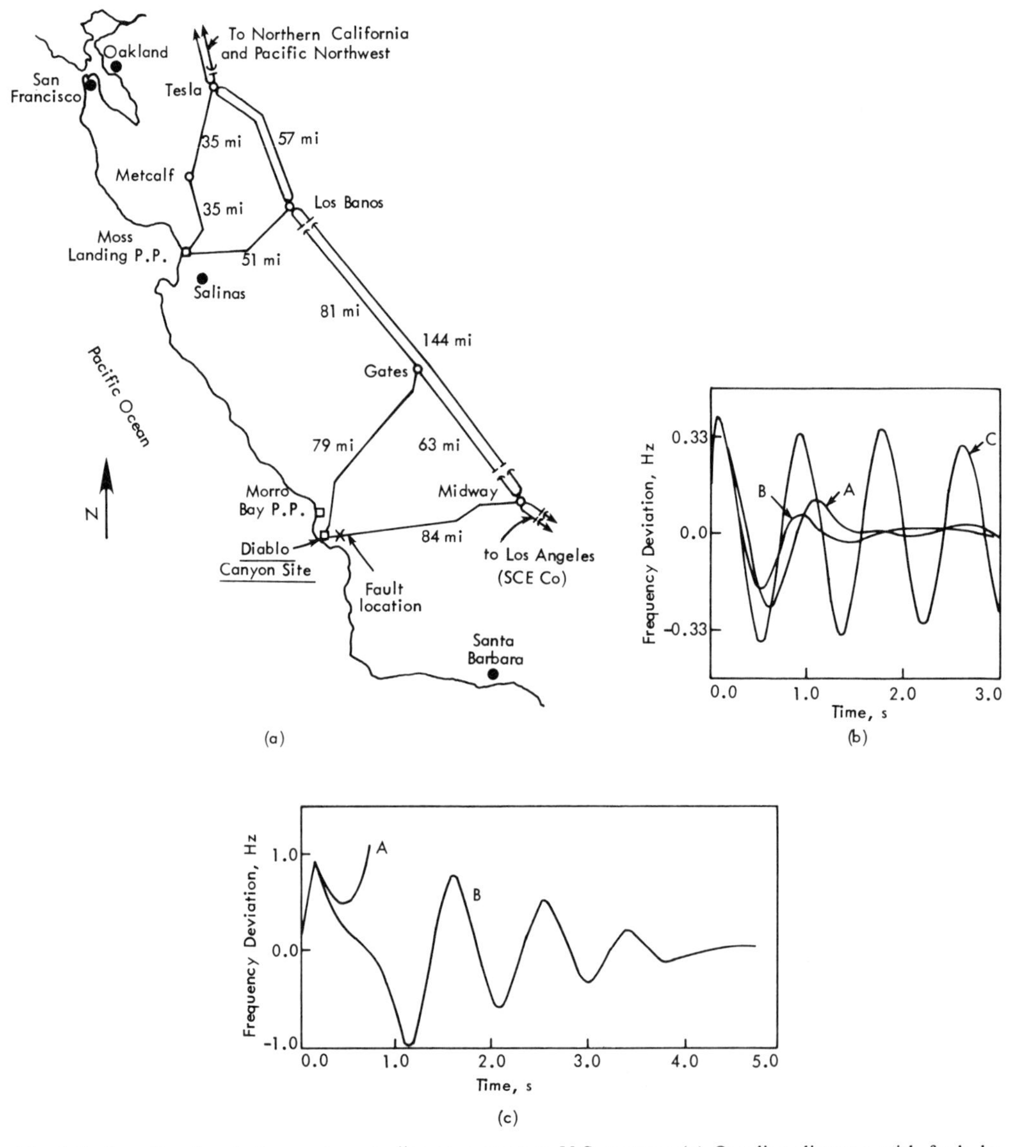

Fig. 8.9 Results of excitation system studies on a western U.S. system: (a) One-line diagram with fault location, (b) frequency deviation comparison for a four-cycle fault, (c) frequency deviation comparison for a 9.6-cycle fault: A = 2.0 ANSI conventional excitation system; B = low time constant excitation system with rate feedback; C = low time constant excitation system without rate feedback. (© IEEE. Reprinted from *IEEE Trans.*, vol. PAS-90, Sept./Oct. 1971.)

means of improving stability [17]. This has prompted significant technological advances in excitation systems.

As an aid to transient stability, the desirable excitation system characteristics are a fast speed of response and a high ceiling voltage. With the help of fast transient forcing of excitation and the boost of internal machine flux, the electrical output of the machine may be increased during the first swing compared to the results obtainable with a slow exciter. This reduces the accelerating power and results in improved transient performance.

Modern excitation systems can be effective in two ways: in reducing the severity of machine swings when subjected to large impacts by reducing the magnitude of the first swing and by ensuring that the subsequent swings are smaller than the first. The latter is an important consideration in present-day large interconnected power systems. Situations may be encountered where various modes of oscillations reinforce each other during later swings, which along with the inherent weak system damping can cause transient instability after the first swing. With proper compensation a modern excitation system can be very effective in correcting this type of problem. However, except for transient stability studies involving faults with long clearing times (or stuck breakers), the effect of the excitation system on the severity of the first swing is relatively small. That is, a very fast, high-response excitation system will usually reduce the first swing by only a few degrees or will increase the generator transient stability power limit (for a given fault) by a few percent.

In a study reported by Perry et al. [18] on part of the Pacific Gas and Electric Company system in northern California, the effect of the excitation system response on the system frequency deviation is studied when a three-phase fault occurs in the network (at the Diablo Canyon site on the Midway circuit adjacent to a 500-kV bus). Some of the results of that study are shown in Figure 8.9. A one-line diagram of the network is shown in Figure 8.9(a). The frequency deviations for 4-cycle and 9.6-cycle faults are shown in Figures 8.9(b) and 8.9(c) respectively. The comparison is made between a 2.0 response ratio excitation system (curve *A*), a modern, low time constant excitation with rate feedback (curve *B*) and without rate feedback (curve *C*). The results of this study support the points made above.

8.3.3 Parametric study

Two recent studies [17, 19] show the effect of the excitation system on "first-swing" transients. Figure 8.10 shows the system studied where one machine is connected to an infinite bus through a transformer and a transmission network. The synchronous machine data is given in Table 8.1.

The transmission network has an equivalent transfer reactance X_e as shown in

Table 8.1. Machine Data for the Studies of Reference [19]

$x_d = 1.72$ pu	$\tau'_{d0} = 6.3$ s
$x'_d = 0.45$ pu	$\tau''_{d0} = 0.033$ s
$x''_d = 0.33$ pu	$\tau'_{q0} = 0.43$ s
$x_q = 1.68$ pu	$\tau''_{q0} = 0.033$ s
$x'_q = 0.59$ pu	$H = 4.0$ s
$x''_q = 0.33$ pu	

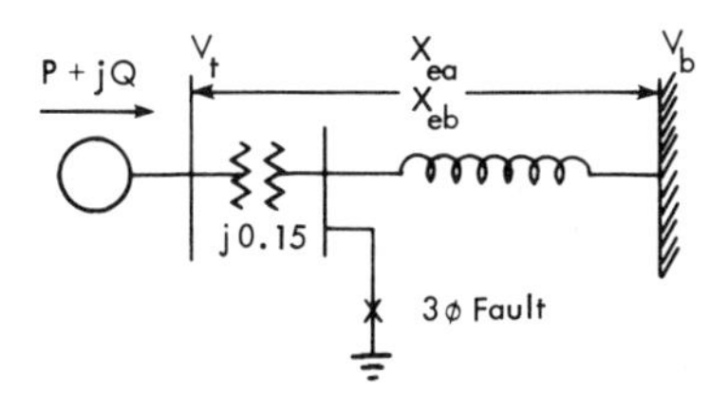

Fig. 8.10 System representation used in a parametric study of the effect of excitation on transient stability. (© IEEE. Reprinted from *IEEE Trans.*, vol. PAS-89, July/Aug. 1970.)

Figure 8.10. A transient is initiated by a three-phase fault on the high-voltage side of the transformer. The fault is cleared in a specified time. After the fault is cleared, the transfer reactance X_e is increased from X_{eb} (the value before the fault) to X_{ea} (its value after the fault is cleared). The machine initial operating conditions are summarized in Table 8.2.

Table 8.2. Prefault Operating Conditions, All Values in pu

X_{eb}	V_t	V_∞	P	Q
0.2	1.0	0.94	0.90	0.39
0.4	1.0	0.90	0.90	0.45
0.6	1.0	0.91	0.90	0.44
0.8	1.0	0.97	0.90	0.44

With the machine operating at approximately rated load and power factor, a three-phase fault is applied at the high-voltage side of the step-up transformer for a given length of time. When the fault is cleared, the transmission system reactance is changed to the postfault reactance X_{ea}, and the simulation is run until it can be determined if the run is stable or unstable. This is repeated for different values of X_{ea} until the maximum value of X_{ea} is found where the system is marginally stable.

Two different excitation system representations were used in the study:

1. A 0.5 pu response alternator-fed diode system shown in Figure 8.11.
2. A 3.0 pu response alternator-fed SCR system with high initial response shown in Figure 8.12. This system has a steady-state gain of 200 pu and a transient gain of 20 pu. An external stabilizer using a signal V_s derived from the shaft speed is also used (see Section 8.7).

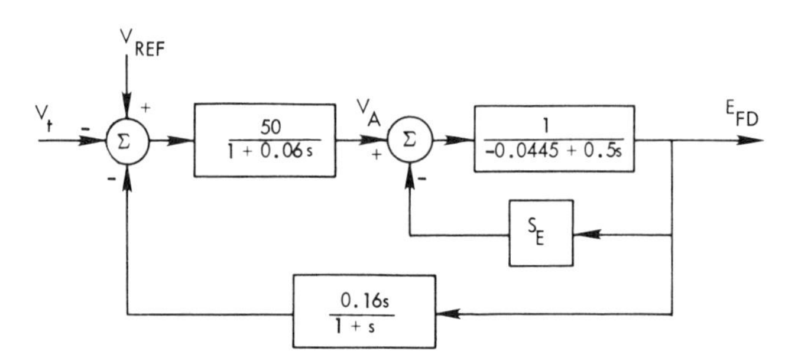

Fig. 8.11 Excitation block diagram for a 0.5 RR alternator-fed diode system. (© IEEE. Reprinted from *IEEE Trans.*, vol. PAS-89, July/Aug. 1970.)

From the data presented in [19], the effect of excitation on the "first-swing" transients is shown in Figure 8.13, where the critical clearing time is plotted against the transmission line reactance for the case where $X_{ea} = X_{eb}$ and for the two different types of excitation system used. The critical clearing time is used as a measure of relative stability for the system under the impact of the given fault. Figure 8.13 shows that for the conditions considered in this study a change in exciter response ratio from 0.5 to 3.0 resulted in a gain of approximately one cycle in critical clearing time.

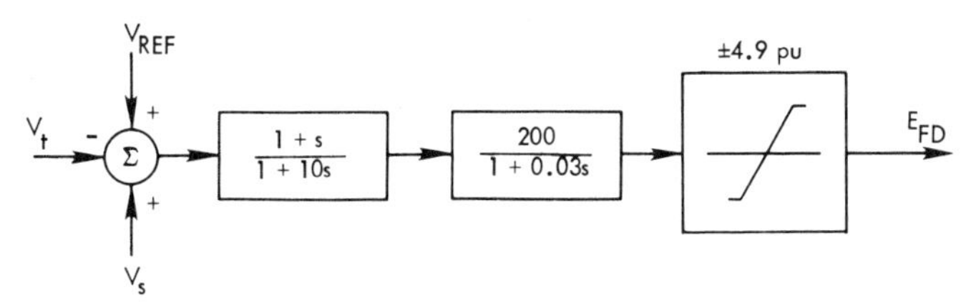

Fig. 8.12 Excitation block diagram for a 3.0 RR alternator-fed SCR excitation system. (© IEEE. Reprinted from *IEEE Trans.*, vol. PAS-89, July/Aug. 1970.)

8.3.4 Reactive power demand during system emergencies

A situation frequently encountered during system emergencies is a high reactive power demand. The capability of modern generators to meet this demand is reduced by the tendency toward the use of higher generator reactances. Modern exciters with high ceiling voltage improve the generator capability to meet this demand. It should be recognized that excitation systems are not usually designed for continuous operation at ceiling voltage and are usually limited to a few seconds of operation at that level. Concordia and Brown [17] recommend that the reactive-power requirement during system emergencies should be determined for a time of from a few minutes to a quarter- or half-hour and that these requirements should be met by the proper selection of the generator rating.

8.4 Effect of Excitation on Dynamic Stability

Modern fast excitation systems are usually acknowledged to be beneficial to transient stability following large impacts by driving the field to ceiling without delay. However, these fast excitation changes are not necessarily beneficial in damping the oscillations that follow the first swing, and they sometimes contribute growing oscillations several seconds after the occurrence of a large disturbance. With proper design and compensation, however, a fast exciter can be an effective means of enhancing stability in the dynamic range as well as in the first few cycles after a disturbance.

Since dynamic stability involves the system response to small disturbances, analysis as a linear system is possible, using the linear generator model derived previously [11]. For simplicity we analyze the problem of one machine connected to an infinite bus

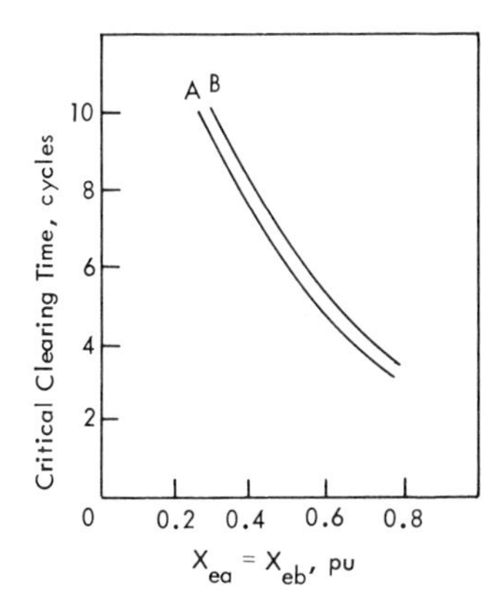

Fig. 8.13 Transient stability studies resulting from studies of [19]: A = 0.5 RR diode excitation system; B = 3.0 pu RR SCR excitation system. (© IEEE. Reprinted from *IEEE Trans.*, vol. PAS-89, July/Aug. 1970.)

through a transmission line. The synchronous machine equations, for small perturbations about a quiescent operating condition, are given by (the subscript Δ is omitted for convenience)

$$T_e = K_1\delta + K_2 E_q' \tag{8.10}$$

$$E_q' = [K_3/(1 + K_3\tau_{d0}'s)]E_{FD} - [K_3K_4/(1 + K_3\tau_{d0}'s)]\delta \tag{8.11}$$

$$V_t = K_5\delta + K_6 E_q' \tag{8.12}$$

$$\tau_j\omega s = T_m - T_e \tag{8.13}$$

where τ_{d0}' is the direct axis open circuit time constant and the constants K_1 through K_6 depend on the system parameters and on the initial operating condition as defined in Chapter 6. In previous chapters it was pointed out that this model is a substantial improvement over the classical model since it accounts for the demagnetizing effects of the armature reaction through the change in E_q' due to change in δ.

We now add to the generator model a regulator-excitation system that is represented as a first-order lag. Thus the change in E_{FD} is related to the change in V_t (again the subscript Δ is dropped) by

$$E_{FD}/V_t = -K_\epsilon/(1 + \tau_\epsilon s) \tag{8.14}$$

where K_ϵ is the regulator gain and τ_ϵ is the exciter-regulator time constant.

8.4.1 Examination of dynamic stability by Routh's criterion

To obtain the characteristic equation for the system described by (8.10)–(8.14), a procedure similar to that used in Section 3.5 is followed. First, we obtain

$$T_e(s) = \left[K_1 - \frac{K_2K_4}{\tau_{d0}'}\frac{s + (1/\tau_\epsilon + K_5K_\epsilon/K_4\tau_\epsilon)}{s^2 + s(1/\tau_\epsilon + 1/K_3\tau_{d0}') + [(1 + K_3K_6K_\epsilon)/K_3\tau_{d0}'\tau_\epsilon]}\right]\delta(s) \tag{8.15}$$

From (8.13) for $T_m = 0$,

$$s^2\delta = -(\omega_R^2/\tau_j)T_e = -(\omega_R/2H)T_e \tag{8.16}$$

By combining (8.15) and (8.16) and rearranging, the following characteristic equation is obtained:

$$s^4 + \alpha s^3 + \beta s^2 + \gamma s + \eta = 0 \tag{8.17}$$

where

$$\alpha = 1/\tau_\epsilon + 1/K_3\tau_{d0}'$$

$$\beta = [(1 + K_3K_6K_\epsilon)/K_3\tau_{d0}'\tau_\epsilon] + K_1(\omega_R/2H)$$

$$\gamma = \frac{\omega_R}{2H}(K_1/\tau_\epsilon + K_1/K_3\tau_{d0}' - K_2K_4/\tau_{d0}')$$

$$\eta = \frac{\omega_R}{2H}\left[\frac{K_1(1 + K_3K_6K_\epsilon)}{K_3\tau_{d0}'\tau_\epsilon} - \frac{K_2K_4}{\tau_{d0}'\tau_\epsilon}\left(1 + \frac{K_5K_\epsilon}{K_4}\right)\right]$$

Applying Routh's criterion to the above system, we establish the array

s^4	1	β	η
s^3	α	γ	0
s^2	a_1	a_2	
s^1	b_1	b_2	
s^0	c_1		

where

$$a_1 = (1/\alpha)(\alpha\beta - \gamma) = \beta - \gamma/\alpha$$
$$a_2 = (1/\alpha)(\alpha\eta - 0) = \eta$$
$$b_1 = (1/a_1)(a_1\gamma - a_2\alpha) = \frac{\gamma(\beta - \gamma/\alpha) - \alpha\eta}{\beta - \gamma/\alpha}$$
$$b_2 = 0 \qquad c_1 = a_2 = \eta \tag{8.18}$$

According to Routh's criterion for stability, the number of changes in sign in the first column ($1, \alpha, a_1, b_1$, and c_1) corresponds to the number of roots of (8.17) with positive real parts. Therefore, for stability the terms α, a_1, b_1, and c_1 must all be greater than zero. Thus the following conditions must be satisfied.

1. $\alpha = 1/\tau_\epsilon + 1/K_3\tau'_{d0} > 0$, and since τ_ϵ and τ'_{d0} are positive,

$$\tau'_{d0}/\tau_\epsilon > -1/K_3 \tag{8.19}$$

K_3 is an impedance factor that is not likely to be negative unless there is an excessive series capacitance in the transmission network. Even then τ'_{d0}/τ_ϵ is usually large enough to satisfy the above criterion.

2. $a_1 = \beta - \gamma/\alpha > 0$

$$\left(\frac{1 + K_3K_6K_\epsilon}{K_3\tau'_{d0}\tau_\epsilon} + K_1\frac{\omega_R}{2H}\right) - \frac{K_3\tau'_{d0}\tau_\epsilon}{K_3\tau'_{d0} + \tau_\epsilon}\frac{\omega_R}{2H}\left[K_1\left(\frac{\tau_\epsilon + K_3\tau'_{d0}}{K_3\tau'_{d0}\tau_\epsilon}\right) - \frac{K_2K_4}{\tau'_{d0}}\right] > 0$$

or

$$K_\epsilon > -\frac{\tau'_{d0}\tau_\epsilon}{K_6}\left[\frac{\omega_R K_2K_3K_4\tau_\epsilon}{2H(K_3\tau'_{d0} + \tau_\epsilon)} + \frac{1}{K_3\tau'_{d0}\tau_\epsilon}\right] \tag{8.20}$$

This inequality is easily satisfied for all values of constants normally encountered in power system operation. Note that negative K_ϵ is not considered feasible. From (8.20) K_ϵ is limited to values greater than some negative number, a constraint that is always satisfied in the physical system.

3. $b_1 = \gamma - \dfrac{\alpha\eta}{\beta - \gamma/\alpha} > 0$

$$\left[K_1\left(\frac{1}{\tau_\epsilon} + \frac{1}{K_3\tau'_{d0}}\right) - \frac{K_2K_4}{\tau'_{d0}}\right]$$
$$-\left\{\frac{(\tau_\epsilon + K_3\tau'_{d0})/K_3\tau'_{d0}\tau_\epsilon}{\dfrac{1 + K_3K_6K_\epsilon}{K_3\tau'_{d0}\tau_\epsilon} + \dfrac{\omega_R}{2H}\dfrac{\tau_\epsilon K_2K_3K_4}{K_3\tau'_{d0} + \tau_\epsilon}}\left[K_1\left(\frac{1 + K_3K_6K_\epsilon}{K_3\tau'_{d0}\tau_\epsilon}\right) - \frac{K_2K_4}{\tau'_{d0}\tau_\epsilon}\left(1 + \frac{K_5K_\epsilon}{K_4}\right)\right]\right\} > 0$$

Rearranging, this expression may be written as

$$\frac{\omega_R K_1K_2K_4}{2H\tau'_{d0}} - \frac{K_2K_4(1 + K_3K_6K_\epsilon)}{K_3\tau'^2_{d0}\tau_\epsilon} - \frac{\omega_R K_2^2K_3K_4^2\tau_\epsilon}{2H\tau'_{d0}(K_3\tau'_{d0} + \tau_\epsilon)}$$
$$+ \left(\frac{K_3\tau'_{d0} + \tau_\epsilon}{K_3\tau'_{d0}\tau_\epsilon}\right)\frac{K_2K_4}{\tau'_{d0}\tau_\epsilon}\left(1 + \frac{K_5K_\epsilon}{K_4}\right) > 0 \tag{8.21}$$

We now recognize the first expression in parentheses in the last term of (8.21) to be the positive constant α defined in (8.17). Making this substitution and rearranging to isolate K_ϵ terms, we have

$$\frac{\omega_R}{2H}\left(K_1 - \frac{K_2K_4}{\alpha\tau'_{d0}}\right) + \frac{1}{\tau_\epsilon^2} > \left(\frac{K_4K_6 - \alpha K_5\tau'_{d0}}{K_4\tau'_{d0}\tau_\epsilon}\right)K_\epsilon \tag{8.22}$$

The expressions in parentheses are positive for any load condition. Equation (8.22) places a maximum value on the gain K_ϵ for stable operation.

4. $c_1 = \eta > 0$

$$K_1\left(\frac{1 + K_3K_6K_\epsilon}{K_3\tau'_{d0}\tau_\epsilon}\right) - \frac{K_2K_4}{\tau'_{d0}\tau_\epsilon}\left(1 + \frac{K_5K_\epsilon}{K_4}\right) > 0$$

$$K_\epsilon(K_1K_6 - K_2K_5) > K_2K_4$$

Since $K_1K_6 - K_2K_5 > 0$ for all physical situations, we have

$$K_\epsilon > K_2K_4/(K_1K_6 - K_2K_5) \tag{8.23}$$

This condition puts a lower limit on the value of K_ϵ.

Example 8.3

For the machine loading of Examples 5.1 and 5.2 and for the values of the constants K_1 through K_6 calculated in Examples 6.6 and 6.7, compute the limitations on the gain constant K_ϵ, using the inequality expressions developed above. Do this for an exciter with time constant $\tau_\epsilon = 0.5$ s.

Solution

In Table 8.3 the values of the constants K_1 through K_6 are given together with the maximum value of K_ϵ from (8.22) and the minimum value of K_ϵ from (8.23). The regulator time constant τ_ϵ used is 0.5 s, $\tau'_{d0} = 5.9$ s, and $H = 2.37$ s. Case 1 is discussed in Examples 5.1 and 6.6; Case 2, in Examples 5.2 and 6.5.

From Table 8.3 it is apparent that the generator operating point plays a significant

Table 8.3. Computed Constants for the Linear Regulated Machine

Constants	Case 1 (Ex. 5.1)	Case 2 (Ex. 5.2)
K_1	1.076	1.448
K_2	1.258	1.317
K_3	0.307	0.307
K_4	1.712	1.805
K_5	−0.041	0.029
K_6	0.497	0.526
α	2.552	2.552
$K_2K_3K_4\tau_\epsilon$	0.331	0.365
$K_3\tau'_{d0} + \tau_\epsilon$	2.313	2.313
$K_3\tau'_{d0}\tau_\epsilon$	0.906	0.906
$K_2K_4/\alpha\tau'_{d0}$	0.143	0.158
K_4K_6	0.851	0.949
$\alpha K_5\tau'_{d0}$	−0.616	0.442
$K_4\tau'_{d0}\tau_\epsilon$	5.051	5.325
$1/\tau_\epsilon^2$	4.000	4.000
$K_\epsilon >$	−2.3	−3.2
$K_\epsilon <$	269.0	1120.2

role in system performance. The loading seems to influence the values of K_1 and K_5 more than the other constants. At heavier loads the values of these constants change such that in (8.22) the left side tends to decrease while the right side tends to increase. This change is in the direction to *lower* the permissible maximum value of exciter-regulator gain K_ϵ. For the problem under study, the heavier load condition of Case 1 allows a lower limit for K_ϵ than that for the less severe Case 2.

Routh's criterion is a feasible tool to use to find the limits of stable operation in a physical system. As shown in Example 8.3, the results are dependent upon both the system parameters and the initial operating point. The analysis here has been simplified to omit the rate feedback loop that is normally an integral part of excitation systems. Rate feedback could be included in this analysis, but the resulting equations become complicated to the point that one is almost forced to find an alternate method of analysis. Computer based methods are available to determine the behavior of such systems and are recommended for the more complex cases [20, 21].

One special case of the foregoing analysis has been extensively studied [11]. This analysis assumes high regulator gain ($K_3 K_6 K_\epsilon \gg 1$) and low exciter time constant ($\tau_\epsilon \ll K_3 \tau'_{d0}$). In this special case certain simplifications are possible. See Problem 8.4.

8.4.2 Further considerations of the regular gain and time constant

At no load the angle δ is zero, and the δ dependence of (8.10)–(8.23) does not apply. For this condition we can easily show that the machine terminal voltage V_t is the same as the voltage E'_q. Changes in this latter voltage follow the changes in E_{FD} with a time lag equal to τ'_{d0}. A block diagram representing the machine terminal voltage at no load is shown in Figure 8.14. From that figure the transfer function for V_t/V_{REF} can be obtained by inspection.

$$V_t/V_{\mathrm{REF}} = K_\epsilon/[(1 + K_\epsilon) + s(\tau_\epsilon + \tau'_{d0}) + \tau'_{d0}\tau_\epsilon s^2] \tag{8.24}$$

Equation (8.24) can be put in the standard form for second-order systems as

$$V_t/V_{\mathrm{REF}} = K/(s^2 + 2\zeta\omega_n s + \omega_n^2) \tag{8.25}$$

where $K = K_\epsilon/\tau'_{d0}\tau_\epsilon$, $\omega_n^2 = (1 + K_\epsilon)/\tau'_{d0}\tau_\epsilon$, $2\zeta\omega_n = (1/\tau_\epsilon + 1/\tau'_{d0})$.

For good dynamic performance, i.e., for good damping characteristics, a reasonable value of ζ is $1/\sqrt{2}$. For typical values of the gains and time constants in fast exciters we usually have $\tau'_{d0} \gg \tau_\epsilon$ and $K_\epsilon \gg 1$. We can show then that for good performance $K_\epsilon \simeq \tau'_{d0}/2\tau_\epsilon$. This is usually lower than the value of gain required for steady-state performance. In [11] de Mello and Concordia point out that the same dynamic performance can be obtained with higher values of K_ϵ by introducing a lead-lag network with the proper choice of transfer function. This is left as an exercise (see Problem 8.5).

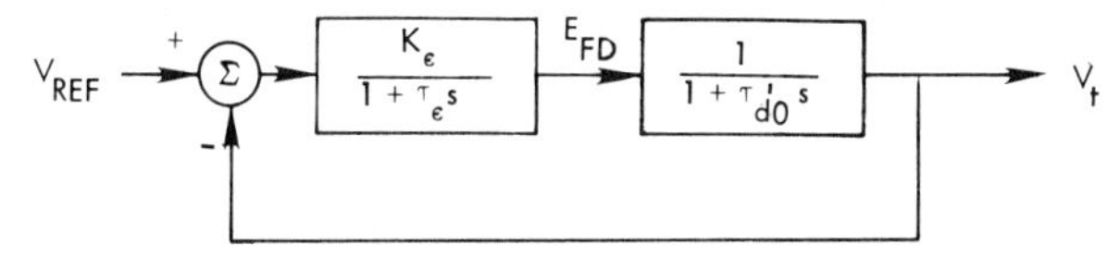

Fig. 8.14 Block diagram representing the machine terminal voltage at no load.

8.4.3 Effect on the electrical torque

The electrical torque for the linearized system under discussion was developed in Chapter 3. With use of the linear model, the electrical torque in pu is numerically equal to the three-phase electrical power in pu. Equation (3.13) gives the change in the electrical torque for the unregulated machine as a function of the angle δ. The same relation for the regulated machine is given by (3.40). From (3.13) we compute the torque as a function of angular frequency to be

$$T_e/\delta = K_1 - [K_2K_3K_4/(1 + \omega^2K_3^2\tau_{d0}'^2)](1 - j\omega K_3\tau_{d0}') \tag{8.26}$$

The real component in (8.26) is the synchronizing torque component, which is reduced by the demagnetizing effect of the armature reaction. At very low frequencies the synchronizing torque T_s is given by

$$T_s \cong K_1 - K_2K_3K_4 \tag{8.27}$$

In the regulated machine there is positive damping introduced by the armature reaction, which is given by the imaginary part of (8.26). This corresponds to the coefficient of the first power of s and is therefore a damping term.

In the regulated machine we may show the effect of the regulator on the electrical torque as follows. From (3.40) the change of the electrical torque with respect to the change in angle is given by

$$\begin{aligned}\frac{T_e}{\delta} &= K_1 - \frac{K_2K_4}{\tau_{d0}'}\,\frac{s + (1/\tau_\epsilon + K_5K_\epsilon/K_4\tau_\epsilon)}{s^2 + s\left(\dfrac{1}{\tau_\epsilon} + \dfrac{1}{K_3\tau_{d0}'}\right) + \left(\dfrac{1 + K_3K_6K_\epsilon}{K_3\tau_{d0}'\tau_\epsilon}\right)}\\ &= K_1 - \frac{K_2K_4(1 + \tau_\epsilon s) + K_2K_5K_\epsilon}{[(1/K_3 + K_6K_\epsilon) + s^2(\tau_{d0}'\tau_\epsilon)] + s(\tau_{d0}' + \tau_\epsilon/K_3)}\end{aligned} \tag{8.28}$$

It can be shown that the effect of the terms $K_2K_4(1 + \tau_\epsilon s)$ in the numerator is very small compared to the term $K_2K_5K_\epsilon$. This point is discussed in greater detail in [11]. Using this simplification, we write the expression for T_e/δ as

$$\frac{T_e}{\delta} \cong K_1 - \frac{K_2K_5K_\epsilon}{[(1/K_3 + K_6K_\epsilon) + \tau_{d0}'\tau_\epsilon s^2] + s(\tau_{d0}' + \tau_\epsilon/K_3)} \tag{8.29}$$

which at a frequency ω can be separated into a real component that gives the synchronizing torque T_s and into an imaginary component that gives the damping torque T_d. These components are given by

$$T_s \cong K_1 - \frac{K_2K_5K_\epsilon[(1/K_3 + K_6K_\epsilon) - \omega^2\tau_{d0}'\tau_\epsilon]}{[(1/K_3 + K_6K_\epsilon) - \omega^2\tau_{d0}'\tau_\epsilon]^2 + \omega^2(\tau_{d0}' + \tau_\epsilon/K_3)^2} \tag{8.30}$$

$$T_d \cong \frac{K_2K_5K_\epsilon(\tau_{d0}' + \tau_\epsilon/K_3)\omega}{[(1/K_3 + K_6K_\epsilon) - \omega^2\tau_{d0}'\tau_\epsilon]^2 + \omega^2(\tau_{d0}' + \tau_\epsilon/K_3)^2} \tag{8.31}$$

Note that the damping torque T_d will have the same sign as K_5. This latter quantity can be negative at some operating conditions (see Example 6.6). In this case the regulator reduces the inherent system damping.

At very low frequencies (8.30) is approximately given by

$$T_s \cong K_1 - K_2K_5/K_6 \tag{8.32}$$

which is higher than the value obtained for the unregulated machine given by (8.27).

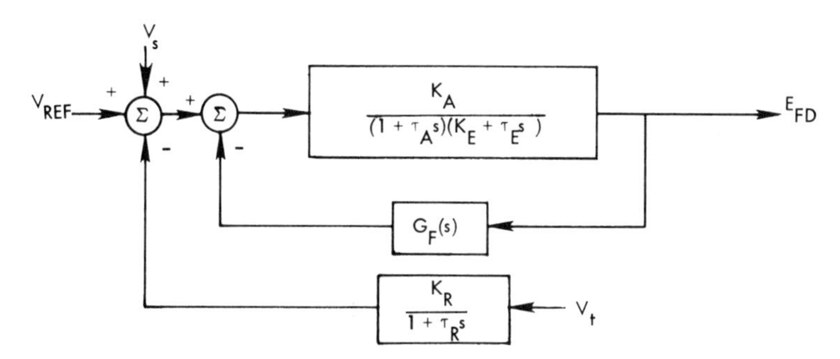

Fig. 8.15 Block diagram of a linearized excitation system model.

Therefore, whereas the regulator improves the synchronizing forces in the machine at low frequencies of oscillation, it reduces the inherent system damping when K_5 is negative, a common condition for synchronous machines operated near rated load.

8.5 Root-Locus Analysis of a Regulated Machine Connected to an Infinite Bus

We have used linear system analysis techniques to study the dynamic response of one regulated synchronous machine. In Section 7.8, while the exciter is represented in detail, a very simple model of the generator is used. In Section 8.4 the exciter model used is a very simple one. In this section a more detailed representation of the exciter is adopted, along with the simplified linear model of the synchronous machine that takes into account the field effects. The excitation system model used here is similar to that in Figure 7.54 except for the omission of the limiter and the saturation function S_E. This model is shown in Figure 8.15. In this figure the function $G_F(s)$ is the rate feedback signal. The signal V_s is the stabilizing signal that can be derived from any convenient signal and processed through a power system stabilizer network to obtain the desired phase relations (see Section 8.7).

The system to be studied is that of one machine connected to an infinite bus through a transmission line. This model used for the synchronous machine is essentially that given in Figure 6.3 and is based on the linearized equations (8.10)–(8.13). To simulate the damping effect of the damper windings and other damping torques, a damping torque component $-D\omega$ is added to the model as shown in Figure 8.16.

The combined block diagram of the synchronous machine and the exciter is given in Figure 8.17 (with the subscript Δ omitted for convenience).

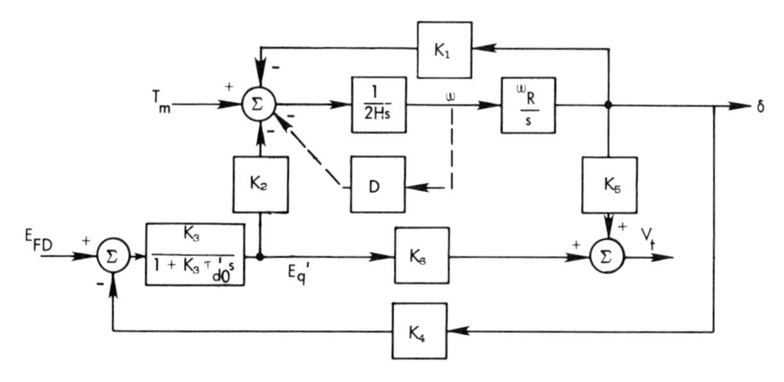

Fig. 8.16 Block diagram of the simplified linear model of a synchronous machine connected to an infinite bus with damping added.

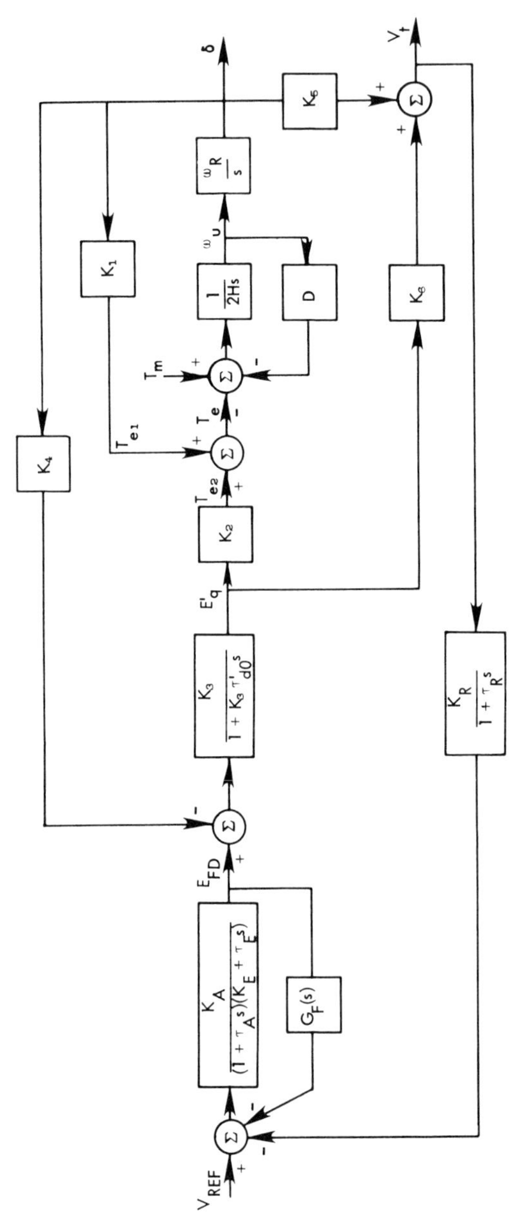

Fig. 8.17 Combined block diagram of a linear synchronous machine and exciter.

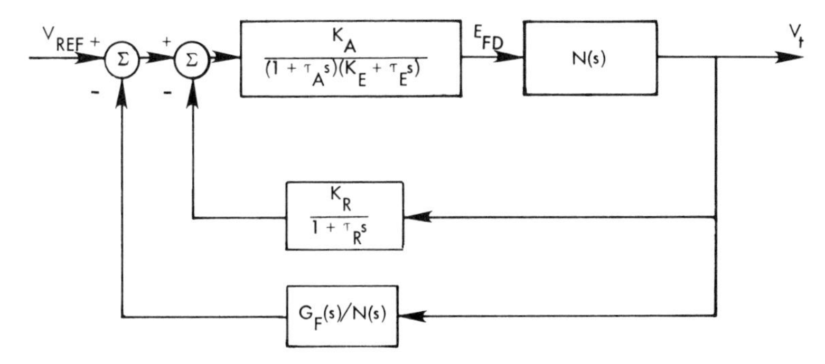

Fig. 8.18 Block diagram with V_t as the takeoff point for feedback loops.

To study the effect of the different feedback loops, we manipulate the block diagram so that all the feedback loops "originate" at the same takeoff point. This is done by standard techniques used in feedback control systems [22]. The common takeoff point desired is the terminal voltage V_t, and feedback loops to be studied are the regulator and the rate feedback $G_F(s)$. The resulting block diagram is shown in Figure 8.18. In that figure the transfer function $N(s)$ is given by

$$N(s) = \frac{K_1 K_6 (2Hs^2 + Ds + K_1 \omega_R) - \omega_R K_2 K_3 K_5}{(1 + K_3 \tau'_{d0} s)(2Hs^2 + Ds + K_1 \omega_R) - \omega_R K_2 K_3 K_4} \tag{8.33}$$

Note that the expression for $N(s)$ can be simplified if the damping D is neglected or if the term containing K_5 is omitted (K_5 is usually very small at heavy load conditions).

The system of Figure 8.18 is solved by linear system analysis techniques, using the digital computer. A number of computer programs are available that are capable of solving very complex linear systems and of displaying the results graphically in several convenient ways or in tabular forms [20, 21]. For a given operating point we can obtain the loci of roots of the open loop system and the frequency response to a sinusoidal input as well as the time response to a small step change in input.

The results of the linear computer analysis are best illustrated by some examples. In the analysis given in this section, the machine discussed in the examples of Chapters 4, 5, and 6 is analyzed for the loading condition of Example 6.7. The exciter data are $K_A = 400$, $\tau_A = 0.05$, $K_E = -0.17$, $\tau_E = 0.95$, $K_R = 1.0$ and $\tau_R = 0$. The machine constants are $2H = 4.74$ s, $D = 2.0$ pu and $\tau'_{d0} = 5.9$ s. The constants K_1 through K_6 in pu for the operating point to be analyzed are

$$K_1 = 1.4479 \qquad K_3 = 0.3072 \qquad K_5 = 0.0294$$
$$K_2 = 1.3174 \qquad K_4 = 1.8052 \qquad K_6 = 0.5257$$

Example 8.4

Use a linear systems analysis program to determine the dynamic response of the system of Figure 8.18 with and without the rate feedback. The following graphical solutions are to be obtained for the above operating conditions:

1. Root-locus plot.
2. Time response of $V_{t\Delta}$ to a step change in V_{REF}.
3. Bode diagram of the closed loop transfer function.
4. Bode diagram of the open loop transfer function.

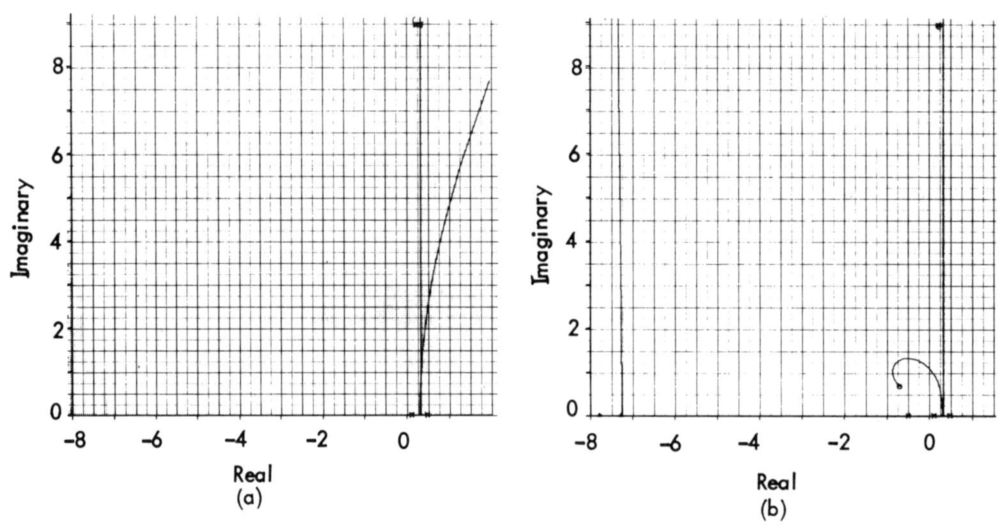

Fig. 8.19 Root locus of the system of Figure 8.17: (a) without rate feedback, (b) with rate feedback.

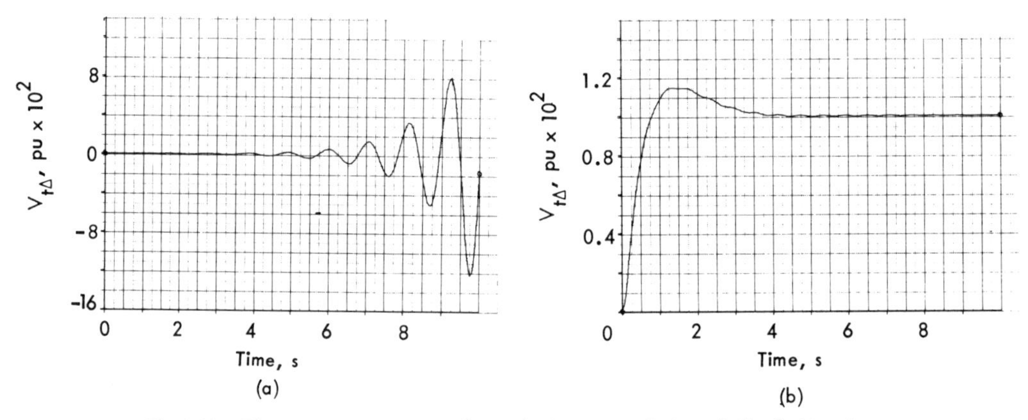

Fig. 8.20 Time response to a step change in V_{REF}: (a) $G_F(s) = 0$, (b) $G_F(s) \neq 0$.

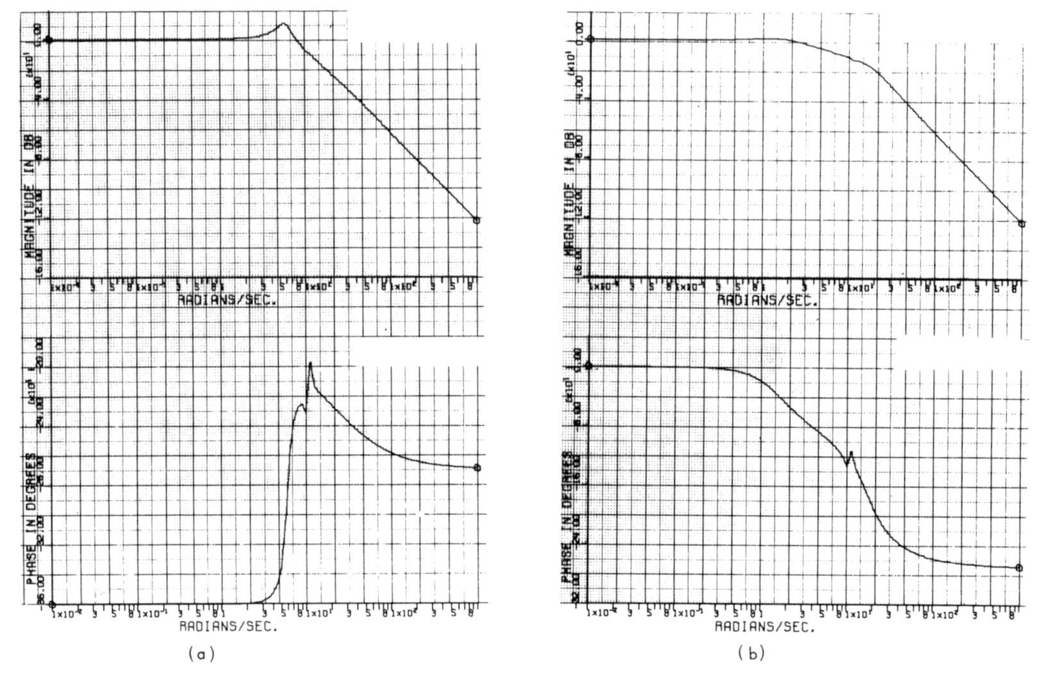

Fig. 8.21 Bode plots of the closed loop transfer function: (a) $G_F = 0$, (b) $G_F \neq 0$.

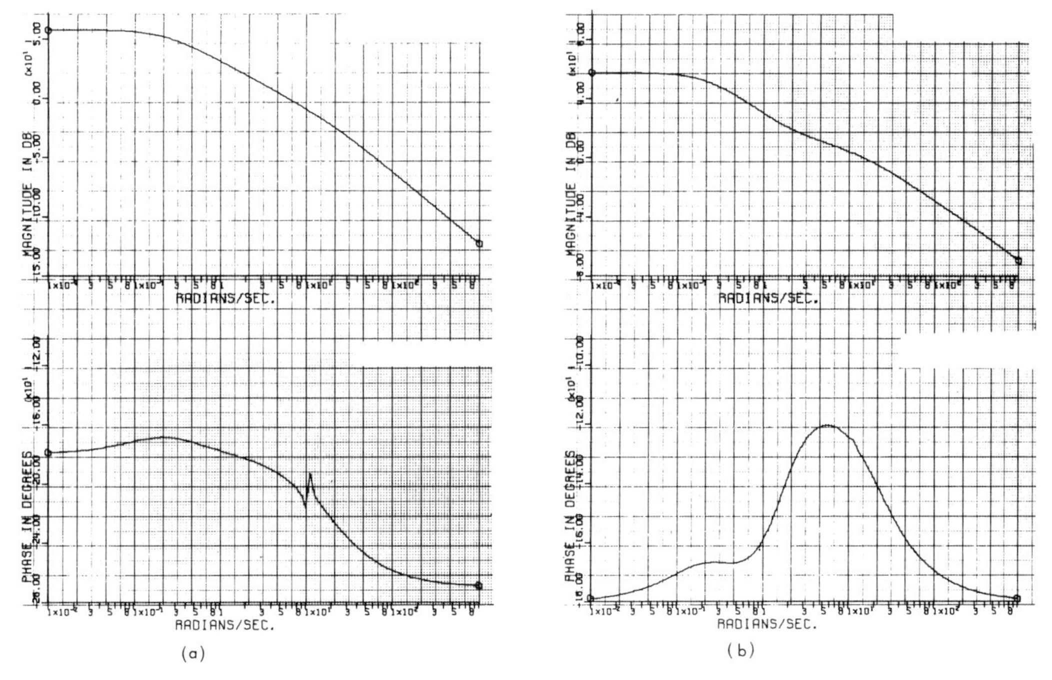

Fig. 8.22 Bode plots of the open loop transfer function: (a) $G_F = 0$, (b) $G_F \neq 0$.

Compute these graphical displays for two conditions:

(a) $G_F(s) = 0$
(b) $G_F(s) = sK_F/(1 + \tau_F s)$, with $K_F = 0.04$, and $\tau_F = 10$ s

Solution

The results of the computer analysis are shown in Figures 8.19–8.22 for the different plots. In each figure, part (a) is for the result without the rate feedback and part (b) is with the rate feedback.

Figures 8.19–8.20 show clearly that the system is unstable for this value of gain without the rate feedback. Note the basic problem discussed in Example 7.7. With $G_F(s) = 0$, the system dynamic response is dominated by two pairs of complex roots near the imaginary axis. The pair that causes instability is determined by the field

Table 8.4. Root-Locus Poles and Zeros of Example 8.4

Condition	Zeros	Poles
(a) $K_F = 0$	−0.21097 + j10.45130	−0.27324
	−0.21097 − j10.45130	−20.00000
		−0.17894
		−0.35020 + j10.72620
		−0.35020 − j10.72620
(b) $K_F = 0.04$	−1.19724 + j0.83244	−20.00000
	−1.19724 − j0.83244	−0.17894
	−0.40337 + j10.69170	−0.27324
	−0.40337 − j10.69170	−0.35021 + j10.72620
		−0.35021 − j10.72620
		−1.00000

winding and exciter parameters. The effect of the pair caused by the torque angle loop is noticeable in the Bode plots of Figures 8.21–8.22. These roots occur near the natural frequency $\omega_n = (1.4479 \times 377/4.74)^{1/2} = 10.73$ rad/s. The rate feedback modifies the root-locus plot in such a way as to make the system stable even with high amplifier gains. The poles and zeros obtained from the computer results are given in Table 8.4.

Example 8.5

Repeat part (b) of Example 8.4 with (a) $D = 0$ and (b) $K_5 = 0$.

Solution

(a) For the case of $D = 0$ it is found (from the computer output) that the poles and zeros affected are only those determined by the torque angle loop. These poles now become $-0.13910 \pm j10.72550$ (instead of $-0.35021 \pm j10.72620$). The net effect is to move the branch of the root locus determined by these poles and zeros to just slightly away from the imaginary axis.

(b) It has been shown that K_5 is numerically small. Except for the situations where K_5 becomes negative, its main effect is to change ω_n to the value

$$\omega^2_n = (\omega_R/2H)(K_1 - K_2K_5/K_6)$$

The computer output for $K_5 = 0$ is essentially the same as that of Example 8.4.

The root-locus plot and the time response to a step change in V_{REF} for the cases of $D = 0$ and $K_5 = 0$ are displayed in Figures 8.23–8.24.

The examples given in this section substantiate the conclusions reached in Section 7.7 concerning the importance of the rate feedback for a stable operation at high values of gain. A very significant point to note about the two pairs of complex roots that dominate the system dynamic response is the nature of the damping associated with them. The damping coefficient D primarily affects the roots caused by the torque angle loop at a frequency near the natural frequency ω_n. The second pair of roots, determined by the field circuit and exciter parameters, gives a somewhat lower fre-

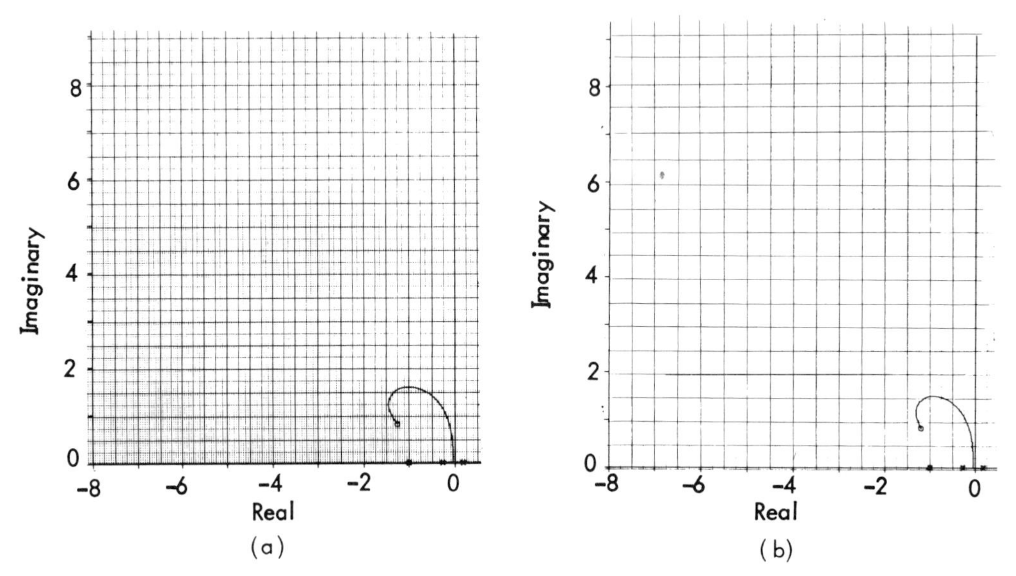

Fig. 8.23 Root locus of the system of Example 8.5: (a) $D = 0$, (b) $K_5 = 0$.

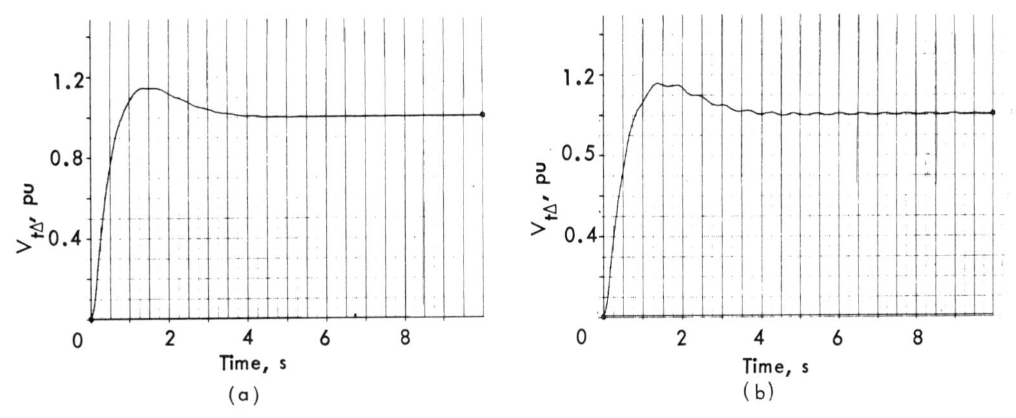

Fig. 8.24 Time response to a step change in V_{REF} for the system of Example 8.5: (a) $D = 0$, (b) $K_5 = 0$.

quency and its damping is inherently poor. This is an important consideration in the study of power system stabilizers.

8.6 Approximate System Representation

In the previous section it is shown that the dynamic system performance is dominated by two pairs of complex roots that are particularly significant at low frequencies. In this frequency range the system damping is inherently low, and stabilizing signals are often needed to improve the system damping (Section 8.7). Here we develop an approximate model for the excitation system that is valid for low frequencies.

We recognize that the effect of the rate feedback $G_F(s)$ in Figure 8.17 is such that it can be neglected at low frequencies ($s = j\omega \rightarrow 0$) or near steady state ($t \rightarrow \infty$). We have already pointed out that K_5 is usually very small and is omitted in this approximate model. The feedback path through K_4 provides a small positive damping component that is usually considered negligible [11]. The resulting reduced system is composed of two subsystems: one representing the exciter-field effects and the other representing the inertial effects. These effects contribute the electrical torque components designated T_{e2} and T_{e1} respectively.

8.6.1 Approximate excitation system representation

The approximate system to be analyzed is shown in Figure 8.25 where the exciter and the generator have been approximated by simple first-order lags [11]. A straightforward analysis of this system gives

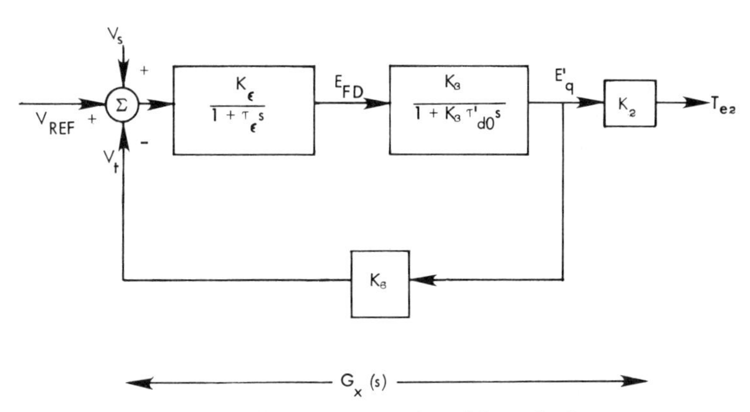

Fig. 8.25 Approximate representation of the excitation system.

$$G_x(s) = \frac{K_2K_3K_\epsilon}{1 + K_3K_6K_\epsilon}\left(1 + \frac{\tau_\epsilon + K_3\tau'_{d0}}{1 + K_3K_6K_\epsilon}s + \frac{K_3\tau'_{d0}\tau_\epsilon}{1 + K_3K_6K_\epsilon}s^2\right) \tag{8.34}$$

Since $K_3K_6K_\epsilon \gg 1$ in all cases of interest, (8.34) can be simplified to

$$\begin{aligned} G_x(s) &= \frac{K_2/K_6}{1 + [(\tau_\epsilon + K_3\tau'_{d0})/K_3K_6K_\epsilon]s + (\tau'_{d0}\tau_\epsilon/K_6K_\epsilon)s^2} \\ &= \frac{K_2K_\epsilon/\tau'_{d0}\tau_\epsilon}{s^2 + [(\tau_\epsilon + K_3\tau'_{d0})/K_3\tau'_{d0}\tau_\epsilon]s + K_6K_\epsilon/\tau'_{d0}\tau_\epsilon} \\ &\triangleq \frac{K_2K_\epsilon/\tau'_{d0}\tau_\epsilon}{s^2 + 2\zeta_x\omega_x s + \omega_x^2} = \frac{K_2K_\epsilon/\tau'_{d0}\tau_\epsilon}{d(s)} \end{aligned} \tag{8.35}$$

where ω_x is the undamped natural frequency and ζ_x is the damping ratio:

$$\omega_x = \sqrt{K_6K_\epsilon/\tau'_{d0}\tau_\epsilon} \qquad \zeta_x = (\tau_\epsilon + K_3\tau'_{d0})/2\omega_x K_3\tau'_{d0}\tau_\epsilon \tag{8.36}$$

We are particularly concerned about the system frequency of oscillation as compared to ω_x. The damping ζ_x is usually small and the system is poorly damped.

The function $G_x(s)$ must be determined either by calculation or by measurement on the physical system. A proven technique for measurement of the parameters of $G_x(s)$ is to monitor the terminal voltage while injecting a sinusoidal input signal at the voltage regulator summing junction [8, 12, 23, 24, 25]. The resulting amplitude and phase (Bode) plot can be used to identify $G_x(s)$ in (8.35).

Lacking field test data, we must estimate the parameters of $G_x(s)$ by calculations derived from a given operating condition. It should be emphasized that this procedure has some serious drawbacks. First, the gains and time constants may not be precisely known, and the use of estimated values may give results that are suspect [10, 12, 24]. Second, the theoretical model based on the constants K_1 through K_6 is not only load dependent but is also based on a one machine–infinite bus system. The use of these constants, then, requires that assumptions be made concerning the proximity of the machine under study with respect to the rest of the system. A procedure based on deriving an *equivalent* infinite bus, connected to the machine under study by a series impedance, is given in Section 8.6.2.

8.6.2 Estimate of $G_x(s)$

The purpose of this section is to develop an approximate method for estimating K_1 through K_6 that can be applied to any machine in the system. These constants can be used in (8.36) to calculate the approximate parameters for $G_x(s)$.

The one machine–infinite bus system assumes that the generator under study is connected to an *equivalent* infinite bus of voltage $V_\infty\underline{/\alpha}$ through a transmission line of impedance $\overline{Z}_e = R_e + jX_e$. This equivalent impedance is assumed to be the Thevenin equivalent impedance as "seen" at the generator terminals. Therefore, if the driving-point short circuit admittance $\overline{Y}_{ii}$ at the generator terminal node i is known, we assume that

$$\overline{Z}_e = 1/\overline{Y}_{ii} \tag{8.37}$$

The equivalent infinite bus voltage $\overline{V}_\infty$ is calculated by subtracting the drop $\overline{I}_i\overline{Z}_e$ from the generator terminal voltage $\overline{V}_{ti}$, where $\overline{I}_i$ is the generator current. The procedure is illustrated by an example.

Example 8.6

Compute the constants K_1 through K_6 for generator 2 of Example 2.6, using the equivalent infinite bus method outlined above. Note that the three-machine system is certainly not considered to have an infinite bus, and the results might be expected to differ from those obtained by a more detailed simulation.

Solution

From Example 2.6 the following data for the machine are known (in pu and s).

$$x_{d2} = 0.8958 \qquad x_{q2} = 0.8645 \qquad x_{\ell 2} = 0.0521 \qquad H_2 = 6.4$$
$$x'_{d2} = 0.1198 \qquad x'_{q2} = 0.1969 \qquad \tau'_{d02} = 6.0$$

We can establish the terminal conditions from the load-flow study of Figure 2.19:

$$\begin{aligned} I_2 \underline{/-\phi_2} &= I_{r2} + jI_{x2} = (P_2 - jQ_2)/V_2 \\ &= (1.630 - j0.066)/1.025 = 1.592 \underline{/-2.339^\circ} \text{ pu} \end{aligned}$$

From Figure 5.6

$$\begin{aligned} \tan(\delta_{20} - \beta_2) &= x_{q2} I_{r2}/(V_2 - x_{q2} I_{x2}) = 1.272 \\ \delta_{20} - \beta_2 &= 51.818^\circ \end{aligned}$$

But from the load flow $\beta_2 = 9.280^\circ$,

$$\delta_{20} = 51.818 + 9.280 = 61.098^\circ$$

Then $\delta_{20} - \beta_2 + \phi_2 = 54.156^\circ$ and

$$\begin{aligned} \overline{V}_2 &= V_2 \underline{/\beta_2 - \delta_2} = 1.025 \underline{/-51.818^\circ} = V_{q2} + jV_{d2} = 0.634 - j0.806 \text{ pu} \\ \overline{I}_2 &= I_2 \underline{/-(\delta_2 - \beta_2 + \phi_2)} = 1.592 \underline{/-54.156^\circ} = I_{q2} + jI_{d2} = 0.932 - j1.290 \text{ pu} \end{aligned}$$

Neglecting the armature resistance, $r = 0$,

$$\begin{aligned} E_{qa0} &= V_{q2} - x_{q2} I_{d2} = 1.749 \text{ pu} \\ E_{20} &= V_{q2} - x_{d2} I_{d2} = 1.789 \text{ pu} \end{aligned}$$

From Table 2.6 the driving-point admittance at the *internal* node of generator 2 is given by

$$\overline{Y}_{22} = 0.420 - j2.724 \text{ pu}$$

The terminal voltage node of generator 2 had been eliminated in the reduction process. However, since it is connected to the internal node by x'_{d2}, $\overline{Z}_e$ can be obtained by using the approximate relation $\overline{Z}_e = 1/\overline{Y}_{22} - jx'_{d2}$. The exact reduction process gives

$$\overline{Z}_e = 0.0550 + j0.2388 = 0.2450 \underline{/77.029^\circ} \text{ pu}$$

Then we compute from (6.56)

$$\begin{aligned} K_I &= 1/[R_e^2 + (x_q + X_e)(x'_d + X_e)] = 1/0.39925 = 2.5084 \\ 1/K_3 &= 1 + K_I(x_d - x'_d)(x_q + X_e) = 3.1476 \\ K_3 &= 0.3177 \end{aligned}$$

We can compute the infinite bus voltage

$$\overline{V}_\infty = V_\infty \angle \alpha = \overline{V}_{t2} - \overline{Z}_e \overline{I}_2$$
$$= 1.025 \angle 9.280^\circ - (0.2450 \angle 77.029^\circ)(1.592 \angle 6.941^\circ)$$
$$= 0.9706 - j0.2226 = 0.9958 \angle -12.914^\circ$$

The angle required in the computations to follow is

$$\gamma = \delta_{20} - \alpha = 61.098 - (-12.914) = 74.012^\circ$$

$$K_1 = K_I V_\infty \{E_{qa0}[R_e \sin\gamma + (x_d' + X_e)\cos\gamma] + I_{q0}(x_q - x_d')[(x_q + X_e)\sin\gamma - R_e \cos\gamma]\}$$
$$= 2.4750$$

$$K_2 = K_I\{R_e E_{qa0} + I_{q0}[R_e^2 + (x_q + X_e)^2]\} = 3.0941$$

$$K_4 = V_\infty K_I (x_d - x_d')[(x_q + X_e)\sin\gamma - R_e \cos\gamma] = 2.0265$$

$$K_5 = (K_I V_\infty / V_{t0})\{x_d' V_{q0}[R_e \cos\gamma - (x_q + X_e)\sin\gamma]$$
$$- x_q V_{d0}[(x_d' + X_e)\cos\gamma + R_e \sin\gamma]\} = 0.0640$$

$$K_6 = (V_{q0}/V_{t0})[1 - K_I x_d'(x_q + X_e)] - (V_{d0}/V_{t0}) K_I x_q R_e = 0.5070$$

Summary:

$$K_1 = 2.475 \qquad K_3 = 0.318 \qquad K_5 = 0.064$$
$$K_2 = 3.094 \qquad K_4 = 2.027 \qquad K_6 = 0.507$$

Note that these constants are in pu on 100-MVA base whereas the machine is a 192-MVA generator. The constants K_1 and K_2 should be divided by 1.92 to convert to the machine base.

Example 8.7

The exciter for generator 2 of the three-machine system has the constants $K_\epsilon = 400$ and $\tau_\epsilon = 0.95$ s. Compute the parameters of $G_x(s)$. For the system natural frequency (see Example 3.4) calculate the excitation control system phase lag. (Here again we emphasize the need for actual measurement of the system parameters. Lacking such measurement, a judgment is made as to which parameters should be used. We use the regulator gain and the exciter time constant. It is judged that the latter is important at the low frequencies of interest. This point is a source of some confusion in the literature. It is sometimes assumed, erroneously, that the regulator time constant is to be used when the excitation system is represented by one time constant. This is not valid for low frequencies.)

Solution

From (8.36) we have

$$\omega_x = \sqrt{(0.507 \times 400)/(6.0 \times 0.95)} = 5.967 \text{ rad/s}$$
$$\zeta_x = (0.95 + 0.318 \times 6.0)/(2 \times 5.967 \times 0.318 \times 6.0 \times 0.95) = 0.132$$

and the excitation system is poorly damped.

From Example 3.4 the dominant frequency of oscillation is approximately 1.4 Hz or $\omega_{osc} \cong 8.8$ rad/s. At any frequency the characteristic equation of $G_x(s)$ is obtained by substituting $s = j\omega$ in the denominator of the first expression in (8.35):

$$d(j\omega) = 1 - 0.0281\omega^2 + j0.0443\omega$$

At the frequency of interest ($\omega = 8.8$ rad/s) we have

$$d(j\omega_{osc}) = -1.1761 + j0.3898$$
$$\phi_{lag} = \tan^{-1}(0.3898/-1.1761) = 161.661^\circ$$

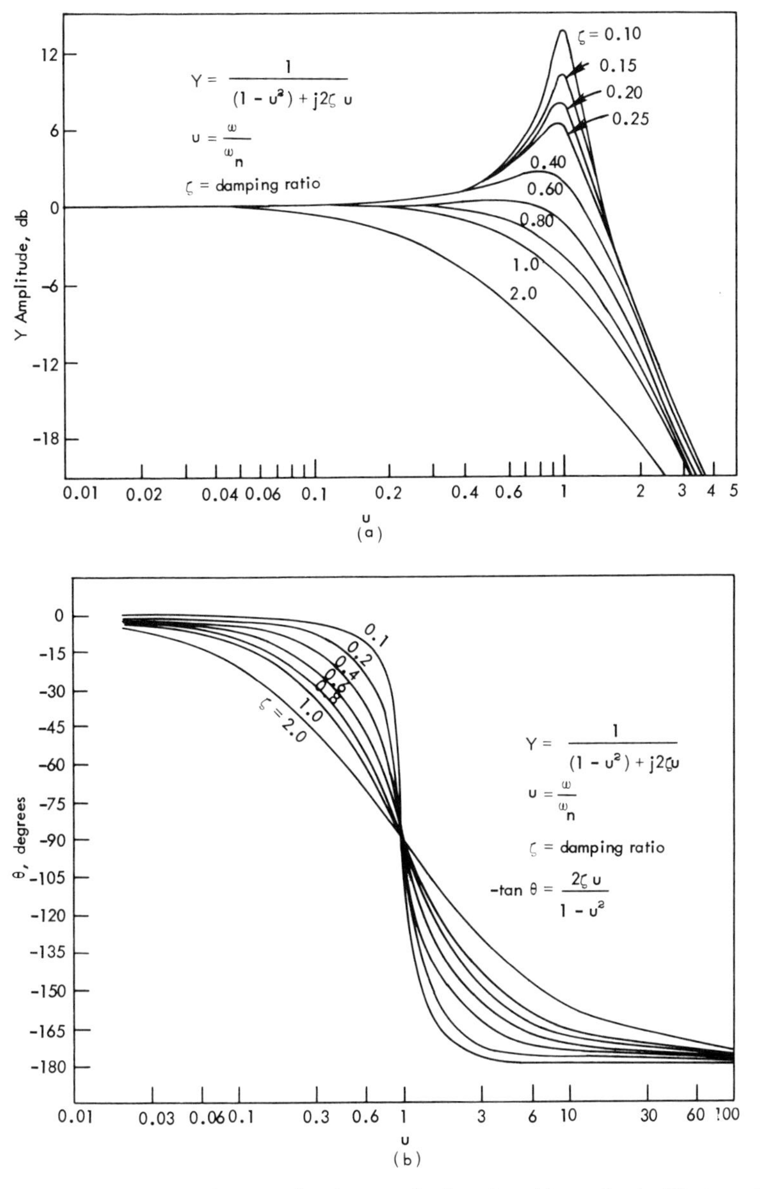

Fig. 8.26 Characteristics of a second-order transfer function: (a) amplitude, (b) phase shift.

The excitation system phase lag in Example 8.7 is rather large, and phase compensation is likely to be required (see Section 8.7). The phase lag is large because $\omega_{osc} > \omega_x$ and ζ_x is small. For small damping the phase changes very fast in the neighborhood of ω_x (where $\phi_{lag} = 90°$). Many textbooks on control systems, such as [22], give curves of phase shift as a function of normalized frequency, $u = \omega/\omega_n$, as shown in Figure 8.26. In the above example, with $u = 8.8/5.967 = 1.47$ and $\zeta = 0.13$, it is apparent from Figure 8.26(b) that the phase lag is great.

8.6.3 The inertial transfer function

The inertial transfer function can be obtained by inspection from Figure 8.17. For the case where damping is present,

$$\frac{-\delta}{T_{e2}} = \frac{\omega_R/2H}{s^2 + \dfrac{D}{2H}s + \dfrac{K_1\omega_R}{2H}} = \frac{\omega_R/2H}{s^2 + 2\zeta_n\omega_n s + \omega_n^2} \tag{8.38}$$

Where ω_n is the natural frequency of the rotating mass and ζ_n is the damping factor,

$$\begin{aligned}\omega_n &= \sqrt{K_1\omega_R/2H}\\ \zeta_n &= D/4H\omega_n = D/2\sqrt{2HK_1\omega_R}\end{aligned} \tag{8.39}$$

The damping of the inertial system is usually very low.

Example 8.8

Compute the characteristic equation, the undamped natural frequency, and the damping factor of the inertial system of generator 2 (Example 2.6). Use $D = 2$ pu.

Solution

From the data of Examples 2.6 and 8.6 we compute

$$\begin{aligned}d(s) &= s^2 + 0.156s + 72.894\\ \omega_n &= \sqrt{72.894} = 8.538 \text{ rad/s}\\ \zeta_n &= 2/[2(12.8 \times 2.975 \times 377)^{1/2}] = 0.009\\ d(j\omega) &= 1 - 0.0137\omega^2 + j0.00214\omega\end{aligned}$$

At the system frequency of oscillation $\omega = \omega_{osc} = 8.8$ rad/s,

$$\phi_{lag} = \tan^{-1}[0.0183/(-0.0222 - 0.0604)] = 163.3°$$

8.7 Supplementary Stabilizing Signals

Equation (8.31) indicates that the voltage regulator introduces a damping torque component proportional to K_5. We noted in Section 8.4.3 that under heavy loading conditions K_5 can be negative. These are the situations in which dynamic stability is of concern. We have also shown in Section 8.6.2 that the excitation system introduces a large phase lag at low system frequencies just above the natural frequency of the excitation system. Thus it can often be assumed that the voltage regulator introduces negative damping. To offset this effect and to improve the system damping in general, artificial means of producing torques in phase with the speed are introduced. These are called "supplementary stabilizing signals" and the networks used to generate these signals have come to be known as "power system stabilizer" (PSS) networks.

Stabilizing signals are introduced in excitation systems at the summing junction where the reference voltage and the signal produced from the terminal voltage are added to obtain the error signal fed to the regulator-exciter system. For example, in the excitation system shown in Figure 7.54 the stabilizing signal is indicated as the signal V_s. To illustrate, the signal usually obtained from speed or a related signal such as the frequency, is processed through a suitable network to obtain the desired phase relationship. Such an arrangement is shown schematically in Figure 8.27.

8.7.1 Block diagram of the linear system

We have previously established the rationale for using linear systems analysis for the study of low-frequency oscillations. For any generator in the system the behavior

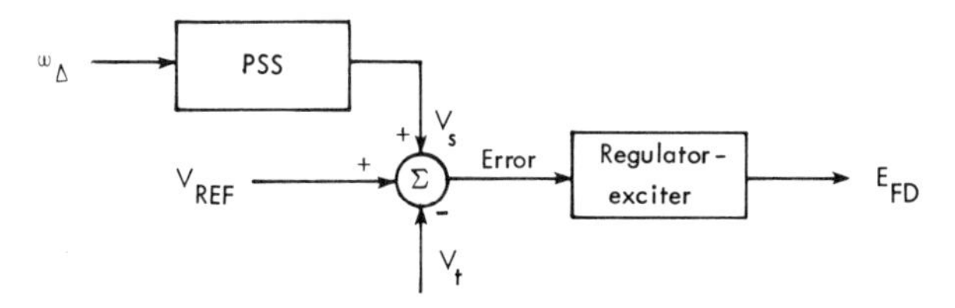

Fig. 8.27 Schematic diagram of a stabilizing signal from speed deviation.

can be conveniently characterized and the unit performance determined, from the linear block diagram of that generator. This block diagram is shown in Figure 8.28.

The constants K_1 through K_6 are load dependent (see Section 8.6 for an approximate method to determine these constants) but may be considered constant for small deviations about the operating point. The damping constant D is usually in the range of 1.0–3.0 pu. The system time constants, gains, and inertia constants are obtained from the equipment manufacturers or by measurement.

The PSS is shown here as a feedback element from the shaft speed and is often given in the form [11]

$$G_S(s) = \frac{K_0 \tau_0 s}{1 + \tau_0 s} \left[\frac{(1 + \tau_1 s)(1 + \tau_3 s)}{(1 + \tau_2 s)(1 + \tau_4 s)} \right] \tag{8.40}$$

The first term in (8.40) is a reset term that is used to "wash out" the compensation effect after a time lag τ_0, with typical values of 4 s [11] to 20 or 30 s [12]. The use of reset control will assure no permanent offset in the terminal voltage due to a prolonged error in frequency, such as might occur in an overload or islanding condition. The second term in $G_S(s)$ is a lead compensation pair that can be used to improve the phase lag through the system from V_{REF} to ω_Δ at the power system frequency of oscillation.

Qualitatively, we can recognize the existence of a potential control problem in the system of Figure 8.28 due to the cascading of several phase lags in the forward loop. In terms of a Bode or frequency analysis (see [22], for example) the system is likely to have inadequate phase margin. This is difficult to show quantitatively in the complete system because of its complexity. We therefore take advantage of the simplified representation developed in Section 8.6 and the results obtained in that section.

8.7.2 Approximate model of the complete exciter-generator system

Having established the complete forward transfer function of the excitation control system and inertia, we may now sketch the complete block diagram as in Figure 8.29.

We note that a common takeoff point is used for the feedback loop, requiring a slight modification of the inertial transfer function using standard block diagram manipulation techniques. We also note that the output in Figure 8.29 is the negative of the speed deviation. The parameters ζ_x, ω_x and ζ_n, ω_n are defined in (8.36) and (8.39) respectively.

Examining Figure 8.29 we can see that to damp speed oscillations, the power system stabilizer must compensate for much of the inherent forward loop phase lag. Thus the PSS network must provide lead compensation.

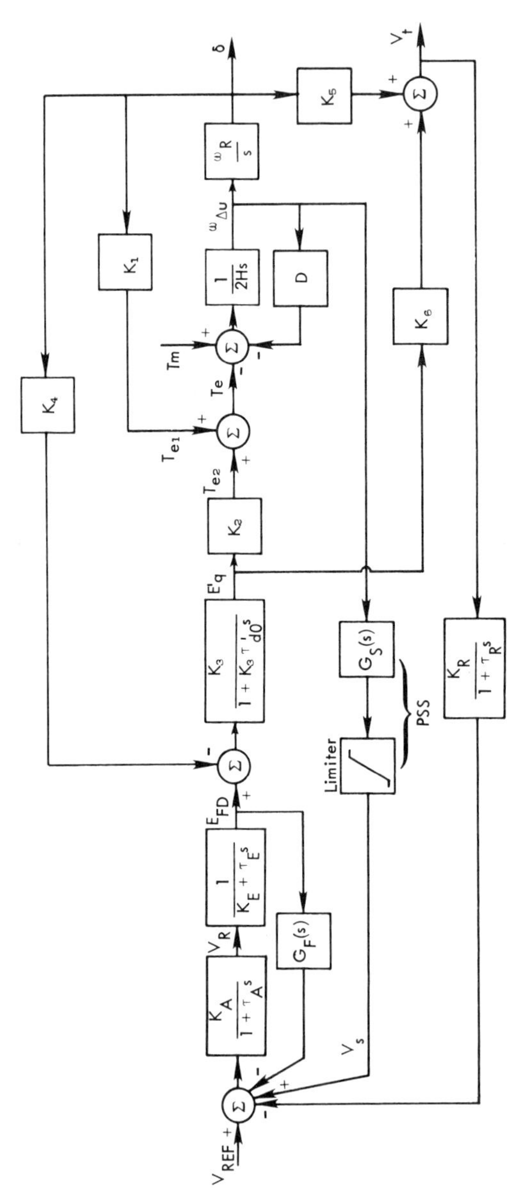

Fig. 8.28 Block diagram of a linear generator with an exciter and power system stabilizer.

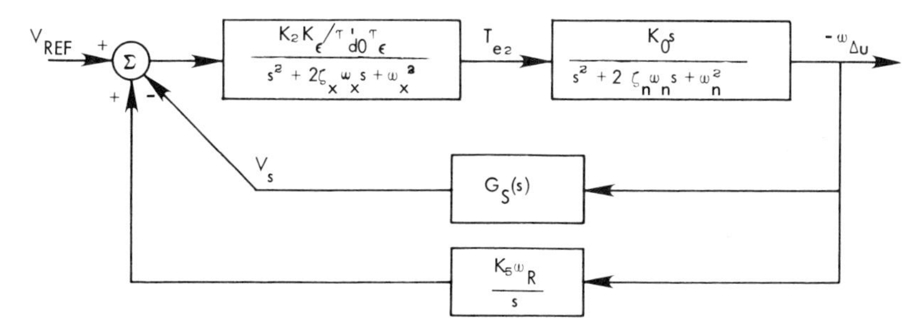

Fig. 8.29 Block diagram of a simplified model of the complete system.

8.7.3 Lead compensation

One method of providing phase lead is with the passive circuit of Figure 8.30(a). If loaded into a high impedance, the transfer function of this circuit is

$$\frac{E_0}{E_i} = \frac{(1/a)(1 + a\tau s)}{1 + \tau s}$$

where

$$a = (R_1 + R_2)/R_2 > 1$$
$$\tau = R_1 R_2 C/(R_1 + R_2) \tag{8.41}$$

The transfer function has the pole zero configuration of Figure 8.30(b), where the zero lies inside the pole to provide phase lead. For this simple network the magnitude of the parameter a is usually limited to about 5.

Another lead network not so restricted in the parameter range is that shown in Figure 8.31 [26]. For this circuit we compute

$$\frac{E_0}{E_i} = \frac{1 + (\tau_A + \tau_B)s}{(1 + \tau_B s)[1 + (\tau_C + \tau_D)s]} \tag{8.42}$$

where $\tau_A = K_1 R C_1$ = lead time constant
$\tau_B = R_1 C_1$ = noise filter time constant $\ll \tau_A$
$\tau_C = K_2 R C_2$ = lag time constant
$\tau_D = R C_F$ = stabilizing circuit time constant $\ll \tau_C$
$K_1 = R_B/(R_A + R_B)$
$K_2 = R_D/(R_C + R_D)$

Approximately, then

$$E_0/E_i = (1 + \tau_A s)/(1 + \tau_C s) = (1 + a\tau s)/(1 + \tau s) \tag{8.43}$$

where $a = K_1 C_1 / K_2 C_2 > 1$.

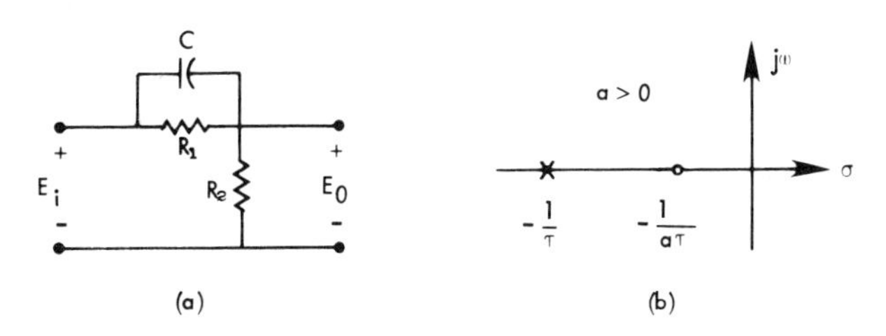

Fig. 8.30 Lead network: (a) passive network, (b) pole zero configuration.

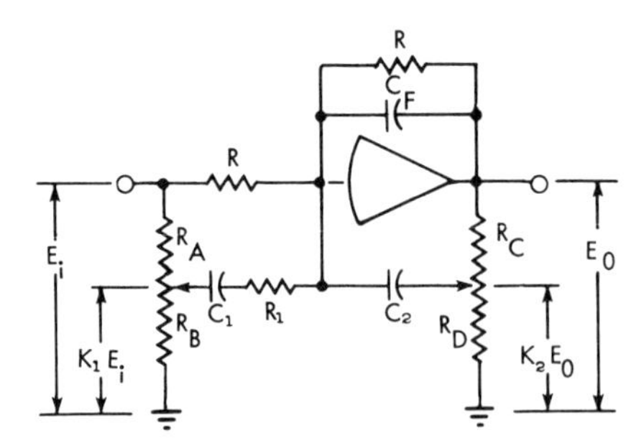

Fig. 8.31 Active lead network.

For any lead network the Bode diagram is that shown in Figure 8.32, where the asymptotic approximation is illustrated [22]. The maximum phase lead ϕ_m occurs at the median frequency ω_m, where ω_m occurs at the geometric mean of the corner frequencies; i.e.,

$$\begin{aligned}\log_{10}\omega_m &= (1/2)[\log_{10}(1/a\tau) + \log_{10}(1/\tau)] \\ &= (1/2)\log_{10}(1/a\tau^2) = \log_{10}(1/\tau\sqrt{a})\end{aligned}$$

Then

$$\omega_m = 1/\tau\sqrt{a} \tag{8.44}$$

The magnitude of the maximum phase lead ϕ_m is computed from

$$\phi_m = \arg[(1 + j\omega_m a\tau)/(1 + j\omega_m\tau)] = \tan^{-1}\omega_m a\tau - \tan^{-1}\omega_m\tau \triangleq x - y \tag{8.45}$$

From trigonometric identities

$$\tan(x - y) = (\tan x - \tan y)/(1 + \tan x \tan y) \tag{8.46}$$

Therefore, using (8.46) in (8.45)

$$\tan\phi_m = (\omega_m a\tau - \omega_m\tau)/[1 + (\omega_m a\tau)(\omega_m\tau)] = \omega_m\tau(a - 1)/(1 + a\omega_m^2\tau^2) \tag{8.47}$$

This expression can be simplified by using (8.44) to compute

$$\tan\phi_m = (a - 1)/2\sqrt{a} \tag{8.48}$$

Now, visualizing a right triangle with base $2\sqrt{a}$, height $(a - 1)$ and hypotenuse b,

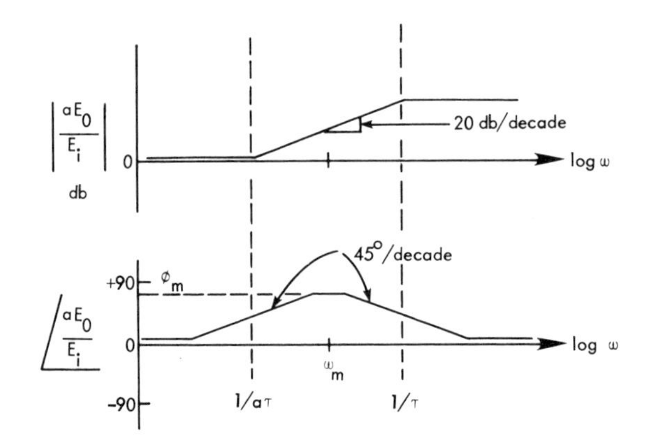

Fig. 8.32 Bode diagram for the lead network $(1 + a\tau s)/(1 + \tau s)$ where $a > 1$.

we compute $b^2 = (a - 1)^2 + 4a = (a + 1)^2$ or

$$\sin \phi_m = (a - 1)/(a + 1) \tag{8.49}$$

This expression can be solved for a to compute

$$a = (1 + \sin \phi_m)/(1 - \sin \phi_m) \tag{8.50}$$

These last two expressions give the desired constraint between maximum phase lead and the parameter a. The procedure then is to determine the desired phase lead ϕ_m. This fixes the parameter a from (8.50). Knowing both a and the frequency ω_m determines the time constant τ from (8.44).

In many practical cases the phase lead required is greater than that obtainable from a single lead network. In this case two or more cascaded lead stages are used. Thus we often write (8.40) as

$$G_S(s) = [K_0 \tau_0 s/(1 + \tau_0 s)][(1 + a\tau s)/(1 + \tau s)]^n \tag{8.51}$$

where n is the number of lead stages (usually $n = 2$ or 3).

Example 8.9

Compute the parameters of the power system stabilizer required to exactly compensate for the excitation control system lag of 161.6° computed in Example 8.7.

Solution

Assume two cascaded lead stages. Then the phase lead per stage is

$$\phi_m = 161.6/2 = 80.8°$$

From (8.50)

$$a = (1 + \sin 80.8)/(1 - \sin 80.8) = 154.48$$

This is a very large ratio, and it would probably be preferable to design the compensator with three lead stages such that $\phi_m = 53.9°$. Then

$$a = (1 + \sin 53.9)/(1 - \sin 53.9) = 9.42$$

which is a reasonable ratio to achieve physically.

The natural frequency of oscillation of the system is $\omega_{osc} = \omega_m = 8.8$ rad/s. Thus from (8.44)

$$\tau = 1/\omega_m\sqrt{a} = 0.037 \qquad a\tau = 0.3488$$

Thus

$$G_S(s) = [K_0\tau_0 s/(1 + \tau_0 s)][(1 + 0.349s)/(1 + 0.037s)]^3$$

A suitable value for the reset time constant is $\tau_0 = 10$ s. The gain K_0 is usually modest [26], say $0.1 < K_0 < 100$, and is usually field adjusted for good response. It is also common to limit the output of the stabilizer, as shown in Figure 8.28, so that the stabilizer output will never dominate the terminal voltage feedback.

Example 8.10

Assume a two-stage lead-compensated stabilizer. Prepare a table showing the phase lead and the compensator parameters as a function of a.

Solution

As before, we assume that $\omega_m = 8.8$ rad/s.

Table 8.5. Lead Compensator Parameters as a Function of a

a	ϕ_m	$2\phi_m$	$\tau = 1/\omega_m\sqrt{a}$	$\omega_{Hi} = 1/\tau$	$a\tau$	$\omega_{LO} = 1/a\tau$
5	41.81	83.62	0.0508	19.68	0.2541	3.935
10	54.90	109.80	0.0359	27.83	0.3593	2.783
15	61.05	122.10	0.0293	34.08	0.4401	2.272
20	64.79	129.58	0.0254	39.35	0.5082	1.968
25	67.38	134.76	0.0227	44.00	0.5682	1.760

These results show that for a large a or large ϕ_m the corner frequencies ω_{Hi} and ω_{LO} must be spread farther apart than for small ϕ_m. See Figure 8.32 and Problem 8.11.

8.8 Linear Analysis of the Stabilized Generator

In previous sections certain simplifying assumptions were made in order to give an approximate analysis of the stabilized generator. In this section the system of Figure 8.28 is solved by linear system analysis techniques using the digital computer (see Section 8.5). The results of the linear computer analysis are best illustrated by an example.

Example 8.11

Use a linear systems analysis program to determine the following graphical solutions for the system of Figure 8.28:

1. Root-locus plot
2. Time response of ω_Δ to a step change in V_{REF}
3. Bode diagram of the closed loop transfer function
4. Bode diagram of the open loop transfer function.

Furthermore, compute these graphical displays for two conditions, (a) no power system stabilizer and (b) a two-stage lead stabilizer with $a = 25$:

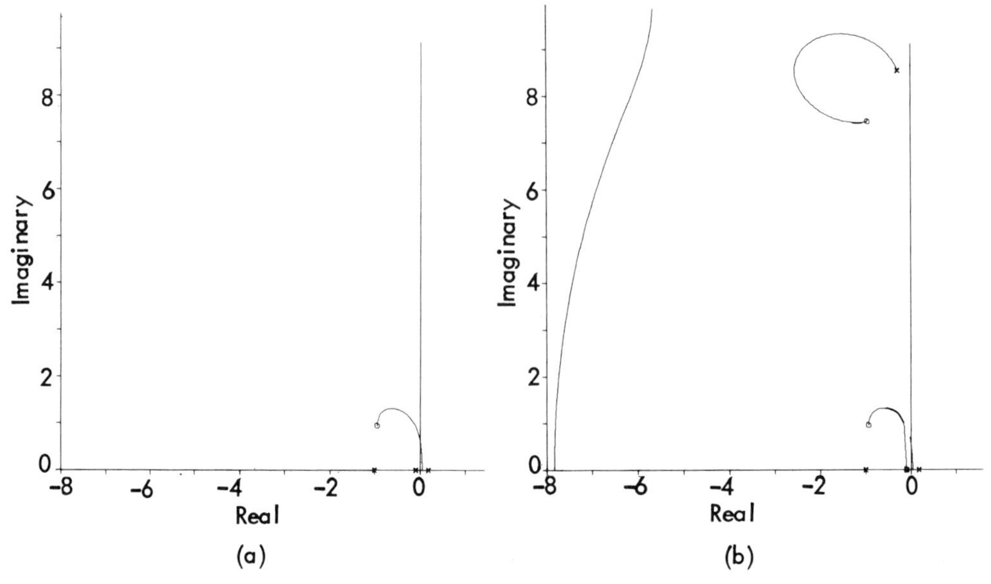

Fig. 8.33 Root locus of the generator 2 system: (a) no PSS, (b) with the PSS having two lead stages with $a = 25$.

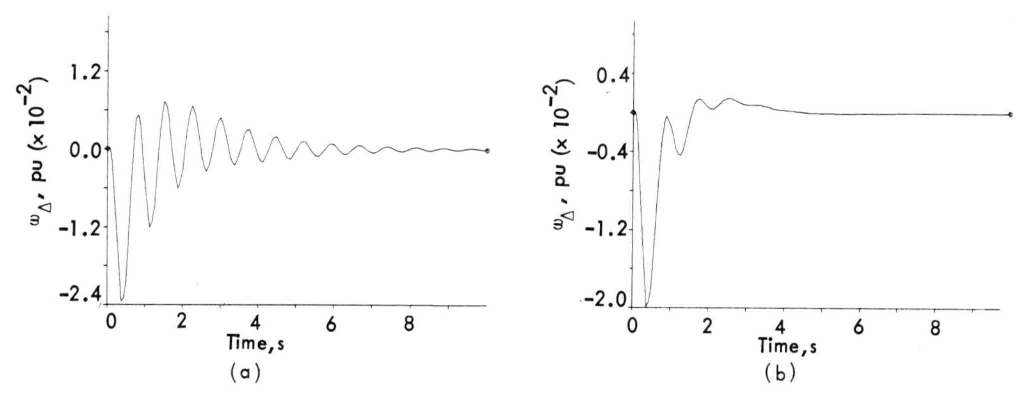

Fig. 8.34 Time response to a step change in V_{REF}: (a) no PSS, (b) with the PSS having two lead stages with $a = 25$.

$$G_S(s) = [10s/(1 + 10s)][(1 + 0.568s)/(1 + 0.0227s)]^2$$

The system constants are the same as Examples 8.7 and 8.8.

Solution

The system to be solved is that of Figure 8.28 except that the PSS limiter cannot be represented in a linear analysis program and is therefore ignored. The results are shown in Figures 8.33–8.36 for the four different plots. In each figure, part (a) is the result without the PSS and part (b) is with the PSS.

In the root-locus plot (Figure 8.33) the major effect of the PSS is to separate the torque-angle zeros from the poles, forcing the locus to loop to the left and downward, thereby increasing the damping. The root locus shows clearly the effect of lead compensation and has been used as a basis for PSS parameter identification [27]. Note that

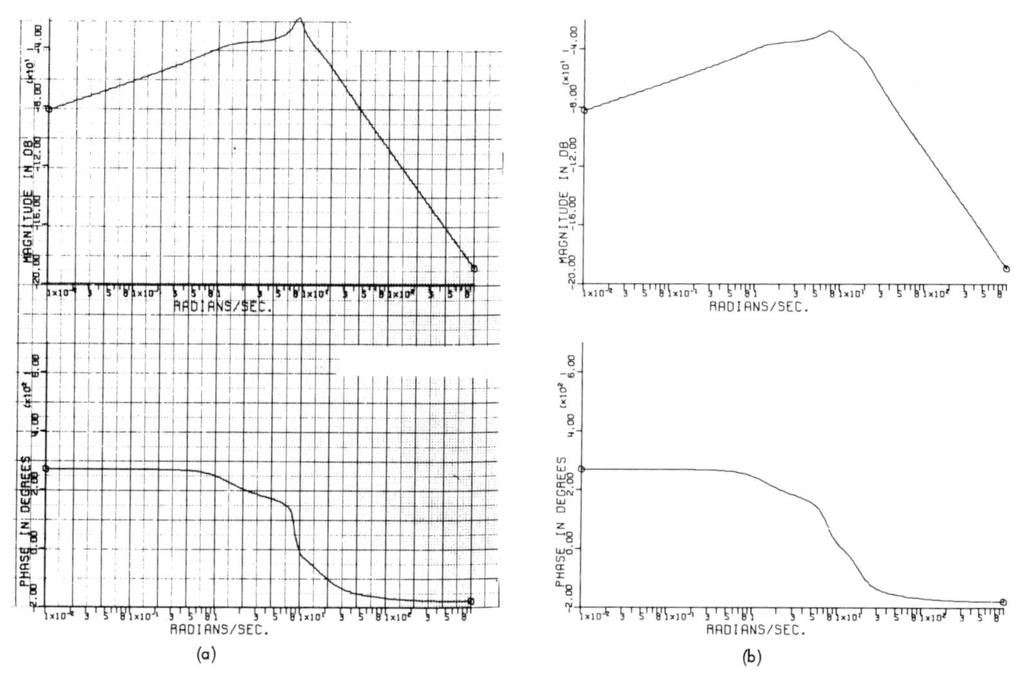

Fig. 8.35 Frequency response (Bode diagram) of the closed loop transfer function: (a) no PSS, (b) with the PSS having two lead stages with $a = 25$.

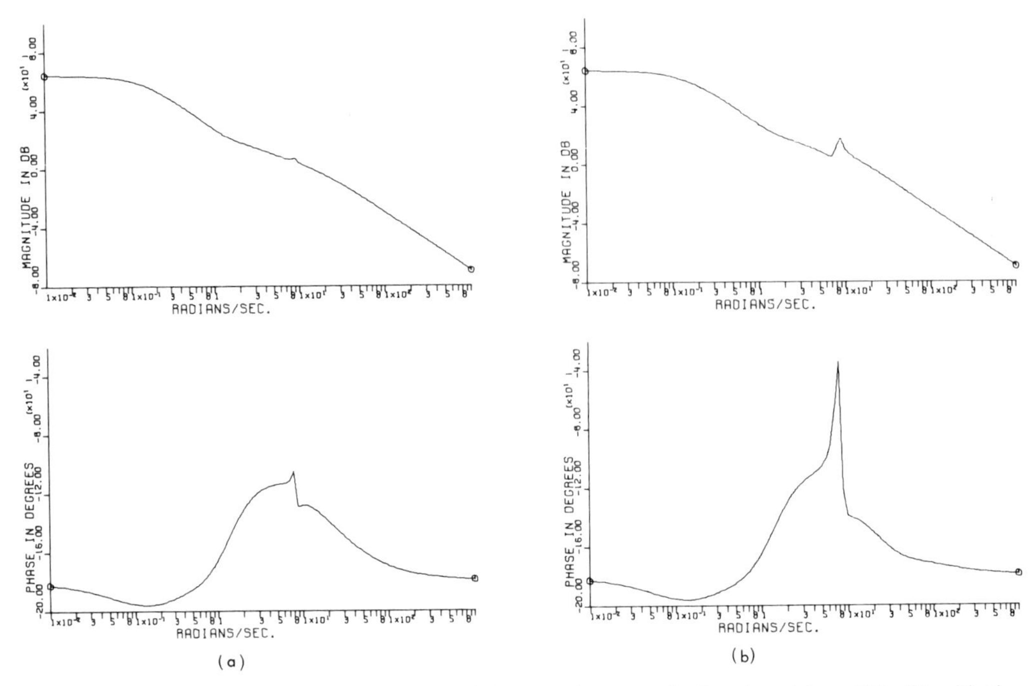

Fig. 8.36 Frequency response (Bode diagram) of the open loop transfer function: (a) no PSS, (b) with the PSS having two lead stages with $a = 25$.

the locus near the origin is unaffected by the PSS, but the locus breaking away vertically from the negative real axis moves closer to the origin as compensation is added [this locus is off scale in 8.33(a)]. From the computer we also obtain the tabulation of poles and zeros given in Table 8.6. From this table we note that the natural radian frequency of oscillation is controlled by the torque-angle poles with a frequency of 8.467 rad/s. This agrees closely with $\omega_n = 8.538$ rad/s computed in Example 8.8 using the approximate model and also checks well with the frequency of δ_{21} in Figure 3.3.

Figure 8.34 shows the substantial improvement in damping introduced by the PSS network. Note the slightly decreased frequency of oscillation in the stabilized response.

Table 8.6. Root-Locus Poles and Zeros

Condition	Poles	Zeros
No PSS	$-20.000 + j0.000$	$-0.944 + j0.955$
	$0.179 + j0.000$	$-0.944 - j0.955$
	$-0.102 + j0.000$	$-0.452 + j8.467$
	$-0.289 + j8.533$	$-0.452 - j8.467$
	$-0.289 - j8.533$	
	$-1.000 + j0.000$	
With PSS $a = 25$	$-20.000 + j0.000$	$-0.100 + j0.000$
	$0.179 + j0.000$	$-0.941 + j0.959$
	$-0.010 + j0.000$	$-0.941 - j0.959$
	$-0.289 + j8.533$	$-0.955 + j7.439$
	$-0.289 - j8.533$	$-0.955 - j7.439$
	$-1.000 + j0.000$	$-45.000 + j24.847$
	$-0.100 + j0.000$	$-45.000 - j24.847$
	$-45.500 + j0.000$	
	$-45.500 - j0.000$	

Figures 8.35 and 8.36 show the frequency response of the closed loop and open loop transfer functions respectively. The uncompensated system has a very sharp drop in phase very near the frequency of oscillation. Lead compensation improves the phase substantially in this region, thus improving gain and phase margins.

8.9 Analog Computer Studies

The analog computer offers a valuable tool to arrive at an optimum setting of the adjustable parameters of the excitation system. With a variety of compensating schemes available to the designer and with each having many adjustable components and parameters, comparative studies of the effectiveness of the various schemes of compensation can be conveniently made. Furthermore, this can be done using the complete nonlinear model of the synchronous machine.

8.9.1 Effect of the rate feedback loop in Type 1 exciter

As a case study, Example 5.8 is extended to include the effect of the excitation system. The synchronous machine used is the same as in the examples of Chapter 4 with the loading condition of Example 5.1. Three IEEE Type 1 exciters (see Section 7.9.1) are used in this study: W TRA, W Brushless, and W Low τ_E Brushless. The parameters for these exciters are given in Table 7.8.

The analog computer representation of the excitation system is shown in Figure 8.37. This system is added to the machine simulation given in Figure 5.18. Note that the output of amplifier 614 (Figure 8.37) connects to the terminal marked E_{FD} in Figure 5.18, and the terminal marked v_t in Figure 5.18 connects with switch 421 in Figure 8.37. The new "free" inputs to the combined diagram are V_{REF} and T_m. The potentiometer settings for the analog computer units are given in Tables 8.7, 8.8 and 8.9 for the three excitation systems described in Table 7.8. Saturation is represented by an analog limiter on V_R in this simulation.

With the generator equipped with a W TRA exciter, the response due to a 10% increase in T_m and 5% change in V_{REF} and the phase plane plot of ω_Δ versus δ_Δ for the initial loading condition of Example 5.1 are shown in Figure 8.38. The results with W Brushless and W Low τ_E Brushless exciters are shown in Figures 7.69 and 8.39 respectively.

Table 8.7. Potentiometer Calculations for a Type 1 Representation of a W TRA Exciter ($a = 20$)

Pot. no.	Amp. no.	Out	L_0	In	L_i	L_0/L_i	C = constant	$(L_0/L_i)C$	Int. cap.	Amp. gain	Pot. set.
600	601	V_{REF}	50	REF	100	0.50	5% of $P601$	0.0250	...	1	0.0250
601	601	V_{REF}	50	REF	100	0.50	$1 + (2.667)(-0.17)/400 = 0.9988$	0.4994	...	1	0.4994
800	800	V_R	1	$-V_e$	50	0.02	$K_A/a\tau_A = 400/(20)(0.05) = 400$	8.0	1.0	10	0.8000
701	800	V_R	1	V_R	1	1.00	$1/a\tau_A = 1/(20)(0.05) = 1$	1.0	1.0	10	0.1000
801	801	$-E_{FD}$	10	V_R	1	10.00	$1/a\tau_E = 1/(20)(0.95) = 0.05263$	0.5263	1.0	1	0.5263
703	801	$-E_{FD}$	10	$-E_{FD}$	10	1.00	$\lvert K_E \rvert /a\tau_E = 0.17/(20)(0.95) = 0.008947$	0.0089	1.0	1	0.0089
802	802	V_z	50	V_y	100	0.50	$1/\tau_F = 1/1.0 = 1.0$	0.5	...	1	0.5000
810	802	V_z	50	$-E_{FD}$	10	5.00	$K_F/\tau_F = 0.04/1.0 = 0.04$	0.2	...	1	0.2000
812	810	$-V_y$	100	V_z	50	2.00	$1/a = 1/20 = 0.05$	0.1	1.0	1	0.1000
803	803	V_x	50	v_t	40	1.25	$1/\sqrt{3} = 0.5773$	0.7217	...	1	0.7217
lim	800	...	...	...	...	...	$V_{R\max} = 3.5$ pu $= 3.5\ v$				
800		...	...	...	...	...	$V_{R\min} = -3.5$ pu $= -3.5\ v$				

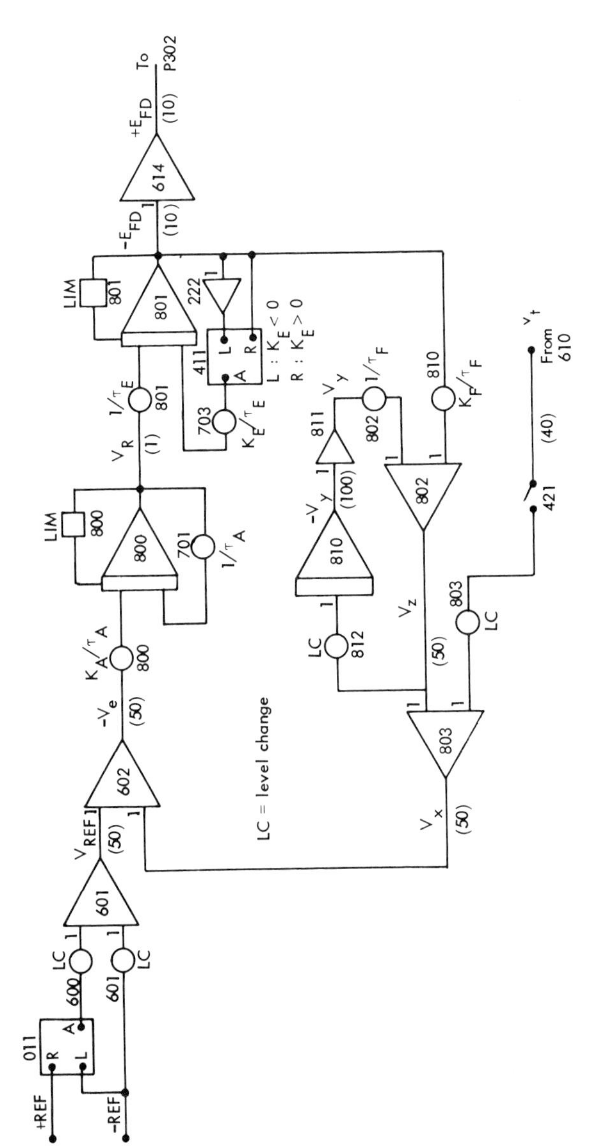

Fig. 8.37 Analog computer representation of a Type 1 excitation system.

Table 8.8. Potentiometer Calculations for a Type 1 Representation of a W Brushless Exciter ($a = 20$)

Pot. no.	Amp. no.	Out	L_0	In	L_i	L_0/L_i	C = constant	$(L_0/L_i)C$	Int. cap.	Amp. gain	Pot. set.
600	601	V_{REF}	50	REF	100	0.50	5% of $P601$	0.0252	...	1	0.0252
601	601	V_{REF}	50	REF	100	0.50	$1 + 2.667/400 = 1.0066$	0.5033	...	1	0.5033
800	800	V_R	1	$-V_e$	50	0.02	$K_A/a\tau_A = 400/(20)(0.02) = 1000$	20.0	0.1	100	0.2000
701	800	V_R	1	V_R	1	1.00	$1/a\tau_A = 1/(20)(0.02) = 2.5$	2.5	0.1	10	0.2500
801	801	$-E_{FD}$	10	V_R	1	10.00	$1/a\tau_E = 1/(20)(0.8) = 0.0625$	0.6250	1.0	1	0.6250
703	801	$-E_{FD}$	10	$-E_{FD}$	10	1.00	$K_E/a\tau_E = 1/(20)(0.8) = 0.0625$	0.0625	1.0	1	0.0625
802	802	V_z	50	V_y	100	0.50	$1/\tau_F = 1/1.0 = 1.0$	0.50	...	1	0.5000
810	802	V_z	50	$-E_{FD}$	10	5.00	$K_F/\tau_F = 0.03/1.0 = 0.03$	0.15	...	1	0.1500
812	810	$-V_y$	100	V_z	50	2.00	$1/a = 1/20 = 0.05$	0.10	1.0	1	0.1000
803	803	V_x	50	v_t	40	1.25	$1/\sqrt{3} = 0.5773$	0.7217	...	1	0.7217
lim	800	...	...	...	...	...	$V_{R\max} = 7.3$ pu $= 7.3\ v$				
800		...	...	...	...	...	$V_{R\min} = -7.3$ pu $= -7.3\ v$				

Comparing the responses shown in Figures 8.38, 7.69, and 8.39 with that of Figure 5.20, we note that without the exciter the slow transient is dominated by the field winding effective time constant. The terminal voltage, the field flux linkage, and the rotor angle are slow in reaching their new steady-state values. From Figures 8.38, 7.69, and 8.39 we can see that the steady-state conditions are reached sooner with the exciter present. At the same time, the response is more oscillatory.

8.9.2 Effectiveness of compensation

A detailed study of the effectiveness of four methods of compensation is given in [28], by comparing the dynamic response due to changes in the mechanical torque T_m and the reference excitation voltage V_{REF} at various machine loadings. The dynamic response comparison is based on observing the rise time, settling time, and percent overshoot of either $P_{e\Delta}$ or $V_{t\Delta}$ in a given transient. For example, a 10% increase in the reference torque is made, and the change in electrical power output $P_{e\Delta}$ is observed. The machine data and loading are essentially those given in the Examples 8.4 and 8.5.

Table 8.9. Potentiometer Calculations for a Type 1 Representation of a W Low τ_E Brushless Exciter ($a = 20$)

Pot. no.	Amp. no.	Out	L_0	In	L_i	L_0/L_i	C = constant	$(L_0/L_i)C$	Int. cap.	Amp. gain	Pot. set.
600	601	V_{REF}	50	REF	100	0.50	5% of $P601$	0.0252	...	1	0.0252
601	601	V_{REF}	50	REF	100	0.50	$1 + 2.667/400 = 1.0066$	0.5033	...	1	0.5033
800	800	V_R	1	$-V_e$	50	0.02	$K_A/a\tau_A = 400/(20)(0.02) = 1000$	20.0	0.1	100	0.2000
701	800	V_R	1	V_R	1	1.00	$1/a\tau_A = 1/(20)(0.02) = 2.5$	2.5	0.1	10	0.2500
801	801	$-E_{FD}$	10	V_R	1	10.00	$1/a\tau_E = 1/(20)(0.015) = 3.3333$	33.333	0.1	100	0.3333
703	801	$-E_{FD}$	10	$-E_{FD}$	10	1.00	$K_E/a\tau_E = 1/(20)(0.015) = 3.3333$	3.3333	0.1	10	0.3333
802	802	V_z	50	V_y	100	0.50	$1/\tau_F = 1/0.5 = 2.0$	1.00	...	10	0.1000
810	802	V_z	50	$-E_{FD}$	10	5.00	$K_F/\tau_F = 0.04/0.5 = 0.08$	0.40	...	1	0.4000
812	810	$-V_y$	100	V_z	50	2.00	$1/a = 1/20 = 0.05$	0.10	1.0	1	0.1000
803	803	V_x	50	v_t	40	1.25	$1/\sqrt{3} = 0.5773$	0.7217	...	1	0.7217
lim	800	...	...	...	...	...	$V_{R\max} = 6.96$ pu $= 6.96\ v$				
800		...	...	...	...	...	$V_{R\min} = -6.96$ pu $= -6.96\ v$				

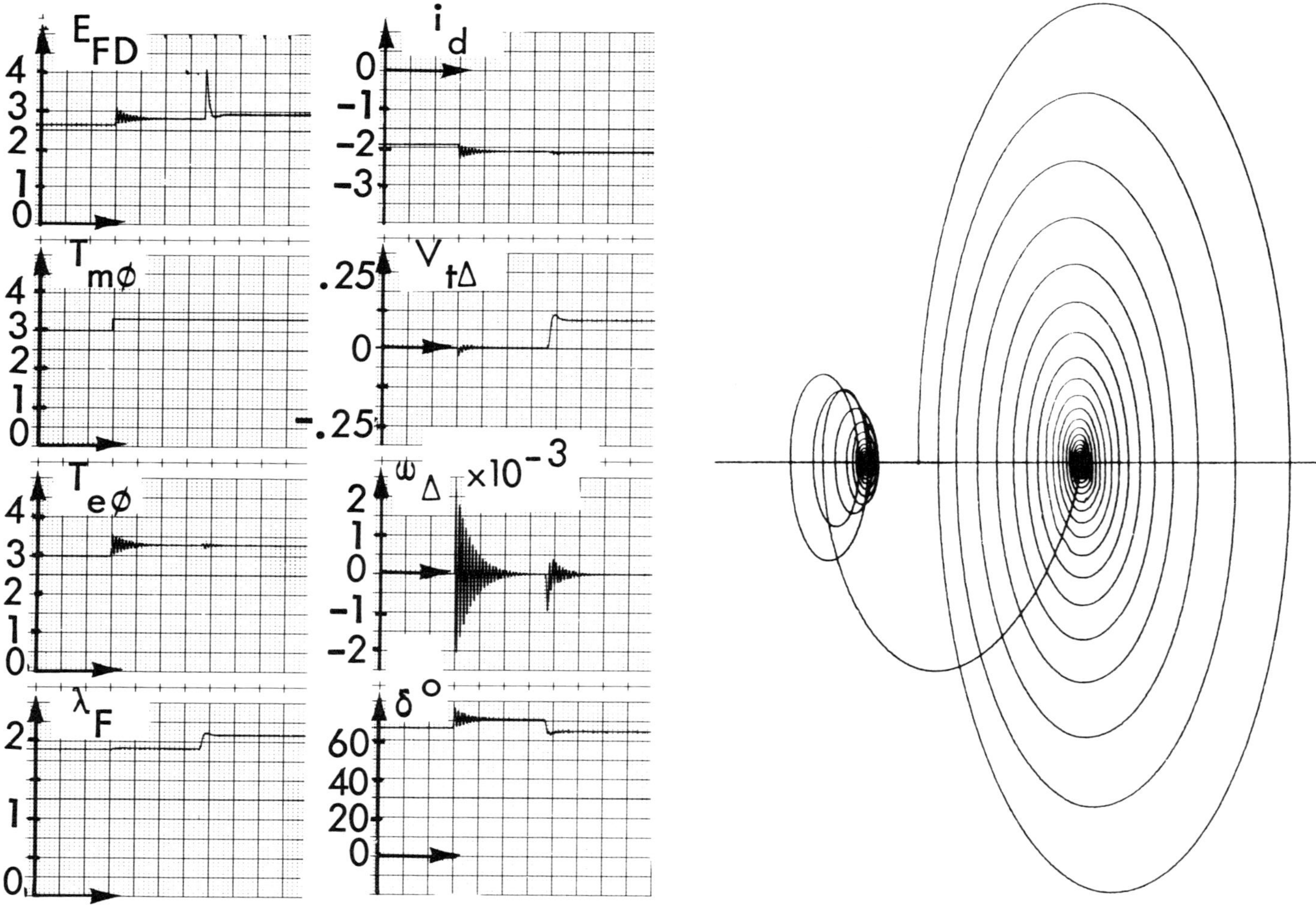

Fig. 8.38 System response to a step change in T_m and V_{REF}, generator equipped with a W TRA exciter.

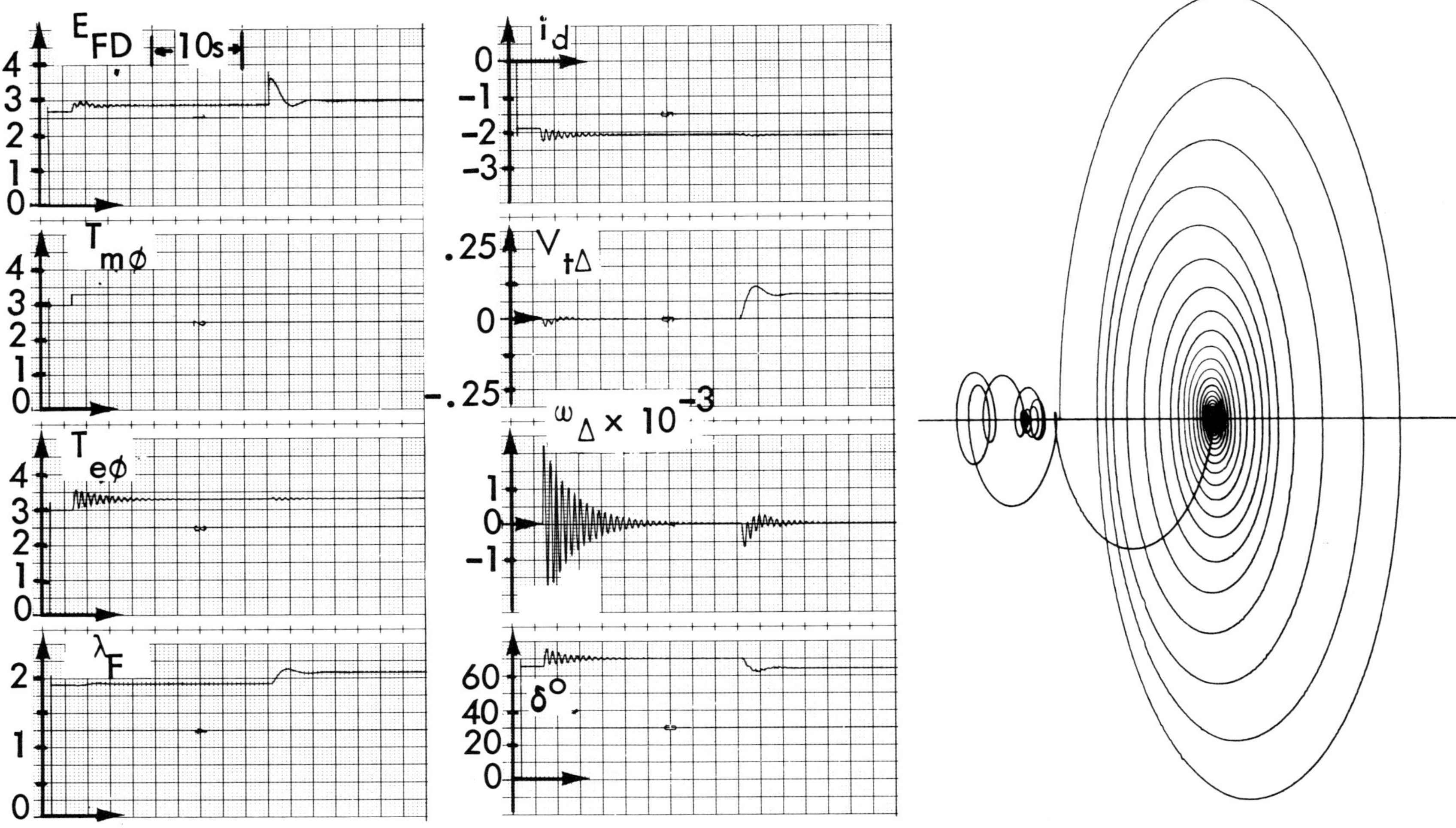

Fig. 8.39 System response to a step change in T_m and V_{REF}, generator equipped with a Low τ_E Brushless exciter.

However, the machine is fully represented on the analog computer. The excitation system used is Type 2, a rotating rectifier system (see Section 7.9.3). The data of the exciter are:

$$K_A = 400 \text{ pu} \qquad K_F = 0.04 \qquad S_{E\max} = 0.86$$
$$\tau_A = 0.02 \text{ s} \qquad \tau_F = 0.05 \text{ s} \qquad S_{E.75\min} = 0.50$$
$$\tau_E = 0.015 \text{ s} \qquad \tau_R = 0.0 \qquad V_{R\max} = 8.26$$
$$K_E = 1.0 \qquad K_R = 1.0 \qquad V_{R\min} = -8.26$$
$$E_{FD\max} = 4.45$$

The methods of compensation used are:

Rate feedback: $sK_F/(1 + \tau_F s)$
Bridge-T filter with transfer function:

$$C/R = (s^2 + rn\,\omega_n s + \omega_n^2)/(s^2 + n\omega_n s + \omega_n^2)$$
$$\omega_n = 21 \text{ rad/s}$$
$$n = 2 \qquad r = 0.1$$

Power system stabilizer:

$$\frac{C}{R} = \frac{Ks}{1 + \tau s}\left(\frac{1 + \tau_1 s}{1 + \tau_2 s}\right)^2$$
$$\tau = 3.0 \text{ s} \qquad \tau_1 = 0.2 \text{ s} \qquad \tau_2 = 0.05 \text{ s}$$

A sample of data given in reference [28] is shown in Table 8.10 for the initial operating condition of $T_{m\phi} = 3.0$ pu at 0.85 PF lagging.

Table 8.10. Comparison of Compensation Schemes

Case	$P_{e\Delta}$			$V_{t\Delta}$		
	Rise time	Settling time	Overshoot %	Rise time	Settling time	Overshoot %
Uncompensated	0.06	0.22	86.6	0.20	0.60	10.0
Excitation rate feedback	0.06	0.22	80.0	0.98	4.20	60.0
Bridged-T only	0.05	0.23	100.0	0.21	0.56	33.0
Bridged-T, two-stage lead-lag and speed	0.04	0.21	73.4	0.28	0.37	5.0
Power system stabilizer	0.05	0.21	82.6	0.23	0.42	5–10

Source: Schroder and Anderson [28].

Other valuable information that can be obtained from analog computer studies is the response of the machine to oscillations originating in the system to which the machine is connected. This can be simulated on the analog representation of one machine connected to an infinite bus by modulating the infinite bus voltage with a signal of the desired frequency. This is particularly valuable in studies to improve the system damping. When growing oscillations occur in large interconnected systems, the frequencies of these oscillations are usually on the order of 0.2–0.3 Hz, with other frequencies superimposed upon them. Thus it is important to know the dynamic response of the synchronous machine under these conditions.

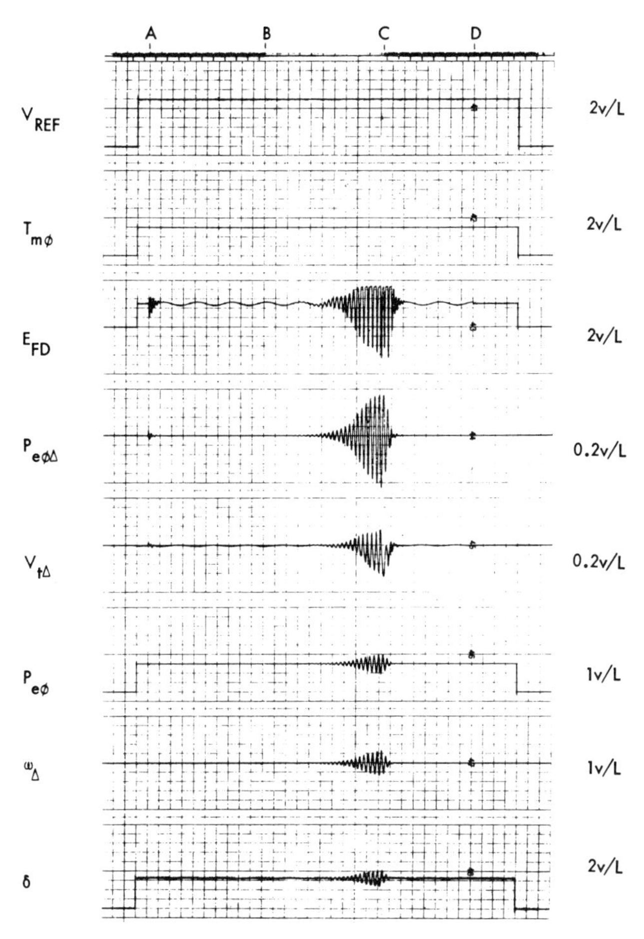

Fig. 8.40 Synchronous machine with PSS operating against an infinite bus whose voltage is being modulated at one-tenth the natural frequency of the machine.

A sample of this type of study, taken from [28], is shown here. The same machine discussed above, but operating under the heavy loading condition of Example 5.1, has its bus voltage modulated by a frequency of one-tenth the natural frequency. The modulating signal varies the infinite bus voltage between 1.02 and 0.98 peak. Figure 8.40 shows the effect of the PSS under these conditions. At time A the modulating signal of 2.1 rad/s is added. The PSS is removed at B, causing growing oscillations to build up especially on $P_{e\Delta}$, which would simulate tie-line oscillations. Note also that the frequency of these oscillations is near the natural frequency of the machine. When the stabilizer is reinstated at point C, the oscillations are quickly damped out. At point D the modulation is removed.

8.10 Digital Computer Transient Stability Studies

To illustrate the effect of the excitation system on transient stability, transient stability studies are made on the nine-bus system used in Section 2.10. The impedance diagram of the system (to 100-MVA base) and the prefault conditions are shown in Figures 2.18 and 2.19 respectively. The generator data are given in Table 2.1. The transient is initiated by a three-phase fault near bus 7 and is cleared by opening the line between bus 5 and bus 7. In this study the loads A, B, and C are represented by

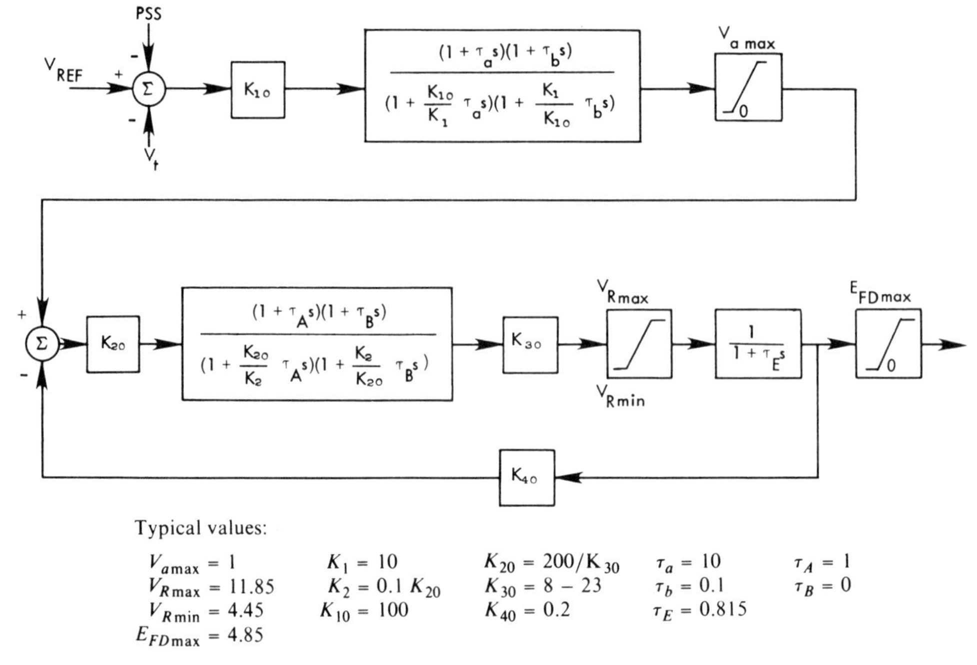

Fig. 8.41 The Brown Boveri Co. alternator diode exciter. (Used with permission of Brown Boveri Co.)

constant impedances; generators 1 and 3 are represented by classical models, i.e., constant voltage behind transient reactance. For generator 2, provision is made for the excitation system representation.

A modified transient stability program was used in this study. (It is based on a program developed by the Philadelphia Electric Co., with modifications to include the required new features.) When the excitation system is represented in detail, the model used for the synchronous machine is the so-called "one-axis model" (see Section 4.15.4) with provision for representing saturation. When the machine EMF E (corresponding to the field current) is calculated, an additional value E_Δ is added due to saturation

Table 8.11. Excitation Systems Data

Parameter	Amplidyne	Mag-A-Stat	SCPT
K_A	25	400	120
K_E	−0.044	−0.17	1.0
K_F	0.0805	0.04	0.02
K_R	1.0	1.0	1.0
K_p	...	...	1.19
K_I	...	...	2.62
$V_{R\max}$	1.0	3.5	1.2
$V_{R\min}$	−1.0	−3.5	−1.2
$V_{B\max}$	...	...	2.78
τ_A	0.20	0.05	0.15
τ_E	0.50	0.95	0.05
τ_F	0.35	1.0	0.60
τ_R	0.06	0	0
A_g	0.0016	0.0039	...
B_g	1.465	1.555	...

Note: See Figure 8.41 for BBC exciter parameters.

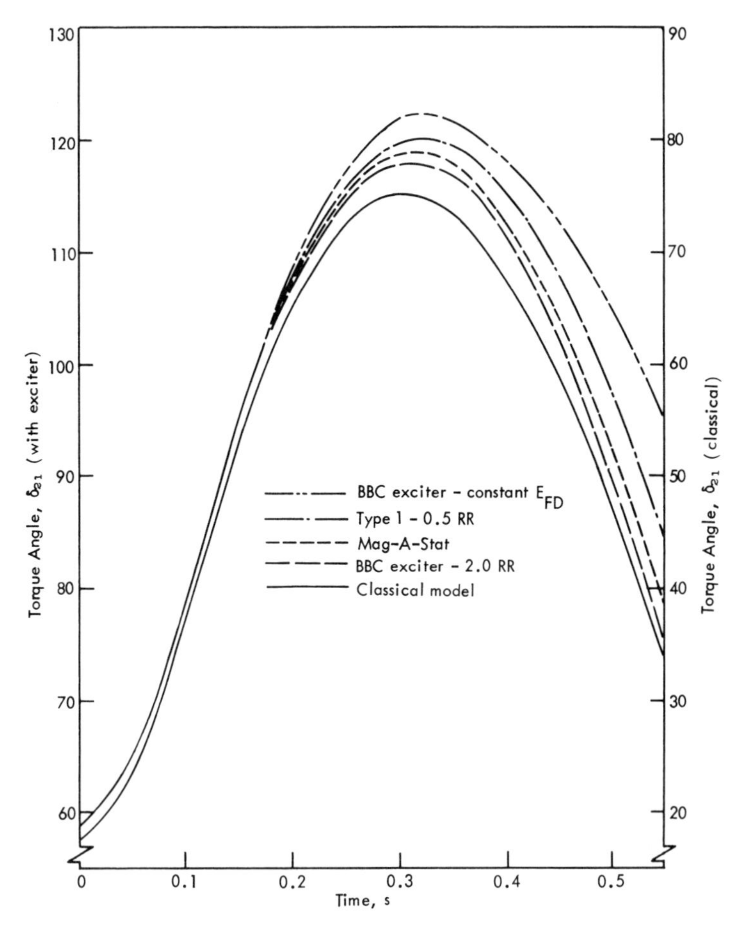

Fig. 8.42 δ_{21} for various exciters with a three-cycle fault.

effect and based on the voltage behind the leakage reactance E_ℓ. This is given by

$$E_\Delta = A_g \exp[B_g(E_\ell - 0.8)] \tag{8.52}$$

The constants A_g and B_g are provided for several exciters [see (4.141)].

The types of field representation used with generator 2 are:

1. Classical model.
2. IEEE Type 1, 0.5 pu response, amplidyne NA101 exciter (see Figure 7.61).
3. IEEE Type 1, 2.0 pu response, Mag-A-Stat exciter (see Figure 7.61).
4. IEEE Type 3, SCPT fast exciter, 2.0 pu response (see Figure 7.66).
5. Brown Boveri Company (BBC) alternator diode exciter (see Figure 8.41).

The excitation system data are given in Table 8.11.

8.10.1 Effect of fault duration

Two sets of runs were made for the same fault location and removal, but for different fault durations. The breaker clearing times used were three cycles and six cycles. For a three-cycle fault, the results of generator 2 data are shown in Figures 8.42–8.46. Similar results for a six-cycle fault are shown in Figures 8.47–8.50.

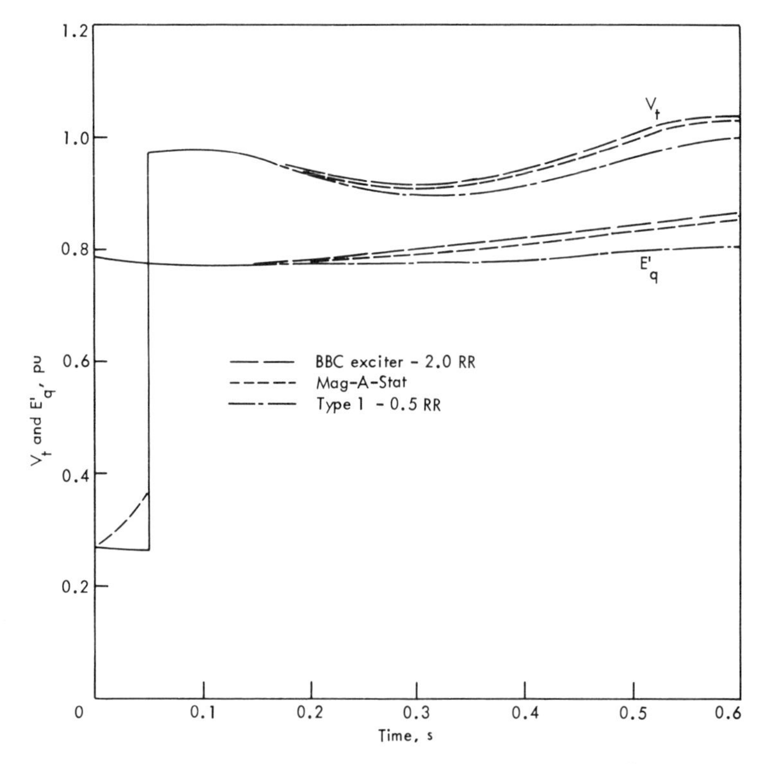

Fig. 8.43 V_t and E'_q for various exciters with a three-cycle fault.

Results with three-cycle fault clearing. Figure 8.42 shows a plot of the first swing of the angle δ_{21} for different field representations. Note that the classical run gives the angle of the voltage behind transient reactance, while all the others give the position of the q axis. A run with constant E_{FD} is also added. We conclude from the results shown in Figure 8.42 that for a three-cycle clearing time the classical model gives approximately the same magnitude of δ_{21} for the first swing as the different exciter representations. When the exciter model was adjusted to give constant E_{FD}, however, a large swing was obtained.

From Figure 8.43 we conclude that the slow exciter gives the nearest simulation of a constant flux linkage in the main field winding (and hence constant E'_q) and minimum variation of the terminal voltage after fault clearing.

The action of the exciter and the armature reaction effects are clearly displayed in Figure 8.44. It is interesting to note that the actual field current, as seen by the EMF E, is hardly affected by the value of E_{FD} for most of the duration of the first swing after the fault is cleared. The effect of the armature reaction is dominant in this period.

Figure 8.45 shows a time plot of P_2 for this transient. Again it can be seen that the different models give essentially the same power swing for this generator. We note, however, that the minimum swing is obtained with the slow exciter while the maximum swing is obtained with the classical model.

In Figure 8.46 the rotor angle δ_{21} is plotted for a period of 2.0 s for the classical model, a slow IEEE Type 1 exciter, and a relatively fast exciter with 2.0 pu response. The plot shows that the first swing is the largest, with the subsequent swings slightly reduced in magnitude.

Figures 8.42–8.46 seem to indicate that for this fault the system is well below the stability limit, since the magnitude of the first swing is on the order of 60°. All generator

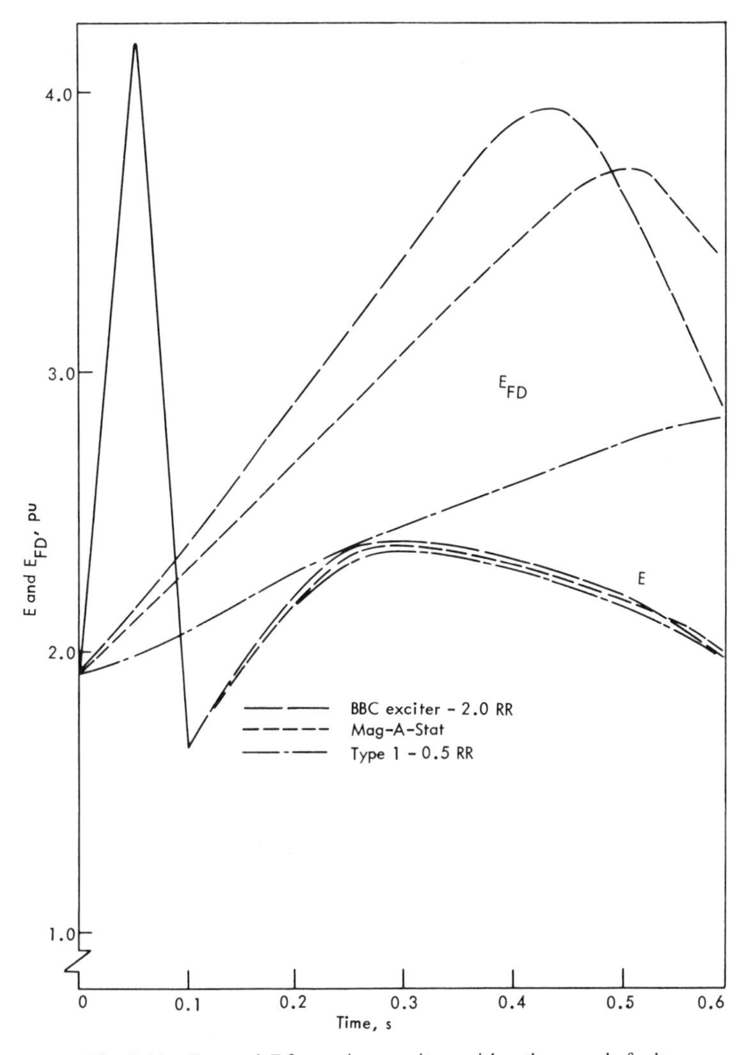

Fig. 8.44 E_{FD} and E for various exciters with a three-cycle fault.

2 models give approximately the same magnitude of rotor angle and power swing and period of oscillation.

Results with six-cycle fault clearing. For the case of a six-cycle clearing time, the plot of the angle δ_{21} is shown in Figure 8.47 for the classical model and for two different types of exciter models. The swing curves indicate that this is a much more severe fault than the previous one, and the system is perhaps close to the transient stability limit. Here the swing curves for the generator with different field representations are quite different in both the magnitude of swings and periods of oscillation. The effect of the 2.0 pu response exciter is pronounced after the first swing. The effect of the power system stabilizer on the response is hardly noticeable until the second swing. The magnitude of the first swing for the cases where the excitation system is represented in detail is significantly larger than for the case of the classical representation. The Type 1 exciter gives the highest swing. Comparing Figures 8.46 and 8.47, we note that for this severe fault the rotor oscillation of generator 2 depends a great deal on the type of excitation system used on the generator. We also note that the classical model does not accurately represent the generator response for this case.

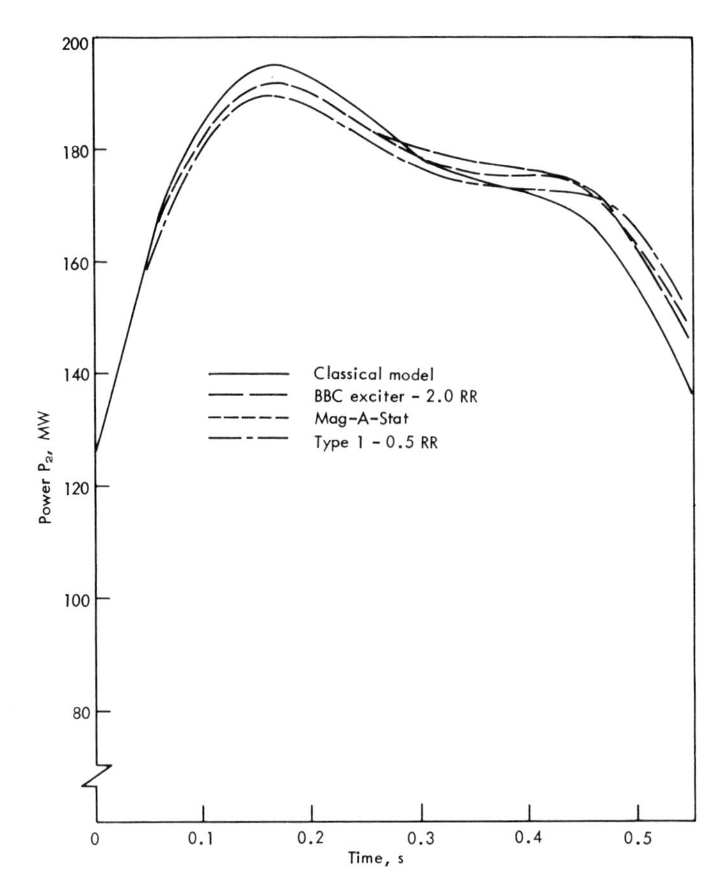

Fig. 8.45 Output power P_2 for various exciters with a three-cycle fault.

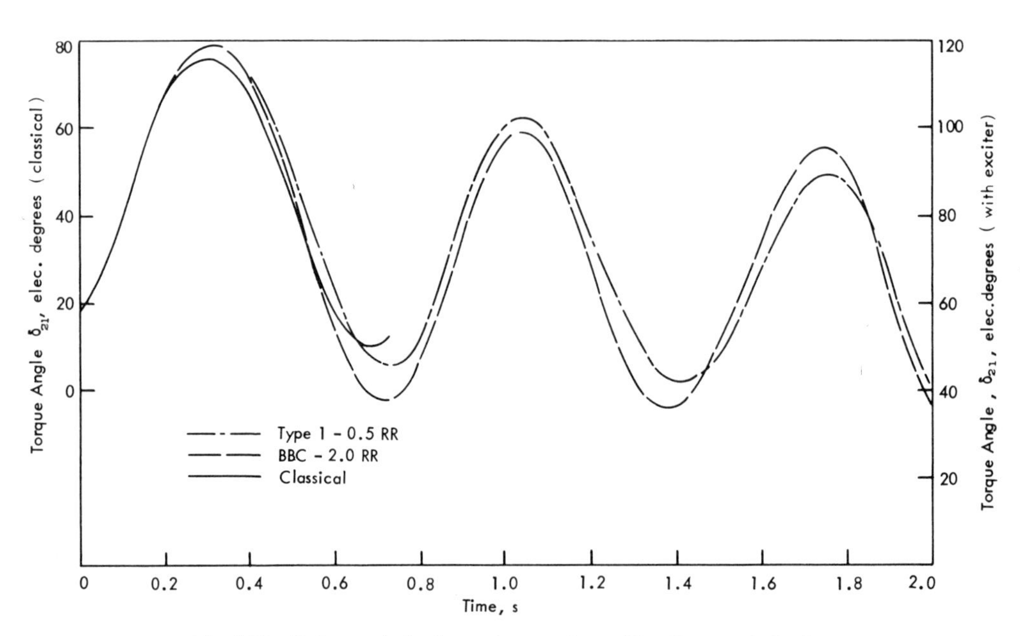

Fig. 8.46 Rotor angle δ_{21} for various exciters with a three-cycle fault.

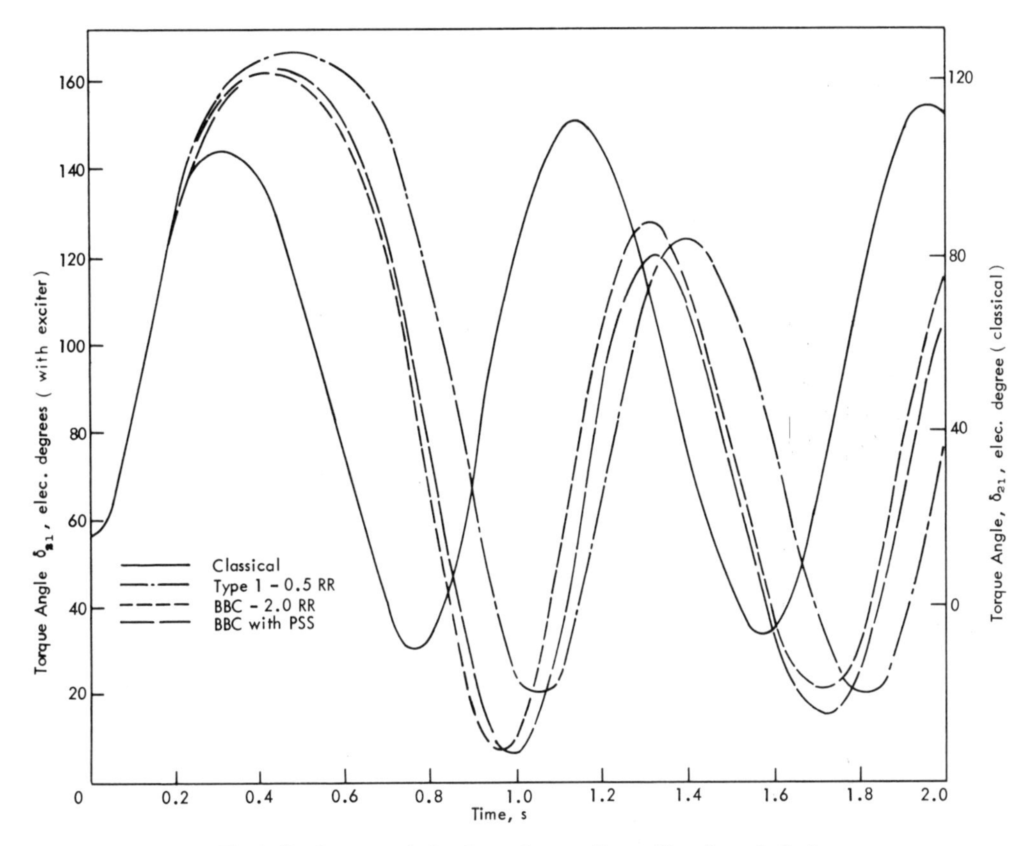

Fig. 8.47 Rotor angle δ_{21} for various exciters with a six-cycle fault.

The output power of generator 2 is shown in Figure 8.48 for different exciter representations. While the general shape of these curves is the same, some significant differences are noted. The excitation system increases the output power of the generator after the first swing. The generator acceleration will thus decrease, causing the rotor swing to decrease appreciably. This effect is not noticed in the classical model. It would appear that for slightly more severe faults the classical model may predict different results concerning stability than those predicted using the detailed representation of the exciter.

Figures 8.49 and 8.50 show plots of the various voltages and EMF's of generator 2 for the case of the 2.0 pu exciter and the Type 1 exciter respectively. The curves for E_q' show that although the fault is near the generator terminal, the flux linkage in the main field winding (reflected in the value of E_q') drops only slightly (by about 5%); and for the duration of the first swing it is fairly constant. The faster recovery occurs with the 2.0 pu exciter, and E_q' reaches a plateau at about 1.1 s and stays fairly constant thereafter. For the Type 1 exciter E_q' recovers slowly and continues to increase steadily. The oscillations of terminal voltage V_t are somewhat complex. The first swing after the fault seems to be dominated by the inertial swing of the rotor, with the action of the exciter dominating the subsequent swings in V_t. Thus after the first voltage dip, the swings in V_t follow the changes in the field voltage E_{FD} with a slight time lag. Again the recovery of the terminal voltage is faster with the 2.0 pu exciter than with the Type 1 exciter. We also note that the excitation system introduces additional frequencies of oscillation, which appear in the V_t response.

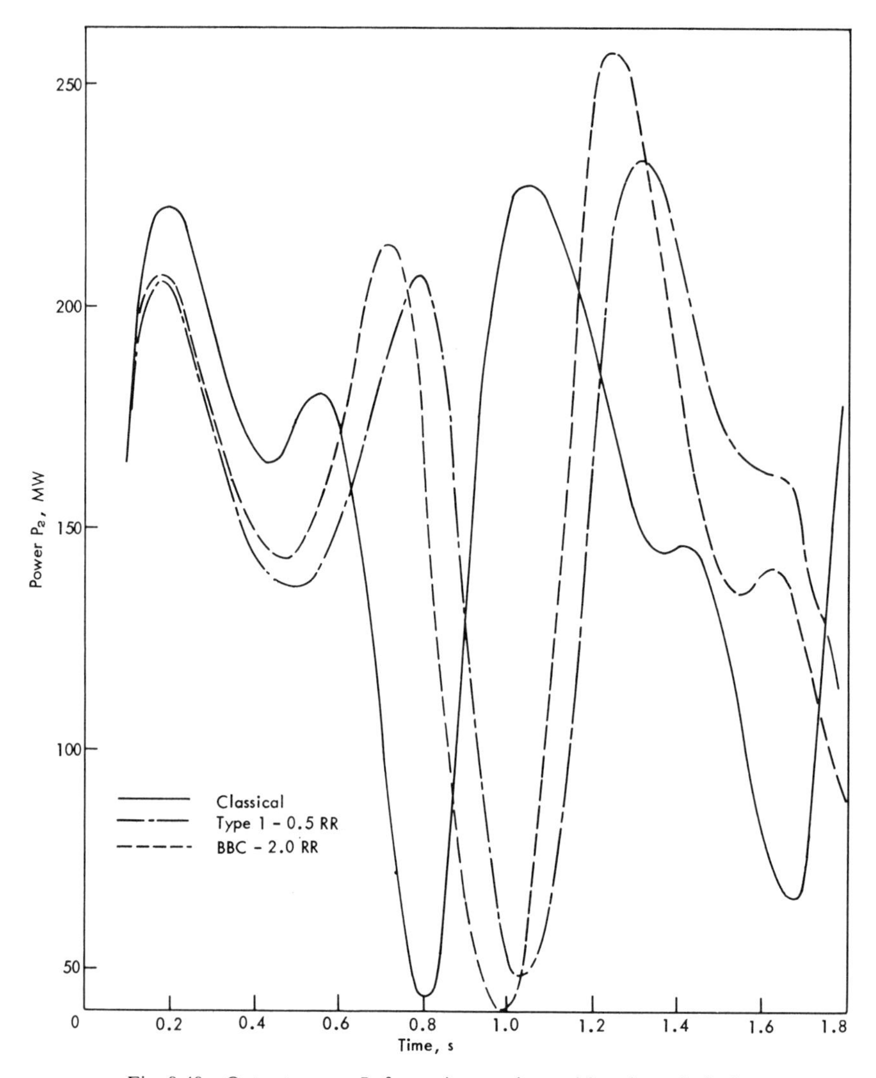

Fig. 8.48 Output power P_2 for various exciters with a six-cycle fault.

The plots of E clearly show the effect of the armature reaction. In the first 0.7 s, for example, the changes in E_{FD} are reflected only in a minor way in the total internal EMF E. The component of E due to the armature reaction seems to be dominant because the field circuit time constant is long. The general shape of the EMF plot, however, is due to the effects of both E_{FD} and the armature reaction.

From the data presented in this study we conclude that for a less severe fault or for fast fault clearing, the excitation representation is not critical in predicting the system dynamic responses. However, for a more severe fault or for studies involving long transient periods, it is important to represent the excitation system accurately to obtain the correct system dynamic response.

8.10.2 Effect of the power system stabilizer

For large disturbances the assumption of linear analysis is not valid. However, the PSS is helpful in damping oscillations caused by large disturbances and can be effective in restoring normal steady-state conditions. Since the initial rotor swing is largely an

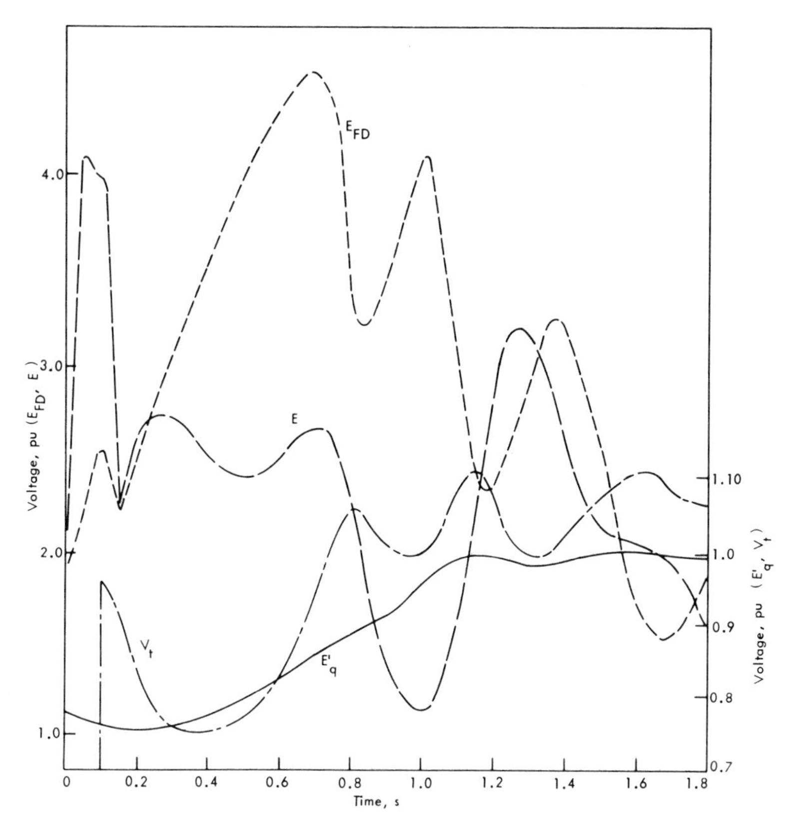

Fig. 8.49 Voltages of generator 2 with BBC exciter.

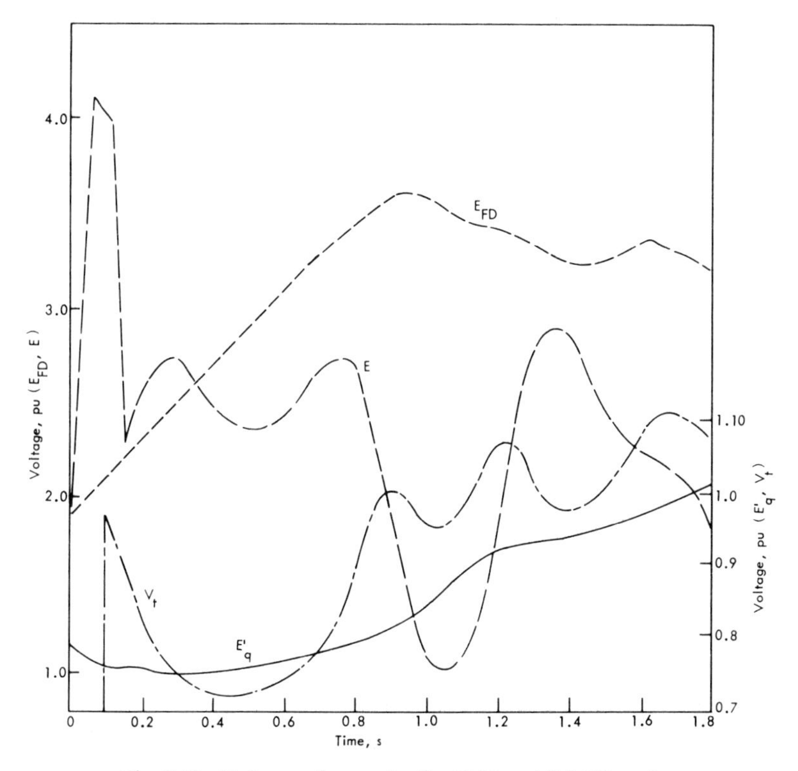

Fig. 8.50 Voltages of generator 2 with Type 1 0.5 RR exciter.

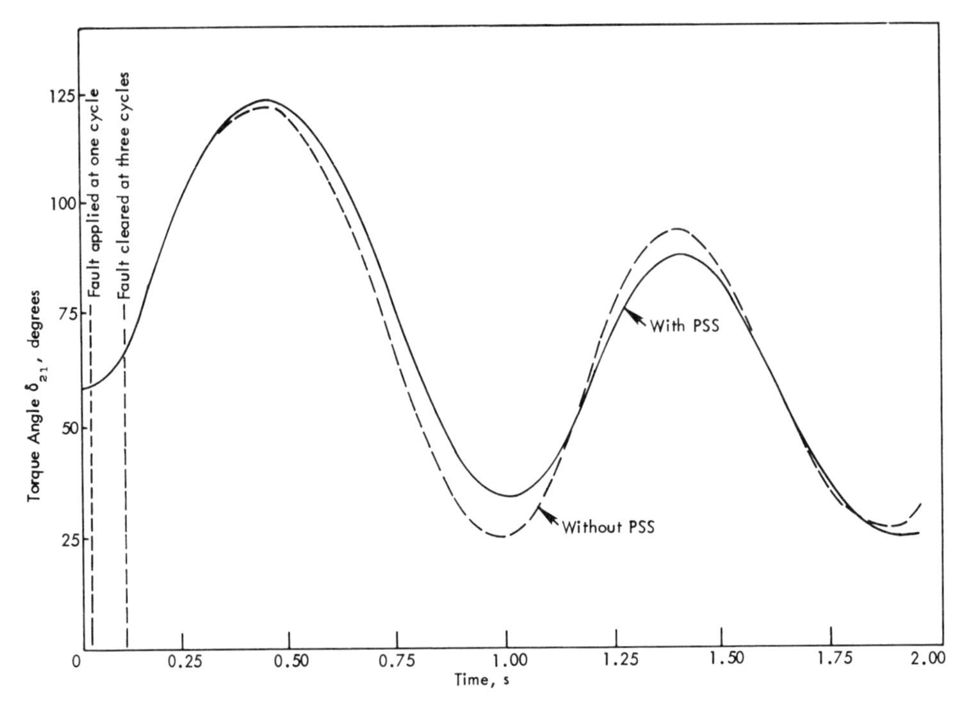

Fig. 8.51 Torque angle δ_{21} for a three-phase fault near generator 2, PSS with $a = 25$, $\omega_{osc} = 8.9$ rad/s.

inertial response to the accelerating torque in the rotor, the stabilizer has little effect on this first swing. On subsequent oscillations, however, the effect of the stabilizer is quite pronounced.

To illustrate the effect of the PSS, some transient stability runs are made for a three-phase fault near bus 7 applied at $t = 0.0167$ s (1 cycle) and cleared by opening line 5–7

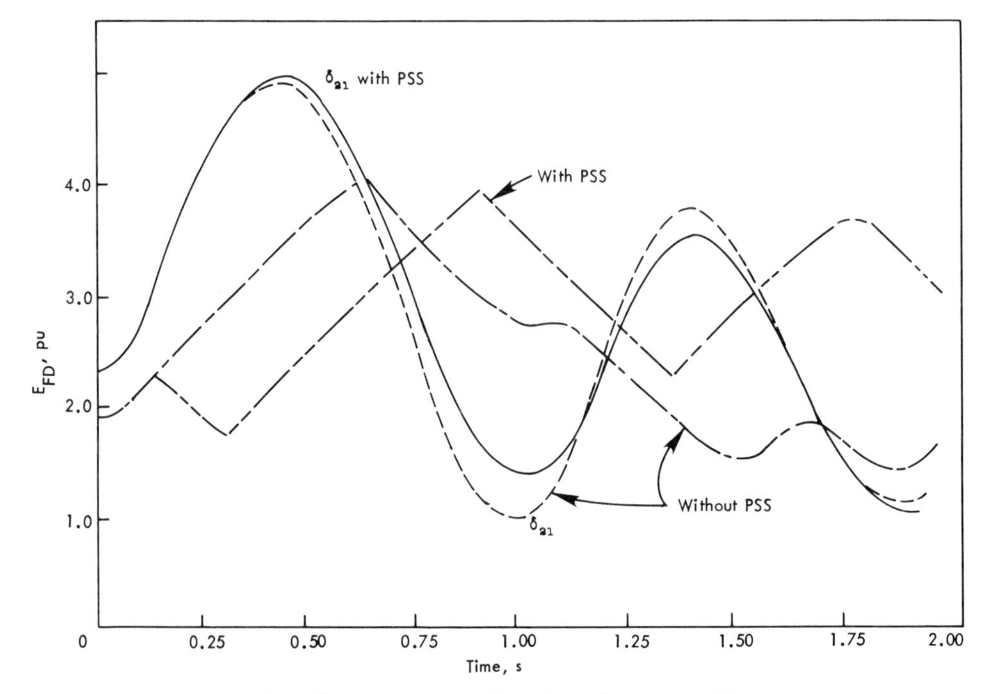

Fig. 8.52 Exciter voltage E_{FD} with and without a PSS.

at $t = 0.10$ s (6 cycles). Generator 2 is equipped with a Type 1 Mag-A-Stat exciter with constants similar to those given in Table 8.11. The PSS constants are the same as in Example 8.12 ($a = 25$) with a limiter included such that the PSS output is limited to ± 0.10 pu. Stability runs were made with and without the PSS. From the stability runs, data for the angle δ_{21} and the voltage E_{FD} are taken with and without the PSS. The results are displayed in Figures 8.51 and 8.52.

From the plot of δ_{21} in Figure 8.51 note that while the change in the first peak (due to the PSS) is very small, the improvement in the peak of the second swing is significant. The comparison in E_{FD}, shown in Figure 8.52, is interesting. Note that this exciter is not particularly fast (RR = 0.5), and the response tends to be a ramp up and then down. The phase of E_{FD} changes when the PSS is applied to produce a field voltage that is almost 180° out of phase with δ_{21}. This results in a delayed E_{FD} ramp as δ swings downward, which tends to limit the downward δ excursion by retarding the building in T_e.

The improvement in the angle δ_{21}, defined as $\delta_{21\Delta} = \delta_{21\,(\text{no PSS})} - \delta_{21\,(\text{PSS})}$, has been investigated for different PSS parameters. It is found that this angle improvement is sensitive to both the amount of lead compensation and to the cutoff level of the PSS limiter. A comparison of several runs is shown in Table 8.12.

Table 8.12. δ_{21} Improvement at Peak of Second Swing

a	Limit = ± 0.10	Limit = ± 0.05
25	5.8°	4.6°
16	5.4°	3.7°

8.11 Some General Comments on the Effect of Excitation on Stability

In the 1940s it was recognized that excitation control can increase the stability limits of synchronous generators. Another way to look at the same problem is to note that fast excitation systems allow operation with higher system reactances. This is felt to be important in view of the trends toward higher capacity generating units with higher reactances. For exciters to perform this function, they need high gain. Series compensation makes it possible to have a high dc gain and at the same time have lower "transient gain" for stable performance.

Modern exciters are faster and more powerful and hence allow for operation with higher series system reactance. Concordia [17], however, warns that "we cannot expect to continue indefinitely to compensate for increases in reactance by more and more powerful excitation systems." A limit may soon be reached when further increases in system reactance should be compensated for by means other than excitation control.

The above summarizes the situation regarding the so-called steady-state stability or power limits. Regarding the dynamic performance, modern excitation systems play an important part in the overall response of large systems to various impacts, both in the so-called transient stability problems and the dynamic stability problems.

The discussion in Section 8.3 and the studies of Section 8.10 seem to indicate that for less severe transients, the effect of modern fast excitation systems on first swing transients is marginal. However, for more severe transients or for transients initiated by faults of longer duration, these modern exciters can have a more pronounced effect. In the first place, for faults near the generator terminals it is important that the synchronous machine be modeled accurately. Also, if the transient study extends beyond the first swing, an accurate representation of the field flux in the machine is needed. If the excitation system is slow and has a low response ratio, optimistic results

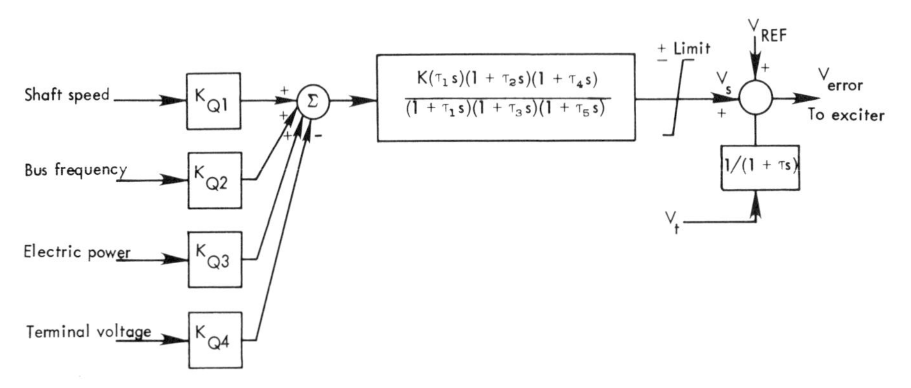

Fig. 8.53 Block diagram of the PSS for the BBC exciter with a 2.0 RR: $K_{Q1} = K_{Q3} = K_{Q4} = 0$, $K_{Q2} = 1$, $\tau_1 = 10$, $\tau_2 = 0.5$, $\tau_3 = 0.05$, $\tau_4 = 0.5$, $\tau_5 = 0.05$, limit $= \pm 0.05$ pu.

may be obtained if the classical machine representation is used. Transient studies are frequently run for a few swings to check on situations where circuit breakers may fail to operate properly and where backup protection is used. It should be mentioned that several transients have been encountered in the systems of North America where subsequent swings were of greater magnitudes than the first, causing eventual loss of synchronism. This is not too surprising in large interconnected systems with numerous modes of oscillations. It is not unlikely that some of the modes may be superimposed at some time after the start of the transient in such a way as to cause increased angle deviation. As shown in Section 8.10, the effect of excitation system compensation on subsequent swings (in large transients) is very pronounced. This has been repeatedly demonstrated in computer simulation studies and by field tests reported upon in the literature [8, 9, 13, 23, 29, 30, 31]. For example, in a stability study conducted by engineers of the Nebraska Public Power District, the effect of the PSS on damping the subsequent swings was found to be quite pronounced, while the effect on the magnitude of the first swing was hardly noticeable. The excitation system used is the Brown Boveri exciter shown in Figure 8.41. The PSS used is shown schematically in Figure 8.53, and the swing curves obtained with and without the PSS (for the same fault) are shown in Figure 8.54.

Voltage regulators can and do improve the synchronizing torques. Their effect on damping torques are small; but in the cases where the system exhibits negative damping characteristics, the voltage regulator usually aggravates the situation by increasing the negative damping. Supplementary signals to introduce artificial damping torques and to reduce intermachine and intersystem oscillations have been used with great success. These signals must be introduced with the proper phase relations to compensate for the excessive phase lag (and hence improve the system damping) at the desired frequencies [32].

Large interconnected power systems experience negative damping at very low frequencies of oscillations. The parameters of the PSS for a particular generator must be adjusted after careful study of the power system dynamic performance and the generator-exciter dynamic response characteristics. As indicated in Section 8.6, to obtain these characteristics, field measurements are preferred. If such measurements are not possible, approximate methods of analysis can be used to obtain preliminary design data, with provision for the adjustment of the PSS parameters to be made on the site after installation. Usually the PSS parameters are optimized over a range of frequencies between the natural mode of oscillation of the machine and the dominant frequency of oscillation of the interconnected power system.

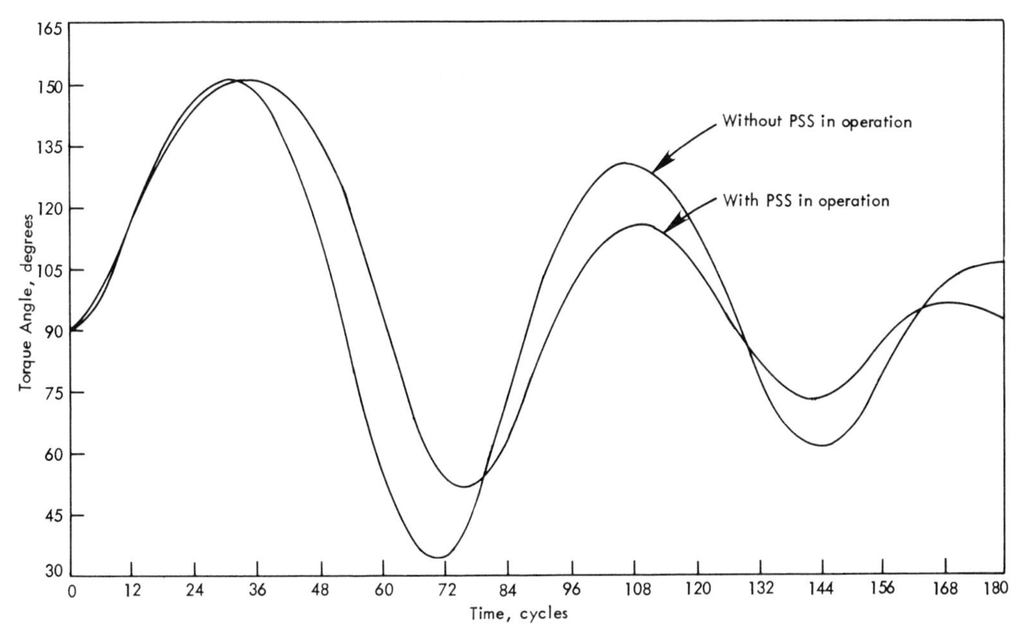

Fig. 8.54 Effect of the PSS on transient stability. (Obtained by private communication and used with permission.)

Recently many studies have been made on the use of various types of compensating networks to meet different situations and stimuli. Most of these studies concentrate on the use of a signal derived from speed or frequency deviation processed through a PSS network to give the proper phase relation to obtain the desired damping characteristic. This approach seems to concentrate on alleviating the problem of growing oscillations on tie lines [11, 13, 14, 24, 26, 30, 33–39]. However, in a large interconnected system it is possible to have a variety of potential problems that can be helped by excitation control. Whether the stabilizing signal derived from speed provides the best answer is an open question. It would seem likely that the principle of "optimal control" theory is applicable to this problem. Here signals derived from the various "states" of the system are fed back with different gains to optimize the system dynamic performance. This optimization is accomplished by assigning a performance index. This index is minimized by a control law described by a set of equations. These equations are solved for the gain constants. This subject is under active investigation by many researchers [40–44].

Problems

8.1 Construct a block diagram for the regulated generator given by (8.10)–(8.14). What is the order of the system?

8.2 Use block diagram algebra to reduce the system of Problem 8.1 to a feed-forward transfer function $KG(s)$ and a feedback transfer function $H(s)$, arranged as in Figure 7.19.

8.3 Determine the open loop transfer function for the system of Problem 8.2, using the numerical data given in Example 8.3. Find the upper and lower limits of the gain K_ϵ for (a) Case 1 and (b) Case 2.

8.4 Repeat the determination of stable operating constraints developed in Section 8.4.1, with the following assumptions (see [11]):

$$K_3 K_6 K_\epsilon \gg 1 \qquad \tau_\epsilon \ll K_3 \tau'_{d0}$$

Recompute the gain limitations, using the numerical constants K_1 through K_6 given in Table 8.3.

8.5 The block diagram shown in Figure 8.14 represents the machine terminal voltage at no load. The s domain equation for V_t/V_{REF} is given by (8.24). It is stated in Section 8.4.2 that a higher value of regulator gain K_ϵ can be used if a suitable lead-lag network is chosen. If the transfer function of such a network is $(1 + \tau_1 s)/(1 + \tau_2 s)$, choose τ_1 and τ_2 such that the value of the gain can be increased eight times.

8.6 In (8.30) and (8.31) assume that $K_6 K_\epsilon \gg 1/K_3$, and $\tau'_{d0} \gg \tau_\epsilon/K_3$. For each of the cases in Example 8.3, plot T_s and T_d as functions of ω between $\omega = 0.1$ rad/s and $\omega = 10$ rad/s (use semilog graph paper).

8.7 Compute the constants K_1 through K_6 for generator 3 of Example 2.6.

8.8 Determine the excitation control system phase lag of Example 8.7 if a low time constant exciter is used where $K_\epsilon = 400$ and $\tau_\epsilon = 0.05$ s.

8.9 Compute the open loop transfer function of the system of Figure 8.28 both with and without the stabilizer. Sketch root loci of each case.

8.10 Analyze the system in Figure 8.29 for a stabilizing signal processed through a bridged T-filter:

$$G_S = (s^2 + rn\omega_n s + \omega_n^2)/(s^2 + n\omega_n s + \omega_n^2),$$

where ω_n is the natural frequency of the machine, $n = 2$ and $r = 0.1$.

8.11 Sketch Bode diagrams of the several lead compensators described in Example 8.10.

8.12 Use a linear systems analysis program (if one is available) to compute root locus, time response to a step change in V_{REF}, and a Bode plot for Example 8.11 with
(a) A dual lead compensator with $a = 15$.
(b) A triple lead compensator with $a = 10$.

8.13 Perform a transient stability run, using a computer library program to verify the results of Section 8.10. Plot E'_q and V_t as functions of time and comment on these results.

8.14 Modify the block diagram of Figure 5.18 showing the analog computer simulation of the synchronous machine to allow modulating the infinite bus voltage.

8.15 With the help of the field voltage equation ($v_F = r_F i_F + \dot{\lambda}_F$), discuss the plots of E_{FD}, E, and E'_q shown in Figures 8.43 and 8.44.

8.16 Explain why the curve for constant E_{FD} in Figure 8.42 shows a larger swing than the other field representation.

References

1. Concordia, C. Steady-state stability of synchronous machines as affected by voltage regulator characteristics. *AIEE Trans.* PAS-63:215–20, 1944.
2. Crary S. B. Long distance power transmission. *AIEE Trans.* 69, (Pt. 2):834–44, 1950.
3. Ellis, H. M., Hardy, J. E., Blythe, A. L., and Skooglund, J. W. Dynamic stability of the Peace River transmission system. *IEEE Trans.* PAS-85:586–600, 1966.
4. Schleif, F. R., and White, J. H. Damping for the northwest-southwest tieline oscillations—An analog study. *IEEE Trans.* PAS-85:1239–47, 1966.
5. Byerly, R. T., Skooglund, J. W., and Keay, F. W. Control of generator excitation for improved power system stability. *Proc. Am. Power Conf.* 29:1011–1022, 1967.
6. Schleif, F. R., Martin, G. E., and Angell, R. R. Damping of system oscillations with a hydrogenating unit. *IEEE Trans.* PAS-86:438–42, 1967.
7. Hanson, O. W., Goodwin, C. J., and Dandeno, P. L. Influence of excitation and speed control parameters in stabilizing intersystem oscillations. *IEEE Trans.* PAS-87:1306–13, 1968.
8. Dandeno, P. L., Karas, A. N., McClymont, K. R., and Watson, W. Effect of high-speed rectifier excitation systems on generator stability limits. *IEEE Trans.* PAS-87:190–201, 1968.
9. Shier, R. M., and Blythe, A. L. Field tests of dynamic stability using a stabilizing signal and computer program verification. *IEEE Trans.* PAS-87:315–22, 1968.
10. Schleif, F. R., Hunkins, H. D., Martin, G. E., and Hattan, E. E. Excitation control to improve power line stability. *IEEE Trans.* PAS-87:1426–34, 1968.
11. de Mello, F. P., and Concordia, C. Concepts of synchronous machine stability as affected by excitation control. *IEEE Trans.* PAS-88:316–29, 1969.
12. Schleif, F. R., Hunkins, H. D., Hattan, E. E., and Gish, W. B. Control of rotating exciters for power system damping: Pilot applications and experience. *IEEE Trans.* PAS-88:1259–66, 1969.

13. Klopfenstein, A. Experience with system stabilizing controls on the generation of the Southern California Edison Co. *IEEE Trans.* PAS-90:698–706, 1971.
14. de Mello, F. P. The effects of control. Modern concepts of power system dynamics. IEEE tutorial course. IEEE Power Group Course Text 70 M 62-PWR, 1970.
15. Young, C. C. The art and science of dynamic stability analysis. IEEE paper 68 CP702-PWR, presented at the ASME-IEEE Joint Power Generation Conference, San Francisco, Calif., 1968.
16. Ramey, D. G., Byerly, R. T., and Sherman, D. E. The application of transfer admittances to the analysis of power systems stability studies. *IEEE Trans.* PAS-90:993–1,000, 1971.
17. Concordia, C., and Brown, P. G. Effects of trends in large steam turbine generator parameters on power system stability. *IEEE Trans.* PAS-90:2211–18, 1971.
18. Perry, H. R., Luini, J. F., and Coulter, J. C. Improved stability with low time constant rotating exciter. *IEEE Trans.* PAS-90:2084–89, 1971.
19. Brown, P. G., de Mello, F. P., Lenfest, E. H., and Mills, R. J. Effects of excitation, turbine energy control and transmission on transient stability. *IEEE Trans.* PAS-89:1247–53, 1970.
20. Melsa, J. L. *Computer Programs for Computational Assistance in the Study of Linear Control Theory.* McGraw-Hill, New York, 1970.
21. Duven, D. J. Data instructions for program LSAP. Unpublished notes, Electrical Engineering Dept., Iowa State University, Ames, 1973.
22. Kuo, Benjamin C. *Automatic Control Systems,* Prentice-Hall, Englewood Cliffs, N.J., 1962.
23. Gerhart, A. D., Hillesland, T., Jr., Luini, J. F., and Rockfield, M. L., Jr. Power system stabilizer: Field testing and digital simulation. *IEEE Trans.* PAS-90:2095–2101, 1971.
24. Warchol, E. J., Schleif, F. R., Gish, W. B. and Church, J. R. Alignment and modeling of Hanford excitation control for system damping. *IEEE Trans.* PAS-90:714–25, 1971.
25. Eilts, L. E. Power system stabilizers: Theoretical basis and practical experience. Paper presented at the panel discussion "Dynamic stability in the western interconnected power systems" for the IEEE Summer Power Meeting, Anaheim, Calif., 1974.
26. Keay, F. W., and South, W. H. Design of a power system stabilizer sensing frequency deviation. *IEEE Trans.* PAS-90:707–14, 1971.
27. Bolinger, K., Laha, A., Hamilton, R., and Harras, T. Power stabilizer design using root-locus methods. *IEEE Trans.* PAS-94:1484–88, 1975.
28. Schroder, D. C., and Anderson, P. M. Compensation of synchronous machines for stability. IEEE paper C 73-313-4, presented at the Summer Power Meeting, Vancouver, B.C., Canada. 1973.
29. Bobo, P. O., Skooglund, J. W., and Wagner, C. L. Performance of excitation systems under abnormal conditions. *IEEE Trans.* PAS-87:547–53, 1968.
30. Byerly, R. T. Damping of power oscillations in salient-pole machines with static exciters. *IEEE Trans.* PAS-89:1009–21, 1970.
31. McClymont, K. R., Manchur, G., Ross, R. J., and Wilson, R. J. Experience with high-speed rectifier excitation systems. *IEEE Trans.* PAS-87:1464–70, 1968.
32. Jones, G. A. Phasor interpretation of generator supplementary excitation control. Paper A75-437-4, presented at the IEEE Summer Power Meeting, San Francisco, Calif., 1975.
33. El-Sherbiny, M. K., and Fouad, A. A. Digital analysis of excitation control for interconnected power systems. *IEEE Trans.* PAS-90:441–48, 1971.
34. Watson, W., and Manchur, G. Experience with supplementary damping signals for generator static excitation systems. *IEEE Trans.* PAS-92:199–203, 1973.
35. Hayes, D. R., and Craythorn, G. E. Modeling and testing of Valley Steam Plant supplemental excitation control system. *IEEE Trans.* PAS-92:464–70, 1973.
36. Marshall, W. K., and Smolinski, W. J. Dynamic stability determination by synchronizing and damping torque analysis. Paper T 73-007-2, presented at the IEEE Winter Power Meeting, New York, 1973.
37. El-Sherbiny, M. K., and Huah, Jenn-Shi. A general analysis of developing a universal stabilizing signal for different excitation controls, which is applicable to all possible loadings for both lagging and leading operation. Paper C74-106-1, presented at the IEEE Winter Power Meeting, New York, 1974.
38. Bayne, J. P., Kundur, P., and Watson, W. Static exciter control to improve transient stability. Paper T74-521-1, presented at the IEEE-ASME Power Generation Technical Conference, Miami Beach, Fla., 1974.
39. Arcidiacono, V., Ferrari, E., Marconato, R., Brkic, T., Niksic, M., and Kajari, M. Studies and experimental results about electromechanical oscillation damping in Yugoslav power system. Paper F75-460-6 presented at the IEEE Summer Meeting, San Francisco, Calif., 1975.
40. Fosha, C. E., and Elgerd, O. I. The megawatt-frequency control problem: A new approach via optimal control theory. *IEEE Trans.* PAS-89:563–77, 1970.
41. Anderson, T. H. The control of a synchronous machine using optimal control theory. *IEEE Trans.* PAS-90:25–35, 1971.
42. Moussa, H. A. M., and Yu, Yao-nan. Optimal power system stabilization through excitation and/or governor control. *IEEE Trans.* PAS-91:1166–74, 1972.
43. Humpage, W. D., Smith, J. R., and Rogers, G. T. Application of dynamic optimization to synchronous generator excitation controllers. *Proc. IEE* (British) 120:87–93, 1973.
44. Elmetwally, M. M., Rao, N. D. and Malik, O. P. Experimental results on the implementation of an optimal control for synchronous machines. *IEEE Trans.* PAS-94:1192–1200, 1974.

chapter 9

Multimachine Systems with Constant Impedance Loads

9.1 Introduction

In this chapter we develop the equations for the load constraints in a multimachine system in the special case where the loads are to be represented by constant impedances. The objective is to give a mathematical description of the multimachine system with the load constraints included.

Representing loads by constant impedance is not usually considered accurate. It has been shown in Section 2.11 that this type of load representation could lead to some error. A more accurate representation of the loads will be discussed in Volume 2 of this work. Our main concern here is to apply the load constraints to the equations of the machines. We choose the constant impedance load case because of its relative simplicity and because with this choice all the nodes other than the generator nodes can be eliminated by network reduction (See Section 2.10.2).

9.2 Statement of the Problem

In previous chapters, mathematical models describing the dynamic behavior of the synchronous machine are discussed in some detail. In Chapter 4 [see (4.103) and (4.138)] it is shown that *each* machine is described mathematically by a set of equations of the form

$$\dot{\mathbf{x}} = \mathbf{f}(\mathbf{x}, \mathbf{v}, T_m, t) \tag{9.1}$$

where $\mathbf{x}$ is a vector of state variables, $\mathbf{v}$ is a vector of voltages, and T_m is the mechanical torque. The dimension of the vector $\mathbf{x}$ depends on the model used. The order of $\mathbf{x}$ ranges from seventh order for the full model (with three rotor circuits) to second order for the classical model where only ω and δ are retained as the state variables.

The vector $\mathbf{v}$ is a vector of voltages that includes v_d, v_q, and v_F. If the excitation system is not represented in detail, v_F is assumed known; but if the excitation system is modeled mathematically, additional state variables, including v_F, are added to the vector $\mathbf{x}$ (see Chapter 7) with a reference quantity such as V_{REF} known. In this chapter we will assume without loss of generality that v_F is known.

Consider the set of equations (9.1). In the current model developed in Chapter 4, it represents a set of seven first-order differential equations *for each machine*. The number of the variables, however, is nine: five currents, ω and δ, and the voltages v_d and v_q. Assuming that there are n synchronous machines in the system, we have a set of $7n$ differential equations with $9n$ unknowns. Therefore, $2n$ additional equations are needed to complete the description of the system. These equations are obtained from the load constraints.

The objective here is to derive relations between v_{di} and v_{qi}, $i = 1, 2, \ldots, n$, and the state variables. This will be obtained in the form of a relation between these voltages, the machine currents i_{qi} and i_{di}, and the angles δ_i, $i = 1, 2, \ldots, n$. In the case of the flux linkage model the currents are linear combinations of the flux linkages, as given in (4.124). For convenience we will use a complex notation defined as follows.

For machine i we define the phasors $\overline{V}_i$ and $\overline{I}_i$ as

$$\overline{V}_i = V_{qi} + \mathrm{j}V_{di} \qquad \overline{I}_i = I_{qi} + \mathrm{j}I_{di} \tag{9.2}$$

where

$$\begin{aligned} V_{qi} &\triangleq v_{qi}/\sqrt{3} \qquad & V_{di} &\triangleq v_{di}/\sqrt{3} \\ I_{qi} &\triangleq i_{qi}/\sqrt{3} \qquad & I_{di} &\triangleq i_{di}/\sqrt{3} \end{aligned} \tag{9.3}$$

and where the axis q_i is taken as the phasor reference in each case. Then we define the complex vectors $\overline{\mathbf{V}}$ and $\overline{\mathbf{I}}$ by

$$\overline{\mathbf{V}} \triangleq \begin{bmatrix} V_{q1} + \mathrm{j}V_{d1} \\ V_{q2} + \mathrm{j}V_{d2} \\ \cdots \\ V_{qn} + \mathrm{j}V_{dn} \end{bmatrix} = \begin{bmatrix} \overline{V}_1 \\ \overline{V}_2 \\ \cdots \\ \overline{V}_n \end{bmatrix}$$

$$\overline{\mathbf{I}} \triangleq \begin{bmatrix} I_{q1} + \mathrm{j}I_{d1} \\ I_{q2} + \mathrm{j}I_{d2} \\ \cdots \\ I_{qn} + \mathrm{j}I_{dn} \end{bmatrix} = \begin{bmatrix} \overline{I}_1 \\ \overline{I}_2 \\ \cdots \\ \overline{I}_n \end{bmatrix} \tag{9.4}$$

Note carefully that the voltage $\overline{V}_i$ and the current $\overline{I}_i$ are referred to the q and d axes of machine i. In other words the different voltages and currents are expressed in terms of different reference frames. The desired relation is that which relates the vectors $\overline{\mathbf{V}}$ and $\overline{\mathbf{I}}$. When obtained, it will represent a set of n complex algebraic equations, or $2n$ real equations. These are the additional equations needed to complete the mathematical description of the system.

9.3 Matrix Representation of a Passive Network

Consider the multimachine system shown in Figure 9.1. The network has n machines and r loads. It is similar to the system shown in Figure 2.17 except that the machines are not represented by the classical model. Thus, the terminal voltages $\overline{V}_i$, $i = 1, 2, \ldots, n$, are shown in Figure 9.1 instead of the internal EMF's in Figure 2.17. Since the loads are represented by constant impedances, the network has only n active sources. Note also that the impedance equivalents of the loads are obtained from the pretransient conditions in the system.

By network reduction the network shown in Figure 9.1 can be reduced to the n-node network shown in Figure 9.2 (see Section 2.10.2). For this network the node currents and voltages expressed in phasor notation are $\overline{I}_1, \overline{I}_2, \ldots, \overline{I}_n$ and $\overline{V}_1, \overline{V}_2, \ldots, \overline{V}_n$ respectively. Again we emphasize that these phasors are expressed in terms of reference frames that are different for each node.

At *steady state* these currents and voltages can be represented by phasors to a com-

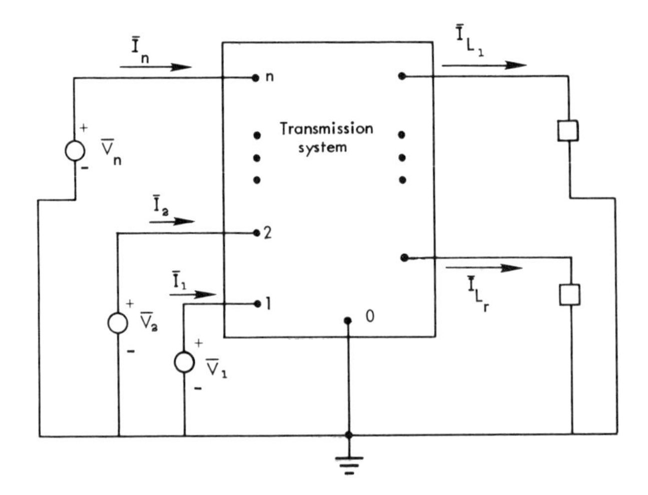

Fig. 9.1. Multimachine system with constant impedance loads.

mon reference frame. To distinguish these phasors from those defined by (9.2), we will use the symbols $\hat{I}_i$ and $\hat{V}_i$, $i = 1, 2, \ldots, n$, to designate the use of a common (network) frame of reference. Similarly, we can form the matrices $\hat{\mathbf{I}}$ and $\hat{\mathbf{V}}$. From the network steady-state equations we write

$$\hat{\mathbf{I}} = \overline{\mathbf{Y}}\hat{\mathbf{V}} \tag{9.5}$$

where

$$\hat{\mathbf{I}} \triangleq \begin{bmatrix} \hat{I}_1 \\ \hat{I}_2 \\ \cdots \\ \hat{I}_n \end{bmatrix} \qquad \hat{\mathbf{V}} \triangleq \begin{bmatrix} \hat{V}_1 \\ \hat{V}_2 \\ \cdots \\ \hat{V}_n \end{bmatrix} \tag{9.6}$$

and $\overline{\mathbf{Y}}$ is the short circuit admittance matrix of the network in Figure 9.2.

9.3.1 Network in the transient state

Consider a branch in the reduced network of Figure 9.2. Let this branch, located between any two nodes in the network, be identified by the subscript k. Let the branch

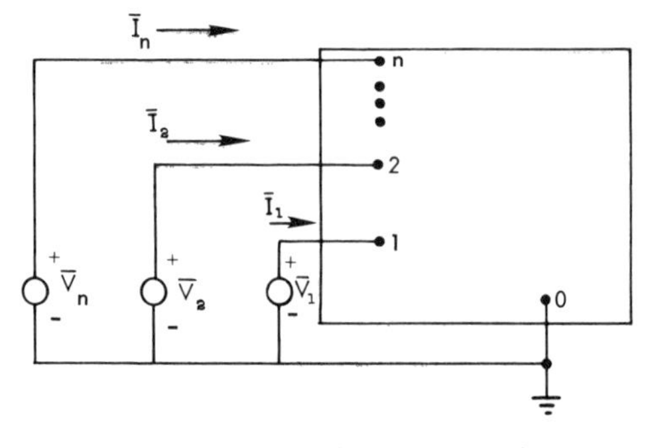

Fig. 9.2. Reduced n-port network.

resistance be r_k, its inductance be ℓ_k, and its impedance be $\bar{z}_k$. The branch voltage drop and current are v_k and i_k.

In the transient state the relation between these quantities is given by

$$v_k = \ell_k \dot{i}_k + r_k i_k \qquad k = 1, 2, \ldots, b \tag{9.7}$$

where b is the number of branches.

Using subscripts abc to denote the phases abc, (9.7) can be written as

$$\mathbf{v}_{abck} = \ell_k \dot{\mathbf{i}}_{abck} + r_k \mathbf{i}_{abck} \qquad k = 1, 2, \ldots, b \tag{9.8}$$

This branch equation could be written with respect to any of the n q-axis references by using the appropriate transformation $\mathbf{P}$. Premultiplying (9.8) by the transformation $\mathbf{P}$ as defined by (4.5),

$$\mathbf{P}\,\mathbf{v}_{abck} = \ell_k \mathbf{P}\,\dot{\mathbf{i}}_{abck} + r_k \mathbf{P}\,\mathbf{i}_{abck} \tag{9.9}$$

Then from (4.31) and (4.32)

$$\mathbf{P}\,\dot{\mathbf{i}}_{abc} = \dot{\mathbf{i}}_{0dq} - \omega \begin{bmatrix} 0 \\ -i_q \\ i_d \end{bmatrix} \tag{9.10}$$

Substituting (9.10) in (9.9) and using (4.7),

$$\mathbf{v}_{0dqk} = \ell_k \left(\dot{\mathbf{i}}_{0dqk} - \omega \begin{bmatrix} 0 \\ -i_{qk} \\ i_{dk} \end{bmatrix} \right) + r_k \mathbf{i}_{0dqk} \tag{9.11}$$

which in the case of balanced conditions becomes

$$\mathbf{v}_{dqk} = \ell_k \left(\dot{\mathbf{i}}_{dqk} + \omega \begin{bmatrix} i_{qk} \\ -i_{dk} \end{bmatrix} \right) + r_k \mathbf{i}_{dqk} \tag{9.12}$$

It is customary to make the following assumptions: (1) the system angular speed does not depart appreciably from the rated speed, or $\omega \simeq \omega_R$ and (2) the terms $\ell \dot{i}$ are negligible compared to the terms $\omega \ell i$. The first assumption makes the term $\omega \ell_k i_k$ approximately equal to $x_k i_k$, and the second assumption suggests that the terms in $\dot{i}_k$ are to be neglected.

Under the above assumptions (9.12) becomes

$$\mathbf{v}_{dqk} = r_k \mathbf{i}_{dqk} + x_k \begin{bmatrix} i_{qk} \\ -i_{dk} \end{bmatrix} \qquad k = 1, 2, \ldots, b \tag{9.13}$$

Equation (9.13) gives a relation between the voltage drop and the current in one branch of the network in the transient state. These quantities are expressed in the q-d frame of reference of any machine. Let the machine associated with this transformation be i. The rotor angle θ_i of this machine is given by

$$\theta_i = \omega_R t + \pi/2 + \delta_i \tag{9.14}$$

where δ_i is the angle between this rotor and a *synchronously rotating* reference frame.

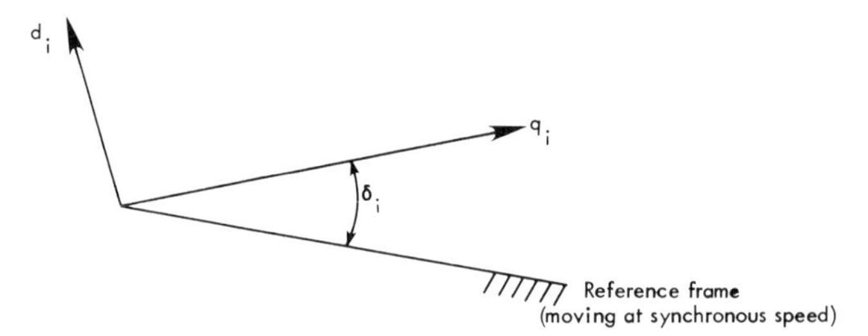

Fig. 9.3. Position of axes of rotor k with respect to reference frame.

From (9.13) multiply both sides by $1/\sqrt{3}$; and using (9.3),

$$V_{qk(i)} = r_k I_{qk(i)} - x_k I_{dk(i)} \qquad V_{dk(i)} = r_k I_{dk(i)} + x_k I_{qk(i)} \tag{9.15}$$

where the subscript i is added to indicate that the rotor of machine i is used as reference.

Expressing (9.15) in phasor notation,

$$\begin{aligned} \overline{V}_{k(i)} &= V_{qk(i)} + \mathrm{j} V_{dk(i)} \\ &= (r_k I_{qk(i)} - x_k I_{dk(i)}) + \mathrm{j}(r_k I_{dk(i)} + x_k I_{qk(i)}) = (r_k + \mathrm{j}x_k)(I_{qk} + \mathrm{j}I_{dk}) \end{aligned}$$

or

$$\overline{V}_{k(i)} = \overline{z}_k \overline{I}_{k(i)} \qquad k = 1, 2, \ldots, b, \tag{9.16}$$

Equation (9.16) expresses, in complex phasor notation, the relation between the voltage drop in branch k and the current in that branch. The reference is the q axis of some (hypothetical) rotor i located at angle δ_i with respect to a synchronously rotating system reference, as shown in Figure 9.3.

9.3.2 Converting to a common reference frame

To obtain general network relationships, it is desirable to express the various branch quantities to the same reference. Let us assume that we want to convert the phasor $\overline{V}_i = V_{qi} + \mathrm{j}V_{di}$ to the common reference frame (moving at synchronous speed). Let the same voltage, expressed in the new notation, be $\hat{V}_i = V_{Qi} + \mathrm{j}V_{Di}$ as shown in Figure 9.4.

From Figure 9.4 by inspection we can show that

$$V_{Qi} + \mathrm{j}V_{Di} = (V_{qi}\cos\delta_i - V_{di}\sin\delta_i) + \mathrm{j}(V_{qi}\sin\delta_i + V_{di}\cos\delta_i)$$

or

$$\hat{V}_i = \overline{V}_i e^{\mathrm{j}\delta_i} \tag{9.17}$$

Now convert the network branch voltage drop equation (9.16) to the system reference frame by using (9.17).

$$\hat{V}_k e^{-\mathrm{j}\delta_i} = \overline{z}_k \hat{I}_k e^{-\mathrm{j}\delta_i}$$

or

$$\hat{V}_k = \overline{z}_k \hat{I}_k \qquad k = 1, 2, \ldots, b \tag{9.18}$$

where b is the number of branches and $\overline{z}_k$ is calculated based on rated angular speed.

Comparing (9.18) and (9.5) under the assumptions stated above, the network in the transient state can be described by equations similar to those describing its steady-state

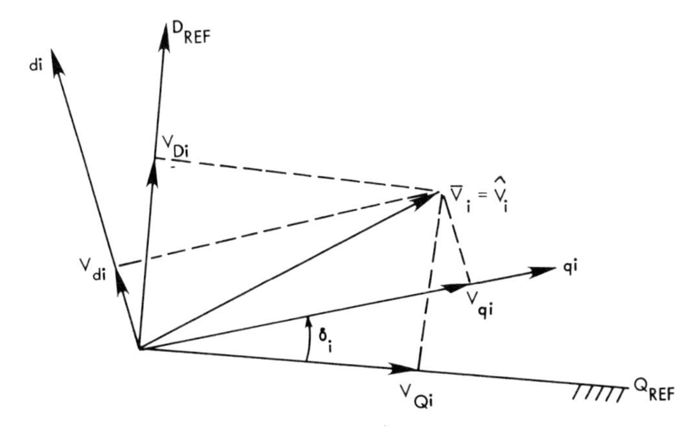

Fig. 9.4. Two frames of reference for phasor quantities for a voltage V_i.

behavior. The network (branch) equations are in terms of quantities expressed to the same frame of reference, conveniently chosen to be moving at synchronous speed (it is also the system reference frame).

Equation (9.18) can be expressed in matrix form

$$\hat{\mathbf{V}}_b = \bar{\mathbf{z}}_b \hat{\mathbf{I}}_b \tag{9.19}$$

where the subscript b is used to indicate a branch matrix. The inverse of the primitive branch matrix $\bar{\mathbf{z}}_b$ exists and is denoted $\bar{\mathbf{y}}_b$, thus

$$\hat{\mathbf{I}}_b = \bar{\mathbf{y}}_b \hat{\mathbf{V}}_b \tag{9.20}$$

Equation (9.20) is expressed in terms of the *primitive admittance matrix* of a passive network. From network theory we learn to construct the node incidence matrix $\mathbf{A}$ which is used to convert (9.20) into a nodal admittance equation

$$\hat{\mathbf{I}} = (\mathbf{A}^t \bar{\mathbf{y}}_b \mathbf{A})\hat{\mathbf{V}} \triangleq \bar{\mathbf{Y}}\hat{\mathbf{V}} \tag{9.21}$$

where $\bar{\mathbf{Y}}$ is the matrix of short circuit driving point and transfer admittances and

$$\begin{aligned} \mathbf{A} = [a_{pq}] &= 1 \text{ if current in branch } p \text{ leaves node } q \\ &= -1 \text{ if current in branch } p \text{ enters node } q \\ &= 0 \text{ if branch } p \text{ is not connected to node } q \end{aligned} \tag{9.22}$$

with $p = 1, 2, \ldots, b$ and $q = 1, 2, \ldots, n$.

Since $\bar{\mathbf{Y}}^{-1} \triangleq \bar{\mathbf{Z}}$ exists,

$$\hat{\mathbf{V}} = \bar{\mathbf{Y}}^{-1}\hat{\mathbf{I}} \triangleq \bar{\mathbf{Z}}\hat{\mathbf{I}} \tag{9.23}$$

where $\bar{\mathbf{Z}}$ is the matrix of the open circuit driving point and transfer impedances of the network. (For the derivation of (9.21)–(9.23), including a discussion of the properties of the $\bar{\mathbf{Y}}$ and $\bar{\mathbf{Z}}$ matrices, see reference [1], Chapter 11.)

9.4 Converting Machine Coordinates to System Reference

Consider a voltage $\mathbf{v}_{abci}$ at node i. We can apply Park's transformation to this voltage to obtain $\mathbf{v}_{dqi}$. From (9.2) this voltage can be expressed in phasor notation as $\bar{V}_i$, using the rotor of machine i as reference. It can also be expressed to the system reference as $\hat{V}_i$, using the transformation (9.17).

Equation (9.17) can be generalized to include all the nodes. Let

$$\mathbf{T} = \begin{bmatrix} e^{j\delta_1} & 0 & \dots & 0 \\ 0 & e^{j\delta_2} & \dots & 0 \\ \dots & \dots & \dots & \dots \\ 0 & 0 & \dots & e^{j\delta_n} \end{bmatrix} \tag{9.24}$$

$$\hat{\mathbf{V}} = \begin{bmatrix} V_{Q1} + jV_{D1} \\ V_{Q2} + jV_{D2} \\ \dots \\ V_{Qn} + jV_{Dn} \end{bmatrix} \qquad \bar{\mathbf{V}} = \begin{bmatrix} V_{q1} + jV_{d1} \\ V_{q2} + jV_{d2} \\ \dots \\ V_{qn} + jV_{dn} \end{bmatrix} \tag{9.25}$$

Then from (9.2), (9.14), (9.17), and (9.25)

$$\hat{\mathbf{V}} = \mathbf{T}\bar{\mathbf{V}} \tag{9.26}$$

Thus $\mathbf{T}$ is a transformation that transforms the d and q quantities of all machines to the system frame, which is a common frame moving at synchronous speed.

We can easily show that the transformation $\mathbf{T}$ is orthogonal, i.e.,

$$\mathbf{T}^{-1} = \mathbf{T}^* \tag{9.27}$$

Therefore, from (9.26) and (9.27)

$$\bar{\mathbf{V}} = \mathbf{T}^*\hat{\mathbf{V}} \tag{9.28}$$

Similarly for the node currents we get

$$\hat{\mathbf{I}} = \mathbf{T}\bar{\mathbf{I}} \qquad \bar{\mathbf{I}} = \mathbf{T}^*\hat{\mathbf{I}} \tag{9.29}$$

9.5 Relation between Machine Currents and Voltages

From (9.22) $\hat{\mathbf{I}} = \bar{\mathbf{Y}}\hat{\mathbf{V}}$. By using (9.29) in (9.22),

$$\mathbf{T}\bar{\mathbf{I}} = \bar{\mathbf{Y}}\mathbf{T}\bar{\mathbf{V}} \tag{9.30}$$

Premultiplying (9.30) oy $\mathbf{T}^{-1}$

$$\bar{\mathbf{I}} = (\mathbf{T}^{-1}\bar{\mathbf{Y}}\mathbf{T})\bar{\mathbf{V}} \triangleq \bar{\mathbf{M}}\bar{\mathbf{V}} \tag{9.31}$$

where

$$\bar{\mathbf{M}} \triangleq (\mathbf{T}^{-1}\bar{\mathbf{Y}}\mathbf{T}) \tag{9.32}$$

and if $\bar{\mathbf{M}}^{-1}$ exists,

$$\bar{\mathbf{V}} = (\mathbf{T}^{-1}\bar{\mathbf{Y}}\mathbf{T})^{-1}\bar{\mathbf{I}} = (\mathbf{T}^{-1}\bar{\mathbf{Z}}\mathbf{T})\bar{\mathbf{I}} \tag{9.33}$$

Equation (9.33) is the desired relation needed between the terminal voltages and currents of the machines. It is given here in an equivalent phasor notation for convenience and compactness. It is, however, a set of algebraic equations between $2n$ real voltages $V_{q1}, V_{d1}, \dots, V_{qn}, V_{dn}$, and $2n$ real currents $I_{q1}, I_{d1}, \dots, I_{qn}, I_{dn}$.

Example 9.1

Derive the expression for the matrix $\bar{\mathbf{M}}$ for an n-machine system.

Solution

The matrix $\overline{\mathbf{Y}}$ of the network is of the form

$$\overline{\mathbf{Y}} = \begin{bmatrix} Y_{11}e^{j\theta_{11}} & Y_{12}e^{j\theta_{12}} & \dots & Y_{1n}e^{j\theta_{1n}} \\ Y_{21}e^{j\theta_{21}} & Y_{22}e^{j\theta_{22}} & \dots & Y_{2n}e^{j\theta_{2n}} \\ \dots & \dots & \dots & \dots \\ Y_{n1}e^{j\theta_{n1}} & Y_{n2}e^{j\theta_{n2}} & \dots & Y_{nn}e^{j\theta_{nn}} \end{bmatrix} \tag{9.34}$$

and from (9.24)

$$\mathbf{T} = \begin{bmatrix} e^{j\delta_1} & & \\ & \ddots & \\ & & e^{j\delta_n} \end{bmatrix} \qquad \mathbf{T}^{-1} = \begin{bmatrix} e^{-j\delta_1} & & \\ & \ddots & \\ & & e^{-j\delta_n} \end{bmatrix} \tag{9.35}$$

From (9.34) and (9.35)

$$\overline{\mathbf{Y}}\mathbf{T} = \begin{bmatrix} Y_{11}e^{j(\theta_{11}+\delta_1)} & Y_{12}e^{j(\theta_{12}+\delta_2)} & \dots & Y_{1n}e^{j(\theta_{1n}+\delta_n)} \\ Y_{21}e^{j(\theta_{21}+\delta_1)} & Y_{22}e^{j(\theta_{22}+\delta_2)} & \dots & Y_{2n}e^{j(\theta_{2n}+\delta_n)} \\ \dots & \dots & \dots & \dots \\ Y_{n1}e^{j(\theta_{n1}+\delta_1)} & Y_{n2}e^{j(\theta_{n2}+\delta_2)} & \dots & Y_{nn}e^{j(\theta_{nn}+\delta_n)} \end{bmatrix}$$

and premultiplying by $\mathbf{T}^{-1}$, we get the desired result

$$\mathbf{T}^{-1}\overline{\mathbf{Y}}\mathbf{T} \triangleq \overline{\mathbf{M}} = \begin{bmatrix} Y_{11}e^{j\theta_{11}} & Y_{12}e^{j(\theta_{12}-\delta_{12})} & \dots & Y_{1n}e^{j(\theta_{1n}-\delta_{1n})} \\ Y_{21}e^{j(\theta_{21}-\delta_{21})} & Y_{22}e^{j\theta_{22}} & \dots & Y_{2n}e^{j(\theta_{2n}-\delta_{2n})} \\ \dots & \dots & \dots & \dots \\ Y_{n1}e^{j(\theta_{n1}-\delta_{n1})} & Y_{n2}e^{j(\theta_{n2}-\delta_{n2})} & \dots & Y_{nn}e^{j\theta_{nn}} \end{bmatrix} \tag{9.36}$$

To simplify (9.36), we note that

$$Y_{ik}e^{j(\theta_{ik}-\delta_{ik})} = (G_{ik}\cos\delta_{ik} + B_{ik}\sin\delta_{ik}) + j(B_{ik}\cos\delta_{ik} - G_{ik}\sin\delta_{ik})$$

Now define

$$\begin{aligned} F_{G+B}(\delta_{ik}) &= F_{G+B} = G_{ik}\cos\delta_{ik} + B_{ik}\sin\delta_{ik} \\ F_{B-G}(\delta_{ik}) &= F_{B-G} = B_{ik}\cos\delta_{ik} - G_{ik}\sin\delta_{ik} \end{aligned} \tag{9.37}$$

Then the matrix $\overline{\mathbf{M}}$ is given by

$$\overline{\mathbf{M}} = \mathbf{H} + j\mathbf{S} \tag{9.38}$$

where $\mathbf{H}$ and $\mathbf{S}$ are real matrices of dimensions $(n \times n)$. Their diagonal and off-diagonal terms are given by

$$h_{ii} = G_{ii} \qquad h_{ik} = F_{G+B}(\delta_{ik}) \qquad s_{ii} = B_{ii} \qquad s_{ik} = F_{B-G}(\delta_{ik}) \tag{9.39}$$

Example 9.2

Derive the relations between the d and q machine voltages and currents for a two-machine system.

Solution

From (9.31) and (9.38)

$$\bar{\mathbf{I}} = (\mathbf{H} + \mathrm{j}\mathbf{S})\begin{bmatrix} V_{q1} + \mathrm{j}V_{d1} \\ \vdots \\ V_{qn} + \mathrm{j}V_{dn} \end{bmatrix} = (\mathbf{H} + \mathrm{j}\mathbf{S})(\mathbf{V}_q + \mathrm{j}\mathbf{V}_d)$$

$$= (\mathbf{HV}_q - \mathbf{SV}_d) + \mathrm{j}(\mathbf{SV}_q + \mathbf{HV}_d) \tag{9.40}$$

For a two-machine system the q axis currents are given by

$$\begin{bmatrix} I_{q1} \\ I_{q2} \end{bmatrix} = \begin{bmatrix} G_{11} & F_{G+B}(\delta_{12}) \\ F_{G+B}(\delta_{21}) & G_{22} \end{bmatrix}\begin{bmatrix} V_{q1} \\ V_{q2} \end{bmatrix} - \begin{bmatrix} B_{11} & F_{B-G}(\delta_{12}) \\ F_{B-G}(\delta_{21}) & B_{22} \end{bmatrix}\begin{bmatrix} V_{d1} \\ V_{d2} \end{bmatrix}$$

and the d axis currents are given by

$$\begin{bmatrix} I_{d1} \\ I_{d2} \end{bmatrix} = \begin{bmatrix} B_{11} & F_{B-G}(\delta_{12}) \\ F_{B-G}(\delta_{21}) & B_{22} \end{bmatrix}\begin{bmatrix} V_{q1} \\ V_{q2} \end{bmatrix} + \begin{bmatrix} G_{11} & F_{G+B}(\delta_{12}) \\ F_{G+B}(\delta_{21}) & G_{22} \end{bmatrix}\begin{bmatrix} V_{d1} \\ V_{d2} \end{bmatrix}$$

We note that a relation between the voltages and currents based upon (9.33) (i.e., giving V_{q1}, V_{q2}, V_{d1}, and V_{d2} in terms of I_{q1}, I_{q2}, I_{d1}, and I_{d2}) can be easily derived. It would be analogous to (9.40) except that the admittance parameters are replaced with the parameters of the $\bar{\mathbf{Z}}$ matrix of the network.

Example 9.3.

Derive the complete system equations for a two-machine system. The machines are to be represented by the two-axis model (see Section 4.15.3), and the loads are to be represented by constant impedances.

Solution

The transient equivalent circuit of each synchronous machine is given in Figure 4.16. A further approximation, commonly used with this model, is that $x_d' \simeq x_q' \triangleq x'$. The network is now shown in Figure 9.5. The representation is similar to that of the classical model except that in Figure 9.5 the voltages $\bar{E}_1'$ and $\bar{E}_2'$ are not constant.

The first step is to reduce the network to the "internal" generator nodes 1 and 2. Thus the transient generator impedances $r_1 + \mathrm{j}x_1'$ and $r_2 + \mathrm{j}x_2'$ are included in the network $\bar{\mathbf{Y}}$ (or $\bar{\mathbf{Z}}$) matrix. The voltages at the nodes are $\bar{E}_1' = E_{q1}' + \mathrm{j}E_{d1}'$ and $\bar{E}_2' = E_{q2}' + \mathrm{j}E_{d2}'$, and the currents are $\bar{I}_1 = I_{q1} + \mathrm{j}I_{d1}$ and $\bar{I}_2 = I_{q2} + \mathrm{j}I_{d2}$. The relation between them is

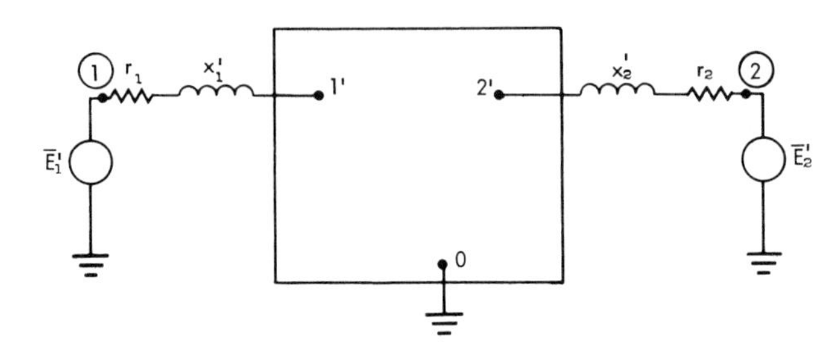

Fig. 9.5. Network of Example 9.3.

given by an equation similar to (9.40). The equations for each machine, under the assumption that $x'_d \simeq x'_q$, are the two axis equations of Section 4.15.3.

$$\begin{aligned}
\tau'_{q0i}\dot{E}'_{di} &= -E'_{di} - (x_{qi} - x'_i)\,I_{qi} \\
\tau'_{d0i}\dot{E}'_{qi} &= E_{FDi} - E'_{qi} + (x_{di} - x'_i)\,I_{di} \\
\tau_{ji}\dot{\omega}_i &= T_{mi} - (I_{di}E'_{di} + I_{qi}E'_{qi}) - D_i\omega_i \\
\dot{\delta}_i &= \omega_i - 1 \qquad i = 1,2
\end{aligned} \tag{9.41}$$

Equations (9.40), with $\overline{V}_i$ replaced with $\overline{E}'_i$, and (9.41) completely describe the system. Each machine represents a fourth-order system, with state variables E'_{qi}, E'_{di}, ω_i, and δ_i.

The complete system equations are given by

$$\begin{aligned}
\tau'_{q01}\dot{E}'_{d1} &= -[1 - (x_{q1} - x'_1)\,B_{11}]\,E'_{d1} - (x_{q1} - x'_1)\,G_{11}E'_{q1} \\
&\quad - (x_{q1} - x'_1)\,F_{G+B}(\delta_{12})\,E'_{q2} + (x_{q1} - x'_1)\,F_{B-G}(\delta_{12})\,E'_{d2} \\
\tau'_{q02}\dot{E}'_{d2} &= -[1 - (x_{q2} - x'_2)\,B_{22}]\,E'_{d2} - (x_{q2} - x'_2)\,G_{22}E'_{q2} \\
&\quad - (x_{q2} - x'_2)\,F_{G+B}(\delta_{21})\,E'_{q1} + (x_{q2} - x'_2)\,F_{B-G}(\delta_{21})\,E'_{d1} \\
\tau'_{d01}\dot{E}'_{q1} &= E_{FD1} - [1 - (x_{d1} - x'_1)\,B_{11}]\,E'_{q1} \\
&\quad + (x_{d1} - x'_1)[G_{11}E'_{d1} + F_{B-G}(\delta_{12})\,E'_{q2} + F_{G+B}(\delta_{12})\,E'_{d2}] \\
\tau'_{d02}\dot{E}'_{q2} &= E_{FD2} - [1 - (x_{d2} - x'_2)\,B_{22}]\,E'_{q2} \\
&\quad + (x_{d2} - x'_2)[F_{B-G}(\delta_{21})\,E'_{q1} + F_{G+B}(\delta_{21})\,E'_{d1} + G_{22}E'_{d2}] \\
\tau_{j1}\dot{\omega}_1 &= T_{m1} - D_1\omega_1 - [G_{11}(E'^2_{d1} + E'^2_{q1}) + F_{B-G}(\delta_{12})(E'_{d1}E'_{q2} - E'_{q1}E'_{d2}) \\
&\quad + F_{G+B}(\delta_{12})(E'_{d1}E'_{d2} + E'_{q1}E'_{q2})] \\
\tau_{j2}\dot{\omega}_2 &= T_{m2} - D_2\omega_2 - [G_{22}(E'^2_{d2} + E'^2_{q2}) + F_{B-G}(\delta_{21})\,(E'_{q1}E'_{d2} - E'_{q2}E'_{d1}) \\
&\quad + F_{G+B}(\delta_{21})\,(E'_{d1}E'_{d2} + E'_{q1}E'_{q2})] \\
\dot{\delta}_1 &= \omega_1 - 1 \qquad \dot{\delta}_2 = \omega_2 - 1
\end{aligned} \tag{9.42}$$

The system given by (9.42) is *not* an eighth-order system since the equations are not independent. This system is actually a seventh-order system with state variables E'_{q1}, E'_{d1}, E'_{q2}, E'_{d2}, ω_1, ω_2, and δ_{12}. The reduction of the order is obtained from the last two equations

$$\dot{\delta}_{12} = \omega_1 - \omega_2$$

Furthermore, if damping is uniform; i.e., if $D_1/\tau_{j1} = D_2/\tau_{j2} = D/\tau_j$ (or if damping is not present) *then* the system is further reduced in order by one, and the two torque equations can be combined in the form

$$\dot{\omega}_{12} = \frac{T_{m1}}{\tau_{j1}} - \frac{T_{m2}}{\tau_{j2}} + f(E'_{q1}, E'_{d1}, E'_{q2}, E'_{d2}, \delta_{12}) - \frac{D}{\tau_j}\,\omega_{12}$$

9.6 System Order

In Example 9.3 it was shown that with damping present the order of the system was reduced by one if the angle of one machine is chosen as reference. It was also pointed out that if damping is uniform, a further reduction of the system order is achieved. We now seek to generalize these conclusions. We consider first the classical model with zero transfer conductances. We can show that the system equations are given by

$$\tau_{ji}\dot{\omega}_i + D_i\omega_i = \sum_{\substack{j=1\\ j\neq i}}^{n} E_iE_jB_{ij}(\sin\delta_{ij}^s - \sin\delta_{ij})$$

$$\dot{\delta}_i = \omega_i \qquad i = 1,2,\ldots,n \tag{9.43}$$

where the superscript s indicates the stable equilibrium angle. Defining the state vector $\mathbf{x}$, the vector $\boldsymbol{\sigma}$, and the function $\mathbf{f}$ by

$$\mathbf{x}^t = [\omega_1, \omega_2, \ldots, \omega_n, (\delta_1 - \delta_1^s), (\delta_2 - \delta_2^s), \ldots, (\delta_n - \delta_n^s)]$$

$$f_k(\sigma_k) = E_pE_qB_{pq}[\sin(\sigma_k + \delta_{pq}^s) - \sin\delta_{pq}^s] \qquad k = 1,2,\ldots,m$$

$$m = n(n-1)/2$$

and $\boldsymbol{\sigma} = \mathbf{C}\,\mathbf{x}$ where $\mathbf{C}$ is a constant matrix. The system (9.43) may then be written in the form

$$\dot{\mathbf{x}} = \mathbf{A}\,\mathbf{x} - \mathbf{B}\,\mathbf{f}(\boldsymbol{\sigma}) \tag{9.44}$$

where $\mathbf{A}$ and $\mathbf{B}$ are constant matrices.

The order of the system (9.44) is determined by examining the transfer function of the linear part (with s the Laplace variable)

$$\mathbf{W}(s) = \mathbf{C}(s\mathbf{I} - \mathbf{A})^{-1}\mathbf{B} \tag{9.45}$$

This has been done in the literature [2, 3]. Expanding (9.45) in partial fractions and examining the ranks of the coefficients obtained, the minimal order of the system is obtained. It is shown that the *minimal order* for this system is $2n - 1$. For the uniform damping case, i.e., for constant D_i/τ_{ji}, the order of the system becomes $2n - 2$ (see also [4]).

The conclusions summarized above for the classical model can be generalized as follows. If the order of the mathematical model describing the synchronous machine i is k_i, $i = 1,2,\ldots,n$, and if damping terms are nonuniform damping, the order of the system is $\sum_{i=1}^{n}(k_i - 1)$. However, if the damping coefficients are uniform or if the damping terms are not present, a further reduction of the order is obtained by referring all the speeds to the speed of the reference machine. The system order then becomes $\sum_{i=1}^{n}(k_i - 2)$.

The above rule should be kept in mind, especially in situations where eigenvalues are obtained such as in the linearized models used in Chapter 6. Unless angle differences are used, the sum of the column of δ's will be zero and a zero eigenvalue will be obtained (see Section 9.12.4).

9.7 Machines Represented By Classical Methods

In the discussion presented above, it is assumed that all the nodes are connected to controlled sources, with all other nodes eliminated by Kron reduction (see Chapter 2, Section 2.10.2). The procedure used to obtain (9.31) assumes that all the machines are represented in detail using Park's transformation. For these machines we seek a relation, such as (9.31), between the currents $\bar{\mathbf{I}}$ and the voltages $\bar{\mathbf{V}}$. The former are either among the state variables if the current model is used, or are derived from the state variables if the flux linkage model is used (see [5]).

If some machines are represented by the classical model, the magnitudes of their internal voltages are known. If machine r is represented by the classical model, the angle δ_r for this machine is the angle between this internal voltage and the system reference axis. In phasor notation the voltage of that node, expressed to the system reference, is given by

$$\hat{V}_r = V_{Qr} + jV_{Dr} = E_r \cos\delta_r + jE_r \sin\delta_r \tag{9.46}$$

At any instant if δ_r is known, V_{Qr} and V_{Dr} are also known.

Since the voltage E_r is considered to be along the q axis of the machine represented by the classical model, we can also express the voltage of this machine in phasor notation as

$$\overline{V}_r = E_r + j0 \qquad r = 1, 2, \ldots, c \tag{9.47}$$

where c is the number of machines represented by the classical model. Also from (4.93) on a per phase base

$$P_{e\phi} = v_d i_d + v_q i_q \quad \text{pu}$$

Dividing both sides by three changes the base power to a three-phase base and divides each voltage and current by $\sqrt{3}$, converting to stator rms equivalent quantities. Thus we have

$$P_e = V_d I_d + V_q I_q \quad \text{pu}(3\phi)$$

and using (9.47),

$$P_{er} = I_{qr} E_r \quad \text{pu}(3\phi) \tag{9.48}$$

Note that E_r is in per unit to a base of rated voltage to neutral.

Assuming that the speed does not deviate appreciably from the synchronous speed, then $T_e \cong P_e$ and from the swing equation (4.90) on a three-phase base

$$\dot{\omega}_r = (1/\tau_{jr})(T_{mr} - E_r I_{qr}) - (D_r/\tau_{jr})\omega_r \qquad \dot{\delta}_r = \omega_r - 1 \tag{9.49}$$

A machine r represented by the classical model will have only ω_r and δ_r as state variables. In (9.49) E_r is known, while I_{qr} is a variable that should be eliminated. To do this we should obtain a relation between I_{qr} and the currents of the machines represented in detail. Similarly the voltages V_{qi} and V_{di} of the machines represented in detail should be expressed in terms of the currents I_{qi} and I_{di} of these machines and the voltages E_r of the machines represented classically. To obtain the above desired relations, the following procedure is suggested.

Let m be the number of machines represented in detail, and c the number of machines represented by the classical model; i.e.,

$$m + c \triangleq n$$

Let the vectors $\overline{\mathbf{I}}$ and $\overline{\mathbf{V}}$ be partitioned as

$$\overline{\mathbf{I}} = \begin{bmatrix} I_{q1} + jI_{d1} \\ \cdots \\ I_{qm} + jI_{dm} \\ \hline I_{qm+1} + jI_{dm+1} \\ \cdots \\ I_{qn} + jI_{dn} \end{bmatrix} \triangleq \begin{bmatrix} \overline{\mathbf{I}}_m \\ \hline \overline{\mathbf{I}}_r \end{bmatrix} \qquad \overline{\mathbf{V}} = \begin{bmatrix} V_{q1} + jV_{d1} \\ \cdots \\ V_{qm} + jV_{dm} \\ \hline E_{m+1} + j0 \\ \cdots \\ E_n + j0 \end{bmatrix} \triangleq \begin{bmatrix} \overline{\mathbf{V}}_m \\ \hline \overline{\mathbf{E}}_r \end{bmatrix} \tag{9.50}$$

Then from (9.50) and (9.31)

$$\begin{bmatrix} \mathbf{I}_m \\ \hline \mathbf{I}_r \end{bmatrix} = \begin{bmatrix} \overline{\mathbf{M}}_{11} & \overline{\mathbf{M}}_{12} \\ \hline \overline{\mathbf{M}}_{21} & \overline{\mathbf{M}}_{22} \end{bmatrix} \begin{bmatrix} \mathbf{V}_m \\ \hline \mathbf{E}_r \end{bmatrix} \tag{9.51}$$

where in (9.51) the complex matrix $\overline{\mathbf{M}}$ is partitioned. Now since $\overline{\mathbf{M}}_{11}^{-1}$ exists, (9.51) can be rearranged with the aid of matrix algebra to obtain

$$\begin{bmatrix} \mathbf{V}_m \\ \hline \mathbf{I}_r \end{bmatrix} = \begin{bmatrix} \overline{\mathbf{M}}_{11}^{-1} & -\overline{\mathbf{M}}_{11}^{-1}\overline{\mathbf{M}}_{12} \\ \hline \overline{\mathbf{M}}_{21}\overline{\mathbf{M}}_{11}^{-1} & \overline{\mathbf{M}}_{22} - \overline{\mathbf{M}}_{21}\overline{\mathbf{M}}_{11}^{-1}\overline{\mathbf{M}}_{12} \end{bmatrix} \begin{bmatrix} \mathbf{I}_m \\ \hline \mathbf{E}_r \end{bmatrix} \tag{9.52}$$

Equation (9.52) is the desired relation between the voltages of the machines represented in detail along with the currents of the machines represented classically, as functions of the current variables of the former machines and the known internal voltages of the latter group. We note that the matrices $\overline{\mathbf{M}}_{11}$, $\overline{\mathbf{M}}_{12}$, $\overline{\mathbf{M}}_{21}$, and $\overline{\mathbf{M}}_{22}$ are functions of the angle differences as well as the admittance parameters.

Example 9.4

Repeat Example 9.2 assuming that machine 1 is represented in detail by the two-axis model and machine 2 by the classical model.

Solution

From (9.37) and using $\overline{Y}_{12} = \overline{Y}_{21}$ and $\delta_{12} = -\delta_{21}$,

$$\overline{\mathbf{M}} = \begin{bmatrix} Y_{11}e^{j\theta_{11}} & Y_{12}e^{j(\theta_{12}-\delta_{12})} \\ \hline Y_{12}e^{j(\theta_{12}+\delta_{12})} & Y_{22}e^{j\theta_{22}} \end{bmatrix} \tag{9.53}$$

and from (9.53) by inspection

$$\overline{M}_{11}^{-1} = \frac{1}{Y_{11}} e^{-j\theta_{11}}$$

$$-\overline{M}_{11}^{-1}\overline{M}_{12} = -\frac{1}{Y_{11}} e^{-j\theta_{11}} \; Y_{12}e^{j(\theta_{12}-\delta_{12})} = -\frac{Y_{12}}{Y_{11}} e^{j(\theta_{12}-\theta_{11}-\delta_{12})}$$

$$\overline{M}_{21}\overline{M}_{11}^{-1} = Y_{12}e^{j(\theta_{12}+\delta_{12})}\frac{1}{Y_{11}} e^{-j\theta_{11}} = \frac{Y_{12}}{Y_{11}} e^{j(\theta_{12}-\theta_{11}+\delta_{12})}$$

$$-\overline{M}_{21}\overline{M}_{11}^{-1}\overline{M}_{12} = -\frac{Y_{12}}{Y_{11}} e^{j(\theta_{12}-\theta_{11}+\delta_{12})} \, Y_{12}e^{j(\theta_{12}-\delta_{12})} = -\frac{Y_{12}^2}{Y_{11}} e^{j(2\theta_{12}-\theta_{11})}$$

$$\overline{M}_{22} - \overline{M}_{21}\overline{M}_{11}^{-1}\overline{M}_{12} = Y_{22}e^{j\theta_{22}} - \frac{Y_{12}^2}{Y_{11}} e^{j(2\theta_{12}-\theta_{11})} \tag{9.54}$$

From (9.50) and (9.52)

$$\begin{bmatrix} V_{q1} + jV_{d1} \\ \hline I_{q2} + jI_{d2} \end{bmatrix} = \begin{bmatrix} \dfrac{1}{Y_{11}} e^{-j\theta_{11}} & -\dfrac{Y_{12}}{Y_{11}} e^{j(\theta_{12}-\theta_{11}-\delta_{12})} \\ \hline \dfrac{Y_{12}}{Y_{11}} e^{j(\theta_{12}-\theta_{11}+\delta_{12})} & Y_{22}e^{j\theta_{22}} - \dfrac{Y_{12}^2}{Y_{11}} e^{j(2\theta_{12}-\theta_{11})} \end{bmatrix} \begin{bmatrix} I_{q1} + jI_{d1} \\ \hline E_2 + j0 \end{bmatrix} \tag{9.55}$$

or

$$\begin{aligned}
V_{q1} &= \left(\frac{1}{Y_{11}}\cos\theta_{11}\right) I_{q1} - \left[\frac{Y_{12}}{Y_{11}}\cos(\theta_{12} - \theta_{11} - \delta_{12})\right]E_2 + \left(\frac{1}{Y_{11}}\sin\theta_{11}\right)I_{d1} \\
V_{d1} &= -\left(\frac{1}{Y_{11}}\sin\theta_{11}\right)I_{q1} + \left(\frac{1}{Y_{11}}\cos\theta_{11}\right)I_{d1} - \left[\frac{Y_{12}}{Y_{11}}\sin(\theta_{12} - \theta_{11} - \delta_{12})\right]E_2 \\
I_{q2} &= \left[\frac{Y_{12}}{Y_{11}}\cos(\theta_{12} - \theta_{11} + \delta_{12})\right]I_{q1} - \left[\frac{Y_{12}}{Y_{11}}\sin(\theta_{12} - \theta_{11} + \delta_{12})\right]I_{d1} \\
&\quad + \left[Y_{22}\cos\theta_{22} - \frac{Y_{12}^2}{Y_{11}}\cos(2\theta_{12} - \theta_{11})\right]E_2 \\
I_{d2} &= \left[\frac{Y_{12}}{Y_{11}}\sin(\theta_{12} - \theta_{11} + \delta_{12})\right]I_{q1} + \left[\frac{Y_{12}}{Y_{11}}\cos(\theta_{12} - \theta_{11} + \delta_{12})\right]I_{d1} \\
&\quad + \left[Y_{22}\sin\theta_{22} - \frac{Y_{12}^2}{Y_{11}}\sin(2\theta_{12} - \theta_{11})\right]E_2
\end{aligned} \tag{9.56}$$

Note that the variables needed to solve for the swing equations are only V_{q1}, V_{d1}, and I_{q2}.

Example 9.5

Repeat Example 9.3, with machine 1 represented mathematically by the two-axis model and machine 2 by the classical model.

Solution

Again the nodes retained are the "internal" generator nodes, and the transient impedances of both generators are included in the network $\overline{\mathbf{Y}}$ (or $\overline{\mathbf{Z}}$) matrix. The equations needed to describe this system are (9.41) for generator 1, (9.49) for generator 2, and an additional set of algebraic equations relating the node currents to the node voltages. Since the two-axis model retains E'_q and E'_d as state variables, it is convenient to use (9.51). For the two-machine system this is the same as (9.40), with $\overline{E}'_1$ replacing $\overline{V}_1$ and $\overline{E}'_2 = E_2 + j0$ replacing $\overline{V}_2$. The system is now fifth order. The state variables for this system are E'_{q1}, E'_{d1}, ω_1, ω_2, and δ_{12}. The complete system equations are given by

$$\begin{aligned}
\tau'_{q01}\dot{E}'_{d1} &= [B_{11}(x_{q1} - x'_1) - 1]E'_{d1} - (x_{q1} - x'_1)[G_{11}E'_{q1} - F_{G+B}(\delta_{12})E_2] \\
\tau'_{d01}\dot{E}'_{q1} &= E_{FD1} + [B_{11}(x_{d1} - x'_1) - 1]E'_{q1} + (x_{d1} - x'_1)[G_{11}E'_{d1} + F_{B-G}(\delta_{12})E_2] \\
\tau_{j1}\dot{\omega}_1 &= T_{m1} - D_1\omega_1 - [G_{11}(E'^2_{d1} + E'^2_{q1}) + F_{B-G}(\delta_{12})E'_{d1}E_2 + F_{G+B}(\delta_{12})E'_{q1}E_2] \\
\tau_{j2}\dot{\omega}_2 &= T_{m2} - D_2\omega_2 - E_2[F_{G+B}(\delta_{21})E'_{q1} - F_{B-G}(\delta_{21})E'_{d1} + G_{22}E_2] \\
\dot{\delta}_{12} &= \omega_1 - \omega_2
\end{aligned} \tag{9.57}$$

9.8 Linearized Model for the Network

From (9.26) $\hat{\mathbf{V}} = \mathbf{T}\overline{\mathbf{V}}$, where $\mathbf{T}$ is defined by (9.24) and $\overline{\mathbf{V}}$ and $\hat{\mathbf{V}}$ are defined by (9.4) and (9.17). Also from (9.31) $\overline{\mathbf{I}} = \overline{\mathbf{M}}\overline{\mathbf{V}}$, where $\overline{\mathbf{M}}$ is given by (9.32). Linearizing (9.31),

$$\overline{\mathbf{I}}_\Delta = \overline{\mathbf{M}}_0\overline{\mathbf{V}}_\Delta + \overline{\mathbf{M}}_\Delta\overline{\mathbf{V}}_0 \tag{9.58}$$

where $\overline{\mathbf{M}}_0$ is evaluated at the initial angles δ_{i0}, $i = 1, 2, \ldots, n$, and $\overline{\mathbf{V}}_0$ is the initial value of the vector $\overline{\mathbf{V}}$.

Let $\delta_i = \delta_{i0} + \delta_{i\Delta}$. Then the matrix $\overline{\mathbf{M}}$ becomes

$$\overline{\mathbf{M}} = \begin{bmatrix} Y_{11}e^{j\theta_{11}} & Y_{12}e^{j(\theta_{12} - \delta_{120} - \delta_{12\Delta})} & \cdots & Y_{1n}e^{j(\theta_{1n} - \delta_{1n0} - \delta_{1n\Delta})} \\ \cdots & \cdots & \cdots & \cdots \\ Y_{n1}e^{j(\theta_{n1} - \delta_{n10} - \delta_{n1\Delta})} & Y_{n2}e^{j(\theta_{n2} - \delta_{n20} - \delta_{n2\Delta})} & \cdots & Y_{nn}e^{j\theta_{nn}} \end{bmatrix} \quad (9.59)$$

The general term $\overline{m}_{ij}$ of the matrix $\overline{\mathbf{M}}$ is of the form $Y_{ij}e^{j(\theta_{ij} - \delta_{ij0} - \delta_{ij\Delta})}$, thus

$$\overline{m}_{ij} = Y_{ij}e^{j(\theta_{ij} - \delta_{ij0})}e^{-j\delta_{ij\Delta}}$$

Using the relation $\cos \delta_{ij\Delta} \cong 1$, $\sin \delta_{ij\Delta} \cong \delta_{ij\Delta}$, we get for the general term

$$\overline{m}_{ij} \cong Y_{ij}e^{j(\theta_{ij} + \delta_{ij0})}(1 - j\delta_{ij\Delta}) \quad (9.60)$$

Therefore the general term in $\overline{\mathbf{M}}_\Delta$ is given by

$$\overline{m}_{ij\Delta} \cong -jY_{ij}e^{j(\theta_{ij} - \delta_{ij0})}\delta_{ij\Delta} \quad (9.61)$$

Thus $\overline{\mathbf{M}}_\Delta$ has off-diagonal terms only, with all the diagonal terms equal to zero.

$$\overline{\mathbf{M}}_\Delta\overline{\mathbf{V}}_0 = -j\begin{bmatrix} 0 & \cdots & Y_{1n}e^{j(\theta_{1n} - \delta_{1n0})}\delta_{1n\Delta} \\ Y_{21}e^{j(\theta_{21} - \delta_{210})}\delta_{21\Delta} & \cdots & Y_{2n}e^{j(\theta_{2n} - \delta_{2n0})}\delta_{2n\Delta} \\ \cdots & \cdots & \cdots \\ Y_{n1}e^{j(\theta_{n1} - \delta_{n10})}\delta_{n1\Delta} & \cdots & 0 \end{bmatrix}\begin{bmatrix} \overline{V}_{10} \\ \overline{V}_{20} \\ \cdots \\ \overline{V}_{n0} \end{bmatrix}$$

$$= -j\begin{bmatrix} \sum_{k=1}^{n} Y_{1k}e^{j(\theta_{1k} - \delta_{1k0})}\overline{V}_{k0}\delta_{1k\Delta} \\ \sum_{k=1}^{n} Y_{2k}e^{j(\theta_{2k} - \delta_{2k0})}\overline{V}_{k0}\delta_{2k\Delta} \\ \cdots \\ \sum_{k=1}^{n} Y_{nk}e^{j(\theta_{nk} - \delta_{nk0})}\overline{V}_{k0}\delta_{nk\Delta} \end{bmatrix} \quad (9.62)$$

and the linearized equation (9.58) becomes

$$\begin{bmatrix} \overline{I}_{1\Delta} \\ \overline{I}_{2\Delta} \\ \cdots \\ \overline{I}_{n\Delta} \end{bmatrix} = \begin{bmatrix} Y_{11}e^{j\theta_{11}} & \cdots & Y_{1n}e^{j(\theta_{1n} - \delta_{1n0})} \\ Y_{21}e^{j(\theta_{21} - \delta_{210})} & \cdots & Y_{2n}e^{j(\theta_{2n} - \delta_{2n0})} \\ \cdots & \cdots & \cdots \\ Y_{n1}e^{j(\theta_{n1} - \delta_{n10})} & \cdots & Y_{nn}e^{j\theta_{nn}} \end{bmatrix}\begin{bmatrix} \overline{V}_{1\Delta} \\ \overline{V}_{2\Delta} \\ \cdots \\ \overline{V}_{n\Delta} \end{bmatrix} - j\begin{bmatrix} \sum_{k=1}^{n} \overline{V}_{k0}Y_{1k}e^{j(\theta_{1k} - \delta_{1k0})}\delta_{1k\Delta} \\ \sum_{k=1}^{n} \overline{V}_{k0}Y_{2k}e^{j(\theta_{2k} - \delta_{2k0})}\delta_{2k\Delta} \\ \cdots \\ \sum_{k=1}^{n} \overline{V}_{k0}Y_{nk}e^{j(\theta_{nk} - \delta_{nk0})}\delta_{nk\Delta} \end{bmatrix} \quad (9.63)$$

The set of equations (9.63) is that needed to complete the description of the system. A similar equation analogous to (9.63) can be derived relating $\overline{\mathbf{V}}_\Delta$ to $\overline{\mathbf{I}}_\Delta$ and $\boldsymbol{\delta}_{ij\Delta}$. The network elements involved in this case are elements of the open circuit impedance matrix $\overline{\mathbf{Z}}$.

We now formulate (9.63) in a more compact form. From (9.24) let $\mathbf{T} = \mathbf{T}_0 + \mathbf{T}_\Delta$ to compute

$$\mathbf{T}_\Delta = \mathrm{j}\mathbf{T}_0\boldsymbol{\delta}_\Delta \qquad \boldsymbol{\delta}_\Delta \triangleq \mathrm{diag}\,(\delta_{1\Delta}, \ldots, \delta_{n\Delta}) \tag{9.64}$$

Similarly, we let $\mathbf{T}^{-1} \triangleq \mathbf{N} = \mathbf{N}_0 + \mathbf{N}_\Delta$ to compute

$$\mathbf{N}_\Delta = (\mathbf{T}^{-1})_\Delta = -\mathrm{j}\mathbf{N}_0\boldsymbol{\delta}_\Delta = -\mathrm{j}\mathbf{T}_0^{-1}\boldsymbol{\delta}_\Delta \tag{9.65}$$

Note carefully that $\mathbf{T}^{-1} \neq \mathbf{T}_0^{-1} + \mathbf{T}_\Delta^{-1}$ and that $(\mathbf{T}_\Delta)^{-1} \neq (\mathbf{T}^{-1})_\Delta = \mathbf{N}_\Delta$. We can show, however, that $(\mathbf{T}_0)^{-1} = (\mathbf{T}^{-1})_0 = \mathbf{N}_0$. Thus from $\overline{\mathbf{M}} = \overline{\mathbf{M}}_0 + \overline{\mathbf{M}}_\Delta$ we compute $\overline{\mathbf{M}}_0 + \overline{\mathbf{M}}_\Delta = (\mathbf{N}_0 + \mathbf{N}_\Delta)\overline{\mathbf{Y}}(\mathbf{T}_0 + \mathbf{T}_\Delta)$. Neglecting second-order terms,

$$\overline{\mathbf{M}}_\Delta = -\mathrm{j}(\mathbf{T}_0^{-1}\boldsymbol{\delta}_\Delta\overline{\mathbf{Y}}\mathbf{T}_0 - \mathbf{T}_0^{-1}\overline{\mathbf{Y}}\mathbf{T}_0\boldsymbol{\delta}_\Delta) \tag{9.66}$$

From matrix algebra we get the following relations,

$$\mathbf{T}_0\boldsymbol{\delta}_\Delta = \begin{bmatrix} e^{\mathrm{j}\delta_{10}} & & \\ & \ddots & \\ & & e^{\mathrm{j}\delta_{n0}} \end{bmatrix} \begin{bmatrix} \delta_{1\Delta} & & \\ & \ddots & \\ & & \delta_{n\Delta} \end{bmatrix} = \begin{bmatrix} e^{\mathrm{j}\delta_{10}}\delta_{1\Delta} & & \\ & \ddots & \\ & & e^{\mathrm{j}\delta_{n0}}\delta_{n\Delta} \end{bmatrix}$$

$$\mathbf{T}_0^{-1}\overline{\mathbf{Y}} = \begin{bmatrix} e^{-\mathrm{j}\delta_{10}} & & \\ & \ddots & \\ & & e^{-\mathrm{j}\delta_{n0}} \end{bmatrix} \begin{bmatrix} Y_{11}e^{\mathrm{j}\theta_{11}} & \cdots & Y_{1n}e^{\mathrm{j}\theta_{1n}} \\ \cdots & \cdots & \cdots \\ Y_{n1}e^{\mathrm{j}\theta_{n1}} & \cdots & Y_{nn}e^{\mathrm{j}\theta_{nn}} \end{bmatrix}$$

$$= \begin{bmatrix} Y_{11}e^{\mathrm{j}(\theta_{11}-\delta_{10})} & \cdots & Y_{1n}e^{\mathrm{j}(\theta_{1n}-\delta_{10})} \\ \cdots & \cdots & \cdots \\ Y_{n1}e^{\mathrm{j}(\theta_{n1}-\delta_{n0})} & \cdots & Y_{nn}e^{\mathrm{j}(\theta_{nn}-\delta_{n0})} \end{bmatrix}$$

$$\mathbf{T}_0^{-1}\overline{\mathbf{Y}}\mathbf{T}_0\boldsymbol{\delta}_\Delta = \begin{bmatrix} Y_{11}e^{\mathrm{j}\theta_{11}} & \cdots & Y_{1n}e^{\mathrm{j}(\theta_{1n}-\delta_{1n0})} \\ Y_{21}e^{\mathrm{j}(\theta_{21}-\delta_{210})} & \cdots & Y_{2n}e^{\mathrm{j}(\theta_{2n}-\delta_{2n0})} \\ \cdots & \cdots & \cdots \\ Y_{n1}e^{\mathrm{j}(\theta_{n1}-\delta_{n10})} & \cdots & Y_{nn}e^{\mathrm{j}\theta_{nn}} \end{bmatrix} \begin{bmatrix} \delta_{1\Delta} & & \\ & \ddots & \\ & & \delta_{n\Delta} \end{bmatrix} = \overline{\mathbf{M}}_0\boldsymbol{\delta}_\Delta \tag{9.67}$$

Also

$$\mathbf{T}_0^{-1}\boldsymbol{\delta}_\Delta = \begin{bmatrix} e^{-j\delta_{10}}\delta_{1\Delta} & & \\ & \ddots & \\ & & e^{-j\delta_{n0}}\delta_{n\Delta} \end{bmatrix}$$

$$\overline{\mathbf{Y}}\mathbf{T}_0 = \begin{bmatrix} Y_{11}e^{j(\theta_{11}+\delta_{10})} & \cdots & Y_{1n}e^{j(\theta_{1n}+\delta_{n0})} \\ \cdots & \cdots & \cdots \\ Y_{n1}e^{j(\theta_{n1}+\delta_{10})} & \cdots & Y_{nn}e^{j(\theta_{nn}+\delta_{n0})} \end{bmatrix}$$

$$\mathbf{T}^{-1}\boldsymbol{\delta}_\Delta\overline{\mathbf{Y}}\mathbf{T}_0 = \begin{bmatrix} Y_{11}e^{j\theta_{11}}\ \delta_{1\Delta} & \cdots & Y_{1n}e^{j(\theta_{1n}-\delta_{1n0})}\delta_{1\Delta} \\ \cdots & \cdots & \cdots \\ Y_{n1}e^{j(\theta_{n1}-\delta_{n10})}\delta_{n\Delta} & \cdots & Y_{nn}e^{j\theta_{nn}}\ \delta_{n\Delta} \end{bmatrix} = \boldsymbol{\delta}_\Delta\overline{\mathbf{M}}_0 \tag{9.68}$$

From (9.66), (9.67), and (9.68)

$$\overline{\mathbf{M}}_\Delta = -j[\boldsymbol{\delta}_\Delta\overline{\mathbf{M}}_0 - \overline{\mathbf{M}}_0\boldsymbol{\delta}_\Delta] \tag{9.69}$$

and the network equation is given by

$$\overline{\mathbf{I}}_\Delta = \overline{\mathbf{M}}_0\overline{\mathbf{V}}_\Delta - j[\boldsymbol{\delta}_\Delta\overline{\mathbf{M}}_0 - \overline{\mathbf{M}}_0\boldsymbol{\delta}_\Delta]\overline{\mathbf{V}}_0 \tag{9.70}$$

Note that (9.70) is the *same* as (9.63).

To obtain a relation between $\overline{\mathbf{V}}_\Delta$ and $\overline{\mathbf{I}}_\Delta$, we can either manipulate (9.70) to obtain

$$\overline{\mathbf{V}}_\Delta = \overline{\mathbf{M}}_0^{-1}\overline{\mathbf{I}}_\Delta - j[\boldsymbol{\delta}_\Delta - \overline{\mathbf{M}}_0^{-1}\boldsymbol{\delta}_\Delta\overline{\mathbf{M}}_0]\overline{\mathbf{V}}_0 \tag{9.71}$$

or follow a procedure similar to the above. Define

$$\overline{\mathbf{Q}} \triangleq \overline{\mathbf{M}}^{-1} = \mathbf{T}^{-1}\overline{\mathbf{Y}}^{-1}\mathbf{T} \tag{9.72}$$

We can then show that

$$\overline{\mathbf{V}}_\Delta = \overline{\mathbf{Q}}_0\overline{\mathbf{I}}_\Delta - j(\boldsymbol{\delta}_\Delta\overline{\mathbf{Q}}_0 - \overline{\mathbf{Q}}_0\boldsymbol{\delta}_\Delta)\overline{\mathbf{I}}_0 \tag{9.73}$$

Example 9.6

Derive the relations between $\overline{\mathbf{V}}_\Delta$ and $\overline{\mathbf{I}}_\Delta$ for a two-machine system.

Solution

From (9.53) we get for $\overline{\mathbf{M}}_0$

$$\overline{\mathbf{M}}_0 = \begin{bmatrix} Y_{11}e^{j\theta_{11}} & Y_{12}e^{j(\theta_{12}-\delta_{120})} \\ Y_{12}e^{j(\theta_{12}+\delta_{120})} & Y_{22}e^{j\theta_{22}} \end{bmatrix} \tag{9.74}$$

$$\boldsymbol{\delta}_\Delta = \begin{bmatrix} \delta_{1\Delta} & 0 \\ 0 & \delta_{2\Delta} \end{bmatrix} \qquad \overline{\mathbf{V}}_\Delta = \begin{bmatrix} V_{q1\Delta} + jV_{d1\Delta} \\ V_{q2\Delta} + jV_{d2\Delta} \end{bmatrix} \tag{9.75}$$

$$\overline{\mathbf{M}}_0\boldsymbol{\delta}_\Delta = \begin{bmatrix} Y_{11}e^{j\theta_{11}}\delta_{1\Delta} & Y_{12}e^{j(\theta_{12}-\delta_{120})}\delta_{2\Delta} \\ Y_{12}e^{j(\theta_{12}+\delta_{120})}\delta_{1\Delta} & Y_{22}e^{j\theta_{22}}\delta_{2\Delta} \end{bmatrix}$$

$$\boldsymbol{\delta}_\Delta\overline{\mathbf{M}}_0 = \begin{bmatrix} Y_{11}e^{j\theta_{11}}\delta_{1\Delta} & Y_{12}e^{j(\theta_{12}-\delta_{120})}\delta_{1\Delta} \\ Y_{12}e^{j(\theta_{12}+\delta_{120})}\delta_{2\Delta} & Y_{22}e^{j\theta_{22}}\delta_{2\Delta} \end{bmatrix}$$

$$\mathrm{j}(\delta_\Delta \overline{\mathbf{M}}_0 - \overline{\mathbf{M}}_0 \delta_\Delta)\overline{\mathbf{V}}_0 = \mathrm{j}\begin{bmatrix} 0 & Y_{12}e^{\mathrm{j}(\theta_{12}-\delta_{120})}\delta_{12\Delta} \\ -Y_{12}e^{\mathrm{j}(\theta_{12}+\delta_{120})}\delta_{12\Delta} & 0 \end{bmatrix}\begin{bmatrix} V_{q10} + \mathrm{j}V_{d10} \\ V_{q20} + \mathrm{j}V_{d20} \end{bmatrix}$$

$$= \begin{bmatrix} \mathrm{j}Y_{12}e^{\mathrm{j}(\theta_{12}-\delta_{120})}(V_{q20} + \mathrm{j}V_{d20}) \\ -\mathrm{j}Y_{12}e^{\mathrm{j}(\theta_{12}+\delta_{120})}(V_{q10} + \mathrm{j}V_{d10}) \end{bmatrix}\delta_{12\Delta} \tag{9.76}$$

$$\overline{\mathbf{M}}_0\overline{\mathbf{V}}_\Delta = \begin{bmatrix} Y_{11}e^{\mathrm{j}\theta_{11}}(V_{q1\Delta} + \mathrm{j}V_{d1\Delta}) + Y_{12}e^{\mathrm{j}(\theta_{12}-\delta_{120})}(V_{q2\Delta} + \mathrm{j}V_{d2\Delta}) \\ Y_{12}e^{\mathrm{j}(\theta_{12}+\delta_{120})}(V_{q1\Delta} + \mathrm{j}V_{d1\Delta}) + Y_{22}e^{\mathrm{j}\theta_{22}}(V_{q2\Delta} + \mathrm{j}V_{d2\Delta}) \end{bmatrix} \tag{9.77}$$

Substituting (9.76), (9.77), in (9.70),

$$\begin{bmatrix} I_{q1\Delta} + \mathrm{j}I_{d1\Delta} \\ I_{q2\Delta} + \mathrm{j}I_{d2\Delta} \end{bmatrix} = \begin{bmatrix} Y_{11}e^{\mathrm{j}\theta_{11}}V_{q1\Delta} + \mathrm{j}Y_{11}e^{\mathrm{j}\theta_{11}}V_{d1\Delta} + Y_{12}e^{\mathrm{j}(\theta_{12}-\delta_{120})}V_{q2\Delta} + \mathrm{j}Y_{12}e^{\mathrm{j}(\theta_{12}-\delta_{120})}V_{d2\Delta} \\ Y_{12}e^{\mathrm{j}(\theta_{12}+\delta_{120})}V_{q1\Delta} + \mathrm{j}Y_{12}e^{\mathrm{j}(\theta_{12}+\delta_{120})}V_{d1\Delta} + Y_{22}e^{\mathrm{j}\theta_{22}}V_{q2\Delta} + \mathrm{j}Y_{22}e^{\mathrm{j}\theta_{22}}V_{d2\Delta} \end{bmatrix}$$
$$- \begin{bmatrix} \mathrm{j}Y_{12}e^{\mathrm{j}(\theta_{12}-\delta_{120})}(V_{q20} + \mathrm{j}V_{d20}) \\ -\mathrm{j}Y_{12}e^{\mathrm{j}(\theta_{12}+\delta_{120})}(V_{q10} + \mathrm{j}V_{d10}) \end{bmatrix}\delta_{12\Delta} \tag{9.78}$$

By separating the real and the imaginary terms in equation (9.78), we get four real equations between $I_{q1\Delta}$, $I_{d1\Delta}$, $I_{q2\Delta}$, and $I_{d2\Delta}$ and $V_{q1\Delta}$, $V_{d1\Delta}$, $V_{q2\Delta}$, $V_{d2\Delta}$, and $\delta_{12\Delta}$. These are given below:

$$\begin{aligned}
I_{q1\Delta} &= G_{11}V_{q1\Delta} - B_{11}V_{d1\Delta} + Y_{12}\cos(\theta_{12} - \delta_{120})V_{q2\Delta} - Y_{12}\sin(\theta_{12} - \delta_{120})V_{d2\Delta} \\
&\quad + Y_{12}[\sin(\theta_{12} - \delta_{120})V_{q20} + \cos(\theta_{12} - \delta_{120})V_{d20}]\delta_{12\Delta} \\
I_{d1\Delta} &= B_{11}V_{q1\Delta} + G_{11}V_{d1\Delta} + Y_{12}\sin(\theta_{12} - \delta_{120})V_{q2\Delta} + Y_{12}\cos(\theta_{12} - \delta_{120})V_{d2\Delta} \\
&\quad + Y_{12}[-\cos(\theta_{12} - \delta_{120})V_{q20} + \sin(\theta_{12} - \delta_{120})V_{d20}]\delta_{12\Delta} \\
I_{q2\Delta} &= Y_{12}\cos(\theta_{12} + \delta_{120})V_{q1\Delta} - Y_{12}\sin(\theta_{12} + \delta_{120})V_{d1\Delta} + G_{22}V_{q2\Delta} - B_{22}V_{d2\Delta} \\
&\quad - Y_{12}[\sin(\theta_{12} + \delta_{120})V_{q10} + \cos(\theta_{12} + \delta_{120})V_{d10}]\delta_{12\Delta} \\
I_{d2\Delta} &= Y_{12}\sin(\theta_{12} + \delta_{120})V_{q1\Delta} + Y_{12}\cos(\theta_{12} + \delta_{120})V_{d1\Delta} + B_{22}V_{q2\Delta} + G_{22}V_{d2\Delta} \\
&\quad + Y_{12}[\cos(\theta_{12} + \delta_{120})V_{q10} - \sin(\theta_{12} + \delta_{120})V_{d10}]\delta_{12\Delta}
\end{aligned} \tag{9.79}$$

Example 9.7

Linearize the two-axis model of the synchronous machine as given by (9.41) and the classical model as given by (9.48).

Solution

From (9.41) we get

$$\begin{aligned}
\tau'_{q0}\dot{E}'_{d\Delta} &= -E'_{d\Delta} - (x_q - x')I_{q\Delta} \\
\tau'_{d0}\dot{E}'_{q\Delta} &= E_{FD\Delta} - E'_{q\Delta} + (x_d - x')I_{d\Delta} \\
\tau_j\dot{\omega}_\Delta &= T_{m\Delta} - D\omega_\Delta - (I_{d0}E'_{d\Delta} + I_{q0}E'_{q\Delta} + E'_{d0}I_{d\Delta} + E'_{q0}I_{q\Delta}) \\
\dot{\delta}_\Delta &= \omega_\Delta
\end{aligned} \tag{9.80}$$

From (9.48) we get

$$\tau_j\dot{\omega}_\Delta = T_{m\Delta} - EI_{q\Delta} - D\omega_\Delta \qquad \dot{\delta}_\Delta = \omega_\Delta \tag{9.81}$$

Example 9.8
Linearize the two-machine system of Example 9.5. One machine is represented by the two-axis model, and the second is represented classically.

Solution
From (9.79), (9.80), and (9.81) and dropping the Δ subscripts for convenience,

$$
\begin{aligned}
\tau'_{q01}\dot{E}'_{d1} &= [(x_{q1} - x'_1)B_{11} - 1]E'_{d1} - G_{11}(x_{q1} - x'_1)E'_{q1} \\
&\quad -[(x_{q1} - x'_1)Y_{12}E_2\sin(\theta_{12} - \delta_{120})]\,\delta_{12} \\
\tau'_{d01}\dot{E}'_{q1} &= E_{FD1} + [(x_{d1} - x'_1)B_{11} - 1]E'_{q1} + (x_{d1} - x'_1)G_{11}E'_{d1} \\
&\quad - [(x_{d1} - x'_1)Y_{12}E_2\cos(\theta_{12} - \delta_{120})]\,\delta_{12} \\
\tau_{j1}\dot{\omega}_1 &= T_{m1} - D_1\omega_1 - \{(E'_{d10}B_{11} + E'_{q10}G_{11} + I_{q10})E'_{q1} + (E'_{d10}G_{11} - E'_{q10}B_{11} + I_{d10})E'_{d1} \\
&\qquad + Y_{12}E_2[E'_{q10}\sin(\theta_{12} - \delta_{120}) - E'_{d10}\cos(\theta_{12} - \delta_{120})]\,\delta_{12}\} \\
\tau_{j2}\dot{\omega}_2 &= T_{m2} - D_2\omega_2 - E_2\{[Y_{12}\cos(\theta_{12} + \delta_{120})]E'_{q1} - [Y_{12}\sin(\theta_{12} + \delta_{120})]E'_{d1} \\
&\qquad - Y_{12}[E'_{q10}\sin(\theta_{12} + \delta_{120}) + E'_{d10}\cos(\theta_{12} + \delta_{120})]\,\delta_{12}\} \\
\dot{\delta}_{12} &= \omega_1 - \omega_2
\end{aligned}
\tag{9.82}
$$

Equation (9.82) is a set of five first-order linear differential equations. It is of the form $\dot{\mathbf{x}} = \mathbf{Ax} + \mathbf{Bu}$, where

$$
\mathbf{x}^t = [E'_{q1}\;E'_{d1}\;\omega_1\;\omega_2\;\delta_{12}] \qquad \mathbf{u} = \begin{bmatrix} E_{FD1} \\ T_{m1} \\ T_{m2} \end{bmatrix}
$$

$$
\mathbf{A} = \begin{array}{c}
\begin{array}{ccccc} E'_{q1} & E'_{d1} & \omega_1 & \omega_2 & \delta_{12} \end{array} \\
\left[\begin{array}{ccccc}
\dfrac{(x_{d1} - x'_1)B_{11} - 1}{\tau'_{d01}} & \dfrac{(x_{d1} - x'_1)G_{11}}{\tau'_{d01}} & 0 & 0 & \dfrac{-(x_{d1} - x'_1)Y_{12}E_2\cos(\theta_{12} - \delta_{120})}{\tau'_{d01}} \\
\dfrac{-(x_{q1} - x'_1)G_{11}}{\tau'_{q01}} & \dfrac{(x_{q1} - x'_1)B_{11} - 1}{\tau'_{q01}} & 0 & 0 & \dfrac{-(x_{q1} - x'_1)Y_{12}E_2\sin(\theta_{12} - \delta_{120})}{\tau'_{q01}} \\
\dfrac{-(E'_{d10}B_{11} + E'_{q10}G_{11} + I_{q10})}{\tau_{j1}} & \dfrac{-(E'_{d10}G_{11} - E'_{q10}B_{11} + I_{d10})}{\tau_{j1}} & -D_1/\tau_{j1} & 0 & \dfrac{-Y_{12}E_2[E'_{q10}\sin(\theta_{12} - \delta_{120}) - E'_{d10}\cos(\theta_{12} - \delta_{120})]}{\tau_{j1}} \\
\dfrac{-E_2Y_{12}\cos(\theta_{12} + \delta_{120})}{\tau_{j2}} & \dfrac{E_2Y_{12}\sin(\theta_{12} + \delta_{120})}{\tau_{j2}} & 0 & -D_2/\tau_{j2} & \dfrac{Y_{12}E_2[E'_{q10}\sin(\theta_{12} + \delta_{120}) + E'_{d10}\cos(\theta_{12} + \delta_{120})]}{\tau_{j2}} \\
0 & 0 & -1 & -1 & 0
\end{array}\right]
\end{array}
\tag{9.84}
$$

From the initial conditions, which determine E'_{d10}, E'_{q10}, E_2, I'_{q10}, I'_{d10}, and δ_{120} and from the network $\overline{\mathbf{Y}}$ matrix all the coefficients of the **A** matrix of (9.84) can be determined. Stability analysis (such as discussed in Chapter 6) can be conducted.

We note again (as per the discussion in Section 9.6) that the order of the mathematical description of machine 1 is four, that of machine 2 is two. The system order, however, is $4 + 2 - 1 = 5$. If the damping terms are not present, the variables ω_1 and ω_2 can be combined in one variable ω_{12}.

9.9 Hybrid Formulation

Where a combination of classical and detailed machine representations exists, a hybrid formulation is convenient. Let m machines be represented in detail, and c machines represented classically, $m + c = n$. Then from (9.58),

$$\overline{\mathbf{V}}_\Delta = \begin{bmatrix} \overline{V}_{1\Delta} \\ \cdots \\ \overline{V}_{m\Delta} \\ \hline \mathbf{0} \end{bmatrix} \triangleq \begin{bmatrix} \overline{\mathbf{V}}_{m\Delta} \\ \hline \mathbf{0} \end{bmatrix}; \qquad \text{also } \overline{\mathbf{I}}_\Delta = \begin{bmatrix} \overline{\mathbf{I}}_{m\Delta} \\ \hline \overline{\mathbf{I}}_{c\Delta} \end{bmatrix}$$

From (9.70)

$$\overline{\mathbf{I}}_\Delta = \left[\begin{array}{c|c} \overline{\mathbf{M}}_{011} & \overline{\mathbf{M}}_{012} \\ \hline \overline{\mathbf{M}}_{021} & \overline{\mathbf{M}}_{022} \end{array}\right] \begin{bmatrix} \overline{\mathbf{V}}_{m\Delta} \\ \hline \mathbf{0} \end{bmatrix} - \mathrm{j}(\boldsymbol{\delta}_\Delta \overline{\mathbf{M}}_0 - \overline{\mathbf{M}}_0 \boldsymbol{\delta}_\Delta)\overline{\mathbf{V}}_0 \tag{9.85}$$

where the subscript m indicates a vector of dimension m.

By comparing (9.85) and (9.63),

$$(\boldsymbol{\delta}_\Delta \overline{\mathbf{M}}_0 - \overline{\mathbf{M}}_0 \boldsymbol{\delta}_\Delta)\overline{\mathbf{V}}_0 = \begin{bmatrix} \sum_{k=1}^{n} Y_{1K} e^{\mathrm{j}(\theta_{1k} + \delta_{1k0})} \overline{V}_{k0} \delta_{1k\Delta} \\ \cdots \\ \sum_{k=1}^{n} Y_{nk} e^{\mathrm{j}(\theta_{nk} + \delta_{nk0})} \overline{V}_{k0} \delta_{nk\Delta} \end{bmatrix} \triangleq \begin{bmatrix} \overline{\mathbf{K}}_m(\boldsymbol{\delta}_\Delta) \\ \hline \overline{\mathbf{K}}_c(\boldsymbol{\delta}_\Delta) \end{bmatrix} \tag{9.86}$$

where $\overline{\mathbf{K}}_m(\boldsymbol{\delta}_\Delta)$ is an $(m \times 1)$ vector and $\overline{\mathbf{K}}_c(\boldsymbol{\delta}_\Delta)$ is a $(c \times 1)$ vector.

From (9.85) and (9.86)

$$\begin{bmatrix} \overline{\mathbf{I}}_{m\Delta} \\ \hline \overline{\mathbf{I}}_{c\Delta} \end{bmatrix} = \left[\begin{array}{c|c} \overline{\mathbf{M}}_{011} & \overline{\mathbf{M}}_{012} \\ \hline \overline{\mathbf{M}}_{021} & \overline{\mathbf{M}}_{022} \end{array}\right] \begin{bmatrix} \overline{\mathbf{V}}_{m\Delta} \\ \hline \mathbf{0} \end{bmatrix} - \mathrm{j} \begin{bmatrix} \overline{\mathbf{K}}_m(\boldsymbol{\delta}_\Delta) \\ \hline \overline{\mathbf{K}}_c(\boldsymbol{\delta}_\Delta) \end{bmatrix} \tag{9.87}$$

Therefore

$$\overline{\mathbf{I}}_{m\Delta} = \overline{\mathbf{M}}_{011} \overline{\mathbf{V}}_{m\Delta} - \mathrm{j}\overline{\mathbf{K}}_m(\boldsymbol{\delta}_\Delta) \qquad \overline{\mathbf{I}}_{c\Delta} = \overline{\mathbf{M}}_{021} \overline{\mathbf{V}}_{m\Delta} - \mathrm{j}\overline{\mathbf{K}}_c(\boldsymbol{\delta}_\Delta) \tag{9.88}$$

from which we get

$$\begin{aligned} \overline{\mathbf{V}}_{m\Delta} &= \overline{\mathbf{M}}_{011}^{-1} \overline{\mathbf{I}}_{m\Delta} + \mathrm{j}\overline{\mathbf{M}}_{011}^{-1} \mathbf{K}_m(\boldsymbol{\delta}_\Delta) \\ \overline{\mathbf{I}}_{c\Delta} &= \overline{\mathbf{M}}_{021} \overline{\mathbf{M}}_{011}^{-1} \overline{\mathbf{I}}_{m\Delta} + \mathrm{j}\overline{\mathbf{M}}_{021} \overline{\mathbf{M}}_{011}^{-1} \overline{\mathbf{K}}_m(\boldsymbol{\delta}_\Delta) - \mathrm{j}\overline{\mathbf{K}}_c(\boldsymbol{\delta}_\Delta) \end{aligned} \tag{9.89}$$

Example 9.9

Obtain the linearized hybrid formulation for the two-machine system in Example 9.4.

Solution

From Example 9.2

$$\overline{\mathbf{M}}_0 = \begin{bmatrix} Y_{11} e^{\mathrm{j}\theta_{11}} & Y_{12} e^{\mathrm{j}(\theta_{12} - \delta_{120})} \\ Y_{12} e^{\mathrm{j}(\theta_{12} + \delta_{120})} & Y_{22} e^{\mathrm{j}\theta_{22}} \end{bmatrix}$$

and from (9.78) and (9.86)

$$\begin{bmatrix} \overline{\mathbf{K}}_m(\boldsymbol{\delta}_\Delta) \\ \hline \overline{\mathbf{K}}_c(\boldsymbol{\delta}_\Delta) \end{bmatrix} = \begin{bmatrix} (V_{q20} + \mathrm{j}V_{d20}) Y_{12} e^{\mathrm{j}(\theta_{12} - \delta_{120})} \delta_{12\Delta} \\ \hline (V_{q10} + \mathrm{j}V_{d10}) Y_{21} e^{\mathrm{j}(\theta_{21} - \delta_{210})} \delta_{21\Delta} \end{bmatrix} = \begin{bmatrix} \overline{V}_{20} Y_{12} e^{\mathrm{j}(\theta_{12} - \delta_{120})} \delta_{12\Delta} \\ \hline -\overline{V}_{10} Y_{12} e^{\mathrm{j}(\theta_{12} + \delta_{120})} \delta_{12\Delta} \end{bmatrix} \tag{9.90}$$

Substituting in (9.89)

$$\overline{V}_{1\Delta} = \frac{1}{Y_{11}} e^{-j\theta_{11}} \overline{I}_{1\Delta} + j \frac{1}{Y_{11}} e^{-j\theta_{11}} \overline{V}_{20} Y_{12} e^{j(\theta_{12}-\delta_{120})} \delta_{12\Delta}$$

or

$$V_{q1\Delta} + jV_{d1\Delta} = \frac{e^{-j\theta_{11}}}{Y_{11}} [(I_{q1\Delta} + jI_{d1\Delta}) + jY_{12}e^{j(\theta_{12}-\delta_{120})}(V_{q20} + jV_{d20})\delta_{12\Delta}] \tag{9.91}$$

and

$$I_{q2\Delta} + jI_{d2\Delta} = \frac{Y_{12}}{Y_{11}} e^{j(\theta_{12}-\theta_{11}+\delta_{120})}[(I_{q1\Delta} + jI_{d1\Delta}) + j(V_{q20} + jV_{d20})Y_{12}e^{j(\theta_{12}-\delta_{120})}\delta_{12\Delta}]$$

$$- j(V_{q10} + jV_{d10})Y_{12}e^{j(\theta_{12}+\delta_{120})}\delta_{12\Delta}$$

or

$$I_{q2\Delta} + jI_{d2\Delta} = \frac{Y_{12}}{Y_{11}} e^{j(\theta_{12}-\theta_{11}+\delta_{120})}(I_{q1\Delta} + jI_{d1\Delta})$$

$$+ j\left[\frac{Y_{12}^2}{Y_{11}} e^{j(2\theta_{12}-\theta_{11})}(V_{q20} + jV_{d20}) - Y_{12}e^{j(\theta_{12}+\delta_{120})}(V_{q10} + jV_{d10})\right]\delta_{12\Delta} \tag{9.92}$$

Equations (9.91) and (9.92) are the desired relations giving $\overline{V}_{1\Delta}$ and $\overline{I}_{2\Delta}$ in terms of $\overline{I}_{1\Delta}$ and $\overline{\delta}_{12\Delta}$. These complex equations represent four real equations:

$$V_{q1\Delta} = \frac{1}{Y_{11}} \cos\theta_{11} I_{q1\Delta} + \frac{1}{Y_{11}} \sin\theta_{11} I_{d1\Delta} + \frac{Y_{12}}{Y_{11}} [\sin(\theta_{11} - \theta_{12} + \delta_{120})V_{q20}$$
$$- \cos(\theta_{11} - \theta_{12} + \delta_{120})V_{d20}]\delta_{12\Delta}$$

$$V_{d1\Delta} = -\frac{1}{Y_{11}} \sin\theta_{11} I_{q1\Delta} + \frac{1}{Y_{11}} \cos\theta_{11} I_{d1\Delta} + \frac{Y_{12}}{Y_{11}} [\cos(\theta_{11} - \theta_{12} + \delta_{120})V_{q20}$$
$$+ \sin(\theta_{11} - \theta_{12} + \delta_{120})V_{d20}]\delta_{12\Delta}$$

$$I_{q2\Delta} = \frac{Y_{12}}{Y_{22}} \cos(\theta_{12} - \theta_{11} + \delta_{120})I_{q1\Delta} - \frac{Y_{12}}{Y_{22}} \sin(\theta_{12} - \theta_{11} + \delta_{120})I_{d1\Delta}$$
$$+ \left\{-\frac{Y_{12}^2}{Y_{22}} [\sin(2\theta_{12} - \theta_{22})V_{q20} + \cos(2\theta_{12} - \theta_{11})V_{d20}]\right.$$
$$\left. + Y_{12}[\sin(\theta_{12} + \delta_{120})V_{q10} + \cos(\theta_{12} + \delta_{120})V_{d10}]\right\}\delta_{12\Delta}$$

$$I_{d2\Delta} = \frac{Y_{12}}{Y_{11}} \sin(\theta_{12} - \theta_{11} + \delta_{120})I_{q1\Delta} + \frac{Y_{12}}{Y_{22}} \cos(\theta_{12} - \theta_{11} + \delta_{120})I_{d1\Delta}$$
$$+ \left\{\frac{Y_{12}^2}{Y_{11}} [\cos(2\theta_{12} - \theta_{11})V_{q20} - \sin(2\theta_{12} - \theta_{11})V_{d20}]\right.$$
$$\left. - Y_{12}[\cos(\theta_{12} + \delta_{120})V_{q10} - \sin(\theta_{12} + \delta_{120})V_{d10}]\right\}\delta_{12\Delta} \tag{9.93}$$

9.10 Network Equations with Flux Linkage Model

The network equation for the flux linkage description is taken from (9.33) and (9.72).

$$\overline{\mathbf{V}} = \overline{\mathbf{Q}}\overline{\mathbf{I}} \tag{9.94}$$

This is a complex equation of order n, or $2n$ real equations.

If the flux linkage model is used, I_q and I_d for the various machines are not state variables. Therefore, auxiliary equations are needed to relate these currents to the flux linkages. These equations are obtained from Section 4.12. For machine i we have

$$
\begin{aligned}
I_{qi} &= \frac{1}{\ell_q}\left(1 - \frac{L_{MQ}}{\ell_q}\right)\Lambda_{qi} - \frac{L_{MQ}}{\ell_q \ell_Q}\Lambda_{Qi} \\
I_{di} &= \frac{1}{\ell_d}\left(1 - \frac{L_{MD}}{\ell_d}\right)\Lambda_{di} - \frac{L_{MD}}{\ell_d \ell_F}\Lambda_{Fi} - \frac{L_{MD}}{\ell_d \ell_D}\Lambda_{Di} \qquad i = 1,2,\ldots,n
\end{aligned} \tag{9.95}
$$

Equations (9.94) and (9.95) are the desired network equations. Together with the machine equations they complete the description of the system. While the above procedure appears to be conceptually simple, it is exceedingly complex to implement. This is illustrated below. To simplify the notation, (9.95) is put in the form

$$
\begin{aligned}
I_{qi} &= \sigma_{qi}\Lambda_{qi} + \sigma_{Qi}\Lambda_{Qi} \\
I_{di} &= \sigma_{di}\Lambda_{di} + \sigma_{Fi}\Lambda_{Fi} + \sigma_{Di}\Lambda_{Di} \qquad i = 1,2,\ldots,n
\end{aligned} \tag{9.96}
$$

The complex vector $\bar{\mathbf{I}}$ thus becomes

$$
\bar{\mathbf{I}} = \begin{bmatrix} I_{q1} + jI_{d1} \\ I_{q2} + jI_{d2} \\ \cdots \\ I_{qn} + jI_{dn} \end{bmatrix}
= \begin{bmatrix} \sigma_{q1}\Lambda_{q1} + \sigma_{Q1}\Lambda_{Q1} \\ \cdots \\ \sigma_{qn}\Lambda_{qn} + \sigma_{Qn}\Lambda_{Qn} \end{bmatrix} + j\begin{bmatrix} \sigma_{d1}\Lambda_{d1} + \sigma_{F1}\Lambda_{F1} + \sigma_{D1}\Lambda_{D1} \\ \cdots \\ \sigma_{dn}\Lambda_{dn} + \sigma_{Fn}\Lambda_{Fn} + \sigma_{Dn}\Lambda_{Dn} \end{bmatrix} \tag{9.97}
$$

Now the matrix $\bar{\mathbf{Q}}$ in (9.94) is of the form

$$
\begin{aligned}
\bar{\mathbf{Q}} &= \begin{bmatrix} Z_{11}e^{j\theta_{11}} & \cdots & Z_{1n}e^{j(\theta_{1n}-\delta_{1n})} \\ \cdots & \cdots & \cdots \\ Z_{n1}e^{j(\theta_{n1}-\delta_{n1})} & \cdots & Z_{nn}e^{j\theta_{nn}} \end{bmatrix} \\
&= \begin{bmatrix} R_{11} & \cdots & Z_{1n}\cos(\theta_{1n} - \delta_{1n}) \\ \cdots & \cdots & \cdots \\ Z_{n1}\cos(\theta_{n1} - \delta_{n1}) & \cdots & R_{nn} \end{bmatrix} + j\begin{bmatrix} X_{11} & \cdots & Z_{1n}\sin(\theta_{1n} - \delta_{1n}) \\ \cdots & \cdots & \cdots \\ Z_{n1}\sin(\theta_{n1} - \delta_{n1}) & \cdots & X_{nn} \end{bmatrix} \\
&= \mathbf{Q}_R + j\mathbf{Q}_I
\end{aligned} \tag{9.98}
$$

Expanding (9.94),

$$
\begin{aligned}
\mathbf{V}_q + j\mathbf{V}_d &= (\mathbf{Q}_R + j\mathbf{Q}_I)(\mathbf{I}_q + j\mathbf{I}_d) \\
&= (\mathbf{Q}_R\mathbf{I}_q - \mathbf{Q}_I\mathbf{I}_d) + j(\mathbf{Q}_I\mathbf{I}_q + \mathbf{Q}_R\mathbf{I}_d)
\end{aligned} \tag{9.99}
$$

and substituting (9.97) into (9.99),

$$
\begin{aligned}
\mathbf{V}_q = &\begin{bmatrix} R_{11} & \cdots & Z_{1n}\cos(\theta_{1n} - \delta_{1n}) \\ \cdots & \cdots & \cdots \\ Z_{n1}\cos(\theta_{n1} - \delta_{n1}) & \cdots & R_{nn} \end{bmatrix}\begin{bmatrix} \sigma_{q1}\Lambda_{q1} + \sigma_{Q1}\Lambda_{Q1} \\ \cdots \\ \sigma_{qn}\Lambda_{qn} + \sigma_{Qn}\Lambda_{Qn} \end{bmatrix} \\
&- \begin{bmatrix} X_{11} & \cdots & Z_{1n}\sin(\theta_{1n} - \delta_{1n}) \\ \cdots & \cdots & \cdots \\ Z_{n1}\sin(\theta_{n1} - \delta_{n1}) & \cdots & X_{nn} \end{bmatrix}\begin{bmatrix} \sigma_{d1}\Lambda_{d1} + \sigma_{F1}\Lambda_{F1} + \sigma_{D1}\Lambda_{D1} \\ \cdots \\ \sigma_{dn}\Lambda_{dn} + \sigma_{Fn}\Lambda_{Fn} + \sigma_{Dn}\Lambda_{Dn} \end{bmatrix}
\end{aligned} \tag{9.100}
$$

$$\mathbf{V}_d = \begin{bmatrix} X_{11} & \cdots & Z_{1n}\sin(\theta_{1n} - \delta_{1n}) \\ \cdots & \cdots & \cdots \\ Z_{n1}\sin(\theta_{n1} - \delta_{n1}) & \cdots & X_{nn} \end{bmatrix} \begin{bmatrix} \sigma_{q1}\Lambda_{q1} + \sigma_{Q1}\Lambda_{Q1} \\ \cdots \\ \sigma_{qn}\Lambda_{qn} + \sigma_{Qn}\Lambda_{Qn} \end{bmatrix}$$

$$+ \begin{bmatrix} R_{11} & \cdots & Z_{1n}\cos(\theta_{1n} - \delta_{1n}) \\ \cdots & \cdots & \cdots \\ Z_{n1}\cos(\theta_{n1} - \delta_{n1}) & \cdots & R_{nn} \end{bmatrix} \begin{bmatrix} \sigma_{d1}\Lambda_{d1} + \sigma_{F1}\Lambda_{d1} + \sigma_{D1}\Lambda_{D1} \\ \cdots \\ \sigma_{dn}\Lambda_{dn} + \sigma_{Fn}\Lambda_{Fn} + \sigma_{Dn}\Lambda_{Dn} \end{bmatrix} \tag{9.101}$$

Equations (9.100) and (9.101) are needed to eliminate V_{qi} and V_{di} in the state-space equations when the flux linkage model, such as given in (4.138), is used.

The above illustrates the complexity of the use of the full-machine flux linkage model together with the network equations. Much of the labor is reduced when some of the simplified synchronous machine models of Section 4.15 are used. For example, if the constant voltage behind subtransient reactance is used, the voltages E''_{qi} and E''_{di} become state variables. The network is reduced to the generator internal nodes. This allows the direct use of a relation similar to (9.31) to complete the mathematical description of the system model. This has been illustrated in some of the examples used in this chapter.

The linearized equations for the flux linkage model are obtained from (9.97), which is linear, and (9.73). Following a procedure similar to that used in deriving (9.100) and (9.101), we expand (9.73) into real and imaginary terms as follows:

$$\begin{aligned} \bar{\mathbf{V}}_\Delta &= \mathbf{V}_{q\Delta} + \mathrm{j}\mathbf{V}_{d\Delta} \\ &= (\mathbf{Q}_{R0} + \mathrm{j}\mathbf{Q}_{I0})(\mathbf{I}_{q\Delta} + \mathrm{j}\mathbf{I}_{d\Delta}) - \mathrm{j}[\boldsymbol{\delta}_\Delta(\mathbf{Q}_{R0} + \mathrm{j}\mathbf{Q}_{I0}) - (\mathbf{Q}_{R0} + \mathrm{j}\mathbf{Q}_{I0})\boldsymbol{\delta}_\Delta](\mathbf{I}_{q0} + \mathrm{j}\mathbf{I}_{d0}) \\ &= [\mathbf{Q}_{R0}\mathbf{I}_{q\Delta} - \mathbf{Q}_{I0}\mathbf{I}_{d\Delta} + (\boldsymbol{\delta}_\Delta\mathbf{Q}_{I0} - \mathbf{Q}_{I0}\boldsymbol{\delta}_\Delta)\mathbf{I}_{q0} + (\boldsymbol{\delta}_\Delta\mathbf{Q}_{R0} - \mathbf{Q}_{R0}\boldsymbol{\delta}_\Delta)\mathbf{I}_{d0}] \\ &\quad + \mathrm{j}[\mathbf{Q}_{I0}\mathbf{I}_{q\Delta} + \mathbf{Q}_{R0}\mathbf{I}_{d\Delta} - (\boldsymbol{\delta}_\Delta\mathbf{Q}_{R0} - \mathbf{Q}_{R0}\boldsymbol{\delta}_\Delta)\mathbf{I}_{q0} + (\boldsymbol{\delta}_\Delta\mathbf{Q}_{I0} - \mathbf{Q}_{I0}\boldsymbol{\delta}_\Delta)\mathbf{I}_{d0}] \end{aligned} \tag{9.102}$$

The terms in $\mathbf{I}_{q\Delta}$, $\mathbf{I}_{d\Delta}$, $\mathbf{I}_{q0}$, and $\mathbf{I}_{d0}$ are substituted for by the linear combinations of the flux linkages given by (9.97).

9.11 Total System Equations

From (4.103) for *each* synchronous machine and hence for each node in Figure 9.2, the following relations apply

$$\begin{aligned} \dot{\mathbf{i}}_k &= -\mathbf{L}_k^{-1}(\mathbf{R}_k + \omega_k\mathbf{N}_k)\mathbf{i}_k - \mathbf{L}_k^{-1}\mathbf{v}_k \\ \dot{\omega}_k &= (1/3\tau_{jk})(-\lambda_{dk}i_{qk} + \lambda_{qk}i_{dk} - 3D_k\omega_k + 3T_{mk}) \\ \dot{\delta}_k &= \omega_k - 1 \qquad k = 1, 2, \ldots, n \end{aligned} \tag{9.103}$$

where $\mathbf{i}_k = [i_{dk}\, i_{Fk}\, i_{Dk}\, i_{qk}\, i_{Qk}]'$, $\mathbf{v}_k = [v_{dk}\ -v_{Fk}\ 0\ v_{qk}\ 0]'$ and the matrices $\mathbf{R}_k$, $\mathbf{L}_k$ and $\mathbf{N}_k$ are defined by (4.74). The whole system is of the form

$$\dot{\mathbf{x}} = \mathbf{f}(\mathbf{x}, \mathbf{v}, \mathbf{T}_m, t) \tag{9.104}$$

(see [5, 6, 7, 8, and 9]). Assuming that v_{Fk} and T_{mk}, $k = 1, 2, \ldots, n$, are known, (9.104) represents a set of $7n$ nonlinear differential equations. The vector $\mathbf{x}$ includes *all the stator and rotor* currents of the machines, and the vector $\mathbf{v}$ includes the stator voltages plus the rotor voltages (which are assumed to be known). The set (9.31) provides a constraint between all the stator voltages and currents (in phasor notation) as functions of the machine angles. These equations are also nonlinear.

By examining (9.103) and (9.31) we note the following: The differential equations describing the changes in the machine currents, rotor speeds, and angles are given in terms of the individual machine parameters only. The voltage-current relationships (9.31) are functions of the angles of all machines. This creates difficulties in the solution of these equations and is referred to in the literature as "the interface problem" [10]. The nature of the system equations forces the solution methods to be performed in two different phases (or cycles). One phase deals with the state of the network, in terms of node voltages and currents, assuming "known" internal machine quantities. The other phase is the solution of the differential equations of (9.103) only. The solution alternates between these two phases. This problem is mentioned here to focus attention on the system and solution complexities. This problem will be discussed further in Volume 2 of this work.

Finally, if the flux linkage model is used (for the case where saturation is neglected), the system equations will be (4.138), (9.100), and (9.101). Again the "interface problem" and the computational difficulties are encountered.

Example 9.10

Give the complete system equations for a two-machine system with the machines represented by the voltage-behind-subtransient-reactance model and the loads represented by constant impedances.

Solution

The network constraints for this system are given in complex notation in (9.31) or in real variables in (9.40), and the machine equations are given in Section 4.15.2. The machine equations are obtained from (4.234) and (4.270). They are

$$
\begin{aligned}
E''_{qi} &= K_{1i}E'_{qi} + K_{2i}\Lambda_{Di} \\
V_{di} &= -r_i I_{di} - I_{qi}X''_i + E''_{di} \\
V_{qi} &= -r_i I_{qi} + I_{di}X''_i + E''_{qi}
\end{aligned}
\tag{9.105}
$$

and

$$
\begin{aligned}
\tau''_{q0i}\dot{E}''_{di} &= -E''_{di} - (x_{qi} - x''_{qi})I_{qi} \\
\tau''_{d0i}\dot{\Lambda}_{Di} &= E'_{qi} - \Lambda_{Di} + (x'_{di} - x_{\ell i})I_{di} \\
\tau'_{d0i}\dot{E}'_{qi} &= E_{FDi} - (1 + K_{di})E'_{qi} + x_{di}I_{di} + K_{di}\Lambda_{Di} \\
\tau_{ji}\dot{\omega}_i &= T_{mi} - I_{qi}E''_{qi} - I_{di}E''_{di} \\
\dot{\delta}_i &= \omega_i - 1 \qquad i = 1,2
\end{aligned}
\tag{9.106}
$$

The network constraints are obtained from (9.40).

The system has ten differential equations, six auxiliary machine equations, and four algebraic equations for the network (or two complex equations). As per the discussion in Section 9.6, some differential equations can be eliminated by using $\delta_1 - \delta_2$ and $\omega_1 - \omega_2$ as state variables.

Some of the computational labor can be reduced if the subtransient reactances of the generators are included in the network $\overline{\mathbf{Y}}$ matrix (or $\overline{\mathbf{Z}}$ matrix). The network equations would then give relations between the currents I_{qi} and I_{di}, $i = 1,2$, and the voltages E''_{qi} and E''_{di}, $i = 1,2$. The auxiliary equations for V_{di} and V_{qi} can be omitted. Also in (9.40), E''_{qi} and E''_{di} should replace V_{qi} and V_{di}.

9.12 Multimachine System Study

The nine-bus system discussed in Section 2.10 is to be examined for dynamic stability at the initial operating point given in Section 2.10. Linearized machine equations are to be used. The loads are to be simulated by constant impedances based on the initial operating conditions.

The system under study comprises three generators and three loads. A one-line impedance diagram is given in Figure 2.18. The initial operating system condition, indicating the power flows and bus voltages, is given in Figure 2.19. Data for the three generators are given in Table 2.1 (some of which are repeated below for convenience).

The synchronous machine models to be used are as follows: classical model for generator 1, and the two-axis model for generators 2 and 3.

9.12.1 Preliminary calculations

Let the generator terminal voltage be $V\underline{/\beta}$, and the q axis be located at angle δ. All angles are measured from reference. The generator current $\bar{I}$ lags the terminal voltage by the power factor angle ϕ. The following relations, derived in Section 5.5, are used ($r \simeq 0$) to obtain the data in Table 9.1:

$$I_r + jI_x = I\underline{/\beta - \phi} = (P - jQ)/V \qquad \tan(\delta - \beta) = x_q I_r/(V - x_q I_x)$$

$$E'_q = V_q - I_d x'_d \qquad E'_d \simeq V_d + I_q x'_d$$

$$I\underline{/-(\delta - \beta + \phi)} = I_q + jI_d \qquad V\underline{/-\delta} = V_q + jV_d$$

Table 9.1. Three-Machine System Data

Quantity	Unit	Generator 1 (classical)	Generator 2 (two-axis)	Generator 3 (two-axis)
H(MW·s/100 MVA)	s	23.6400	6.4000	3.0100
$\tau_j = 2H\omega_B$	pu	17824.1400	4825.4863	2269.4865
$x_d - x'_d$	pu	0.0852	0.7760	1.1312
$x_q - x'_d$	pu	0.0361	0.7447	1.0765
τ'_{q0}	s	0	0.5350	0.6000
τ'_{q0}	pu	0	201.6900	226.1900
τ'_{d0}	s	8.9600	6.0000	5.8900
τ'_{d0}	pu	3377.8404	2261.9467	2220.4777
E'_{q0}	pu	1.0558	0.7882	0.7679
E'_{d0}	pu	−0.0419	−0.6940	−0.6668
I_{q0}	pu	0.6780	0.9320	0.6194
I_{d0}	pu	−0.2872	−1.2902	−0.5615
V_{q0}	pu	1.0392	0.6336	0.6661
V_{d0}	pu	−0.0412	−0.8057	−0.7791
δ_0	elec deg	2.2717°	61.0975°	54.1431°
E'	pu	1.0566	...	...

9.12.2 Linearized network equations

The network is assumed to include the transient reactances of the generators. The network is reduced to the generator internal nodes. At these nodes the voltages are $\bar{E}'_1$, $\bar{E}'_2$, and $\bar{E}'_3$.

From (9.63) with $\bar{V}$ replaced with $\bar{E}'$ and for a three-machine system (using $\delta_{12} = -\delta_{21}, \delta_{13} = -\delta_{31}$),

$$\begin{bmatrix} \bar{I}_{1\Delta} \\ \bar{I}_{2\Delta} \\ \bar{I}_{3\Delta} \end{bmatrix} = \begin{bmatrix} Y_{11}e^{j\theta_{11}} & Y_{12}e^{j(\theta_{12}-\delta_{120})} & Y_{13}e^{j(\theta_{13}-\delta_{130})} & -j\bar{E}'_{20}Y_{12}e^{j(\theta_{12}-\delta_{120})} & -j\bar{E}'_{30}Y_{13}e^{j(\theta_{13}-\delta_{130})} & 0 \\ Y_{21}e^{j(\theta_{21}-\delta_{210})} & Y_{22}e^{j\theta_{22}} & Y_{23}e^{j(\theta_{23}-\delta_{230})} & j\bar{E}'_{10}Y_{21}e^{j(\theta_{21}-\delta_{210})} & 0 & -j\bar{E}'_{30}Y_{23}e^{j(\theta_{23}-\delta_{230})} \\ Y_{31}e^{j(\theta_{31}-\delta_{310})} & Y_{32}e^{j(\theta_{32}-\delta_{320})} & Y_{33}e^{j\theta_{33}} & 0 & j\bar{E}'_{10}Y_{31}e^{j(\theta_{31}-\delta_{310})} & j\bar{E}'_{20}Y_{32}e^{j(\theta_{32}-\delta_{320})} \end{bmatrix} \begin{bmatrix} \bar{E}'_{1\Delta} \\ \bar{E}'_{2\Delta} \\ \bar{E}'_{3\Delta} \\ \delta_{12\Delta} \\ \delta_{13\Delta} \\ \delta_{23\Delta} \end{bmatrix} \tag{9.107}$$

With generator 1 represented classically, $\bar{E}'_{1\Delta} = 0$ and $\bar{E}'_{10} = \bar{E}'_1$; and with node 1 as the arbitrary reference node $\bar{E}'_1 = E_1 + j0 = E_1$ (a constant). Substituting in (9.107) and using $\delta_{23} = \delta_{13} - \delta_{12}$,

$$\begin{bmatrix} \bar{I}_{1\Delta} \\ \bar{I}_{2\Delta} \\ \bar{I}_{3\Delta} \end{bmatrix} = \begin{bmatrix} Y_{12}e^{j(\theta_{12}-\delta_{120})} & Y_{13}e^{j(\theta_{13}-\delta_{130})} & -j\bar{E}'_{20}Y_{12}e^{j(\theta_{12}-\delta_{120})} & -j\bar{E}'_{30}Y_{13}e^{j(\theta_{13}-\delta_{130})} \\ Y_{22}e^{j\theta_{22}} & Y_{23}e^{j(\theta_{23}-\delta_{230})} & \begin{array}{l} j[E_1Y_{21}e^{j(\theta_{12}+\delta_{120})} \\ \quad + \bar{E}'_{30}Y_{23}e^{j(\theta_{23}-\delta_{230})}] \end{array} & -j\bar{E}'_{30}Y_{23}e^{j(\theta_{23}-\delta_{230})} \\ Y_{32}e^{j(\theta_{23}+\delta_{230})} & Y_{33}e^{j\theta_{33}} & -j\bar{E}'_{20}Y_{23}e^{j(\theta_{23}+\delta_{230})} & \begin{array}{l} j[E_1Y_{13}e^{j(\theta_{13}+\delta_{130})} \\ \quad + \bar{E}'_{20}Y_{23}e^{j(\theta_{23}+\delta_{230})}] \end{array} \end{bmatrix} \begin{bmatrix} \bar{E}'_{2\Delta} \\ \bar{E}'_{3\Delta} \\ \delta_{12\Delta} \\ \delta_{13\Delta} \end{bmatrix} \tag{9.108}$$

Separating real and imaginary parts and dropping the subscript Δ for convenience,

$$\begin{array}{c} \\ \begin{bmatrix} I_{q1} \\ I_{d1} \\ I_{q2} \\ I_{d2} \\ I_{q3} \\ I_{d3} \end{bmatrix} \end{array} = \begin{array}{c} \begin{array}{cccccc} E'_{q2} & E'_{d2} & E'_{q3} & E'_{d3} & \delta_{12} & \delta_{13} \end{array} \\ \begin{bmatrix} Y_{12}\cos(\theta_{12}-\delta_{120}) & -Y_{12}\sin(\theta_{12}-\delta_{120}) & Y_{13}\cos(\theta_{13}-\delta_{130}) & -Y_{13}\sin(\theta_{13}-\delta_{130}) & \begin{array}{l} Y_{12}[E'_{d20}\cos(\theta_{12}-\delta_{120}) \\ + E'_{q20}\sin(\theta_{12}-\delta_{120})] \end{array} & \begin{array}{l} Y_{13}[E'_{d30}\cos(\theta_{13}-\delta_{130}) \\ + E'_{q30}\sin(\theta_{13}-\delta_{130})] \end{array} \\ Y_{12}\sin(\theta_{12}-\delta_{120}) & Y_{12}\cos(\theta_{12}-\delta_{120}) & Y_{13}\sin(\theta_{13}-\delta_{130}) & Y_{13}\cos(\theta_{13}-\delta_{130}) & \begin{array}{l} Y_{12}[E'_{d20}\sin(\theta_{12}-\delta_{120}) \\ - E'_{q20}\cos(\theta_{12}-\delta_{120})] \end{array} & \begin{array}{l} Y_{13}[E'_{d30}\sin(\theta_{13}-\delta_{130}) \\ - E'_{q30}\cos(\theta_{13}-\delta_{130})] \end{array} \\ G_{22} & -B_{22} & Y_{23}\cos(\theta_{23}-\delta_{230}) & -Y_{23}\sin(\theta_{23}-\delta_{230}) & \begin{array}{l} -E_1Y_{12}\sin(\theta_{12}+\delta_{120}) \\ - E'_{d30}Y_{23}\cos(\theta_{23}-\delta_{230}) \\ - E'_{q30}Y_{23}\sin(\theta_{23}-\delta_{230}) \end{array} & \begin{array}{l} Y_{23}[E'_{d30}\cos(\theta_{23}-\delta_{230}) \\ + E'_{q30}\sin(\theta_{23}-\delta_{230})] \end{array} \\ B_{22} & G_{22} & Y_{23}\sin(\theta_{23}-\delta_{230}) & Y_{23}\cos(\theta_{23}-\delta_{230}) & \begin{array}{l} E_1Y_{12}\cos(\theta_{12}+\delta_{120}) \\ - E'_{d30}Y_{23}\sin(\theta_{23}-\delta_{230}) \\ + E'_{q30}Y_{23}\cos(\theta_{23}-\delta_{230}) \end{array} & \begin{array}{l} Y_{23}[E'_{d30}\sin(\theta_{23}-\delta_{230}) \\ - E'_{q30}\cos(\theta_{23}-\delta_{230})] \end{array} \\ Y_{23}\cos(\theta_{23}+\delta_{230}) & -Y_{23}\sin(\theta_{23}+\delta_{230}) & G_{33} & -B_{33} & \begin{array}{l} Y_{23}[E'_{d20}\cos(\theta_{23}+\delta_{230}) \\ + E'_{q20}\sin(\theta_{23}+\delta_{230})] \end{array} & \begin{array}{l} -E_1Y_{13}\sin(\theta_{13}+\delta_{130}) \\ - Y_{23}E'_{d20}\cos(\theta_{23}+\delta_{230}) \\ - Y_{23}E'_{q20}\sin(\theta_{23}+\delta_{230}) \end{array} \\ Y_{23}\sin(\theta_{23}+\delta_{230}) & Y_{23}\cos(\theta_{23}+\delta_{230}) & B_{33} & G_{33} & \begin{array}{l} Y_{23}[E'_{d20}\sin(\theta_{23}+\delta_{230}) \\ - E'_{q20}\cos(\theta_{23}+\delta_{230})] \end{array} & \begin{array}{l} E_1Y_{13}\cos(\theta_{13}+\delta_{130}) \\ - E'_{d20}Y_{23}\sin(\theta_{23}+\delta_{230}) \\ + E'_{q20}Y_{23}\cos(\theta_{23}+\delta_{230}) \end{array} \end{bmatrix} \end{array} \begin{array}{c} \\ \begin{bmatrix} E'_{q2} \\ E'_{d2} \\ E'_{q3} \\ E'_{d3} \\ \delta_{12} \\ \delta_{13} \end{bmatrix} \end{array} \tag{9.109}$$

Equation (9.109) is the desired linearized network equation. It relates the incremental currents to the incremental state variables E'_{q2}, E'_{d2}, E'_{q3}, E'_{d3}, δ_{12}, and δ_{13}.

9.12.3 Generator equations

From Example 9.7 we obtain the following generator equations (again the subscript Δ is omitted):

Generator 1 (classical)

$$\begin{aligned} \tau_{j1}\dot{\omega}_1 &= T_{m1} - E_1 I_{q1} - D_1\omega_1 \\ \dot{\delta}_1 &= \omega_1 \end{aligned} \tag{9.110}$$

Generators 2 and 3 (two-axis model)

$$\begin{aligned}
\tau'_{q0i}\dot{E}'_{di} &= -E'_{di} - (x_{qi} - x'_i)I_{qi} \\
\tau'_{d0i}\dot{E}'_{qi} &= E_{FDi} - E'_{qi} + (x_{di} - x'_i)I_{di} \\
\tau_{ji}\dot{\omega}_i &= T_{mi} - D_i\omega_i - I_{di0}E'_{di} - I_{qi0}E'_{qi} - E'_{di0}I_{di} - E'_{qi0}I_{qi} \\
\dot{\delta}_i &= \omega_i \qquad i = 2, 3
\end{aligned} \tag{9.111}$$

Again we recall that, to obtain an independent set, the last equations in (9.110) and (9.111) are combined to give

$$\dot{\delta}_{1i} = \omega_1 - \omega_i \qquad i = 2, 3 \tag{9.112}$$

By using (9.109), I_{q1}, I_{d1}, I_{q2}, I_{d2}, I_{q3}, and I_{d3} are eliminated from (9.110) and (9.111). The resulting system comprises nine linear first-order differential equations. The state variables are $E'_{q2}, E'_{d2}, E'_{q3}, E'_{d3}; \omega_1, \omega_2, \omega_3, \delta_{12}$, and δ_{13}.

9.12.4 Development of the A matrix

The $\bar{\mathbf{Y}}$ matrix of the network, reduced to the internal generator nodes and including the generator transient reactance, is given in Table 2.6 as the prefault $\bar{\mathbf{Y}}$ matrix. It is repeated here in Table 9.2. Data for the terms in (9.109) are calculated and given in Table 9.3.

Table 9.2. Reduced $\bar{\mathbf{Y}}$ Matrix for a Three-Machine System

Node	1	2	3
1	0.8455 − j2.9883	0.2871 + j1.5129 = 1.5399 $\angle 79.25°$	0.2096 + j1.2256 = 1.2434 $\angle 80.30°$
2	0.2871 + j1.5129 = 1.5399 $\angle 79.25°$	0.4200 − j2.7238	0.2133 + j1.0879 = 1.1086 $\angle 78.91°$
3	0.2096 + j1.2256 = 1.2434 $\angle 80.30°$	0.2133 + j1.0879 = 1.1086 $\angle 78.91°$	0.2770 − j2.3681

The coefficients of (9.109) and (9.111) are then calculated. The main system equations are given below. The incremental currents I_{qi} and I_{di} are calculated from (9.109).

$$\begin{bmatrix} I_{q1} \\ I_{d1} \\ I_{q2} \\ I_{d2} \\ I_{q3} \\ I_{d3} \end{bmatrix} = \begin{bmatrix} -1.1458 & -1.0288 & -0.8347 & -0.9216 & 1.6062 & 1.2642 \\ 1.0288 & -1.1458 & 0.9216 & -0.8347 & 0.1891 & 0.0265 \\ 0.4200 & 2.7239 & 0.3434 & -1.0541 & -1.1484 & 0.5805 \\ -2.7239 & 0.4200 & 1.0541 & 0.3434 & 2.4914 & -0.9666 \\ 0.0800 & -1.1058 & 0.2770 & 2.3681 & 0.8160 & -1.4414 \\ 1.1058 & 0.0800 & -2.3681 & 0.2770 & -0.8305 & 1.9859 \end{bmatrix} \begin{bmatrix} E'_{q2} \\ E'_{d2} \\ E'_{q3} \\ E'_{d3} \\ \delta_{12} \\ \delta_{13} \end{bmatrix} \tag{9.113}$$

The generator differential equations are:

Generator 1 (classical)

$$\begin{aligned}
\dot{\omega}_1 &= 5.6104 \times 10^{-5}T_{m1} - 5.6104 \times 10^{-5}D_1\omega_1 - 5.9279 - 10^{-5}I_{q1} \\
\dot{\delta}_1 &= \omega_1
\end{aligned}$$

Table 9.3. Preliminary Calculations for Three-Machine System

Nodes	1–2	1–3	2–3
Y_{ij}	1.5399	1.2434	1.1086
θ_{ij}	79.2544	80.2952	78.9084
δ_{ij0}	−58.8259	−51.8714	6.9545
$\theta_{ij} - \delta_{ij0}$	138.0802	132.1666	71.9540
$Y_{ij}\cos(\theta_{ij} - \delta_{ij0})$	−1.1458	−0.8347	0.3434
$Y_{ij}\sin(\theta_{ij} - \delta_{ij0})$	1.0288	0.9216	1.0541
$\theta_{ij} + \delta_{ij0}$	20.4285	28.4238	85.8629
$Y_{ij}\cos(\theta_{ij} + \delta_{ij0})$	1.4431	1.0935	0.0800
$Y_{ij}\sin(\theta_{ij} + \delta_{ij0})$	0.5375	0.5919	1.1058

Generator 2 (two-axis)

$$
\begin{aligned}
\dot{E}'_{d2} &= -4.9581 \times 10^{-3} E'_{d2} - 3.6923 \times 10^{-3} I_{q2} \\
\dot{E}'_{q2} &= 4.4210 \times 10^{-4} E_{FD2} - 4.4210 \times 10^{-4} E'_{q2} + 3.4307 \times 10^{-4} I_{d2} \\
\dot{\omega}_2 &= 2.0723 \times 10^{-4} T_{m2} - 2.0723 \times 10^{-4} D_2\omega_2 - 1.9314 \times 10^{-4} E'_{q2} \\
&\quad + 2.6736 \times 10^{-4} E'_{d2} + 1.4383 \times 10^{-4} I_{d2} - 1.6334 \times 10^{-4} I_{q2} \\
\dot{\delta}_2 &= \omega_2
\end{aligned}
$$

Generator 3 (two-axis)

$$
\begin{aligned}
\dot{E}'_{d3} &= -4.4210 \times 10^{-3} E'_{d3} - 4.7592 \times 10^{-3} I_{q3} \\
\dot{E}'_{q3} &= 4.5035 \times 10^{-4} E_{FD3} - 4.5035 \times 10^{-4} E'_{q3} + 5.0944 \times 10^{-4} I_{d3} \\
\dot{\omega}_3 &= 4.4063 \times 10^{-4} T_{m3} - 4.4063 \times 10^{-4} D_3\omega_3 + 2.4741 \times 10^{-4} E'_{d3} \\
&\quad - 2.7292 \times 10^{-4} E'_{q3} + 2.9380 \times 10^{-4} I_{d3} - 3.3836 \times 10^{-4} I_{q3} \\
\dot{\delta}_3 &= \omega_3
\end{aligned}
\tag{9.114}
$$

By using (9.113), the currents are then eliminated in (9.114). Combining terms and using the relation $\delta_{ij} = \delta_i - \delta_j$, we obtain the linearized differential equations for the three-machine system. The results are shown in (9.115), which is of the form

$$
\begin{bmatrix} \dot{\omega}_1 \\ \dot{E}'_{q2} \\ \dot{E}'_{d2} \\ \dot{\omega}_2 \\ \dot{E}'_{q3} \\ \dot{E}'_{d3} \\ \dot{\omega}_3 \\ \dot{\delta}_{12} \\ \dot{\delta}_{13} \end{bmatrix}
= 10^{-4}
\begin{array}{c}
\begin{matrix} \omega_1 & E'_{q2} & E'_{d2} & \omega_2 & E'_{q3} & E'_{d3} & \omega_3 & \delta_{12} & \delta_{13} \end{matrix} \\
\begin{bmatrix}
-0.5610D_1 & 0.6793 & 0.6099 & 0 & 0.4948 & 0.5463 & 0 & -0.9520 & -0.7494 \\
0 & -13.7658 & 1.4409 & 0 & 3.6163 & 1.1781 & 0 & 8.5472 & -3.3161 \\
0 & -15.5076 & -150.1554 & 0 & -12.6793 & 38.9205 & 0 & 42.4023 & -21.4333 \\
0 & -6.5352 & -1.1714 & -2.0723D_2 & 0.9552 & 2.2156 & 0 & 5.4592 & -2.3385 \\
0 & 5.6334 & 0.4076 & 0 & -16.5675 & 1.4111 & 0 & -4.2309 & 10.1170 \\
0 & -3.8073 & 52.6270 & 0 & -13.1829 & -156.9117 & 0 & -38.8349 & 68.5987 \\
0 & 2.9781 & 3.9766 & 0 & -10.6238 & -4.7247 & -4.4063D_3 & -5.2010 & 10.7116 \\
10000 & 0 & 0 & -10000 & 0 & 0 & 0 & 0 & 0 \\
10000 & 0 & 0 & 0 & 0 & 0 & -10000 & 0 & 0
\end{bmatrix}
\end{array}
\begin{bmatrix} \omega_1 \\ E'_{q2} \\ E'_{d2} \\ \omega_2 \\ E'_{q3} \\ E'_{d3} \\ \omega_3 \\ \delta_{12} \\ \delta_{13} \end{bmatrix}
$$

$$
+ 10^{-4} \begin{bmatrix} 0.5610T_{m1} \\ 4.4210E_{FD2} \\ 0 \\ 2.0723T_{m2} \\ 4.5035E_{FD3} \\ 0 \\ 4.4063T_{m3} \\ 0 \\ 0 \end{bmatrix}
\tag{9.115}
$$

$$\dot{\mathbf{x}} = \mathbf{A}\mathbf{x} + \mathbf{B}\mathbf{u}$$

where $\mathbf{x}^t = [\omega_1\ E'_{q2}\ E'_{d2}\ \omega_2\ E'_{q3} E'_{d3}\ \omega_3\ \delta_{12}\ \delta_{13}]$

$\mathbf{u}^t = [T_{m1}\ E_{FD2}\ T_{m2}\ E_{FD3}\ T_{m3}]$

The eigenvalues of the $\mathbf{A}$ matrix are obtained for the case of $D_1 = D_2 = D_3 =$ 1.0 pu, using a library computer program. They are

$\lambda_1 = -0.002664 + j0.034648$ $\quad\lambda_6 = -0.010373$
$\lambda_2 = -0.002664 - j0.034648$ $\quad\lambda_7 = -0.000455$
$\lambda_3 = -0.000622 + j0.022984$ $\quad\lambda_8 = -0.000199 + j0.000129$
$\lambda_4 = -0.000622 - j0.022984$ $\quad\lambda_9 = -0.000199 - j0.000129$
$\lambda_5 = -0.016644$

All the eigenvalues have negative real parts, and the system is stable for the operating point under study. The dominant frequencies are about 2.1 Hz and 1.4 Hz respectively. These frequencies are the rotor electromechanical oscillations and should be very similar to the frequencies obtained in Example 3.4. Thus if we plot P_{12} from the data of Figure 3.8, we find that the dominant frequency is about 1.4 Hz, which checks with the data obtained here.

A similar run was obtained for the same data except for $D_1 = D_2 = D_3 = 0$. The eigenvalues are

$\lambda_1 = -0.000458$ $\quad\lambda_6 = -0.000529 + j0.022983$
$\lambda_2 = -0.000281$ $\quad\lambda_7 = -0.000529 - j0.022983$
$\lambda_3 = -0.010366$ $\quad\lambda_8 = -0.002459 + j0.034636$
$\lambda_4 = -0.016659$ $\quad\lambda_9 = -0.002459 - j0.034636$
$\lambda_5 = 0$

Since this is a special case of uniform damping ($D/\tau_j = 0$), the system order is reduced by one. The frequencies corresponding to the electromechanical oscillations are almost unchanged, while the long period frequency has disappeared.

Problems

9.1 If the $\overline{\mathbf{Y}}$ matrix of the network, reduced to the generator nodes, is such that $\theta_{ij} = 90°$, $i \neq j$, derive the general form of the matrix $\overline{\mathbf{M}}$.

9.2 For the conditions of Problem 9.1, obtain the real matrices for $\mathbf{I}_q$ and $\mathbf{I}_d$ in terms of $\mathbf{V}_q$ and $\mathbf{V}_d$. Compare with (9.40) for a two-machine system with $G_{12} = G_{21} = 0$.

9.3 Repeat Example 9.3, using the synchronous machine model called the one-axis model (see Section 4.15.4).

9.4 Repeat Example 9.5, neglecting the amortisseur effects for the synchronous machine represented in detail (Section 4.15.1).

9.5 Linearize the voltage-behind-subtransient-reactance model of the synchronous machine.

9.6 Repeat Example 9.8, using the results of Problem 9.5.

9.7 Develop (9.89) for a three-machine system with zero transfer conductances.

9.8 For the nine-bus system of Section 2.10 the dynamic stability of the postfault system (with line 5–7 open) is to be examined. The generator powers are the same as those of prefault conditions.

a. From a load-flow study obtain the system flows, voltages, and angles.
b. Calculate the initial position of the q axes; $I_{q0}, I_{d0}, V_{q0}, V_{d0}, E'_{q0}$, and E'_{d0} for each machine; and the angles δ_{120} and δ_{130}.
c. Obtain the $\mathbf{A}$ matrix and examine the system eigenvalues for stability.

References

1. Anderson, Paul M. *Analysis of Faulted Power Systems*. Iowa State University Press, Ames, 1973.
2. Pai, M. A., and Murthy, P. G. New Liapunov functions for power systems based on minimal realizations. *Int. J. Control* 19:401–15, 1974.
3. Willems, J. L. A partial stability approach to the problem of transient power system stability. *Int. J. Control* 19:1–14, 1974.
4. Pal, M. K. State-space representation of multimachine power systems. IEEE Paper C 74 396-8, presented at the Summer Power Meeting, Anaheim, Calif, 1974.
5. Prabhashankar, K., and Janischewskyj, W. Digital simulation of multimachine power systems for stability studies. *IEEE Trans.* PAS-87:73–80, 1968.
6. Undrill, J. M. Dynamic stability calculations for an arbitrary number of interconnected synchronous machines. *IEEE Trans.* PAS-87:835–44, 1968.
7. Janischewskyj, W., Prabhashankar, K., and Dandeno, P. Simulation of the nonlinear dynamic response of interconnected synchronous machines (in two parts). *IEEE Trans.* PAS-91:2064–77, 1972.
8. Van Ness, J. E., and Goddard, W. F. Formation of the coefficient matrix of a large dynamic system. *IEEE Trans.* PAS-87:80–84, 1968.
9. Laughton, M. A. Matrix analysis of dynamic stability in synchronous multi-machine systems. *Proc. IEE* (British) 113:325–36, 1966.
10. Tinney, W. F. Evaluation of concepts for studying transient stability. IEEE Power Engineering Society Tutorial, Spec. Publ. 70 M62-PWR, pp. 53–60, 1970.

appendix A

Trigonometric Identities for Three-Phase Systems

In solving problems involving three-phase systems, the engineer encounters a large number of trigonometric functions involving the angles $\pm 120°$. Some of these are listed here to save the time and effort of computing these same quantities over and over. Although the symbol (°) has been omitted from angles $\pm 120°$, it is always implied.

$$\sin(\theta \pm 120) = -1/2 \sin\theta \pm \sqrt{3}/2 \cos\theta \tag{A.1}$$

$$\cos(\theta \pm 120) = -1/2 \cos\theta \mp \sqrt{3}/2 \sin\theta \tag{A.2}$$

$$\begin{aligned}\sin^2(\theta \pm 120) &= 1/4 \sin^2\theta + 3/4 \cos^2\theta \mp \sqrt{3}/2 \sin\theta \cos\theta \\ &= 1/2 + 1/4 \cos 2\theta \mp \sqrt{3}/4 \sin 2\theta\end{aligned} \tag{A.3}$$

$$\begin{aligned}\cos^2(\theta \pm 120) &= 1/4 \cos^2\theta + 3/4 \sin^2\theta \pm \sqrt{3}/2 \sin\theta \cos\theta \\ &= 1/2 - 1/4 \cos 2\theta \pm \sqrt{3}/4 \sin 2\theta\end{aligned} \tag{A.4}$$

$$\begin{aligned}\sin\theta \sin(\theta \pm 120) &= -1/2 \sin^2\theta \pm \sqrt{3}/2 \sin\theta \cos\theta \\ &= -1/4 + 1/4 \cos 2\theta \pm \sqrt{3}/4 \sin 2\theta\end{aligned} \tag{A.5}$$

$$\begin{aligned}\cos\theta \cos(\theta \pm 120) &= -1/2 \cos^2\theta \mp \sqrt{3}/2 \sin\theta \cos\theta \\ &= -1/4 - 1/4 \cos 2\theta \mp \sqrt{3}/4 \sin 2\theta\end{aligned} \tag{A.6}$$

$$\begin{aligned}\sin\theta \cos(\theta \pm 120) &= -1/2 \sin\theta \cos\theta \mp \sqrt{3}/2 \sin^2\theta \\ &= -1/4 \sin 2\theta \pm \sqrt{3}/4 \cos 2\theta \mp \sqrt{3}/4\end{aligned} \tag{A.7}$$

$$\begin{aligned}\cos\theta \sin(\theta \pm 120) &= -1/2 \sin\theta \cos\theta \pm \sqrt{3}/2 \cos^2\theta \\ &= -1/4 \sin 2\theta \pm \sqrt{3}/4 \cos 2\theta \pm \sqrt{3}/4\end{aligned} \tag{A.8}$$

$$\begin{aligned}\sin(\theta + 120)\cos(\theta + 120) &= -1/2 \sin\theta \cos\theta - \sqrt{3}/4 \cos^2\theta + \sqrt{3}/4 \sin^2\theta \\ &= -1/4 \sin 2\theta - \sqrt{3}/4 \cos 2\theta\end{aligned} \tag{A.9}$$

$$\sin(\theta + 120)\cos(\theta - 120) = \sin\theta \cos\theta - \sqrt{3}/4 = 1/2 \sin 2\theta - \sqrt{3}/4 \tag{A.10}$$

$$\sin(\theta - 120)\cos(\theta + 120) = \sin\theta \cos\theta + \sqrt{3}/4 = 1/2 \sin 2\theta + \sqrt{3}/4 \tag{A.11}$$

$$\begin{aligned}\sin(\theta - 120)\cos(\theta - 120) &= -1/2 \sin\theta \cos\theta + \sqrt{3}/4 \cos^2\theta - \sqrt{3}/4 \sin^2\theta \\ &= -1/4 \sin 2\theta + \sqrt{3}/4 \cos 2\theta\end{aligned} \tag{A.12}$$

$$\sin(\theta + 120)\sin(\theta - 120) = 1/4 \sin^2\theta - 3/4 \cos^2\theta = -1/4 - 1/2 \cos 2\theta \tag{A.13}$$

$$\cos(\theta + 120)\cos(\theta - 120) = 1/4\cos^2\theta - 3/4\sin^2\theta = -1/4 + 1/2\cos 2\theta \quad (A.14)$$

$$\sin(2\theta \pm 120) = -1/2\sin 2\theta \pm \sqrt{3}/2\cos 2\theta \quad (A.15)$$

$$\cos(2\theta \pm 120) = -1/2\cos 2\theta \mp \sqrt{3}/2\sin 2\theta \quad (A.16)$$

$$\sin\theta + \sin(\theta - 120) + \sin(\theta + 120) = 0 \quad (A.17)$$

$$\cos\theta + \cos(\theta - 120) + \cos(\theta + 120) = 0 \quad (A.18)$$

$$\sin^2\theta + \sin^2(\theta - 120) + \sin^2(\theta + 120) = 3/2 \quad (A.19)$$

$$\cos^2\theta + \cos^2(\theta - 120) + \cos^2(\theta + 120) = 3/2 \quad (A.20)$$

$$\sin\theta\cos\theta + \sin(\theta - 120)\cos(\theta - 120) + \sin(\theta + 120)\cos(\theta + 120) = 0 \quad (A.21)$$

In addition to the above, the following commonly used identities are often required:

$$\sin^2\theta + \cos^2\theta = 1$$

$$\sin\theta\cos\theta = 1/2\sin 2\theta$$

$$\cos^2\theta - \sin^2\theta = \cos 2\theta$$

$$\cos^2\theta = (1 + \cos 2\theta)/2$$

$$\sin^2\theta = (1 - \cos 2\theta)/2$$

appendix **B**

Some Computer Methods for Solving Differential Equations

The solution of dynamic systems of any kind involves the integration of differential equations. Some physical systems, such as power systems, are described by a large number of differential equations. Hand computation of such large systems of equations is exceedingly cumbersome, and computer solutions are usually called for.

Computer solutions fall into two categories, analog and digital, with hybrid systems as a combination of the two. The purpose of this appendix is to reinforce the material of the text by providing some of the fundamentals of computer solutions. This material is divided into two parts: analog computer fundamentals and digital computer solutions of ordinary differential equations. A short bibliography of references on analog and digital solutions is included at the end of this appendix.

B.1 Analog Computer Fundamentals

The analog computer is a device designed to solve differential equations. This is done by means of electronic components that perform the functions usually required in such problems. These include summation, integration, multiplication, division, multiplication by a constant, and other special functions.

The purpose of this appendix is to acquaint the beginner with the basic fundamentals of analog computation. As such it may be a valuable aid to the understanding of some of the text material and may be helpful in attempting an actual analog simulation. It should be used as a supplement to the many excellent books on the subject. In particular, the engirieer who attempts an actual simulation will surely need the instruction manual for the computer actually used.

B.1.1 Analog computer components

Here we consider the most important analog computer components. Later, we will connect several components to solve a simple differential equation. We discuss these components using the common symbolic language of analog computation and omit entirely the electronic means of accomplishing these ends.

The summer. The first important component is the summer or summing amplifier shown in Figure B.1, where both the analog symbol and the mathematical operation are indicated. Note that the amplifier inverts (changes the sign) of the input sum and multiplies each input voltage by a gain constant k_i selected by the user. On most computers k_i may have values of 1 or 10, but some models have other gains available. Usually V_4 is limited to 100 V (10 V on some computers).

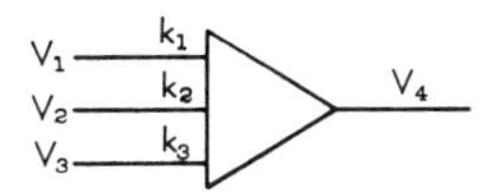

Fig. B.1. The summer; $V_4 = -(k_1 V_1 + k_2 V_2 + k_3 V_3)$.

The integrator. It is necessary to be able to perform integration if differential equations are to be solved. Fortunately, integration may be done rapidly and very reliably by electronic means, as shown in Figure B.2, where V_0 is the initial value (at $t = 0$) of the output variable V_4. Gain constants k_i are chosen by the operator and are restricted to values available on the computer, usually 1 and 10. The output voltage is limited, usually to 100 V.

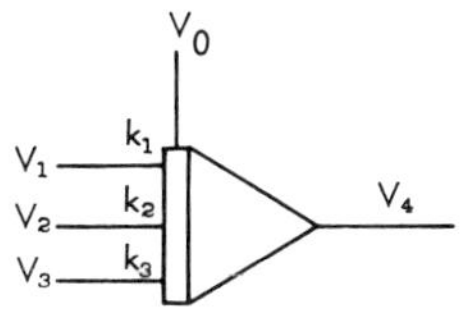

Fig. B.2. The integrator; $V_4 = -[V_0 + \int_0^t (k_1 V_1 + k_2 V_2 + k_3 V_3)dt]$.

The potentiometer. The potentiometer is used to scale down a voltage by an exact amount as shown in Figure B.3, where the signal is implied as going from left to right. Potentiometers are usually 10-turn pots and can be reliably set to three decimals with excellent accuracy.

Fig. B.3. The potentiometer; $V_2 = kV_1, 0 \leq k \leq 1$.

The function generator. The function generator is a device used to simulate a nonlinear function by straight-line segments. Function generators are represented by the "pointed box" shown in Figure B.4 where the function f is specified by the user, and this function is set according to the instructions for the particular computer used. This feature makes it possible to simulate with reasonable accuracy certain nonlinear functions such as generator saturation. The function f must be single valued.

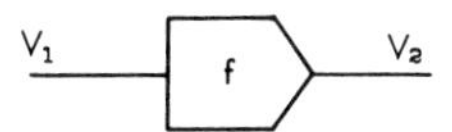

Fig. B.4. The function generator; $V_2 = f(V_1)$.

The high-gain amplifier. On some analog computers it is necessary to use high-gain amplifiers to simulate certain operations such as multiplication. The symbol usually used for this is shown in Figure B.5, although it should be mentioned that this symbol is not used by all manufacturers of analog equipment. Note that the gain of the amplifiers is very high, usually being greater than 10^4 and often greater than 10^6. This

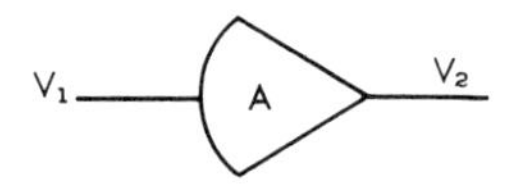

Fig. B.5. The high-gain amplifier; $V_2 = -AV_1, A > 10^4$.

means that the input voltage of such amplifiers is essentially zero since the output is always limited to a finite value (often 100 V).

The multiplier. The multiplier used on modern analog computers is an electronic quarter-square multiplier that operates on the following principle. Suppose v and i are to be multiplied to find the instantaneous power; i.e., $p = vi$. To do this, we begin with two voltages, one proportional to v, the other proportional to i. Then we form sum and difference signals, which in turn are squared and subtracted; i.e.,

$$M = (v + i)^2 - (v - i)^2 = (v^2 + 2vi + i^2) - (v^2 - 2vi + i^2) = 4vi$$

and $p = (1/4)M$, or one *quarter* of the difference of the *squared* signals.

The symbol used for multiplication varies with the actual components present in the computer multiplier section, but in its simplest form it may be represented as shown in Figure B.6. Note that it is usually necessary to supply both the positive and negative of one signal, say V_1. The multiplier inverts and divides the result by 100 (on a 100-V computer).

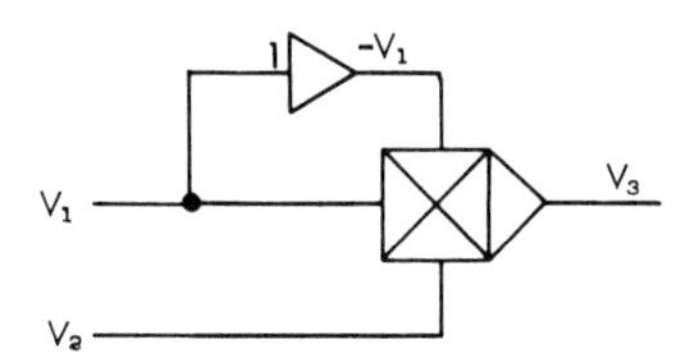

Fig. B.6. The multiplier; $V_3 = -V_1 V_2/100 \text{ V} = -V_1 V_2$ pu.

Other components. Most full-scale analog computers have other components not described here, including certain logical elements to control the computer operation. These specialized devices are left for the interested reader to discover for himself.

B.1.2 Analog computer scaling

Two kinds of scaling are necessary in analog computation, time scaling and amplitude scaling. Time scaling can be illustrated by means of a simple example. Consider the first-order equation

$$T\frac{dv}{dt} = f(v, t) \tag{B.1}$$

where v is the dependent variable that is desired, T is a constant, and f is a nonlinear function of v and t. The constant T would appear to be merely an amplitude scale factor, but such is not the case. Suppose we write

$$T\frac{dv}{dt} = \frac{dv}{d(t/T)} = \frac{dv}{d\tau} = f(v, t) \tag{B.2}$$

where $\tau = t/T$. Thus replacing the constant T by unity as in (B.2) amounts to time scaling the equation. In an analog computation the integration time must be chosen so that the computed results may be conveniently plotted or displayed. For example, if the output plotter has a frequency limit of 1.0 kHz, the computer should be time scaled to plot the results more slowly than this limit.

Analog computers must also be amplitude scaled so that no variables will exceed the rating of the computer amplifiers (usually 100 V). This requires that the user estimate the maximum value of all variables to be represented and scale the values of these variables so that the maximum excursion is well below the computer rating.

Actually, it is convenient to scale time and amplitude simultaneously. One reason for this is that the electronic integrator is unable to tell the difference between the two scale factors. Moreover, this makes one equation suffice for both kinds of scaling. We begin with the following definitions. Let the time scaling constant a be defined as follows:

$$\tau = \text{computer time} \qquad t = \text{real time} \qquad a = \frac{\tau}{t} = \frac{\text{computer time}}{\text{real time}} \tag{B.3}$$

For example, if $a = 100$, this means that it will take the computer 100 times as long to solve the problem as the real system would require. It also means that 100 s on the output plotter corresponds to 1 s of real time.

Also define L as the *level* of a particular variable in volts, corresponding to 1.0 pu of that variable. For example, suppose the variable v in (B.1) ordinarily does not go above 5.0 pu. If the computer is rated 100 V, we could set $L = 20$ V on the amplifier supplying v. Then if v goes to 5.0 pu, the amplifier would reach 100 V, its maximum safe value. The scaling procedure follows:

1. Choose a time scale a that is compatible with plotting equipment and will give reasonable computation times (a few minutes at most).
2. Choose levels for all variables at the output of all summers and integrators.

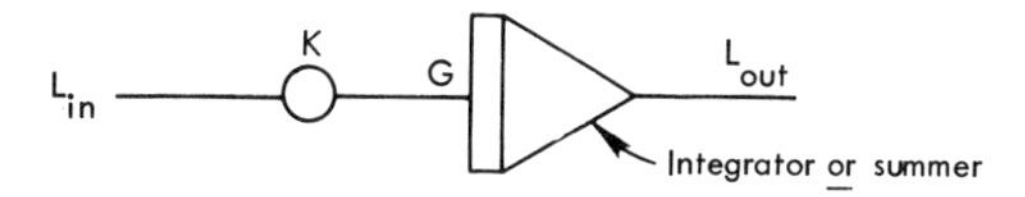

Fig. B.7. Time and amplitude scaling.

3. Apply the following formula to all potentiometer settings (see Figure B.7):

$$PG = KL_{out}/aL_{in} \tag{B.5}$$

where a = time scale factor
P = potentiometer setting, $0 \leq P \leq 1$
G = amplifier or integrator gain
K = physical constant computed for this potentiometer
L_{out} = assigned output level, V
L_{in} = assigned input level, V

B.1.3 Analog computation

Example B.1

Suppose the integrator in Figure B.7 is to integrate $-\dot{\delta}$ (in pu) to get the torque angle δ in radians. Then we write

$$\delta = -\delta_0 - \omega_R \int_0^t \dot{\delta}\, dt \tag{B.6}$$

Thus the constant K in Figure B.7 and (B.5) is ω_R, which is required to convert from $\dot{\delta}$ in pu to $\dot{\delta}$ in rad/s. In our example let $\omega_R = 377$.

Solution

Let $a = 50$. Then the levels are computed as follows: $\delta_{max} = 100° = 1.745$ rad, so let $L_{out} = 50$ V, $(1.745 \times 50 < 100)$. Also estimate $\dot{\delta}_{max} = 1.25$ pu, so let $L_{in} = 75$ V, $(1.25 \times 75 < 100)$. Then compute

$$PG = KL_{out}/aL_{in} = (377 \times 50)/(50 \times 75) = 5.03$$

Since $0 \leqq P \leqq 1$ let $G = 10$ = gain of integrator and $P = 0.503$ = potentiometer setting.

Example B.2

Compute the buildup curve of a dc exciter by analog computer and compare with the method of formal integration used in Chapter 7. Use numerical data from Examples 7.4, 7.5, and 7.6.

Solution

For this problem we have the first-order differential equation

$$\dot{v}_F = (v - Ri)/T \tag{B.7}$$

where $v = v_p$ when separately excited
$= v_F$ when self-excited
$= v_R + v_F$ when boost-buck excited

where both v_p and v_R are constants. Thus the analog computer diagram is that shown in Figure B.8, where $v_{F0} = v_F(0)$.

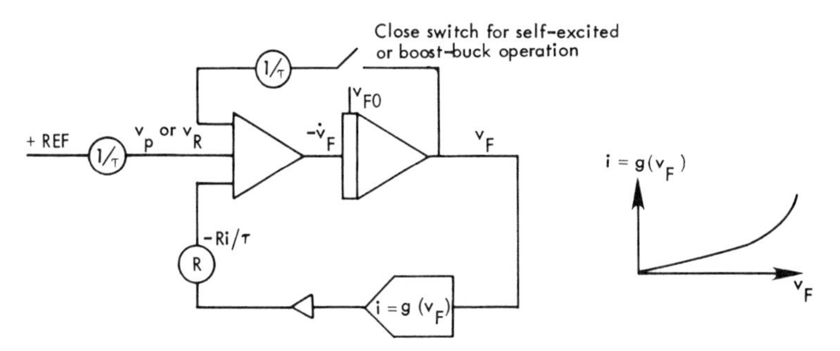

Fig. B.8. Solution diagram for dc exciter buildup.

An alternate solution utilizing the Frohlich approximation to the magnetization curve is described by the equation

$$\tau_E \dot{v}_F = v - \frac{bRv_f}{a - v_f} \tag{B.8}$$

Solving this equation should exactly duplicate the results of Chapter 7 where this same equation was solved by formal integration.

Using numerical data from Example 7.4 we have

$$\tau_E = 0.25 \text{ s} \qquad a = 279.9 \qquad b = 5.65$$

The values of R and v depend upon the type of buildup curve being simulated. From Examples 7.4, 7.5, and 7.6 we have

Separately excited: $v = v_p = 125$ V $\quad R = 34\ \Omega$
Self excited: $v = v_F \quad R = 30\ \Omega$
Boost-buck excited: $v = v_F + 50$ V $\quad R = 43.6\ \Omega$

and these values will give a ceiling of 110.3 V in all cases. Also, from Table 7.5 we note that the derivative of v_F can be greater than 100 V/s. This will help us scale the voltage level of $\dot{v}_F$.

Rewriting equation (B.8) with numerical values, we have

$$0.25\,\dot{v}_F = v - 5.65\,Rv_F/(279.9 - v_F)\ \text{V} \tag{B.9}$$

where R and v depend on the type of system being simulated. Suppose we choose a base voltage of 100 V. Then dividing (B.9) by the base voltage we have the pu equation

$$0.25\,\dot{v}_F = v - 0.0565\,Rv_F/(2.799 - v_F) \tag{B.10}$$

where v_F and v are now in pu.

A convenient time scale factor is obtained by writing

$$\tau_E \dot{v}_F = \tau_E \frac{dv_F}{dt} = \frac{dv_F}{d(t/\tau_E)} = \frac{dv_F}{d\tau}$$

or $a = \tau/t = 1/\tau_E = 4.0\ \text{s}^{-1}$
Then the factor 0.25 in front of (B.10) becomes unity, and 4 s on the computer corresponds to 1.0 s of real time.

The analog computer solution for (B.10) is shown in Figure B.9, and the potentiometer settings are given in Table B.1. By moving the three switches simultaneously to positions R, C, and L, the same computer setup solves the separately excited, self-excited, and boost-buck buildup curves respectively. Voltage levels are assumed for

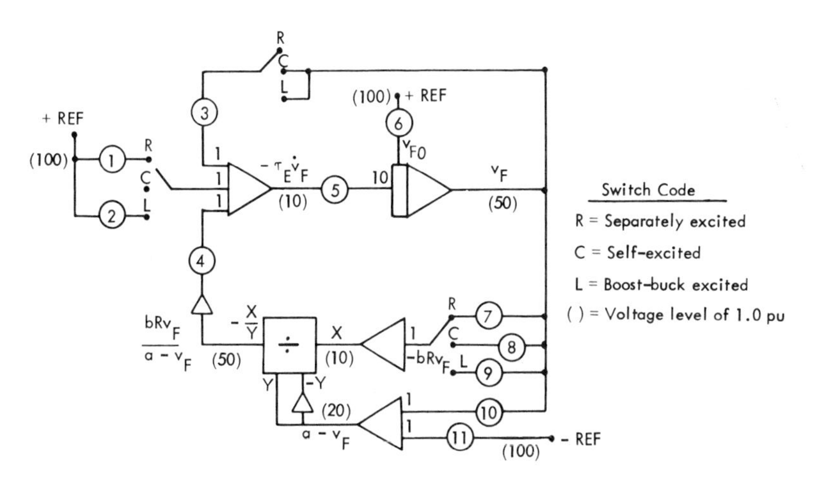

Fig. B.9. Solution diagram for Frohlich approximated buildup.

each amplifier and are noted in parentheses. These values are substituted into (B.5) to compute the PG products given in Table B.1. For example, for potentiometer 5

$$PG = (K/a)(L_{out}/L_{in}) = (1.0/4)(50/10) = 1.25 = 0.125 \times 10$$

or for potentiometer 7

$$PG = (1.92/1)(10/50) = 0.384 = 0.384 \times 1$$

Other table entries are similarly computed.

Table B.1. Potentiometer and Gain Calculations for Figure B.9

Potentiometer	Function	K	PG	P	G
1	v_p	1.25	0.125	0.125	1
2	v_R	0.50	0.050	0.050	1
3	scale	1.0	0.20	0.20	1
4	scale	1.0	0.20	0.20	1
5	time scale	1.0	1.25	0.125	10
6	initial value, v_{F0}	0.45	...	0.45	...
7	bR (separately)	1.92	0.384	0.384	1
8	bR (self)	1.695	0.339	0.339	1
9	bR (boost-buck)	2.46	0.492	0.492	1
10	scale	1.0	0.40	0.40	1
11	a	2.799	0.56	0.56	1

The computed results are shown in Examples 7.4, 7.5, and 7.6.

B.2 Digital Computer Solution of Ordinary Differential Equations

The purpose of this section is to present a brief introduction to the solution of ordinary differential equations by numerical techniques. The treatment here is simple and is intended to introduce the subject of numerical analysis to the reader who wishes to see how equations can be solved numerically.

One effective method of introducing a subject is to turn immediately to a simple example that can be solved without getting completely immersed in details. We shall use this technique. Our sample problem is the dc exciter buildup equation from Chapter 7, which was solved by integration in Examples 7.4–7.6. Since the solution is known, our numerical exercise will serve as a check on the work of Chapter 7. However, the real reason for choosing this example is that it is a scalar (one-dimensional) system that we can solve numerically with relative ease. Larger n-dimensional systems of equations are more challenging, but the principles are the same. The nonlinear differential equation here is

$$\dot{v}_F = \frac{dv_F}{dt} = \frac{1}{\tau_E}(v - Ri) \tag{B.11}$$

which we will solve by numerical techniques using a digital computer. Such problems are generally called "initial value problems" because the dependent variable v_F is known to have the initial value (at $t = 0$) of $v_F(0) = v_{F0}$.

B.2.1 Brief survey of numerical methods

There are several well-documented methods for solving the initial value problem by numerical integration. All methods divide the time domain into small segments Δt long and solve for the value of v_F at the end of each segment. In doing this there are three problems: getting the integration started, the speed of computation, and the generation of errors. Some methods are self-starting and others are not; therefore, a given computation scheme may start the integration using one method and then change to another method for increased speed or accuracy. Speed is important because, although the digital computer may be fast, any process that generates a great deal of computation may be expensive. Thus, for example, choosing Δt too small may greatly increase the cost of a computed result and may not provide enough improvement in accuracy to be worth the extra cost.

A brief outline of some known methods of numerical integration is given in Table B.2. Note that the form of equation is given in each case as an nth-order equation. However, it is easily shown that any nth-order equation can be written as n first-order equations. Thus instead of

$$v^{(n)} = f(v, t) \tag{B.12}$$

we may write

$$\begin{aligned} \dot{x}_1 &= f_1(v, t) \\ \dot{x}_2 &= f_2(v, t) \\ &\cdots\cdots \\ \dot{x}_n &= f_n(v, t) \end{aligned}$$

or in matrix form

$$\dot{\mathbf{x}} = \mathbf{f}(\mathbf{x}, t) \tag{B.13}$$

Thus we concern ourselves primarily with the solution of a first-order equation.

Table B.2. Some Methods of Numerical Integration of Differential Equations

Method	Form of equation	Order of errors	Remarks
Direct integration, trapezoidal rule, Simpson's rule	$v^{(n)} = f(t)$	Δt	Must know $n - 1$ derivatives to solve for $v^{(n)}$
Euler	$v^{(n)} = f(v, t)$	$(\Delta t)^2$	Self-starting
Modified Euler (Heun)	$v^{(n)} = f(v, t)$	$(\Delta t)^3$	Self-starting predictor-corrector
Runge-Kutta	$v^{(n)} = f(v, t)$	$(\Delta t)^3$	Self-starting, slow
Milne	$v^{(n)} = f(v, t)$	$(\Delta t)^5$	Start by Runge-Kutta or Taylor series
Hamming	$v^{(n)} = f(v, t)$	...	Imposes maximum condition on Δt for stable solution
Crane	$v^{(n)} = f(v, t)$	...	Varies size of Δt to control error

A complete analysis of every method in Table B.2 is beyond the scope of this appendix and the interested reader is referred to the many excellent references on the subject. Instead, we will investigate only the modified Euler method in enough detail to be able to work a simple problem.

B.2.2 Modified Euler method

Consider the first-order differential equation

$$\dot{v} = f(v, t) \tag{B.14}$$

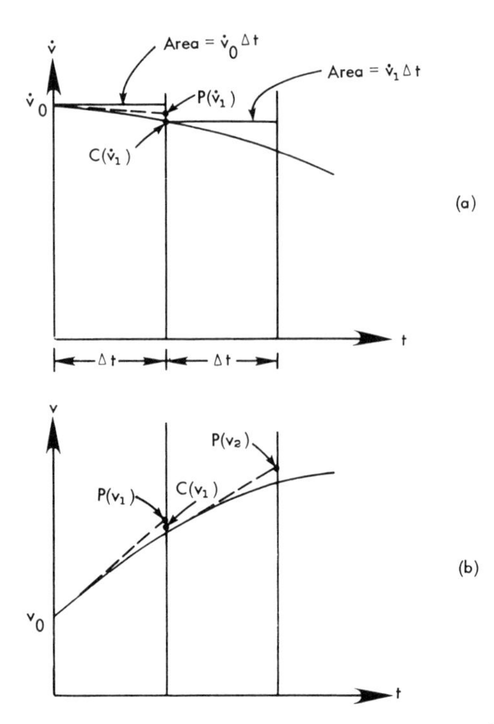

Fig. B.10. Graphical interpretation of the predictor-corrector routine: (a) $\dot{v}$ versus t, (b) v versus t.

where v is known for $t = 0$ (the initial value). Suppose the curves for v and $\dot{v}$ are as shown in Fig. B.10, where the time base has been divided into finite intervals Δt wide. Now define

$$\dot{v}_0 = f(v_0, 0) \tag{B.15}$$

which gives the initial slope of the v versus t curve. Next a *predicted* value for v at the end of the first interval is computed. If we define $v = v_1$ when $t = \Delta t$, we compute the *predicted* value v_1 as

$$P(v_1) = v_0 + \dot{v}_0 \Delta t \tag{B.16}$$

which is an extension of the initial slope out to the end of the first interval, as shown in Figure B.10(b). But $\dot{v}_0 \Delta t$ is the rectangular area shown in Figure B.10(a) and is obviously larger than the true area under the $\dot{v}$ versus t curve, so we conclude that $P(v_1)$ is too large [also see Figure B.10(b)].

Suppose we now approximate the value of $\dot{v}_1$ by substituting $P(v_1)$ into the given differential equation (B.14). Calling this value $P(\dot{v}_1)$, we compute

$$P(\dot{v}_1) = f[P(v_1), \Delta t] \tag{B.17}$$

Now approximate the true area under the $\dot{v}$ versus t curve between 0 and Δt by a trapezoid whose top is the straight line from $\dot{v}_0$ to $P(\dot{v}_1)$, as shown by the dashed line in Figure B.10(a). Using this area rather than the rectangular area, we compute a corrected value of v_1, which we call $C(v_1)$,

$$C(v_1) = v_0 + \{[\dot{v}_0 + P(\dot{v}_1)]/2\} \Delta t \tag{B.18}$$

We call (B.18) the corrector equation. Now we substitute the corrected value of v_1, $C(v_1)$, into the original equation to get a corrected $\dot{v}_1$.

$$C(\dot{v}_1) = f[C(v_1), \Delta t] \tag{B.19}$$

We now repeat this operation, using $C(\dot{v}_1)$ in (B.18) rather than $P(\dot{v}_1)$ to obtain an even better value for $C(v_1)$. This is done over and over again until successive values of $C(v_1)$ differ from one another by less than some prescribed precision index or until

$$C(v_1)^k - C(v_1)^{k-1} \leqq \epsilon \tag{B.20}$$

where k is the iteration number and ϵ is some convenient, small precision index (10^{-6}, for example). Once v_1 is determined as above, we use it as the starting point to find v_2 by the same method.

The general form of predictor and corrector equations is

$$P(v_{i+1}) = v_i + \dot{v}_i(\Delta t) \tag{B.21}$$

$$C(v_{i+1}) = v_i + \{[\dot{v}_i + P(\dot{v}_{i+1})]/2\}\, \Delta t \tag{B.22}$$

B.2.3 Use of the modified Euler method

Example B.3

Solve the separately excited buildup curve by the predictor-corrector method of numerical integration. Use numerical values from Example 7.4.

Solution

The equation requiring solution is

$$\tau_E \dot{v}_F = v_p - Ri \tag{B.23}$$

where i as a function of v_F is known from Table 7.3. We could proceed in two different ways at this point. We could store the data of Table 7.3 in the computer and use linear (or other means) interpolation to compute values of i for v_F between given data points. Thus using linear interpolation, we have for any value of v between v_1 and v_2

$$i = i_1 + (i_2 - i_1)(v - v_1)/(v_2 - v_1) \tag{B.24}$$

In this way we can compute the value of i corresponding to any v_F and substitute in (B.23) to find $\dot{v}_F$. An alternative method is to use an approximate formula to represent the nonlinear relationship between v_F and i. Thus, by the Frohlich equation,

$$i = bv_F/(a - v_F) \tag{B.25}$$

where a and b may be found as in Example 7.2.

Let us proceed using the latter of the two methods, where from Example 7.2 we have

$$a = 279.9 \qquad b = 5.65$$

Thus (B.23) becomes

$$\dot{v}_F = \frac{v_p}{\tau_E} - \frac{Rbv_F}{\tau_E(a - v_F)} \tag{B.26}$$

or

$$\dot{v}_F = 500 - 282.5\, v_F/(279.9 - v_F) \tag{B.27}$$

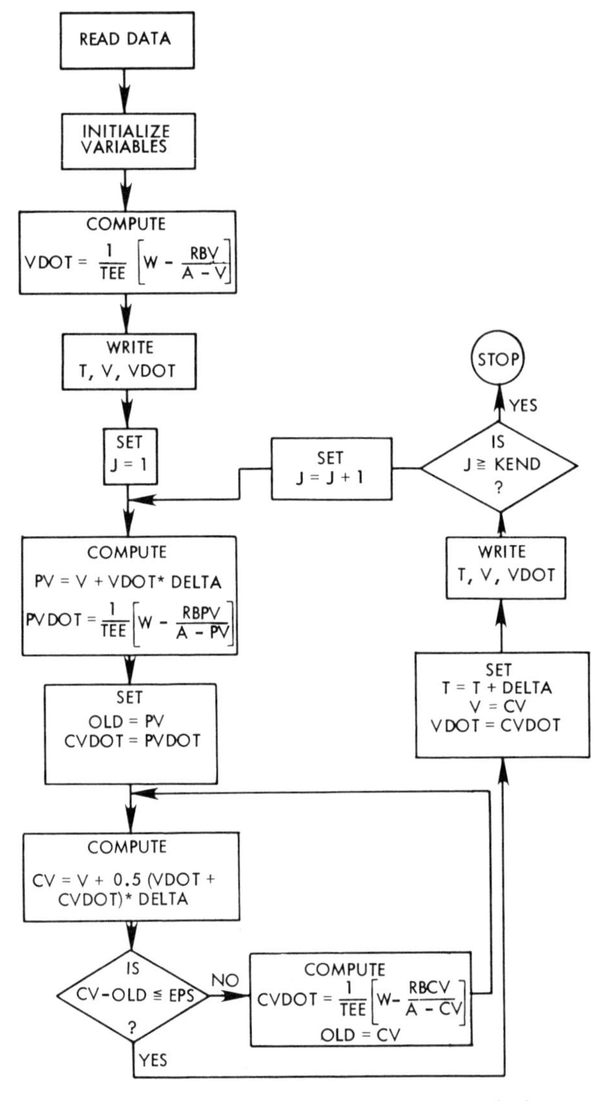

Fig. B.11. Computer flow diagram, separately excited case.

To avoid confusion in programming, we drop the subscript on v_F, represent v_p by a constant W, and replace τ by T to write

$$\dot{v} = W/T - (Rb/T)[v/(a - v)] \tag{B.28}$$

The data that must be input to begin the solution is shown in Table B.3 with certain additional variables that must be defined.

The computer flow diagram is shown in Figure B.11 for the separately excited case. The FORTRAN coding is given in Figure B.12. The solution is printed in tabular form in Table B.4 for values of t from 0 to 0.8 s. Note that both v_F and $\dot{v}_F$ are given. The derivative may not be needed, but it is known and can just as well be printed. The computed results agree almost exactly with the results of Example 7.4 and are therefore not plotted.

```
      VDOT1(W,V)=(W-R*B*V/(A-V))/TEE
      READ(1,101)W,TEE,R,B,A,VO,DELTA,KEND,EPS
  101 FORMAT (F5.2,F4.3,F5.2,F5.3,F6.3,F5.2,F5.4,I3,F7.7)
      V=VO
      VDOT=0.0
      PV=0.0
      CV=0.0
      PVDOT=0.0
      CVDOT=0.0
      T=0.0
      VDOT=VDOT1 (W,V)
      WRITE(3,110)T,V,VDOT
      DO 200 J=1, KEND
  105 PV=V+VDOT*DELTA
      PVDOT=VDOT1 (W,PV)
  102 OLD=PV
  103 CVDOT=PVDOT
  104 CV=V+0.5*(VDOT+CVDOT)*DELTA
      IF(CV-OLD-EPS) 107,107,106
  106 CVDOT=VDOT1(W,CV)
      OLD=CV
      GO TO 104
  107 T=T+DELTA
      V=CV
      VDOT=CVDOT
      WRITE(3,110)T,V,VDOT
  110 FORMAT(' ',F10.3,F10.2,F10.2)
  200 CONTINUE
      STOP
      END
```

Fig. B.12. FORTRAN coding for the separately excited case.

Table B.3. Data and Variable Symbols, Names, and Formats

Symbol	Name	Format	Constant	Variable
v_p	W	F5.2	x	
T	TEE	F4.3	x	
R	R	F5.2	x	
b	B	F5.3	x	
a	A	F6.3	x	
$v(0)$	VO	F5.2	x	
Δt	DELTA	F5.4	x	
—	KEND	I3	x	
ϵ	EPS	F7.7	x	
v	V	F5.2		x
$\dot{v}$	VDOT	F6.2		x
$P(v_{i+1})$	PVDOT			x
$C(v_{i+1})$	CVDOT			x
$P(\dot{v}_{i+1})$	PV			x
$C(\dot{v}_{i+1})$	CV			x
t	T	F5.3		x

Table B.4. Separately Excited Results in Tabular Form

t	v_F	$\dot{v}_F$	t	v_F	$\dot{v}_F$
			0.400	158.55	130.72
0.0	40.00	452.90	0.410	159.82	123.82
0.010	44.50	446.55	0.420	161.02	117.15
0.020	48.93	440.10	0.430	162.16	110.72
0.030	53.30	433.50	0.440	163.24	104.52
0.040	57.60	426.75	0.450	164.26	98.58
0.050	61.83	419.84	0.460	165.21	92.87
0.060	66.00	412.78	0.470	166.11	87.42
0.070	70.09	405.57	0.480	166.96	82.20
0.080	74.11	398.20	0.490	167.76	77.23
0.090	78.05	390.69	0.500	168.51	72.50
0.100	81.92	383.03	0.510	169.21	68.00
0.110	85.71	375.23	0.520	169.87	63.73
0.120	89.42	367.29	0.530	170.49	59.68
0.130	93.06	359.21	0.540	171.06	55.85
0.140	96.61	351.01	0.550	171.60	52.23
0.150	100.08	342.68	0.560	172.11	48.82
0.160	103.46	334.24	0.570	172.58	45.59
0.170	106.76	325.70	0.580	173.02	42.56
0.180	109.97	317.05	0.590	173.43	39.71
0.190	113.10	308.32	0.600	173.82	37.03
0.200	116.14	299.52	0.610	174.17	34.51
0.210	119.09	290.65	0.620	174.51	32.15
0.220	121.95	281.74	0.630	174.82	29.94
0.230	124.72	272.79	0.640	175.11	27.87
0.240	127.41	263.82	0.650	175.38	25.93
0.250	130.00	254.84	0.660	175.63	24.12
0.260	132.50	245.88	0.670	175.86	22.43
0.270	134.92	236.94	0.680	176.08	20.85
0.280	137.24	228.05	0.690	176.28	19.37
0.290	139.48	219.21	0.700	176.46	18.00
0.300	141.63	210.46	0.710	176.64	16.72
0.310	143.69	201.80	0.720	176.80	15.52
0.320	145.66	193.26	0.730	176.95	14.41
0.330	147.56	184.84	0.740	177.09	13.38
0.340	149.36	176.57	0.750	177.22	12.41
0.350	151.09	168.45	0.760	177.34	11.52
0.360	152.73	160.51	0.770	177.45	10.68
0.370	154.30	152.76	0.780	177.55	9.91
0.380	155.79	145.20	0.790	177.65	9.19
0.390	157.20	137.85	0.800	177.73	8.52

References

Analog Computation

Ashley, J. R. *Introduction to Analog Computation*. Wiley, New York, 1963.

Blum, J. J. *Introduction to Analog Computation*. Harcourt, Brace and World, New York, 1969.

Hausner, A. *Analog and Analog/Hybrid Computer Programming*. Prentice-Hall, Englewood Cliffs, N.J., 1971.

James, M. L., Smith, G. M., and Wolford, J. C. *Analog and Digital Computer Methods in Engineering Analysis*. International Textbook Co., Scranton, Pa., 1964.

———. *Analog Computer Simulation of Engineering Systems*, 2nd ed. Intext Educational Publ., Scranton, Pa., 1971.

Jenness, R. R. *Analog Computation and Simulation*. Allyn and Bacon, Boston, 1965.

———. *Analog Computation and Simulation: Laboratory Approach*. Allyn and Bacon, Boston, 1965.

Johnson, C. L. *Analog Computer Techniques*. McGraw-Hill, New York, 1963.

Digital Computation

Hildebrand, F. B. *Introduction to Numerical Analysis.* McGraw-Hill, New York, 1956.

James, M. L., Smith, G. M., and Wolford, J. C. *Analog and Digital Computer Methods in Engineering Analysis.* International Textbook Co., Scranton, Pa., 1964.

Korn, G. A., and Korn, T. M. *Mathematics Handbook for Scientists and Engineers.* McGraw-Hill, New York, 1968.

Pennington, R. H. *Introductory Computer Methods and Numerical Analysis.* Macmillan, New York, 1965.

Pipes, L. A. *Matrix Methods for Engineering.* Prentice-Hall, Englewood Cliffs, N.J., 1963.

Stagg, G. W., and El-Abiad, A. H. *Computer Methods in Power System Analysis.* McGraw-Hill, New York, 1968.

Stephenson, R. E. *Computer Simulation for Engineers.* Harcourt Brace Jovanovich, New York, 1971.

Wilf, H. S. *Mathematics for the Physical Sciences.* Wiley, New York, 1962.

appendix C

Normalization

There are many ways that equations can be normalized, and no one system is clearly superior to the others [1, 2, 3]. For the study of system dynamic performance it is important to choose a normalization scheme that provides a convenient *simulation* of the equations. At the same time it is also important to consider the traditions that have been established over the years [1, 2] and either comply wholly or provide a clear transition to a new system.

Having carefully considered a number of normalization schemes for synchronous machines and weighed the merits of each, the authors have adopted the following guidelines against which any normalization system should be measured.

1. *The system voltage equations must be exactly the same whether the equations are in* pu *or* MKS *units.* This means that the equations are symbolically always the same and no normalization constants are required in the pu equations.
2. *The system power equation must be exactly the same whether the equation is in* pu *or* MKS *units.* This means that power is invariant in undergoing the normalization. Thus both before and after normalization we may write

$$p = kv^t i \tag{C.1}$$

   and k is the same both before and after normalization.
3. *All mutual inductances must be capable of representation as tee circuits after normalization.* This requirement is included to simplify the *simulation* of the pu equations.
4. *The major* pu *impedances traditionally provided by the manufacturers must be maintained in the adopted system for the convenience of the users.* Other pu impedances must be related to and easily derived from the data supplied by the manufacturer.

The normalization scheme used by U.S. manufacturers does not satisfy requirement 2. The manufacturers use the original Park's transformation, as given by (4.22), which is different from the transformation used in this book, as given by (4.5). However, the pu system is to be developed so that *the same* pu *stator and rotor impedance values are obtained.*

C.1 Normalization of Mutually Coupled Coils

Consider the ideal transformer shown in Figure C.1. First we write the equations in MKS quantities, i.e., volts, amperes, ohms, and henrys.

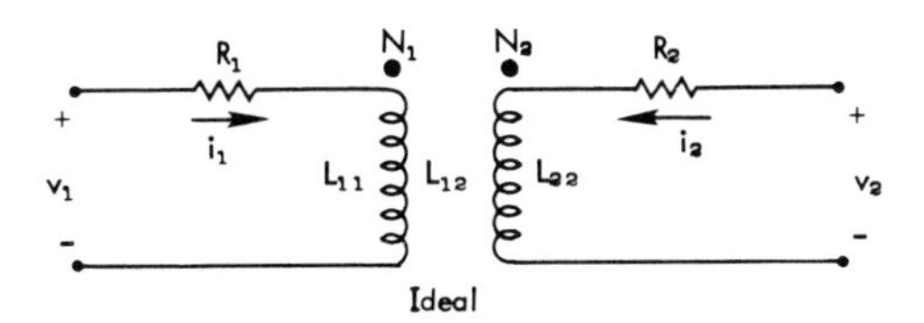

Fig. C.1. Schematic diagram of an ideal transformer.

$$v_1 = R_1 i_1 + L_{11}\frac{di_1}{dt} + L_{12}\frac{di_2}{dt} \text{ V}$$
$$v_2 = R_2 i_2 + L_{22}\frac{di_2}{dt} + L_{21}\frac{di_1}{dt} \text{ V} \tag{C.2}$$

where, in terms of the mutual permeance $\mathcal{P}_m$ and the coil turns N, $L_{jk} = \mathcal{P}_m N_j N_k$ for $j, k = 1, 2$. Now choose base values for voltage, current, and time in each circuit, i.e.,

For circuit 1: $V_{1B} I_{1B} t_{1B}$

For circuit 2: $V_{2B} I_{2B} t_{2B}$

Then since any quantity is the product of its per unit and base quantities, we have, using the subscript u to clearly distinguish pu quantities,

$$V_{1B} v_{1u} = R_1 I_{1B} i_{1u} + \frac{L_{11} I_{1B}}{t_{1B}}\frac{di_{1u}}{dt_{1u}} + \frac{L_{12} I_{2B}}{t_{2B}}\frac{di_{2u}}{dt_{2u}} \text{ V}$$
$$V_{2B} v_{2u} = R_2 I_{2B} i_{2u} + \frac{L_{22} I_{2B}}{t_{2B}}\frac{di_{2u}}{dt_{2u}} + \frac{L_{21} I_{1B}}{t_{1B}}\frac{di_{1u}}{dt_{1u}} \text{ V} \tag{C.3}$$

Dividing each equation by its base voltage, we have the pu (normalized) voltage equations

$$v_{1u} = \frac{R_1}{V_{1B}/I_{1B}} i_{1u} + \frac{L_{11} I_{1B}}{V_{1B} t_{1B}}\frac{di_{1u}}{dt_{1u}} + \frac{L_{12} I_{2B}}{V_{1B} t_{2B}}\frac{di_{2u}}{dt_{2u}}$$
$$v_{2u} = \frac{R_2}{V_{2B}/I_{2B}} i_{2u} + \frac{L_{22} I_{2B}}{V_{2B} t_{2B}}\frac{di_{2u}}{dt_{2u}} + \frac{L_{21} I_{1B}}{V_{2B} t_{1B}}\frac{di_{1u}}{dt_{1u}} \tag{C.4}$$

We can define

$$R_{1u} = \frac{R_1}{V_{1B}/I_{1B}} = \frac{R_1}{R_{1B}} \qquad R_{2u} = \frac{R_2}{V_{2B}/I_{2B}} = \frac{R_2}{R_{2B}}$$
$$L_{11u} = \frac{L_{11}}{V_{1B} t_{1B}/I_{1B}} = \frac{L_{11}}{L_{1B}} \qquad L_{22u} = \frac{L_{22}}{V_{2B} t_{2B}/I_{2B}} = \frac{L_{22}}{L_{2B}} \tag{C.5}$$

Now examine the mutual inductance coefficients. To preserve reciprocity, we require that

$$L_{12} I_{2B}/V_{1B} t_{2B} = L_{21} I_{1B}/V_{2B} t_{1B}$$

and since $L_{12} = L_{21}$ H, we compute

$$V_{1B} I_{1B}/t_{2B} = V_{2B} I_{2B}/t_{1B}$$

or

$$S_{1B}/t_{2B} = S_{2B}/t_{1B} \tag{C.6}$$

The ideal transformer is also characterized as having the following constraints on primary and secondary quantities:

$$n = i_2/i_1 = v_1/v_2 \tag{C.7}$$

where $n = N_1/N_2$. Rewriting in terms of base and pu values, we have

$$n = I_{2B}i_{2u}/I_{1B}i_{1u} = V_{1B}v_{1u}/V_{2B}v_{2u}$$

Thus the pu turns ratio n_u must be

$$n_u = i_{2u}/i_{1u} = nI_{1B}/I_{2B} = v_{1u}/v_{2u} = nV_{2B}/V_{1B} \tag{C.8}$$

and base quantities are often chosen to make $n_u = 1$. From (C.8) we compute

$$I_{1B}/I_{2B} = V_{2B}/V_{1B}$$

or

$$V_{1B}I_{1B} = V_{2B}I_{2B} \qquad S_{1B} = S_{2B} \triangleq S_B \tag{C.9}$$

Combining with (C.6), it is apparent that we must have

$$t_{1B} = t_{2B} \triangleq t_B \tag{C.10}$$

and the mutual inductance terms of the voltage equation (C.4) become

$$L_{12u} = \frac{L_{12}}{V_{1B}t_B/I_{2B}} = \frac{L_{12}}{L_{12B}} \qquad L_{21u} = \frac{L_{21}}{V_{2B}t_B/I_{1B}} = \frac{L_{21}}{L_{21B}} = \frac{L_{12}}{L_{12B}} \tag{C.11}$$

Then the voltage equation is exactly the same in pu as in volts, and the first requirement is satisfied. Furthermore, if this identical relationship exists between currents and voltages, the power is also invariant and the second requirement is also met.

C.2 Equal Mutual Flux Linkages

To adapt the voltage equations to a pu tee circuit, we divide the coil inductances into a leakage and a magnetizing inductance; i.e.,

$$L_{11} = \ell_1 + L_{m1} \qquad L_{22} = \ell_2 + L_{m2} \text{ H} \tag{C.12}$$

From the flux linkage equations we write (in MKS units)

$$\begin{bmatrix} \lambda_1 \\ \lambda_2 \end{bmatrix} = \begin{bmatrix} L_{11} & L_{12} \\ L_{21} & L_{22} \end{bmatrix} \begin{bmatrix} i_1 \\ i_2 \end{bmatrix} = \begin{bmatrix} \ell_1 & 0 \\ 0 & \ell_2 \end{bmatrix} \begin{bmatrix} i_1 \\ i_2 \end{bmatrix} + \begin{bmatrix} L_{m1} & L_{12} \\ L_{21} & L_{m2} \end{bmatrix} \begin{bmatrix} i_1 \\ i_2 \end{bmatrix} \tag{C.13}$$

Injecting a base current in circuit 1 with circuit 2 open, i.e., with $i_1 = I_{1B}$ and $i_2 = 0$, gives the following mutual flux linkages

$$\lambda_{m1} = L_{m1}I_{1B} \qquad \lambda_{m2} = L_{21}I_{1B} \text{ Wb turns} \tag{C.14}$$

In pu these flux linkages are

$$\lambda_{m1u} = \lambda_{m1}/\lambda_{1B} = L_{M1}I_{1B}/L_{1B}I_{1B} = L_{m1}/L_{1B} \tag{C.15}$$

$$\lambda_{m2u} = \lambda_{m2}/\lambda_{2B} = L_{21}I_{1B}/L_{2B}I_{2B} \tag{C.16}$$

Equal pu mutual flux linkages require that

$$\lambda_{m1u} = \lambda_{m2u} \tag{C.17}$$

or

$$L_{m1}/L_{1B} = L_{m1u} = L_{21}I_{1B}/L_{2B}I_{2B} \tag{C.18}$$

Following a similar procedure, we can show that injecting a base current in circuit 2 with circuit 1 open (i.e., with $i_2 = I_{2B}$ and $i_1 = 0$) gives the following pu flux linkages:

$$\lambda_{m1u} = L_{12}I_{2B}/L_{1B}I_{1B} \qquad \lambda_{m2u} = L_{m2}/L_{2B} \tag{C.19}$$

Again equal pu flux linkages give

$$L_{m2}/L_{2B} = L_{m2u} = L_{12}I_{2B}/L_{1B}I_{1B} \tag{C.20}$$

From $S_{1B} = S_{2B}$

$$I_{1B}^2 L_{1B} = I_{2B}^2 L_{2B} \tag{C.21}$$

and from (C.20) and (C.21)

$$L_{m2}/L_{2B} = L_{m2u} = L_{12}I_{1B}/L_{2B}I_{2B} \tag{C.22}$$

Comparing (C.18) and (C.22),

$$L_{m1u} = L_{m2u} \triangleq L_{mu} \tag{C.23}$$

Now using (C.12), (C.20), (C.22), and (C.23) in the voltage equation (C.4),

$$\begin{aligned} v_{1u} &= R_{1u}i_{1u} + \ell_1 \dot{i}_{1u} + L_{mu}(\dot{i}_{1u} + \dot{i}_{2u}) \\ v_{2u} &= R_{2u}i_{2u} + \ell_2 \dot{i}_{2u} + L_{mu}(\dot{i}_{1u} + \dot{i}_{2u}) \end{aligned} \tag{C.24}$$

which is represented schematically by the tee circuit shown in Figure C.2. Thus the third requirement is satisfied.

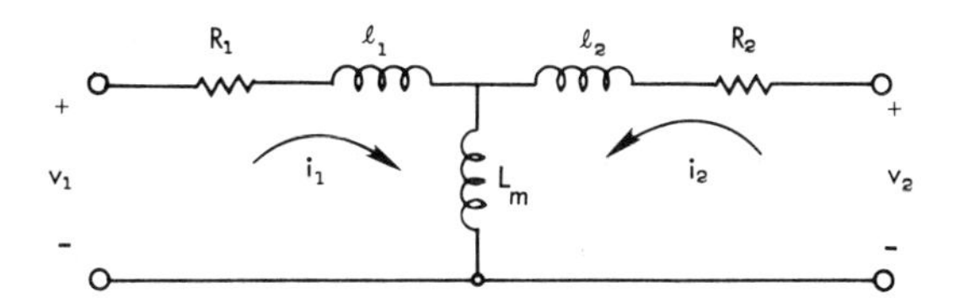

Fig. C.2. Tee circuit representation of a transformer.

An interesting point to be made here is that the requirement for equal pu mutual flux linkages is the same as equal base MMF's.

$$S_B(L_{m1}/L_{1B}) = S_B(L_{m2}/L_{2B})$$

or

$$\begin{aligned} (L_{m1}/L_{1B})(I_{1B}^2 L_{1B}) &= (L_{m2}/L_{2B})(I_{2B}^2 L_{2B}) \\ L_{m1}I_{1B}^2 &= L_{m2}I_{2B}^2 \end{aligned} \tag{C.25}$$

or in terms of the mutual permeance $\mathcal{P}_m$

$$\mathcal{P}_m N_1^2 I_{1B}^2 = \mathcal{P}_m N_2^2 I_{2B}^2 \tag{C.26}$$

or

$$N_1^2 I_{1B}^2 = N_2^2 I_{2B}^2 \tag{C.27}$$

or in terms of MMF

$$F_{1B} = F_{2B} \tag{C.28}$$

C.2.1 Summary

The first three normalization specifications require that

1. All circuits must have the same VA base (C.9).
2. All circuits must have the same time base (C.6), (C.9), and (C.10).
3. The requirement of a common pu tee circuit means equal pu magnetizing inductance in all circuits (C.23). This requires equal pu mutual flux linkages (C.17), which in turn requires that the base MMF be the same in all circuits (C.28).

C.3 Comparison with Manufacturers' Impedances

We now select the base stator and rotor quantities to satisfy the fourth requirement, namely, to give the same pu impedances as those supplied by the manufacturers.

The choice of the stator base voltage V_{1B} and the stator base current I_{1B} determines the base stator impedance. Because of a certain awkwardness in the original Park's transformation resulting from the fact that the transformation is not power invariant, a system of stator base quantities is used by U.S. manufacturers that facilitates the choice of rotor base quantities. For this reason it is customary to use a stator base voltage equal to the *peak* line-to-neutral voltage and a stator base current equal to the *peak* line current. Such a choice, along with the requirement of equal base ampere turns (or equal pu mutuals), leads to a rotor VA base equal to the three-phase stator VA base.

Since the transformation used in this book is power invariant, the awkwardness referred to above is not encountered. A variety of possible stator base quantities can be chosen to satisfy the condition of having the same pu stator impedances as supplied by the manufacturers. For example, among the possible choices for the stator base: peak line-to-neutral voltage and peak line current (same as the manufacturers), rms line-to-neutral voltage and rms line current, or rms line voltage and $\sqrt{3}$ times rms line current. Note that in all these choices the base stator impedance is the same. However, the other three requirements stated in the previous sections may not be satisfied.

To illustrate, it would appear that adoption of stator base quantities of rated rms line voltage and $\sqrt{3}$ times line current would be attractive. The factor of $\sqrt{3}$ appearing in the d and q axis equations of Chapter 4 would be eliminated. Careful examination, however, would reveal that the requirement of having the same identical equation hold for the MKS and the pu systems would be violated. For example, if the phase voltage $v_a = \sqrt{2}V\cos(\omega_R t + \alpha)$, the d and q axis voltages are obtained by a relation similar to that of (4.146)

$$v_d = -\sqrt{3}V\sin(\delta - \alpha) \qquad v_q = \sqrt{3}V\cos(\delta - \alpha) \text{ V} \tag{C.29}$$

where V = rms voltage to neutral. Choosing $V_{1B} = \sqrt{3}V_{LN}$ (rated), we get

$$\begin{aligned} v_{du} &= -\frac{\sqrt{3}V\sin(\delta - \alpha)}{\sqrt{3}V_{LN}} = -(V/V_{LN})\sin(\delta - \alpha) \text{ pu} \\ v_{qu} &= (V/V_{LN})\cos(\delta - \alpha) \text{ pu} \end{aligned} \tag{C.30}$$

Note that (C.29) and (C.30) are not identical, and hence this choice of stator base quantities does not meet requirement number 1.

In this book the stator base quantities selected to meet the requirements stated above are

$$
\begin{aligned}
S_{1B} &= \text{rated per phase voltampere, VA}\\
V_{1B} &= \text{rated rms voltage to neutral, V}\\
I_{1B} &= \text{rated rms line current, A}\\
t_{1B} &= 1/\omega_R,\ \text{s}
\end{aligned}
\tag{C.31}
$$

The rotor base quantities are selected to meet the conditions of equal S_B, t_B, and F_B (or λ_m). Equal VA base gives

$$V_{1B}I_{1B} = V_{2B}I_{2B} \quad \text{VA} \tag{C.32}$$

(The subscript 2 is used to indicate *any* rotor circuit. The same derivation applies to a field circuit or to an amortisseur circuit.) Equal mutual flux linkages require that the mutual flux linkage in the d axis stator produced by a base stator current would be the same as the d axis stator flux linkage produced by a d axis rotor base current. Thus in MKS units,

$$I_{1B}L_{m1} = I_{2B}\mathrm{k}M_F \qquad \mathrm{k} = \sqrt{3/2}$$

or

$$I_{2B} = (L_{m1}/\mathrm{k}M_F)I_{1B} = (1/k_F)I_{1B} \quad \text{A} \tag{C.33}$$

where $k_F = \mathrm{k}M_F/L_{m1}$.

From (C.32) and (C.33) we obtain for the rotor circuit base voltage

$$V_{2B} = V_{1B}I_{1B}/I_{2B} = k_F V_{1B} \tag{C.34}$$

From (C.33) and (C.34) for the rotor resistance base

$$R_{2B} = V_{2B}/I_{2B} = k_F^2(V_{1B}/I_{1B}) = k_F^2 R_{1B} \quad \Omega \tag{C.35}$$

The inductance base for the rotor circuit is then given by

$$L_{2B} = V_{2B}t_B/I_{2B} = (\mathrm{k}M_F/L_{m1})^2(V_{1B}/I_{1B})\omega_R = k_F^2 L_{1B} \tag{C.36}$$

The base for the mutual inductance is obtained from (C.11) and (C.33)

$$L_{12B} = \frac{V_{1B}t_B}{I_{2B}} = \frac{V_{1B}\omega_R}{(L_{m1}/\mathrm{k}M_F)I_{1B}} = k_F L_{1B} \tag{C.37}$$

The pu d axis mutual inductance is then given by

$$\mathrm{k}M_{Fu} = \frac{\mathrm{k}M_F}{L_{12B}} = \frac{\mathrm{k}M_F}{(\mathrm{k}M_F/L_{m1})L_{1B}} = \frac{L_{m1}}{L_{1B}} = L_{m1u} \tag{C.38}$$

Thus the value of the pu d axis mutual inductance of *any* rotor circuit is the same as the pu magnetizing inductance of the stator.

$$\mathrm{k}M_{Fu} = \mathrm{k}M_{Du} = M_{Ru} = L_{m1u} \tag{C.39}$$

A comparison between the pu system derived in this book and that used by U.S. manufacturers is given in the Table C.1. Note that the base inductances and resistances are the same in both systems.

Table C.1. Comparison of Base Quantities

Quantity/system	Per unit system used: In this book	By U.S. manufacturers*
V_{1B}	V_{LN}	$\sqrt{2}V_{LN}$
I_{1B}	I_L	$\sqrt{2}I_L$
L_{1B}	$V_{LN}/\omega_R I_L$	$V_{LN}/\omega_R I_L$
$S_{1B} = S_{2B}$	$V_{LN}I_L$	$(3/2)V_{1B}I_{1B} = 3V_{LN}I_L$
I_{2B}	$(L_{m1}/\sqrt{3/2}M_F)I_L$	$\sqrt{2}(L_{m1}/M_F)I_L$
V_{2B}	$(\sqrt{3/2}M_F/L_{m1})V_{LN}$	$(3/\sqrt{2})(M_F/L_{m1})V_{LN}$
R_{2B}	$(3/2)(M_F/L_{m1})^2(V_{LN}/I_L)$	$(3/2)(M_F/L_{m1})^2(V_{LN}/I_L)$
L_{2B}	$(3/2)(M_F/L_{m1})^2L_{1B}$	$(3/2)(M_F/L_{m1})^2L_{1B}$
L_{12B}	$(\sqrt{3/2}M_F/L_{m1})L_{1B}$	$(M_F/L_{m1})L_{1B}$
Per unit mutual inductances	$kM_{Fu} = L_{mu} = L_{ADu}$	$M_{Fu} = L_{mu} = L_{ADu}$
$v_{qu}^2 + v_{du}^2$	$3V_u^2$	V_u^2

*See reference [4].

C.4 Complete Data for Typical Machine

To complement the discussion on normalization given in this appendix, we provide a consistent set of data for a typical synchronous generator. Starting with the pu impedances supplied by the manufacturer, the base quantities are derived and all the impedance values are calculated.

The machine used for this data is the 160-MVA, two-pole machine that is used in many of the text examples. The method used is that of Section 5.8 of the text. The data given and results computed are the same as in Example 5.5. Computations here are carried to about eight significant figures using a pocket "slide rule" calculator.

The following data is provided by the manufacturer (this is actual data on an actual machine with data from the manufacturers bid or "guaranteed" data).

Ratings:

$$160\ \text{MVA} \qquad 136\ \text{MW} \qquad 0.85\ \text{PF} \qquad 15\ \text{kV} \tag{C.40}$$

Unsaturated reactances in pu:

$$\begin{aligned} x_d &= 1.70 & x_q' &= 0.380 & x_q'' &= 0.185 \\ x_q &= 1.64 & x_\ell &= 0.150 & x_2 &= 0.185 \\ x_d' &= 0.245 & x_d'' &= 0.185 & x_0 &= 0.100 \end{aligned} \tag{C.41}$$

Time constants in seconds:

$$\tau_{d0}' = 5.9 \qquad \tau_d'' = 0.023 \qquad \tau_a = 0.24 \qquad \tau_{q0}'' = 0.075 \tag{C.42}$$

Excitation at rated load:

$$v_F = 345\ \text{V} \qquad i_F = 926\ \text{A} \tag{C.43}$$

Resistances in ohms at 25°C:

$$r_a = 0.001113 \qquad r_F = 0.2687 \tag{C.44}$$

Computations are given in Example 5.5. One problem not mentioned there is that of finding the correct value of field resistance to use in the generator simulation. There are three possibilities:

1. Compute from (C.43), at operating temperature,

$$r_F = 345/926 = 0.37257 \ \Omega \tag{C.45}$$

2. Compute from (C.44) at an assumed operating temperature of 125°C:

$$r_F = 0.2687[(234.5 + 125.0)/(234.5 + 25.0)] = 0.372245 \ \Omega \tag{C.46}$$

3. Compute from (5.59), using L_F from Table C.3

$$r_F = L_F/\tau'_{d0} = 2.189475/5.9 = 0.371097 \ \Omega$$

The value computed from L_F/τ'_{d0} must be used if the correct time constant is to result. Working backward to compute the corresponding operating temperature, we have

$$0.2687[(234.5 + \theta)/(234.5 + 25)] = 0.371097 \tag{C.48}$$

or the operating temperature is $\theta = 123.8$ C, which is a reasonable result.

The base quantities for all circuits are given in Table C.2. Stator base values are derived from nameplate data for voltamperes, voltage, and frequency. The method of relating stator to field base quantities through the constant k_F is shown in Example 4.1 where we compute

$$k_F = \mathrm{k}M_F/L_{md} = 109.0102349 \ \mathrm{mH}/5.781800664 \ \mathrm{mH} = 18.85402857 \tag{C.49}$$

Note that a key element in determining the factor k_F, and hence all the rotor base quantities, is the value of M_F (in H). This is obtained from the air gap line of the magnetization curve provided by the manufacturer. Unfortunately, no such data is given for any of the amortisseur circuits. Thus, while the pu values of the various amortisseur elements can be determined, their corresponding MKS data are not known.

Using the base values from Table C.2 and the pu values from Example 5.5, we may construct Table C.3 of d axis parameters and Table C.4 of q axis parameters. The given values are easily identified since they are written to three decimals.

Table C.2. Base Values in MKS Units

Circuit	Base quantity	Formula	Numerical value	Units
Stator	S_B	$S_{B3}/3$	53.333 333 333	MVA/phase
	V_B	$V_{LL}/\sqrt{3}$	8.660 254 036	kV_{LN}
	t_B	$1/2\pi 60$	2.652 582 384	ms
	I_B	S_B/V_B	6158.402 872	A
	R_B	V_B/I_B	1.406 250	Ω
	λ_B	$V_B t_B$	22.972 0373	Wb
	L_B	λ_B/I_B	3.730 193 98	mH
Field	S_{FB}	S_B	53.333 333 333	MVA/phase
	V_{FB}	S_B/I_{FB}	163 280.677	V
	t_{FB}	t_B	2.652 582 384	ms
	I_{FB}	I_B/k_F	326.635 915	A
	R_{FB}	V_{FB}/I_{FB}	499.885 8653	Ω
	λ_{FB}	$V_{FB}t_B$	433.115 4475	Wb
	L_{FB}	λ_{FB}/I_{FB}	1.325 988 441	H
	M_{FB}	$\sqrt{L_B L_{FB}}$	0.070 329 184	H

Table C.3. Direct Axis Parameters in pu and MKS

Symbol	pu value	MKS value	Units
L_d	1.700	6.341 329 761	mH
L'_d	0.245		
L''_d	0.185		
L_{md}	1.550	5.781 800 664	mH
ℓ_d	0.150	0.559 529 097	mH
L_F	1.651 202 749	2.189 475 759	H
L_{mF}	1.550	2.055 282 084	H
ℓ_F	0.101 202 749	0.134 193 675	H
L_D	1.605 416 667		
L_{mD}	1.550		
ℓ_D	0.055 416 667		
M_F	1.265 5697	0.089 006 484	H
kM_F	1.550	0.109 010 235	H
M_D	1.265 5697		
kM_D	1.550		
M_R	1.550		
L_{MD}	0.028 378 3784		
r_a 25°C	$0.791\ 607\ 397 \times 10^{-3}$	1.113	mΩ
r_a 125°C	$1.096\ 463\ 455 \times 10^{-3}$	1.541 901 734	mΩ
r_F 25°C	. . .	0.2687 (not used)	Ω
r_F Hot	$0.742\ 364\ 295 \times 10^{-3}$	0.371 097 586	Ω
r_D	$13.099\ 135\ \ 90 \times 10^{-3}$		
τ_a	90.477 868 44	0.24	s
τ'_{d0}	2224.247 599	5.90	s
τ'_d	320.442 450 7	0.85	s
τ''_{d0}	11.482 945 69	0.030 459	s
τ''_d	8.670 795 726	0.023	s

Table C.4. Quadrature Axis Parameters in pu and MKS

Symbol	pu value	MKS value	Units
L_q	1.640	6.117 518 122	mH
L'_q	0.380 (not used)		
L''_q	0.185		
L_{mq}	1.490	5.557 989 025	mH
ℓ_q	0.150	0.559 529 097	mH
L_Q	1.525 808 581		
L_{mQ}	1.490		
ℓ_Q	0.035 808 581		
M_Q	1.216 579 905		
kM_Q	1.490		
L_{MQ}	0.028 357 4715		
r_a 25°C	$0.791\ 607\ 397 \times 10^{-3}$	1.113	mΩ
r_a 125°C	$1.096\ 463\ 455 \times 10^{-3}$	1.541 901 734	mΩ
r_Q	0.053 955 165		
τ'_{q0}	203.575 204	0.54	s
τ''_{q0}	28.274 333 89	0.075	s
τ''_q	3.189 482 785	8.460 365 85	ms

References

1. Rankin, A. W. Per unit impedances of synchronous machines. *AIEE Trans.* 64:569–841, 1945.
2. Lewis, W. A. A basic analysis of synchronous machines, Pt. 1. *AIEE Trans.* 77:436–56, 1958.
3. Harris, M. R., Lawrenson, P. J., and Stephenson, J. M. *Per Unit Systems: With Special Reference to Electrical Machines.* IEE Monogr. Ser. 4, Cambridge Univ. Press, 1970.
4. General Electric Co. Power system stability. Electric Utility Engineering Seminar, Section on Synchronous Machines. Schenectady, N.Y., 1973.

appendix D

Typical System Data

In studying system control and stability, it is often helpful to have access to typical system constants. Such constants help the student or teacher become acquainted with typical system parameters, and they permit the practicing engineer to estimate values for future installations.

The data given here were chosen simply because they were available to the authors and are probably typical. A rather complete set of data is given for various sizes of machines driven by both steam and hydraulic turbines. In most cases such an accumulation of information is not available without special inquiry. For example, data taken from manufacturers' bids are limited in scope, and these are often the only known data for a machine. Thus it is often necessary for the engineer to estimate or calculate the missing information.

Data are also provided that might be considered typical for certain prime mover systems. This is helpful in estimating simulation constants that can be used to represent other typical medium to large units. Finally, data are provided for typical transmission lines of various voltages. (See Tables D.1–D.8 at the end of this appendix.)

D.1 Data for Generator Units

Included here are all data normally required for dynamic simulation of the synchronous generator, the exciter, the turbine-governor system, and the power system stabilizer. The items included in the tabulations are specified in Table D.1.

Certain items in Table D.1 require explanation. Table references on these items are given in parentheses following the identifying symbol. An explanation of these referenced items follows.

(1) Short circuit ratio

The SCR is the "short circuit ratio" of a synchronous machine and is defined as the ratio of the field current required for rated open circuit voltage to the field current required for rated short circuit current [1]. Referring to Figure D.1, we compute

$$\mathrm{SCR} = I_B/I_S \ \ \mathrm{pu} \tag{D.1}$$

It can be shown that

$$\mathrm{SCR} \simeq 1/x_d \ \ \mathrm{pu} \tag{D.2}$$

where x_d is the saturated d axis synchronous reactance.

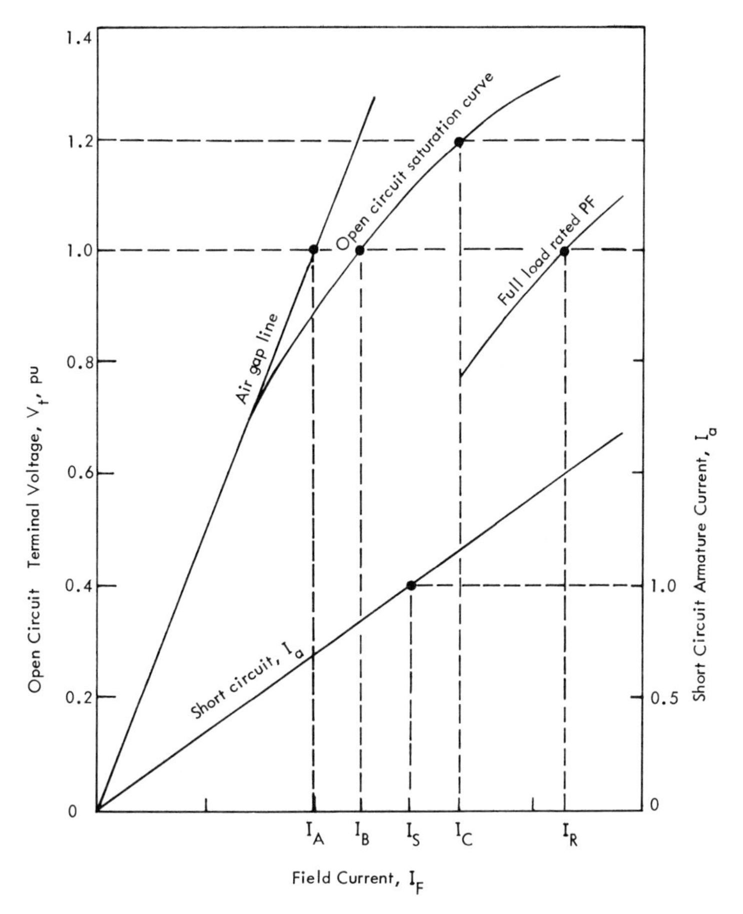

Fig. D.1. Open circuit, full load, and short circuit characteristics of a synchronous generator.

(2) Generator saturation

Saturation of the generator is often specified in terms of a pu saturation function S_G, which is defined in terms of the open circuit terminal voltage versus field current characteristic shown in Figure D.2. We compute

$$S_G \text{ at } V_{t1} = (I_{F2} - I_{F1})/I_{F1} \tag{D.3}$$

where (D.3) is valid for any point V_{t1} [2, 3]. With use of this definition, it is common to specify two values of saturation at $V_t = 1.0$ and 1.2 pu. These values are given under open circuit conditions so that V_t is actually the voltage behind the leakage reactance and is the voltage across L_{AD}, the pu saturated magnetizing inductance. Thus we can easily determine two saturation values from the generator saturation curve to use as the basis for defining a saturation function. From Figure D.1 we arbitrarily define

$$S_{G1.0} = (I_B - I_A)/I_A \tag{D.4}$$
$$S_{G1.2} = (I_C - 1.2I_A)/1.2I_A \tag{D.5}$$

and will use these two values to generate a saturation function.

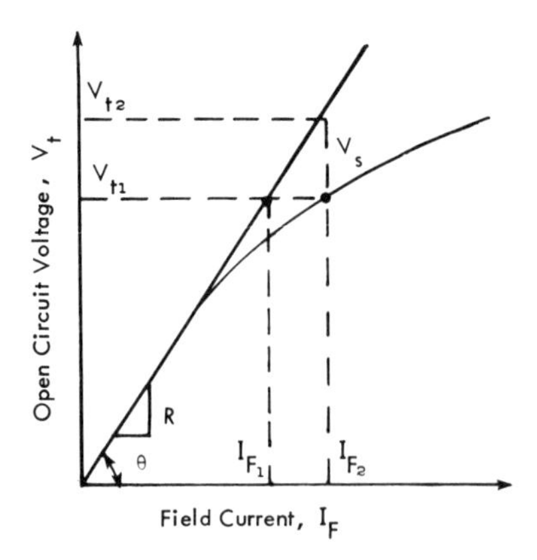

Fig. D.2. Construction used for computing saturation.

There are several ways to define a saturation function, one of which is given in Section 5.10.1 where we define

$$S_G = A_G e^{B_G V_\Delta} \tag{D.6}$$

where

$$V_\Delta = V_t - 0.8 \tag{D.7}$$

is the difference between the open circuit terminal voltage and the assumed saturation threshold of 0.8 pu. Since (D.6) contains two unknowns and the quantities S_G and V_Δ are known at two points, we can solve for A_G and B_G explicitly.

From the given data we write

$$S_{G1.0} = A_G e^{0.2B_G} \qquad 1.2S_{G1.2} = A_G e^{0.4B_G} \tag{D.8}$$

Rearranging and taking logarithms,

$$\ln(S_{G1.0}/A_G) = 0.2\,B_G \qquad \ln(1.2S_{G1.2}/A_G) = 0.4\,B_G \tag{D.9}$$

Then,

$$(S_{G1.0}/A_G)^2 = 1.2S_{G1.2}/A_G$$

or

$$A_G = S_{G1.0}^2/1.2S_{G1.2} \qquad B_G = 5\ln(1.2S_{G1.2}/S_{G1.0}) \tag{D.10}$$

Example D.1

Suppose that measurements on a given generator saturation curve provide the following data:

$$S_{G1.0} = 0.20 \qquad S_{G1.2} = 0.80$$

Then we compute, using (D.10),

$$A_G = (0.20)^2/1.2(0.80) = 0.04167 \qquad B_G = 5\ln(1.2 \times 0.8/0.20) = 7.843$$

This gives an idea of the order of magnitude of these constants; A_G is usually less than 0.1 and B_G is usually between 5 and 10.

The value of S_G determined above may be used to compute the open circuit voltage (or flux linkage) in terms of the *saturated* value of field current (or MMF). Referring again to Figure D.1, we write the voltage on the air gap line as

$$V_t = RI_F \tag{D.11}$$

Refer to Figure D.2. When saturation is present, current I_{F2} does not give $V_{t2} = RI_{F2}$ but only produces V_{F1}, or

$$V_{t1} = V_{t2} - V_s = RI_{F2} - V_s \tag{D.12}$$

where V_s is the drop in voltage due to saturation. But from Figure D.2

$$\tan\theta = R = V_s/(I_{F2} - I_{F1}) \tag{D.13}$$

From (D.3) we write

$$S_G = (I_{F2} - I_{F1})/I_{F1} = V_s/RI_{F1} = V_s/V_{t1} \tag{D.14}$$

Then from (D.12)

$$V_{t1} = RI_{F2} - S_G V_{t1} \tag{D.15}$$

where S_G is clearly a function of V_{t1}. Equation (D.15) describes how V_{t1} is reduced by saturation below its air gap value RI_{F2} at no load. Usually, we assume a similar reduction occurs under load.

Note that the exponential saturation function does not satisfy the definition (D.3) in the neighborhood of $V_t = 0.8$, where we assume that saturation begins. The computed saturation function has the shape shown in Figure D.3. Note that $S_G > 0$ for any V_Δ. The error is small, however, and the approximation solution is considered adequate in the neighborhood of 1.0 pu voltage. Note that A_G is usually a very small number, so the saturation computed for $V_t < 0.8$ is negligible.

Other methods of treating saturation are found in the literature [1, 2, 4, 5, 6, 7].

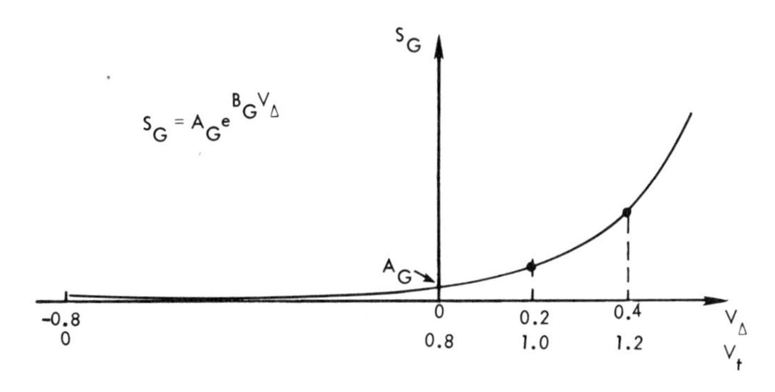

Fig. D.3. The approximate saturation function, S_G.

(3) Damping

It is common practice in stability studies to provide a means of adding damping that is proportional to speed or slip. This concept is discussed in Sections 2.3, 2.4, 2.9, 4.10, and 4.15 and is treated in the literature [8–12]. The method of introducing the damping is by means of a speed or slip feedback term similar to that shown in Figure 3.4, where D is the pu damping coefficient used to compute a damping torque T_d

defined as

$$T_d = D\omega_{\Delta u} \text{ pu} \tag{D.16}$$

where all quantities are in pu. The value used for D depends greatly on the kind of generator model used and particularly on the modeling of the amortisseur windings. For example, a damping of 1–3 pu is often used to represent damping due to turbine windage and load effects [2]. A much higher value, up to 25 pu is sometimes used as a representation of amortisseur damping if this important source of damping is omitted from the machine model.

The value of D also depends on the units of (D.16). In some simulations the torque is computed in megawatts. Then with the slip ω_Δ in pu

$$T_d = (S_{B3}D)\omega_{\Delta u} \text{ MW} \tag{D.17}$$

It is also common to see the slip computed in hertz, i.e., f_Δ Hz. Then (D.17) becomes

$$T_d = (S_{B3}D/f_R)f_\Delta = D'f_\Delta \text{ MW} \tag{D.18}$$

where S_{B3} is the three-phase MVA base, f_R is the base frequency in Hz, and f_Δ is the slip in Hz. A value sometimes used for D' in (D.18) is

$$D' = P_G/f_R \text{ MW/Hz} \tag{D.19}$$

wnere P_G is the scheduled power generated in MW for this unit. This corresponds to $D = P_G/S_{B3}$ pu.

(4) Voltage regulator type

The type of voltage regulator system is tabulated using an alphabetical symbol that corresponds to the block diagrams shown in Figures D.4–D.11. Excitation systems have undergone significant changes in the past decade, both in design and in the models for representing the various designs. The models proposed by the IEEE committee in 1969 [3] have been largely superseded by newer systems and alternate models for certain older systems. The approach used here is the alphabetic labeling adopted by the Western Systems Coordinating Council (WSCC), provided through private communication. The need for expanded modeling and common format for exchange of modeling data is under study by an IEEE working group at the time of publication of this book.

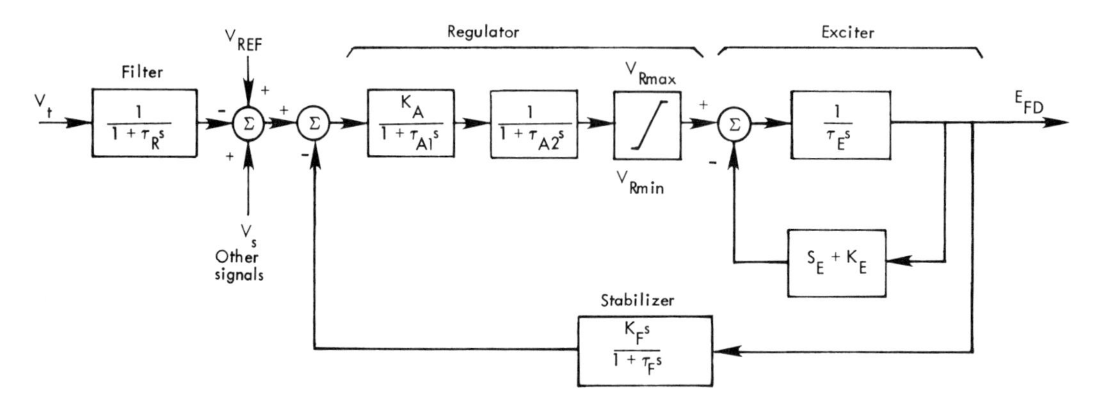

Fig. D.4. Type A—continuously acting dc rotating excitation system. Representative systems: (1) $T_R = 0$: General Electric NA143, NA108; Westinghouse Mag-A-Stat, WMA; Allis Chalmers Regulux; (2) $T_R \neq 0$: General Electric NA 101; Westinghouse Rototrol, Silverstat, TRA.

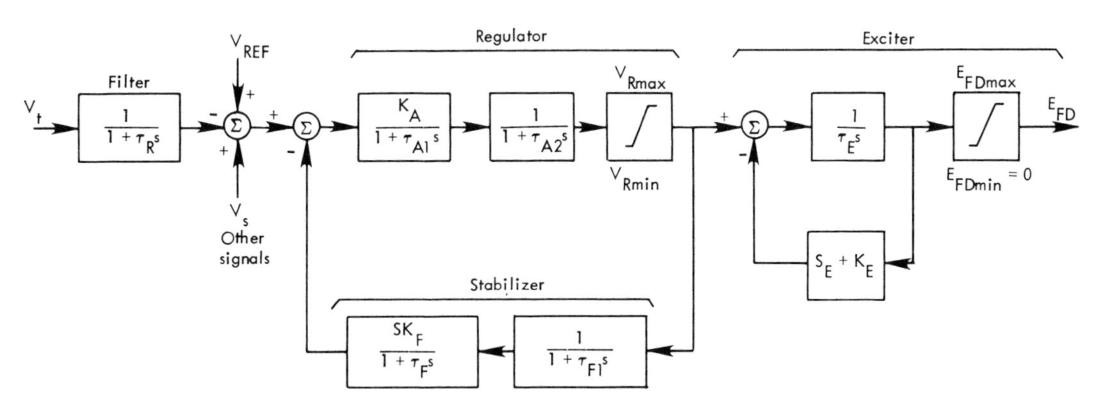

Fig. D.5. Type B—Westinghouse pre-1967 brushless.

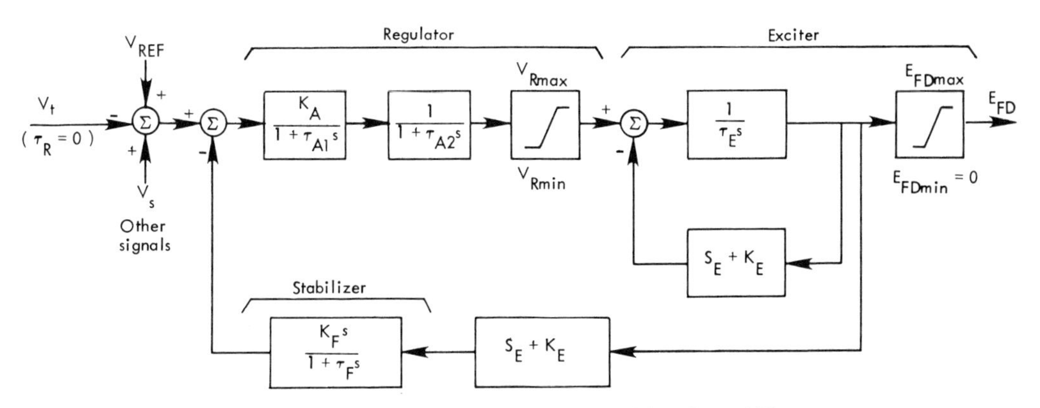

Fig. D.6. Type C—Westinghouse brushless since 1966.

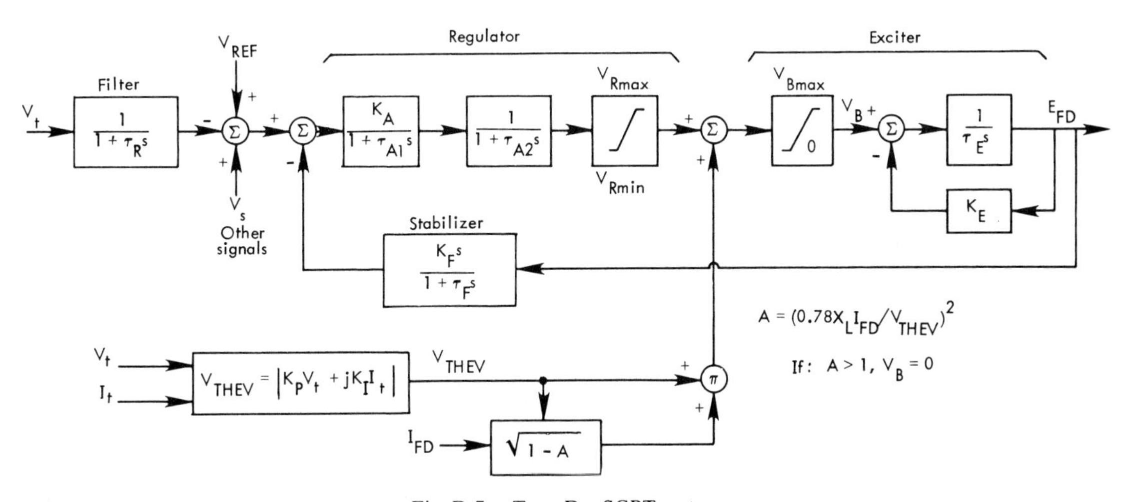

Fig. D.7. Type D—SCPT system.

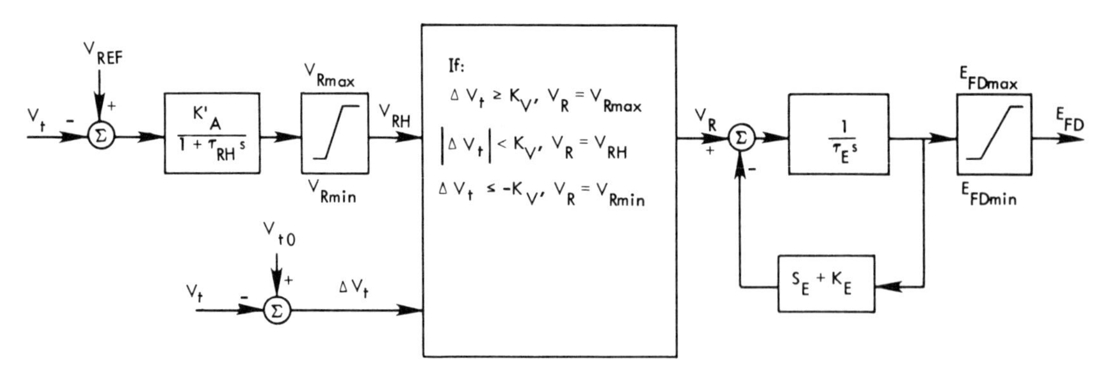

Fig. D.8. Type E—noncontinuously acting rheostatic excitation system. Representative systems: General Electric GFA4, Westinghouse BJ30.

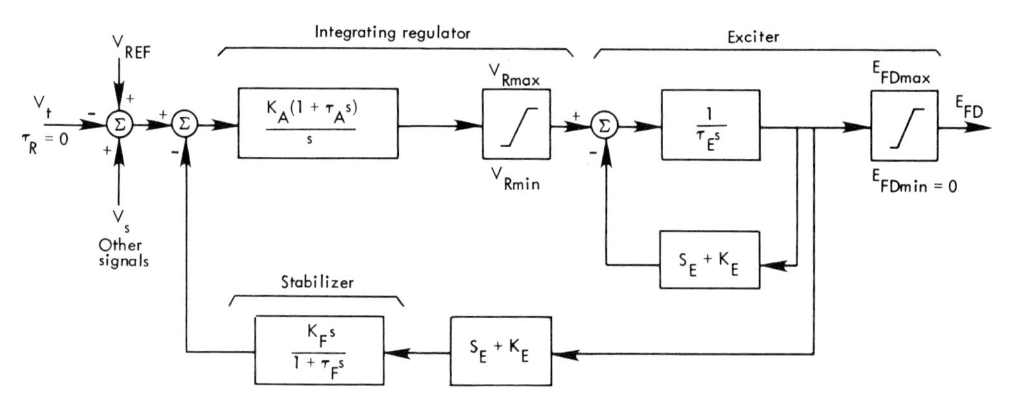

Fig. D.9. Type F—Westinghouse continuously acting brushless rotating alternator excitation system.

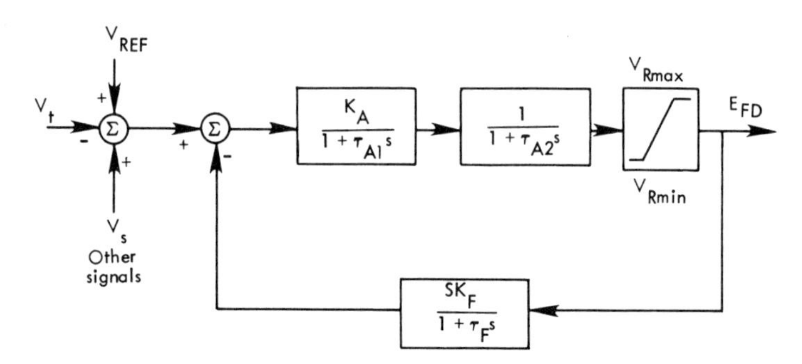

Fig. D.10. Type G—General Electric SCR excitation system.

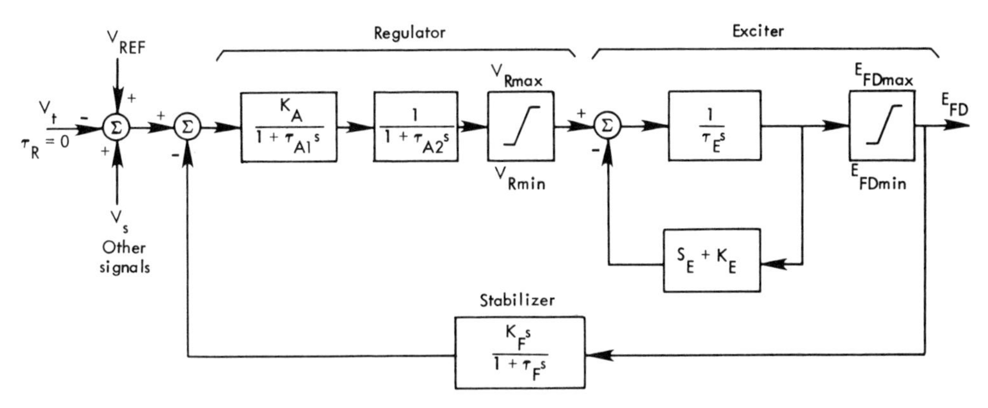

Fig. D.11. Type K—General Electric Alterrex.

Note that the regulator base voltage used to normalize V_R may be chosen arbitrarily. Since the exciter input signal is usually $V_R - (S_E + K_E)E_{FD}$, choosing a different base affects the constant S_E and K_E and also the gain K_A.

(5) Exciter saturation

The saturation of dc generator exciters is represented by an exponential model derived to fit the actual saturation curve at the exciter ceiling (max) voltage (zero field rheostat setting) and at 75% of ceiling. Referring to Figure D.12, we define the following constants at ceiling, 0.75 of ceiling and full load.

$$S_{E\max} = (A - B)/B \qquad S_{E.75\max} = (E - F)/F \qquad S_{EFL} = (C - D)/D \qquad \text{(D.20)}$$

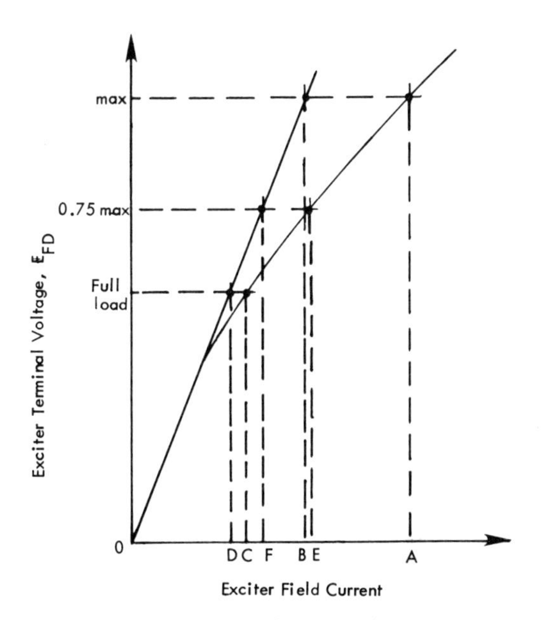

Fig. D.12. A dc exciter saturation curve.

Then in pu with E_{FDFL} as a base (actually, any convenient base may be used),

$$E_{FD\max} = E_{FD\max}(\mathrm{V})/E_{FDFL}(\mathrm{V}) = B/D \text{ pu}$$

or

$$B = DE_{FD\max} \tag{D.21}$$

We can also compute

$$B/F = 4/3 = DE_{FD\max}/F$$

or

$$F = 0.75DE_{FD\max} \tag{D.22}$$

Combining (D.20)–(D.22) we can write

$$S_{E\max} = (A - B)/B = (A - B)/DE_{FD\max}$$
$$S_{E.75\max} = (E - F)/F = (E - F)/0.75DE_{FD\max} = (4/3)(E - F)/DE_{FD\max} \tag{D.23}$$

Now define the saturation function

$$S_E \triangleq A_{EX}e^{B_{EX}E_{FD}} \tag{D.24}$$

which gives the approximate saturation for any E_{FD}. Suppose we are given the numerical values of saturation at $E_{FD\max}$ and $0.75E_{FD\max}$. These values are called $S_{E\max}$ and $S_{E.75\max}$ respectively. Using these two saturation values, we compute the two unknowns A_{EX} and B_{EX} as follows. At $E_{FD} = E_{FD\max}$

$$S_E = S_{E\max} = (A - B)/DE_{FD\max} = A_{EX}e^{B_{EX}E_{FD\max}} \tag{D.25}$$

and at $E_{FD} = 0.75E_{FD\max}$

$$S_E = S_{E.75\max} = (4/3)(E - F)/DE_{FD\max} = A_{EX}e^{B_{EX}(0.75E_{FD\max})} \tag{D.26}$$

We then solve (D.25) and (D.26) simultaneously to find

$$A_{EX} = S^4_{E.75\max}/S^3_{E\max} \qquad B_{EX} = (4/E_{FD\max})\ln(S_{E\max}/S_{E.75\max}) \tag{D.27}$$

(6) Governor representation

Three types of governor representation are specified in this appendix: a general governor model that can be used for both steam and hydro turbines, a cross-compound governor model, and a hydraulic governor model. The appropriate model is identified by the letters G, C, and H in the tabulation. The governor block diagrams are given in Figures D.13–D.15. The regulation R is the steady-state regulation or droop and is usually factory set at 5% for U.S. units.

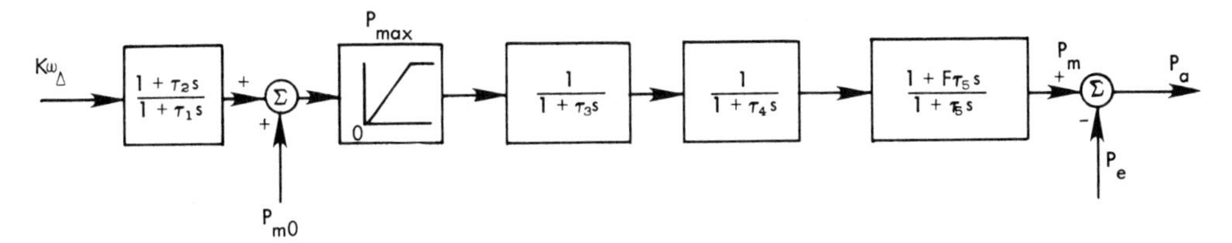

Fig. D.13. General purpose governor block diagram.

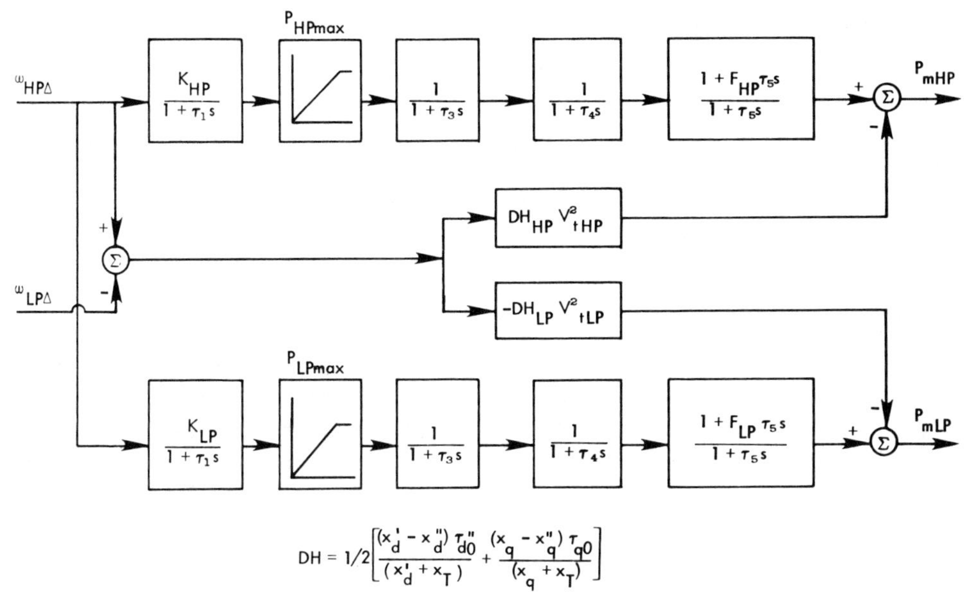

$$DH = 1/2\left[\frac{(x_d' - x_d'')\tau_{d0}''}{(x_d' + x_T)} + \frac{(x_q - x_q'')\tau_{q0}}{(x_q + x_T)}\right]$$

Fig. D.14. Cross-compound governor block diagram.

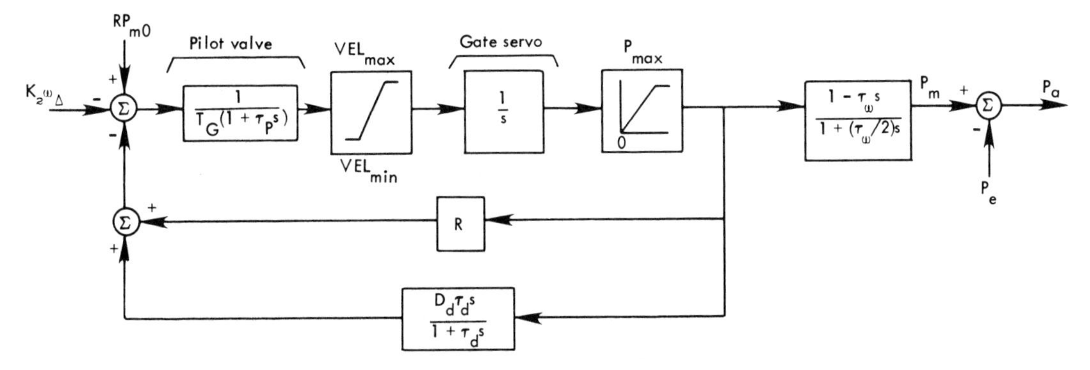

Fig. D.15. Hydroturbine governor block diagram.

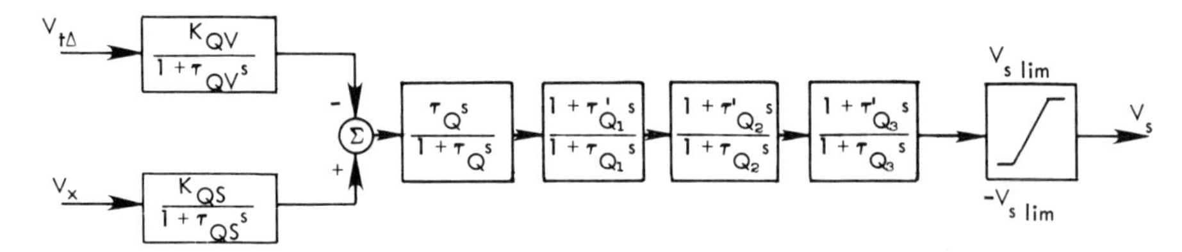

Fig. D.16. Power system stabilizer block diagram. Stabilizer types: (1) V_x = rotor slip = ω_Δ, (2) V_x = frequency deviation = f_Δ, (3) V_x = accelerating power = P_a.

(7) Power system stabilizer

The constants used for power system stabilizer (PSS) settings will always depend on the location of a unit electrically in the system, the dynamic characteristics of the system, and the dynamic characteristics of the unit. Still there is some merit in having approximate data that can be considered typical of stabilizer settings. Values given in Tables D.2–D.5 are actual settings used at certain locations and may be used as a rough estimate for stabilizer adjustment studies. The PSS block diagram is given in Figure D.16.

D.2 Data for Transmission Lines

Data are provided in Table D.8 for estimating the impedance of transmission lines. Usually, accurate data are available for transmission circuits, based on actual utility line design information. Table D.8 provides data for making rough estimates of transmission line impedances for a variety of common 60-Hz ac transmission voltages.

References

1. Fitzgerald, A. E., Kingsley, C., Jr., and Kusko, A. *Electric Machinery*, 3rd ed. McGraw Hill, New York, 1971.
2. Byerly, R. T., Sherman, D. E., and McCauley, T. M. Stability program data preparation manual. Westinghouse Electric Corp. Rept. 70–736. 1970. (Rev. Dec. 1972.)
3. IEEE Working Group. Computer representation of excitation systems. *IEEE Trans.* PAS-87:1460–64, 1968.
4. Prubhashankar, K., and Janischewdkyj, W. Digital simulation of multi-machine power systems for stability studies. *IEEE Trans.* PAS-87:73–80, 1968.
5. Crary, S. B., Shildneck, L. P., and March, L. A. Equivalent reactance of synchronous machines. *Electr. Eng.* Jan.: 124–32; discussions, Mar.: 484–88; Apr.: 603–7, 1934.
6. Kingsley, C., Jr. Saturated synchronous reactance. *Electr. Eng.* Mar.: 300–305, 1935.
7. Kilgore, L. A. Effects of saturation on machine reactances. *Electr. Eng.* May: 545–50, 1935.
8. Concordia, C. Effect of steam-turbine reheat on speed-governor performances. *ASME J. Eng. Power.* Apr.: 201–6, 1950.
9. Kirchmayer, L. K. *Economic Control of Interconnected Systems*. Wiley, New York, 1959.
10. Young, C. C., and Webler, R. M. A new stability program for predicting dynamic performance of electric power systems. *Proc. Am. Power Conf.* 29:1126–39, 1967.
11. Crary, S. B. *Power System Stability*, Vol. 2. Wiley, New York, 1947.
12. Concordia, C. Synchronous machine damping and synchronizing torques. *AIEE Trans.* 70:731–37, 1951.

Table D.1. Definitions of Tabulated Generator Unit Data

GENERATOR		
Unit no.		Arbitrary reference number
Rated MVA		Machine-rated MVA; base MVA for impedances
Rated kV		Machine-rated terminal voltage in kV; base kV for impedances
Rated PF		Machine-rated power factor
SCR	(1)	Machine short circuit ratio
x_d''	pu	Unsaturated d axis subtransient reactance
x_d'	pu	Unsaturated d axis transient reactance
x_d	pu	Unsaturated d axis synchronous reactance
x_q''	pu	Unsaturated q axis subtransient reactance
x_q'	pu	Unsaturated q axis transient reactance
x_q	pu	Unsaturated q axis synchronous reactance
r_a	pu	Armature resistance
x_ℓ or x_p	pu	Leakage or Potier reactance
r_2	pu	Negative-sequence resistance
x_2	pu	Negative-sequence reactance
x_0	pu	Zero-sequence reactance
τ_d''	s	d axis subtransient short circuit time constant
τ_d'	s	d axis transient short circuit time constant
τ_{d0}''	s	d axis subtransient open circuit time constant
τ_{d0}'	s	d axis transient open circuit time constant
τ_q''	s	q axis subtransient short circuit time constant
τ_q'	s	q axis transient short circuit time constant
τ_{q0}''	s	q axis subtransient open circuit time constant
τ_{q0}'	s	q axis transient open circuit time constant
τ_a	s	Armature time constant
W_R	MW·s	Kinetic energy of turbine + generator at rated speed in MJ or MW·s
r_F	Ω	Machine field resistance in Ω
$S_{G1.0}$	(2)	Machine saturation at 1.0 pu voltage in pu
$S_{G1.2}$	(2)	Machine saturation at 1.2 pu voltage in pu
E_{FDFL}	(2)	Machine full load excitation in pu
D	(3)	Machine load damping coefficient

EXCITER		
VR Type	(4)	Excitation system type
Name		Excitation system name
RR	(4)	Exciter response ratio (formerly ASA response)
τ_R	s	Regulator input filter time constant
K_A	pu	Regulator gain (continuous acting regulator) or fast raise-lower contact setting (rheostatic regulator)
τ_A or τ_{A1}	s	Regulator time constant (#1)
τ_{A2}	s	Regulator time constant (#2)

EXCITER (*continued*)		
$V_{R\max}$	pu (4)	Maximum regulator output, starting at full load field voltage
$V_{R\min}$	pu (4)	Minimum regulator output, starting at full load field voltage
K_E	pu	Exciter self-excitation at full load field voltage
τ_E	s	Exciter time constant
$S_{E.75\max}$	(5)	Rotating exciter saturation at 0.75 ceiling voltage, or K_I for SCPT exciter
$S_{E\max}$	(5)	Rotating exciter saturation at ceiling voltage, or K_p for SCPT exciter
A_{EX}	(5)	Derived saturation constant for rotating exciters
B_{EX}	(5)	Derived saturation constant for rotating exciters
$E_{FD\max}$	pu (5)	Maximum field voltage or ceiling voltage, pu
$E_{FD\min}$	pu	Minimum field voltage
K_F	pu	Regulator stabilizing circuit gain
τ_F or τ_{F1}	s	Regulator stabilizing circuit time constant (#1)
τ_{F2}	s	Regulator stabilizing circuit time constant (#2)

TURBINE-GOVERNOR		
GOV	(6)	Governor type: G = general, C = cross-compound, H = hydraulic
R	(6)	Turbine steady-state regulation setting or droop
$P_{\max}$	MW	Maximum turbine output in MW
τ_1	s	Control time constant (governor delay) or governor response time (type H)
τ_2	s	Hydro reset time constant (type G) or pilot valve time (type H)
τ_3	s	Servo time constant (type G or C), or hydro gate time constant (type G) or dashpot time constant (type H)
τ_4	s	Steam valve bowl time constant (zero for type G hydrogovernor) or ($\tau_W/2$ for type H)
τ_5	s	Steam reheat time constant or 1/2 hydro water starting time constant (type C or G) or minimum gate velocity in MW/s (type H)
F	(6)	pu shaft output ahead of reheater or −2.0 for hydro units (types C or G), or maximum gate velocity in MW/s (type H)

STABILIZER		
PSS	(7)	PSS feedback: F = frequency, S = speed, P = accelerating power
K_{QV}	(7)	PSS voltage gain, pu
K_{QS}	(7)	PSS speed gain, pu
τ_Q	s	PSS reset time constant
τ_{Q1}'	s	First lead time constant
τ_{Q1}	s	First lag time constant
τ_{Q2}'	s	Second lead time constant
τ_{Q2}	s	Second lag time constant
τ_{Q3}'	s	Third lead time constant
τ_{Q3}	s	Third lag time constant
$V_{S\lim}$	pu	PSS output limit setting, pu

Table D.2. Typical Data for Hydro (H) Units

GENERATOR										
Unit no.		$H1$	$H2$	$H3$	$H4$	$H5$	$H6$	$H7$	$H8$	$H9$
Rated MVA		9.00	17.50	25.00	35.00	40.00	54.00	65.79	75.00	86.00
Rated kV		6.90	7.33	13.20	13.80	13.80	13.80	13.80	13.80	13.80
Rated PF		0.90	0.80	0.95	0.90	0.90	0.90	0.95	0.95	0.90
SCR	(1)	1.250	...	2.280	1.167	1.180	1.18	1.175	2.36	1.18
x''_d	pu	0.329	0.330	0.310	0.235	0.288	0.340	0.240	0.140	0.258
x'_d	pu	0.408	...	...	0.260	0.318	0.380	0.260	0.174	0.320
x_d	pu	0.911	1.070	1.020	1.000	0.990	1.130	0.900	0.495	1.050
x''_q	pu	...	...	...	0.264	0.306	0.340	...	0.135	0.306
x'_q	pu	0.580	0.660	0.650	0.620	0.615	0.680	0.540	...	0.670
x_q	pu	0.580	0.660	0.650	0.620	0.615	0.680	0.540	0.331	0.670
r_a	pu	...	0.003	0.0032	0.004	0.0029	0.0049	0.0022	0.0041	0.0062
x_ℓ or x_p	pu	...	0.310	0.924	0.170	0.224	0.2100	...	0.120	0.140
r_2	pu	...	0.030	0.030	0.040	...	...	0.014	...	0.060
x_2	pu	...	0.490	0.460	0.270	0.297	0.340	0.260	0.130	0.312
x_0	pu	...	0.200	0.150	0.090	0.125	0.180	0.130	0.074	0.130
τ''_d	s	...	0.035	0.035	0.035	...	...	...	...	0.044
τ'_d	s	...	1.670	2.190	2.300	1.700	3.000	1.600	1.850	2.020
τ''_{d0}	s	...	...	...	...	...	...	...	...	0.051
τ'_{d0}	s	4.200	5.400	7.200	7.100	5.300	8.500	5.500	8.400	4.000
τ''_q	s	...	0.035	0.035	0.035	...	...	...	...	0.017
τ'_q	s	...	0.835	1.100	1.150	...	...	...	...	...
τ''_{q0}	s	...	...	...	...	...	...	...	...	0.033
τ'_{q0}	s	...	...	...	...	...	...	...	...	...
τ_a	s	0.1800	...	...	...	...	...	...	...	0.286
W_R	MW·s	23.50	117.00	183.00	254.00	107.90	168.00	176.00	524.00	233.00
r_F	Ω	...	...	...	...	0.269	0.301	0.199	0.155	0.332
$S_{G1.0}$	(2)	0.160	0.064	0.064	0.064	0.194	0.3127	0.1827	0.170	0.245
$S_{G1.2}$	(2)	0.446	1.018	1.018	1.018	0.685	0.7375	0.507	0.440	0.770
E_{FDFL}	(2)	2.080	2.130	2.130	2.130	2.030	2.320	1.904	1.460	2.320
D	(3)	2.000	2.000	2.000	2.000	2.000	2.000	2.000	2.000	2.000
EXCITER										
VR type	(4)	E	E	E	E	A	A	A	A	A
Name		RHEO	AJ23	GFA4	WMA	NA108	REGULUX	WMA	NA108	NA143
RR	(4)	0.88	0.5	0.5	0.5	0.5	0.5	1.85	0.5	0.5
τ_R	s	0.000	0.000	0.000	0.000	0.000	0.000	0.000	0.000	0.000
K_A	pu	0.050	0.050	0.050	0.050	65.200	25.000	37.300	180.000	242.000
τ_A or τ_{A1}	s	20.000	20.000	20.000	20.000	0.200	0.200	0.120	1.000	0.060
τ_{A2}	s	0.000	0.000	0.000	0.000	0.000	0.000	0.012	0.000	0.000
$V_{R\max}$	pu (4)	4.320	5.940	4.390	5.940	2.607	1.000	1.410	3.000	5.320
$V_{R\min}$	pu (4)	0.000	1.210	0.000	1.210	−2.607	−1.000	−1.410	−3.000	−5.320
K_E	pu	1.000	1.000	1.000	1.000	−0.111	−0.057	−0.137	−0.150	−0.1219
τ_E	s	2.019	0.760	1.970	0.760	1.930	0.646	0.560	2.000	2.700
$S_{E.75\max}$	(5)	0.099	0.220	0.096	0.220	0.176	0.0885	0.328	0.623	0.450
$S_{E\max}$	(5)	0.385	0.950	0.375	0.950	0.610	0.3480	0.687	1.327	1.500
A_{EX}	(5)	0.0017	0.0027	0.0016	0.0027	0.0042	0.0015	0.0357	0.0645	0.0121
B_{EX}	(5)	1.7412	1.9185	1.7059	1.9185	0.9488	1.5738	1.1507	1.1861	1.3566
$E_{FD\max}$	pu (5)	3.120	3.050	3.195	3.050	5.240	3.480	2.570	2.550	3.550
$E_{FD\min}$	pu	0.000	1.210	0.000	1.210	−5.240	−3.480	−2.570	−2.550	−3.550
K_F	pu	0.000	0.000	0.000	0.000	0.120	0.103	0.055	0.150	0.100
τ_F or τ_{F1}	s	0.000	0.000	0.000	0.000	1.000	1.000	1.000	1.000	1.000
τ_{F2}	s	0.000	0.000	0.000	0.000	0.000	0.000	0.000	0.000	0.000

Table D.2 *(continued)*

TURBINE-GOVERNOR										
GOV	(6)	*G*	*G*	*G*	*G*	*G*	*G*	*G*	*G*	*G*
R	(6)	0.050	0.050	0.050	0.050	0.056	0.050	0.050	0.050	0.050
P_{max}	MW	8.60	14.00	23.80	40.00	40.00	52.50	65.50	90.00	86.00
τ_1	s	48.440	16.000	16.000	16.000	0.000	0.000	25.600	20.000	12.000
τ_2	s	4.634	2.400	2.400	2.400	0.000	0.000	2.800	4.000	3.000
τ_3	s	0.000	0.920	0.920	0.920	0.500	0.000	0.500	0.500	0.500
τ_4	s	0.000	0.000	0.000	0.000	0.000	0.000	0.000	0.000	0.000
τ_5	s	0.579	0.300	0.300	0.300	0.430	0.785	0.350	0.850	1.545
F	(6)	−2.000	−2.000	−2.000	−2.000	−2.000	−2.000	−2.000	−2.000	−2.000
STABILIZER										
PSS	(7)	...	...	...	...	...	...	*F*	*F*	*F*
K_{QV}	(7)	...	...	...	...	...	...	0.000	0.000	0.000
K_{QS}	(7)	...	...	...	...	...	...	1.000	4.000	3.150
τ_Q	s	...	...	...	...	...	...	30.000	30.000	10.000
τ'_{Q1}	s	...	...	...	...	...	...	0.500	0.700	0.758
τ_{Q1}	s	...	...	...	...	...	...	0.030	0.100	0.020
τ'_{Q2}	s	...	...	...	...	...	...	0.500	0.700	0.758
τ_{Q2}	s	...	...	...	...	...	...	0.030	0.050	0.020
τ'_{Q3}	s	...	...	...	...	...	...	0.000	0.000	0.000
τ_{Q3}	s	...	...	...	...	...	...	0.000	0.000	0.000
$V_{s\,lim}$	pu	...	...	...	...	...	...	0.100	0.100	0.095

Table D.2. (*cont.*)

Table D.2 *(continued)*

GENERATOR										
Unit no.		*H*10	*H*11	*H*12	*H*13	*H*14	*H*15	*H*16	*H*17	*H*18
Rated MVA		100.10	115.00	125.00	131.00	145.00	158.00	231.60	250.00	615.00
Rated kV		13.80	12.50	13.80	13.80	14.40	13.80	13.80	18.00	15.00
Rated PF		0.90	0.85	0.90	0.90	0.90	0.90	0.95	0.85	0.975
SCR	(1)	1.20	1.05	1.155	1.12	1.20	...	1.175	1.050	...
x''_d	pu	0.280	0.250	0.205	0.330	0.273	0.220	0.245	0.155	0.230
x'_d	pu	0.314	0.315	0.300	0.360	0.312	0.300	0.302	0.195	0.2995
x_d	pu	1.014	1.060	1.050	1.010	0.953	0.920	0.930	0.995	0.8979
x''_q	pu	0.375	0.287	0.221	0.330	0.402	0.290	0.270	0.143	0.2847
x'_q	pu	0.770	0.610	0.686	0.570	0.573	0.510	...	0.568	0.646
x_q	pu	0.770	0.610	0.686	0.570	0.573	0.510	0.690	0.568	0.646
r_a	pu	0.0049	0.0024	0.0023	0.004	...	0.002	0.0021	0.0014	...
x_ℓ or x_p	pu	0.163	0.147	0.218	0.170	0.280	0.130	0.340	0.160	0.2396
r_2	pu	...	0.027	0.008	...	...	0.045	...	...	...
x_2	pu	0.326	0.269	0.211	0.330	...	0.255	0.258	...	...
x_0	pu	...	0.161	0.150	0.150	...	0.120	0.135	...	...
τ''_d	s	0.035	...	...	0.030	...	0.024	0.020	...	...
τ'_d	s	1.810	2.260	1.940	2.700	...	1.600	3.300	...	...
τ''_{d0}	s	0.039	0.040	...	0.030	0.041	0.029	0.030	...	...
τ'_{d0}	s	6.550	8.680	6.170	7.600	7.070	5.200	8.000	9.200	7.400
τ''_q	s	...	...	...	0.030	...	0.028	0.020	...	...
τ'_q	s	...	...	...	...	...	...	...	...	...
τ''_{q0}	s	0.071	0.080	...	0.040	0.071	0.034	0.060	...	...
τ'_{q0}	s	...	...	...	...	...	...	...	...	...
τ_a	s	0.278	0.330	...	0.180	...	0.360	0.200	...	...
W_R	MW·s	312.00	439.00	392.09	458.40	469.00	502.00	786.00	1603.00	3166.00
r_F	Ω	0.332	0.156	0.379	0.182	...	0.206	0.181	...	...
$S_{G1.0}$	(2)	0.219	0.178	0.200	0.113	0.220	0.1642	0.120	0.0769	0.180
$S_{G1.2}$	(2)	0.734	0.592	0.612	0.478	0.725	0.438	0.400	0.282	0.330
E_{FDFL}	(2)	2.229	2.200	2.220	1.950	2.230	1.990	1.850	1.88	...
D	(3)	2.000	2.000	2.000	2.000	2.000	2.000	2.000	2.000	2.000
EXCITER										
VR type	(4)	*A*	*A*	*A*	*G*	*A*	*A*	*A*	*A*	*J*
Name		WMA	WMA	NA143A	SCR	WMA	NA143	SIEMEN	ASEA	...
RR	(4)	1.0	1.5	1.5	0.5	1.0	0.5	1.0	1.0	...
τ_R	s	0.000	0.000	0.000	0.000	0.000	0.000	0.000	0.000	0.000
K_A	pu	400.000	276.000	54.000	272.000	400.000	17.800	50.000	100.000	200.000
τ_A or τ_{A1}	s	0.050	0.060	0.105	0.020	0.050	0.200	0.060	0.020	0.020
τ_{A2}	s	0.000	0.000	0.011	0.000	0.000	0.000	0.000	0.000	0.000
$V_{R\max}$	pu (4)	4.120	1.960	3.850	2.730	4.120	0.710	1.000	5.990	7.320
$V_{R\min}$	pu (4)	−4.120	−1.960	−3.850	−2.730	−4.120	−0.710	−1.000	−5.990	0.000
K_E	pu	−0.243	−0.184	−0.062	1.000	−0.243	−0.295	−0.080	−0.020	1.000
τ_E	s	0.950	1.290	0.732	0.000	0.950	0.535	0.405	0.100	0.000
$S_{E.75\max}$	(5)	0.484	0.270	0.410	0.000	0.480	0.333	0.200	0.127	0.000
$S_{E\max}$	(5)	1.308	0.560	1.131	0.000	1.310	0.533	0.407	0.300	0.000
A_{EX}	(5)	0.0245	0.0303	0.0195	0.000	0.0236	0.0812	0.0237	0.0096	0.000
B_{EX}	(5)	1.0276	0.5612	1.1274	0.000	1.0377	0.6303	0.9227	1.1461	0.000
$E_{FD\max}$	pu (5)	3.870	5.200	3.600	2.730	3.870	2.985	3.080	3.000	7.320
$E_{FD\min}$	pu	−3.870	−5.200	−3.600	0.000	−3.870	−2.985	−3.080	−3.000	0.000
K_F	pu	0.040	0.0317	0.140	0.0043	0.040	0.120	0.0648	0.000	0.010
τ_F or τ_{F1}	s	1.000	0.480	1.000	0.060	1.000	1.000	1.000	0.000	1.000
τ_{F2}	s	0.000	0.000	0.000	0.000	0.000	0.000	0.000	0.000	0.000

Table D.2 *(continued)*

TURBINE-GOVERNOR										
GOV	(6)	*G*	*G*	*G*	*G*	*G*	*G*	*G*	*G*	*G*
R	(6)	0.030	0.051	0.050	0.050	0.038	0.050	0.050	0.050	0.050
P_{max}	MW	133.00	115.00	171.00	120.00	160.00	155.00	267.00	250.00	603.30
τ_1	s	52.100	...	31.00	27.500	65.300	...	124.470	30.000	36.000
τ_2	s	4.800	...	4.120	3.240	6.200	...	8.590	3.500	6.000
τ_3	s	0.500	...	0.393	0.500	0.500	...	0.250	0.520	0.000
τ_4	s	0.000	...	0.000	0.000	0.000	...	0.000	0.000	0.000
τ_5	s	0.498	...	0.515	0.520	0.650	...	0.740	0.415	0.900
F	(6)	−2.000	−2.000	−2.000	−2.000	−2.000	−2.000	−2.000	−2.000	−2.000
STABILIZER										
PSS	(7)	*F*	*F*	*F*	...	...	...	*F*	*F*	*F*
K_{QV}	(7)	0.000	0.000	0.000	...	...	...	0.000	0.000	0.000
K_{QS}	(7)	1.000	0.300	8.000	...	...	...	4.000	10.000	5.000
τ_Q	s	10.000	10.000	30.000	...	...	...	55.000	15.000	10.000
τ'_{Q1}	s	0.700	0.431	0.600	...	...	...	1.000	0.000	0.380
τ_{Q1}	s	0.020	0.020	0.100	...	...	...	0.020	0.053	0.020
τ'_{Q2}	s	0.700	0.431	0.600	...	...	...	1.000	0.000	0.380
τ_{Q2}	s	0.020	0.020	0.040	...	...	...	0.020	0.053	0.020
τ'_{Q3}	s	0.000	0.000	0.000	...	...	...	0.000	0.000	0.000
τ_{Q3}	s	0.000	0.000	0.000	...	...	...	0.000	0.000	0.000
V_{slim}	pu	0.050	0.100	0.100	...	...	...	0.090	0.050	0.050

Table D.3. Typical Data for Fossil Steam (F) Units

GENERATOR												
Unit no.		$F1$	$F2$	$F3$	$F4$	$F5$	$F6$	$F7$	$F8$	$F9$	$F10$	$F11$
Rated MVA		25.00	35.29	51.20	75.00	100.00	125.00	147.10	160.00	192.00	233.00	270.00
Rated kV		13.80	13.80	13.80	13.80	13.80	15.50	15.50	15.00	18.00	20.00	18.00
Rated PF		0.80	0.85	0.80	0.80	0.80	0.85	0.85	0.85	0.85	0.85	0.85
SCR	(1)	0.80	0.80	0.90	1.00	0.90	0.90	0.64	0.64	0.64	0.64	0.6854
x_d''	pu	0.120	0.118	0.105	0.130	0.145	0.134	0.216	0.185	0.171	0.249	0.185
x_d'	pu	0.232	0.231	0.209	0.185	0.220	0.174	0.299	0.245	0.232	0.324	0.256
x_d	pu	1.250	1.400	1.270	1.050	1.180	1.220	1.537	1.700	1.651	1.569	1.700
x_q''	pu	0.120	...	0.116	0.130	0.145	0.134	0.216	0.185	0.171	0.248	0.147
x_q'	pu	0.715	...	0.850	0.360	0.380	0.250	0.976	0.380	0.380	0.918	0.245
x_q	pu	1.220	1.372	1.240	0.980	1.050	1.160	1.520	1.640	1.590	1.548	1.620
r_a	pu	0.0014	...	...	0.0031	0.0035	0.004	0.0034	0.0031	0.0026	0.0016	0.0016
x_ℓ or x_p	pu	0.134	...	0.108	0.070	0.075	0.078	0.333	0.110	0.102	0.204	0.155
r_2	pu	0.0082	...	...	0.016	0.020	0.017	0.0284	0.016	0.023	...	...
x_2	pu	0.120	0.118	0.105	0.085	0.095	0.134	0.216	0.115	0.171	0.248	0.140
x_0	pu	0.0215	0.077	0.116	0.070	0.065	...	0.093	0.100	...	0.143	0.060
τ_d''	s	0.035	...	...	...	...	0.023	0.035	...	0.023	0.350	0.027
τ_d'	s	0.882	...	0.882	...	...	1.280	...	...	0.829	0.950	0.620
τ_{d0}''	s	0.059	...	...	0.038	0.042	0.033	0.0484	0.033	0.033	0.0437	...
τ_{d0}'	s	4.750	5.500	6.600	6.100	5.900	8.970	4.300	5.900	5.900	5.140	4.800
τ_q''	s	0.035	...	...	...	...	0.023	0.0072	...	0.023	...	...
τ_q'	s	...	...	...	...	...	0.640	...	...	0.415	...	...
τ_{q0}''	s	0.210	...	...	0.099	0.092	0.070	0.218	0.076	0.078	0.141	...
τ_{q0}'	s	1.500	...	...	0.300	0.300	0.500	1.500	0.540	0.535	1.500	0.500
τ_a	s	0.177	...	...	0.140	0.140	0.390	0.470	0.240	0.254	0.420	0.297
W_R	MW·s	125.40	154.90	260.00	464.00	498.50	596.00	431.00	634.00	634.00	960.50	1115.00
r_F	Ω	0.375	...	0.295	0.290	0.215	...	...	0.370	...	...	0.166
$S_{G1.0}$	(2)	0.279	0.210	0.2067	0.100	0.0933	0.1026	0.057	0.1251	0.105	0.0987	0.125
$S_{G1.2}$	(2)	0.886	0.805	0.724	0.3928	0.4044	0.4320	0.364	0.7419	0.477	0.303	0.450
E_{FDFL}	(2)	2.500	3.000	2.310	2.120	2.292	2.220	2.670	2.680	2.640	2.580	2.300
D	(3)	2.000	2.000	2.000	2.000	2.000	2.000	2.000	2.000	2.000	2.000	2.000
EXCITER												
VR type		E	A	A	E	A	A	A	A	A	C	A
Name	(4)	BJ30	NA143A	WMA	GFA4	NA101	NA101	WMA	NA101	NA101	BRLS	BBC
RR	(4)	0.50	0.50	1.50	0.50	0.50	0.50	0.50	0.50	0.50	0.50	0.50
τ_R	s	0.000	0.000	0.000	0.000	0.060	0.060	0.000	0.060	0.060	0.000	0.000
K_A	pu	0.050	57.140	400.000	0.050	25.000	25.000	175.000	25.000	25.000	250.000	30.000
τ_A or τ_{A1}	s	20.000	0.050	0.050	20.000	0.200	0.200	0.050	0.200	0.200	0.060	0.400

τ_{A2}	s	0.000	0.000	0.000	0.000	0.000	0.000	0.000	0.000	0.000	0.000	0.000
$V_{R\max}$	pu (4)	6.812	1.000	0.6130	4.380	1.000	1.000	3.120	1.000	1.000	4.420	4.590
$V_{R\min}$	pu (4)	1.395	−1.000	−0.6130	0.000	−1.000	−1.000	−3.120	−1.000	−1.000	−4.420	−4.590
K_E	pu	1.000	−0.0445	−0.0769	1.000	−0.0582	−0.0601	−0.170	−0.0497	−0.0505	1.000	−0.020
τ_E	s	0.700	0.500	1.370	1.980	0.6544	0.6758	0.952	0.560	0.5685	0.613	0.560
$S_{E.75\max}$	(5)	0.414	0.0684	0.1120	0.0967	0.0895	0.0924	0.220	0.0765	0.0778	0.010	0.730
$S_{E\max}$	(5)	0.908	0.2667	0.2254	0.3774	0.349	0.3604	0.950	0.2985	0.303	0.270	1.350
A_{EX}	(5)	0.0392	0.0012	0.0137	0.0016	0.0015	0.0016	0.0027	0.0013	0.0013	0.000	0.1154
B_{EX}	(5)	0.8807	1.2096	0.6774	1.7128	1.5833	1.6349	1.4628	1.3547	1.3733	3.7884	0.7128
$E_{FD\max}$	pu (5)	3.567	4.500	4.130	3.180	3.438	3.330	4.000	4.020	3.960	3.480	3.450
$E_{FD\min}$	pu	1.417	−4.500	−4.130	0.000	−3.438	−3.330	−4.000	−4.020	−3.960	0.000	−3.450
K_F	pu	0.000	0.080	0.040	0.000	0.105	0.108	0.030	0.0896	0.091	0.053	0.050
τ_F or τ_{F1}	s	0.000	1.000	1.000	0.000	0.350	0.350	1.000	0.350	0.350	0.330	1.300
τ_{F2}	s	0.000	0.000	0.000	0.000	0.000	0.000	0.000	0.000	0.000	0.000	0.000
TURBINE GOVERNOR												
GOV	(6)	*G*	*G*	*G*	*G*	*G*	*G*	*G*	*G*	*G*	*G*	*G*
R	(6)	0.050	0.050	0.078	0.050	0.050	0.050	0.050	0.050	0.050	0.050	0.050
$P_{\max}$	MW	22.50	36.10	53.00	75.00	105.00	132.00	121.00	142.30	175.00	210.00	230.00
τ_1	s	0.200	0.200	0.200	0.090	0.090	0.083	0.200	0.100	0.083	0.150	0.100
τ_2	s	0.000	0.000	0.000	0.000	0.000	0.000	0.000	0.000	0.000	0.000	0.000
τ_3	s	0.300	0.300	0.300	0.200	0.200	0.200	0.300	0.200	0.200	0.100	0.259
τ_4	s	0.090	0.200	0.090	0.300	0.300	0.050	0.090	0.050	0.050	0.300	0.100
τ_5	s	0.000	0.000	0.000	0.000	0.000	5.000	10.000	8.000	8.000	10.000	10.000
F	(6)	1.000	1.000	1.000	1.000	1.000	0.280	0.250	0.300	0.271	0.237	0.272
STABILIZER												
PSS	(7)	...	...	*F*	...	...	...	...	...	*S*	...	...
K_{QV}	(7)	...	...	0.000	...	...	...	...	...	0.000	...	...
K_{QS}	(7)	...	...	0.700	...	...	...	...	...	15.000	...	...
τ_Q	s	...	...	10.000	...	...	...	...	...	10.000	...	...
τ'_{Q1}	s	...	...	0.300	...	...	...	...	...	1.000	...	...
τ_{Q1}	s	...	...	0.020	...	...	...	...	...	0.020	...	...
τ'_{Q2}	s	...	...	0.300	...	...	...	...	...	0.750	...	...
τ_{Q2}	s	...	...	0.020	...	...	...	...	...	0.020	...	...
τ'_{Q3}	s	...	...	0.000	...	...	...	...	...	0.000	...	...
τ_{Q3}	s	...	...	0.000	...	...	...	...	...	0.000	...	...
V_{slim}	pu	...	...	0.100	...	...	...	...	...	0.050	...	...

Table D.3 *(continued)*

GENERATOR											
Unit no.		$F12$	$F13$	$F14$	$F15$	$F16$	$F17$	$F18$	$F19$	$F20$	$F21$
Rated MVA		330.00	384.00	410.00	448.00	512.00	552.00	590.00	835.00	896.00	911.00
Rated kV		20.00	24.00	24.00	22.00	24.00	24.00	22.00	20.00	26.00	26.00
Rated PF		0.90	0.85	0.90	0.85	0.90	0.90	0.95	0.90	0.90	0.90
SCR	(1)	0.580	0.580	0.580	0.580	0.580	0.580	0.500	0.500	0.52	0.64
x''_d	pu	...	0.260	0.2284	0.205	0.200	0.198	0.215	0.339	0.180	0.193
x'_d	pu	0.317	0.324	0.2738	0.265	0.270	0.258	0.280	0.413	0.220	0.266
x_d	pu	1.950	1.798	1.7668	1.670	1.700	1.780	2.110	2.183	1.790	2.040
x''_q	pu	...	0.255	0.2239	0.205	...	0.172	0.215	0.332	...	0.191
x'_q	pu	1.120	1.051	1.0104	0.460	0.470	0.247	0.490	1.285	0.400	0.262
x_q	pu	1.920	1.778	1.7469	1.600	1.650	1.770	2.020	2.157	1.715	1.960
r_a	pu	...	0.0014	0.0019	0.0043	0.004	0.0047	0.0046	0.0019	0.001	0.001
x_ℓ or x_p	pu	0.199	0.1930	0.1834	0.150	0.160	...	0.155	0.246	0.135	0.154
r_2	pu	...	0.0054	...	0.023	...	0.013	0.026	...	0.019	...
x_2	pu	...	0.2374	0.2261	0.175	...	0.167	0.215	0.309	0.135	0.192
x_0	pu	...	0.1320	0.1346	0.140	...	0.112	0.150	0.174	0.130	0.105
τ''_d	s	...	0.035	...	0.023	...	0.030	0.0225	...	0.035	...
τ'_d	s	...	0.159	...	1.070	...	0.550	...	...	0.596	...
τ''_{d0}	s	...	0.042	0.042	0.032	...	...	0.032	0.041	0.032	...
τ'_{d0}	s	6.000	5.210	5.432	3.700	3.800	3.650	4.200	5.690	4.300	6.000
τ''_q	s	...	0.035	...	...	...	...	0.0225	...	0.035	...
τ'_q	s	...	0.581	...	...	...	...	...	...	0.298	...
τ''_{q0}	s	...	0.042	0.158	0.060	...	...	0.062	0.144	...	...
τ'_{q0}	s	1.500	1.500	1.500	0.470	0.480	1.230	0.565	1.500	...	0.900
τ_a	s	...	0.450	...	0.150	...	...	0.140	...	0.160	...
W_R	MW·s	992.00	1006.50	1518.70	1190.00	1347.20	3010.00	1368.00	2206.40	2625.00	2265.00
r_F	Ω	...	0.1245	...	0.1357	...	0.0711	0.1094	...	...	...
$S_{G1.0}$	(2)	0.082	0.162	0.2632	0.0910	0.090	0.111	0.079	0.134	0.090	0.340
$S_{G1.2}$	(2)	0.290	0.508	0.5351	0.400	0.400	0.518	0.349	0.617	0.402	1.120
E_{FDFL}	(2)	...	3.053	2.7895	2.870	2.700	3.000	2.980	3.670	3.330	3.670
D	(3)	2.000	2.000	2.000	2.000	2.000	2.000	2.000	2.000	2.000	2.000
EXCITER											
VR type		A	C	C	A	G	A	G	C	G	A
Name	(4)	WMA	BRLS	BRLS	NA143A	ALTHYREX	BBC	ALTHYREX	WTA	ALTHYREX	BBC
RR	(4)	0.50	0.50	0.50	0.50	1.50	0.50	3.50	2.00	2.50	0.50
τ_R	s	0.000	0.000	0.000	0.000	0.000	0.000	0.000	0.000	0.000	0.000
K_A	pu	400.000	400.000	400.000	50.000	200.000	30.000	200.000	400.000	250.000	50.000
τ_A or τ_{A1}	s	0.050	0.020	0.020	0.060	0.3950	0.400	0.3575	0.020	0.200	0.060
τ_{A2}	s	0.000	0.000	0.000	0.000	0.000	0.000	0.000	0.000	0.000	0.000

$V_{R\max}$	pu (4)	3.810	8.130	5.270	1.000	3.840	5.990	5.730	18.300	5.150	1.000
$V_{R\min}$	pu (4)	−3.810	−8.130	−5.270	−1.000	−3.840	−5.990	−5.730	−18.300	−5.150	−1.000
K_E	pu	−0.170	1.000	1.000	−0.0465	1.000	−0.020	1.000	1.000	1.000	−0.0393
τ_E	s	0.950	0.812	0.920	0.520	0.000	0.560	0.000	0.942	0.000	0.440
$S_{E.75\max}$	(5)	0.220	0.459	0.435	0.071	0.000	0.730	0.000	0.813	0.000	0.064
$S_{E\max}$	(5)	0.950	0.656	0.600	0.278	0.000	1.350	0.000	2.670	0.000	0.235
A_{EX}	(5)	0.0027	0.1572	0.1658	0.0012	0.000	0.1154	0.000	0.023	0.000	0.0013
B_{EX}	(5)	0.3857	0.2909	0.3910	1.2639	0.000	0.5465	0.000	0.9475	0.000	1.1562
$E_{FD\max}$	pu (5)	4.890	4.910	3.290	4.320	3.840	4.500	5.730	5.020	5.150	4.500
$E_{FD\min}$	pu	−4.890	0.000	0.000	−4.320	−3.840	−4.500	−5.730	0.000	−5.150	−4.500
K_F	pu	0.040	0.060	0.030	0.0832	0.0635	0.050	0.0529	0.030	0.036	0.070
τ_F or τ_{F1}	s	1.000	1.000	1.000	1.000	1.000	1.300	1.000	1.000	1.000	1.000
τ_{F2}	s	0.000	0.000	0.000	0.000	0.000	0.000	0.000	0.000	0.000	0.000
TURBINE GOVERNOR											
GOV	(6)	G	G	G	G	G	G	G	G	G	G
R	(6)	0.050	0.050	0.050	0.050	0.050	0.050	0.050	0.050	0.050	0.050
$P_{\max}$	MW	347.00	360.00	367.00	390.00	460.00	497.00	553.00	766.29	810.00	820.00
τ_1	s	0.100	0.220	0.180	0.100	0.150	0.100	0.080	0.180	0.100	0.100
τ_2	s	0.000	0.000	0.000	0.000	0.050	0.000	0.000	0.030	0.000	0.000
τ_3	s	0.400	0.200	0.040	0.300	0.300	0.300	0.150	0.200	0.200	0.200
τ_4	s	0.050	0.250	0.250	0.050	0.260	0.100	0.050	0.000	0.100	0.100
τ_5	s	8.000	8.000	8.000	10.000	8.000	10.006	10.000	8.000	8.720	8.720
F	(6)	0.250	0.270	0.267	0.250	0.270	0.300	0.280	0.300	0.300	0.300
STABILIZER											
PSS	(7)	...	...	...	S	S	...	S	F	S	...
K_{QV}	(7)	...	...	...	0.000	0.000	...	0.000	0.000	0.000	...
K_{QS}	(7)	...	...	...	4.000	26.000	...	24.400	0.400	24.000	...
τ_Q	s	...	...	...	10.000	3.000	...	3.000	10.000	10.000	...
τ'_{Q1}	s	...	...	...	0.230	0.150	...	0.150	0.650	0.200	...
τ_{Q1}	s	...	...	...	0.020	0.050	...	0.050	0.020	0.060	...
τ'_{Q2}	s	...	...	...	0.230	0.150	...	0.150	0.650	0.150	...
τ_{Q2}	s	...	...	...	0.020	0.050	...	0.050	0.020	0.020	...
τ'_{Q3}	s	...	...	...	0.000	0.000	...	0.000	0.000	0.000	...
τ_{Q3}	s	...	...	...	0.000	0.000	...	0.000	0.000	0.000	...
$V_{s\lim}$	pu	...	...	...	0.100	0.050	...	0.050	0.100	0.050	...

Table D.4. Typical Data for Cross-Compound Fossil Steam (*CF*) Units

GENERATOR											
Unit no.		*CF1-HP*	*CF1-LP*	*CF2-HP*	*CF2-LP*	*CF3-HP*	*CF3-LP*	*CF4-HP*	*CF4-LP*	*CF5-HP*	*CF5-LP*
Rated MVA		128.00	128.00	192.00	192.00	278.30	221.70	445.00	375.00	483.00	426.00
Rated kV		13.80	13.80	18.00	18.00	20.00	20.00	22.00	22.00	22.00	22.00
Rated PF		0.85	0.85	0.85	0.85	0.90	0.90	0.90	0.90	0.90	0.90
SCR	(1)	0.64	0.64	0.64	0.64	0.58	0.58	0.64	0.64	0.604	0.645
x_d''	pu	0.171	0.250	0.225	0.225	0.231	0.252	0.205	0.180	0.220	0.205
x_d'	pu	0.232	0.369	0.315	0.315	0.311	0.380	0.260	0.250	0.285	0.285
x_d	pu	1.680	1.660	1.670	1.670	1.675	1.581	1.650	1.500	1.800	1.750
x_q''	pu	0.171	0.250	0.224	0.224	0.229	0.248	0.205	0.181	0.220	0.205
x_q'	pu	0.320	0.565	0.958	0.958	0.979	0.955	0.460	0.440	0.490	0.485
x_q	pu	1.610	1.590	1.640	1.640	1.648	1.531	1.590	1.400	1.720	1.580
r_a	pu	0.0024	0.003	0.0036	0.0036	0.0043	0.0039	0.0043	0.0045	0.0027	0.0036
x_ℓ or x_p	pu	0.095	0.140	0.186	0.186	0.304	0.291	0.150	0.140	0.160	0.155
r_2	pu	0.026	0.020	0.028	0.028	0.029	0.028	0.022	0.022	0.025	0.025
x_2	pu	0.171	0.250	0.224	0.224	0.229	0.249	0.175	0.145	0.220	0.205
x_0	pu	...	...	0.101	0.101	...	...	0.140	0.135	0.150	0.150
τ_d''	s	0.023	0.023	...	...	0.023	0.023	0.020	0.020	0.023	0.023
τ_d'	s	0.815	1.130	0.820	0.820	1.000	1.292	...	...	0.586	1.360
τ_{d0}''	s	0.034	0.037	0.043	0.043	0.047	0.053	0.032	0.036	0.032	0.035
τ_{d0}'	s	5.890	5.100	5.000	5.000	5.400	5.390	4.800	8.000	3.700	8.400
τ_q''	s	0.023	0.023	...	...	0.023	0.023	0.020	0.020	0.023	0.023
τ_q'	s	0.410	0.570	...	...	0.500	0.650	...	...	0.293	0.680
τ_{q0}''	s	0.080	0.070	0.150	0.150	0.150	0.135	0.060	0.070	0.060	0.070
τ_{q0}'	s	0.600	0.326	1.500	1.500	1.500	1.500	0.470	0.410	0.480	0.460
τ_a	s	0.171	0.205	0.390	0.390	0.390	0.330	0.150	0.110	0.150	0.110
W_R	MW·s	305.00	787.00	596.70	650.70	464.00	1418.00	639.50	3383.50	633.00	2539.00
r_F	Ω	...	...	0.141	0.141	...	...	0.1357	0.3958	0.1259	0.343
$S_{G1.0}$	(2)	0.121	0.1122	0.0982	0.0982	0.1249	0.0905	0.0926	0.1333	0.0866	0.177
$S_{G1.2}$	(2)	0.610	0.433	0.4161	0.4161	0.500	0.345	0.4139	0.5555	0.410	0.532
E_{FDFL}	(2)	2.640	2.640	2.840	2.840	2.570	2.500	2.730	2.560	2.900	2.915
D	(3)	2.000	2.000	2.000	2.000	2.000	2.000	2.000	2.000	2.000	2.000
EXCITER											
VR type		*A*	*A*	*A*	*A*	*A*	*A*	*A*	*A*	*G*	*G*
Name	(4)	NA101	NA101	WMA	WMA	WMA	WMA	NA143A	NA143A	ALTHYREX	ALTHYREX
RR	(4)	0.50	0.50	0.50	0.50	0.50	0.50	2.00	2.00	2.50	2.50
τ_R	s	0.060	0.060	0.000	0.000	0.000	0.000	0.000	0.000	0.000	0.000
K_A	pu	25.000	25.000	275.000	275.000	245.000	245.000	592.000	312.000	250.000	250.000

τ_A or τ_{A1}	s	0.200	0.200	0.060	0.060	0.050	0.050	0.053	0.050	0.140	0.060
τ_{A2}	s	0.000	0.000	0.000	0.000	0.000	0.000	0.000	0.000	0.000	0.000
$V_{R\max}$	pu (4)	1.000	1.000	0.984	0.984	2.780	2.780	13.050	10.770	5.150	4.910
$V_{R\min}$	pu (4)	−1.000	−1.000	−0.984	−0.984	−2.780	−2.780	−13.050	−10.770	−5.150	−4.910
K_E	pu	−0.051	−0.051	−0.0667	−0.0667	−0.170	−0.170	−0.591	−0.4035	1.000	1.000
τ_E	s	0.5685	0.5685	1.230	1.230	1.370	1.370	0.512	1.080	0.000	0.000
$S_{E.75\max}$	(5)	0.0778	0.0778	0.1688	0.1688	0.220	0.220	1.094	0.647	0.000	0.000
$S_{E\max}$	(5)	0.3035	0.3035	0.2978	0.2978	0.950	0.950	3.048	2.545	0.000	0.000
A_{EX}	(5)	0.0013	0.0013	0.0307	0.0307	0.0027	0.0027	0.0506	0.0106	0.000	0.000
B_{EX}	(5)	1.3750	1.3750	0.5331	0.5331	1.639	1.639	0.7719	1.0891	0.000	0.000
$E_{FD\max}$	pu (5)	3.960	3.960	4.260	4.260	3.570	3.570	5.310	5.030	5.150	4.910
$E_{FD\min}$	pu	−3.960	−3.960	−4.260	−4.260	−3.570	−3.570	−5.310	−5.030	−5.150	−4.910
K_F	pu	0.091	0.091	0.033	0.033	0.040	0.040	0.070	0.090	0.062	0.025
τ_F or τ_{F1}	s	0.350	0.350	0.330	0.330	1.000	1.000	1.880	2.250	1.000	1.000
τ_{F2}	s	0.000	0.000	0.000	0.000	0.000	0.000	0.000	0.000	0.000	0.000
TURBINE GOVERNOR											
GOV	(6)	*G*	*G*	*G*	*G*	*G*	*G*	*G*	*G*	*G*	*G*
R	(6)	0.050	0.050	0.050	0.050	0.050	0.050	0.050	0.050	0.050	0.050
$P_{\max}$	MW	107.50	107.50	172.50	172.50	267.00	213.00	411.00	339.00	436.00	382.00
τ_1	s	0.100	0.100	0.100	0.100	0.250	0.250	0.100	0.100	0.100	0.100
τ_2	s	0.000	0.000	0.000	0.000	0.000	0.000	0.000	0.000	0.000	0.000
τ_3	s	0.150	0.150	0.150	0.150	0.000	0.000	0.200	0.200	0.300	0.300
τ_4	s	0.300	0.300	0.300	0.300	0.050	0.300	0.100	0.100	0.050	0.050
τ_5	s	10.000	10.000	4.160	4.160	12.000	12.000	8.720	8.720	14.000	14.000
F	(6)	0.606	0.000	0.560	0.000	0.549	0.000	0.540	0.000	0.580	0.000
STABILIZER											
PSS	(7)	*S*	*S*	*F*	*F*	*S*	*S*	*F*	*F*	*S*	*S*
K_{QV}	(7)	0.000	0.000	0.000	0.000	0.000	0.000	0.000	0.000	0.000	0.000
K_{QS}	(7)	12.000	8.000	0.600	0.600	10.000	10.000	1.170	1.170	24.000	24.000
τ_Q	s	10.000	10.000	10.000	10.000	10.000	10.000	10.000	10.000	10.000	10.000
τ'_{Q1}	s	1.000	1.000	0.490	0.455	0.250	0.700	0.265	0.640	0.200	0.200
τ_{Q1}	s	0.020	0.020	0.020	0.020	0.020	0.020	0.020	0.020	0.050	0.070
τ'_{Q2}	s	0.750	0.250	0.490	0.455	0.400	0.450	0.265	0.640	0.200	0.300
τ_{Q2}	s	0.020	0.020	0.020	0.020	0.020	0.020	0.020	0.020	0.020	0.020
τ'_{Q3}	s	0.000	0.000	0.000	0.000	0.000	0.000	0.000	0.000	0.000	0.000
τ_{Q3}	s	0.000	0.000	0.000	0.000	0.000	0.000	0.000	0.000	0.000	0.000
$V_{s\lim}$	pu	0.050	0.050	0.080	0.080	0.050	0.050	0.060	0.080	0.050	0.050

Table D.5. Typical Data for Nuclear Steam (N) Units

GENERATOR									
Unit no.		$N1$	$N2$	$N3$	$N4$	$N5$	$N6$	$N7$	$N8$
Rated MVA		76.80	245.5	500.00	920.35	1070.00	1280.00	1300.00	1340.00
Rated kV		13.8	14.4	18.00	18.00	22.00	22.00	25.00	25.00
Rated PF		0.85	0.85	0.90	0.90	0.90	0.95	0.90	0.90
SCR	(1)	0.650	0.640	0.580	0.607	0.500	0.500	0.480	0.480
x''_d	pu	0.190	0.210	0.283	0.275	0.312	0.237	0.315	0.281
x'_d	pu	0.320	0.320	0.444	0.355	0.467	0.358	0.467	0.346
x_d	pu	1.660	1.710	1.782	1.790	1.933	2.020	2.129	1.693
x''_q	pu	0.120	0.210	0.277	0.275	...	0.237	0.308	0.281
x'_q	pu	0.470	0.510	1.201	0.570	1.144	0.565	1.270	0.991
x_q	pu	1.580	1.630	1.739	1.660	1.743	1.860	2.074	1.636
r_a	pu	...	0.0032	0.0041	0.0048	0.360	0.0019	0.0029	0.0021
x_ℓ or x_p	pu	0.150	0.125	0.275	0.215	...	0.205	0.251	0.228
r_2	pu	...	0.025	0.029	0.028	...	0.029	...	...
x_2	pu	0.125	0.160	0.280	0.230	0.284	0.215	...	0.228
x_0	pu	0.450	0.110	...	0.195	...	0.195	...	...
τ''_d	s	...	0.230	0.035	...	...	...	...	...
τ'_d	s	...	...	1.512	...	...	...	...	...
τ''_{d0}	s	0.032	0.038	0.055	0.032	...	0.034	0.052	0.043
τ'_{d0}	s	4.780	7.100	6.070	7.900	6.660	9.100	6.120	6.580
τ''_q	s	...	...	0.035	...	...	...	...	...
τ'_q	s	...	...	0.756	...	...	...	...	...
τ''_{q0}	s	...	0.073	0.152	0.055	...	0.059	0.144	0.124
τ'_{q0}	s	...	0.380	1.500	0.41	...	0.460	1.500	1.500
τ_a	s	...	0.210	0.310	0.19	...	0.180	...	...
W_R	MW·s	281.70	1136.00	1990.00	3464.00	3312.00	4690.00	4580.00	4698.00
r_F	Ω	...	0.217	...	0.0901	...	0.0979	0.0576	0.0576
$S_{G1.0}$	(2)	0.0857	0.1309	0.0900	0.0816	...	0.0779	0.0714	0.0769
$S_{G1.2}$	(2)	0.3244	0.5331	0.3520	0.3933	...	0.3055	0.3100	0.4100
E_{FDFL}	(2)	2.587	2.730	2.710	2.870	...	2.945	3.340	2.708
D	(3)	2.000	...	2.000	...	2.000	2.000	2.000	2.000
EXCITER									
VR type		A	A	A	A	C	A	C	C
Name	(4)	NA101	NA101	WMA	NA143	BRLS	EA210	BRLS	BRLS
RR	(4)	0.50	0.50	0.50	0.50	2.00	1.50	2.23	2.00
τ_R	s	0.060	0.060	0.000	0.000	0.000	0.000	0.000	0.000
K_A	pu	25.000	25.000	256.000	25.000	400.000	50.000	400.000	400.000
τ_A or τ_{A1}	s	0.200	0.200	0.050	0.200	0.020	0.020	0.020	0.020
τ_{A2}	s	0.000	0.000	0.000	0.000	0.000	0.000	0.000	0.000
$V_{R\max}$	pu (4)	1.000	1.000	2.858	1.000	10.650	1.000	6.960	6.020
$V_{R\min}$	pu (4)	−1.000	−1.000	−2.858	−1.000	−10.650	−1.000	−6.960	−6.020
K_E	pu	−0.0516	−0.0489	−0.170	−0.0464	1.000	−0.0244	1.000	1.000
τ_E	s	0.579	0.550	2.150	0.522	1.000	0.1455	0.015	0.015
$S_{E.75\max}$	(5)	0.0794	0.0752	0.2200	0.0714	0.375	0.0863	0.3400	0.3900
$S_{E\max}$	(5)	0.3093	0.2932	0.9500	0.2784	1.220	0.2148	0.5600	0.5630
A_{EX}	(5)	0.0013	0.0016	0.0027	0.0016	...	0.0056	0.0761	0.1296
B_{EX}	(5)	1.4015	1.6120	1.5966	1.5330	...	0.6818	0.4475	0.3814
$E_{FD\max}$	pu (5)	3.881	4.090	3.665	4.310	4.800	5.350	4.460	3.850
$E_{FD\min}$	pu	−3.881	−4.090	−3.665	−4.310	0.000	0.000	0.000	0.000
K_F	pu	0.093	0.088	0.040	0.084	0.060	0.0233	0.040	0.040
τ_F or τ_{F1}	s	0.350	0.350	1.000	1.000	1.000	0.7750	0.050	0.050
τ_{F2}	s	0.000	0.000	0.000	0.000	0.000	0.000	0.000	0.000

Table D.5. *(continued)*

TURBINE GOVERNOR									
GOV	(6)	*G*	*G*	*G*	*G*	*G*	*G*	*G*	*G*
R	(6)	0.050	0.050	0.050	0.050	0.050	0.050	0.050	0.050
P_{max}	MW	65.00	208.675	450.00	790.18	951.00	1216.00	1090.00	1205.00
τ_1	s	...	...	0.250	...	0.180	0.150	0.180	0.180
τ_2	s	...	...	0.000	...	0.030	0.000	0.000	0.000
τ_3	s	...	...	0.000	...	0.100	0.210	0.040	0.040
τ_4	s	...	...	0.300	...	0.200	0.814	0.200	0.200
τ_5	s	...	...	5.000	...	6.280	2.460	5.000	5.000
F	(6)	...	...	0.320	...	0.330	0.340	0.300	0.300
STABILIZER									
PSS	(7)	*S*	...	...	...	*F*	*S*	*F*	*F*
K_{QV}	(7)	0.000	...	...	...	0.000	0.000	0.000	0.000
K_{QS}	(7)	0.200	...	...	...	10.000	1.530	20.000	20.000
τ_Q	s	10.000	...	...	...	10.000	3.000	10.000	10.000
τ'_{Q1}	s	1.330	...	...	...	0.080	0.150	0.300	0.300
τ_{Q1}	s	0.020	...	...	...	0.020	0.050	0.020	0.020
τ'_{Q2}	s	1.330	...	...	...	0.080	0.150	0.000	0.000
τ_{Q2}	s	0.020	...	...	...	0.020	0.050	0.000	0.000
τ'_{Q3}	s	0.000	...	...	...	0.000	0.000	0.000	0.000
τ_{Q3}	s	0.000	...	...	...	0.000	0.000	0.000	0.000
V_{slim}	pu	0.100	...	...	...	0.100	0.100	0.100	0.100

Table D.6. Typical Data for Synchronous Condensor (*SC*) Units

GENERATOR						
Unit no.		*SC*1	*SC*2	*SC*3	*SC*4	*SC*5
Rated MVA		25.00	40.00	50.00	60.00	75.00
Rated kV		13.80	13.80	12.70	13.80	13.80
Rated PF		0.00	0.00	0.00	0.00	0.00
SCR	(1)	...	0.558	1.004	0.477	0.800
x_d''	pu	0.2035	0.231	0.141	0.257	0.170
x_d'	pu	0.304	0.343	0.244	0.385	0.320
x_d	pu	1.769	2.373	1.083	2.476	1.560
x_q''	pu	0.199	...	0.170	0.261	0.200
x_q'	pu	0.5795	1.172	0.720	1.180	1.000
x_q	pu	0.855	1.172	0.720	1.180	1.000
r_a	pu	0.0025	...	0.006	0.0024	0.0017
x_ℓ or x_p	pu	0.1045	0.132	...	0.146	0.0987
r_2	pu	0.0071	...	0.160	...	0.180
x_2	pu	0.177	...	...	0.225	0.185
x_0	pu	0.115	...	0.058	0.165	0.128
τ_d''	s	...	0.035	0.041	0.035	0.041
τ_d'	s	...	...	0.858	...	3.230
τ_{d0}''	s	0.0525	0.058	0.050	0.058	0.039
τ_{d0}'	s	8.000	11.600	6.000	12.350	16.000
τ_q''	s	...	...	...	...	0.0473
τ_q'	s	...	...	...	...	...
τ_{q0}''	s	0.0151	0.201	...	0.188	0.235
τ_{q0}'	s	...	...	0.150	...	...
τ_a	s	...	0.159	0.200	0.290	0.288
W_R	MW·s	30.00	60.80	105.00	60.60	89.98
r_F	Ω	0.4407	...	0.0631	0.274	0.279
$S_{G1.0}$	(2)	0.304	0.295	0.0873	0.180	0.150
$S_{G1.2}$	(2)	0.666	0.776	0.310	0.708	0.500
E_{FDFL}	(2)	3.560	4.180	2.338	4.224	3.730
D	(3)	...	...	...	...	...
EXCITER						
VR type		*A*	*A*	*A*	*A*	*A*
Name	(4)	WMA	WMA	...	WMA	NA143
RR	(4)	0.50	1.00	3.85	1.00	2.00
τ_R	s	0.000	0.000	0.000	0.000	0.000
K_A	pu	400.000	400.000	200.000	400.000	18.000
τ_A or τ_{A1}	s	0.050	0.050	0.050	0.050	0.200
τ_{A2}	s	0.000	0.000	0.000	0.000	0.000
$V_{R\max}$	pu (4)	4.407	6.630	11.540	5.850	1.000
$V_{R\min}$	pu (4)	−4.407	−6.630	−11.540	−5.850	−1.000
K_E	pu	−0.170	−0.170	−0.170	−0.170	−0.0138
τ_E	s	0.950	0.950	1.000	0.950	0.0669
$S_{E.75\max}$	(5)	0.220	0.220	0.220	0.220	0.0634
$S_{E\max}$	(5)	0.950	0.950	0.950	0.950	0.1512
A_{EX}	(5)	0.0027	0.0027	0.0027	0.0027	0.0047
B_{EX}	(5)	1.0356	0.6884	0.3956	0.7802	0.4782
$E_{FD\max}$	pu (5)	5.650	8.500	14.790	7.500	7.270
$E_{FD\min}$	pu	−5.650	−8.500	−14.790	−7.500	−7.270
K_F	pu	0.040	0.040	0.070	0.040	0.0153
τ_F or τ_{F1}	s	1.000	1.000	1.000	1.000	1.000
τ_{F2}	s	0.000	0.000	0.000	0.000	0.000

Table D.7. Typical Data for Combustion Turbine (CT) Units

GENERATOR			
Unit no.		CT1	CT2
Rated MVA		20.65	62.50
Rated kV		13.80	13.80
Rated PF		0.85	0.85
SCR	(1)	0.580	0.580
x''_d	pu	0.155	0.102
x'_d	pu	0.225	0.159
x_d	pu	1.850	1.640
x''_q	pu	...	0.100
x'_q	pu	...	0.306
x_q	pu	1.740	1.575
r_a	pu	...	0.034
x_ℓ or x_p	pu	...	0.113
r_2	pu	...	0.352
x_2	pu	...	0.102
x_0	pu	...	0.051
τ''_d	s	...	0.035
τ'_d	s	...	0.730
τ''_{d0}	s	...	0.054
τ'_{d0}	s	4.610	7.500
τ''_q	s	...	0.035
τ'_q	s	...	0.188
τ''_{q0}	s	...	0.107
τ'_{q0}	s	...	1.500
τ_a	s	...	0.350
W_R	MW·s	183.30	713.50
r_F	Ω	...	0.261
$S_{G1.0}$	(2)	...	0.0870
$S_{G1.2}$	(2)	...	0.2681
E_{FDFL}	(2)	2.640	2.4348
D	(3)	...	2.000

EXCITER			
VR type		*D*	*C*
Name	(4)	SCPT	BRLS
RR	(4)	...	0.50
τ_R	s	0.000	0.000
K_A	pu	120.000	400.000
τ_A or τ_{A1}	s	0.050	0.020
τ_{A2}	s	0.000	0.000
$V_{R\max}$	pu (4)	1.200	7.300
$V_{R\min}$	pu (4)	−1.200	−7.300
K_E	pu	1.000	1.000
τ_E	s	0.500	0.253
$S_{E.75\max}$	(5)	...	0.500
$S_{E\max}$	(5)	...	0.860
A_{EX}	(5)	...	0.0983
B_{EX}	(5)	...	0.2972
$E_{FD\max}$	pu (5)	...	7.300
$E_{FD\min}$	pu	...	0.000
K_F	pu	0.020	0.030
τ_F or τ_{F1}	s	0.461	1.000
τ_{F2}	s	...	0.000
K_P		1.19	
K_I		2.32	

TURBINE GOVERNOR				
GOV	(6)	*G*		*G*
R	(6)	0.050		0.040
$P_{\max}$	MW	17.55		82.00
τ_1	s	0.000		0.500
τ_2	s	0.000		1.250
		Fuel: Oil	Gas	
τ_3	s	0.025	0.100	0.700
τ_4	s	0.000	0.000	0.700
τ_5	s	0.025	0.100	0.000
F	(6)	0.5	0.0	1.000

Table D.8. Typical 60-Hz Transmission Line Data

Line-to-line voltage (kV)	Conductors per phase @ 18 in. spacing	ACSR Conductor area (or diam) kCM (in.)	Flat phase spacing (ft)	Geometric mean distance (ft)	60-Hz inductive reactance Ω/mi			60-Hz capacitive reactance MΩ·mi			Surge impedance z_0 (Ω)	Surge impedance loading (MVA)
					x_a	x_d	$x_a + x_d$	x'_a	x'_d	$x'_a + x'_d$		
69	1	226.8	12	15.1	0.465	0.3294	0.7944	0.1074	0.0805	0.1879	386.4	12
115	1	336.4	14	17.6	0.451	0.3480	0.7990	0.1039	0.0851	0.1890	388.6	34
138	1	397.5	16	20.1	0.441	0.3641	0.8051	0.1015	0.0890	0.1905	391.6	49
161	1	477.0	18	22.7	0.430	0.3789	0.8089	0.0988	0.0926	0.1914	393.5	66
230	1	556.5	22	27.7	0.420	0.4030	0.8230	0.0965	0.0985	0.1950	400.6	132
345	1	(1.750)	28	35.3	0.3336	0.4325	0.7761	0.0777	0.1057	0.1834	374.8	318
345	2	(1.246)	28	35.3	0.1677	0.4325	0.6002	0.0379	0.1057	0.1436	293.6	405
500	1	(2.500)	38	47.9	0.2922	0.4694	0.7616	0.0671	0.1147	0.1818	372.1	672
500	2	(1.602)	38	47.9	0.1529	0.4694	0.6223	0.0341	0.1147	0.1488	304.3	822
500	3	(1.165)	38	47.9	0.0988	0.4694	0.5682	0.0219	0.1147	0.1366	278.6	897
500	4	(0.914)	38	47.9	0.0584	0.4694	0.5278	0.0126	0.1147	0.1273	259.2	965
735	3	(1.750)	56	70.6	0.0784	0.5166	0.5950	0.0179	0.1263	0.1442	292.9	1844
735	4	(1.382)	56	70.6	0.0456	0.5166	0.5622	0.0096	0.1263	0.1359	276.4	1955

appendix E

Excitation Control System Definitions

There are two important recently published documents dealing with excitation control system definitions. The first [1] appeared in 1961 under the title "Proposed excitation system definitions for synchronous machines" and provided many definitions of basic system elements. The second report [2] was published in 1969 under the same title and, using the first report as a starting point, added the new definitions required by technological change and attempted to make all definitions agree with accepted language of the automatic control community. The definitions that follow are those proposed by the 1969 report.[1]

Reference is also made to the definitions given in ANSI Standard C42.10 on rotating machines [3], ANSI Standard C85.1 on automatic control [4], and the supplement to C85.1 [5]. Finally, reference is made to the IEEE Committee Report "Computer representation of excitation systems" [6], which defines certain time constants and gain factors used in excitation control systems.

Proposed IEEE Definitions

1.0 Systems

1.01 Control system, feedback. A control system which operates to achieve prescribed relationships between selected system variables by comparing functions of these variables and using the difference to effect control.

1.02 Control system, automatic feedback. A feedback control system which operates without human intervention.

1.03 Excitation system [1, definition 4]. The source of field current for the excitation of a synchronous machine and includes the exciter, regulator, and manual control.

1.04 Excitation control system (new). A feedback control system which includes the synchronous machine and its excitation system.

1.05 High initial response excitation system (new). An excitation system having an excitation system voltage response time of 0.1 second or less.

1. © IEEE. Reprinted with permission from *IEEE Trans.*, vol. PAS-88, 1969.

2.0 Components

2.01 Adjuster [1, definition 40]. An element or group of elements associated with a feedback control system by which adjustment of the level of a controlled variable can be made.

2.02 Amplifier. A device whose output is an enlarged reproduction of the essential features of an input signal and which draws power therefore from a source other than the input signal.

2.03 Compensator [1, definition 44]. A feedback element of the regulator which acts to compensate for the effect of a variable by modifying the function of the primary detecting element.

Notes:

1. Examples are reactive current compensator and active current compensator. A reactive current compensator is a compensator that acts to modify the functioning of a voltage regulator in accordance with reactive current. An active current compensator is a compensator that acts to modify the functioning of a voltage regulator in accordance with active current.
2. Historically, terms such as "equalizing reactor" and "cross-current compensator" have been used to describe the function of a reactive compensator. These terms are deprecated.
3. Reactive compensators are generally applied with generator voltage regulators to obtain reactive current sharing among generators operating in parallel. They function in the following two ways.
   a. Reactive droop compensation is the more common method. It creates a droop in generator voltage proportional to reactive current and equivalent to that which would be produced by the insertion of a reactor between the generator terminals and the paralleling point.
   b. Reactive differential compensation is used where droop in generator voltage is not wanted. It is obtained by a series differential connection of the various generator current transformer secondaries and reactive compensators. The difference current for any generator from the common series current creates a compensating voltage in the input to the particular generator voltage regulator that acts to modify the generator excitation to reduce to minimum (zero) its differential reactive current.
4. Line drop compensators modify generator voltage by regulator action to compensate for the impedance drop from the machine terminals to a fixed point. Action is accomplished by insertion within the regulator input circuit of a voltage equivalent to the impedance drop. The voltage drops of the resistance and reactance portions of the impedance are obtained respectively in pu quantities by an "active compensator" and a "reactive compensator."

2.04 Control, manual (new). Those elements in the excitation control system which provide for manual adjustment of the synchronous machine terminal voltage by open loop (human element) control.

2.05 Elements, feedback. Those elements in the controlling system which change the feedback signal in response to the directly controlled variable.

2.06 Elements, forward. Those elements situated between the actuating signal and the controlled variable in the closed loop being considered.

2.07 Element, primary detecting. That portion of the feedback elements which first either utilizes or transforms energy from the controlled medium to produce a signal which is a function of the value of the directly controlled variable.

2.08 Exciter [1, definition 5]. The source of all or part of the field current for the excitation of an electric machine.

2.09 Exciter, main [1, definition 5]. The source of all or part of the field current for the excitation of an electric machine, exclusive of another exciter.

2.09.1 DC generator commutator exciter. An exciter whose energy is derived from a dc generator. The exciter includes a dc generator with its commutator and brushes. It is exclusive of input control elements. The exciter may be driven by a motor, prime mover, or the shaft of the synchronous machine.

2.09.2 Alternator rectifier exciter. An exciter whose energy is derived from an alternator and converted to dc by rectifiers. The exciter includes an alternator and power rectifiers which may be either noncontrolled or controlled, including gate circuitry. It is exclusive of input control elements. The alternator may be driven by a motor, prime mover, or by the shaft of the synchronous machine. The rectifiers may be stationary or rotating with the alternator shaft.

2.09.3 Compound rectifier exciter. An exciter whose energy is derived from the currents and potentials of the ac terminals of the synchronous machine and converted to dc by rectifiers. The exciter includes the power transformers (current and potential), power reactor, power rectifiers which may be either noncontrolled or controlled, including gate circuitry. It is exclusive of input control elements.

2.09.4 Potential source rectifier exciter. An exciter whose energy is derived from a stationary ac potential source and converted to dc by rectifiers. The exciter includes the power potential transformers, where used, power rectifiers which may be either noncontrolled or controlled, including gate circuitry. It is exclusive of input control elements.

2.10 Exciter, pilot [1, definition 7]. The source of all or part of the field current for the excitation of another exciter.

2.11 Limiter [1, definition 43]. A feedback element of the excitation system which acts to limit a variable by modifying or replacing the function of the primary detector element when predetermined conditions have been reached.

2.12 Regulator, synchronous machine [1, definition 8]. A synchronous machine regulator couples the output variables of the synchronous machine to the input of the exciter through feedback and forward controlling elements for the purpose of regulating the synchronous machine output variables.

Note: In general, the regulator is assumed to consist of an error detector, preamplifier, power amplifier, stabilizers, auxiliary inputs, and limiters. As shown in Figure 7.20, these regulator components are assumed to be self-explanatory, and a given regulator may not have all the items included. Functional regulator definitions describing types of regulators are listed below. The term "dynamic-type" regulator has been omitted as a classification [1, Definition 15].

2.12.1 Continuously acting regulator [1, definition 10]. One that initiates a corrective action for a sustained infinitesimal change in the controlled variable.

2.12.2 Noncontinuously acting regulator [1, definition 11]. One that requires a sustained finite change in the controlled variable to initiate corrective action.

2.12.3 Rheostatic type regulator [1, definition 12]. One that accomplishes the regulating function by mechanically varying a resistance.

Note [1, Definitions 13, 14]: Historically, rheostatic type regulators have been further defined as direct-acting and indirect-acting. An indirect-acting type of regulator is a rheostatic type that controls the excitation of the exciter by acting on an intermediate device not considered part of the regulator or exciter.

A direct-acting type of regulator is a rheostatic type that directly controls the excitation of an exciter by varying the input to the exciter field circuit.

2.13 Stabilizer, excitation control system (new). An element or group of elements which modifies the forward signal by either series or feedback compensation to improve the dynamic performance of the excitation control system.

2.14 Stabilizer, power system (new). An element or group of elements which provides an additional input to the regulator to improve power system dynamic performance. A number of different quantities may be used as input to the power system stabilizer such as shaft speed, frequency, synchronous machine electrical power and other.

3.0 Characteristics and performance

3.01 Accuracy, excitation control system (new). The degree of correspondence between the controlled variable and the ideal value under specified conditions such as load changes, ambient temperature, humidity, frequency, and supply voltage variations. Quantitatively, it is expressed as the ratio of difference between the controlled variable and the ideal value.

3.02 Air gap Line. The extended straight line part of the no-load saturation curve.

3.03 Ceiling voltage, excitation system [1, definition 26]. The maximum dc component system output voltage that is able to be attained by an excitation system under specified conditions.

3.04 Ceiling voltage, exciter [1, definition 24]. Exciter ceiling voltage is the maximum voltage that may be attained by an exciter under specified conditions.

3.05 Ceiling voltage, exciter nominal [1, definition 25]. Nominal exciter ceiling voltage is the ceiling voltage of an exciter loaded with a resistor having an ohmic value equal to the resistance of the field winding to be excited and with this field winding at a temperature of

1. 75°C for field windings designed to operate at rating with a temperature rise of 60°C or less.
2. 100°C for field windings designed to operate at rating with a temperature rise greater than 60°C.

3.06 Compensation. A modifying or supplementary action (also, the effect of such action) intended to improve performance with respect to some specified characteristics.

Note: In control usage this characteristic is usually the system deviation. Compensation is frequently qualified as "series," "parallel," "feedback," etc., to indicate the relative position of the compensating element.

3.07 Deviation, system. The instantaneous value of the ultimately controlled variable minus the command.

3.08 Deviation, transient. The instantaneous value of the ultimately controlled variable minus its steady-state value.

3.09 Disturbance. An undesired variable applied to a system which tends to affect adversely the value of a controlled variable.

3.10 Duty, excitation system (new). Those voltage and current loadings imposed by the synchronous machine upon the excitation system including short circuits and all conditions of loading. The duty cycle will include the action of limiting devices to maintain synchronous machine loading at or below that defined by ANSI C50.13-1965.

3.11 Duty, excitation system (new). An initial operating condition and a subsequent sequence of events of specified duration to which the excitation system will be exposed.

Note: The duty cycle usually involves a three-phase fault of specified duration located electrically close to the synchronous generator. Its primary purpose is to specify the duty that the excitation system components can withstand without incurring maloperation or specified damage.

3.12 Drift [1, definition 36]. An undesired change in output over a period of time, which change is unrelated to input, environment, or load.

Note: The change is a plus or minus variation of short periods that may be superimposed on plus or minus variations of a long time period. On a practical system, drift is determined as the change in output over a specified time with fixed command and fixed load, with specified environmental conditions.

3.13 Dynamic. Referring to a state in which one or more quantities exhibit appreciable change within an arbitrarily short time interval.

3.14 Error. An indicated value minus an accepted standard value, or true value.

Note: ANSI C85 deprecates use of the term as the negative of deviation. See also accuracy, precision in ANSI C85.1.

3.15 Excitation system voltage response [1, definition 21]. The rate of increase or decrease of the excitation system output voltage determined from the excitation system voltage-time response curve, which rate if maintained constant, would develop the same voltage-time area as obtained from the curve for a specified period. The starting point for determining the rate of voltage change shall be the initial value of the excitation system voltage time response curve. Referring to Fig. E-1, the excitation system voltage response is illustrated by line *ac*. This line is determined by establishing the area *acd* equal to area *abd*.

Notes:

1. Similar definitions can be applied to the excitation system major components such as the exciter and regulator.
2. A system having an excitation system voltage response time of 0.1 s or less is defined as a high initial response excitation system (Definition 1.05).

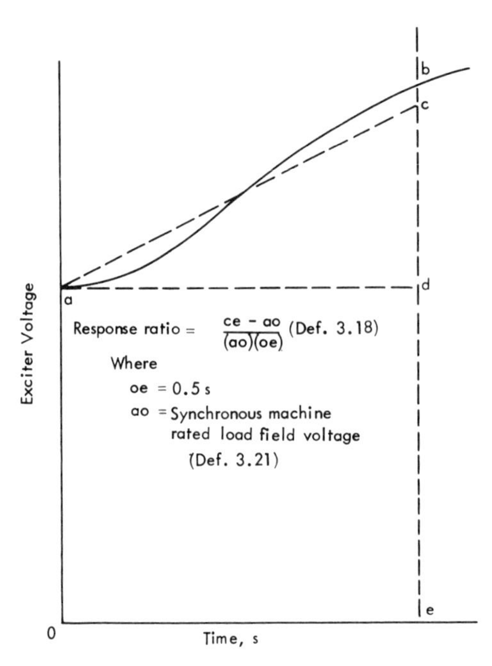

Fig. E.1. Exciter or synchronous machine excitation system voltage response (Def. 3.15).

3.16 Excitation system voltage response time (**new**). The time in seconds for the excitation voltage to reach 95 percent of ceiling voltage under specified conditions.

3.17 Excitation system voltage time response [**1, definition 19**]. The excitation system output voltage expressed as a function of time, under specified conditions.

Note: A similar definition can be applied to the excitation system major components: the exciter and regulator separately.

3.18 Excitation system voltage response ratio [**1, definition 23**]. The numerical value which is obtained when the excitation system voltage response in volts per second, measured over the first half-second interval unless otherwise specified, is divided by the rated-load field voltage of the synchronous machine. Unless otherwise specified, the excitation system voltage response ratio shall apply only to the increase in excitation system voltage. Referring to Fig. E.1 the excitation system voltage response ratio = $(ce - ao)/(ao)(oe)$, where ao = synchronous machine rated load field voltage (Definition 3.21) and oe = 0.5 second, unless otherwise specified.

3.19 Exciter main response ratio; formerly nominal exciter response. The main exciter response ratio is the numerical value obtained when the response, in volts per second, is divided by the rated-load field voltage; which response, if maintained constant, would develop, in one half-second, the same excitation voltage-time area as attained by the actual exciter.

Note: The response is determined with no load on the exciter, with the exciter voltage initially equal to the rated-load field voltage, and then suddenly establishing circuit conditions that would be used to obtain nominal exciter ceiling voltage. For a rotating exciter, response should be determined at rated speed. This definition does not apply to main exciters having one or more series fields (except a light differential series field) nor to electronic exciters.

3.20 Field voltage, base (new). The synchronous machine field voltage required to produce rated voltage on the air gap line of the synchronous machine at field temperatures.

1. 75°C for field windings designed to operate at rating with a temperature rise of 60°C or less.
2. 100°C for field windings designed to operate at rating with a temperature rise greater then 60°C.

Note: This defines one pu excitation system voltage for use in computer representation of excitation systems [6].

3.21 Field voltage, rated-load [1, definition 38]; formerly nominal collector ring voltage. Rated-load field voltage is the voltage required across the terminals of the field winding or an electric machine under rated continuous-load conditions with the field winding at one of the following.

1. 75°C for field windings designed to operate at rating with a temperature rise of 60°C or less.
2. 100°C for field windings designed to operate at rating with a temperature rise greater than 60°C.

3.22 Field voltage, no-load [1, definition 39]. No-load field voltage is the voltage required across the terminals of the field winding of an electric machine under conditions of no load, rated speed, and terminal voltage and with the field winding at 25°C.

3.23 Gain, proportional. The ratio of the change in output due to proportional control action to the change in input. Illustration: $Y = \pm PX$ where P = proportional gain, X = input transform, and Y = output transform.

3.24 Limiting. The intentional imposition or inherent existence of a boundary on the range of a variable, e.g., on the speed of a motor.

3.25 Regulation, load. The decrease of controlled variable (usually speed or voltage) from no load to full load (or other specified limits).

3.26 Regulated voltage, band of [1, definition 37]. Band of regulated voltage is the band or zone, expressed in percent of the rated value of the regulated voltage, within which the excitation system will hold the regulated voltage of an electric machine during steady or gradually changing conditions over a specified range of load.

3.27 Regulated voltage, nominal band of. Nominal band of regulated voltage is the band of regulated voltage for a load range between any load requiring no-load field voltage and any load requiring rated-load field voltage with any compensating means used to produce a deliberate change in regulated voltage inoperative.

3.28 Signal, actuating. The reference input signal minus the feedback signal (Figure 7.19).

3.29 Signal, error. In a closed loop, the signal resulting from subtracting a particular return signal from its corresponding input signal (Figure 7.19).

3.30 Signal, feedback. That return signal which results from the reference input signal (Figure 7.19).

3.31 Signal, input. A signal applied to a system or element.

3.32 Signal, output. A signal delivered by a system or element.

3.33 Signal, rate (new). A signal that is responsive to the rate of change of an input signal.

3.34 Signal, reference input. One external to a control loop which serves as the standard of comparison for the directly controlled variable.

3.35 Signal, return. In a closed loop, the signal resulting from a particular input signal, and transmitted by the loop and to be subtracted from that input signal.

3.36 Stability. For a feedback control system or element, the property such that its output is asymptotic, i.e., will ultimately attain a steady-state, within the linear range and without continuing external stimuli. For certain nonlinear systems or elements, the property that the output remains bounded, e.g., in a limit cycle of continued oscillation, when the input is bounded.

3.37 Stability limit. A condition of a linear system or one of its parameters which places the system on the verge of instability.

3.38 Stability, excitation system. The ability of the excitation system to control the field voltage of the principal electric machine so that transient changes in the regulated voltage are effectively suppressed and sustained oscillations in the regulated voltage are not produced by the excitation system during steady-load conditions or following a change to a new steady-load condition.

Note: It should be recognized that under some system conditions it may be necessary to use power system stabilizing signals as additional inputs to excitation control systems to achieve stability of the power system including the excitation system.

3.39 Steady state. That in which some specified characteristic of a condition, such as value, rate, periodicity, or amplitude, exhibits only negligible change over an arbitrarily long interval of time.

Note: It may describe a condition in which some characteristics are static, others dynamic.

3.40 Transient. In a variable observed during transition from one steady-state operating condition to another that part of the variation which ultimately disappears.

Note: ANSI C85 deprecates using the term to mean the total variable during the transition between two steady states.

3.41 Variable, directly controlled. In a control loop, that variable whose value is sensed to originate a feedback signal.

References

1. AIEE Committee Report. Proposed excitation system definitions for synchronous machines. *AIEE Trans.* PAS-80:173–180, 1961.
2. IEEE Committee Report. Proposed excitation system definitions for synchronous machines. *IEEE Trans.* PAS-88:1248–58, 1969.
3. ANSI Standard C42.10. Definitions of electrical terms, rotating machinery (group 10). American National Standards Institute, New York, 1957.
4. ANSI Standard C85.1-1963. Terminology for automatic control. American National Standards Institute, New York, 1963.
5. ANSI Standard C85.1a-1966. Supplement to terminology for automatic control C85.1-1963. American National Standards Institute, New York, 1963.
6. IEEE Committee Report. Computer representation of excitation systems. *IEEE Trans.* PAS-87:1460–64, 1968.

Index

POWER SYSTEM CONTROL AND STABILITY

By P. M. Anderson and A. A. Fouad

CORRECTIONS

Chapter 2

Page No.	Place	Error	Correction
15	Line above (2-14)	N.m.	Kg.m^2
23	ℓ. 3B	$\underline{/8.06°}$	$\underline{/8.13°}$
19	Line below (2.26)	... machine VA base	
	(2.30)	$\Sigma S_B/\Sigma S_{SB}$	$\Sigma S_{SB}/\Sigma S_B$
25	Equation (2.38)	$\sqrt{M/P_s}$	$\sqrt{M/P_s S_{B3}}$
33	ℓ. 7 of Sec. 2.8.1	P_m	P_M
	(2.51)	$\cos^{-1}[.][.]$	$\cos^{-1}\{[.][.]\}$
35	ℓ. 3 of Sec. 2.8.3	(See Problem 2.10)	(See Problem 2.14)

Chapter 3

Page No.	Place	Error	Correction
57	Numerator of second term in (3.12)	K_3	K_2K_3
61	General Formula for A's in Eq. 3.32		

$$A_{ii} = -\sum_{\substack{j=1\\ j\neq i}}^{n} \frac{\omega_R}{2H_i} P_{sij} - \frac{\omega_R}{2H_n} P_{sni}$$

$$A_{ij} = \frac{\omega_R}{2H_i} P_{sij} - \frac{\omega_R}{2H_n} P_{snj}$$

where n is the number of machines and machine n is the reference.

Page No.	Place	Error	Correction
62	First equation	$\delta_{31\Delta}$	$\delta_{13\Delta}$
	Fourth equation	$\frac{\omega_R}{2H_1}$ (in two places)	$\frac{\omega_R}{2H_2}$

Page	Location	Printed	Should read
63	Near Bottom	$\alpha_{21} = \frac{\omega_R}{2}\left(\frac{P_{s31}}{H_3} - \frac{P_{s21}}{H_1}\right)$ $= 68.21$	$\alpha_{21} = \frac{\omega_R}{2}\left(\frac{P_{s31}}{H_3} - \frac{P_{s21}}{H_2}\right)$ $= 33.841$
		$\alpha_{22} = \_\_\_\_ = 119.065$	$\alpha_{22} = \_\_\_\_ = 153.460$
	Last equation should read	$\lambda^2 = -(1/2)[-257.56 \pm 66336 - 55841]$ $= -77.56$ or -180.0	
64	Table 3.2 should read	$\pm j8.807$ $\pm j13.416$ 8.807 13.416 1.402 2.135 0.713 0.462	
	Table 3.2	Ts 0.912 0.474	Ts 0.713 0.468
	line below Table 3.2	... 1.1 Hz ...	... 1.4 Hz ...
67	Last line	(see Problem 3.3)	(see Problem 3.7)
73	Caption of Fig. 3.8	10 pu	10 MW
75	Caption of Fig. 3.9	10 pu	10 MW
	ℓ. 8B	$\frac{d\bar{\omega}}{dt}$	$d\bar{\omega}_\Delta/dt$
77	Figure 3.12, reactance of line 3 - 4	j0.800	j1.800
78	ℓ. 4 of (c)	$P_{s21} = \ldots = 0.509$	$P_{s21} = \ldots = 0.485$
	ℓ. 8 of (d)	$\bar{Y}_{14} = -0.451 + j1.042$	$\bar{Y}_{14} = -0.103 + j0.533$
	ℓ. 12 of (d)	= (1.009) (1.042 cos 10.532 + 0.451 sin 10.532) = 1.117	= (1.009) (0.533 cos 10.532 + 0.103 sin 10.532) = 0.548
	ℓ. 16 of (d)	$P_{1\Delta}(0^+) = (0.493)(0.2) = 0.0986$	= (0.323) (0.2) = 0.0646
	ℓ. 17 of (d)	$P_{2\Delta}(0^+) = (0.507)(0.2) = 0.1014$	= (0.677) (0.2) = 0.1354

78,79 Paragraph at the end of example should read:

'In this example the synchronizing power coefficient P_{s14} is smaller than P_{s24}, while the inertia of area 1 is greater than that of area 2. Thus while initially area 1 picks up only about one third of the load $P_{4\Delta}$, at a later time $t = t_1$, it picks up about 59% of the load and area 2 picks up the remaining 41%.'

Chapter 4

Page	Location	Error	Correction
105	ℓ. 5	$\frac{2H}{\omega_B}\omega = T_{au}$	$\frac{2H}{\omega_B}\dot{\omega} = T_{au}$
107	Equation (4.101)	$-\frac{D}{3\tau_j}$ term	$-\frac{D}{\tau_j}$
	Equation (4.103)	$-\frac{D}{3\tau_j}, \frac{T_m}{3\tau_j}$ terms	$-\frac{D}{\tau_j}, \frac{T_m}{\tau_j}$
109	Fig. 4.6	Voltage in Q branch is v_q	v_Q
112	(4.138)	Column of variables has $\dot{\lambda}_q$	λ_q

113 1. The last five terms in the preliminary calculations should be corrected as follows:

$$\frac{L_{MD}}{3\tau_j \ell_d^2} = 0.000235, \quad \frac{L_{MD}}{3\tau_j \ell_d \ell_F} = 0.000349, \quad \frac{L_{MD}}{3\tau_j \ell_d \ell_D} = 0.000642$$

$$\frac{L_{MQ}}{3\tau_j \ell_q \ell_Q} = 0.000980, \quad \frac{L_{MQ}}{3\tau_j \ell_q^{2}} = 0.000235$$

2. Line below should read:

   and we get the state-space equation for $D = 0$

3. Changes in the matrix:

   (a) in the column of variables: remove ω

4. Line 1 of matrix: ω change to $\omega \times 10^3$

113 (b) modify the last row as shown

$$\begin{bmatrix} \\ \\ \\ \end{bmatrix} = 10^{-3} \begin{bmatrix} & & & & \\ & & & & \\ -0.235\lambda_q & -0.349\lambda_q & -0.642\lambda_q & 0.235\lambda_d & 0.980\lambda_d \end{bmatrix} \begin{bmatrix} \\ \\ \omega \end{bmatrix}$$

remove (ω)

118 First two terms in ℓ. 2 of (4.156) — L_{MD} is missing in $\dfrac{\omega L_e L_{MD}}{\ell_d \ell_F} \lambda_F + \dfrac{\omega L_e L_{MD}}{\ell_d \ell_D} \lambda_D$

Equation (4.157) — First term on R.H.S. should have − sign

120 Last matrix should read

$$\hat{\underline{L}}^{-1} \begin{bmatrix} -K \sin\gamma \\ -v_F \\ 0 \\ K\cos\gamma \\ 0 \end{bmatrix} = \begin{bmatrix} -1.71K \sin\gamma + 0.591\, v_F \\ 0.591K \sin\gamma - 6.67\, v_F \\ 1.08K \sin\gamma + 5.87\, v_F \\ 1.71\, K\cos\gamma \\ -1.67\, K\cos\gamma \end{bmatrix}$$

121 1. The second part of the top matrix should be modified as shown above.

2. The first part of the top matrix should read

$$\begin{bmatrix} \dot{i}_d \\ \dot{i}_F \\ \dot{i}_D \\ \hline \dot{i}_q \\ \dot{i}_Q \\ \hline \dot{\omega} \\ \dot{\delta} \end{bmatrix} = \left[\begin{array}{ccccc|cc} -0.00187 & 0.00044 & 0.0141 & -3.487\omega & 2.547\omega & 0 & 0 \\ 0.00065 & -0.00495 & 0.0769 & 1.206\omega & 0.881\omega & 0 & 0 \\ 0.00118 & 0.00436 & -0.0960 & 2.202\omega & 1.609\omega & 0 & 0 \\ 3.590\omega & 2.65\omega & 2.65\omega & -0.0019 & 0.0901 & 0 & 0 \\ -3.506\omega & -2.588\omega & -2.588\omega & 0.00183 & 0.12332 & 0 & 0 \\ \hline -0.00032 i_q & -0.00029 i_q & -0.00029 i_q & 0.00031 i_d & 0.00028 i_d & -0.000559D & 0 \\ 0 & 0 & 0 & 0 & 0 & 1 & 0 \end{array}\right] \begin{bmatrix} i_d \\ i_F \\ i_D \\ i_q \\ i_Q \\ \hline \omega \\ \delta \end{bmatrix}$$

122	The top line of the second part of the final matrix should read		

$$+\begin{bmatrix} 0.316\ K \sin\gamma + 0.236\ v_F \\ \\ \\ \\ \end{bmatrix}$$

123	ℓ. below (4.169)	... reactance	... inductance
128	First term in (4.205)	Minus sign missing	$-(r/L_q)\lambda_q$
129	(4.217) should read		

$$T_{e\emptyset} = \lambda_q(\lambda_d/L_q - \lambda_d/L_d') + (L_{AD}\lambda_F/L_d'L_F)\ \lambda_q$$

137	Expression for $X_{xd} = \dfrac{(\quad)(x_d'' - x_\ell)}{(\quad)}$		
139	Equ. (4.286)		$e_d = -L_{AQ}\ i_Q$
	Equ. (4.287)	$x_q''\ I_q$	$x_q'\ I_q$
143	Equation (4.301), denom. for I_d, I_q	$R^2 + X''^2$	$\hat{R}^2 + \hat{X}''^2$
	(4.300)	$V_{\infty d} = \sqrt{3}\,V_\infty \sin(\delta-\alpha)$	$V_{\infty d} = -V_\infty \sin(\delta-\alpha)$
		$V_{\infty q} = \sqrt{3}\,V_\infty \cos(\delta-\alpha)$	$V_{\infty q} = V_\infty \cos(\delta-\alpha)$
147	ℓ. 5 of Problem 4.25	A statement is missing	Add: Let $\lambda_{ADT}=0.8\sqrt{3}$.
	ℓ. 2 of Problem 4.26		Omit: Let $\lambda_{ADT}=0.8\sqrt{3}$.
148	Part (b) of Problem 4.32	λ_d	$\dot{\lambda}_d$
	ℓ. 2 of Problem 4.33	A statement is missing	... in Figure P4.33 if the rotor has two circuits on the q-axis.
149	Ref. 18	Prubhaskankar	Prabhashankar
159	ℓ. 4B	From (5.42) and Figure 5.7	... and Figure 5.6
160	ℓ. 4	From (5.42)	From (5.43)

161	Mid-page	... similar to (5.32, viz ...	... similar to (5.42), viz, ...
161	In Ex. 5.3 it is preferable to solve for V_a directly by writing the voltage drop equation <u>and</u> solving the resulting quadratic equation, with the result V_a = 1.2056+j0.3876 = 1.2664 $\lfloor$ 17.823°		
164	Mid-page	$F_p = \tan^{-1}(Q/P)=0.859$	$F_p = \cos 30.746° = 0.859$
165	Mid-page	$V_{\infty d} = V_\infty \sin(\delta-\alpha)=$___	$V_{\infty d} = -V_\infty \sin(\delta-\alpha)=$___
168	(5.52)		$/[L_{AD}-\ell_F-L_F(L_d''-\ell_d)]$
170	Near end of Example 5.5	$\tau_q''=(L_q''/L_q)\tau_{qo}''$	$\tau_q''=(L_q''/L_q)\tau_{qo}''$
175	Figure 5.18 lower R.H. corner	connection between Ampl. 611 and line from v_t should be removed	
178	Pot. no. 700, 702	C = 1.667	C = 1.0
206	ℓ. 4 of Problem 5.1	δ	ϕ
	ℓ. 7 of Problem 5.4	$L_e i_d = L_e i_q \simeq 0$	$L_e i_d = L_e i_q \simeq 0$

Chapter 6

211	Equation (6.22)	First term on R.H. S.	Should have minus sign.
212	K matrix	Remove minus sign outside matrix	
214	(6.29)	[ ] [ ]	-[ ] [ ]
215	Terms above matrix K	$\frac{1}{3}(-kM_F i_{qo})$	$\frac{1}{3}(-kM_D i_{qo})$
		$\frac{1}{3}(-kM_D i_{qo})$	$\frac{1}{3}(-\lambda_{do} + L_q i_{do})$
216	Top A matrix, last row	1.0	1000
	Bottom A matrix fifth row	-1233.20	-123.320
	Last row	1.0	1000
221	Last row of <u>A</u> and <u>C</u>	1.0	1000
222	A matrix last row	1.0	1000
223	(6.57)	E_{FD}	$E_{FD\Delta}$
223	(6.60)	E_q'	$E_{q\Delta}'$

224	expression for K_2	T_e	$T_{e\Delta}$
226	Data from Ex. 5.1 used in Ex. 6.6 gives generator d-q values for non-zero r_a. This violates one of the assumptions for the linear model, p. 222, but should cause neglible error here.		
230	Solution, second line of P_s	$= \frac{1.3164 \times 1.0}{0.4162}$ (−)	$= \frac{1.3186 \times 1.0}{0.4164}$ (+)
231	Figure 6.4 summer at left	Move + sign to left of summer	
232	Ref. 1.	Hefron	Heffron

Chapter 7

265	ℓ. 2 below (7.38)	(7.76)	(7.26)
266	ℓ. 2 above 7.6.4.	respectable τ_{do}'	reasonable τ_{do}'
274	ℓ. 2 of the exciter	Figure 7.44	Figure 7.45
		Figure 7.48	Figure 7.50
285	ℓ. 2 of (7.67)	E_{FD}	E_{FD}
	ℓ. 4 of (7.67)	(S_E+K_F)	(S_E+K_E)
	Paragraph above (7.68):	Arbitrarily, we set x_8, x_9, x_{10} and x_{11} to correspond to the variables v_1, v_3, v_R and E_{FD}. In rewriting (7.67) to eliminate E_{FD} in the second equation we observe that, when per unit time is used, the product $(\tau_F\tau_E)$ must be divided by ω_R for the system of units to be consistent. The following equations are obtained.	
285 and 286	ℓ. 2 of (7.68)	$K_F/\tau_F\tau_E$	$\omega_R K_F/\tau_F\tau_E$
	Bottom matrix, fifth row	$-\frac{K_F(S_E'+K_E)}{\tau_F\tau_E}$	$-\frac{\omega_R K_F(S_E'+K_E)}{\tau_F\tau_E}$
287	(7.78)	$d_o v_q$	$d_o v_d$
288	K matrix: ℓ. 1	$-\sqrt{3}\ V_{\infty do}$	$-\sqrt{3}\ V_{\infty do}$
	ℓ. 4	$-\sqrt{3}\ V_{\infty qo}$	$-\sqrt{3}\ V_{\infty do}$
	ℓ. 9	Multiply last two terms by ω_R	

289 ℓ. 8B should be:

$$K_{9-10} = -\frac{\omega_R K_F}{\tau_F \tau_E} = -\frac{0.04 \times 377}{(269.5 \times 188.5)} = -2.967 \times 10^{-4}$$

ℓ. 7B

$$K_{9-11} = \frac{\omega_R K_F (S_E' + K_E)}{\tau_F \tau_E} = 2.967 \times 10^{-4} \times 0.26 = 7.7 \times 10^{-5}$$

290 K matrix: Divide elements of row 6 by 3: -0.014, -0.362, -0.362, -1.428, -0.790

Row 8: Change 0.056 to -0.056

Last 2 elements of row 9 should be: -2.967×10^{-4}, 7.7×10^{-5}

M matrix, row 6: -1786.94

A matrix:

Row 2: Change -0.6057 to -605.7

Row 7: Change 1.0 to 1000

Row 9: Last 2 elements should be: 0.2967, -0.077

Row 11: Change 53.052 to 5.3052

291		$\lambda_9 = -0.0002 + j0.0064$ $\lambda_{10} = -0.0002 - j0.0064$	$\lambda_9 = -0.00125 + j0.00297$ $\lambda_{10} = -0.00125 - j0.00297$
292	ℓ. 2, 3	Example 6.3	Example 6.4
297	ℓ. 5 of (7.99)	$1/\tau_R$	$1/\tau_E$
298	Fig. 7.66	Arrow from V_t to V_{TH} is reversed	
303	Fig. 7.69 scale of $V_{t\Delta}$	-25	0.25

Chapter 8

309	ℓ. 2 of second paragraph	... lead to ...	... led to ...
313	Line below (8.5)	$\delta = \delta_1 - \delta_2$	$\delta = \delta_1 + \delta_2$
	Line above (8.5)	arc of length	arc of radius
317	Table in Fig. 8.8, middle column heading	Clearing Time	τ_e

324	ℓ7	$K_\varepsilon(K_1K_6-K_2K_5)>K_2K_4$	$K_\varepsilon(K_1K_6-K_2K_5)>K_2K_4-K_1/K_3$
	making (8.23)	$K_6 > \dfrac{K_2K_4-K_1/K_3}{K_1K_6-K_2K_5}$	
325	Title of 8.4.2.	regular	regulator
326	Line below (8.27)	In the regulated...	In the unregulated...
329	Numerator of (8.33)	$K_1K_6(\ \ )$	$K_3K_6(\ \ )$
331	Item (b)	$\tau_F = 10s$	$\tau_F = 1.0s$
334	(8.34)	$G_x(s) = \dfrac{K_2K_3K_4}{1+K_3K_6K_\varepsilon}(\ \ )$	$G_x = \dfrac{K_2K_3K_4}{1+K_3K_6K_\varepsilon} \Big/ (\ \ )$
335	Middle of page	$V_2 \angle \beta_2 - \delta_2$	$V_2 \angle \beta_2 - \delta_{20}$
	Middle of page	$I_2 \angle -(\delta_2 - \beta_2 + \phi_2)$	$I_2 \angle -(\delta_{20} - \beta_2 + \phi_2)$
362	Figure 8.51	Fault cleared at <u>three</u> cycles	... at six ...
367	ℓ. 2 of ref. 41	PAS-90:	Proc. IEEE-59:
372	Equation (9.15)	$\underline{V}_{qk(i)}$ (bold face)	$V_{qk(i)}$ (not bold face)
378	2nd paragraph below (9.45)	$\sum_{i=1}^{m} (k_i - 1)$	$\sum_{i=1}^{m} k_i - 1$
		$\sum_{i=1}^{m} (k_i - 2)$	$\sum_{i=1}^{m} k_i - 2$
387	Equ. (9.86)	$j(\theta_{1k} + \delta_{1ko})$	$j(\theta_{1k} - \delta_{1ko})$
		$j(\theta_{nk} + \delta_{nko})$	$j(\theta_{nk} - \delta_{nko})$
388	Equ. (9.93) First, second term in $I_{q2\Delta}$, Second term in $I_{d\Delta}$	y_{12}/y_{22}	y_{12}/y_{11}
392	Equations in 9.12.1:		
	ℓ. 1	$I \angle \beta - \phi$	$I \angle - \phi$
	ℓ. 3	$V \angle - \delta$	$V \angle \beta - \delta$
	ℓ. 2 of Table 9.1	$\tau_j = 2H\omega_B$	$\tau_j = 2H\omega_R$

392	Table 9.1	gen 1	
	E'_{qo}	1.0558	—
	E'_{do}	-0.0419	—
415	Last 2 equations	Change t_{1B} to t_{2B}, and t_{2B} to t_{1B}	
419	(c.36) - (c.37)	ω_R	$1/\omega_R$
428	Section (4), ℓ. 5	1969	1968